Metalloids and Nonmetals

Noble Gases

			Group IIIA	Group IVA	Group VA	Group VIA	Group VIIA (Halogens)	Group 0 (Noble Gases)
			+3	+4 / -3	+5 / -2	-1		
								2 **He** helium 4.00260
			5 **B** boron 10.81	6 **C** carbon 12.011	7 **N** nitrogen 14.0067	8 **O** oxygen 15.9994	9 **F** fluorine 18.998403	10 **Ne** neon 20.179
	Group IB	Group IIB	13 **Al** aluminum 26.98154	14 **Si** silicon 28.0855	15 **P** phosphorus 30.97376	16 **S** sulfur 32.06	17 **Cl** chlorine 35.453	18 **Ar** argon 39.948
28 **Ni** nickel 58.70	29 **Cu** copper 63.546	30 **Zn** zinc 65.38	31 **Ga** gallium 69.72	32 **Ge** germanium 72.59	33 **As** arsenic 74.9216	34 **Se** selenium 78.96	35 **Br** bromine 79.904	36 **Kr** krypton 83.80
46 **Pd** palladium 106.4	47 **Ag** silver 107.868	48 **Cd** cadmium 112.41	49 **In** indium 114.82	50 **Sn** tin 118.69	51 **Sb** antimony 121.75	52 **Te** tellurium 127.60	53 **I** iodine 126.9045	54 **Xe** xenon 131.30
78 **Pt** platinum 195.09	79 **Au** gold 196.9665	80 **Hg** mercury 200.59	81 **Tl** thallium 204.37	82 **Pb** lead 207.2	83 **Bi** bismuth 208.9804	84 **Po** polonium (209)	85 **At** astatine (210)	86 **Rn** radon (222)

Noble

+4 SnS₂

64 **Gd** gadolinium 157.25	65 **Tb** terbium 158.9254	66 **Dy** dysprosium 162.50	67 **Ho** holmium 164.9304	68 **Er** erbium 167.26	69 **Tm** thulium 168.9342	70 **Yb** ytterbium 173.04	71 **Lu** lutetium 174.97
96 **Cm** curium (247)	97 **Bk** berkelium (247)	98 **Cf** californium (251)	99 **Es** einsteinium (254)	100 **Fm** fermium (257)	101 **Md** mendelevium (258)	102 **No** nobelium (255)	103 **Lr** lawrencium (260)

☐ metals ☐ nonmetals ☐ metalloids ☐ noble gases

General Chemistry

B. Richard Siebring

Mary Ellen Schaff

University of Wisconsin, Milwaukee

With assistance in rewriting the final draft by

Douglas Vaughan
Science writer

Wadsworth Publishing Company
Belmont, California
A division of Wadsworth, Inc.

Wadsworth Series in Chemistry

General Chemistry
B. Richard Siebring and Mary Ellen Schaff

Organic Chemistry
James D. Morrison

Chemistry: A Basic Introduction
G. Tyler Miller, Jr.

Chemistry: A Contemporary Approach
G. Tyler Miller, Jr.

Chemistry: Principles and Applications
G. Tyler Miller, Jr.

Energy and Environment: Four Energy Crises
G. Tyler Miller, Jr.

Dedicated to the memory of Professor Elsa Shipman, who was a warm, kind, and gentle person.

© 1980 by Wadsworth, Inc. All rights reserved. No part of this book may be reproduced, stored in a retrieval system, or transcribed, in any form or by any means, electronic, mechanical, photocopying, recording, or otherwise, without the prior written permission of the publisher, Wadsworth Publishing Company, Belmont, California 94002, a division of Wadsworth, Inc.

Printed in the United States of America

1 2 3 4 5 6 7 8 9 10—84 83 82 81 80

Library of Congress Cataloging in Publication Data

Siebring, B. Richard
   General chemistry.

   Includes index.
   1. Chemistry.   I. Schaff, Mary Ellen, joint author.
II. Title
QD31.2.S552   540   79-24947
ISBN 0-534-00802-X

Chemistry Editor: Jack C. Carey

Production: Greg Hubit Bookworks

Technical Illustrator: Scientific Illustrators, Inc.

Cover photo (a crystal of potassium acid phthalate, magnified 33-1/3 times):
   Suzanne Groet/Stock, Boston

Preface

This book is designed for the standard one-year general chemistry course that enrolls not only preprofessional students and chemistry and other science majors but also humanities and social science majors who want a substantial background in chemistry. Some of the book's important features are summarized below.

Sensitivity to Students' Needs and Backgrounds This text assumes that students have some reasonable facility with algebra, but it makes no use of calculus. Although most general chemistry students have a high school chemistry background, their degree of understanding is so variable that we assume no specific knowledge or skills in chemistry. Hence this text begins each topic at a low level and develops it step by step to the appropriate degree of sophistication. To help make this text as clear as possible, the manuscript (including all problems and questions) was class-tested in each of its five successive drafts. More than 5000 students read the manuscript altogether, and where they had trouble understanding any point, we rewrote. Furthermore, each draft of the manuscript was reviewed in detail by many professors who teach the course.

A Professional Writer with a Ph.D. in Chemistry Douglas Vaughan, a professional science writer who has a doctorate in chemistry from the University of California at Berkeley, helped the authors rewrite the final draft. His contribution, as the final-draft reviewers enthusiastically agreed, has made the book even clearer and more readable.

Sequence of Topics The order of topics was determined primarily by pedagogical considerations—by asking those who teach the course what they want. Wadsworth sent out a series of questionnaires to thousands of general chemistry professors, asking which of the organizational changes made in successful texts over the past decade or so have met with their approval. When a substantial majority of the respondents preferred a certain order, we listened. As a result, we believe that our chapter sequence will be comfortable and workable for most professors.

Balance The text maintains a sensible balance between exposition of principles and descriptive chemistry. Many worked-out problems clarify the principles, and applications and relevant material enliven the descriptive chemistry, making both it and the chemical principles more meaningful.

Learning by Involvement: Study Aids The key to learning is getting the learners actively involved. To this end we have incorporated the following study aids—which add up to an unusually complete and integrated system—into the book and its supplements. Such a system is essential in a field as complex as chemistry.

Problems and questions This text includes many more worked-out problems, end-of-section questions, and end-of-chapter problems and questions than most texts. The worked-out problems make clear the reasoning involved.

Dimensional analysis method for problem-solving This method is emphasized throughout the text.

Varied level of challenge in exercises The most challenging ones have an asterisk (*) beside them.

End-of-section questions Questions follow most sections within each chapter to encourage students to think briefly about what they have just read. However, the questions do not require so much time and effort that the continuity of discussions is interrupted.

End-of-chapter problems and questions organized by chapter section The exercises at the end of the chapters are organized by section so that instructors can, if they wish, assign some questions and problems before students have finished the whole chapter. Also, students can easily refer to the part of the chapter that would be most helpful in completing the exercises. Most important, this section-by-section arrangement encourages the students to test their understanding of one section before moving on to another (providing immediate feedback) and allows students to approach each chapter in manageable segments.

Discussion-starters Discussion-starters covering the entire chapter follow the problems and questions.

Study Guide We have written our own Study Guide. It contains objectives, key terms, and more worked-out problems for appropriate chapters, in that order. There are also many additional questions and unworked problems. All of these items are arranged by chapter section so that students can zero in on sections where they need help and skip the sections they have mastered.

Answers and solutions Answers to many of the questions and problems appear in the text. For those questions and problems that have answers, the numbers appear in color. The Solutions Manual gives complete solutions for all numerical problems as well as answers to all nonnumerical questions.

Key terms and glossary All key terms appear in color. To help students review such terms in context, the page number on which each key term is defined appears in boldface type in that term's index entry. All key terms are also defined in a comprehensive glossary that appears in the student study guide. A list of key terms by chapter also appears in that guide.

Correlation with laboratory work In organizing this book and our Laboratory Manual, we kept in mind the difficulty of providing meaningful laboratory experience at the beginning of a general course. For example, although we delay the chapter on solutions until midway in the text (for sound pedagogical reasons), we discuss molarity of solutions in Chapter 3 (Stoichiometry), making possible meaningful experiments early in the semester.

Laboratory Manual The manual we have written for our book has been class-tested for ten years by at least 10,000 students, including students at colleges and universities other than our own. It has the following format: (1) Objectives, (2) Principles, (3) Procedures, (4) Pre-laboratory questions that test students on the principles and procedures, (5) Report sheets, and (6) Post-laboratory exercises. The pre-laboratory questions are on a separate perforated sheet so they can be handed in before the start of the laboratory session. The report sheets and post-laboratory exercises are also on separate perforated pages so they can be handed in separately when the lab is completed. A separate Instructor's Manual for the Laboratory Manual contains the following items: (1) estimate of time allotment, (2) supply list, (3) safety considerations, (4) suggestions for instruction, (5) sources of error, (6) summary of students' results, (7) answers to pre-laboratory questions, and (8) answers to post-laboratory exercises.

Summary of Supplementary Material We have either authored or coauthored all of the following supplements with Jacqlynn Behnke and Marguerite Fulton:
1. Instructor's Manual with hundreds of multiple-choice questions and problems arranged by chapter, all of which have answers. For each chapter, there is also a chapter quiz.
2. Detailed Solutions Manual.
3. Student Study Guide.
4. Laboratory Manual.
5. Instructor's Manual for Laboratory Manual.
6. Transparency masters of the most important illustrations in the text.

Reviewing Process In the course of its development from a series of lectures to a book, the manuscript was reviewed in several drafts by over 40 chemistry professors who are listed on page vii. These reviewers not only pointed out errors but also suggested additions, deletions, better approaches, more effective language, and other valuable changes.

Acknowledgments The book is dedicated to the memory of Professor Elsa Shipman, of the Department of English at the University of Wisconsin, Milwaukee, who read the entire manuscript in the early stages of its development and made numerous suggestions about language and sentence structure—gleefully pouncing on our dangling participles.

After Elsa, we would like to recognize the approximately 5000 students who, in effect, proofread the book in its various stages. Students enjoy pointing out instructor's errors, and of course we profited from their enjoyment by correcting errors that ranged from typos to misstatements. Sometimes we completely revised a section that was obviously just not working for the students.

Several people contributed the sketches upon which the final art was based. Outstanding among these was Tom Lesperance, who prepared fully 75 percent of the sketches. Others whose drawings were used are Tony Garcia, James Iserman, and Barbara Parkman.

Finally, we want to thank Jack Carey, Chemistry Editor at Wadsworth, who directed the entire process. With several successful books already to his credit, we were glad to have him lead us through the long, laborious, sometimes frustrating but ultimately rewarding experience of writing a book.

Extensive Manuscript Review

Hundreds of teachers responded to an initial questionnaire concerning their needs in this basic chemistry course. The detailed outline for this book was then reviewed by over a hundred teachers from many kinds of schools. The following reviewers from universities, colleges, and community colleges read the original and revised manuscripts.

David L. Adams
North Shore Community College

John J. Alexander
University of Cincinnati

Ronald M. Backus
American River College

Paul A. Barks
North Hennepin Junior College

Luther K. Brice, Jr.
Virginia Polytechnic Institute

Joseph H. Cecil
Kent State University

Ulrich O. dela Camp
California State University, Dominquez Hills

John S. DiYorio
Wytheville Community College

Norman V. Duffy
Kent State University

Gordon G. Evans
Tufts University

Thomas H. Epps
Virginia State College

Pat Garvey
Des Moines Area Community College

Melvin E. Gleiter
University of Wisconsin, Eau Claire

Peter J. Hansen
Northwestern College

William E. Hatfield
University of North Carolina, Chapel Hill

H. Fred Henneike
Georgia State University

Forrest C. Hentz, Jr.
North Carolina State University

William J. Husa, Jr.
Middle Georgia College, Raleigh

J. S. Johar
Wayne State College

Delwin D. Johnson
Forrest Park Community College

Richard L. Kiefer
College of William and Mary

Craig K. Kellogg
Georgia Southern College

C. Richard Kistner
University of Wisconsin, La Crosse

Soter G. Kokalis
William Rainey Harper College

Lawrence F. Koons
Tuskegee Institute

Andrew G. Lang
Miami University

Ralph H. Marking
University of Wisconsin, Eau Claire

Robert L. McNeely
University of Tennessee, Chattanooga

Lester R. Morss
Rutgers University, The State University of New Jersey

Bruce Murray
University of Wisconsin, River Falls

Jack E. Powell
Iowa State University

Patricia A. Redden
Saint Peter's College

S. Rosen
Olney Central College

Kenneth H. Russell
California State University, Fresno

Donald E. Sands
University of Kentucky

Donald W. Sink
Appalachian State University

Jimmy C. Stokes
West Georgia College

Bruce Storhoff
Ball State University

Benjamin P. Stormer
Ulster County Community College

Judith A. Strong
Moorhead State University

Donald C. Taylor
Slippery Rock State College

Forrest D. Thomas II
University of Montana

Brief Contents

1	Introductory Concepts	1
2	Some Basic Concepts of Chemistry	33
3	Equations and Stoichiometry	54
4	Atomic Structure	76
5	Electron Configurations and the Periodic Table	96
6	Chemical Bonding—Electron Formulas	145
7	Covalent Bonding—Molecular Structure	183
8	Gases	214
9	Liquids and Solids—The Condensed Phases	244
10	Solutions	307
11	Thermodynamics	340
12	Chemical Kinetics	373
13	Molecular Equilibria	400
14	Acids and Bases	426
15	Acid-Base Equilibria	454
16	Precipitation and Solubility Products	493
17	Oxidation and Reduction	513
18	Electrochemistry	540
19	The Chemistry of the Nonmetals of Groups VIA and VIIA	573
20	The Chemistry of the Nonmetals of Groups IVA and VA and the Noble Gases	611
21	The Chemistry of Metals	646
22	Organic Chemistry	674
23	The Structure and Properties of Biomolecules	733
24	Coordination Compounds	764
25	Nuclear Chemistry	796
	Four-Place Logarithms	831
	Answers to Selected Questions and Problems	833
	Index	

Detailed Contents

1	**Introductory Concepts**		**1**
	1.1	Concerns of Chemistry	1
	1.2	The Classification of Matter	8
	1.3	Units	10
	1.4	Exponential Numbers	12
	1.5	Significant Figures	15
	1.6	Unit Conversion	19
	1.7	Density and Specific Gravity	22
	1.8	Heat	25
2	**Some Basic Concepts of Chemistry**		**33**
	2.1	The Atomic Concept	33
	2.2	Names and Symbols of the Elements	38
	2.3	Atoms and Molecules	39
	2.4	Formulas	40
	2.5	The Mole Concept	42
	2.6	Percent Composition from the Formula	45
	2.7	Establishing the Formula of a Compound from Experimental Data	46
3	**Equations and Stoichiometry**		**54**
	3.1	Chemical Equations	54
	3.2	Stoichiometry	57
	3.3	Stoichiometry in Solutions—Molarity	67
4	**Atomic Structure**		**76**
	4.1	Electrical Energy and Chemical Change	76
	4.2	Subatomic Particles	78
	4.3	Radioactivity and the Nuclear Atom	86
	4.4	Atomic Weights and Isotopes	91
5	**Electron Configurations and the Periodic Table**		**96**
	5.1	Electromagnetic Radiation	96
	5.2	Atomic Spectra	101

	5.3	Electrons, Quantum Numbers, and Energy States	106
	5.4	Periodic Classification of the Elements	114
	5.5	Electron Configurations and the Periodic Table	119
	5.6	Categories of Elements and Periodic Table Position	128
	5.7	Atomic Properties	133
	5.8	Electron Formulas for Atoms	138
6	**Chemical Bonding—Electron Formulas**		**145**
	6.1	The Ionic Bond	146
	6.2	The Covalent Bond	149
	6.3	Electronegativities and Bond Polarity	151
	6.4	Multiple Covalent Bonding	154
	6.5	Metal Ions That Deviate from Noble Gas Electron Configurations	155
	6.6	Molecules That Deviate from Noble Gas Electron Configurations	160
	6.7	Resonance	162
	6.8	The Coordinate Covalent Bond	163
	6.9	Summary of Generalizations for Deducing Electron Formulas	165
	6.10	Oxidation Numbers	170
	6.11	Introduction to Chemical Nomenclature	172
7	**Covalent Bonding—Molecular Structure**		**183**
	7.1	Molecular Shape and Polarity	183
	7.2	Orbitals and Bonding	185
	7.3	Hybridization of Orbitals	188
	7.4	Geometry of Multiple Bonds	198
	7.5	Resonance and Delocalized Electrons	201
	7.6	Summary of Rules for Predicting Molecular Geometries	204
8	**Gases**		**214**
	8.1	Air, the Atmosphere, and Atmospheric Pressure	215
	8.2	Qualitative Descriptions of Gases	218
	8.3	The Gas Laws	220
	8.4	Combining Volumes and the General Gas Law Equation	227
	8.5	Stoichiometry and Gas Volumes	231
	8.6	Dalton's Law of Partial Pressures	231
	8.7	The Kinetic Molecular Theory	233
	8.8	Molecular Velocities	235
	8.9	Deviations from the Gas Laws	237
9	**Liquids and Solids—The Condensed Phases**		**244**
	9.1	Condensation	244
	9.2	Evaporation	245
	9.3	Liquid-Vapor Equilibrium	247
	9.4	Collecting Gas Over Water	251
	9.5	Boiling Points	253

	9.6	The Solid-Liquid Equilibrium	253
	9.7	Phase Diagrams	255
	9.8	Intermolecular Forces	258
	9.9	The Kinetic Theory of Liquids	264
	9.10	Properties of Solids	265
	9.11	Molecular Solids	274
	9.12	Metallic Solids	280
	9.13	Ionic Solids	283
	9.14	Network Solids	292
	9.15	Graphite	298
	9.16	Ice	301
	9.17	Pseudosolids	301
10	**Solutions**		**307**
	10.1	Types of Solutions and Colloids	307
	10.2	Gas-in-Liquid Solutions	309
	10.3	Liquid-in-Liquid Solutions	312
	10.4	Solid-in-Liquid Solutions	313
	10.5	Concentrations	316
	10.6	Saturation	322
	10.7	Physical Properties of Solutions	323
	10.8	Osmosis	330
11	**Thermodynamics**		**340**
	11.1	Heat and Work	341
	11.2	Thermodynamic Systems	344
	11.3	The First Law of Thermodynamics	345
	11.4	Thermochemical Equations	350
	11.5	Bond Enthalpies	356
	11.6	Entropy	359
	11.7	Free Energy	365
12	**Chemical Kinetics**		**373**
	12.1	Rates of Reactions	373
	12.2	Factors Affecting Rates: Chemical and Physical Condition of Reactants	377
	12.3	Factors Affecting Rates: Temperature	377
	12.4	Factors Affecting Rates: Concentrations	380
	12.5	Reaction Mechanisms	387
	12.6	Theory of the Activated Complex	389
	12.7	Chain Reactions	393
	12.8	Factors Affecting Rates: Catalysts	395
13	**Molecular Equilibria**		**400**
	13.1	The Equilibrium Constant	400
	13.2	Le Chatelier's Principle and Chemical Equilibrium	404

13.3	Quantitative Aspects of Chemical Equilibrium	407
13.4	K_p, the Equilibrium Constant in Terms of Partial Pressures	413
13.5	Free Energy and Equilibrium	413
13.6	Reaction Rates and Equilibrium	416
13.7	The Ammonia Synthesis	417

14 Acids and Bases — 426

14.1	Autoionization of Water	426
14.2	The Arrhenius Concept	427
14.3	Characterization and Classification of Arrhenius Bases	428
14.4	Characterizations and Classification of Arrhenius Acids	430
14.5	Some Typical Reactions of Arrhenius Acids and Bases	432
14.6	Hydrolysis	435
14.7	The Brønsted-Lowry Acid-Base Concept	437
14.8	Relative Strengths of Brønsted-Lowry Acids and Bases	438
14.9	The Lewis Acid-Base Concept	441
14.10	Acid-Base Trends in the Periodic Table	444
14.11	Equivalents and Normality	447

15 Acid-Base Equilibria — 454

15.1	The Water Equilibrium	454
15.2	Strong Acids and Strong Bases	455
15.3	pH	457
15.4	Weak Acids and Weak Bases	458
15.5	Common-Ion Effect in Acid-Base Reactions	466
15.6	Buffers	469
15.7	Polyprotic Acids	472
15.8	Hydrolysis	476
15.9	Indicators	480
15.10	Titrations	482

16 Precipitation and Solubility Products — 493

16.1	Equations for Precipitation Reactions	493
16.2	Separation of Ions by Precipitation Reactions	495
16.3	Stoichiometry of Precipitation Reactions	498
16.4	Solubility Equilibria and Solubility Products	499
16.5	Solubility Products and the Common-Ion Effect	506
16.6	Predicting the Formation of Precipitates	507

17 Oxidation and Reduction — 513

17.1	Reactions Involving Oxygen	513
17.2	Oxidation and Reduction	514
17.3	Oxidation Numbers	516
17.4	Half-Reactions	521
17.5	Predicting the Direction of Oxidation-Reduction Reactions	524

	17.6	The Preparation of Hydrogen	526
	17.7	Writing Equations for Redox Reactions	529
	17.8	Equivalent Weights in Redox Reactions	533

18 Electrochemistry — 540

	18.1	Galvanic Cells	540
	18.2	Commercial Cells	548
	18.3	Electrode Potentials	554
	18.4	The Nernst Equation	559
	18.5	Electrolysis	562
	18.6	Applications of Electrolysis	567
	18.7	Quantitative Aspects of Electrolysis	569

19 The Chemistry of the Nonmetals of Groups VIA and VIIA — 573

	19.1	The Halogens	573
	19.2	Hydrogen	586
	19.3	Oxygen	592
	19.4	Sulfur	597

20 The Chemistry of the Nonmetals of Groups IVA and VA and the Noble Gases — 611

	20.1	Nitrogen	611
	20.2	Phosphorus	623
	20.3	Carbon	633
	20.4	Silicon	637
	20.5	The Noble Gases	639

21 The Chemistry of Metals — 646

	21.1	Abundance, Availability, and Usefulness	646
	21.2	The Metals of Groups IA and IIA	648
	21.3	Transition Elements	655
	21.4	Rare Earths: Inner Transition Metals	662
	21.5	The Representative Metals of Groups IIIA, IVA, and VA	663
	21.6	The Metalloids	666
	21.7	Corrosion	669

22 Organic Chemistry — 674

	22.1	Structure and Nomenclature of Alkanes	676
	22.2	Optical Isomerism	685
	22.3	Structure and Nomenclature of Alkenes	688
	22.4	Geometric Isomerism	689
	22.5	Structure and Nomenclature of Alkynes	690
	22.6	Structure and Nomenclature of Cycloalkanes	691
	22.7	Structure and Nomenclature of Cycloalkenes	692
	22.8	Structure and Nomenclature of Aromatic Hydrocarbons	693

22.9	Physical Properties of Hydrocarbons	696
22.10	Combustion of Hydrocarbons	697
22.11	Substitution Reactions	698
22.12	Reactions of Unsaturated Hydrocarbons	699
22.13	Polymers	701
22.14	Aromatic Substitution	704
22.15	Petroleum	704
22.16	Other Sources of Hydrocarbons	706
22.17	Air Pollution from Hydrocarbon Fuels	707
22.18	Hydrocarbon Derivatives	709
22.19	Halides	709
22.20	Alcohols	713
22.21	Ethers	716
22.22	Carbonyl Compounds	717
22.23	Carboxylic Acids	720
22.24	Esters	722
22.25	Synthetic Fibers	724

23 The Structure and Properties of Biomolecules — 733

23.1	Lipids	734
23.2	Carbohydrates	739
23.3	Amino Acids	747
23.4	Proteins	751
23.5	Medicinal Chemistry and Enzymes	758

24 Coordination Compounds — 764

24.1	Coordination	764
24.2	Terminology	767
24.3	Nomenclature	769
24.4	The Stability of Coordination Complexes	773
24.5	Geometry	776
24.6	Magnetic Properties and Orbitals	777
24.7	Splitting of d Orbitals	779
24.8	Isomerism	782
24.9	Color and Some Interesting Complexes	786

25 Nuclear Chemistry — 796

25.1	Theories of Nuclear Structure	796
25.2	Natural Radioactivity	797
25.3	Detection of Radiation	799
25.4	Nuclear Equations	802
25.5	Nuclear Stability	804
25.6	Neutron-to-Proton Ratio	809
25.7	Half-Life	812
25.8	Induced Nuclear Transformations	814

25.9	Accelerators	816
25.10	Artificial Radioactivity	818
25.11	Nuclear Fission	819
25.12	Nuclear Fusion	823
25.13	Energy from the Nucleus	824
25.14	Applications of Nuclear Chemistry	826

Four-Place Logarithms — 831

Answers to Selected Questions and Problems — 833

Index

Introductory Concepts

One of the most pressing questions on the mind of anyone about to take a chemistry course is, "What is chemistry all about?" Formal definitions abound but they do not really answer the question. They are vaguely unsatisfying—like saccharin when you expect sugar. Chemistry is best understood by experience, not by definition. Therefore, we shall forego a formal definition and lead you directly into the experience of chemistry, first by way of some things chemistry is concerned with and then by introducing some of the concepts we shall be using. You would be mistaken, of course, to assume that definitions are not a part of studying chemistry, but understanding the underlying concepts is always more important than merely memorizing words.

1.1 Concerns of Chemistry

Chemistry is concerned with matter and its properties. Nearly everyone has an intuitive notion of matter. It is the "stuff" of the earth, the things on the surface of the earth, the earth's atmosphere, indeed all the universe. Formally, matter is defined as *everything that has mass and occupies space*. It thus includes stone, flesh, water, air, and smoke but not heat, electricity, light, wisdom, or friendship.

Since we have defined matter in terms of mass, it should be useful to understand the difference between **mass** and **weight.** We commonly use the two terms interchangeably, but they do not have the same meaning. *Mass* can be thought of as measuring the amount of matter in a substance, and it does not vary with location; mass is a constant. *Weight,* on the other hand, is the measure of the mutual attractive force between two masses, and it is dependent on locations of the masses; weight is a variable. We can emphasize this difference by observing that matter attracts matter with a force that is inversely proportional to the distance between the bodies of matter and directly proportional to the masses of the bodies. This attraction is called *gravitational force*.

As long as we do not venture too far into space, the attraction between the earth and other masses completely overshadows all other gravitational attractions. Therefore we can give a practical definition of weight as the measure of the attraction of the earth for bodies of mass on or near its surface, and we can see that the weight of a body will vary with its distance from the earth. As an illustration, consider an 81.8-kg (180-lb) astronaut bound for the moon. As he leaves the surface of the earth, his weight decreases and he approaches weightlessness. As he nears the moon, his weight gradually increases again, as the attraction of the moon is substituted for the attraction of the earth, but his weight on the moon is only 13.6 kg, since the moon has a mass only one-sixth that of the earth. However, the mass of the astronaut is constant—on the earth's surface, in space, or on the moon.

Weight can be measured with a spring scale as illustrated in Figure 1.1: The greater the elongation of the spring, the greater the weight. The elongation

Figure 1.1 Spring scale is used to measure weight.

of the spring depends on the gravitational force, so the reading on the moon will be different from that on the earth. Indeed, an object weighed with a spring scale will weigh more at sea level than on a nearby mountain peak. Mass, on the other hand, is measured with a balance: An unknown object must be balanced with a quantity of matter that has a known mass (Figure 1.2). In this case, the gravitational force affects both sides of the balance equally; hence the result is independent of the location.

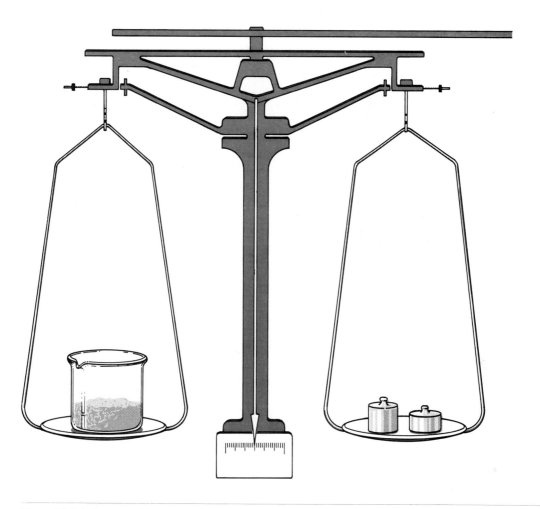

Figure 1.2 A balance used to measure mass.

The balance used in the modern chemistry laboratory is more likely to look like those shown in Figures 1.3 and 1.4 than the one in Figure 1.2, but the principle is the same. In weighing an object with a balance similar to the one shown in Figure 1.3, the object with an unknown mass is placed on the pan

Introductory Concepts 3

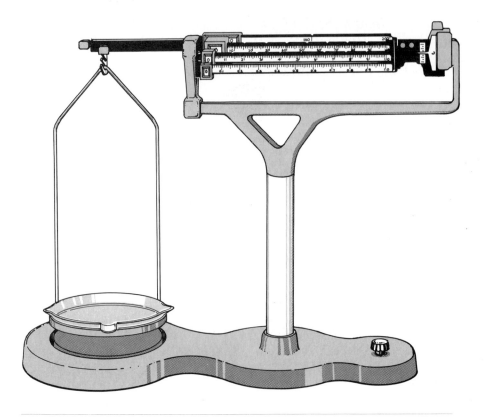

Figure 1.3 A triple-beam balance.

and positions of the weights on the three beams are changed until balance is achieved. Similarly in Figure 1.4 the unknown mass is placed on the pan and small metal "weights" (of known masses) are adjusted within the balance by turning knobs on the outside of the instrument. When the mass on the pan is balanced by the internal weights, the mass is read from an illuminated dial. This measurement of mass is generally, if somewhat loosely, called "weighing."

As we said at the beginning of this section, chemistry is concerned with the *properties* of matter. We shall now try to define more precisely just what these properties are that distinguish one kind of matter from another. **Physical properties** can be described without reference to any other matter. Physical properties thus describe how a substance looks and acts by itself. For example, one can distinguish copper from magnesium by observing such physical properties as color, electrical conductivity, thermal conductivity, hardness, and tensile strength, each of which can be measured for each metal alone. **Chemical properties,** however, cannot be described without reference to other matter. For example, one might distinguish copper from magnesium by observing how each interacts with oxygen. Magnesium burns rapidly in oxygen with a brilliant hot flame, yielding a white powder, whereas copper changes slowly to a black

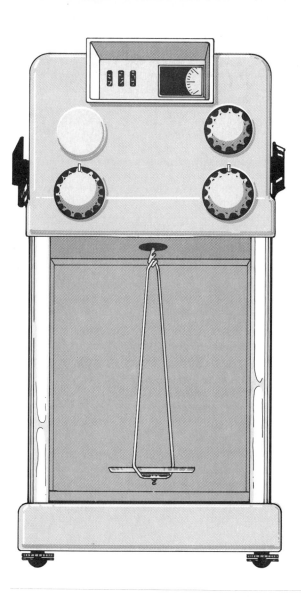

Figure 1.4 A modern balance with a direct reading scale.

powder, even when heated to very high temperatures in oxygen. This example also illustrates that when investigating chemical properties we must often change the substance being investigated into another substance altogether. No amount of physical manipulation (cooling, melting, crushing, and so forth) will regenerate magnesium and copper from the products of their combustion. The metals, with their distinctive properties, have been replaced with new substances having entirely different physical and chemical properties. Table 1.1 contrasts physical and chemical properties and gives examples of each.

Table 1.1 Comparison of Physical and Chemical Properties

Physical	Chemical
Can be observed without reference to other material	Can be observed only in relation to other materials
Can be observed without changing the composition of the matter	Often *cannot* be observed without changing the composition of the matter in question
Examples of Physical Properties of Water	Examples of Chemical Properties of Water
Colorless	Forms an acid with carbon dioxide
Odorless	Produces a colorless gas (hydrogen) on contact with calcium
Does not conduct current appreciably	Does not produce any new material on contact with gold even at high temperatures
Expands on freezing	
Boiling Point = 100°C or 212°F	Produces carbon monoxide and hydrogen upon contact with carbon at high temperatures
Freezing Point = 0°C or 32°F	Can be decomposed into hydrogen and oxygen

Chemistry is concerned with changes that matter undergoes. Since matter has both physical and chemical properties, it should not surprise us to find that it undergoes both physical and chemical changes. A **physical change** never changes one substance into another. The melting of ice is therefore a physical change, since solid water is merely changing to liquid water. Other examples of physical change are evaporation, freezing, crystallization, and condensation.

In a **chemical change,** on the other hand, there are always changes in the indentity and composition of the matter undergoing change: One or more substances are converted into one or more new substances. The burning of magnesium that we described previously is an example. The properties of the matter are entirely different following a chemical change. Rusting of iron, fermentation of fruit juices, souring of milk, decay of animal and vegetable materials, and digestion of food are common chemical changes.

Unfortunately, it is not always easy to determine whether a change in composition (a chemical change) has occurred or not. For example, the solution process is often regarded as a physical change. When white copper sulfate (which is often used to prevent the growth of algae in swimming pools) is added to water, however, the solution is blue—suggesting a chemical change. Indeed, when the solution is evaporated, beautiful blue crystals are formed, indicating that a new substance has resulted from the interaction of copper sulfate with water. This observation well illustrates that the more scientists learn about phenomena, the more "gray" areas of uncertainty they uncover. You should be prepared to find that many of the definitions and classifications in chemistry have similar vague gray outlines. All of us find these ambiguities frustrating at times, but they can also be great sources of fascination.

Chemistry is concerned with conditions necessary to cause or to prevent changes in matter. We normally think of metals as substances that do not burn. If your firewood has iron nails in it, you will find them in the ashes when you clean the fireplace. We have seen, however, that magnesium will burn if it gets hot enough. Iron, too, burns if finely divided, heated to high temperature, and immersed in oxygen. The same is true for nearly all metals. Understanding these and other conditions for chemical change is an important part of chemistry.

Chemistry is concerned with the relative amounts of matter and energy involved in changes in matter. Energy is often defined as *the capacity to do work.* It can take several forms, including potential energy (energy of position or composition), kinetic energy (energy of motion), electrical energy, heat energy, and radiant energy. Changes from one form of energy to another are taking place around us constantly. The water at the top of a waterfall has energy of position, or potential energy. As the water flows over the precipice, this potential energy is converted into kinetic energy. If a power plant corrals the rushing water and directs it through turbines, this kinetic energy may be converted to electrical energy. Electrical energy in turn might be converted to light (radiant energy) in an incandescent light bulb or to heat energy in an electric stove or hair dryer.

Both chemical and physical changes usually involve such transformations of energy. However, the energy changes accompanying chemical changes are usually larger than those accompanying physical changes. Trees transform the radiant energy of sunlight into chemical energy (a form of potential energy) through complex chemical changes, and the storage battery in your automobile transforms chemical energy to electrical energy (and vice versa) by means of simpler chemical changes.

The chemist must be able to deal quantitatively with physical and chemical changes. As an example, a given amount of oxygen will not burn an unlimited quantity of magnesium. A chemist can determine how much magnesium will be burned by a given amount of oxygen, how much energy will be liberated, and how much of the new material will be formed in the process. Soon you will learn how to calculate these quantities.

This brief discussion has been an attempt to sketch a rough answer to the questions, "What is chemistry all about?" However, we have said nothing about the practical scope of the subject or about the state of our chemical knowledge. We know a great deal about the makeup of our universe. We have analyzed moon rocks, investigated the composition of the sun, and identified the components of the earth's atmosphere as well as its crust. We know the composition of the foods we eat and the drugs we use; in fact, their components are usually listed on the containers in which we buy them. Our clothing, so often made from synthetic fibers, usually bears labels listing the kinds and amounts of the various fibers used in each garment. We also know a great deal about the composition of our own bodies: Our blood and body wastes are routinely analyzed to help determine the state of our health.

So complex and numerous are the substances that surround us, indeed, of which we are made, that much remains unknown. What Lincoln Steffens (American journalist, 1866–1936) said in 1936 is still true: "Chemistry is little more than a sparkling mass of beckoning questions."[1]

Questions

1. Can the weight of an object be changed without changing its mass? How?

2. Can the mass of an object be changed without changing its weight? How?

3. Distinguish between physical and chemical properties.

1.2   The Classification of Matter

Chemistry is concerned with all the millions of different species of matter. Hence it seems reasonable that we begin to classify matter at the outset. We shall introduce new classifications and refine the old throughout the text; however, we shall take the first steps here. As shown in Figure 1.5, all materials can be divided into two categories—heterogeneous mixtures and homogeneous materials. **Heterogeneous mixtures** not only vary from sample to sample but also lack uniformity within a single sample. Any heterogeneous mixture will therefore exhibit a variety of properties. Granite, for example, is a heterogeneous mixture of igneous rock and is a blend of pinks, grays, and tans. **Homogeneous materials** are uniform and can be further classified as either *homogeneous mixtures* (usually called *solutions*) or *pure substances*. Solutions have a uniform composition throughout a given sample, but other samples might have widely different compositions. Sugar dissolved in water is a homogeneous mixture, and within a given container the composition is uniform; however, there are many homogeneous mixtures of sugar and water with different compositions and properties.

The components of mixtures, both heterogeneous and homogeneous, can be separated by physical changes. Chemical changes are not required. We can separate table salt from sand by putting the heterogeneous mixture in water so that the salt dissolves and then filtering out the insoluble sand. We can recover the salt from the new homogeneous mixture (salt solution) simply by evaporating the water.

In principle, all forms of matter can be classified as either homogeneous or (more frequently) heterogeneous. Again, however, we encounter a gray area. The distinction between these two categories is not always simple. Suppose we pour equal quantities of red marbles and white marbles into a

---

[1] Lincoln Steffens, *Lincoln Steffens Speaking*, Harcourt Brace Jovanovich, New York, 1936, p. 148.

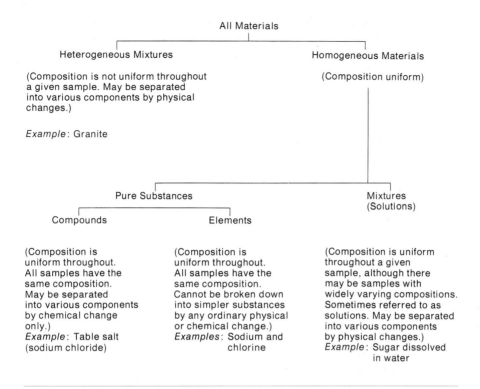

Figure 1.5 Classification of matter.

large barrel and commence mixing them. How thoroughly must they be mixed before we can consider the mixture homogeneous? Suppose we cannot look into the barrel but must rely upon samples withdrawn for observation. The most obvious answer would be that *the mixture is homogeneous when every sample contains the same ratio of red marbles to white marbles.* This depends on the size of the sample. If we withdraw samples by the bucketful, we will pronounce the mixture homogeneous long before we could do so if we sample the mixture with a teacup. Just how homogeneous is homogeneous?

Unlike samples of the mixtures we have considered so far, every sample of a pure substance has the same composition. Sodium chloride, a *compound* used on the dinner table and commonly called salt, has a constant composition (by mass) of 39.33 percent sodium and 60.67 percent chlorine. These percentages never vary. Whether we obtain it from the salt flats of Utah, the Dead Sea, the Atlantic Ocean, the coast of China, or the deep-well brines of Michigan, sodium chloride is always the same. In general, therefore, compounds differ from mixtures in three ways: Compounds are always homogeneous, whereas mixtures can be heterogeneous; compounds always have a constant composition; and compounds cannot be broken down into simpler substances by physical changes. Compounds can, however, be broken down into simpler substances by chemical changes. For example, if we pass an electric current through melted salt, we obtain sodium and chlorine. These and other simple substances that make up the millions of known compounds are called *elements*.

Elements cannot be broken down into simpler substances by ordinary physical or chemical changes and are therefore the basic substances with which the chemist is concerned. There are only slightly more than 100 elements, of which 88 are known to occur naturally. But as you will discover in the following pages, the variety of compounds that can be made from these few elements is inexhaustible.

Questions

4. Distinguish between
   a. solution and heterogeneous mixture
   b. solution and compound
   c. compound and element

5. List several examples of homogeneous and heterogeneous materials other than those discussed in the chapter.

1.3 Units

Since chemistry is a quantitative science, the chemist needs a system of units for measuring amounts of matter and energy. In our daily lives, most of us still use the English system (foot-pound-quart). The scientist, however, works with a system known as the International System of Units, abbreviated SI. It was established and is maintained by the International Bureau of Weights and Measures, at Sèvres, France, and is supported by 40 countries, including the United States. Representatives of the supporting nations meet occasionally to revise definitions of the units.

Although the SI system was not established until 1960, it overlaps considerably with the much older metric system, which has been used by scientists for more than a hundred years. In this text we shall not adhere rigidly to the SI system but shall use some familiar units from the metric system that are not a part of the SI system. The trend is toward the SI system, however, and you should be aware that today in certain types of scientific literature the SI system is used exclusively.

There are seven base units in the SI system. These units, along with their symbols and definitions, are listed in Table 1.2. The definitions of these base units have been selected so that the units can be standardized precisely and reproducibly. Over the years, base units and their definitions have changed from time to time, and they are likely to be modified in the future. You may find some of the definitions mystifying and unintelligible now, but they will mean more to you as you study further. In the meantime, you may find it useful to compare the SI units with their more familiar English equivalents.

Table 1.2 International System of Units

Quantity	Name of Base Unit	Symbol	Definition*	English Equivalent
Length	meter	m	The length equal to 1,650,763.73 wavelengths in vacuum of the radiation corresponding to the transition between the levels $2p^{10}$ and $5d^5$ of the krypton-86 atom.	1.00 m = 39.4 in.
Mass	kilogram	kg	The mass of the international prototype of the kilogram. (A cylinder of platinum-iridium alloy stored at the International Bureau of Weights and Measures at Sèvres, France.)	1.00 kg = 2.20 lb
Temperature	kelvin	K	1/273.16 of the thermodynamic temperature of the triple point of water. [The origin of the Kelvin scale is absolute zero (the point at which all atomic vibration ceases).]	See pp. 25–27 for temperature conversion.
Time	second	s	The duration of 9,192,631,770 periods of the radiation corresponding to the transition between two hyperfine levels of the fundamental state of the atom of cesium-133.	—
Amount of substance	mole	mol	The amount of substance which contains as many elementary entities as there are atoms in 0.012 kilogram of carbon-12 (6.02×10^{23}).	—
Electric current	ampere	A	The constant current which, if maintained in two straight parallel conductors of infinite length, of negligible circular cross section, and placed 1 meter apart in vacuum, would produce between these conductors a force equal to 2×10^{-7} newton per meter of length.	—
Luminous intensity	candela	cd	The luminous intensity, in the perpendicular direction, of a surface of 1/600,000 square meter of a blackbody at the temperature of freezing platinum under a pressure of 101,325 newtons per square meter.	—

* "The International System of Units (SI)," C. H. Page, P. Vigoureux, Eds., *National Bureau of Standards Special Publication 330*, U.S. Government Printing Office, Washington, D.C., 1972. Translation approved by the International Bureau of Weights and Measures of its 1970 publication, "Le Système International d'Unités (SI)."

Table 1.3 lists the prefixes that define the decimal multiples and submultiples of SI units. The prefixes most commonly used by scientists are *kilo*, *centi*, *milli*, and *micro*. These prefixes greatly extend the flexibility of the system. You will notice that one of these prefixes is already incorporated in *kilogram*, the base unit for mass. Therefore, multiples and submultiples of *kilogram* are formed by combining the appropriate prefix with the word *gram* (1 g = 0.001 kg or 1 kg = 1000 g).

Introductory Concepts   11

Table 1.3 SI Prefixes

Multiples and Submultiples*	Prefix	Symbols
$1{,}000{,}000{,}000{,}000 = 10^{12}$	tera	T
$1{,}000{,}000{,}000 = 10^{9}$	giga	G
$1{,}000{,}000 = 10^{6}$	mega	M
$1000 = 10^{3}$	kilo	k
$100 = 10^{2}$	hecto	h
$10 = 10^{1}$	deka	da
Base Unit (except for kilogram) $1 = 10^{0}$		
$0.1 = 10^{-1}$	deci	d
$0.01 = 10^{-2}$	centi	c
$0.001 = 10^{-3}$	milli	m
$0.000{,}001 = 10^{-6}$	micro	μ
$0.000{,}000{,}001 = 10^{-9}$	nano	n
$0.000{,}000{,}000{,}001 = 10^{-12}$	pico	p
$0.000{,}000{,}000{,}000{,}001 = 10^{-15}$	femto	f
$0.000{,}000{,}000{,}000{,}000{,}001 = 10^{-18}$	atto	a

* Perhaps exponential notation is unclear to you now, but an explanation of exponential numbers follows in Section 1.4.

In addition to the base units, along with their multiples and submultiples, there are numerous derived units. Among these are units of volume. For example, volume might be measured in cubic meters (m^3) or cubic centimeters (cm^3).

Several units in common usage among scientists lie outside the SI system but can be exactly defined in terms of SI units. These include units of volume such as the *liter* and *milliliter*. One liter is equal to 0.001 m^3 or 1000 cm^3. The milliliter is the same as the cubic centimeter. In English units, 1.00 liter is 1.06 qt.

Questions

6. What is the base unit in the International System of Units for
   a. length?   b. temperature?   c. time?   d. mass?

7. What are the proper prefixes for the following?
   a. 1/100 of a meter   b. 1/1,000,000 of a gram
   c. 1/1000 of a liter   d. 1000 grams

8. Indicate the correct abbreviations for the units above.

1.4 Exponential Numbers

The chemist often finds it necessary to work with very large or very small numbers. These numbers are cumbersome to work with when expressed in the usual manner, so we write them in *exponential form*. For example, the number

602,000,000,000,000,000,000,000 expressed in exponential form is 6.02×10^{23}. This means that 6.02 must be multiplied by 10 twenty-three times, which is the same as moving the decimal twenty-three places to the right. Similarly, 0.000,000,000,000,000,000,000,205 is written as 2.05×10^{-25}. This means that 2.05 must be divided by 10 twenty-five times, or the decimal moved twenty-five places to the left.

It is customary to write exponential numbers with only one digit to the left of the decimal; thus 1 million would be written 1×10^6 rather than 100×10^4 or 0.1×10^7. When adding or subtracting two exponential numbers, however, both numbers must have the same exponent. Consider the following examples.

Example 1.1 Add 1.400×10^{11} and 1.900×10^{13}.

Solution

a. In addition problems, all numbers must be expressed with the same exponent. In this case we shall write both numbers with 11 as the exponent; however, we could just as well have selected 13.

$$1.900 \times 10^{13} = 190.0 \times 10^{11}$$

b. Add the nonexponential portions of the numbers, leaving the exponent unchanged:

$$\begin{array}{r} 190.0 \times 10^{11} \\ +1.400 \times 10^{11} \\ \hline 191.4 \times 10^{11} \end{array}$$

c. Rewrite the sum in the customary form with one digit to the left of the decimal:

$$191.4 \times 10^{11} = \mathbf{1.914 \times 10^{13}}$$

Example 1.2 Subtract 4.00×10^{-23} from 5.70×10^{-22}.

Solution

a. Both numbers must be expressed with the same exponent:

$$4.00 \times 10^{-23} = 0.400 \times 10^{-22}$$

b. Subtract the nonexponential portions of the numbers, leaving the exponent unchanged:

$$\begin{array}{r} 5.70 \times 10^{-22} \\ -0.400 \times 10^{-22} \\ \hline \mathbf{5.30 \times 10^{-22}} \end{array}$$

In multiplication, the nonexponential portions of the numbers are multiplied, and the exponents are added algebraically. The exponents need not be the same.

Example 1.3 Multiply 3.00×10^{-10} by 5.00×10^{14}.

Solution

a. Multiply 3.00 by 5.00, and then add -10 and 14 to obtain the exponent:

$$(3.00 \times 10^{-10})(5.00 \times 10^{14}) = (3.00 \times 5.00)(10^{-10+14})$$
$$= 15.0 \times 10^4$$

b. Rewrite the result with one digit to the left of the decimal:

$$15.0 \times 10^4 = \mathbf{1.50 \times 10^5}$$

In division, we divide the nonexponential portions and then subtract the exponent of the denominator from the exponent of the numerator. The exponents in the denominator and numerator need not be the same.

Example 1.4 Divide 5.00×10^{14} by 3.00×10^{-10}.

Solution Divide 5.00 by 3.00 and then subtract -10 from 14 algebraically to obtain the exponent:

$$\frac{5.00 \times 10^{14}}{3.00 \times 10^{-10}} = \left(\frac{5.00}{3.00}\right) \times [10^{14-(-10)}] = \mathbf{1.67 \times 10^{24}}$$

To extract the square root of a number, we divide the exponent by two; to extract the cube root, we divide it by three; and so forth. In each case, however, we must move the decimal so that the division results in a whole number. Finally, we extract the root of the nonexponential part of the number.

Example 1.5 Take the fifth root of 3.00×10^{18}.

Solution

a. Rewrite the number so that the exponent is evenly divisible by 5:

$$\sqrt[5]{3.00 \times 10^{18}} = \sqrt[5]{3000 \times 10^{15}}$$

b. Divide the exponent by 5:

$$\sqrt[5]{3000 \times 10^{15}} = \sqrt[5]{3000} \times 10^3$$

c. Calculate the fifth root of 3000. (Use either your calculator or a logarithm table to obtain the logarithm of 3000, divide by 5, and then obtain the antilog.)

$$\sqrt[5]{3000} \times 10^3 = \mathbf{4.96 \times 10^3}$$

To compute the power of an exponential number, multiply the exponent by the desired power and then raise the nonexponential portion to the same power.

Example 1.6 Raise 2.00×10^{17} to the fourth power.

Solution

a. Multiply the exponent by 4:

$$(2.00 \times 10^{17})^4 = (2.00)^4 \times 10^{68}$$

b. Calculate the fourth power of 2.00:

$$(2.00)^4 \times 10^{68} = 16.0 \times 10^{68}$$

c. Express the result with one digit to the left of the decimal:

$$16.0 \times 10^{68} = \mathbf{1.60 \times 10^{69}}$$

Questions

9. Write the following in exponential form:
  a. 1,000,000
  b. 36,000
  c. 0.000,034,8
  d. 10.0
  e. 0.000,000,000,000,830
  f. 91,200,000,000
  g. 0.100
  h. 413
  i. 317,000
  j. 0.000,670

10. Write the following exponential numbers as ordinary numbers:
  a. 1.00×10^{-15}
  b. 3.00×10^4
  c. 6.11×10^9
  d. 6.43×10^{-14}
  e. 7.88×10^{-5}
  f. 3.36×10^8
  g. 4.79×10^{-13}
  h. 3.00×10^2
  i. 8.63×10^{-4}
  j. 1.90×10^{-9}

1.5 Significant Figures

In most chemical calculations, the numbers we deal with are approximate rather than exact, since quantitative measurements are limited by the reliability

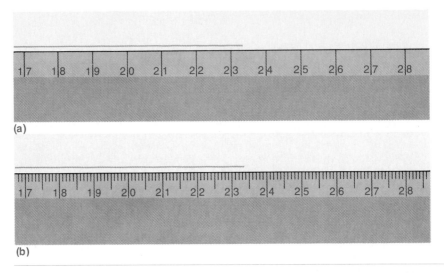

Figure 1.6 (a) Measuring the length of a line with a ruler graduated in centimeters (23.3 cm). **(b)** Measuring the length of a line with a ruler graduated in millimeters (23.35 cm).

of our observations. An example is illustrated in Figure 1.6. When measuring the length of the line with a ruler graduated in centimeters, we find that the end of the line appears to be about three-tenths of the way between graduations marked 23 and 24; hence, we report the line as 23.3 cm long. In doing this we are using the concept of significant figures. We understand that there is some uncertainty about the last digit in the number but that all other digits are definitely known. The final 3 is merely an estimate as to how far beyond the 23-cm mark the line extends. In measuring the same line with a ruler graduated in tenths of centimeters (millimeters) we find the end of the line about halfway between 23.3 and 23.4. Therefore, we report the length of the line with a number containing four digits, 23.35 cm, where the 5 is the doubtful and least significant digit.

Zeros may or may not be significant. If they serve only to locate the decimal point, as in 0.0002335 (the length of the line in Figure 1.6 in kilometers), they are not significant. On the other hand, final zeros after the decimal point are significant. If the length of our line had been determined as 23.30 (instead of 23.35) cm, the zero would be significant, since it indicates that we have made an effort to measure to the nearest hundredth of a centimeter and that we are certain of the other three digits.

Final zeros in a whole number present a problem. If the population of a city is reported as 1,430,000, the zeros function only to locate the decimal point. If a student has earned 100 academic credit hours, however, the record shows 100 academic credit hours exactly, and the zeros are significant. In order to avoid this ambiguity, large numbers are better expressed in exponential form. Thus if a city's population is written 1.430×10^6, it is clear that this quantity is known to only four significant figures.

With practice you will be able to recognize very quickly the number of significant figures in different quantities. The following are examples:

Quantity	Number of Significant Figures
103.22 g	5
0.036 m	2
3431.00 liter	6
1.63×10^7 cm	3
0.003 mg	1
400 ml	?

When performing arithmetical procedures, it is important that the result not imply more precision than is justified. In addition or subtraction, the result *must not* go beyond the last digit position common to all the numbers being added or subtracted. For example, when 3.213 cm is added to 5.63 cm, the sum is 8.84 cm and *not* 8.843 cm. In multiplication or division, the answer must contain no more digits than the least exact item used in the calculation. The product of 2.89 cm and 4.70 cm is 13.6 cm^2 and not 13.583 cm^2.

When rounding off, the last significant figure is

1. Increased by 1 if the discarded digit is greater than 5.

2. Left unchanged if the discarded digit is less than 5.

3. Brought to the nearest *even* number if the discarded digit is 5.

Thus the following numbers are rounded off to three significant figures as shown.

1.532	1.53
1.537	1.54
1.535	1.54
1.545	1.54

It is important to remember that the rules above are not applicable to *exact numbers*. There are 12 items to a dozen. This information is not subject to uncertainties of measurement, and we may add a decimal and as many zeros to the 12 as we like. Thus the number of oranges in 133.0 dozen is 1596 and not 1.6×10^3. The following examples illustrate the principles governing significant figures.

Example 1.7 Add 4.66 g, 0.089 g, and 16.4 g.

Solution

a. Last digit position common to all numbers

```
  4.66  g
  0.089 g
 16.4   g
 ───────
 21.149 g
```

b. Since 4 is less than 5, round off the sum to **21.1 g.** (Note that digits are *not* rounded off one at a time. If they were, we would obtain the following *incorrect* result: 21.149 ⟶ 21.15 ⟶ 21.2.)

Example 1.8 Subtract 12.35 cm³ from 14.945 cm³.

Solution

a. Last digit position common to both numbers
↓
$$\begin{array}{r} 14.945 \text{ cm}^3 \\ 12.35 \text{ cm}^3 \\ \hline 2.595 \text{ cm}^3 \end{array}$$

b. Since the result must be rounded off, we discard the final 5 and round off the answer to the nearest even number, giving **2.60.** The zero is significant and must be retained. This example shows that in subtraction it is possible to obtain a result with fewer significant figures than either of the original numbers.

Example 1.9 Find the volume of a metal bar with the following dimensions:

$$14.69 \text{ cm} \times 11.21 \text{ cm} \times 2.09 \text{ cm}$$

Perform the multiplication with an electronic calculator.

Solution

a. 14.69 cm × 11.21 cm × 2.09 cm = 344.17054 cm³
b. For all the virtues of the electronic calculator, it cannot make judgments about significant figures. Here it reports a result apparently accurate to eight digits of which only the first three are significant. Since one of the dimensions is expressed in only three significant figures, the volume may likewise be expressed in only three significant figures. Hence we report the volume of the metal bar as **344 cm³.**

Example 1.10 Divide 1.400 × 10³ g by 5283 ml. Use an electronic calculator.

Solution

a. $\dfrac{1.400 \times 10^3 \text{ g}}{5283 \text{ ml}} = 0.2650009 \text{ g/ml}$

b. Again we must use the result carefully. Both divisor and dividend are expressed in four significant figures, so we must report the quotient as **0.2650 g/ml.** (Note the difficulty we would have faced if the dividend had been written 1400 instead of 1.400 × 10³.)

Example 1.11 Find the average of the following numbers:

24.0 g,   33.11 g,   49.653 g,   38.6 g

Solution

a.   Last digit common to all numbers
              ↓
    24.0   g
    33.11 g
    49.653 g
    38.6   g
    ―――――――
    145.363 g

b. Round off the sum to 145.4 g.

c. $\dfrac{145.4 \text{ g}}{4} = 36.35 \text{ g}$

d. Since 4 is an exact number (it indicates exactly how many items are being averaged), the quotient must be reported in the same number of significant figures as in the dividend.

Questions

11. When are zeros significant? When are zeros not significant?

12. List two numbers containing significant zeros. List two numbers containing non-significant zeros.

1.6 Unit Conversion

If this chapter is your first encounter with SI units, exponential notation, and the concept of significant figures, you may be somewhat confused by it all. A firm logical basis underlies each of these scientific tools, however, and you will soon be more comfortable with them. Indeed, both the new SI units and the metric units are far easier to learn and use than the English system we often use outside the laboratory. In this section we shall present several solved problems to help you learn the relationships among the three systems of units. We hope you will think of these problems as practical ones, since it is planned that the United States will soon abandon the cumbersome English system in favor of the units from the SI and metric systems. Thus, we shall be buying sugar by the kilogram, gasoline by the liter, and cloth by the meter, and we shall be measuring our automobile speed in kilometers per hour.

    To convert a measurement from one unit to another we need an equation called a *conversion equation* to relate the two units. The equations relating SI units to English units are listed in Table 1.2. You should also remember that 1 liter = 1 × 10⁻³ m³ and 1.00 liter = 1.06 qt, and you should recall the meanings of the prefixes micro, milli, centi, and kilo from Table 1.3. You will then be ready to do the most important conversions you will encounter in studying chemistry. Others may be necessary later, but they will logically follow what you learn here.

To use a conversion equation it is usually convenient to change it into a *conversion factor*. We do this by dividing *both* sides of the equation by one side or the other. There are hence two conversion factors for each conversion equation. Consider the equation

$$12 \text{ in.} = 1 \text{ ft}$$

Dividing both sides of this equation by 1 ft produces one of the two conversion factors:

$$\frac{12 \text{ in.}}{1 \text{ ft}} = \frac{1 \text{ ft}}{1 \text{ ft}} = 1$$

This first conversion factor can be read as *12 inches per foot*. The second is produced by dividing both sides of the original equation by 12 in. and can be read as *1 foot per 12 inches*:

$$\frac{12 \text{ in.}}{12 \text{ in.}} = \frac{1 \text{ ft}}{12 \text{ in.}} = 1$$

Notice that all conversion factors are equal to unity: When multiplied by a quantity, they change only the units, never the value of the quantity. If 3.00 ft is to be converted to inches, the first conversion factor will eliminate feet from the expression:

$$(3.00 \text{ ft}) \left(\frac{12 \text{ in.}}{1 \text{ ft}} \right) = 36.0 \text{ in.}$$

To convert 36.0 in. to feet, we use the second conversion factor:

$$(36.0 \text{ in.}) \left(\frac{1 \text{ ft}}{12 \text{ in.}} \right) = 3.00 \text{ ft}$$

In general, the proper conversion factor will have the given units in the denominator and the desired units in the numerator. Choosing the wrong (inverted) conversion factor, on the other hand, will yield a meaningless result:

$$(36.0 \text{ in.}) \left(\frac{12 \text{ in.}}{1 \text{ ft}} \right) = 432 \frac{\text{in.}^2}{\text{ft}}$$

Example 1.12 A rod-shaped bacterium is 1.18×10^{-4} in. long. Express the length of this bacterium in micrometers.

Solution

a. In this problem two conversion factors are necessary. We can consider the solution as a series of steps, each of which requires an appropriate conversion factor:

$$\text{inches} \longrightarrow \text{meters} \longrightarrow \text{micrometers}$$

b. From the equation relating English and SI linear measure, we obtain the first conversion factor:

$$1.00 \text{ m} = 39.4 \text{ in.} \longrightarrow \frac{1.00 \text{ m}}{39.4 \text{ in.}}$$

c. From the definition of the prefix micro, we find the second:

$$1 \text{ μm} = 10^{-6} \text{ m} \longrightarrow \frac{1 \text{ μm}}{10^{-6} \text{ m}}$$

d. Using both of these conversion factors, we are prepared to convert inches to micrometers:

$$(1.18 \times 10^{-4} \text{ in.}) \left(\frac{1.00 \text{ m}}{39.4 \text{ in.}}\right) \left(\frac{1 \text{ μm}}{10^{-6} \text{ m}}\right) = 2.99 \text{ μm}$$

Example 1.13 A sample of blood is 15.0 percent hemoglobin (the iron-containing protein that carries oxygen to the body tissues) by mass. Calculate the number of grams of hemoglobin in 2.00 lb of blood.

Solution

a. Again we need more than one conversion factor. Since we know the English–metric relationship between kilograms and pounds, we can represent the step-by-step solution of the problem as follows:

$$\text{pounds} \longrightarrow \text{kilograms} \longrightarrow \text{grams} \longrightarrow 15 \text{ percent of total}$$

b. For the first step, the conversion factor comes from the equation relating pounds and kilograms:

$$1.00 \text{ kg} = 2.20 \text{ lb} \longrightarrow \frac{1.00 \text{ kg}}{2.20 \text{ lb}}$$

c. The second conversion factor changes kilograms to grams:

$$1 \text{ kg} = 1000 \text{ g} \longrightarrow \frac{1000 \text{ g}}{1 \text{ kg}}$$

d. We could use these two conversion factors to find the mass of the blood in grams. What we are seeking is 15 percent of that total. Hence we have

$$(2.00 \text{ lb blood}) \left(\frac{1.00 \text{ kg blood}}{2.20 \text{ lb blood}}\right) \left(\frac{1000 \text{ g blood}}{1 \text{ kg blood}}\right) \left(\frac{15.0 \text{ g hemoglobin}}{100 \text{ g blood}}\right)$$

$$= 136 \text{ g hemoglobin}$$

Questions

13. Distinguish between a conversion factor and a conversion equation.

14. How can one judge if the correct conversion factor has been used in solving a problem?

15. What conversion factors are used to make the following conversions?
   a. 10.3 m to feet
   b. 4.00 ft to millimeters
   c. 4.31 gal to cubic meters
   d. 5.91 lb to grams
   e. 4.31 kg to milligrams
   f. 791 tons to kilograms
   g. 3.12 liters to cubic inches

1.7 Density and Specific Gravity

We have already outlined the differences between chemical and physical properties. Now we must distinguish between *properties of a substance* and *properties of an object*. Gold is very malleable (that is, we can shape it by hammering). This is a property of the substance. But the fact that a piece of gold foil is only 0.001 cm thick is a property of the object (not all pieces of gold are 0.001 cm thick). Similarly, the luster of gold is a property of the substance. But if a gold ornament is shaped to reveal and emphasize the luster of the gold, its shape is a property of the object.

Mass, despite being a universal property of matter, cannot be used as a property to distinguish one substance from another. It does not make sense to talk about the mass of lead—one lead object will have one mass; a second object, another mass. When people say that lead is heavier (more massive) than cork, what they really mean is, for a given volume, lead is more massive than cork. They are talking about *density*, which is a property of the substance. **Density** is mass per unit volume ($D = M/V$) and is usually expressed in grams per cubic centimeter or grams per milliliter (Table 1.4). The density of lead is 11.34 g/cm^3, regardless of its shape or volume.

Table 1.4 Densities of Some Common Substances at Room Temperature (20°C)

Substance	Density (g/cm³)
Gold	19.30
Mercury	13.55
Lead	11.34
Copper	8.89
Iron	7.90
Aluminum	2.70
Water	0.998
Alcohol	0.79
Gasoline	0.69
Cork wood	0.22
Carbon dioxide gas*	0.00184
Hydrogen gas*	0.0000838
Air*	0.001205

* At 1.00 atm of pressure. See Chapter 8.

Density does, however, vary slightly with temperature. Water has a density of 1.000 g/cm^3 at $3.98°C$: This is its maximum density. Above or below this temperature, the density of water decreases. For example, at $0°C$ it is 0.9999 g/cm^3, and at $80°C$ it is 0.9718 g/cm^3. Water is a substance almost unique in this respect. For most substances, the density decreases as the temperature increases; that is, it expands as it is heated.

A concept related to density is *specific gravity*. The **specific gravity** of a liquid or solid is the ratio of its density to that of water at $3.98°C$ (1.000 g/cm^3). The specific gravity of a substance will therefore be numerically equal to its density, if the density is expressed in grams per cubic centimeter (or kilograms per liter). Since specific gravity is a ratio, however, it has no units. It is specific gravity that we measure to determine how much antifreeze our radiator contains or to indicate the concentration of sulfuric acid in a lead-acid automobile battery.

Example 1.14 If the average adult human body contains 5.5×10^3 ml of blood weighing 13.5 lb, what is the density of blood in grams per milliliter?

Solution

a. Density is mass per unit volume; the ratio supplied is pounds per milliliter.

$$\frac{13.5 \text{ lb}}{5.5 \times 10^3 \text{ ml}}$$

b. Since the problem asks for density in grams per milliliter, we need the following conversion equations:

$$1.00 \text{ kg} = 2.20 \text{ lb}$$
$$1 \text{ kg} = 1000 \text{ g}$$

c. Using the correct conversion factors, we obtain

$$\left(\frac{13.5 \text{ lb}}{5.5 \times 10^3 \text{ ml}}\right)\left(\frac{1.00 \text{ kg}}{2.20 \text{ lb}}\right)\left(\frac{1000 \text{ g}}{1 \text{ kg}}\right) = 1.1 \text{ g/ml}$$

Example 1.15 The density of glucose (sometimes called blood sugar because it is the form in which carbohydrates are transported to body tissues) is 1.56 g/ml. Calculate the volume of 1.00 kg of glucose in kiloliters.

Solution

a. This problem illustrates the use of density as a conversion factor. For glucose we have

$$\frac{1.56 \text{ g}}{1.00 \text{ ml}} \quad \text{and} \quad \frac{1.00 \text{ ml}}{1.56 \text{ g}}$$

When converting from volume to mass, we would use the first factor. In this problem, we need the second factor.

b. We also need conversion equations relating grams to kilograms and milliliters to kiloliters:

$$1 \text{ kg} = 1000 \text{ g}$$
$$1 \text{ liter} = 1000 \text{ ml}$$
$$1 \text{ kl} = 1000 \text{ liters}$$

c. Using all four conversion factors, we write the answer:

$$(1.00 \text{ kg})\left(\frac{1000 \text{ g}}{1 \text{ kg}}\right)\left(\frac{1.00 \text{ ml}}{1.56 \text{ g}}\right)\left(\frac{1 \text{ liter}}{1000 \text{ ml}}\right)\left(\frac{1 \text{ kl}}{1000 \text{ liters}}\right) = 6.41 \times 10^{-4} \text{ kl}$$

Questions

16. Distinguish between density and specific gravity.
17. Criticize the following statement: "Lead is heavier than water."
18. What conversion factor is used to make each of the following conversions?
    a. 133 lb to grams   b. 14.0 qt to liters
    c. 309 ml to liters   d. 127 in. to meters

1.8 Heat

In the opening pages of this chapter, we mentioned that the chemist was often concerned with the energy involved when matter undergoes changes. Most often, though by no means always, this energy takes the form of heat. The chemist, therefore, must often measure both the intensity and the quantity of *heat energy*. Heat intensity, the "hotness" or "coldness" of an object, is measured by the *temperature*. This of course is an idea with which we are all familiar. We are also well acquainted with another important observation: Relative temperature determines the direction of heat flow; that is, heat will flow from the hotter of two objects to the colder, and the rate of flow depends on the difference in temperature.

The measurement of temperature is usually based upon the measurement of some other physical property that varies with temperature. Often this property is the thermal expansion of mercury. If mercury is confined to a narrow tube with a uniform bore, connected to a thin-walled bulb, an increase in temperature causes the mercury to expand and rise in the tube. By measuring the distance the mercury rises we can therefore measure the temperature.

There are three temperature scales (thermometers) in common use today: the familiar Fahrenheit scale and the more logical Celsius and Kelvin scales. The Celsius (or centigrade) thermometer is the one most often used by chemists. This thermometer is based upon the freezing and boiling points of water at 1 atmosphere of pressure. The height of the mercury column at the freezing point of water is labeled 0°; the height of the column at the boiling point of water is labeled 100°. The distance between these two points is then divided into 100 equal divisions, each of which represents 1°C (Celsius).

The kelvin (K), which is the SI unit for heat intensity, is the same size as the Celsius degree; that is, there are also 100 K between the freezing and boiling points of water (see Figure 1.7). Therefore to convert a Celsius reading to a kelvin reading we need only consider the different origins of the two scales. Since zero on the kelvin scale is 273°C *below* the zero on the Celsius scale, we convert Celsius to kelvin simply by adding 273.

Converting Celsius to Fahrenheit is not so simple, since the two kinds of degrees are not the same size. By referring to Figure 1.7, you will note that there are 180° between the freezing and boiling points of water on the Fahrenheit thermometer. Thus there are $\frac{9}{5}$ as many degrees between any two points on the Fahrenheit scale as on the Celsius scale. Furthermore, there is a difference in the origins of the two thermometers. The Fahrenheit scale reads 32° at the freezing point of water, whereas the Celsius scale reads 0°. This means that after we convert the number of Celsius degrees to Fahrenheit degrees by multiplying by $\frac{9}{5}$, we must add 32. This relationship between the two scales is stated algebraically as

$$°F = \tfrac{9}{5}°C + 32$$

To convert the other way (from °F to °C), we must first subtract 32 from the

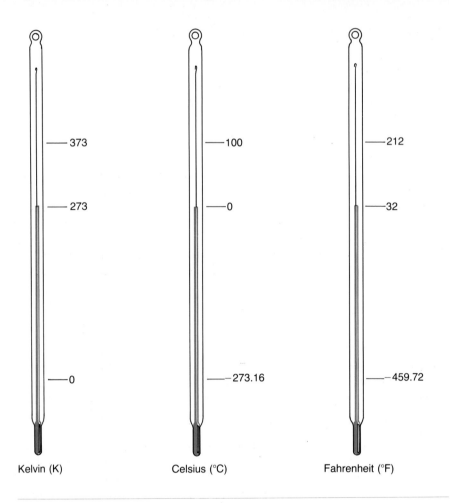

Figure 1.7 Relationships among three temperature scales.

Fahrenheit reading and then multiply by $\frac{5}{9}$. Stated algebraically,

$$°C = \tfrac{5}{9}(°F - 32)$$

Example 1.16 Body temperature is 98.6°F. Express this temperature in degrees Celsius and in kelvins.

Solution

a. First we show the conversion equations:

$$°C = \tfrac{5}{9}(°F - 32°)$$
$$K = °C + 273$$

b. Since the conversion equation for kelvins involves Celsius, solve for Celsius first:

$$°C = \tfrac{5}{9}(98.6 - 32) = 37.0°$$

c. Then complete the conversion by adding 273:

$$K = 37 + 273 = 310$$

When measuring the quantity of heat that an object contains, we must consider not only its temperature but also its mass and its ability to absorb heat. A kilogram of water at a given temperature contains more heat than does a gram of water at the same temperature. Moreover a kilogram of water contains more heat than a kilogram of mercury at the same temperature, since water has a greater ability to absorb heat than mercury does. Clearly then the temperature scales will not do for measuring *quantities* of heat. The unit of heat quantity in the SI system is the *joule* (J). It will be defined and used in a later chapter. The unit most commonly used for this purpose is the *calorie* (cal), which is defined as the heat necessary to raise the temperature of 1 gram of water from 14.5°C to 15.5°C. In general, it is sufficient to remember the calorie as the quantity of heat necessary to raise the temperature of 1 gram of water 1° Celsius. Another common unit, the *large Calorie* (used to measure the energy value of food) is 1000 calories, or a *kilocalorie* (kcal). Thus an 1800-Calorie meal actually releases 1,800,000 calories, equal to the energy needed to raise 1 gram of water 1.8×10^6 °C or to raise 1.8×10^6 g of water 1° Celsius! Note that the large Calorie is capitalized, but the small calorie (the chemist's calorie) is always written with a small *c*.

Equipped with a unit for measuring quantities of heat, we can now consider the differing abilities of substances to absorb heat. This ability is most usefully expressed as the amount of heat required to raise the temperature of 1 gram of the material 1° Celsius and is referred to as the *specific heat* of the material. The specific heat of water, for example, among the highest of all specific heats, is $\dfrac{1.00 \text{ cal}}{(1.00 \text{ g})(1.00°C)}$ and that of iron is $\dfrac{0.108 \text{ cal}}{(1.00 \text{ g})(1.00°C)}$. Although the specific heat of a substance varies with the temperature, the change is slight and can usually be ignored.

Example 1.17 Calculate the number of calories required to raise the temperature of 1.00×10^2 g of iron from 22.0°C to 82.0°C.

Solution

a. To determine the quantity of heat required we must consider temperature change, mass, and ability to absorb heat (specific heat). Thus the number of calories required equals the temperature change × mass × specific heat.

b. $(82.0 - 22.0°C)(1.00 \times 10^2 \text{ g}) \left[\dfrac{0.108 \text{ cal}}{(1.00 \text{ g})(1.00°C)} \right] = 648 \text{ cal}$

Example 1.18 A 47.3-g sample of metal shot at a temperature of 98.0°C, was placed into 34.3 g of water at a temperature of 18.0°C. The temperature of the thoroughly stirred mixture was 28.4°C. Assuming that all heat lost by the metal was absorbed by the water, calculate the specific heat of the metal.

Solution

a. The metal cooled from 98.0°C to 28.4°C and the water was heated from 18.0°C to 28.4°C. We can therefore calculate the temperature change for each:

Water	Metal
28.4°C	98.0°C
−18.0	−28.4
10.4°C	69.6°C

b. Using the value for the specific heat of water, we then calculate the amount of heat necessary to raise the temperature of 34.3 g of water 10.4°C.

$$(34.3 \text{ g})(10.4°C) \frac{(1.00 \text{ cal})}{(1.00 \text{ g})(1.00°C)} = 357 \text{ cal}$$

c. We are assuming that all the heat absorbed by the water was lost by the metal.

heat lost by hotter body = heat gained by cooler body

Hence the 47.3 g of metal, in cooling from 98.0°C to 28.4°C, must have given off 357 calories. Therefore the specific heat of the metal (the number of calories necessary to change the temperature of 1.00 g of the metal 1.00°C) is

$$\frac{357 \text{ cal}}{(47.3 \text{ g})(69.6°C)} = \frac{0.108 \text{ cal}}{(1.00 \text{ g})(1.00°C)}$$

Questions

19. What is an example of a unit of heat intensity? A unit of quantity of heat?

20. Compare the magnitude of the following:
    a. Celsius degree, Fahrenheit degree, kelvin   b. Large Calorie and calorie

Questions and Problems

1.1 Concerns of Chemistry

21. Classify each of the following as a physical or chemical property:
    a. Silver conducts electricity.   b. Milk sours.   c. Eggs decay.
    d. Ice floats.   e. A metal corrodes.   f. Grape juice ferments.

g. Food is digested. h. Steam condenses. i. Gasoline evaporates.
j. Milk freezes. k. Gold is malleable.

1.2 Classification of Matter

22. Classify each of the following as an element, a compound, or a mixture:
    a. coke
    b. aluminum wire
    c. air
    d. table salt
    e. water
    f. granite
    g. earth
    h. wood
    i. fudge
    j. water from the Atlantic Ocean
    k. oxygen
    l. magnesium
    m. spinach
    n. a dog

1.3 Units

23. How many
    a. meters in a kilometer?
    b. liters in a milliliter?
    c. cubic centimeters in a milliliter?
    d. meters in 100 centimeters?
    e. grams in a microgram?

1.4 and 1.5 Exponential Numbers and Significant Figures

24. Perform the following arithmetic operations:
    a. $1.03 \times 10^4 + 6.70 \times 10^6$
    b. $7.83 \times 10^8 - 4.30 \times 10^6$
    c. $9.00 \times 10^{-11} + 13.0 \times 10^{-15}$
    d. $5.00 \times 10^{-13} - 6.70 \times 10^{-14}$
    e. $3.00 \times 10^{11} + 9.10 \times 10^{12} - 4.10 \times 10^{10}$
    f. $369.78 \times 10^{-7} + 9.73 \times 10^{-4}$
    g. $9.38 \times 10^{14} + 608 \times 10^{11} - 301 \times 10^9$

25. Perform the following arithmetic operations:
    a. $\sqrt{(1.00 \times 10^{16})(3.00 \times 10^{-11})}$
    b. $(7.20 \times 10^{-3})(17.0 \times 10^{-9})$
    c. $\sqrt{(309 \times 10^{12})(4.36 \times 10^{-8})}$
    d. $(7.30 \times 10^{-12})(4.90 \times 10^{-6})$
    e. $(7.30 \times 10^{-10} + 7.60 \times 10^{-13})(4.00 \times 10^8)$
    f. $\sqrt{\dfrac{9.00 \times 10^{-13}}{4.00 \times 10^7}}$
    g. $\sqrt{\dfrac{4.00 \times 10^{18}}{3.20 \times 10^{-9}}}$
    h. $\dfrac{3.80 \times 10^{-11}}{9.60 \times 10^{-27}}$
    i. $\dfrac{42.0 \times 10^{13}}{812 \times 10^{19}}$
    j. $\sqrt{\dfrac{3.90 \times 10^{-28}}{4.70 \times 10^6}}$

26. Perform the following arithmetic operations:
    a. $\sqrt{9.00 \times 10^{-12}}$
    b. $(7.00 \times 10^{13})^2$
    c. $\sqrt{11.0 \times 10^7}$
    d. $(17.0 \times 10^{-4})^2$
    e. $(3.00 \times 10^{12})^2$
    f. $\sqrt[3]{27.0 \times 10^{18}}$
    g. $\sqrt{144 \times 10^7}$
    h. $(7.80 \times 10^{-13})^3$
    i. $\sqrt[3]{613 \times 10^{146}}$
    j. $\sqrt[3]{4.63 \times 10^8}$

27. Perform the following arithmetic operations:
    a. $\dfrac{(7.00 \times 10^{12})(4.00 \times 10^{-8})}{(121 \times 10^{18})(7.00 \times 10^{-21})}$
    b. $\dfrac{(5.00 \times 10^{13} + 6.00 \times 10^{14})(3.00 \times 10^{-11})}{7.00 \times 10^{14}}$

c. $\sqrt{\dfrac{(4.00 \times 10^{19})(7.00 \times 10^{13})^2}{4.00 \times 10^{11}}}$

1.6 Unit Conversions

28. Add the following:
    a. 1093 g + 1.03 × 10⁴ mg + 1.93 kg   b. 3.09 m + 0.00813 km + 9451 mm

29. How many gallons of gasoline are contained in a 100-liter tank? (Assume the 100-liter measurement is known to three significant figures.)

30. If an automobile will go 13.4 miles to a gallon, how many kilometers will it travel per liter?

31. A football field is 3.00×10^2 ft by 1.60×10^2 ft. Calculate its area in square meters.

32. Our language contains many idiomatic expressions that involve units of measurement. In many cases the units are understood but not expressed. For each of the following, make an appropriate, parallel statement using SI units.
    a. He must have been doing 90!
    b. I wouldn't touch it with a 10-ft pole.
    c. The fullback was six-four and weighed at least 220.
    d. In what year did they first break 200 at Indianapolis?
    e. Her fever was over 105.
    f. He was built like a 10-ton truck.
    g. It was over 100 in the shade.

33. Determine the final value for each of the following:
    a. 3 ml of water is added to a flask containing 1.27×10^4 ml of water.
    b. 30.9 g of sand is added to a container holding 1.61 kg of sand.
    c. 0.91 ml of solution is spilled in transfer of 1.26 liters of solution from one flask to another.

34. If steak costs $3.60 per pound, what is its cost per kilogram?

1.7 Density and Specific Gravity

35. Determine the values missing from the following table by making the necessary conversions.

	Density	Mass	Volume
a		10.0 g	5.00 liters
b	2.31 g/cm³	15.9 g	
c		5.10×10^2 g	1212 ml
d	4.67 lb/ft³	3.00 tons	
e		6.00×10^3 kg	8.32×10^4 liters
f	339 g/liter		9.02×10^3 ml
g	1.23 lb/ft³		889 ml

36. A rectangular bar of metal weighs 39.1 lb. Its dimensions are 1.32 ft by 23.0 cm by 213 mm. What is the specific gravity of the metal?

37. Calculate the density of water in
  a. grams per liter
  b. kilograms per liter
  c. kilograms per cubic centimeter
  d. pounds per cubic foot
  e. pounds per gallon

38. A cubic box with an edge length of 10.0 cm is packed with 4 cubes of iron, each 5.00 cm on a side; 25 cubes of cork wood, each 2.00 cm on a side; 25 cubes of lead, also 2.00 cm on a side; and 100 cubes of gold, each 1.00 cm on a side. Calculate the average density of the contents of the box.

1.8 Heat

39. Determine the values missing from the following table by making the necessary conversions.

	°C	°F	kelvins
a		−10	
b	9.00×10^2		
c			48
d	−112		
e		104	

40. Room temperature is 68°F. What is this temperature in kelvins?

41. Which is hotter, 113 K or −312°F?

42. Which is colder, 311°F or 150°C?

43. A pizza is baked at 425°F. Convert this temperature to Celsius.

* **44.** At what temperature do the Celsius and Fahrenheit scales read the same?

* **45.** At what Celsius reading would the Fahrenheit reading be twice the Celsius reading? Five times the Celsius reading?

* **46.** Assume a temperature scale is developed on which the melting point of benzene (5.5°C) and the boiling point of benzene (80.1°C) are taken as 0.0° and 100.0°, respectively. Develop an equation relating a reading on this scale to a reading on the Celsius scale.

47. Calculate the quantity of heat (in calories) necessary to change the temperature of 1.00 qt of water from 20.0°C to its boiling point. (Use the value of 1.00 g/ml for the density of water.)

48. Determine the final values for each of the following:
  a. The temperature of a sample of iron at 4981.1 K is raised by 0.05 K.
  b. The temperature of an oven at 415°F drops by 0.2°C.

49. The British thermal unit (BTU) is defined as the quantity of heat necessary to raise the temperature of 1.00 lb of water 1°F. Calculate the number of calories in 1.00 BTU.

50. How much heat is necessary to raise the temperature of 1.00 cup of water from 20.0°C to 100.0°C? How much is needed to raise the temperature of the same volume of aluminum by the same amount? Facts you need: 1.00 cup = 236 ml; the density of water is 1.00 g/ml; the density of aluminum is 2.70 g/ml; the specific heat of aluminum is 0.212 cal/(g)(°C); the specific heat of water is 1.00 cal/(g)(°C).

* **51.** A student heated 34.0 g of copper shot to 98.5°C and then placed it into 43.0 ml of water at 23.0°C. Assuming that no heat was lost to the surroundings, what temperature should the mixture reach? The specific heat of copper is 0.0924 cal/(g)(°C).

Discussion Starters

52. Are systems of measurement purely arbitrary? If not, what are some of the factors that influence the choice of a standard of measurement?

53. Account for the fact that a liter is 1×10^{-3} m^3 but it is also 1×10^3 cm^3.

54. The great Greek mathematician Archimedes, seeking to determine the composition of the king's crown, purportedly placed it and equal weights of gold and silver in separate containers brimful of water. He then measured the overflow from each. What was he able to report to the king on the basis of his data?

Some Basic Concepts of Chemistry

In this chapter we shall continue to explore the basic concepts of chemistry. These include the concepts of the atom and the mole and the use of chemical formulas. The ideas discussed in Chapter 1 were for the most part tools necessary for the study of chemistry; here we begin that study in earnest by laying the basic foundation. In later chapters we shall study these fundamental concepts in more detail, building upon them at the same time.

2.1 The Atomic Concept

Suppose we were to divide a sheet of copper foil in half and then divide the halves into quarters, the quarters into eighths, the eighths into sixteenths, and so on. When would this process of division end? Physically, of course, it would end when the pieces of copper became so small that we could no longer handle them. But if we ignore the practical problem, would we reach a point beyond which further division would be impossible? Or might we reach a point where further division would result in a product other than copper?

These questions, of course, have been asked for centuries. The ancient Greek philosophers speculated a good deal about them. Democritus (460–370 B.C.), who promulgated the concept of the atomic nature of matter in the early Greek civilization, argued that there was a point beyond which a piece of copper, or any other substance, could not be divided. He referred to the indivisible particle as the *atom*, describing it as a small, incompressible, indivisible, homogeneous particle. There were as many kinds of atoms as there were kinds of matter. Thus, there were atoms of wood, stone, blood, flesh, water, air, and so on. Democritus attributed many of the properties of a material to the shape of its atoms. For example, he thought that the atoms of a liquid were smooth and spherical and therefore glided easily over one another, whereas atoms of solids were rough and angular. In some cases he pictured atoms with hooks, which were used to connect them to one another.

The ideas of Democritus and other atomists were not widely accepted. Instead, the ideas of two other noted Greek philosophers, Aristotle and Plato, were accepted for centuries by most learned men. Because of the pervasive influence of Aristotle and Plato, their arguments against the postulates of the atomists caused the idea of the atom to be sidetracked. In fact, their philosophies led to the development of the pseudoscience, *alchemy*, one of whose objectives was to turn common metals into gold.

The foundation for the modern atomic theory was finally laid by the French chemist, Antoine Lavoisier (1743–1794). Using precise balances, Lavoisier was able to demonstrate what other investigators had suspected: *The quantity of matter does not change during a chemical reaction.* This is the **law of conservation of mass** which, stated another way, says that the mass of the *products* (substances produced in a chemical reaction) is equal to the mass of the *reactants* (substances consumed in the reaction). This law is not always easy to establish, especially when gases are involved in the chemical change. Nonetheless, Lavoisier demonstrated this law by studying the reaction of mercury with a gas. He showed that mercury could be made to combine with a certain portion of the air, to form a red powder, which we now know as mercury(II) oxide. Lavoisier identified this portion of air as an element and gave it the name oxygen. When the red powder was heated, it decomposed to yield the same quantities of mercury and "air" that had originally combined to form it; that is, there was no loss or gain of weight in the formation and decomposition of mercuric oxide.

As another example of the law of conservation of mass, consider the flashbulb used in photography. It contains magnesium (or zirconium) wires sealed in oxygen gas. When the camera shutter is released, a small electric current ignites the magnesium wire, producing a brilliant white light. The law of conservation of mass can be demonstrated by weighing the bulb before and after the chemical change, as shown in Figure 2.1.

Building upon Lavoisier's law of conservation of mass, other chemists were able to show that *if two or more elements combine to form a certain compound, they always do so in the same mass ratio.* The magnesium oxide formed in the flashbulb always contains 60.3 percent magnesium and 39.7 percent oxygen by mass. Pure water, wherever we find it, is always 11.2 percent hydrogen and 88.8 percent oxygen. Notice that this principle, known generally as the **law of definite proportions,** does not demand that hydrogen and oxygen never combine in a different ratio to form *another* compound. Hydrogen peroxide, for example, is always 5.9 percent hydrogen and 94.1 percent oxygen by mass. In describing this principle, the word *law* is perhaps unfortunate—but has long been accepted. A natural law is nothing more than a *general (logical) conclusion, based on observation of natural phenomena, which always or very nearly always proves to be true.*

At the beginning of the nineteenth century John Dalton, an English schoolmaster and professor, proposed to explain the law of conservation of mass and the law of definite proportions by assuming that each element is composed of simple units of specific mass. For each of these simple units he adopted the word *atom*. The essentials of Dalton's atomic theory can be summarized as follows:

1. The smallest part of an element that takes part in a chemical reaction is an atom.

2. Atoms are permanent and unchanging; they cannot be divided, created, or destroyed.

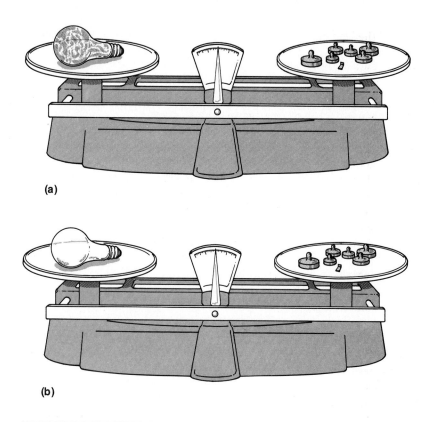

Figure 2.1 The flashbulb and contents weigh the same before **(a)** and after **(b)** the magnesium and oxygen combine to form magnesium oxide.

3. All atoms of a single element are identical in size, mass, and all other properties, whereas atoms of different elements are dissimilar.

4. Compounds are formed by the union of the atoms of elements into the simplest chemical unit having the appropriate composition. (Dalton called this unit a *compound atom* but it later came to be known as a molecule.)

Another useful aspect of Dalton's theory was that it could be used to explain the mass relationships involved when two elements combine to form more than one compound. Familiar to us all, for example, are the two compounds of carbon and oxygen, carbon monoxide and carbon dioxide. A third law, the **law of multiple proportions,** is based upon observation of compounds such as these. It states: *If two elements combine to form more than one compound, the masses of one element which combine with a fixed mass of the other element are in ratios which can be expressed by small, whole numbers.* In other words, if two elements, A and B, combine to form more than one compound, the mass of B that combines with a certain mass of A in the first compound will form some whole-number ratio with the mass of B that combines with the same mass of A in the second compound. This law is not as complicated as it appears, and to clarify it we offer two examples. In each case we have chosen to compare the amounts of element B that combine with 1 gram of element A. The selection of 1 gram is arbitrary but convenient.

Example 2.1 In carbon monoxide, 1.00 g of carbon combines with 1.33 g of oxygen. In carbon dioxide, 1.00 g of carbon combines with 2.67 g of oxygen. The ratio of the masses of oxygen that combine with 1.00 g of carbon in the two compounds is

$$\frac{2.67}{1.33} = \frac{2}{1}$$

Example 2.2 There are both a yellow lead oxide and a red lead oxide. In the yellow oxide, 1.00 g of lead combines with 0.0772 g of oxygen. In the red oxide, 1.00 g of lead combines with 0.116 g of oxygen. The ratio of the masses of oxygen that combine with 1.00 g of lead is therefore

$$\frac{0.116}{0.0772} = \frac{3}{2}$$

Using Dalton's theory we can explain the law of multiple proportions simply by saying that in carbon dioxide there are twice as many atoms of oxygen for each carbon atom as there are in carbon monoxide.

There were, of course, certain parts of John Dalton's theory that are no longer accepted. We have since found that the atom is not the ultimate particle of matter and that there can be small differences in mass among different atoms of the same element (these *isotopes* will be discussed in detail in Chapter 4). These discoveries, however, have not lessened the importance of Dalton's contribution. Of great significance was the emphasis he placed upon the relative masses of atoms. Since he was not able to measure the absolute weight of each atom, Dalton proposed assigning a *number to each element to represent the relative mass of its atom*. He called this number the *atomic weight*. The atomic weight of 1 was originally assigned to hydrogen, the lightest element. To see how this leads to a set of atomic weights, we shall look at several examples.

In the simpler of the two carbon-oxygen compounds, 1.00 g of carbon combines with 1.33 g of oxygen. If we assume that this represents a one-atom-to-one-atom ratio, it follows that the atom of oxygen has a mass four-thirds that of the carbon atom. Thus, if carbon is assigned an atomic weight of 12, oxygen must be assigned an atomic weight of 16. Of course this reasoning depends on the critical assumption that the atoms exist in the compound in a one-to-one ratio. In this compound, carbon monoxide, the assumption is correct. Similar reasoning, however, led the early investigators to assign many erroneous atomic weights. For example, accepting the atomic weight of hydrogen as 1 and assuming that water is composed of hydrogen and oxygen in a one-atom-to-one-atom ratio, early investigators assigned oxygen an atomic weight of 8 instead of the presently accepted value of 16. We now know (from evidence we shall study in Chapter 8) that there is not a one-atom-to-one-atom ratio in water but rather a ratio of two atoms of hydrogen to one atom of oxygen. In hydrogen peroxide, a compound not known in Dalton's time, where 1.00 g of hydrogen combines with 16.0 g of oxygen, the ratio is one to one.

Early chemists were uncertain about other atomic weights as well, not because they were unable to measure the combining weights of elements precisely but because they often had no way of knowing how many atoms of each element were combining to form each "compound atom" or molecule. They knew from the earlier example of water that they could not assume that the simplest weight ratio always represented a one-atom-to-one-atom ratio. For instance, consider the two lead oxides of Example 2.2. If the simpler of the two ratios—that found in the yellow oxide—is the result of a one-atom-to-one-atom ratio, we conclude that if 0.0772 g of oxygen combines with 1.00 g of lead, then 16.0 g of oxygen would combine with 207 g of lead:

$$\frac{1.00 \text{ g lead}}{0.0772 \text{ g oxygen}} = \frac{x \text{ g lead}}{16.0 \text{ g oxygen}}$$

$$x = \frac{16.0}{0.0772} = 207 \text{ g}$$

If we accept 16.0 as the atomic weight of oxygen, this would establish the atomic weight of lead as 207. On the other hand, if we assume that in this compound two atoms of lead combine with each atom of oxygen, the same 207 g of lead would represent two relative weights of lead, and we would calculate the atomic weight as 104. If we assume a one-to-one ratio for the red oxide, where 0.116 g of oxygen combines with 1.00 g of lead, we calculate an atomic weight for lead of 138:

$$\frac{1.00 \text{ g lead}}{0.116 \text{ g oxygen}} = \frac{x \text{ g lead}}{16.0 \text{ g oxygen}}$$

$$x = \frac{16.0}{0.116} = 138 \text{ g}$$

This riddle was neatly solved by the **law of Dulong and Petit,** which is yet another product of experimental observations. It says that when the atomic weight of an element in the solid state is multiplied by its specific heat, the product is *approximately* 6. Let us see how this works with the data on the lead oxides.

The specific heat of lead is 0.029 cal/(1.00 g)(1.00°C). If each of the suggested values for the atomic weight is multiplied by 0.029,

$$(104)(0.029) = 3.0$$

$$(138)(0.029) = 4.0$$

$$(207)(0.029) = 6.0$$

We quickly see that the value of 207 is the correct one and that there is a one-atom-to-one-atom ratio in the yellow oxide. This approximation helped remove the uncertainty about the atomic weights of several elements.

Hydrogen did not serve long as the reference standard for atomic weights. Oxygen (with an atomic weight of 16) was found to be more convenient, and the set of atomic weights based upon oxygen was used—and refined—by chemists for more than 100 years. In 1961 the standard changed again: The most abundant of the several kinds of naturally occurring carbon atoms now serves as the reference standard for atomic weights. Recall that Dalton was wrong when he postulated that all atoms of a given element are identical. In fact, atoms of a given element may differ slightly in mass. These different kinds of atoms of a given element are called *isotopes* (Chapter 4), and it is the most abundant isotope of carbon that has now been assigned an atomic weight of 12, *exactly*. Thus all atomic weights are relative weights compared to the standard carbon atom. When magnesium is assigned an atomic weight of 24.3, we know that the average magnesium atom (accounting for the natural abundance of the several isotopes) is 24.3/12.0 or 2.02 times as heavy as the standard carbon atom. Similarly, when we say that the atomic weight of titanium is 47.9, we mean that the average titanium atom (again, weighted in terms of natural abundance) is 47.9/12.0 or 3.99 times as heavy as the standard carbon atom.

The modern method of atomic weight determination does not involve combining measured portions of elements to form compounds but instead depends on the *mass spectrometer*. This instrument will be described in Chapter 4.

Questions

1. Compare and contrast the atom as described by Democritus with the atom as described by Dalton.

2. What are the differences between a natural law and one passed by the state legislature?

3. What aspects of Dalton's atomic theory are no longer accepted?

4. Dalton believed that the ammonia molecule consisted of one atom of nitrogen and one of hydrogen. If this belief had been correct, what would be the atomic weight of nitrogen, assuming we assign hydrogen an atomic weight of 1? What would have been the atomic weight of hydrogen if instead nitrogen had been assigned an atomic weight of 14?

2.2 Names and Symbols of the Elements

The names of the elements have been given to them through the years for various reasons. Some were assigned in honor of a geographical area. For example, polonium was named to honor Poland, the native country of its discoverer, Marie Curie; other elements named for countries are francium and germanium. Elements named for towns and cities include berkelium (Berkeley, California), hafnium (Latin for Copenhagen), and strontium (Strontian, Scotland). Americium and europium were named for continents, whereas mercury, neptunium, and plutonium were named for planets. Other elements were named for scientists; among them are curium (Marie and Pierre Curie), einsteinium (Albert Einstein), and mendelevium (Dmitri Mendeleev). Still others were given names that describe prominent properties: chlorine (from the Greek meaning green), iodine (from the French meaning violet), and oxygen (from the French meaning "former of acids," since it was once believed that all acids contain oxygen).

Early in the nineteenth century, chemists adopted the practice of assigning symbols to elements. John Dalton introduced the first systematic symbols, consisting of marks or designs inside circles. For example, he represented hydrogen by a circle with a dot in it and copper by a circle around the letter C. Later, when Jons Berzelius introduced the present system, these clumsy and confusing symbols were abandoned. Each element is now represented either by the first letter of its name or by the first letter plus one other letter of its name. Some elements, however, have symbols taken from Latin names. They are elements which have been known for at least several hundred years—since the days, in fact, when Latin was the language of scientists. These elements are listed below, along with the Latin names and symbols.

English	Latin	Symbol
Antimony	Stibium	Sb
Copper	Cuprum	Cu
Gold	Aurum	Au
Iron	Ferrum	Fe
Lead	Plumbum	Pb
Mercury	Hydrargyrum	Hg
Potassium	Kalium	K
Silver	Argentum	Ag
Sodium	Natrium	Na
Tin	Stannum	Sn
Tungsten	Wolfram (German)	W

Note that only the first letter in a symbol is capitalized. This convention is necessary to distinguish, for example, between the metal cobalt (Co) and the poisonous gas carbon monoxide (CO).

Questions

5. Consult an appropriate reference book for origins of the names of the following elements.
   a. cerium   b. dysprosium   c. helium   d. nickel   e. rhenium

6. Why do the symbols of some elements appear to be unrelated to their names?

2.3 Atoms and Molecules

For reasons discussed in Chapter 6, we rarely encounter isolated atoms in nature. All the noble gases (He, Ne, Ar, Kr, Xe, Rn) exist in the monatomic state. Most elements and, by definition, all compounds consist of structural units other than atoms. If a substance is a gas, other than a noble gas, or a liquid, the structural unit will usually be a molecule in which atoms are bound tightly enough together that they exist as a single particle. These molecules are the smallest particles of the substance (element or compound) that can exist alone and exhibit the composition and properties of the substance. To illuminate this statement we might ask the same question about a compound such as sugar as we asked previously about the element copper. To what extent might we divide a sugar cube and still have sugar? The answer is: until we reduce it to a single molecule. Further division would yield substances that have properties considerably different from those of sugar. Of course it is difficult to talk about some properties in terms of a single molecule. For example, it makes no sense to discuss the boiling point of a single molecule. One molecule exhibits essentially the same chemical properties, however, as a kilogram of molecules.

Identifying molecules of a compound is rarely difficult, but distinguishing between the atoms and molecules of an element can sometimes be troublesome. In the case of the noble gases, each molecule is a single atom: Atoms and molecules are identical; hence we call the noble gases *monoatomic* gases. A number of other elements exist as diatomic molecules. These include hydrogen (H_2), oxygen (O_2), fluorine (F_2), chlorine (Cl_2), bromine (Br_2), iodine (I_2), and nitrogen (N_2). When a chemist talks about an atom of oxygen, he is referring to the smallest particle of oxygen that can combine to form a compound: the particle that combines with hydrogen to form water or with carbon to form carbon dioxide or with sulfur to form sulfur dioxide. When a chemist speaks of molecular oxygen, on the other hand, he is referring to diatomic oxygen, the form in which it occurs in the atmosphere. Often the molecules of elements contain more than two atoms. For example, the phosphorus molecule is tetratomic (P_4) and the sulfur molecule is octatomic (S_8).

Question

7. Distinguish between an atom of an element and a molecule of an element.

2.4 Formulas

Since the composition of a compound is exact and unvarying, it should be possible to express its composition in terms of a *formula*. A **formula** consists of the symbol for each element in the compound, together with subscripts to indicate the relative numbers of atoms of each of these elements. The formula for water is H_2O. This means that water consists of hydrogen and oxygen and that the ratio of hydrogen atoms to oxygen atoms is two to one. Note that the subscript 1 is not written. The formula for table salt, NaCl, tells us that salt contains sodium and chlorine in a one-to-one ratio. Some more complex formulas contain parentheses. For example, the formula for ammonium sulfate, a common lawn fertilizer, is $(NH_4)_2SO_4$. Here the subscript following the parentheses refers to everything within the parentheses. In this case the ratio of atoms is two nitrogen atoms, eight hydrogen atoms, and four oxygen atoms for each sulfur atom. There are good reasons (which we shall discuss in Chapter 6) for writing the formula $(NH_4)_2SO_4$ rather than $N_2H_8SO_4$. Both formulas indicate the ratio of atoms, but the former also tells us something about the structure or arrangement of atoms.

As we saw earlier, atoms are arranged on a scale of atomic weights according to their masses. This scale is based on one atom—currently the most abundant isotope of carbon—chosen as the standard, with the other atoms falling in place according to their relative masses. Also, every compound can be characterized by its *formula weight*, which is obtained simply by adding the atomic weights of the constituent elements, counting each atomic weight as many times as the symbol for the element appears in the formula. Thus formula weights are based on the same standard used for atomic weights.

Example 2.3 One of the additives used in gasoline to increase the efficiency of the internal combustion engine is tetraethyl lead. Its formula is $Pb(C_2H_5)_4$. The lead compounds emitted as combustion products of this compound are serious air pollutants. Calculate the formula weight of $Pb(C_2H_5)_4$.

Solution

a. Atomic weights are

C	12.011
H	1.0079
Pb	207.2

b. The subscript 4 in this formula refers to everything enclosed within the parentheses. Thus each molecule of tetraethyl lead contains 8 carbon atoms and 20 hydrogen atoms. (Note that 8 and 20 are exact numbers.)

$$\begin{aligned} 1 \times 207.2 &= 207.2 \\ 8 \times 12.011 &= 96.088 \\ 20 \times 1.0079 &= \underline{20.158} \\ &\ 323.446 \end{aligned}$$

c. Since only one place beyond the decimal is common to all numbers being added, we must round off the result to **323.4**. Figures beyond the 4 are not significant.

Questions

8. Identify the elements present in each of the following compounds.
    a. C_6H_{14}
    b. $Pb(NO_3)_2$
    c. $Al_2(SO_4)_3$
    d. $K_3Fe(CN)_6$
    e. $(NH_4)_3PO_4$

9. For each of the formulas listed in Question 8, determine the relative numbers of atoms of each element.

10. Calculate the formula weight for each formula listed in Question 8.

2.5 The Mole Concept

As we have stressed, atomic weights and formula weights are merely relative measures of atomic and molecular masses. Scientists developed these concepts long before they had the means to measure absolute numbers or masses of atoms. Chemists also needed a relative unit of mass *proportional* to the number of atoms in a sample. This would allow them to combine samples having a known *ratio* of atoms, even though they were unable to measure the absolute number of atoms. Thus we have the concept of the *gram atomic weight*, which is the mass of an element in grams equal to its atomic weight. A gram atomic weight of any element contains the same number of atoms as one gram atomic weight of another. For example, consider the elements hydrogen and oxygen. The atomic weight of hydrogen is 1.01, whereas that of oxygen is 16.0. Since atomic weights are proportional to absolute atomic masses, each oxygen atom must weigh approximately 16 times as much as each hydrogen atom. Therefore, in any two samples of oxygen atoms and hydrogen atoms containing the same *number of atoms*, the sample of oxygen will always weigh 16 times as much as the sample of hydrogen. Thus one gram atomic weight of oxygen (16.0 g) contains the same number of atoms as one gram atomic weight of hydrogen (1.01 g).

We now know that 1.00 gram atomic weight of any element contains 6.02×10^{23} atoms. This number is known as **Avogadro's number,** so called in honor of Amadeo Avogadro (1776–1856), an Italian physicist, and it is represented by the letter N. It has been adopted into the SI system as a basic unit, and the word *mole* (abbreviated as *mol*) has been coined to refer to 6.02×10^{23} entities of any kind. Thus the word *mole* can be thought of as a scientific analog to the word dozen, inasmuch as they both express the number of units in a sample. We buy eggs by the dozen and count atoms and molecules by the mole. To give you an idea of the different sizes of the two numbers, however, consider that if each person consumed 3 eggs a day, it would take the entire population of the world 143 billion years to eat 1 mole of eggs! Notice also that since a mole is merely a number, it is important always to make clear what you are counting. There is an important difference between a mole of hydrogen atoms and a mole of hydrogen molecules.

We can easily extend the concept of gram atomic weight to *gram molecular weight* and *gram formula weight*. Just as one gram atomic weight is the mass in grams of a mole of atoms, one gram molecular weight is the mass in grams of a mole of molecules, and one gram formula weight is the mass in grams of a mole of formula units, where *formula unit* is a general term used when we cannot be sure that the formula accurately represents the molecule. That is, the formula may be an *empirical formula* and simply give the ratio of atoms without indicating the actual number of each atom in each molecule. The gram formula weight is the mass in grams of 6.02×10^{23} of whatever appears in the formula.

To summarize, a gram atomic weight is the weight (mass) in grams of 6.02×10^{23} atoms; a gram atomic weight is the weight in grams of 1.00 mol of atoms; a gram atomic weight is also the atomic weight in grams. The relationships among these terms for hydrogen (atomic weight 1.008) are

$$\begin{aligned}\text{one gram atomic weight of hydrogen} &= \text{mass in grams of } 6.02 \times 10^{23} \text{ hydrogen atoms} \\ &= \text{mass in grams of one mole of H} = 1.008 \text{ g}\end{aligned}$$

A gram molecular weight is the weight (mass) in grams of 6.02×10^{23} molecules; a gram molecular weight is the weight in grams of 1.00 mol of molecules; a gram molecular weight is also the molecular weight in grams. The relationships among these terms for diatomic hydrogen gas (molecular weight 2.016) are

$$\begin{aligned}\text{one gram molecular weight of hydrogen} &= \text{mass in grams of } 6.02 \times 10^{23} \text{ hydrogen molecules} \\ &= \text{mass in grams of one mole of } H_2 = 2.016 \text{ g}\end{aligned}$$

A gram formula weight is the weight (mass) in grams of 6.02×10^{23} formula units; a gram formula weight is the weight in grams of 1.00 mol of formula units; a gram formula weight is also the formula weight in grams. The relationships among these terms for NaCl (formula weight 58.4) are

$$\begin{aligned}\text{one gram formula weight of NaCl} &= \text{mass in grams of } 6.02 \times 10^{23} \text{ NaCl formula units} \\ &= \text{mass in grams of one mole of NaCl} = 58.4 \text{ g}\end{aligned}$$

Although all of these terms are helpful in certain circumstances, the word *mole* is the most consistently useful in discussions of atoms, molecules, or formula units.

Example 2.4 Calculate the mass in grams of a single atom of chlorine.

Solution Using the appropriate conversion factors,

$$\left(\frac{35.45 \text{ g Cl}}{1.00 \text{ mol Cl}}\right)\left(\frac{1.00 \text{ mol Cl}}{6.02 \times 10^{23} \text{ Cl atoms}}\right) = 5.89 \times 10^{-23} \text{ g/Cl atom}$$

Some Basic Concepts of Chemistry

Example 2.5 How many moles and how many atoms are contained in 57.5 g of sodium? What is the mass of a single atom of sodium?

Solution

a. One mole of sodium weighs 23.0 g. The conversion factor

$$\frac{1.00 \text{ mol Na}}{23.0 \text{ g Na}}$$

allows us to calculate the number of moles in 57.5 g of sodium:

$$(57.5 \text{ g Na}) \left(\frac{1.00 \text{ mol Na}}{23.0 \text{ g Na}} \right) = 2.50 \text{ mol Na}$$

b. Since every mole contains 6.02×10^{23} atoms, we use the conversion factor

$$\frac{6.02 \times 10^{23} \text{ Na atoms}}{1.00 \text{ mol Na}}$$

to calculate the number of atoms in 2.50 mol of Na:

$$(2.50 \text{ mol Na}) \left(\frac{6.02 \times 10^{23} \text{ Na atoms}}{1.00 \text{ mol Na}} \right) = 1.50 \times 10^{24} \text{ Na atoms}$$

c. One mole of sodium weighs 23.0 g and contains 6.02×10^{23} atoms. We can find the mass of a single atom as we did in Example 2.4:

$$\left(\frac{23.0 \text{ g Na}}{1.00 \text{ mol Na}} \right) \left(\frac{1.00 \text{ mol Na}}{6.02 \times 10^{23} \text{ Na atoms}} \right) = 3.82 \times 10^{-23} \text{ g/Na atom}$$

Questions

11. A mole is a unit of what?

12. Which of the following is (are) mole(s)?
    a. 6.02×10^{23} atoms
    b. 6.02×10^{23} molecules
    c. 6.02×10^{23} students
    d. 6.02×10^{23} formula units
    e. 6.02×10^{23} books

13. Which weighs more:
    a. a mole of oxygen atoms or a mole of oxygen molecules?
    b. a gram atomic weight of hydrogen or a gram molecular weight of hydrogen?
    c. a gram atomic weight of chlorine or a mole of chlorine molecules?

2.6 Percent Composition from the Formula

One of the useful pieces of information that we can derive from a formula is the percent composition, that is, the contribution of each element to the total mass of the substance. We can calculate the percent composition using only the formula and a table of atomic weights. In general the percent by mass of one element in a compound is obtained by dividing the number of grams of that element in a sample by the total mass of the sample and then multiplying by 100. If the sample is one mole, for example, the percent by mass of an element is

$$\frac{\text{mass of element in one mole of compound}}{\text{mass of one mole of compound}} \times 100$$

Use the table inside the back of this book cover for the following problems.

Example 2.6 Calculate the mass of a mole of water in three significant figures. Also calculate the percent composition of water.

Solution

a. We first find the formula weight by adding the atomic weights of the constituent elements, counting each the number of times indicated by the subscript. The formula weight of H_2O is

$$2(1.01) + 1(16.0) = 18.0$$

The mass of a mole of a compound is its formula weight in grams. Hence a mole of water weighs **18.0 g.**

b. $\% \text{ H} = \left(\dfrac{2.02 \text{ g H}}{18.0 \text{ g H}_2\text{O}}\right) \times 100 = \mathbf{11.2\%}$

$\% \text{ O} = \left(\dfrac{16.0 \text{ g O}}{18.0 \text{ g H}_2\text{O}}\right) \times 100 = \mathbf{88.9\%}$

c. Using the appropriate conversion factors would produce the same result:

$$\left(\frac{2 \text{ mol H}}{1 \text{ mol H}_2\text{O}}\right)\left(\frac{1.01 \text{ g H}}{1.00 \text{ mol H}}\right)\left(\frac{1.00 \text{ mol H}_2\text{O}}{18.0 \text{ g H}_2\text{O}}\right) \times 100 = 11.2\% \text{ hydrogen}$$

$$\left(\frac{1 \text{ mol O}}{1 \text{ mol H}_2\text{O}}\right)\left(\frac{16.0 \text{ g O}}{1.00 \text{ mol O}}\right)\left(\frac{1.00 \text{ mol H}_2\text{O}}{18.0 \text{ g H}_2\text{O}}\right) \times 100 = 88.9\% \text{ oxygen}$$

Example 2.7 Calculate the percent composition of glucose (blood sugar) from its formula, $C_6H_{12}O_6$.

Solution

a. Calculate the formula weight:

$$12(1.01) + 6(16.0) + 6(12.0) = 180.1$$

Hence 1.00 mol of glucose weighs 180.1 g.

b. $\% \text{ H} = \left(\dfrac{12.1 \text{ g H}}{180.1 \text{ g C}_6\text{H}_{12}\text{O}_6}\right) \times 100 = 6.72\%$

$\% \text{ O} = \left(\dfrac{96.0 \text{ g O}}{180.1 \text{ g C}_6\text{H}_{12}\text{O}_6}\right) \times 100 = 53.3\%$

$\% \text{ C} = \left(\dfrac{72.0 \text{ g C}}{180.1 \text{ g C}_6\text{H}_{12}\text{O}_6}\right) \times 100 = 40.0\%$

Questions

14. How many moles of hydrogen atoms are there in 1.00 mol of each of the following compounds?
   a. $C_{12}H_{22}O_{11}$
   b. $Al(OH)_3$
   c. $Ca(H_2PO_4)_2$

15. Calculate the percent hydrogen in each of the compounds above.

2.7 Establishing the Formula of a Compound from Experimental Data

In Section 2.6 we calculated percent composition from formulas. Of even greater practical importance, however, is determining the identity of a compound (that is, its empirical formula) from information about its percent composition. This data is the result of work called *chemical analysis*, which usually comprises three phases. First the analyst must obtain a pure sample. Second he performs a *qualitative analysis* of the substance, determining which elements are present. Finally he must do a *quantitative analysis*, which reveals the percent by mass of the constituent elements. We shall not consider the details of chemical analysis here, but we shall look at examples of the calculations that permit us to deduce formulas from experimental data.

Example 2.8 Analysis of a reddish brown gas found in smog revealed that it had the following composition:

$$\% \text{ N} = 30.4\% \qquad \% \text{ O} = 69.6\%$$

Determine the simplest formula of this compound.

Solution

a. In a sense, the percent composition of a compound is a formula in terms of grams, that is, every 100 g of this compound contains 30.4 g of nitrogen and 69.6 g of oxygen. Since the formulas used by chemists give the relative numbers of moles of elements rather than the relative numbers of grams, however, we need to convert grams to moles.

$$(30.4 \text{ g N}) \left(\frac{1.00 \text{ mol N}}{14.0 \text{ g N}} \right) = 2.17 \text{ mol N}$$

$$(69.6 \text{ g O}) \left(\frac{1.00 \text{ mol O}}{16.0 \text{ g O}} \right) = 4.35 \text{ mol O}$$

b. The relative numbers of moles of each element are also the relative numbers of atoms, since one mole of any element contains 6.02×10^{23} atoms. Thus the numbers 2.17 and 4.35 tell us the atomic ratio of nitrogen and oxygen for this compound. We now need only to convert this ratio to a ratio of small whole numbers:

$$\frac{4.35 \text{ atoms O}}{2.17 \text{ atoms N}} = \frac{2 \text{ atoms O}}{1 \text{ atom N}}$$

c. Therefore the simplest formula of this compound is NO_2.

In Example 2.8 the atomic ratio is not complicated and is easily determined. Arriving at the atomic ratio for more complex compounds, however, often involves more arithmetic. Consider example 2.9.

Example 2.9 Boric acid, a white crystalline compound, was formerly used as a mild antiseptic. Its percent composition is indicated below. What is its formula?

% H = 4.89%   % B = 17.5%   % O = 77.7%

Solution

a. In 1.00×10^2 g (an arbitary, but convenient quantity—any size sample could be used) of boric acid there are 4.89 g of hydrogen, 17.5 g of boron, and 77.7 g of oxygen. Each of these masses divided by the mass of one mole of the element yields the relative number of moles of each element.

$$(4.89 \text{ g H}) \left(\frac{1.00 \text{ mol H}}{1.01 \text{ g H}} \right) = 4.84 \text{ mol H}$$

$$(17.5 \text{ g B}) \left(\frac{1.00 \text{ mol B}}{10.8 \text{ g B}} \right) = 1.62 \text{ mol B}$$

$$(77.7 \text{ g O}) \left(\frac{1.00 \text{ mol O}}{16.0 \text{ g O}} \right) = 4.86 \text{ mol O}$$

b. To express the mole relationship in small whole numbers, we must divide each of the two larger numbers of moles (those for hydrogen and oxygen) by the smallest number of moles (that for boron).

$$\frac{4.84 \text{ mol H}}{1.62 \text{ mol B}} = \frac{2.99 \text{ mol H}}{1.00 \text{ mol B}}$$

$$\frac{4.86 \text{ mol O}}{1.62 \text{ mol B}} = \frac{3.00 \text{ mol O}}{1.00 \text{ mol B}}$$

c. Taking into account the slight deviations from whole numbers that resulted from rounding off to three significant figures, we conclude that the formula for boric acid is **H_3BO_3**.

Example 2.10 Cortisone is a hormone that affects carbohydrate and protein metabolism. It is now available synthetically and is used in the treatment of acute arthritis. If its percent composition is that indicated below, what is its formula?

$$\% \text{ C} = 70.0\% \qquad \% \text{ H} = 7.86\% \qquad \% \text{ O} = 22.2\%$$

Solution

a. Again the percent composition indicates the relative numbers of grams of each element in the compound. To determine the relative numbers of moles, we need only the atomic weights of the elements.

$$(70.0 \text{ g C}) \left(\frac{1.00 \text{ mol C}}{12.0 \text{ g C}} \right) = 5.83 \text{ mol C}$$

$$(7.86 \text{ g H}) \left(\frac{1.00 \text{ mol H}}{1.01 \text{ g H}} \right) = 7.78 \text{ mol H}$$

$$(22.2 \text{ g O}) \left(\frac{1.00 \text{ mol O}}{16.0 \text{ g O}} \right) = 1.39 \text{ mol O}$$

b. Again we divide each of the two larger numbers of moles (those for carbon and hydrogen) by the smallest number of moles (that for oxygen).

$$\frac{5.83 \text{ mol C}}{1.39 \text{ mol O}} = \frac{4.19 \text{ mol C}}{1.00 \text{ mol O}}$$

$$\frac{7.78 \text{ mol H}}{1.39 \text{ mol O}} = \frac{5.60 \text{ mol H}}{1.00 \text{ mol O}}$$

Here we do not yet have small whole numbers. If we multiply each by five, however, we obtain nearly whole numbers. (Again, the slight deviation from a whole number in the case of carbon is a result of rounding to three significant figures.)

$$\left(\frac{4.19 \text{ mol C}}{1.00 \text{ mol O}}\right)\left(\frac{5}{5}\right) = \frac{20.95 \text{ mol C}}{5.00 \text{ mol O}}$$

$$\left(\frac{5.60 \text{ mol H}}{1.00 \text{ mol O}}\right)\left(\frac{5}{5}\right) = \frac{28.0 \text{ mol H}}{5.00 \text{ mol O}}$$

If in problems of this sort it is not obvious which integer factor you need to obtain whole numbers, simply start with 2, then if necessary try 3, and so forth. Usually it is not necessary to go beyond 5.

c. The formula of cortisone is $C_{21}H_{28}O_5$.

The formulas calculated on the basis of chemical analyses alone are *empirical formulas*. As we saw in Section 2.5, empirical formulas indicate merely the relative numbers of atoms present in a compound, but they are often sufficient to identify a compound. However, there can often be several compounds with the same empirical formula. In such a case we would need to know the actual number of atoms of each element in one molecule to identify the compound. For example, there are two compounds with the empirical formula calculated in Example 2.8. One has one nitrogen atom and two oxygen atoms in each molecule; the other has two nitrogen atoms and four oxygen atoms in each molecule.

Actually the example has specified the compound, since only the compound with one nitrogen atom and two oxygen atoms is reddish brown. The other compound is colorless. Without such information, however, the identification of the compound often depends on determining the *molecular formula*, which expresses the actual number of atoms of each element in each molecule and need not be the same as the empirical formula. The molecular formulas for the two nitrogen-oxygen compounds we have been discussing, both with empirical formula NO_2, are NO_2 (reddish brown) and N_2O_4 (colorless).

To determine the molecular formula from the empirical formula, we must know the molecular weight of the compound. The molecular weight is always a small integral multiple of the empirical formula weight. In the examples above the molecular weights are 46.0 for NO_2 and 92.0 for N_2O_4, whereas the empirical formula weight for each is 46.

We shall defer until a later chapter discussion of experimental methods of molecular weight determination, but the following example will illustrate how to determine the molecular formula if the molecular weight is already known.

Example 2.11 One of the compounds present in the rocket fuel for the Apollo-11 flight had a molecular weight of 60.1 and had the following composition by mass: 46.6% N, 13.4% H, and 39.9% C. What is the molecular formula of this compound?

Solution

a. First we convert the percent composition to numbers expressing the relative numbers of moles of each element:

$$(46.6 \text{ g N}) \left(\frac{1.00 \text{ mol N}}{14.0 \text{ g N}} \right) = 3.33 \text{ mol N}$$

$$(13.4 \text{ g H}) \left(\frac{1.00 \text{ mol H}}{1.01 \text{ g H}} \right) = 13.3 \text{ mol H}$$

$$(39.9 \text{ g C}) \left(\frac{1.00 \text{ mol C}}{12.0 \text{ g C}} \right) = 3.32 \text{ mol C}$$

b. We then reduce these numbers to small whole numbers, hence obtaining the empirical formula.

$$\frac{3.33 \text{ mol N}}{3.32 \text{ mol C}} = \frac{1.00 \text{ mol N}}{1.00 \text{ mol C}}$$

$$\frac{13.3 \text{ mol H}}{3.32 \text{ mol C}} = \frac{4.01 \text{ mol H}}{1.00 \text{ mol C}}$$

The empirical formula of the compound is H_4CN.

c. The empirical formula weight is

$$4(1.01) + 12.0 + 14.0 = 30.0$$

Since the molecular weight is 60.1, each molecule must contain twice the number of atoms indicated in the empirical formula. Therefore the molecular formula is $H_8C_2N_2$. This compound is called unsymmetrical dimethylhydrazine.

Questions

16. What are the three phases of chemical analysis?

17. Distinguish between
   a. molecular formula and empirical formula.
   b. qualitative analysis and quantitative analysis.

18. What information is needed to determine the molecular formula that is not needed to determine the empirical formula?

Questions and Problems

2.1   **The Atomic Concept**

19. The analysis of 10.0-g samples of three different oxides of nitrogen produced the following results:

Sample A contained 3.05 g of nitrogen.
Sample B contained 3.70 g of nitrogen.
Sample C contained 2.61 g of nitrogen.

Determine from these values the three ratios that illustrate the law of multiple proportions.

20. The weight analysis of three chromium oxides yielded the data given below. The specific heat of chromium is 0.107 cal/(g)(°C). Assume that the atomic weight of oxygen has been established as 16.0. Use this information together with the law of Dulong and Petit to calculate the atomic weight of chromium.

Oxide A: 61.9% chromium
Oxide B: 76.5% chromium
Oxide C: 68.4% chromium

2.4 Formulas

21. Calculate the formula weight of each of the following compounds:
    a. table salt   NaCl
    b. carbon tetrachloride CCl_4
    c. marble $CaCO_3$
    d. lime   CaO
    e. slaked lime $Ca(OH)_2$
    f. sucrose $C_{12}H_{22}O_{11}$
    g. ethyl alcohol C_2H_5OH

22. Calculate the percent composition of each element in the compounds in Question 21.

23. Calculate the number of grams of hydrogen that combine with 1.00 g of carbon in each of the following:
    a. benzene C_6H_6
    b. toluene C_7H_8
    c. xylene C_8H_{10}
    d. naphthalene $C_{10}H_8$

2.5 The Mole Concept

24. Calculate the number of atoms in each of the following samples:
    a. 74.3 g of fluorine
    b. 3.45 gram atomic weights of carbon
    c. 1.24×10^{-2} gram atomic weights of sodium

25. Calculate the number of gram atomic weights (moles) in each of the following samples:
    a. 0.768 g of barium
    b. 6.35×10^{-8} g of K
    c. 2.41×10^{20} atoms of Mg

26. Calculate the number of grams in each sample.
    a. 3.45 mol of oxygen gas
    b. 1.94 gram atomic weights of copper
    c. 0.0340 gram formula weight of K_3PO_4

27. Calculate the number of moles in the following:
   a. 1.02 g of KCl
   b. 0.348 g of K_3PO_4
   c. 0.0197 g of $(NH_4)_3PO_4$

28. Calculate the number of gram atomic weights in the following:
   a. 12.0×10^{23} atoms of Na
   b. 73.8 g of K
   c. 6.05 mol of O_2

29. Calculate the number of atoms in the following:
   a. 0.0160 mol of water
   b. 48.0 g of water
   c. 3.01×10^{22} molecules of water

30. Which of the following has the largest mass?
   a. 14.2 g of chromium
   b. 1.70×10^{23} atoms of Cr
   c. 0.236 gram atomic weight of Cr

31. If one molecule of a compound weighs 2.33×10^{-22} g, what is the molecular weight of the compound?

32. Calculate the number of atoms of copper in 6.00 in. of a copper wire with a diameter of 0.00100 in. The density of copper is 8.96 g/cm^3.

2.6 Percent Composition from the Formula

33. Calculate the percent composition of carbon, hydrogen, and oxygen in vitamin C (ascorbic acid) from its molecular formula $C_6H_8O_6$.

34. What weight in grams of CO_2 will be produced if wood containing 1.34 lb of carbon is burned? (Assume that all the carbon is converted to CO_2.)

35. In which of these compounds is the percentage of carbon the highest? In which is it the lowest?

$$CH_4 \quad CH_3Cl \quad CH_2Cl_2 \quad CHCl_3 \quad CCl_4$$

36. How many grams of aluminum are in 1.00 lb of an alum with the formula $NaAl(SO_4)_2 \cdot 12\,H_2O$?

37. There are three common oxides of iron. The weight of oxygen that combines in each with 1.00 g of iron is given below. Calculate the relative numbers of atoms of Fe and O in each compound, and then write the formulas.
   a. 1.00 g of Fe and 0.381 g of O
   b. 1.00 g of Fe and 0.278 g of O
   c. 1.00 g of Fe and 0.429 g of O

2.7 Establishing the Formula of a Compound from Experimental Data

38. Calculate the empirical formulas for the compounds with the following percent compositions:
   a. 80.0% C, 20.0% H
   b. 26.6% K, 35.4% Cr, 38.0% O

c. 12.7% Al, 19.7% N, 67.6% O
d. 21.8% Mg, 27.9% P, 50.3% O
e. 72.7% Sb, 28.3% S
f. 79.9% Cu, 20.1% S
g. 23.1% Al, 15.4% C, 61.5% O
h. 28.2% N, 8.0% H, 20.8% P, 43.0% O
i. 32.4% Na, 21.9% P, 45.1% O, 0.7% H

39. Determine the molecular formulas for the compounds with the following empirical formulas and molecular weights:

Empirical Formula	Molecular Weight
HO	34
CH_2O	180
CH_3	30
CH_2	70
HF	40
$NaSO_4$	238
BH_3	27.6
NaS_2	174

40. The percent composition of morphine is 71.6% C, 6.7% H, 4.9% N, and 16.8% O. Calculate the empirical formula of morphine.

41. The percent composition of nicotine is 74.1% C, 8.8% H, and 17.3% N. Determine the empirical formula of nicotine.

42. Benzopyrene is a carcinogen (an agent that causes cancerous growths in living tissue). Its percent composition is 95.2% C and 4.8% H. If its molecular weight is 252, what is its molecular formula?

43. The yellowish material found in carrots and tomatoes is β-carotene. Its percent composition is 89.5% C and 10.5% H, and it has a molecular weight of 536. Determine its molecular formula.

Equations and Stoichiometry

Perhaps the most fundamental principle of chemistry is the *law of conservation of mass*, which we first encountered in Chapter 2. This natural law allows us to study chemical changes, confident that there can be no loss or gain of mass during the reactions. All the atoms present somewhere in the substances reacting (the reactants) must be present in the substances produced (the products). Therefore each chemical change can be thought of as merely a rearrangement of atoms. This is an especially useful way of understanding **chemical equations**, which are simply shorthand expressions of chemical rearrangements using the symbols and formulas we discussed in Chapter 2. The formulas or symbols of the reactants are placed to the left of an arrow, which indicates the direction of the reaction, and the formulas or symbols of the products are placed to the right.

The most important feature of a chemical equation, aside from identifying the products and reactants, is expressing the relative quantities of the substances involved. Again we are relying on the law of conservation of mass. When we study these quantitative features of a chemical reaction, we are studying the **stoichiometry** of the reaction. An understanding of stoichiometry underlies practically every quantitative calculation a chemist makes and is therefore of paramount importance in understanding chemistry.

3.1 Chemical Equations

The first step in writing a chemical equation is identifying the reactants and products. Since we have just begun our study of chemical transformations, we have studied few chemical reactions; however, we do know that hydrogen (H_2) and oxygen (O_2) will combine under certain conditions to form water (H_2O). Using these chemical formulas we can represent this reaction by writing

$$H_2 + O_2 \longrightarrow H_2O$$

There is more to writing an accurate chemical equation, however, because a chemical equation, like an algebraic equation, must be balanced. Again we invoke the law of conservation of mass: There must be as many atoms of each element in the products as were present in the reactants. Note that the expression above is not balanced, since we have two oxygen atoms on the left but only one on the right. We might balance the expression by removing the subscript from the O_2 and writing

$$H_2 + O \longrightarrow H_2O$$

but this misrepresents the chemical facts since we know that the reactants are both diatomic molecules. The equation must not suggest that the reaction involves *atomic* oxygen. In general, *once we have established the chemical facts, we must not alter formulas and symbols to balance the equation*. Instead, we must use *coefficients*, which are numbers placed in front of the formulas of the reactants and products to balance the chemical equation. In this example, if we place a coefficient of 2 in front of H_2 and H_2O, we shall have a balanced expression:

$$2\,H_2 + O_2 \longrightarrow 2\,H_2O$$

Let us now consider some expressions that are more difficult to balance. For the time being, you should accept the chemical facts as we present them. Later you will often be able to predict the products from a given set of reactants, but our immediate objective is to learn to balance equations. Consider the expression

$$SiO_2 + HF \longrightarrow SiF_4 + H_2O$$

As a starting point it is usually best to choose the substance containing the largest number of atoms per formula, in this case SiF_4. Since in the products of this expression the ratio of fluorine (F) atoms to silicon (Si) atoms is 4 to 1, the ratio of fluorine atoms to silicon atoms in the reactants must also be 4 to 1. Indeed the absolute numbers of silicon and fluorine atoms must be the same on both sides of the equation. So we place a coefficient of 4 in front of HF.

$$SiO_2 + 4\,HF \longrightarrow SiF_4 + H_2O$$

The ratio of hydrogen atoms to oxygen atoms is now the same on both sides of the expression (2 H to 1 O), but there are twice as many of each on the left as on the right. Therefore, we can balance the expression simply by placing a coefficient of 2 in front of H_2O. Our equation is thus

$$SiO_2 + 4\,HF \longrightarrow SiF_4 + 2\,H_2O$$

If in doubt, you should check the expression by counting the number of atoms of each element on both sides of the arrow.

Example 3.1 Balance the following expression:

$$C_3H_8 + O_2 \longrightarrow CO_2 + H_2O$$

Solution

a. Again we select the substance containing the most atoms per formula:

$$C_3H_8$$

b. The ratio of hydrogen atoms to carbon atoms is 8:3. We obtain this same ratio between the hydrogen and carbon atoms of the products by placing a coefficient of 3 in front of CO_2 and a coefficient of 4 in front of H_2O.

$$C_3H_8 + O_2 \longrightarrow 3\,CO_2 + 4\,H_2O$$

c. Since there are 10 oxygen atoms on the right (6 from 3 CO_2 and 4 from 4 H_2O), we balance the equation with a coefficient of 5 in front of O_2. The result is

$$C_3H_8 + 5\,O_2 \longrightarrow 3\,CO_2 + 4\,H_2O$$

Example 3.2 Balance the following expression:

$$NH_3 + O_2 \longrightarrow NO_2 + H_2O$$

Solution

a. The formula with the most atoms is NH_3.

b. The ratio of three hydrogen atoms to one nitrogen atom must be maintained in the products. We might do this by placing a coefficient of 1.5 in front of water, but as a matter of convenience and convention we usually use only integral coefficients. Consequently, to preserve the 3:1 ratio of hydrogen and nitrogen atoms and to eliminate fractional coefficients, we assign H_2O a coefficient of 3 and give NH_3 and NO_2 coefficients of 2. At this point we have

$$2\,NH_3 + O_2 \longrightarrow 2\,NO_2 + 3\,H_2O$$

c. There are 7 oxygen atoms on the right, thus a coefficient of 3.5 is required for O_2. Again, to avoid fractional coefficients, we assign O_2 a coefficient of 7 and double the other three coefficients as well:

$$4\,NH_3 + 7\,O_2 \longrightarrow 4\,NO_2 + 6\,H_2O$$

In the equations we have seen so far, there has been no attempt to specify the physical states of the reactants and products. Sometimes, however, it is useful to have this information. Therefore physical states are often indicated using the symbols (s), (l), (g), and (aq), which signify solid, liquid, gas, and aqueous (from Latin *aqua*, meaning water—indicating that the species is dissolved in water), respectively. For example, the equation

$$Zn(s) + H_2SO_4(aq) \longrightarrow H_2(g) + ZnSO_4(aq)$$

states that solid zinc reacts with an aqueous solution of sulfuric acid to form hydrogen gas and zinc sulfate solution.

Question

1. Distinguish between a subscript and a coefficient. Which of these may be changed when an unbalanced expression is to be made into a chemical equation?

3.2 Stoichiometry

We can now introduce the study of stoichiometry by looking more closely at the information contained in a balanced chemical equation. As an example, consider the burning of propane gas (C_3H_8) (see Example 3.1) a common fuel used by farmers for drying crops and by nonurban dwellers for cooking and heating. It is commonly called "bottled gas." When propane is burned under ideal conditions, the reaction can be represented by the equation written in Example 3.1:

$$C_3H_8(g) + 5\ O_2(g) \longrightarrow 3\ CO_2(g) + 4\ H_2O(g)$$

The coefficients indicate that 1 molecule of C_3H_8 reacts with 5 molecules of O_2 to produce 3 molecules of CO_2 and 4 molecules of H_2O. But these numbers fix only the ratios among the four substances. It would be equally correct, for example, to say that 2 molecules of C_3H_8 react with 10 molecules of O_2 to produce 6 molecules of CO_2 and 8 molecules of H_2O. Or we might multiply each coefficient by 6.02×10^{23}, the number of molecules in a mole, and then say that 1 mol of C_3H_8 reacts with 5 mol of O_2 to produce 3 mol of CO_2 and 4 mol of H_2O. Finally, recalling that each mole is one gram molecular weight, we can convert the ratio in moles to the ratio in grams by using the gram molecular weight of each substance. Hence

$$(1\ \text{mol}\ C_3H_8)\left(\frac{44.1\ \text{g}\ C_3H_8}{1.00\ \text{mol}\ C_3H_8}\right) \text{ or } 44.1\ \text{g}\ C_3H_8$$

react with

$$(5\ \text{mol}\ O_2)\left(\frac{32.0\ \text{g}\ O_2}{1.00\ \text{mol}\ O_2}\right) \text{ or } 160\ \text{g}\ O_2$$

to produce

$$(3 \text{ mol CO}_2) \left(\frac{44.0 \text{ g CO}_2}{1.00 \text{ mol CO}_2} \right) \text{ or } 132 \text{ g CO}_2$$

and

$$(4 \text{ mol H}_2\text{O}) \left(\frac{18.0 \text{ g H}_2\text{O}}{1.00 \text{ mol H}_2\text{O}} \right) \text{ or } 72.0 \text{ g H}_2\text{O}$$

These quantitative relationships for the combustion of propane are tabulated below.

$$C_3H_8 + 5 O_2 \longrightarrow 3 CO_2 + 4 H_2O$$

Molecule ratio:

1 molecule C_3H_8 + 5 molecules O_2
$\longrightarrow$ 3 molecules CO_2 + 4 molecules H_2O

Mole ratio:

1 mol C_3H_8 + 5 mol O_2 $\longrightarrow$ 3 mol CO_2 + 4 mol H_2O

Mass ratio:

44.1 g C_3H_8 + 160 g O_2 $\longrightarrow$ 132 g CO_2 + 72.0 g H_2O

Note that (within the limitations of significant digits) the sum of the masses of the reactants equals the sum of the masses of the products.

Example 3.3 How many molecules of CO_2 are produced if 6.02×10^{23} molecules of oxygen are used in burning propane?

Solution

a. The chemical equation discussed above shows several different relationships (conversion equations) between products and reactants—relationships involving numbers of molecules, numbers of moles, and mass. It is therefore necessary to select the conversion equation applicable to the problem at hand. In this case, we need only the conversion equation that relates the number of molecules of carbon dioxide to the number of molecules of oxygen. In fact, once we have the appropriate balanced equation, we can ignore the propane and water altogether. The ratio of CO_2 produced to O_2 consumed is

$$\frac{3 \text{ molecules } CO_2}{5 \text{ molecules } O_2}$$

b. This conversion factor, multiplied by the number of O_2 molecules consumed, yields the number of CO_2 molecules produced:

$$(6.02 \times 10^{23} \text{ molecules } O_2) \left(\frac{3 \text{ molecules } CO_2}{5 \text{ molecules } O_2} \right)$$

$$= 3.61 \times 10^{23} \text{ molecules } CO_2$$

The solution to this example illustrates the typical process of solving a stoichiometric problem. First, write the appropriate balanced chemical equation. Second, identify the quantity given and the quantity to be calculated. Third, derive a ratio (a conversion factor) between these quantities in terms of the coefficients from the chemical equation. Finally, multiply the quantity given by the conversion factor:

quantity given × ratio (conversion factor) = quantity requested

It is often useful to check the units before doing the arithmetic. For example, in Example 3.3, if we had mistakenly inverted the conversion factor, we would have obtained nonsense units in the answer:

$$(6.02 \times 10^{23} \text{ molecules } O_2) \left(\frac{5 \text{ molecules } O_2}{3 \text{ molecules } CO_2} \right)$$

$$= \frac{1.00 \times 10^{24} \, (\text{molecules } O_2)^2}{\text{molecules } CO_2}$$

Of course, molecules are cumbersome units to work with; hence Example 3.3 is not a very practical one. On the other hand, the mole is a very practical unit for measuring the quantities of substances in the laboratory. Consider the following mole-mole problem (where both the quantity given and the quantity requested are in moles).

Example 3.4 When a quantity of propane is burned, 4.46 mol of CO_2 are produced. How many moles of water are produced at the same time?

Solution

a. Equation:

$$C_3H_8(g) + 5\,O_2(g) \longrightarrow 3\,CO_2(g) + 4\,H_2O(g)$$

b. Quantity given: 4.46 mol of CO_2
Quantity requested: moles of H_2O

c. From the equation, note that the ratio of moles of water produced to moles of CO_2 produced is

$$\frac{4 \text{ mol } H_2O}{3 \text{ mol } CO_2}$$

d. Finally, multiply the given quantity by the conversion factor:

$$(4.46 \text{ mol } CO_2)\left(\frac{4 \text{ mol } H_2O}{3 \text{ mol } CO_2}\right) = 5.95 \text{ mol } H_2O$$

Example 3.4 illustrates that mole-mole problems can be solved in a one-step operation. Of course there are many other practical units in which to express quantities of reactants or products. In such cases, the problems can be slightly more complicated since all units must *first be converted to moles*. Consider the following mole-mass problem.

Example 3.5 How many grams of C_3H_8 could be burned by 213 mol of O_2?

Solution

a. The solution of this problem or any stoichiometric problem must begin with a balanced equation:

$$C_3H_8(g) + 5\,O_2(g) \longrightarrow 3\,CO_2(g) + 4\,H_2O(g)$$

b. Quantity given: 213 mol O_2
Quantity requested: grams of C_3H_8

c. The ratio of moles of C_3H_8 reacting to moles of O_2 reacting is

$$\frac{1 \text{ mol } C_3H_8}{5 \text{ mol } O_2}$$

d. The number of moles of C_3H_8 consumed is equal to the given quantity multiplied by this conversion factor:

$$(213 \text{ mol } O_2)\left(\frac{1 \text{ mol } C_3H_8}{5 \text{ mol } O_2}\right)$$

e. Since the quantity requested is grams of C_3H_8, we need a second conversion factor relating moles of C_3H_8 to grams of C_3H_8:

$$1.00 \text{ mol } C_3H_8 = 44.1 \text{ g } C_3H_8$$

$$(213 \text{ mol } O_2)\left(\frac{1 \text{ mol } C_3H_8}{5 \text{ mol } O_2}\right)\left(\frac{44.1 \text{ g } C_3H_8}{1.00 \text{ mol } C_3H_8}\right) = 1.88 \times 10^3 \text{ g } C_3H_8$$

Example 3.6 Carbon monoxide, one of the chief pollutants of the atmosphere, can be detected using the following reaction:

$$I_2O_5(s) + 5 \text{ CO}(g) \longrightarrow I_2(s) + 5 \text{ CO}_2(g)$$

How many grams of CO will react with an excess of I_2O_5 to form 312 g of iodine?

Solution

a. We could solve this problem by constructing a mass ratio between CO (quantity requested) and I_2O_5 (quantity given); however, since moles are simpler to work with, we usually solve problems of this sort using a mole ratio. *All stoichiometric problems should be solved on the basis of moles.*

b. Quantity given: 312 g of I_2
Quantity requested: grams of CO

c. Convert grams of I_2 to moles of I_2:

$$254 \text{ g } I_2 = 1.00 \text{ mol } I_2$$

$$(312 \text{ g } I_2)\left(\frac{1.00 \text{ mol } I_2}{254 \text{ g } I_2}\right)$$

d. From the equation, determine the ratio of moles of CO to moles of I_2:

$$\frac{5 \text{ mol CO}}{1 \text{ mol } I_2}$$

e. The number of moles of CO consumed is therefore

$$(312 \text{ g } I_2)\left(\frac{1.00 \text{ mol } I_2}{254 \text{ g } I_2}\right)\left(\frac{5 \text{ mol CO}}{1 \text{ mol } I_2}\right)$$

f. Finally, convert moles of CO to grams of CO:

$$28.0 \text{ g CO} = 1.00 \text{ mol CO}$$

$$(312 \text{ g } I_2)\left(\frac{1.00 \text{ mol } I_2}{254 \text{ g } I_2}\right)\left(\frac{5 \text{ mol CO}}{1 \text{ mol } I_2}\right)\left(\frac{28.0 \text{ g CO}}{1.00 \text{ mol CO}}\right) = 172 \text{ g CO}$$

So far we have dealt only with examples involving one chemical reaction; however, many chemical preparations require a series of reactions. Example 3.7

illustrates how we can quickly solve for the weight of a reactant when given the quantity of product, even though more than one chemical step intervenes.

Example 3.7 Phosphoric acid (H_3PO_4) has a number of applications in medicine. Two examples are its use as an astringent (a substance that contracts body tissue or blood vessels and restricts the flow of blood) and an antipyretic (a fever reducer). It can be produced by the following reactions:

$$P_4(s) + 5\ O_2(g) \longrightarrow P_4O_{10}(s)$$
$$P_4O_{10}(s) + 6\ H_2O(l) \longrightarrow 4\ H_3PO_4(l)$$

How many grams of O_2 are required to produce 112 g of H_3PO_4?

Solution

a. First determine the number of moles of H_3PO_4 formed:

$$1.00\ \text{mol}\ H_3PO_4 = 98.0\ \text{g}\ H_3PO_4$$

$$(112\ \text{g}\ H_3PO_4)\left(\frac{1.00\ \text{mol}\ H_3PO_4}{98.0\ \text{g}\ H_3PO_4}\right)$$

b. Since P_4O_{10} is obtained by the reaction of P_4 with O_2 (the quantity to be determined), we must calculate next the ratio of the number of moles of P_4O_{10} consumed to the number of moles of H_3PO_4 produced:

$$\frac{1\ \text{mol}\ P_4O_{10}}{4\ \text{mol}\ H_3PO_4}$$

c. Therefore the number of moles of P_4O_{10} required to produce 112 g of H_3PO_4 is

$$(112\ \text{g}\ H_3PO_4)\left(\frac{1.00\ \text{mol}\ H_3PO_4}{98.0\ \text{g}\ H_3PO_4}\right)\left(\frac{1\ \text{mol}\ P_4O_{10}}{4\ \text{mol}\ H_3PO_4}\right)$$

d. From the first equation, we now determine the ratio of the number of moles of O_2 consumed to the number of moles of P_4O_{10} produced:

$$\frac{5\ \text{mol}\ O_2}{1\ \text{mol}\ P_4O_{10}}$$

e. Hence the number of moles of O_2 required to produce 112 g of H_3PO_4 is

$$(112\ \text{g}\ H_3PO_4)\left(\frac{1.00\ \text{mol}\ H_3PO_4}{98.0\ \text{g}\ H_3PO_4}\right)\left(\frac{1\ \text{mol}\ P_4O_{10}}{4\ \text{mol}\ H_3PO_4}\right)\left(\frac{5\ \text{mol}\ O_2}{1\ \text{mol}\ P_4O_{10}}\right)$$

f. Finally, convert moles of O_2 to grams of O_2:

$$32.0 \text{ g } O_2 = 1.00 \text{ mol } O_2$$

$$(112 \text{ g } H_3PO_4)\left(\frac{1.00 \text{ mol } H_3PO_4}{98.0 \text{ g } H_3PO_4}\right)\left(\frac{1 \text{ mol } P_4O_{10}}{4 \text{ mol } H_3PO_4}\right)$$
$$\left(\frac{5 \text{ mol } O_2}{1 \text{ mol } P_4O_{10}}\right)\left(\frac{32.0 \text{ g } O_2}{1.00 \text{ mol } O_2}\right) = 45.7 \text{ g } O_2$$

We are now prepared to deal with a practical problem that chemists often face. In the previous examples we assumed that a given quantity of a reactant was completely consumed. We took it for granted that there was more than enough of the other substances to react completely with the specified reactant. More frequently, you will know the amounts of more than one reactant and will have to determine which reactant is in excess and which reactant limits the quantity of product.

Example 3.8 Sulfur dioxide is one of the major industrial pollutants of the atmosphere. It is formed wherever sulfur-containing fuels are burned. Some conservationists have suggested that industrial pollution might be reduced by the reaction

$$\underset{8.4 \text{ g}}{16 \text{ H}_2S(g)} + \underset{7.82 \text{ g}}{8 \text{ SO}_2(g)} \longrightarrow \underset{X \text{ g}}{3 \text{ S}_8(s)} + 16 \text{ H}_2O(g)$$

At least part of the cost of the pollution-control process might then be recovered by selling the sulfur produced. If a sample of gas contains 8.40 g of H_2S and 7.82 g of SO_2, how much sulfur will be formed when the reaction is complete?

Solution

a. In this problem the quantities of *two* reactants are given. If one of these compounds is present in a greater amount than is required to react completely with the other, then it is said to be "present in *excess*." Obviously the *excess* does not react. The quantity of product formed is determined solely by the amount of the reactant *not* in excess. This reagent is called the *limiting reagent*. The first step, therefore, must be to determine which reactant is the *limiting reagent*.

b. Determine the number of moles of each gas present in the reaction mixture:

$$34.1 \text{ g } H_2S = 1.00 \text{ mol } H_2S$$

$$64.1 \text{ g } SO_2 = 1.00 \text{ mol } SO_2$$

$$(8.40 \text{ g } H_2S)\left(\frac{1.00 \text{ mol } H_2S}{34.1 \text{ g } H_2S}\right) = 0.246 \text{ mol } H_2S$$

$$(7.82 \text{ g } SO_2)\left(\frac{1.00 \text{ mol } SO_2}{64.1 \text{ g } SO_2}\right) = 0.122 \text{ mol } SO_2$$

c. Determine the ratio of moles of H_2S to moles of SO_2 in the reaction mixture:

$$\frac{0.246 \text{ mol } H_2S}{0.122 \text{ mol } SO_2} = \frac{2.02 \text{ mol } H_2S}{1.00 \text{ mol } SO_2}$$

d. From the equation, now determine the ratio of moles of H_2S to moles of SO_2 that actually react.

$$\frac{16 \text{ mol } H_2S}{8 \text{ mol } SO_2} = \frac{2 \text{ mol } H_2S}{1 \text{ mol } SO_2}$$

e. We can now see that H_2S is present in excess. To see how much excess, determine the number of moles of H_2S that react with 0.122 mol SO_2:

$$\left(\frac{2 \text{ mol } H_2S}{1 \text{ mol } SO_2}\right)(0.122 \text{ mol } SO_2) = 0.244 \text{ mol } H_2S$$

Hence 0.246 mol H_2S − 0.244 mol H_2S = 0.002 mol H_2S remains unreacted. We shall now use the limiting reactant, SO_2, to calculate the amount of sulfur produced.

f. From the equation, determine the ratio between moles of sulfur produced and moles of SO_2 consumed:

$$\frac{3 \text{ mol } S_8}{8 \text{ mol } SO_2}$$

g. Calculate the number of moles of sulfur produced:

$$(0.122 \text{ mol } SO_2)\left(\frac{3 \text{ mol } S_8}{8 \text{ mol } SO_2}\right)$$

h. Determine the number of grams of sulfur produced:

$$1.00 \text{ mol } S_8 = 256 \text{ g } S_8$$

$$(0.122 \text{ mol } SO_2)\left(\frac{3 \text{ mol } S_8}{8 \text{ mol } SO_2}\right)\left(\frac{256 \text{ g } S_8}{1.00 \text{ mol } S_8}\right) = 11.7 \text{ g } S_8$$

A second and less direct approach to the example would have been to calculate the number of grams of S_8 that can be prepared from 8.40 g of H_2S, assuming an excess of SO_2, and then the number of grams of S_8 that can be produced from 7.82 g of SO_2, assuming an excess of H_2S. The smaller of these results would be the quantity produced from the specified reaction mixture.

Another practical problem we have not yet faced is that reactions rarely take place under ideal conditions. For example, many reactions do not go to completion; that is, they do not continue until one or more of the reactants is exhausted. Instead they reach a state of equilibrium with appreciable quantities of reactants remaining. (We shall discuss chemical equilibrium in Chapter 13.) In other cases, more than one reaction takes place at the same time, forming more than one set of products. For example, when we burn gasoline in our automobiles, the ideal situation would be to burn the hydrocarbons (compounds of carbon and hydrogen) completely to form carbon dioxide and water. In fact, some of the hydrocarbons in gasoline react to form carbon monoxide and water or carbon and water. We can illustrate these parallel reactions using octane (C_8H_{18}), which is one of the major constituents of gasoline:

$$2\ C_8H_{18} + 25\ O_2 \longrightarrow 16\ CO_2 + 18\ H_2O$$

$$2\ C_8H_{18} + 17\ O_2 \longrightarrow 16\ CO + 18\ H_2O$$

$$2\ C_8H_{18} + 9\ O_2 \longrightarrow 16\ C + 18\ H_2O$$

The desired reaction (burning to carbon dioxide and water) is referred to as the *main reaction* and the other reactions are called *side reactions*. In this case both side reactions are particularly harmful. Carbon monoxide from automobile exhaust is a major cause of air pollution in large cities, and the carbon formed in the other side reaction deposits in the engine, reducing its efficiency.

In preparing a compound in the laboratory or in industry, the chemist must isolate the desired product from the reaction mixture, which may contain other substances produced by the main reaction, products from the side reactions, and unreacted reactants. In the separation and purification processes, some of the product is often lost. Consequently, the *actual* quantities isolated may be far less than the *theoretical* quantities calculated. To see how much less, it is necessary to calculate *percent yield*, which is defined as

$$\left(\frac{\text{actual yield}}{\text{theoretical yield}}\right) \times 100 = \text{percent yield}$$

To increase percentage yield the chemist will try to control the reaction conditions, especially the temperature. The experimenter may also ensure that inexpensive or easily available reactants are present in excess, hoping to convert the expensive or less readily available reactants more completely to the desired product.

Example 3.9 Phenolphthalein ($C_{20}H_{14}O_4$), a white powder, is the active ingredient in most candy-type laxatives. It is prepared from phthalic anhydride ($C_8H_4O_3$) and phenol (C_6H_6O).

$$C_8H_4O_3(s) + 2\ C_6H_6O(s) \longrightarrow C_{20}H_{14}O_4(s) + H_2O(l)$$

If 5.0 g of pure phenolphthalein is isolated from a reaction mixture consisting of 10.0 g of phthalic anhydride and 20.0 g of phenol, what is the percent yield?

Solution

a. The first step in this problem is to determine which reactant is the limiting one. To do this we must first calculate the number of moles of each reactant in the reaction mixture and then compute the ratio of moles of phenol to moles of phthalic anhydride. We can then compare this ratio to that obtained from the balanced chemical equation.

$$148 \text{ g } C_8H_4O_3 = 1.00 \text{ mol } C_8H_4O_3$$

$$94.1 \text{ g } C_6H_6O = 1.00 \text{ mol } C_6H_6O$$

$$(10.0 \text{ g } C_8H_4O_3) \left(\frac{1.00 \text{ mol } C_8H_4O_3}{148 \text{ g } C_8H_4O_3} \right) = 0.0676 \text{ mol } C_8H_4O_3$$

$$(20.0 \text{ g } C_6H_6O) \left(\frac{1.00 \text{ mol } C_6H_6O}{94.1 \text{ g } C_6H_6O} \right) = 0.212 \text{ mol } C_6H_6O$$

The equation states that two moles of phenol react with one mole of phthalic anhydride. Since the number of moles of phenol in the reaction mixture is clearly greater than twice the number of moles of phthalic anhydride, phenol is in excess, and phthalic anhydride is the limiting reagent.

b. Since each mole of phthalic anhydride produces one mole of phenolphthalein, the theoretical yield is 0.0676 mol.

c. We can now calculate the theoretical yield in grams:

$$318 \text{ g phenolphthalein} = 1.00 \text{ mol phenolphthalein}$$

$$(0.0676 \text{ mol phenolphthalein}) \left(\frac{318 \text{ g phenolphthalein}}{1.00 \text{ mol phenolphthalein}} \right)$$

$$= 21.5 \text{ g phenolphthalein}$$

d. The percent yield is

$$\left(\frac{\text{actual yield}}{\text{theoretical yield}} \right) \times 100 = \left(\frac{5.00 \text{ g phenolphthalein}}{21.5 \text{ g phenolphthalein}} \right) \times 100$$

$$= 23.3\%$$

Questions

2. Distinguish between
   a. reactant and product
   b. percent yield and actual yield

3. List several factors that may cause the actual yield to be less than the theoretical yield.

3.3 Stoichiometry in Solutions—Molarity

In Chapter 1 we introduced solutions as homogeneous mixtures. Now we want to consider stoichiometry in solutions, since solutions are so often involved in chemical reactions.

Each component of a solution is usually referred to as either a *solute* or the *solvent*. The component that is usually present in the largest amount and that was originally in the same physical state as the resulting solution is the dispersing medium and is called the **solvent**. All other components are the dispersed substances and are called **solutes**. The most common solvent is water.

Since the compositions of solutions often vary, we need units to describe their composition. For example, the terms *concentrated* and *dilute* are often used to describe acid solutions. The former usually refers to the acid as it is purchased from the supply house, whereas the latter indicates that the commercial acid has been diluted with water. Although these terms are useful, they are only relative. We know only that the ratio of solute to solvent is smaller in a dilute acid solution than in a concentrated solution of the same acid. Clearly we need to be able to express the compositions of solutions more precisely. This is done by specifying the *concentration*, that is, the amount of solute in a given amount of solvent or solution. The most commonly used method of expressing concentration is *molarity* (M), which is defined as the number of moles of solute per liter of *solution*.

$$M = \frac{\text{number of moles of solute}}{\text{liter of solution}}$$

Note that the volume of the *solution*, not the volume of the *solvent*, is specified. Thus to prepare a 1.00 M solution of sugar, a chemist would weigh out 1.00 mol of sugar and then add enough water to make 1.00 liter of solution.

Example 3.10 What is the molarity of a solution that contains 227 g of H_2SO_4 in 2.00 liters of solution?

Solution

a. The problem requires changing grams per 2.00 liters to moles per liter.

b. Calculate grams per liter:

$$\frac{227 \text{ g } H_2SO_4}{2.00 \text{ liter}}$$

c. Calculate moles per liter:

98.1 g H_2SO_4 = 1.00 mol H_2SO_4

$$\left(\frac{227 \text{ g } H_2SO_4}{2.00 \text{ liter}}\right)\left(\frac{1.00 \text{ mol } H_2SO_4}{98.1 \text{ g } H_2SO_4}\right) = 1.16 \text{ mol } H_2SO_4/\text{liter} = \mathbf{1.16\ } \boldsymbol{M}$$

Example 3.11 How many grams of NaOH are required to prepare 4.00×10^2 ml of $0.600\ M$ NaOH solution?

Solution

a. Molarity is defined as the number of moles per liter. Hence 1.00 liter contains 0.600 mol or

$$\frac{0.600 \text{ mol NaOH}}{1.00 \text{ liter}}$$

b. Convert 4.00×10^2 ml to liters:

$$1000 \text{ ml} = 1 \text{ liter}$$

$$(4.00 \times 10^2 \text{ ml}) \left(\frac{1 \text{ liter}}{1000 \text{ ml}} \right)$$

c. Calculate the number of moles in 400 ml:

$$(4.00 \times 10^2 \text{ ml}) \left(\frac{1 \text{ liter}}{1000 \text{ ml}} \right) \left(\frac{0.600 \text{ mol NaOH}}{1.00 \text{ liter}} \right)$$

d. Convert moles to grams:

$$40.0 \text{ g NaOH} = 1 \text{ mol NaOH}$$

$$(4.00 \times 10^2 \text{ ml}) \left(\frac{1 \text{ liter}}{1000 \text{ ml}} \right) \left(\frac{0.600 \text{ mol NaOH}}{1.00 \text{ liter}} \right) \left(\frac{40.0 \text{ g NaOH}}{1.00 \text{ mol NaOH}} \right)$$

$$= 9.60 \text{ g NaOH}$$

When solutions are involved in stoichiometric calculations, quantities are usually given in volume and concentration rather than in mass. Otherwise the examples are similar to the stoichiometric examples that we have already worked.

Example 3.12 How many liters of $0.600\ M$ NaOH are needed to react with 1.00×10^2 ml of $1.16\ M$ H_2SO_4?

$$2\ NaOH(aq) + H_2SO_4(aq) \longrightarrow Na_2SO_4(aq) + 2\ H_2O(l)$$

Solution

a. Quantity given: 1.00×10^2 ml of $1.16\ M$ H_2SO_4
Quantity requested: volume of $0.600\ M$ NaOH

b. Convert the volume of 1.16 M H_2SO_4 to moles of H_2SO_4:

$$(1.00 \times 10^2 \text{ ml}) \left(\frac{1 \text{ liter}}{1000 \text{ ml}} \right) \left(\frac{1.16 \text{ mol } H_2SO_4}{1.00 \text{ liter}} \right)$$

c. From the equation, determine the ratio of moles of NaOH to moles of H_2SO_4:

$$\frac{2 \text{ mol NaOH}}{1 \text{ mol } H_2SO_4}$$

d. Calculate the moles of NaOH required:

$$(1.00 \times 10^2 \text{ ml}) \left(\frac{1 \text{ liter}}{1000 \text{ ml}} \right) \left(\frac{1.16 \text{ mol } H_2SO_4}{1.00 \text{ liter}} \right) \left(\frac{2 \text{ mol NaOH}}{1 \text{ mol } H_2SO_4} \right)$$

e. Convert moles of NaOH to liters of 0.600 M NaOH:

$$(1.00 \times 10^2 \text{ ml}) \left(\frac{1 \text{ liter}}{1000 \text{ ml}} \right) \left(\frac{1.16 \text{ mol } H_2SO_4}{1.00 \text{ liter}} \right)$$

$$\left(\frac{2 \text{ mol NaOH}}{1 \text{ mol } H_2SO_4} \right) \left(\frac{1.00 \text{ liter}}{0.600 \text{ mol NaOH}} \right) = 0.387 \text{ liter}$$

Example 3.13 Hydrogen peroxide (H_2O_2) is the only compound other than water that consists exclusively of hydrogen and oxygen. Since it is unstable, it is usually stored as a dilute aqueous solution. For example, hydrogen peroxide purchased at the drugstore is a 3 percent (by weight) solution. One of the more interesting uses of hydrogen peroxide is to restore old oil paintings. Some of the paint pigments used by artists several centuries ago were lead compounds. When these compounds react with the H_2S in the polluted air of industrial cities, one of the products is black lead sulfide (PbS), which darkens the original colors. These paintings can be at least partially restored, however, by washing them with a solution of H_2O_2. The hydrogen peroxide reacts with black lead sulfide to form white lead sulfate, according to the following equation:

$$PbS(s) + 4 H_2O_2(aq) \longrightarrow PbSO_4(s) + 4 H_2O(l)$$

A. What is the molarity of a 3.00 percent solution of H_2O_2? (Assume the specific gravity of a 3.00 percent solution of H_2O_2 is 1.00.)

B. How many liters of a 3.00 percent H_2O_2 solution are required to react with 32.1 g of PbS?

Solution

A. a. Since the specific gravity of the solution is 1.00, it has the same density as water, that is, 1.00 g/ml. We also know that each 100 g of solution contains 3.00 g of H_2O_2.

b. Calculate the mass of H_2O_2 in each liter of solution:

$$\left(\frac{1000 \text{ ml solution}}{1 \text{ liter solution}}\right)\left(\frac{1.00 \text{ g solution}}{1.00 \text{ ml solution}}\right)\left(\frac{3.00 \text{ g } H_2O_2}{100 \text{ g solution}}\right)$$

$$= \frac{30.0 \text{ g } H_2O_2}{1.00 \text{ liter solution}}$$

c. Convert grams per liter to moles per liter:

$$1.00 \text{ mol } H_2O_2 = 34.0 \text{ g } H_2O_2$$

$$\left(\frac{30.0 \text{ g } H_2O_2}{1.00 \text{ liter } 3.00\% \text{ solution}}\right)\left(\frac{1.00 \text{ mol } H_2O_2}{34.0 \text{ g } H_2O_2}\right)$$

$$= \frac{0.882 \text{ mol } H_2O_2}{1.00 \text{ liter } 3.00\% \text{ solution}} = 0.882 \ M$$

B. a. Quantity given: 32.1 g PbS
Quantity requested: liters of 3.00 percent H_2O_2 solution
Starting with the quantity given, we can evaluate the quantity requested according to the following scheme:

$$\text{g PbS} \longrightarrow \text{mol PbS} \longrightarrow \text{mol } H_2O_2$$
$$\longrightarrow \text{liters of 3.00 percent } H_2O_2 \text{ solution}$$

b. Convert g PbS to mol PbS:

$$1.00 \text{ mol PbS} = 239 \text{ g PbS}$$

$$(32.1 \text{ g PbS})\left(\frac{1.00 \text{ mol PbS}}{239 \text{ g PbS}}\right)$$

c. From the chemical equation, extract the ratio of moles of H_2O_2 to moles of PbS:

$$(32.1 \text{ g PbS})\left(\frac{1.00 \text{ mol PbS}}{239 \text{ g PbS}}\right)\left(\frac{4 \text{ mol } H_2O_2}{1 \text{ mol PbS}}\right)$$

d. Finally, use the answer from part A to relate moles of H_2O_2 to liters of solution:

$$(32.1 \text{ g PbS})\left(\frac{1.00 \text{ mol PbS}}{239 \text{ g PbS}}\right)\left(\frac{4 \text{ mol } H_2O_2}{1 \text{ mol PbS}}\right)\left(\frac{1.00 \text{ liter } 3.00\% \text{ solution } H_2O_2}{0.882 \text{ mol } H_2O_2}\right)$$

$$= 0.609 \text{ liter of } 3.00\% \ H_2O_2 \text{ solution}$$

Questions

4. Distinguish among solute, solvent, and solution.

5. A solution is prepared by weighing out 1.00 mol of $CaCl_2$ and adding 1.00 liter of water. Is this a 1.00 M solution? Explain.

Questions and Problems

3.1 Chemical Equations

6. Form equations by balancing the following expressions:
- a. $S_8 + O_2 \longrightarrow SO_2$
- b. $C + O_2 \longrightarrow CO_2$
- c. $C_2H_4 + H_2 \longrightarrow C_2H_6$
- d. $I_2 + H_2S \longrightarrow HI + S_8$
- e. $SO_3 + H_2O \longrightarrow H_2SO_4$
- f. $P_4 + O_2 \longrightarrow P_4O_{10}$
- g. $H_2 + Br_2 \longrightarrow HBr$
- h. $P_4 + O_2 \longrightarrow P_4O_6$
- i. $CS_2 + Cl_2 \longrightarrow CCl_4 + S_2Cl_2$
- j. $SiH_4 + O_2 \longrightarrow SiO_2 + H_2O$
- k. $N_2H_4 + H_2O_2 \longrightarrow N_2 + H_2O$
- l. $H_2O_2 \longrightarrow H_2O + O_2$
- m. $H_2O \longrightarrow H_2 + O_2$
- n. $H_2 + Cl_2 \longrightarrow HCl$
- o. $SO_2 + O_2 \longrightarrow SO_3$
- p. $SiO_2 + HF \longrightarrow SiF_4 + H_2O$
- q. $H_2S + O_2 \longrightarrow H_2O + S_8$
- r. $C + S_8 \longrightarrow CS_2$
- s. $H_2 + S_8 \longrightarrow H_2S$
- t. $B_2H_6 + O_2 \longrightarrow B_2O_3 + H_2O$
- u. $CS_2 + O_2 \longrightarrow CO_2 + SO_2$
- v. $P_4O_{10} + H_2O \longrightarrow H_3PO_4$
- w. $PCl_3 + H_2O \longrightarrow H_3PO_3 + HCl$
- x. $P_4O_6 + H_2O \longrightarrow H_3PO_3$
- y. $HClO_3 \longrightarrow ClO_2 + O_2 + H_2O$
- z. $N_2H_4 + O_2 \longrightarrow NO_2 + H_2O$
- aa. $C_2H_2 + O_2 \longrightarrow CO_2 + H_2O$
- bb. $P_4 + O_2 + H_2O \longrightarrow H_3PO_4$
- cc. $C_2H_4O + O_2 \longrightarrow CO_2 + H_2O$
- dd. $C_{10}H_{14}N_2 + O_2 \longrightarrow CO_2 + H_2O + N_2$
- ee. $C_4H_{10} + O_2 \longrightarrow CO_2 + H_2O$

3.2 Stoichiometry

7. Consider the following equation:

$$2\,C_2H_6(g) + 7\,O_2(g) \longrightarrow 4\,CO_2(g) + 6\,H_2O(g)$$

a. How many molecules of oxygen are required to react with 6 molecules of C_2H_6?
b. How many molecules of water are produced when 12 molecules of CO_2 are produced?
c. If 20.0 mol of oxygen gas react, how many moles of water are produced?
d. If 15.0 mol of CO_2 are produced, how many moles of C_2H_6 react?
e. How many grams of CO_2 are formed when 90.0 g of C_2H_6 react with an excess of oxygen?
f. How many grams of O_2 are needed to burn 90.0 g of C_2H_6?
g. How many grams of O_2 are necessary to produce 9.03×10^{21} molecules of CO_2?

8. How many moles of SO_2 can be prepared from 10.0 g of sulfur and excess oxygen?

$$S_8(s) + 8\,O_2(g) \longrightarrow 8\,SO_2(g)$$

9. When arsenic reacts with excess oxygen, tetraarsenic decaoxide (As_4O_{10}) is formed as the only product. Write the equation for this reaction, and solve for the number of grams of As_4O_{10} which are produced when 90.0 g of arsenic react with excess oxygen.

10. How many grams of CO are required to produce 9.00×10^{24} molecules of phosgene ($COCl_2$)?

$$CO + Cl_2 \longrightarrow COCl_2$$

11. How many moles of O_2 are needed to combine with 57.0 g of nitrogen to form NO_2?

$$N_2(g) + 2\,O_2(g) \longrightarrow 2\,NO_2(g)$$

12. Octane (C_8H_{18}) is a liquid with a specific gravity of 0.703. How many grams of water are produced when 91.0 ml of octane are burned in excess oxygen (O_2) to form carbon dioxide and water?

13. A gas is known to be either CH_4 or C_2H_6. A 1.00-g sample of this gas produces 1.80 g of water upon reacting with oxygen. What is the formula of the gas? Assume complete combustion to CO_2 and H_2O.

14. How many grams of potassium chlorate must be heated to provide enough oxygen to burn 48.6 mol of sulfur?

$$2\,KClO_3(s) \longrightarrow 2\,KCl(s) + 3\,O_2(g)$$
$$S_8(s) + 8\,O_2(g) \longrightarrow 8\,SO_2(g)$$

15. How many grams of oxygen are required to react with 4.00×10^{16} molecules of C_2H_4?

$$C_2H_4(g) + 3\,O_2(g) \longrightarrow 2\,CO_2(g) + 2\,H_2O(g)$$

16. How many tons of carbon are required to reduce 3.39 tons of Fe_2O_3?

$$2\,Fe_2O_3(s) + 3\,C(s) \longrightarrow 3\,CO_2(g) + 4\,Fe(s)$$

17. How many molecules of nitrogen can be obtained by the decomposition of 4.60×10^{-3} g of ammonium nitrite? This compound decomposes on heating to form N_2 and H_2O.

18. How many grams of aluminum hydroxide will react with 313 g of H_2SO_4? The products of this reaction are aluminum sulfate [$Al_2(SO_4)_3$] and water.

19. How many grams of CO_2 can be produced by reacting 129 g of charcoal (carbon) with an excess of oxygen?

20. How many kilograms of oxygen are needed to react with 1.00 kg of benzene (C_6H_6) to form CO_2 and H_2O?

21. Hydrogen sulfide reacts with oxygen (O_2) to form sulfur dioxide (SO_2) and water. If 146 g of H_2S react, determine
    a. the number of grams of SO_2 produced.
    b. the number of grams of O_2 that react.
    c. the number of moles of H_2O produced.

22. How many grams of CO_2 can be prepared from 19.0 g CS_2 and 16.0 g of O_2?

$$CS_2 + 3\ O_2 \longrightarrow CO_2 + 2\ SO_2$$

23. If a mixture of 18.0 g of O_2 and 1.10 g of hydrogen reacts, how many milliliters of water at 25°C are produced?

$$2\ H_2(g) + O_2(g) \longrightarrow 2\ H_2O(g)$$

24. The following equation represents the fermentation of glucose to form ethyl alcohol (C_2H_5OH). What volume of 60.0 percent (by weight) ethyl alcohol can be prepared from 112 lb of glucose? The density of 60.0 percent ethyl alcohol is 0.894 g/ml.

$$C_6H_{12}O_6(s) \longrightarrow 2\ CO_2(g) + 2\ C_2H_5OH(l)$$

25. How many grams of oxygen are required to burn 3.00×10^9 molecules of C_2H_4 to CO_2 and H_2O?

26. An impure sample of aluminum is treated with an excess of sulfuric acid. If 0.0852 g of hydrogen is obtained from 0.780 g of the sample, what is the percent purity of the sample? Assume that none of the impurities will react with sulfuric acid and that all the aluminum reacts.

$$2\ Al(s) + 3\ H_2SO_4(aq) \longrightarrow Al_2(SO_4)_3(aq) + 3\ H_2(g)$$

27. Consider the following equations:

$$4\ NH_3(g) + 5\ O_2(g) \longrightarrow 4\ NO(g) + 6\ H_2O(g)$$
$$2\ NO(g) + O_2(g) \longrightarrow 2\ NO_2(g)$$
$$3\ NO_2(g) + H_2O(l) \longrightarrow 2\ HNO_3(aq) + NO(g)$$
$$3\ Cu(s) + 8\ HNO_3(aq) \longrightarrow 3\ Cu(NO_3)_2(aq) + 2\ NO(g) + 4\ H_2O(l)$$

a. How many molecules of HNO_3 can be synthesized from 19.0 moles of NH_3?
b. How many grams of copper nitrate can be prepared from 38.6 g of ammonia?
c. How many grams of ammonia are required to prepare 14.6 g of $Cu(NO_3)_2$?
d. How many molecules of ammonia are required to prepare 3.86×10^2 g of $Cu(NO_3)_2$?

28. Washing soda (Na_2CO_3) is produced commercially by the thermal decomposition of sodium bicarbonate ($NaHCO_3$) as indicated by the following equation:

$$2\ NaHCO_3(s) \longrightarrow Na_2CO_3(s) + CO_2(g) + H_2O(g)$$

How many pounds of washing soda can be prepared from a 1.00×10^2-lb sample of $NaHCO_3$, if it contains 4.56 percent inert impurities?

29. The thermal decomposition of a 2.56 g impure sample of $KClO_3$ resulted in the production of 0.940 g of O_2. Assuming the impurities of the sample were stable at the temperature at which the decomposition took place, what was the percent of $KClO_3$ in the sample?

3.3 Stoichiometry in Solutions

30. How many grams of solute are dissolved in each of the following solutions?
    a. 10.0 ml of 3.00 M $CaCl_2$
    b. 36.7 ml of 2.82 M glucose ($C_6H_{12}O_6$)
    c. 9.31 liters of 0.913 M NaCl

31. Calculate the amount of A required to make B.

A	B
a. sucrose ($C_{12}H_{22}O_{11}$)	5.50 liters of 3.00 M sucrose solution
b. $Pb(NO_3)_2$	0.560 liter of 0.250 M $Pb(NO_3)_2$ solution
c. $MgCl_2$	10.0 liters of 3.60 M $MgCl_2$ solution

32. What volume of A can be made from B?

A	B
a. 4.00 M $NaNO_3$ solution	10.0 g $NaNO_3$
b. 2.25 M $Al_2(SO_4)_3$ solution	5.00 g $Al_2(SO_4)_3$
c. 0.582 M $KMnO_4$ solution	121 g $KMnO_4$

33. The specific gravity and percent by weight of commercial acids are listed below. Calculate the molarity of each acid.

Acid	Percent	Specific Gravity
a. HCl	37.0	1.19
b. HNO_3	70.0	1.42
c. H_2SO_4	98.0	1.84
d. H_3PO_4	85.0	1.69
e. $HC_2H_3O_2$	99.5	1.05

34. A sample of vinegar is labeled as 4.2 percent by weight acetic acid ($HC_2H_3O_2$). What is the molarity of this vinegar sample? (Assume that the density of vinegar is 1.0 g/ml.)

35. How many grams of NaOH would be required to react with 2.00×10^2 ml of the vinegar in Question 32?

    $$NaOH(aq) + HC_2H_3O_2(aq) \longrightarrow NaC_2H_3O_2(aq) + H_2O(l)$$

36. Human blood usually contains 0.10 g of glucose ($C_6H_{12}O_6$) per 100 ml. Express this concentration in molarity.

37. How many grams of hydrogen can be prepared from 2.00×10^2 ml of 6.00 M H_2SO_4?

    $$Zn(s) + H_2SO_4(aq) \longrightarrow ZnSO_4(aq) + H_2(g)$$

38. How many milliliters of 3.12 M HCl are required to react with 10.0 g of $Al(OH)_3$?

    $$Al(OH)_3(s) + 3\ HCl(aq) \longrightarrow AlCl_3(aq) + 3\ H_2O(l)$$

*39. A metallurgical firm wishes to dispose of 1.53×10^3 gal of waste sulfuric acid. The firm has determined its concentration to be $1.23\ M$. To reduce water pollution, this waste sulfuric acid must be neutralized with slaked lime [$Ca(OH)_2$] before being disposed of in a local river. If slaked lime costs 2¢ per pound, how much will this process cost?

$$H_2SO_4(aq) + Ca(OH)_2(s) \longrightarrow CaSO_4(s) + 2\ H_2O(l)$$

Atomic Structure

4.1 Electrical Energy and Chemical Change

The study of energy and its interaction with matter takes us into that gray area where physics and chemistry overlap. We must undertake this study, even at this early stage, however, since an understanding of energy is a prerequisite to an understanding of the structure and reactions of matter. Earlier we defined energy as the ability to do work. Like all simple definitions of fundamental concepts, this statement was necessarily vague; however, we can now expand on our rudimentary definition.

An object can possess the ability to do work, thus have energy, because of its position, motion, temperature, or electrical charge, or it might have energy bound up in its chemical or nuclear structure. When that energy is not being expended and is present by virtue of the object's state or position, it is known as **potential energy.** When energy is being expended or is present by virtue of the object's motion, it is **kinetic energy.** Consider, for example, a stone resting on the edge of a precipice. Initially, it possesses potential energy, which is converted to kinetic energy if the rock falls. When the stone lands, it may do work (say, by compacting the soil beneath it), or its energy may simply be transformed into more subtle forms of kinetic energy: It may heat the ground slightly (*heat energy*) or it may emit *sound energy*. If we assume that our stone is a lump of coal, we can convert its considerable chemical (potential) energy to heat and light by burning it. If on the other hand it were a lump of uranium, we would find it an even greater reservoir of potential (in this case, nuclear) energy.

The form of energy that has had the greatest historical impact on our understanding of the structure of matter, however, is electrical energy. Electricity has been a source of wonder for centuries. It has manifested itself not only as awesome displays of lightning but also as such a modest phenomenon as the little spark that jumps from your finger to a doorknob after you walk across a plush carpet. For centuries the explanation of such phenomena was the business of magicians and mystics; only in the past 300 years have experimental scientists taken over the job of putting together a theory of electricity.

An understanding of electricity began with elementary observations. A piece of wax, when rubbed with wool, will attract hair, feathers, or small bits of cloth. A glass rod rubbed with silk will do the same. Two such *charged* pieces of wax will repel each other, as will two pieces of charged glass; however, a charged piece of wax will attract a charged piece of glass. It appears, therefore, that there must be two kinds of charge, each of which attracts the other but repels a like charge. By convention, the wax is said to have acquired a *negative charge* and the glass, a *positive charge*. Furthermore, if the charged glass and the charged wax are brought close together but not allowed to touch, a spark will jump between them. The direction of the discharge is from the wax to the glass, that is, from negative to positive. These phenomena are characteristic of what we now call *static electricity*.

The kind of electricity that has become indispensable to modern civilization differs from static electricity in that there is a continual discharge, a current, rather than an intermittent spark from the area of excess negative charge to the area of positive charge. Early experiments with *current electricity* led to the discovery that some substances decompose when a current passes through them. This process is called *electrolysis* and, in the case of water, yields hydrogen and oxygen gases. From this observation and from similar experiments, scientists concluded that there must be some intrinsic relationship between electricity and the forces that bind atoms to one another.

Michael Faraday (1791–1867) pursued this idea and was the first to observe a quantitative relationship between electrical energy and chemical change. With an instrument he developed to measure electrical current he found that: *For a given substance, the weight of product resulting from electrolysis is directly proportional to the amount of electrical charge used.*

Much careful work was done during Faraday's time on the percentage composition of many compounds. For example, investigators discovered that 1.01 g of hydrogen combine with 8.00 g of oxygen in water and that 31.8 g of copper combine with 8.00 g of oxygen in the black oxide of copper. Thus, although there was no compound containing only copper and hydrogen, 31.8 g of copper were said to be "equivalent" to 1.01 g of hydrogen. Using a series of *equivalent weights* developed on this basis, Faraday drew a second important conclusion: *Masses of different substances produced by the same quantity of electrical charge are proportional to the equivalent weights of these substances.* That is, if a certain quantity of electrical charge deposits 31.8 g of copper from a solution of copper sulfate, the same quantity of charge would produce 8.00 g of oxygen (or 1.01 g of hydrogen) from a solution containing an oxygen compound (or hydrogen compound).

Faraday's observations can be thought of as going hand in hand with the atomic theory. Both point the way to a quantitative understanding of the structure of matter: one establishing a relationship between combining masses and the other between combining masses and electrical charge. Since the atomic theory postulated distinct atoms of each element, it seemed logical that there be some minimum "atom" of electricity. Although several years were to elapse before any attempt to measure this unit of electricity was successful, the term *electron* was used for the theoretical unit of electrical charge. Thus scientists had a particulate theory of electricity as well as a particulate theory of matter, and a conviction arose that the two were inherently related.

Question

1. Water at the brink of a waterfall is an example of a substance with a great deal of potential energy. Into what forms might this energy be converted?

4.2 Subatomic Particles

Among the units of measurement we have been working with so far, we find nothing that might be used to measure quantities of electricity. We shall therefore base a new set of units on the *coulomb*, which is defined as the charge that, at a distance of one meter in a vacuum, will repel an identical charge with a force of one newton. (Units of force are defined independently, but we must defer a complete discussion until Chapter 8.) With this definition, we can now easily provide a definition of current, which is merely charge in motion, for example through a copper wire. The *ampere*, which is one of the seven base units in the SI system then becomes

$$1 \text{ ampere} = \frac{1 \text{ coulomb}}{1 \text{ second}}$$

To make these units clearer, we might compare current to the flow of water in a pipe. The quantity of water passing a given point in the pipe might be measured in liters. The analogous unit for electrical charge is the coulomb. The rate of the flow of water would then be measured in liters per second: likewise, the flow of electricity is measured in amperes (coulombs per second). We can extend this analogy by observing that the flow rate of the water depends on the difference in pressure at the two ends of the pipe. We might measure this pressure difference in pounds per square inch. Similarly, the rate of flow of electricity depends on the difference in intensity of charge at the two ends of the wire. This difference is called the *electrical potential* and is measured in *volts*. The amount of current which flows through a conductor is directly proportional to the potential applied across it. This relationship is expressed through a proportionality constant called the *resistance*. If the potential is measured in volts and the current in amperes, the unit of resistance is the *ohm*.

$$\text{potential} = \text{current} \times \text{resistance}$$
$$\text{volts} = \text{amperes} \times \text{ohms}$$

$$1 \text{ ohm} = \frac{1 \text{ volt}}{1 \text{ ampere}}$$

Some of the most important insights into the structure of the atom have been gained through studies of electrical discharges through gas. Such discharges, in contrast to the ease with which electricity passes through solid conductors and many aqueous solutions, require high voltages. Nonetheless such discharges are now common phenomena.

A glass tube such as the one shown in Figure 4.1 is known as a *Crookes tube* or a *cathode-ray tube*. In its simplest form, it contains a pair of **electrodes** (small metal discs attached to a source of electric potential) sealed into opposite ends of the tube and a connection to a vacuum pump. When the tube is filled with air, a high potential is necessary between the electrodes before electricity will flow from the *cathode* (the negative electrode) to the *anode* (the positive electrode). When it does flow, it is a sudden and violent discharge. This is because the air in the tube has a high resistance, thus acting as an *insulator*. As the air is gradually withdrawn, however, a lower and lower voltage is required for discharge. Finally when the air pressure in the tube has dropped below one-thousandth of atmospheric pressure, a sufficiently high voltage causes electricity to flow continuously. At the same time the glass walls glow with a greenish light. If a fluorescent screen is placed between the electrodes, bright lines will appear, extending outward from the cathode (Figure 4.1).

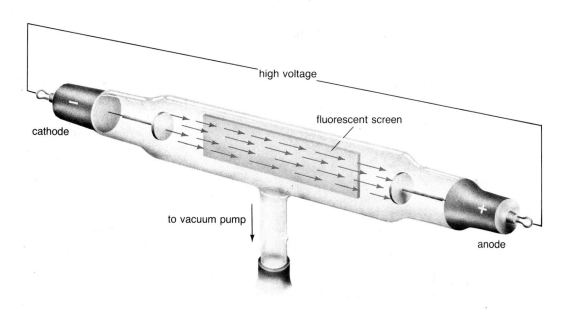

Figure 4.1 Cathode-ray tube.

This continuous flow of electricity was given the descriptive name *cathode rays*; however, it was soon identified as a stream of particles with a negative charge, which had already been given the name **electrons.** We can demonstrate some of the properties of cathode rays by illustrating several modifications to our original tube. Figure 4.2, where a solid object placed in the path of the

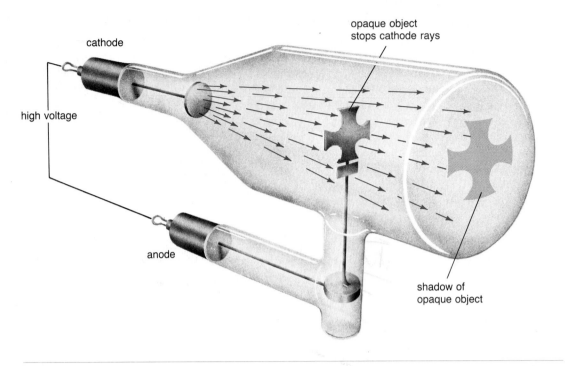

Figure 4.2 Cathode-ray tube with a solid object in the path of the rays.

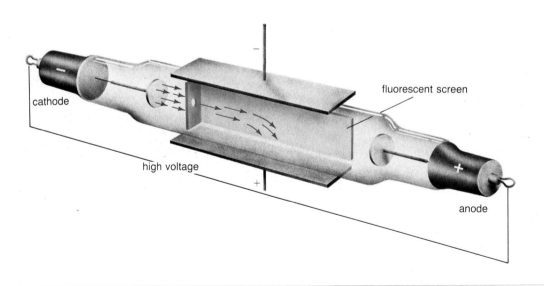

Figure 4.3 Cathode-ray tube in an electrical field.

electrons casts a distinct shadow, suggests that the rays are traveling in straight lines from the cathode. In Figures 4.3 and 4.4 the cathode rays respond characteristically to electric and magnetic fields. These deflections demonstrate conclusively that cathode rays consist of particles, since neither a magnetic nor an electrical field will deflect radiant energy. Early observations also demonstrated that the properties of the cathode rays are independent of the material from which the cathode is made and of the gas within the tube.

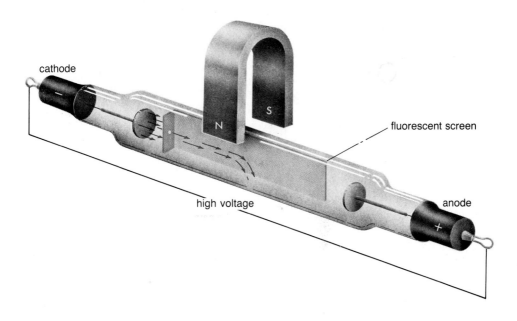

Figure 4.4 Cathode-ray tube in a magnetic field.

J. J. Thomson (1856–1940) was the first to measure the properties of cathode rays quantitatively. He determined the charge-to-mass ratio (e/m) of the electron by balancing the deflection of cathode rays caused by a magnetic field of known strength against the deflection caused by an electrical field of known strength (Figure 4.5). The value Thomson obtained for the e/m ratio was 1.76×10^{11} coul/kg. Therefore, as soon as the charge on each electron could be measured, the mass would be known as well.

Before long the challenge of determining the electronic charge was met. In 1911 an American, Robert A. Millikan (1868–1953), measured it to be 1.60×10^{-19} coul. His experimental apparatus, diagrammed in Figure 4.6, permitted him to observe the rate of fall of electrically charged oil drops between two charged plates. The oil drops were produced by an atomizer at the top of the apparatus and then entered the chamber between the charged plates through

Atomic Structure

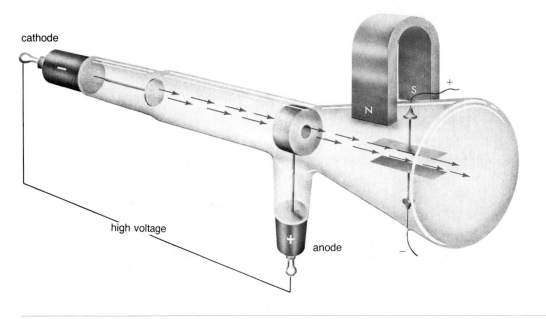

Figure 4.5 Diagram of apparatus used by Thomson in determining the charge-to-mass ratio for the electron.

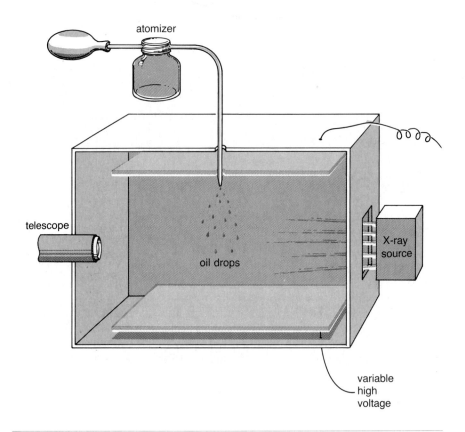

Figure 4.6 Diagram of the apparatus used by Millikan in determining the charge on the electron.

a small opening. In the chamber the drops absorbed one or more charged particles (electrons), which had been produced by irradiating the air in the chamber with X rays. Millikan then observed the behavior of the charged drops with a telescope.

Sidelight
An "Accidental" Discovery

Crookes tubes were electrical curiosities in the 1890s. Many scientists experimented with them. A German physicist named Wilhelm Conrad Roentgen (1845–1903) made an accidental discovery that eventually won him the Nobel Prize (1901). One day, in the course of his experiments, he had wrapped a Crookes tube in black paper. Then, having drawn the window shade, he turned on the high voltage switch to see if there was any light leaking from the tube. He could see no leak, but out of the corner of his eye he saw a greenish glow in a different part of the room. The glow disappeared when he turned off the high voltage but reappeared when he turned it on. Investigating, he discovered that the origin of the glow was some crystals of barium platinocyanide, a fluorescent substance, on a table some distance from the tube. Some kind of ray was penetrating the paper wrapping, reaching the crystals, and making them glow. Roentgen placed one obstruction after another between tube and crystals and found the effect undiminished, whether he used cardboard, wood, glass, rubber, or even metals. He found only lead and platinum impenetrable. He also discovered that the mysterious rays—which he called X rays, because he could not identify them—exposed a photographic plate. He took the first X-ray photograph when he placed his wife's hand on a photographic plate and exposed it to the tube. Seldom has a photograph had a greater or more immediate impact on the world than did this "picture" of the bones of a human hand. Physicians quickly seized upon the X rays for help in locating broken bones (Figure 4.7), but, in a Victorian world, others had different reactions.

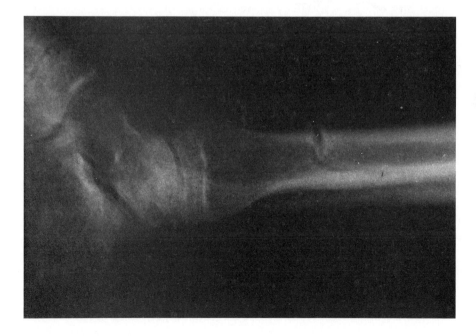

Figure 4.7 An X-ray photograph of the bones of a leg locates the fracture.

Atomic Structure

> Fearing that the X rays might be used "to look through the very clothes that people wore," New Jersey state legislators met the problem by proposing a ban on the use of X rays in opera glasses!

When the plates were not charged, the rate of fall of the oil drops did not vary. When the plates were charged, however, the rate of fall of each drop depended on how many electrons the drop had absorbed. The more it had accumulated, the more strongly it was repelled by the negatively charged plate. Millikan measured the velocities of the drops with the plates charged and uncharged, and from the difference he was able to calculate how much charge each drop had picked up. He found that the charge on each drop was either 1.60×10^{-19} coul or some small whole-number multiple of this value. Since this was the smallest charge he observed, he correctly assumed it to be the basic unit of electric charge, as well as the charge on the electron. By using this value and the e/m ratio for the electron, we can now easily calculate the mass of the electron:

$$\left(\frac{1.60 \times 10^{-19} \text{ coul}}{1 \text{ electron}}\right)\left(\frac{1.00 \text{ kg}}{1.76 \times 10^{11} \text{ coul}}\right) = 9.09 \times 10^{-31} \text{ kg/electron}$$

Even following these discoveries the cathode-ray tube had further contributions to make. Many scientists argued that if the cathode rays were streams of negative particles, there might be a beam of positive particles somewhere else within the tube. Eugen Goldstein (1850–1930) was the first to find these *positive rays*, or *canal rays*. He used a cathode-ray tube with a large perforated cathode plate such as the one shown in Figure 4.8. In addition to the usual

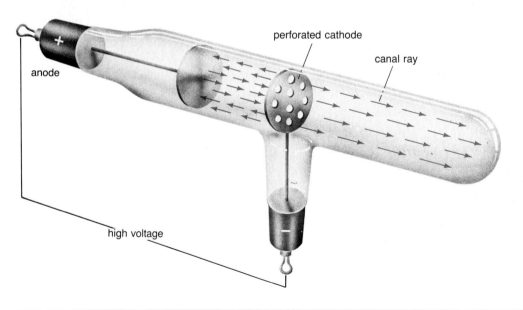

Figure 4.8 Canal rays in a cathode-ray tube with a perforated cathode.

stream of electrons drawn to the anode, Goldstein observed rays on the opposite side of the cathode. These canal rays differed from cathode rays in a number of ways. For one thing, they were luminous. (Cathode rays are never visible themselves and can be traced only by the luminosity they produce in other materials.) When the canal rays were tested for charge and mass with methods similar to those used to characterize the electron, they were found to consist of positive particles. But the e/m ratios for these particles depended on the gas in the tube. This suggested that these canal rays were made up of the positively charged residues of an interaction between the gas in the tube and the cathode rays. The cathode rays collide with molecules of the gas within the tube, knock off one or more electrons, and leave behind positively charged particles, which are then attracted to the cathode.

The least massive positively charged particle is observed when hydrogen is used in the vacuum tube. The e/m ratio for this particle is 9.60×10^7 coul/kg. This particle bears the smallest unit of positive charge ever measured. It is equal but opposite in sign to that of the electron; in other words, it is the fundamental unit of positive charge. We can calculate its mass as follows:

$$\left(\frac{1.60 \times 10^{-19} \text{ coul}}{1 \text{ particle}}\right)\left(\frac{1.00 \text{ kg}}{9.60 \times 10^7 \text{ coul}}\right) = 1.67 \times 10^{-27} \text{ kg/particle}$$

This mass is about 1840 times as great as the mass of an electron and identifies the particle we now know as the **proton.**

The principles used by Thomson for determining the e/m ratio of the electron were the basis for developing an instrument known as the mass spectrograph (Figure 4.9). With this instrument Thomson and F. W. Aston (1877–1946), an English physicist, conducted a number of experiments in which they passed the positively charged particles of the positive rays at high speeds through electric

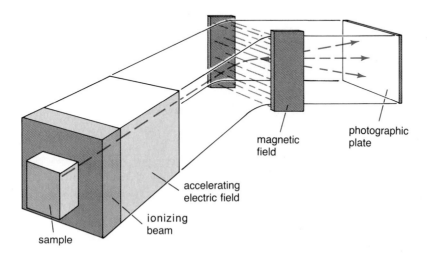

Figure 4.9 Diagram of a mass spectrograph.

and magnetic fields. The fields deflected the charged particles, and the amount of deflection depended on the mass of the particles. Light particles were deflected more than heavier ones. When positively charged particles from a single gaseous element were examined, Thomson and Aston often found that the positive rays contained particles with slightly different masses. For example, they produced two positively charged particles with different masses from the gas neon. This meant that natural neon consists of at least two kinds of atoms. The same is true of most other elements. Atoms of the same element having different masses are known as **isotopes.**

Question

2. Contrast cathode rays and positive rays with regard to
   a. the effect of an electric field.
   b. the effect of a magnetic field.
   c. the origins of the particles that make up the rays.
   d. the effect of the kind of metal in the electrodes on the nature of the particles that make up the rays.
   e. the size of the particles that make up the rays.
   f. the effect of the nature of the residual gas in the tube on particle size.

4.3 Radioactivity and the Nuclear Atom

During the same period that many scientists were busy investigating the electron, Antoine-Henri Becquerel (1852–1908), a French physicist and engineer, was studying *fluorescent* and *phosphorescent* minerals. (Fluorescent substances emit radiation, often visible light, immediately upon absorbing energy. Phosphorescent materials, on the other hand, emit radiation only after a considerable delay following the absorption of energy.) Among the samples with which he was working was a uranium salt. One day in 1896, when his work was interrupted, Becquerel put a sample of the salt in a drawer, inadvertently placing it atop a photographic plate. A few days later, when the plate was developed, it showed halos produced by the salts, even though the plate had been completely wrapped in a protective covering. Apparently the salt had spontaneously emitted an invisible penetrating ray that had exposed the photographic plate. As a result of this incident, Becquerel is now credited with discovering *radioactivity*. Later experiments determined that radioactive elements can, in fact, emit three kinds of radiation. One type is easily absorbed and consists of streams of particles with four times the mass and twice the positive charge of the proton. Each particle is thus equivalent to a helium atom with two electrons removed. These are called *alpha* (α) *rays*. A second type is highly penetrating and consists of streams of negative particles with charge and mass identical to the electron. These are called *beta* (β) *rays*. The third type of radiation has neither mass nor charge but has exceedingly high penetrating power. These are called *gamma* (γ) *rays*. These natural radiations were quickly seized upon by physicists as tools for further investigation of the atom.

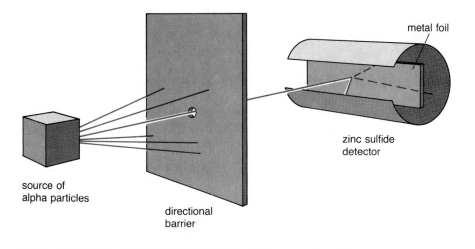

Figure 4.10 Schematic representation of an alpha-scattering experiment. (The actual experiment is carried out in a vacuum.)

The most important use of this newly discovered radiation was directed at understanding how each atom is put together. Since each atom is neutral, it must contain the same number of positive and negative charges, but at first no one knew how these charges were distributed or how the mass was dispersed within each atom. J. J. Thomson had advanced a theory of atomic structure that came to be called the "Thomson raisin-pudding model." He described the atom as a sphere of "uniform positive electrification" with the much smaller negative electrons "moving about inside it." But the Thomson model was upset by a series of experiments performed by Ernest Marsden (1889–1970) and Hans Geiger (1882–1945) under the direction of Ernest Rutherford (1871–1937). In these experiments a beam of alpha particles was directed at a very thin metal foil in a vacuum (Figure 4.10). But even when gold, one of the densest substances known, was used for the foil, most of the alpha particles passed through undeflected. Still more surprising was the observation that when the alpha particles *were* deflected, they were often deflected through angles much greater than could be explained by encounters with particles having the charge and mass of a proton (Figure 4.11). Ernest Rutherford attempted to explain these results with a new theory of atomic structure:

> Considering the evidence as a whole, it seems simplest to suppose that the atom contains a central charge distributed through a very small volume, and that the large single deflexions are due to the central charge as a whole, and not to its constituents.[1]

This suggestion explained the scattering experiments, whereas Thomson's model could not. If most of the mass of each atom and all of its positive charge

[1] E. Rutherford, "The Scattering of α and β Particles by Matter and the Structure of the Atom," *Philosophical Magazine*, **21**, 669, (1911).

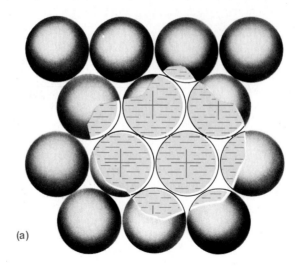

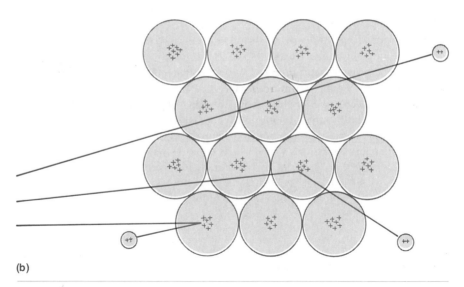

Figure 4.11 (a) The "raisin-pudding model" of Thomson which pictured the atom as a sphere of "uniform positive electrification" with the much smaller electrons "moving about inside it." Since any metal foil, no matter how thin, would be at least several hundred atoms thick, this model predicted that bombarding alpha particles would probably not penetrate very far into the foil until they would be stopped. **(b)** Scattering of alpha particles observed by Geiger and Marsden.

are concentrated in a very small volume (called the **nucleus**), we would expect most alpha particles to pass through the thin foil without deflection. Furthermore, since each nucleus is many times more massive than an alpha particle, occasional encounters will produce deflections as large as 180°, whereas encounters between alpha particles and lone protons would not.

From the measured angles of deflection of the alpha particles, Rutherford calculated the relative size of the nucleus of an atom. (The radius of an atom was already known to be about 1×10^{-10} m.) He calculated the radius of the nucleus to be about 1×10^{-15} m. This means that the nucleus occupies only

$$\frac{\frac{4}{3}\pi(10^{-15} \text{ m})^3}{\frac{4}{3}\pi(10^{-10} \text{ m})^3} = 10^{-15} \quad \text{or} \quad \frac{1}{1,000,000,000,000,000}$$

of the volume of the atom! It is little wonder that most of the alpha particles had passed through the gold foil without deflection.

Shortly after this, through the work of H. G. J. Moseley (1887–1915), the concept of **atomic number** was established. The atomic number of an element indicates the number of (unit) positive charges on the nucleus of the atom. Since each atom is electrically neutral, the atomic number is also equal to the number of electrons it contains. But a mystery still remained: The mass of an atom is typically more than *twice* the total mass of its electrons and protons. Then in 1932 James Chadwick (1891–1974) identified the **neutron,** a particle with a mass approximately equal to that of the proton but with no charge. Thus the total mass of the atom was accounted for. Each nucleus contains not only a number of protons equal to the atom's atomic number but also a number of neutrons. The total number of these *nucleons* (particles in the nucleus) is the **mass number** of the atom. Therefore the number of neutrons in an atomic nucleus is equal to the difference between its mass number and its atomic number (see Table 4.1).

Sidelight
Discoveries

Sometimes, in the midst of careful, routine, repetitive procedures in scientific research, comes an unexpected event that revolutionizes theory. The large deflections of some alpha particles by the metal foil was such an event. Rutherford was quoted as saying, "It was about as credible as if you had fired a 15-inch shell at a piece of tissue paper and it came back and hit you."

Sidelight
H. G. J. Moseley (1887–1915)

Moseley's work had established the concept of atomic number before he was twenty-seven. His promising career was cut short tragically when Great Britain entered the war. He immediately joined the army, left for overseas duty, and was killed by a sniper's bullet within two months.

> **Table 4.1** Relative Numbers of Particles in the Atom
>
> Atomic number = number of protons in the nucleus
> Mass number = number of nucleons = number of protons plus number of neutrons
> In an atom (which is neutral):
>   Number of protons = number of electrons

The neutron also explained the existence of isotopes, those atoms of the same element with different mass. Different isotopes of an element contain the same number of protons (hence have the same atomic number) but different numbers of neutrons (hence have different masses).

Example 4.1 How many protons, neutrons, nucleons, and electrons are there in each atom of a copper isotope with atomic number 29 and mass number 64?

Solution

a. The atomic number is the number of positive charges, or the number of protons, in the nucleus, and for an atom the number of protons equals the number of electrons:

$$\text{number of protons} = 29$$
$$\text{number of electrons} = 29$$

b. The mass number is the sum of the numbers of protons and neutrons; thus the number of neutrons is found by subtracting the number of protons from the mass number:

$$64 - 29 = \mathbf{35} \text{ neutrons}$$

c. Nucleons are particles that reside in the nucleus. The number of nucleons is the sum of the numbers of protons and neutrons.

$$29 + 35 = \mathbf{64} \text{ nucleons}$$

This is, of course, also the mass number.

We use the symbol Z for atomic number and A for mass number. To include this information with an atomic symbol, we write the values for Z and A as a subscript and a superscript, respectively, on the left side of the symbol. Thus the isotope of copper (Z = 29) with a mass number 64 (A) has the symbol:

$$^{64}_{29}\text{Cu}$$

Question

3. Compare isotopes of an element with respect to
   a. atomic number
   b. mass number
   c. number of protons per atom
   d. number of electrons per atom
   e. number of neutrons per atom
   f. number of nucleons per atom

4.4 Atomic Weights and Isotopes

We can now see the relationship between atomic weight, which we discussed in Section 2.1, and the mass number of an element. The atomic weight of each element represents the relative mass of its atom. This number is determined experimentally from the element's interactions with other elements, however; hence it can represent only the weighted average of the atomic weights of the stable isotopes naturally present. Therefore each element has an atomic weight that reflects the natural abundance of its isotopes. In addition each isotope has an atomic weight that depends on the number of nucleons in its nucleus. The arbitrary standard for all these values is the atomic weight of the $^{12}_{6}C$ isotope which is set equal to 12. On this basis the neutron and the proton have atomic weights of 1.0087 and 1.0074, respectively (see Table 4.2). (It will be clear only much later how 12 nucleons with atomic weights greater than 1.000 can combine in a nucleus with an atomic weight of 12.000.) Since each of these values is so close to 1.00, the mass number of an isotope (which is the *number* of neutrons and protons it contains) is always very close to the atomic weight of the isotope. Since the atomic weights of the elements are weighted averages of their stable isotopes, however, they may deviate considerably from whole numbers. For example, the two naturally occurring isotopes of boron, with mass numbers of 10 and 11, have atomic weights of 10.0129 and 11.00931, respectively. However,

Table 4.2 Comparison of the Properties of the Electron, Proton, and Neutron

	Charge	Relative Charge
Electron	$(-)1.60 \times 10^{-19}$ coul	-1
Proton	$(+)1.60 \times 10^{-19}$ coul	$+1$
Neutron	0	0

	Mass	Relative Mass	Atomic Weight
Electron	9.09×10^{-31} kg	1	0.00054
Proton	1.67×10^{-27} kg	1837	1.0074
Neutron	1.67×10^{-27} kg	1837	1.0087

the two isotopes occur naturally in such a ratio that the atomic weight of the element boron is 10.811. (Atomic weights and mass numbers for the elements and for each natural isotope can be found in the *CRC Handbook of Chemistry and Physics*.)

Example 4.2 Calculate the atomic weight of copper from the following facts: One isotope of copper has a mass number of 63, an atomic weight of 62.9298, and a relative abundance of 69.09 percent; a second isotope has a mass number of 65, an atomic weight of 64.9278, and a relative abundance of 30.91 percent.

Solution

a. To deal with a specific amount rather than with percentages, we shall consider an arbitrary sample of N atoms. This sample will contain $0.6909N$ atoms with a mass number of 63 and $0.3091N$ atoms with a mass number of 65.

b. To find the total atomic weight of the N atoms, multiply the number of each kind by its respective atomic weight and add:

$$(0.6909)(N)(62.9298) + (0.3091)(N)(64.9278)$$

c. Since the atomic weight of an element is the weighted average of its isotopes, we must now divide the total by the number of atoms:

$$\frac{(0.6909)(N)(62.9298) + (0.3091)(N)(64.9278)}{N} = (0.6909)(62.9298) + (0.3091)(64.9278)$$

$$= 63.55$$

d. Naturally our arbitrary number N cancels out. This must be true, since the answer must be the same for *any* sample.

Example 4.3 What percent error would be introduced in the answer to Example 4.2 if we used the mass numbers as approximations to the atomic weights of the isotopes?

Solution

a. If we use the mass numbers in place of the atomic weights, the number of significant figures in the answer is determined by the percent data:

$$\frac{(0.6909)(N)(63) + (0.3091)(N)(65)}{N} = (0.6909)(63) + (0.3091)(65)$$

$$= 63.62$$

b. The percent error is

$$\frac{63.62 - 63.55}{63.55} \times 100\% = 0.11\%$$

Example 4.4 Natural chlorine consists of two isotopes of mass numbers 35 and 37. If the atomic weight of chlorine is 35.45, calculate the percent abundance. (Use the mass numbers as close approximations to the atomic weights of the isotopes, but report the percent abundance to only three significant figures.)

Solution

a. Let X = the percent abundance of isotope-35. Then $100 - X$ = the percent abundance of isotope-37.

b. In a sample of N atoms, $(X/100)N$ atoms have a mass number of 35 and $[(100 - X)/100]N$ atoms have a mass number of 37.

c. The atomic weight is the weighted average of the isotopes. Therefore we must set the expression for the weighted average equal to the given value for the atomic weight and then solve for X:

$$\frac{X/100(N)(35) + [(100 - X)/100](N)(37)}{N} = 35.45$$

$$35X + 3700 - 37X = 3545$$

$$2X = 155$$

$$X = 77.5\% \text{ (for isotope-35)}$$

$$100 - X = 22.5\% \text{ (for isotope-37)}$$

Questions and Problems

4.1 Electrical Energy and Chemical Change

4. Compare lightning to the passage of current in a cathode-ray tube.

5. If a piece of glass rubbed with fabric is positively charged, what happens to the fabric used to rub the glass? Explain how the phenomenon occurs in terms of the modern concept of the atom.

4.2 Subatomic Particles

6. Consider the following statements in relation to both cathode rays and positive rays. Answer *true* or *false* in the appropriate columns.

	Cathode Rays	Positive Rays
a. They travel in straight lines.	_____	_____
b. They originate at the cathode.	_____	_____
c. They originate at the anode.	_____	_____
d. They originate from the gas in the tube.	_____	_____
e. They travel from the cathode to the anode.	_____	_____
f. The nature of the ray depends on the kind of gas in the tube.	_____	_____
g. The nature of the ray depends on the type of cathode metal.	_____	_____
h. The nature of the ray depends on the type of anode metal.	_____	_____

4.3 Radioactivity and the Nuclear Atom

7. Calculate the e/m ratio for the alpha particle.

8. Complete the following table.

Symbol	Atomic Number	Mass Number	Number of Protons	Number of Electrons	Number of Neutrons	Number of Nucleons
		52	24			
					74	
				56		136
	36	84				

9. How do the two isotopes of chlorine of mass numbers 35 and 37 differ?

10. What is the charge in coulombs on the nucleus of the nitrogen atom?

11. The radius of an atom is about 10^5 times the radius of its nucleus. If the nucleus had a radius of 1 in., how large would the radius of the atom be? Express this answer in miles.

4.4 Atomic Weights and Isotopes

12. Gallium consists of two isotopes with mass numbers 69 and 71 and with relative abundances of 60.1 and 39.9 percent, respectively. Calculate the atomic weight of gallium, using the mass numbers as close approximations to the atomic weights of the isotopes.

*__13.__ There are three common isotopes of magnesium, having mass numbers 24, 25, and 26. If the relative abundances of these isotopes are 78.7, 10.1, and 11.1 percent, respectively, calculate the atomic weight for magnesium. (Use mass numbers for the atomic weights of the isotopes.)

14. The element boron has two common isotopes of mass numbers 10 and 11. Calculate the relative abundances of these isotopes.

15. The element rubidium has two common isotopes of mass numbers 85 and 87. Calculate the relative abundances.

16. Calculate Avogadro's number from the fact that an atom (weighted average atom) of hydrogen weighs 1.67×10^{-24} g.

17. What is the mass in grams of a weighted average atom of uranium? How many times heavier is this atom than the average atom of hydrogen?

18. If carbon is the reference standard for atomic weights and is given the value 12, exactly, why is the atomic weight of carbon listed on the back cover as 12.011?

19. If the value of 50 instead of 12 had been assigned to the reference carbon isotope, what would be the atomic weight of the chromium isotope that is now given the atomic weight 52.0?

Discussion Starters

20. The ratio between number of neutrons and number of protons in the nucleus of an atom varies with the atomic number. Determine this ratio for the noble gases and then graph the ratio against atomic number. (Use as the mass number the whole number closest to the atomic weight.) State in words a conclusion which may be drawn from your figures and graph.

21. Explain or account for the following:
  a. More static electricity develops on dry days than on humid days.
  b. The mass of an atom is generally more than twice the total mass of its electrons and protons.

5

Electron Configurations and the Periodic Table

In the previous chapters we began discussing the internal structure of atoms and, in so doing, laid the basis for a logical ordering of the elements. We shall now see that we can indeed arrange the elements in such a way that the resulting table yields useful information about trends in atomic properties. Before embarking on a discussion of this periodic arrangement, however, we must take care of some important preliminaries.

5.1 Electromagnetic Radiation

Electromagnetic radiation, or radiant energy, takes several forms, of which visible light is the most familiar. It was natural, therefore, that studies of light were the first important contributions to our understanding of electromagnetic radiation. Since the middle of the nineteenth century it has been apparent that we can best explain many of the characteristics of light by considering it to be a *wave phenomenon*. But not until scientists also accepted the *particle nature* of light did we arrive at a fully satisfactory understanding of radiant energy. The development of our understanding of many phenomena had similar histories. For example, the experiments described in Chapter 4 established the electron as having mass; in other words, they established its particle nature. Today scientists recognize that the electron possesses characteristics of both a particle and a wave.

Any wave phenomenon is characterized by its *wavelength, amplitude, frequency*, and *velocity*. To discuss these properties, let us consider a familiar example—the concentric ripples produced when a stone is dropped into a pool of water (Figure 5.1). In Figure 5.2 we have abstracted the wave form and placed it on a set of rectangular coordinates where we recognize it as a sine curve. The *wavelength* in this example is the distance from a to b; the *amplitude* of the wave is the maximum height of the curve above the axis (c to d); the *frequency* of the wave is the number of wavelengths that pass a given spot in a unit of time; and the *velocity* is the distance traveled by any point on the wave in a unit of time. Wavelengths are usually measured in meters or nanometers; velocities, in meters per second or angstroms per second; and frequencies, in hertz, which are wavelengths per second. Conveniently, all light, in fact all

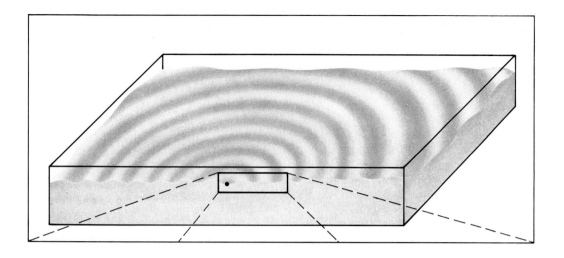

Figure 5.1 Water waves.

forms of radiant energy, travels in a vacuum with the same velocity: 3.00×10^8 m/s (which is 1.86×10^5 mile/s). Therefore, we can relate the constant velocity of light (c), its wavelength (λ, Greek lambda), and its frequency (v, Greek nu) as follows:

$$\lambda = \frac{c}{v}$$

Using this equation, we can now characterize all electromagnetic radiation according to its wavelength (or frequency). If we arrange all radiation

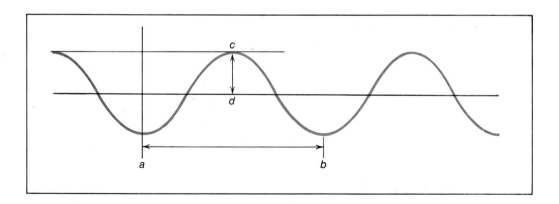

Figure 5.2 Wave form.

Electron Configurations and the Periodic Table 97

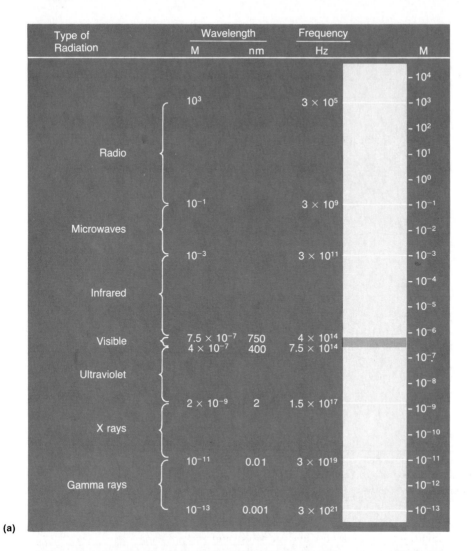
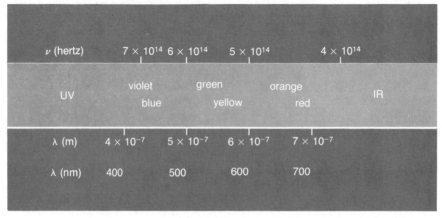

Figure 5.3 (a) The electromagnetic spectrum. (b) The visible spectrum.

along a *continuum* (a continuous series whose parts cannot be precisely separated) of increasing wavelength (or decreasing frequency), we can describe any radiation according to its place in the *electromagnetic spectrum*. Figure 5.3 shows the entire electromagnetic spectrum (a) and, on a larger scale, that portion visible to the human eye (b).

Example 5.1 Calculate the frequency of yellow-orange light having a wavelength (in a vacuum) of 6.00×10^{-7} m.

Solution

a. The relationship between frequency and wavelength, where c is the speed of light in a vacuum, is

$$\lambda = \frac{c}{\nu} \quad \text{or} \quad \nu = \frac{c}{\lambda}$$

b. Substituting values for c and λ,

$$\nu = \left(\frac{3.00 \times 10^8 \text{ m}}{1.00 \text{ s}}\right)\left(\frac{1.00 \text{ wavelength}}{6.00 \times 10^{-7} \text{ m}}\right)$$

$$= \frac{5.00 \times 10^{14} \text{ wavelengths}}{1.00 \text{ s}} \quad \text{or } \mathbf{5.00 \times 10^{14} \text{ Hz}}$$

Isaac Newton (1642–1727) made the first important modern studies of visible light. More than a hundred years elapsed, however, before another form of radiant energy was discovered. Then in 1800 *infrared radiation*, with frequencies lower than light at the red end of the visible spectrum, was discovered, and in 1801 *ultraviolet radiation*, with frequencies higher than light at the violet end of the visible spectrum, was first detected. In the last decades of the nineteenth century, scientists identified radio waves and X rays, and only during the twentieth century has the rest of the spectrum been filled in. In fact, scientists are still a long way from a complete understanding of the extremely high-frequency radiation called *cosmic rays*.

These different kinds of electromagnetic radiation have different sources and are detected with different instruments; however, all share the same velocity in a given medium and all can be described by the same theory. (Indeed we shall sometimes use the term *light* to describe the entire spectrum, reserving the phrase *visible light* for that small portion we can detect with our eyes.)

One of the most important features of the theory of electromagnetic radiation is that it accounts for the characteristics of light that cannot be explained by a simple wave theory. One of these is the distribution of frequencies emitted by a hot glowing body. A piece of metal changes from red hot to white hot as it absorbs energy. An analysis of its radiation spectrum at each temperature shows that at higher temperatures a greater proportion of the radiation consists of light of the higher frequencies. To explain these observations, Max Planck (1858–1947) proposed that the radiating body has only certain discrete energy states, not the predicted continuum of energy states, and that the energies of these states are directly related to the frequency of the radiation emitted. He predicted, therefore, that energy can only be emitted or absorbed during transitions between definite states. The details of Planck's proposal are complicated, but when the calculations are done, his theory correctly explains the distribution of frequencies emitted by a glowing body. The crux of his argument was that radiation must be absorbed and emitted in discrete units and that the energy of the radiation is related to its frequency:

$$E = h\nu$$

where h is *Planck's constant* and is equal to 6.63×10^{-34} joule-second. The joule is defined in terms of SI base units as a kilogram-meter^2 per second^2. The joule, like the calorie, is an energy unit, and 1 cal = 4.1840 J.

A second phenomenon that could not be accounted for by regarding radiant energy as simple waves was the *photoelectric effect*. Heinrich Hertz (1857–1894) produced the first evidence of the effect in 1887 when he found that a lower voltage was required to produce an electric spark between two electrodes if the electrodes were exposed to ultraviolet radiation. Further investigation uncovered the reason: Irradiation of a metal surface with ultraviolet light or X rays often causes the metal to emit electrons. Not all frequencies, however, cause all metals to respond this way. For each metal, in fact, there is a certain minimum frequency, called a *threshold frequency*, at which electrons will be ejected; radiant energy of lower frequency will not produce the photoelectric effect. Thus for zinc, with a threshold frequency of 8.6×10^{14} Hz, only ultraviolet (or higher frequency) radiation will produce electrons; visible light will have no effect. Sodium metal, on the other hand, with a threshold frequency of 4.6×10^{14} Hz, will exhibit the photoelectric effect for all visible light except very-long-wavelength red light. In addition, increasing the intensity of the radiation upon the metal (intensity is a measure of the *amount* of light and is not related to frequency) does *not* increase the energy or the velocity of the electrons emitted; it only increases their number.

Albert Einstein (1879–1955) offered an explanation of the photoelectric effect by carrying Planck's assumption one step further. He proposed that energy is *quantized*—comes in bundles—at all times, whereas Planck had said only that energy is absorbed or emitted in quanta. Einstein suggested that each quantum of energy, which was soon called the **photon,** carries an amount of energy equal to $h\nu$. Quanta, or photons, of different frequencies therefore have different amounts of energy, and photons of higher frequencies have higher energies. Einstein's theory explained the photoelectric effect by assuming that an ejected electron (called a *photoelectron*) receives its energy by absorbing a single photon; thus light below the threshold frequency, regardless of its intensity, does not have enough energy to remove electrons. Likewise, increasing the intensity of light above the threshold frequency merely produces more, rather than more energetic, electrons.

Questions

1. Which light has the longer wavelength, red light or blue light? Which has the higher frequency?

2. Are radio waves of relatively high or low frequency? Of long or short wavelengths? Of great or small energy?

3. Which type of rays in Figure 5.3(a) has the greatest energy?

Sidelight
The Photoelectric Effect and the Garage Door

The alkali metals, such as sodium or cesium, produce the photoelectric effect with visible light. Thus the headlights of a car, focused on a cesium cathode in an "electric eye," cause the metal to eject electrons. A second electrode within the tube of the electric eye has a slight positive charge, which is maintained by a battery. This electrode attracts the electrons, and current flows through the circuit, which in turn triggers an electric motor and the garage door opens.

5.2 Atomic Spectra

When white light, which contains light of all visible frequencies, passes through a prism (Figure 5.4), it is expanded into the familiar rainbow of colors. Thus

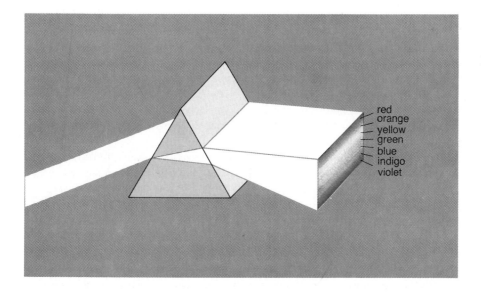

Figure 5.4 White light is resolved into its component wavelengths by a prism.

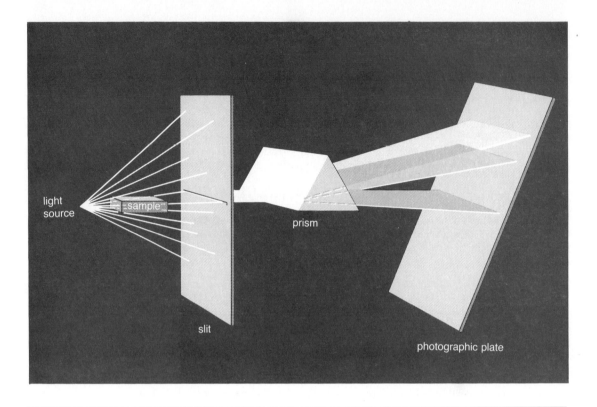

Figure 5.5 Diagrammatic representation of a spectrometer.

when white light is analyzed in a *spectrometer*, which is an instrument that analyzes the frequency of light, the result is a *continuous spectrum* (Figure 5.5). In contrast, when the radiation from a heated sample of a single element in the vapor state passes through the slit and prism of a spectrometer, the result is a discontinuous or *line spectrum* (a spectrum consisting of discrete lines). This indicates that the radiation contains only certain specific wavelengths of light and that energy is emitted in discrete amounts (quanta). The *emission spectra* for atomic hydrogen and helium are shown in Figure 5.6. *Absorption spectra* can be thought of as the inverse of emission spectra. They are generated when white light is first passed through the sample of hydrogen or helium and then analyzed in the spectrometer. Absorption spectra are continuous except for black lines that occur at the same frequencies as the emission lines for the element being analyzed. This indicates that for a given element energy is absorbed only in the same discrete quanta as it is emitted. Absorption and emission spectra for a given element will show identical patterns.

Niels Bohr (1885–1962), a Danish physicist, used the evidence of spectra like these to formulate a **quantum theory** of the atom. His theory incorporated the hypotheses of Planck and Einstein, and although it has been modified and extended, it remains the basis of our current understanding of the atom. Bohr

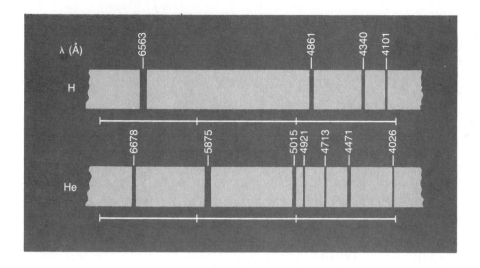

Figure 5.6 Emission spectra for hydrogen and helium.

built on Rutherford's model of the atom, with its small positively charged nucleus and its diffuse distribution of electrons, by assuming that the electrons can exist only in certain discrete *energy levels*. He theorized that the atom does not radiate energy when the electrons are *in* these energy levels but that discrete bundles of energy (photons) are emitted when electrons jump from one energy level to a lower one. Energy is absorbed only when an electron is raised (*excited*) to a higher level. Therefore the quantity of energy emitted or absorbed is equal to the difference between the energies of the atom before and after the electron changes energy levels.

To do as Bohr did—to take the mass of available emission spectra and to deduce the energy levels of the atom—required a stroke of genius. We shall find beginning with the theory and working backward considerably easier. Let us suppose that, in a hypothetical atom, electrons can exist in energy states E_1, E_2, E_3, and E_4, with E_1 the lowest energy level and E_4 the highest. Recall that when the atom absorbs energy, an electron will move to a higher energy level and that when the energy source is removed, the excited electron will relax to its original level and energy will be emitted. In our hypothetical atom, if the electron moves from E_3 to E_2, the energy emitted will be

$$E = E_3 - E_2$$

and will be detected as radiation with a frequency

$$\nu = \frac{E}{h} = \frac{E_3 - E_2}{h}$$

Electron Configurations and the Periodic Table

Electrons, of course, will not be restricted to one-step changes, so we might expect emissions at any of the following frequencies:

$$\frac{E_2 - E_1}{h} \quad \frac{E_3 - E_1}{h} \quad \frac{E_4 - E_1}{h} \quad \frac{E_3 - E_2}{h} \quad \frac{E_4 - E_2}{h} \quad \frac{E_4 - E_3}{h}$$

Each of these emissions will, therefore, appear as a separate line in the emission spectrum of the element.

As a confirmation of this kind of analysis, Figure 5.7 illustrates the emission spectrum of the hydrogen atom. It contains several distinct series of lines, each

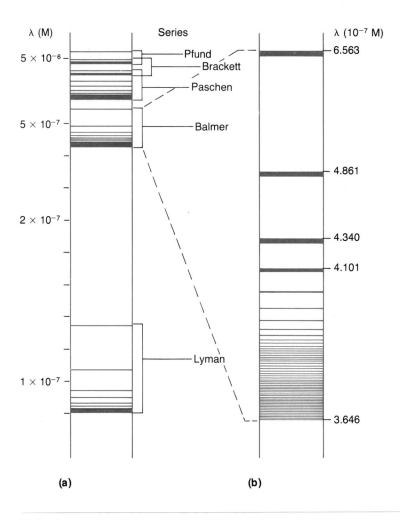

Figure 5.7 (a) The emission spectrum of the hydrogen atom. (b) The Balmer series (visible range) at greater magnification. (Note that the wavelength scale is a logarithmic rather than a linear scale.)

named for its discoverer. In Figure 5.8 we can see the origin of each series. The Lyman series, which is the series of highest energy, is produced when electrons drop from higher energy levels to level E_1. When the electron is in the lowest available energy level (E_1), the atom is said to be in the *ground state*. Since relaxations to the ground state of the hydrogen atom are relatively high-energy transitions, the emissions have high frequencies and the Lyman series appears in the ultraviolet region of the spectrum. Next in energy and thus in frequency is the Balmer series, which appears in the visible region. These emissions are produced by electrons dropping to level E_2. The next two series, the Paschen series and the Brackett series, appear at still lower energies in the infrared region of the spectrum.

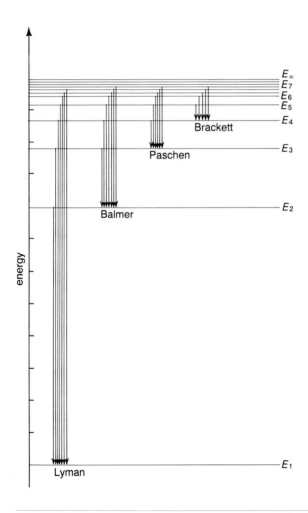

Figure 5.8 Diagram of energy levels for an electron on the hydrogen atom. Energy is expressed as that amount necessary to remove an electron in the energy level from an atom.

Emission spectra of only the first two or three elements are as simple and readily analyzable as the hydrogen spectrum. The line spectra of heavier elements are characteristic and thus provide a reliable means for identification, but they can be extremely complex. Furthermore, only gaseous atoms yield line spectra. In liquids and solids, the interactions among the electrons of the many closely packed atoms introduce so many different energy levels that emission spectra are essentially continuous.

Question

4. What everyday things other than a triangular prism will separate white light into a continuous spectrum?

5.3 Electrons, Quantum Numbers, and Energy States

The energy of an electron is manifested primarily in its velocity. The higher the energy level, the higher the average velocity. Furthermore, the faster an electron travels, the farther it tends to move from the nucleus to which it is bound. In Bohr's original theory, he described electrons as rotating about the nucleus in circular paths called *orbits*. Thus each discrete energy level of the atom became a concentric *shell* of several orbits. In mathematical terms, his theory required an electron in an energy level to conform to the relationship:

$$mvr = \frac{nh}{2\pi}$$

where m and v are, respectively, the mass and the velocity of the electron, h is Planck's constant, r is the radius of the electron's orbit (the Bohr orbit), and n is any positive integer (1, 2, 3, ...). If we write the equation as

$$2\pi r = \frac{nh}{mv}$$

we can see more clearly that the circumference ($2\pi r$) of the orbit of an electron increases with increasing values of n. This integer, n, is called the *principal quantum number*.

A number of modifications have been made to Bohr's original revolutionary theory. For one thing, Bohr's theory could not account for the energies of atoms with more than one electron. In addition, with the development of spectrometers that gave better resolution of the spectral lines, scientists found that what had appeared to be single lines were in fact clusters of lines. More lines meant more energy transitions, and that implied more energy levels. Since the lines came in clusters, the new energy levels were described as *sublevels* within the principal energy levels. The slight variations in energy that yielded these new lines were first explained as the result of deviations from the original circular orbits. Thus each sublevel of a principal energy level could be thought of as corresponding to a discrete elliptical orbit. This required a second quantum number l, called the *azimuthal quantum number*, to describe these elliptical paths or sublevels. Within a level with principal quantum number n, l may have the values $0, 1, 2, \ldots, (n-1)$. In other words, when $n = 1$, $l = 0$, which says that the first energy level consists of only a single sublevel. But when $n = 2$, then $l = 0$ and $l = 1$; thus the second energy level contains two sublevels. For $n = 3$, $l = 0, 1$, and 2, thus producing three sublevels, and so on. In Bohr's theory, n was related to the size of the orbit and l, to its shape. It should also be clear that the principal quantum number, which specifies the energy level, also specifies the *number of* sublevels in that energy level. The spectral lines corresponding to these different sublevels were first classified as sharp, principle, diffuse, or fundamental. This obsolete system survives in the symbols s, p, d, and f for the sublevels with $l = 0, 1, 2$, and 3, respectively.

Further spectroscopic study revealed that many emission lines were split into three or more lines when a magnetic field was applied to the sample. This meant that many energy sublevels in fact contained several distinct electronic orbits, whose energies differed *only* in the presence of a magnetic field. Therefore, a third quantum number m, the *magnetic quantum number*, was introduced to account for these variations. Within a sublevel l, m may have the following values: $-l, (-l+1), \ldots, 0, \ldots, (l-1), l$. We can think of the values of m as corresponding to different *orientations* of the orbits within a given sublevel. Thus the first energy level $(n = 1)$ has only one sublevel $(l = 0)$, which in turn has only one orientation $(m = 0)$. The second energy level $(n = 2)$ has two sublevels $(l = 0$ and $l = 1)$. The s sublevel $(l = 0)$ within the second energy level has only one orientation $(m = 0)$, but the p sublevel $(l = 1)$ has three orientations $(m = -1, 0, +1)$. We now designate these three orientations by the symbols p_x, p_y, p_z to indicate orientations relative to an arbitrary set of three coordinate axes. The third energy level $(n = 3)$ consists of three sublevels $(l = 0, 1,$ and $2)$. Of these, the s sublevel has one orientation $(m = 0)$, the p sublevel has three orientations $(m = -1, 0, +1)$, and the d sublevel $(l = 2)$ has five orientations $(m = -2, -1, 0, +1, +2)$.

In 1927 still another complexity was added. Otto Stern (1888–1969) and W. Gerlach (b. 1899) made careful measurements of the spectra of several elements with an odd number of electrons per atom. They found, for example, that when they subjected a beam of vaporized sodium atoms (atomic number $= 11$) to a magnetic field, the beam split into two equivalent beams. They interpreted this to mean that the electron, while traveling around the nucleus, spins on its own axis. Ten of the electrons in each sodium atom have paired spins, and only the eleventh has a spin that is not canceled out. Thus half the sodium atoms have the eleventh electron spinning one way, and half have it spinning the other. Each of these two populations responds differently to a magnetic field. Thus we need a fourth and final quantum number s, the *spin quantum number*, to describe the energy state of an atom. The spin quantum number may have the value $+\frac{1}{2}$ or $-\frac{1}{2}$.

Table 5.1 Allowed Values for the Quantum Numbers

$n = 1, 2, 3, \ldots$

$l = 0, 1, 2, \ldots, (n - 1)$

$m = -l, (-l + 1), \ldots, 0, \ldots, (l - 1), l$

$s = +\frac{1}{2}, -\frac{1}{2}$

Table 5.1 summarizes the allowed values for the four quantum numbers, and Table 5.2 the symbols for the **orbitals** (the Bohr term *orbit* has been discarded as no longer descriptive) available in the first three energy levels, along with the symbol for each orbital. (The subscripts to the symbols for the $3d$ orbitals, as with the p orbitals, designate different orientations of the orbitals with respect to an arbitrary set of coordinate axes.) Note that there is no mention in Table 5.2 of the spin quantum number; we simply assume that an electron in any of the energy states listed may have either $s = +\frac{1}{2}$ or $s = -\frac{1}{2}$. Two electrons in the same orbital, but with different values for s, differ only in regard to their spin and have identical energies. Electrons in the same sublevel, having only different values for m, have different energies only in the presence of a magnetic field.

Table 5.2 Available Orbitals and Orbital Symbols for the First Three Energy Levels

	n Value	l Value	m Value	Symbol for Orbital
First energy level	1	0 (s)	0	$1s$
Second energy level	2	0 (s)	0	$2s$
	2	1 (p)	1	$2p_x$
	2	1 (p)	0	$2p_y$
	2	1 (p)	-1	$2p_z$
Third energy level	3	0 (s)	0	$3s$
	3	1 (p)	1	$3p_x$
	3	1 (p)	0	$3p_y$
	3	1 (p)	-1	$3p_z$
	3	2 (d)	2	$3d_{xy}$
	3	2 (d)	1	$3d_{xz}$
	3	2 (d)	0	$3d_{z^2}$
	3	2 (d)	-1	$3d_{yz}$
	3	2 (d)	-2	$3d_{x^2-y^2}$

Nothing we have said so far suggests that Bohr's picture of the atom as a positive nucleus surrounded by orbiting electrons is inadequate. But, in fact, there are several reasons that an atom cannot be accurately represented in such a simple way. Most important among them is the difficulty we encounter when we attempt to determine the position or velocity of an orbiting electron. The way we locate an object is by some interaction of it with a measuring device. Ordinary objects are perceived and measured by the light they reflect. In the case of an electron, whose dimensions are smaller than the wavelength of visible light, we must use X rays or γ rays. The shorter the wavelength of the radiation being used, however, the more energetic must be its interaction with the object being observed. Thus when we attempt to "see" an electron, we necessarily perturb it. Indeed, the smaller the object being observed, the more energetic the radiation that must be used and the greater the perturbation. It is therefore impossible to obtain a complete description of the velocity and the whereabouts of an electron at any instant. Werner Heisenberg (1901–1976) first enunciated this principle, which now bears his name—the **Heisenberg uncertainty principle.** The principle has a mathematical formulation, but in words it says that it is impossible even in theory to know both the position and the velocity of an electron at the same time.

To account for this and other objections to the precise orbits of the Bohr atom, we must now think of electrons as occupying orbitals, and we must replace the circular path of each electron with a fuzzy *electron cloud*. Each orbital may, of course, be unambiguously defined by a set of quantum numbers (and may also be described mathematically), but the configuration of its electron cloud, or *probability density pattern*, provides a more concrete image.

Figure 5.9 is a cross section of such a density pattern for an electron in a 1s orbital. You might think of it as being produced by superimposing several thousand snapshots of the electron (snapshots that would certainly simplify things if they were not merely imaginary). If the electron's position in each snapshot is represented by a dot, the density of dots in any region is related to the probability of the electron being found in that region. There is no outer limit to the diagram, because there is some probability that the electron will be found at any position. But an arbitrary boundary, drawn to include 95 percent of the dots, gives a shape that is called the *orbital shape*. Such *boundary surfaces* for the 2s, $2p_x$, $2p_y$, and $2p_z$ orbitals are given in Figure 5.10. The 2s orbital is

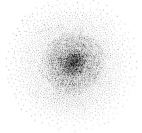

Figure 5.9 Cross section of a density pattern for an electron in a 1s orbital.

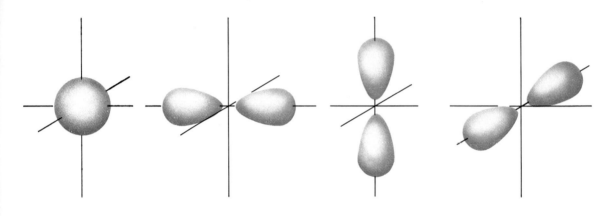

Figure 5.10 Boundary surfaces for the 2s, $2p_x$, $2p_y$, and $2p_z$ orbitals.

spherical, as is the 1s orbital, but the p orbitals are two-lobed distorted ellipsoids. The lobes of the p_x orbitals extend along the x axis; those of the p_y orbitals, along the y axis; and those of the p_z orbitals, along the z axis of an arbitrary set of Cartesian coordinates. In Figure 5.11 the three p orbitals are shown as they might be visualized on one atom, and in Figure 5.12, with some difficulty, we show the 1s, 2s, $2p_x$, $2p_y$, and $2p_z$ orbitals, superimposed. Boundary surfaces for d orbitals are even more complicated, as shown by Figure 5.13. In addition, recall that even though these orbitals are shown singly in the illustration, they are actually superimposed upon each other and upon underlying s and p orbitals.

Sidelight
Recognition

The scientific world was quick to recognize the fundamental importance of the work done by the men mentioned on these pages. Many were awarded the Nobel Prize:

Max Planck, 1918: "For his hypothesis that all radiation was emitted in units called quanta."
Albert Einstein, 1921: "For services to physics and the law of the photoelectric effect."
Niels Bohr, 1922: "For his atomic theory that laid the groundwork for later atomic research."
Werner Heisenberg, 1932: "For the theory that began modern quantum mechanics."

We have seen that the four quantum numbers describe the size, shape, and orientation of the orbitals and the relative spin on the electron. They may therefore be looked upon as an *ordered quadruple* that identifies a certain electron much as an *ordered pair* locates a certain point on a plane. In a given atom, electrons are assigned quantum numbers according to the following principles, along with the restrictions outlined in Table 5.1.

1. The Pauli exclusion principle: *No two electrons on one atom may have the same set of four quantum numbers.* Since the spin quantum number s has only two values, this principle limits the population of each orbital to two electrons, which must be spinning in opposite directions.

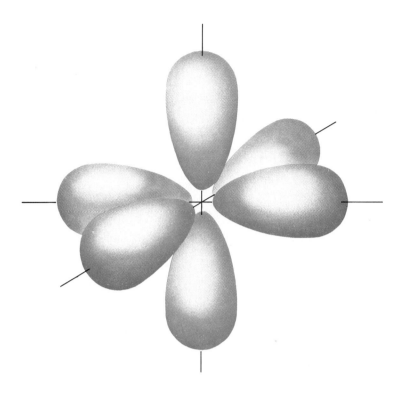

Figure 5.11 Three p orbitals on one atom.

Figure 5.12 The $1s$, $2s$, $2p_x$, $2p_y$, and $2p_z$ orbitals on one atom.

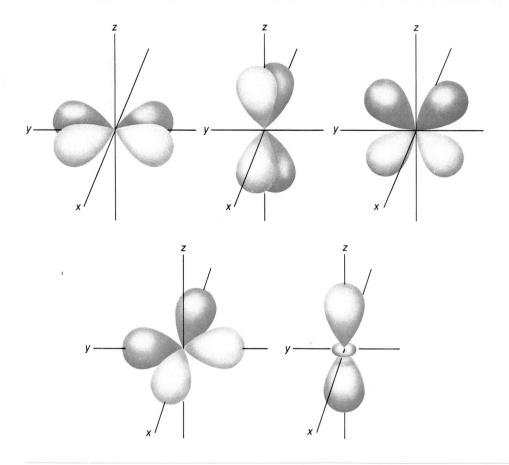

Figure 5.13 Boundary surfaces for d orbitals.

2. **The Aufbau (building-up) principle:** *When not excited, the electrons of an atom will be in the lowest energy states allowed by the Pauli exclusion principle.* This principle allows us to determine the arrangement or *configuration* of the electrons on any atom by assigning all electrons to the lowest available energy levels or sublevels.

3. **Hund's rule:** *In any sublevel, each orbital will accommodate only one electron until every orbital in that sublevel has one electron.* Furthermore, all of these electrons will have the same ("parallel") spin. The fourth electron to enter a *p* sublevel or the sixth electron to enter a *d* sublevel or the eighth to enter an *f* sublevel begins the pairing of electron spins.

Using these principles, we can extend Table 5.2 to show in Table 5.3 the quantum numbers for all electrons allowed in the first three energy levels. In addition we can tabulate the numbers of sublevels and orbitals in each energy level, as well as the maximum number of electrons allowed in each energy level (Table 5.4).

Table 5.3 Allowable Quantum Numbers in the First Three Energy Levels

First Energy Level $n = 1;\ l = 0;$ $m = 0$	Second Energy Level $n = 2;\ l = 0, 1;$ $m = 1, 0, -1$	Third Energy Level $n = 3;\ l = 0, 1, 2;$ $m = 2, 1, 0, -1, -2$
$(1, 0, 0, +\frac{1}{2})$	$(2, 0, 0, +\frac{1}{2})$	$(3, 0, 0, +\frac{1}{2})$
$(1, 0, 0, -\frac{1}{2})$	$(2, 0, 0, -\frac{1}{2})$	$(3, 0, 0, -\frac{1}{2})$
	$(2, 1, 1, +\frac{1}{2})$	$(3, 1, 1, +\frac{1}{2})$
	$(2, 1, 0, +\frac{1}{2})$	$(3, 1, 0, +\frac{1}{2})$
	$(2, 1, -1, +\frac{1}{2})$	$(3, 1, -1, +\frac{1}{2})$
	$(2, 1, 1, -\frac{1}{2})$	$(3, 1, 1, -\frac{1}{2})$
	$(2, 1, 0, -\frac{1}{2})$	$(3, 1, 0, -\frac{1}{2})$
	$(2, 1, -1, -\frac{1}{2})$	$(3, 1, -1, -\frac{1}{2})$
		$(3, 2, 2, +\frac{1}{2})$
		$(3, 2, 1, +\frac{1}{2})$
		$(3, 2, 0, +\frac{1}{2})$
		$(3, 2, -1, +\frac{1}{2})$
		$(3, 2, -2, +\frac{1}{2})$
		$(3, 2, 2, -\frac{1}{2})$
		$(3, 2, 1, -\frac{1}{2})$
		$(3, 2, 0, -\frac{1}{2})$
		$(3, 2, -1, -\frac{1}{2})$
		$(3, 2, -2, -\frac{1}{2})$

Table 5.4 Sublevels, Orbitals, and Electrons in Each Energy Level

Sublevel	Number of Orbitals
$l = 0\ (s)$	1
$= 1\ (p)$	3
$= 2\ (d)$	5
$= 3\ (f)$	7

Energy Level	Number of Sublevels	Number of Orbitals	Maximum Number of Electrons
$n = 1$	1	1	2
$= 2$	2	4	8
$= 3$	3	9	18
$= 4$	4	16	32
$= 5$	5	25	50

In general each energy level, characterized by the principal quantum number n, contains n sublevels, n^2 orbitals, and a maximum of $2n^2$ electrons.

Note especially the numbers in the right-hand column of Table 5.4: 2, 8, 18, 32, 50. We shall soon find that these same numbers arise in a totally different context—from a consideration of the properties of the elements.

Question

5. Point out at least two fundamental differences between Bohr's concept of the atom and the modern concept.

5.4 Periodic Classification of the Elements

We now know enough about atomic structure to see that one important difference among the elements is the difference in electron configurations. It remains to show that this difference can be used as the basis for a logical ordering of the elements according to their properties. A knowledge of the properties of matter is important to a chemist. He must know that oxygen, nitrogen, and chlorine are gases at room conditions, that bromine and mercury are liquids, and that sulfur, iodine, magnesium, copper, and phosphorus are solids. He must know that fluorine is so active that it requires specially lined containers and that potassium and sodium metals must be stored under kerosene to prevent their spontaneous ignition. For his own safety, he needs to know that bromine causes severe burns and that mercury vapor is dangerously toxic. Clearly it would be useful to organize these apparently unrelated observations by classifying the elements according to their properties.

As soon as the atomic weights were established, chemists sought to organize observations about the elements in terms of atomic weights, however, a cursory examination showed no obvious similarity among elements with similar atomic weights. In fact, chemists found striking similarities among elements of quite different atomic weights. For instance, the *halogens* ("salt formers")—fluorine, chlorine, bromine, and iodine—have many similar characteristics, although their atomic weights are quite dissimilar: 19.0, 35.5, 79.9, and 127, respectively. The so-called *noble metals*, silver (Ag), gold (Au), and platinum (Pt), are similar in their durability and resistance to corrosion, and the *light metals*, lithium (Li), sodium (Na), and potassium (K), all combine with oxygen so readily that they are never found uncombined in nature. However, no relationship among atomic weights is obvious in either of these cases.

One of the first noteworthy attempts to classify the elements according to their atomic weights was that of J. A. R. Newlands (1837–1898) in 1865. He observed that when the elements were arranged in order of increasing atomic weights, every eighth element was similar. A partial illustration of the arrangement suggested by Newlands is given in Table 5.5. In the first column appear hydrogen, fluorine, and chlorine, all gases at room temperature and all nonmetals. The second column contains the very light metals, lithium, sodium, and potassium, each extremely reactive. The similarity of beryllium to mag-

Table 5.5 Newlands' Classification of the Elements

H	Li	Be	B	C	N	O
F	Na	Mg	Al	Si	P	S
Cl	K	Ca	Cr	Ti	Mn	Fe

nesium and calcium is not so apparent, but the latter two, both active metals, share many physical and chemical properties. Boron and aluminum are both elements that have some properties of metals and some of nonmetals, and silicon and carbon are both nonmetals, as are nitrogen and phosphorus and oxygen and sulfur.

Newlands called his classification the *law of octaves*. He said: "The eighth element starting from a given one is a kind of a repetition of the first, like the eighth note of an octave of music." Since he failed to anticipate that many elements had not yet been discovered, he was forced to misplace several elements, for example, grouping iron with sulfur and oxygen.

Although we now recognize that Newland's attempt was a step in the right direction, it failed to gain the support of scientists. That support was reserved for Dmitri Mendeleev (1834–1907), a Russian chemist, who published his organization of the elements—a "periodic table" (Figure 5.14)—around 1870. Unlike Newlands, Mendeleev left spaces for undiscovered elements rather than group obviously dissimilar elements together. Furthermore, on the basis of the known elements in a group, he predicted the properties of the undiscovered elements. When the elements were subsequently found, many of his predictions were shown to be correct. For example, Mendeleev designated positions for the unknown element between calcium and titanium and for two elements between zinc and arsenic and gave them the temporary names of eka-boron, eka-aluminum, and eka-silicon. He used the prefix *eka* (from Sanskrit), meaning one or first, to denote the similarity of the undiscovered element to the known element. Today we are able to compare his remarkable predictions with the properties of present-day elements 21, 31, and 32—scandium, gallium, and germanium. Table 5.6 compares the predicted properties for eka-aluminum with those of gallium. Of course not all Mendeleev's predictions were as clearly substantiated.

The **periodic law** upon which Mendeleev based his table was *the chemical and physical properties of the elements are periodic functions of their atomic weights*. This law led ultimately to the acceptance of the periodic table; however, there were still difficulties. In a number of places, it was necessary to depart from an order based strictly upon atomic weights. For example, if tellurium and iodine were placed in the table in strict order of increasing atomic weights, both elements would be in the wrong places according to other properties. Thus, on the basis of the work of Moseley, the law was modified some years later to read, "*The chemical and physical properties of the elements are periodic functions of their atomic number.*" (Review Chapter 4 for the distinction between atomic weight and atomic number.) As with the insights of Dalton and Bohr, later modifications have not lessened the importance of the original concept.

Group	I	II	III	IV	V	VI	VII	VIII	0
Period I	H 1								(He) 2
Period II	Li 3	Be 4	B 5	C 6	N 7	O 8	F 9		(Ne) 10
Period III	Na 11	Mg 12	Al 13	Si 14	P 15	S 16	Cl 17		(Ar) 18
Period IV	K 19	Ca 20	(Sc) 21	Ti 22	V 23	Cr 24	Mn 25	Fe Co Ni 26 27 28	(Kr) 36
	Cu 29	Zn 30	(Ga) 31	(Ge) 32	As 33	Se 34	Br 35		
Period V	Rb 37	Se 38	Y 39	Zr 40	Nb 41	Mo 42	(Tc) 43	Ru Rh Pd 44 45 46	(Xe) 54
	Ag 47	Cd 48	In 49	Sn 50	Sb 51	Te 52	I 53		
Period VI	Cs 55	Ba 56	57 ↓ 71	(Hf) 72	Ta 73	W 74	(Re) 75	Os Ir Pt 76 77 78	(Rn) 86
	Au 79	Hg 80	Tl 81	Pb 82	Bi 83	(Po) 84	(At) 85		
Period VII	(Fr) 87	(Ra) 88	89 ↓ 103						

Figure 5.14 A modern form of Mendeleev's periodic table. (*Note:* Circled elements are those discovered since Mendeleev wrote his law.)

Table 5.6 Comparison of the Properties Predicted by Mendeleev for Eka-Aluminium with the Measured Properties of Gallium

Property	Eka-Aluminum	Gallium
Atomic weight	Approx. 68	69.9
Density	5.9	5.93
Melting point	low	30.1°C
Formula of oxide	Ea_2O_3	Ga_2O_3

Periodic tables in use today, such as the one on the inside of the front cover of this book, have been developed from Mendeleev's original classification. They provide an organization of the properties of the elements, without which chemistry would be a tedious undertaking. With the proper use of the periodic table, however, we can quite easily learn much about the properties of elements—and deduce more. The following generalizations about the periodic table will make this observation clearer. (Refer to the table on the inside of the front cover as you read.)

Sidelight
Coincidental Discoveries and Ideas

In the middle of the nineteenth century, communication within the scientific community was limited, and scientists working independently often worked in ignorance of each other's efforts. A German scientist, J. Lothar Meyer (1830–1895), discovered a periodic relationship among the properties of the elements almost simultaneously with Mendeleev. He also predicted properties for unknown elements and, in some respects, his predictions were better than those of Mendeleev. Meyer represented the periodic law graphically by plotting atomic weights against atomic volumes (g at. wt. divided by the density). The resulting curve is shown in Figure 5.15. This plot is characterized by "waves" in which similar elements occupy analogous positions. This method can be used to illustrate the periodicity of other properties as a function of atomic weight. For example, when boiling point, melting point, and hardness are plotted against atomic weight, the same kind of recurrent waves are observed. Meyer and Mendeleev were both recognized for their work in their own lifetimes. They were jointly awarded the Davy Medal by the British Royal Society in 1882.

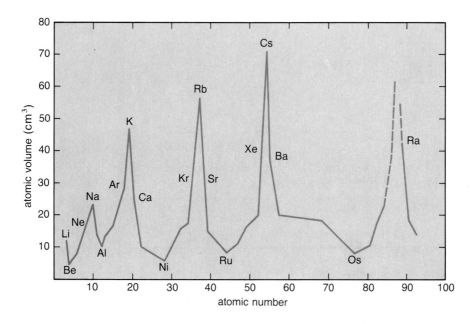

Figure 5.15 A modern representation of Lothar Meyer's atomic volume curve—an example of the periodic law.

IA																	0
1 H	IIA											IIIA	IVA	VA	VIA	VIIA	2 He
3 Li	4 Be											5 B	6 C	7 N	8 O	9 F	10 Ne
11 Na	12 Mg	IIIB	IVB	VB	VIB	VIIB	⎯ VIIIB ⎯			IB	IIB	13 Al	14 Si	15 P	16 S	17 Cl	18 Ar
19 K	20 Ca	21 Sc	22 Ti	23 V	24 Cr	25 Mn	26 Fe	27 Co	28 Ni	29 Cu	30 Zn	31 Ga	32 Ge	33 As	34 Se	35 Br	36 Kr
37 Rb	38 Sr	39 Y	40 Zr	41 Nb	42 Mo	43 Tc	44 Ru	45 Rh	46 Pd	47 Ag	48 Cd	49 In	50 Sn	51 Sb	52 Te	53 I	54 Xe
55 Cs	56 Ba	57 La	72 Hf	73 Ta	74 W	75 Re	76 Os	77 Ir	78 Pt	79 Au	80 Hg	81 Tl	82 Pb	83 Bi	84<:Po	85 At	86 Rn
87 Fr	88 Ra	89 Ac	104	105	106												

58 Ce	59 Pr	60 Nd	61 Pm	62 Sm	63 Eu	64 Gd	65 Tb	66 Dy	67 Ho	68 Er	69 Tm	70 Yb	71 Lu
90 Th	91 Pa	92 U	93 Np	94 Pu	95 Am	96 Cm	97 Bk	98 Cf	99 Es	100 Fm	101 Md	102 No	103 Lr

Figure 5.16 The complete periodic table.

1. With the elements arranged in order of increasing atomic number, the trend from left to right and from top to bottom is from lighter to heavier atoms (with a few exceptions such as iodine and tellurium).

2. In the form most commonly used, the table places similar elements in vertical columns called *groups*. Note that there are seven A groups, a zero group, and eight B groups. Note further that the B groups start with IIIB, that Group VIIIB contains three columns, and that the last two B groups are numbered IB and IIB. The seven horizontal rows are called *periods*.

3. Portions of the sixth and seventh periods are set off below. These portions actually lie between elements 57 and 72 in the sixth period and between elements 89 and 104 in the seventh period. Figure 5.16 shows the table in its complete—but far less practical—form. This latter form does, however, permit us to note the numbers of elements in each row: 2, 8, 8, 18, 18, 32. The seventh row would also contain 32 elements if it were complete. Note that these are the same numbers we obtained by applying the restrictions on quantum numbers (p. 114). There they represented the maximum electron populations of the successive energy levels in the atom; here they represent the number of elements in successive periods of the periodic table. It remains now to explain this correspondence between the properties of the elements and their electron configurations.

Question

6. What is meant by a "periodic" property? Look up this mathematical term if its meaning is not clear to you. What common mathematical functions have periodic properties?

5.5 Electron Configurations and the Periodic Table

In the periodic table on the inside of the front cover, the atomic number for each element appears above the symbol. Since the atomic number is the number of protons in the nucleus, the difference between any two adjacent elements is one proton. The nucleus of the calcium atom (atomic number 20) contains one more proton than the nucleus of the potassium atom (atomic number 19), and the nucleus of the scandium atom (atomic number 21) contains one more proton than the nucleus of the calcium atom. Since we are considering only neutral atoms, the calcium atom contains one more electron than the potassium atom, and the scandium atom contains one more electron than the calcium atom, and so on. We shall see that a description of the energy level, sublevel, orbital, and spin of this one electron, called the *differentiating electron* for each element, is directly related to the chemical and physical properties of that element.

Let us now describe, by using the quantum number rules, the electron configurations of some of the elements. We shall consider all atoms to be in an unexcited state—the ground state—so that each electron will occupy whichever available orbital has the lowest energy.

Hydrogen

The hydrogen atom contains one proton and one electron. The electron is in the first energy level, which consists only of a single sublevel (s) and a single orbital. The quantum numbers for the electron are $(1, 0, 0, +\frac{1}{2})$.

Helium

The helium atom contains two protons and two electrons. The two electrons can occupy the same orbital if they have opposite spins. The quantum numbers for the two electrons are $(1, 0, 0, +\frac{1}{2})$ and $(1, 0, 0, -\frac{1}{2})$.

Lithium

The lithium atom contains three electrons. The first energy level has a capacity of two electrons, so the third electron is found in the second energy level. There are two sublevels in the second energy level and the electron is located in an orbital of the sublevel having lower energy (s). The quantum numbers for the three electrons are $(1, 0, 0, +\frac{1}{2})$, $(1, 0, 0, -\frac{1}{2})$ for the electrons in the first energy level, and $(2, 0, 0, +\frac{1}{2})$ or $(2, 0, 0, -\frac{1}{2})$ for the electron in the second energy level.

Note that any unpaired electron may have either $s = +\frac{1}{2}$ or $s = -\frac{1}{2}$; henceforth we shall arbitrarily assign unpaired electrons a spin quantum number of $+\frac{1}{2}$.

Obviously it would become quite cumbersome to list the quantum numbers for every electron in atoms much larger than lithium. Therefore we shall introduce three shorthand notations for designating electron configurations. We shall refer to these as (1) the electron configuration in terms of one quantum number (or in terms of energy levels); (2) the electron configuration in terms of two quantum numbers (or in terms of sublevels); and (3) the electron configuration in terms of four quantum numbers (or in terms of orbital and spin). These three notations are shown below for hydrogen, helium, and lithium.

(1) Element	Atomic Number	Electron Configuration in Terms of One Quantum Number
Hydrogen	1	1
Helium	2	2
Lithium	3	2, 1

An atom of hydrogen has one electron in the first energy level, an atom of helium has two electrons in the first energy level, and an atom of lithium has two electrons in the first and one in the second energy level.

(2) Element	Atomic Number	Electron Configuration in Terms of Two Quantum Numbers
Hydrogen	1	$1s^1$
Helium	2	$1s^2$
Lithium	3	$1s^2, 2s^1$

Each superscript indicates the number of electrons in each sublevel, and the number preceding the symbol for the sublevel corresponds to the principal quantum number, n.

(3) Element	Atomic Number	Electron Configuration in Terms of Four Quantum Numbers
Hydrogen	1	1s [↑]
Helium	2	1s [↑↓]
Lithium	3	1s 2s [↑↓] [↑]

Again the number and the letter designate the principal and azimuthal quantum numbers. Each box represents an individual orbital with a capacity of two electrons, each of which is symbolized by an arrow. Two electrons in the same orbital (with their opposite spins indicated by the opposite directions of the arrows) are said to be paired. For example, there are three electrons in the lithium atom, two with paired spins in the $1s$ orbital and a third unpaired in the $2s$ orbital.

Using these shorthand notations, we can now continue with our description of the elements.

Beryllium

4 Be

The fourth electron in the beryllium atom fills the single orbital in the s sublevel of the second energy level.
Quantum numbers of differentiating electron: $(2, 0, 0, -\frac{1}{2})$
Electron configuration in terms of one quantum number: 2, 2
Electron configuration in terms of two quantum numbers: $1s^2, 2s^2$
Electron configuration in terms of four quantum numbers: 1s   2s
[↑↓] [↑↓]

Boron

5 B

The first four electrons in the boron atom have the same configuration as the electrons in the beryllium atom. The differentiating (fifth) electron occupies an orbital in the p sublevel of the second energy level.
Quantum numbers of differentiating electron: $(2, 1, +1, +\frac{1}{2})$
Electron configuration in terms of energy levels: 2, 3
Electron configuration in terms of sublevels: $1s^2, 2s^2, 2p^1$
Electron configuration in terms of orbital and spin: 1s   2s   2p
[↑↓] [↑↓] [↑][ ][ ]

(Note that in representing the electron configuration in terms of orbital and spin, we depict all orbitals of a given sublevel, even when some of them are empty.)

Configurations for the next elements are examples of Hund's rule. The fifth and sixth electrons in the carbon atom have parallel spins, as do the fifth, sixth, and seventh electrons in the nitrogen atom. Only with oxygen will we begin to pair the 2p electrons, leaving two unpaired. In fluorine one unpaired electron remains; in neon, none. As we shall see later, the number of unpaired electrons in an atom has a significant influence on the properties of the element.

Carbon

6 C

Quantum numbers of differentiating electron: $(2, 1, 0, +\frac{1}{2})$
Electron configuration: 2, 4
$1s^2, 2s^2, 2p^2$
1s   2s   2p
[↑↓] [↑↓] [↑|↑|]

Nitrogen

7 N

Quantum numbers of differentiating electron: $(2, 1, -1, +\frac{1}{2})$
Electron configuration: 2, 5
$1s^2, 2s^2, 2p^3$
1s   2s   2p
[↑↓] [↑↓] [↑|↑|↑]

Oxygen

8 O

Quantum numbers of differentiating electron: $(2, 1, +1, -\frac{1}{2})$
Electron configuration: 2, 6
$1s^2, 2s^2, 2p^4$
1s   2s   2p
[↑↓] [↑↓] [↑↓|↑|↑]

Fluorine

9 F

Quantum numbers of differentiating electron: $(2, 1, 0, -\frac{1}{2})$
Electron configuration: 2, 7
$1s^2, 2s^2, 2p^5$
1s   2s   2p
[↑↓] [↑↓] [↑↓|↑↓|↑]

Neon

10 Ne

Quantum numbers of differentiating electron: $(2, 1, -1, -\frac{1}{2})$
Electron configuration: 2, 8
$1s^2, 2s^2, 2p^6$
1s   2s   2p
[↑↓] [↑↓] [↑↓|↑↓|↑↓]

Sodium

11 Na

Quantum numbers of differentiating electron: $(3, 0, 0, +\frac{1}{2})$
Electron configuration: 2, 8, 1
$1s^2, 2s^2, 2p^6, 3s^1$
1s   2s   2p   3s
[↑↓] [↑↓] [↑↓|↑↓|↑↓] [↑]

The first two energy levels can accommodate only 10 electrons. Thus the sodium atom, with 11 electrons, contains a single electron in the third energy level. Recall that lithium, which is directly above sodium in the periodic table, has a single electron in the second energy level. If we were to continue our detailed atom-by-atom discussion, we would find that potassium $(Z = 19)$, rubidium $(Z = 37)$, cesium $(Z = 55)$, and francium $(Z = 87)$ are the first elements with electrons in the fourth, fifth, sixth, and seventh energy levels, respectively. Thus we may draw our first general conclusion about the correspondence between electron configurations and the periodic table: *The period in which an element is located corresponds to the number of energy levels containing at least one electron.* Another generalization from the observations above, one that will become more apparent as we examine the electron configurations of more elements, might be stated as follows: *Among the A groups and Group 0 (except for He), all elements in a given group have the same number of electrons in their outermost energy levels.*

The electron configurations in terms of two quantum numbers are given below for the elements sodium through argon (11 through 18). The differentiating electron in each case is in an orbital of the third energy level.

Na	(sodium) atomic number 11	$1s^2, 2s^2, 2p^6, 3s^1$
Mg	(magnesium) atomic number 12	$1s^2, 2s^2, 2p^6, 3s^2$
Al	(aluminum) atomic number 13	$1s^2, 2s^2, 2p^6, 3s^2, 3p^1$
Si	(silicon) atomic number 14	$1s^2, 2s^2, 2p^6, 3s^2, 3p^2$
P	(phosphorus) atomic number 15	$1s^2, 2s^2, 2p^6, 3s^2, 3p^3$
S	(sulfur) atomic number 16	$1s^2, 2s^2, 2p^6, 3s^2, 3p^4$
Cl	(chlorine) atomic number 17	$1s^2, 2s^2, 2p^6, 3s^2, 3p^5$
Ar	(argon) atomic number 18	$1s^2, 2s^2, 2p^6, 3s^2, 3p^6$

Beginning with the next element, potassium, the fourth energy level is occupied:

K	(potassium) atomic number 19	$1s^2, 2s^2, 2p^6, 3s^2, 3p^6, 4s^1$
Ca	(calcium) atomic number 20	$1s^2, 2s^2, 2p^6, 3s^2, 3p^6, 4s^2$

It is logical at this point to ask why electrons occupy the $4s$ orbital before any of the $3d$ orbitals. Before answering this question, however, we shall tabulate the electron configurations for elements 21 to 30. We need these in the discussion that follows.

Element	Symbol	Atomic Number	Electron Configuration
			$1s$ $\;$ $2s$ $\;$ $2p$ $\;$ $3s$ $\;$ $3p$ $\;$ $3d$ $\;$ $4s$
Scandium	Sc	21	[↑↓] [↑↓] [↑↓][↑↓][↑↓] [↑↓] [↑↓][↑↓][↑↓] [↑] [ ] [ ] [ ] [ ] [↑↓]
Titanium	Ti	22	[↑↓] [↑↓] [↑↓][↑↓][↑↓] [↑↓] [↑↓][↑↓][↑↓] [↑] [↑] [ ] [ ] [ ] [↑↓]
Vanadium	V	23	[↑↓] [↑↓] [↑↓][↑↓][↑↓] [↑↓] [↑↓][↑↓][↑↓] [↑] [↑] [↑] [ ] [ ] [↑↓]
Chromium	Cr	24	[↑↓] [↑↓] [↑↓][↑↓][↑↓] [↑↓] [↑↓][↑↓][↑↓] [↑] [↑] [↑] [↑] [↑] [↑]
Manganese	Mn	25	[↑↓] [↑↓] [↑↓][↑↓][↑↓] [↑↓] [↑↓][↑↓][↑↓] [↑] [↑] [↑] [↑] [↑] [↑↓]
Iron	Fe	26	[↑↓] [↑↓] [↑↓][↑↓][↑↓] [↑↓] [↑↓][↑↓][↑↓] [↑↓] [↑] [↑] [↑] [↑] [↑↓]
Cobalt	Co	27	[↑↓] [↑↓] [↑↓][↑↓][↑↓] [↑↓] [↑↓][↑↓][↑↓] [↑↓] [↑↓] [↑] [↑] [↑] [↑↓]
Nickel	Ni	28	[↑↓] [↑↓] [↑↓][↑↓][↑↓] [↑↓] [↑↓][↑↓][↑↓] [↑↓] [↑↓] [↑↓] [↑] [↑] [↑↓]
Copper	Cu	29	[↑↓] [↑↓] [↑↓][↑↓][↑↓] [↑↓] [↑↓][↑↓][↑↓] [↑↓] [↑↓] [↑↓] [↑↓] [↑↓] [↑]
Zinc	Zn	30	[↑↓] [↑↓] [↑↓][↑↓][↑↓] [↑↓] [↑↓][↑↓][↑↓] [↑↓] [↑↓] [↑↓] [↑↓] [↑↓] [↑↓]

Table 5.7 Electron Configurations for the Elements

Energy Level	1	2		3			4				5					6			7
Sublevel	1s	2s	2p	3s	3p	3d	4s	4p	4d	4f	5s	5p	5d	5f	5g	6s	6p	6d	7s
1. H	1																		
2. He	2																		
3. Li	2	1																	
4. Be	2	2																	
5. B	2	2	1																
6. C	2	2	2																
7. N	2	2	3																
8. O	2	2	4																
9. F	2	2	5																
10. Ne	2	2	6																
11. Na	2	2	6	1															
12. Mg	2	2	6	2															
13. Al	2	2	6	2	1														
14. Si	2	2	6	2	2														
15. P	2	2	6	2	3														
16. S	2	2	6	2	4														
17. Cl	2	2	6	2	5														
18. Ar	2	2	6	2	6														
19. K	2	2	6	2	6		1												
20. Ca	2	2	6	2	6		2												
21. Sc	2	2	6	2	6	1	2												
22. Ti	2	2	6	2	6	2	2												
23. V	2	2	6	2	6	3	2												
24. Cr	2	2	6	2	6	5	1												
25. Mn	2	2	6	2	6	5	2												
26. Fe	2	2	6	2	6	6	2												
27. Co	2	2	6	2	6	7	2												
28. Ni	2	2	6	2	6	8	2												
29. Cu	2	2	6	2	6	10	1												
30. Zn	2	2	6	2	6	10	2												
31. Ga	2	2	6	2	6	10	2	1											
32. Ge	2	2	6	2	6	10	2	2											
33. As	2	2	6	2	6	10	2	3											
34. Se	2	2	6	2	6	10	2	4											
35. Br	2	2	6	2	6	10	2	5											
36. Kr	2	2	6	2	6	10	2	6											
37. Rb	2	2	6	2	6	10	2	6			1								
38. Sr	2	2	6	2	6	10	2	6			2								
39. Y	2	2	6	2	6	10	2	6	1		2								
40. Zr	2	2	6	2	6	10	2	6	2		2								
41. Nb	2	2	6	2	6	10	2	6	4		1								
42. Mo	2	2	6	2	6	10	2	6	5		1								
43. Tc	2	2	6	2	6	10	2	6	(5)		(2)								
44. Ru	2	2	6	2	6	10	2	6	7		1								
45. Rh	2	2	6	2	6	10	2	6	8		1								
46. Pd	2	2	6	2	6	10	2	6	10										
47. Ag	2	2	6	2	6	10	2	6	10		2								
48. Cd	2	2	6	2	6	10	2	6	10		2								
49. In	2	2	6	2	6	10	2	6	10		2	1							
50. Sn	2	2	6	2	6	10	2	6	10		2	2							

Table 5.7 (continued)

Energy Level	1	2		3			4				5					6			7
Sublevel	1s	2s	2p	3s	3p	3d	4s	4p	4d	4f	5s	5p	5d	5f	5g	6s	6p	6d	7s
51. Sb	2	2	6	2	6	10	2	6	10		2	3							
52. Te	2	2	6	2	6	10	2	6	10		2	4							
53. I	2	2	6	2	6	10	2	6	10		2	5							
54. Xe	2	2	6	2	6	10	2	6	10		2	6							
55. Cs	2	2	6	2	6	10	2	6	10		2	6				1			
56. Ba	2	2	6	2	6	10	2	6	10		2	6				2			
57. La	2	2	6	2	6	10	2	6	10		2	6	1			2			
58. Ce	2	2	6	2	6	10	2	6	10	2	2	6				2			(?)
59. Pr	2	2	6	2	6	10	2	6	10	3	2	6				2			
60. Nd	2	2	6	2	6	10	2	6	10	4	2	6				2			
61. Pm	2	2	6	2	6	10	2	6	10	5	2	6				2			
62. Sm	2	2	6	2	6	10	2	6	10	6	2	6				2			
63. Eu	2	2	6	2	6	10	2	6	10	7	2	6				2			
64. Gd	2	2	6	2	6	10	2	6	10	7	2	6	1			2			
65. Tb	2	2	6	2	6	10	2	6	10	8	2	6	1			2			
66. Dy	2	2	6	2	6	10	2	6	10	9	2	6	1			2			
67. Ho	2	2	6	2	6	10	2	6	10	10	2	6	1			2			
68. Er	2	2	6	2	6	10	2	6	10	11	2	6	1			2			
69. Tu	2	2	6	2	6	10	2	6	10	13	2	6				2			
70. Yb	2	2	6	2	6	10	2	6	10	14	2	6				2			
71. Lu	2	2	6	2	6	10	2	6	10	14	2	6	1			2			
72. Hf	2	2	6	2	6	10	2	6	10	14	2	6	2			2			
73. Ta	2	2	6	2	6	10	2	6	10	14	2	6	3			2			
74. W	2	2	6	2	6	10	2	6	10	14	2	6	4			2			
75. Re	2	2	6	2	6	10	2	6	10	14	2	6	5			2			
76. Os	2	2	6	2	6	10	2	6	10	14	2	6	6			2			
77. Ir	2	2	6	2	6	10	2	6	10	14	2	6	7			2			
78. Pt	2	2	6	2	6	10	2	6	10	14	2	6	9			1			
79. Au	2	2	6	2	6	10	2	6	10	14	2	6	10			1			
80. Hg	2	2	6	2	6	10	2	6	10	14	2	6	10			2			
81. Tl	2	2	6	2	6	10	2	6	10	14	2	6	10			2	1		
82. Pb	2	2	6	2	6	10	2	6	10	14	2	6	10			2	2		
83. Bi	2	2	6	2	6	10	2	6	10	14	2	6	10			2	3		
84. Po	2	2	6	2	6	10	2	6	10	14	2	6	10			2	4		
85. At	2	2	6	2	6	10	2	6	10	14	2	6	10			2	5		
86. Rn	2	2	6	2	6	10	2	6	10	14	2	6	10			2	6		
87. Fr	2	2	6	2	6	10	2	6	10	14	2	6	10			2	6		1
88. Ra	2	2	6	2	6	10	2	6	10	14	2	6	10			2	6		2
89. Ac	2	2	6	2	6	10	2	6	10	14	2	6	10			2	6	1	2
90. Th	2	2	6	2	6	10	2	6	10	14	2	6	10			2	6	2	2
91. Pa	2	2	6	2	6	10	2	6	10	14	2	6	10	2		2	6	1	2
92. U	2	2	6	2	6	10	2	6	10	14	2	6	10	3		2	6	1	2
93. Np	2	2	6	2	6	10	2	6	10	14	2	6	10	5		2	6		2 (?)
94. Pu	2	2	6	2	6	10	2	6	10	14	2	6	10	6		2	6		2
95. Am	2	2	6	2	6	10	2	6	10	14	2	6	10	7		2	6		2
96. Cm	2	2	6	2	6	10	2	6	10	14	2	6	10	7		2	6	1	2
97. Bk	2	2	6	2	6	10	2	6	10	14	2	6	10	9		2	6		2 (?)
98. Cf	2	2	6	2	6	10	2	6	10	14	2	6	10	10		2	6		2 (?)
99. Es	2	2	6	2	6	10	2	6	10	14	2	6	10	11		2	6		2 (?)
100. Fm	2	2	6	2	6	10	2	6	10	14	2	6	10	12		2	6		2 (?)
101. Mv	2	2	6	2	6	10	2	6	10	14	2	6	10	13		2	6		2 (?)
102. No	2	2	6	2	6	10	2	6	10	14	2	6	10	14		2	6		2 (?)
103. Lr	2	2	6	2	6	10	2	6	10	14	2	6	10	14		2	6	1	2 (?)

The elements beginning with scandium illustrate that, after filling the 4s orbital, electrons begin occupying the 3d sublevel. This order of filling could not have been predicted from the material we have studied so far, but it turns out to be an extension of a principle we have already encountered: Electrons tend to fill the available orbitals with the lowest energies. It can be shown by experiment or by detailed calculations that the 4s orbital lies at a lower energy than the 3d orbitals, which in turn lie at lower energies than the 4p orbitals.

The relative energies of the sublevels, thus the filling order of the orbitals, become even more complicated in larger atoms; therefore, a memory device is a useful tool. One such device can be outlined as follows:

1. Arrange the designations for the sublevels needed for the known elements as follows:

```
1s
2s  2p
3s  3p  3d
4s  4p  4d  4f
5s  5p  5d  5f
6s  6p  6d  6f
7s
```

2. Draw parallel diagonal lines from upper right to lower left. Notice that the first two lines cross one term each; the second pair of lines cross two terms each; the third pair cross three terms each; and the next line crosses four terms.

3. Read the filling order by following each arrow in turn from upper right to lower left.

4. Thus the filling order is 1s, 2s, 2p, 3s, 3p, 4s, 3d, 4p, 5s, 4d, 5p, 6s, 4f, 5d, 6p, 7s.

You will probably need this memory device for only a short time. When the periodic table is well understood, it usually provides all the help needed to arrive at the electron configuration for an atom. Table 5.7 provides complete electron configurations of the elements.

Equipped with a general rule, we must now be prepared to confront exceptions. Fortunately, among the first 40 elements, there are only two elements that do not strictly obey the filling order above. They are chromium and copper. Each of these atoms has only one electron in the 4s sublevel, even though the rest of the elements in the fourth row of the periodic table have two. Since one of our basic principles is that every atom, in its ground state, is in the lowest available energy state, we must conclude from the example of chromium that a half-filled 3d sublevel plus one electron in the 4s sublevel (thus also half-filled) represents a lower energy state than four electrons in the 3d orbitals and two in the 4s. From the example of copper, we deduce that a completely filled 3d sublevel plus one electron in the 4s sublevel is a lower energy state than nine electrons in the 3d and two in the 4s. This extra stability, which a half-filled or completely filled d sublevel gives an atom, also manifests itself in the electron configurations of elements in the fifth and sixth periods.

Example 5.2 Write the electron configuration in terms of one and in terms of two quantum numbers for the element with atomic number 23. Then draw the electron configuration in terms of four quantum numbers for the outer two energy levels.

Solution

a. It is easiest to write the electron configuration in terms of two quantum numbers first. To do this we need only to list the sublevels in the correct filling order, each with a superscript indicating the sublevel's capacity of electrons, until we have accounted for 23 electrons. The first 20 electrons are in the following sublevels:

$1s^2, 2s^2, 2p^6, 3s^2, 3p^6, 4s^2$

b. The next sublevel in the filling order is 3d. Thus the electron configuration in terms of two quantum numbers for element 23 is

$1s^2, 2s^2, 2p^6, 3s^2, 3p^6, 3d^3, 4s^2$

c. To write the electron configuration in terms of one quantum number, we simply add the number of electrons in each energy level. Therefore, the electron configuration in terms of one quantum number for element 23 is

2, 8, 11, 2

d. The electron configuration for the outer two energy levels in terms of all four quantum numbers is

3s	3p	3d	4s
↑↓	↑↓ ↑↓ ↑↓	↑ ↑ ↑		↑↓

Example 5.3 Answer the following questions about the element in Example 5.2: In the ground state of the atom, how many energy levels contain electrons?

How many electrons are in its outermost energy level? What are the first two quantum numbers of the differentiating electron?

Solution

a. How many energy levels contain electrons? four

b. Which level is the outermost? the fourth

c. How many electrons are in the fourth energy level? two

d. In what sublevel is the twenty-third electron located? 3d

e. What are the n and l values for the twenty-third electron? $n = 3, l = 2$

Question

7. Which sublevel in an atom
    a. contains only one orbital?
    b. is first filled in the third energy level?
    c. has a capacity of 14 electrons?

5.6 Categories of Elements and Periodic Table Position

The elements can be classified in several ways. In this section we shall look at the two most useful classifications: (1) that based upon electron configurations, in which the elements are described as **representative, transition,** or **rare earth** elements, and (2) that based upon properties in which the elements are described as **metals, nonmetals, metalloids,** or **noble gases.**

The classification based upon electron configurations is depicted in Figure 5.17. The representative elements, those in the seven A groups and Group 0, have their differentiating electron in an *s* or a *p* sublevel. The *noble gases*, Group 0, are those elements whose atoms contain eight electrons (or two in the case of helium) in the outermost energy level. The transition elements, those in the B groups and Group VIIIB, are located between Groups IIA and IIIA. For most of these elements, the differentiating electron is in a *d* sublevel, although there are exceptions such as manganese and zinc. The rare earths (also called the *inner transition elements*) are set off at the bottom of the table and can usually be characterized as having the differentiating electron in an *f* sublevel. We sometimes call the 14 elements in the first row of rare earths the *lanthanides* and those in the second row the *actinides*. There are no group numbers assigned to this portion of the table, because similarities within each period are more important than any vertical relationships.

As we have already noted, the period number of an atom corresponds to the number of occupied energy levels. We may now ask what each group number indicates. *For Group A elements the group number corresponds to the total number of electrons in the outermost energy level.* Atoms in Group 0 have a filled outermost

Representative Elements	Transition Elements	Representative Elements
s Sublevels Being Filled	*d* Sublevels Being Filled	*p* Sublevels Being Filled

IA	IIA	IIIB	IVB	VB	VIB	VIIB	VIIIB	VIIIB	VIIIB	IB	IIB	IIIA	IVA	VA	VIA	VIIA	0
1 **H** 1.0079 (1)																	2 **He** 4.003 (2)
3 **Li** 6.941 (2, 1)	4 **Be** 9.01218 (2, 2)											5 **B** 10.81 (2, 3)	6 **C** 12.011 (2, 4)	7 **N** 14.007 (2, 5)	8 **O** 15.9994 (2, 6)	9 **F** 18.9984 (2, 7)	10 **Ne** 20.179 (2, 8)
11 **Na** 22.9898 (2, 8, 1)	12 **Mg** 24.305 (2, 8, 2)				90							13 **Al** 26.982 (2, 8, 3)	14 **Si** 28.0855 (2, 8, 4)	15 **P** 30.974 (2, 8, 5)	16 **S** 32.06 (2, 8, 6)	17 **Cl** 35.453 (2, 8, 7)	18 **Ar** 39.948 (2, 8, 8)
19 **K** 39.0983 (2, 8, 8, 1)	20 **Ca** 40.08 (2, 8, 8, 2)	21 **Sc** 44.9559 (2, 8, 9, 2)	22 **Ti** 47.90 (2, 8, 10, 2)	23 **V** 50.941 (2, 8, 11, 2)	24 **Cr** 51.996 (2, 8, 13, 1)	25 **Mn** 54.9380 (2, 8, 13, 2)	26 **Fe** 55.847 (2, 8, 14, 2)	27 **Co** 58.9332 (2, 8, 15, 2)	28 **Ni** 58.70 (2, 8, 16, 2)	29 **Cu** 63.546 (2, 8, 18, 1)	30 **Zn** 65.38 (2, 8, 18, 2)	31 **Ga** 69.72 (2, 8, 18, 3)	32 **Ge** 72.59 (2, 8, 18, 4)	33 **As** 74.922 (2, 8, 18, 5)	34 **Se** 78.96 (2, 8, 18, 6)	35 **Br** 79.904 (2, 8, 18, 7)	36 **Kr** 83.80 (2, 8, 18, 8)
37 **Rb** 85.4678 (2, 8, 18, 8, 1)	38 **Sr** 87.62 (2, 8, 18, 8, 2)	39 **Y** 88.9059 (2, 8, 18, 9, 2)	40 **Zr** 91.22 (2, 8, 18, 10, 2)	41 **Nb** 92.906 (2, 8, 18, 12, 1)	42 **Mo** 95.94 (2, 8, 18, 13, 1)	43 **Tc** (97) (2, 8, 18, 13, 2)	44 **Ru** 101.07 (2, 8, 18, 15, 1)	45 **Rh** 102.9055 (2, 8, 18, 16, 1)	46 **Pd** 106.4 (2, 8, 18, 18)	47 **Ag** 107.87 (2, 8, 18, 18, 1)	48 **Cd** 112.41 (2, 8, 18, 18, 2)	49 **In** 114.82 (2, 8, 18, 18, 3)	50 **Sn** 118.69 (2, 8, 18, 18, 4)	51 **Sb** 121.75 (2, 8, 18, 18, 5)	52 **Te** 127.60 (2, 8, 18, 18, 6)	53 **I** 126.90 (2, 8, 18, 18, 7)	54 **Xe** 131.30 (2, 8, 18, 18, 8)
55 **Cs** 132.9054 (2, 8, 18, 18, 8, 1)	56 **Ba** 137.33 (2, 8, 18, 18, 8, 2)	57 **La** 138.906 (2, 8, 18, 18, 9, 2)	72 **Hf** 178.49 (2, 8, 18, 32, 10, 2)	73 **Ta** 180.948 (2, 8, 18, 32, 11, 2)	74 **W** 183.85 (2, 8, 18, 32, 12, 2)	75 **Re** 186.207 (2, 8, 18, 32, 13, 2)	76 **Os** 190.2 (2, 8, 18, 32, 14, 2)	77 **Ir** 192.22 (2, 8, 18, 32, 15, 2)	78 **Pt** 195.09 (2, 8, 18, 32, 17, 1)	79 **Au** 196.97 (2, 8, 18, 32, 18, 1)	80 **Hg** 200.59 (2, 8, 18, 32, 18, 2)	81 **Tl** 204.37 (2, 8, 18, 32, 18, 3)	82 **Pb** 207.2 (2, 8, 18, 32, 18, 4)	83 **Bi** 208.98 (2, 8, 18, 32, 18, 5)	84 **Po** (209) (2, 8, 18, 32, 18, 6)	85 **At** (210) (2, 8, 18, 32, 18, 7)	86 **Rn** (222) (2, 8, 18, 32, 18, 8)
87 **Fr** (223) (2, 8, 18, 32, 18, 8, 1)	88 **Ra** 226.0254 (2, 8, 18, 32, 18, 8, 2)	89 **Ac** (227) (2, 8, 18, 32, 18, 9, 2)	104	105	106												

Rare Earths

f Sublevels Being Filled

58 **Ce** 140.12 (2, 8, 18, 20, 8, 2)	59 **Pr** 140.9077 (2, 8, 18, 21, 8, 2)	60 **Nd** 144.24 (2, 8, 18, 22, 8, 2)	61 **Pm** (145) (2, 8, 18, 23, 8, 2)	62 **Sm** 150.4 (2, 8, 18, 24, 8, 2)	63 **Eu** 151.96 (2, 8, 18, 25, 8, 2)	64 **Gd** 157.25 (2, 8, 18, 25, 9, 2)	65 **Tb** 158.92 (2, 8, 18, 26, 9, 2)	66 **Dy** 162.50 (2, 8, 18, 27, 9, 2)	67 **Ho** 164.9304 (2, 8, 18, 28, 9, 2)	68 **Er** 167.26 (2, 8, 18, 29, 9, 2)	69 **Tm** 168.93 (2, 8, 18, 31, 8, 2)	70 **Yb** 173.04 (2, 8, 18, 32, 8, 2)	71 **Lu** 174.97 (2, 8, 18, 32, 9, 2)
90 **Th** 232.04 (2, 8, 18, 32, 18, 10, 2)	91 **Pa** 231.0359 (2, 8, 18, 32, 20, 9, 2)	92 **U** 238.029 (2, 8, 18, 32, 21, 9, 2)	93 **Np** 237.0482 (2, 8, 18, 32, 23, 8, 2)	94 **Pu** (244) (2, 8, 18, 32, 24, 8, 2)	95 **Am** (243) (2, 8, 18, 32, 25, 8, 2)	96 **Cm** (247) (2, 8, 18, 32, 25, 9, 2)	97 **Bk** (247) (2, 8, 18, 32, 27, 8, 2)	98 **Cf** (251) (2, 8, 18, 32, 28, 8, 2)	99 **Es** (254) (2, 8, 18, 32, 29, 8, 2)	100 **Fm** (257) (2, 8, 18, 32, 30, 8, 2)	101 **Md** (258) (2, 8, 18, 32, 31, 8, 2)	102 **No** (255) (2, 8, 18, 32, 32, 8, 2)	103 **Lr** (260) (2, 8, 18, 32, 32, 9, 2)

Figure 5.17 Periodic table with electron configurations.

energy level. For transition elements, however, no such relationship holds, since the numbering of the B groups developed historically rather than logically. In the periodic table of Figure 5.17 the configuration for each element in terms of one quantum number is given in the vertical column immediately preceding the symbol of the element.

Example 5.4 Consider the three elements whose electron configurations are given below. Classify each as representative, transition, or rare earth, and give its group and period numbers. When you have completed the problem, check your answer with the periodic table.

Element A: $1s^2, 2s^2, 2p^6, 3s^2, 3p^6, 3d^7, 4s^2$
Element B: $1s^2, 2s^2, 2p^6, 3s^2, 3p^6$
Element C: $1s^2, 2s^2, 2p^6, 3s^2, 3p^5$

Solution

a. The sublevel containing the differentiating electron determines the classification of the element. In element A, the differentiating electron is in the $3d$ sublevel. Therefore A is a transition element. In elements B and C the differentiating electrons are in the $3p$ sublevels, so both B and C are representative elements. More specifically, on B, the differentiating electron is the sixth electron to enter that sublevel, so element B is a noble gas.

b. The period number of each element is the same as the number of energy levels in the atom that contain electrons. Thus element A is in the fourth period, and B and C are in the third.

c. Group numbers are sometimes more difficult to determine. Since the element B has a filled outermost energy level, it must be in Group 0. For the other representative element, C, the group number is given by the number of electrons in the outermost (third) energy level. Thus it is in Group VIIA. For a transition element, however, we must calculate the group number from the number of d electrons the atom contains, keeping in mind the exceptions we have already discussed. (Recall that the B groups start with IIIB and that Group VIIIB contains three columns of elements.) Element A, with seven d electrons, is therefore in the second column of Group VIIIB.

Figure 5.18 again illustrates the periodic arrangement of the elements; this time, however, they are classified according to their properties. Many of the metals are already familiar to us: as aluminum pots and pans and wrought-iron fences, on chromium-embellished motorcycles, and in silver jewelry. The characteristic physical properties of the metals are *luster, malleability, ductility,* high melting points, and thermal and electrical conductivity. All clean metal surfaces are lustrous. Some metals, like chromium, retain their luster, whereas many corrode or tarnish, some slowly (aluminum) and others rapidly (sodium).

Figure 5.18 Four classes of elements.

All metals are malleable; that is, they can be rolled or hammered into thin sheets. Gold, for example, can be hammered into sheets less than a hundredth of a centimeter thick. Gold is also the most ductile metal, allowing it to be drawn into thin threads or wires. The ductility of another metal, copper, makes it suitable for the millions of miles of wire upon which modern communications depend. As evidence of the high melting points of metals, note that all except one are solids at room temperature and that approximately two-thirds of them melt above 800°C. Mercury (melting point = −39°C) is the single liquid among the metals, and gallium, whose melting point is 29.8°C, will melt in your hand. Finally, we hardly need to mention the thermal and electrical conductivities of metals: Cast-iron skillets illustrate the former; copper and aluminum power lines, the latter.

About 75 percent of the elements are metals. The rest are the noble gases, metalloids, and the nonmetals. The physical properties of the nonmetals may at first seem more varied than those of the metals: Some are gases at ordinary temperatures and pressures; one (bromine) is a liquid; and some are solids. However, we can think of this variability as merely a manifestation of the low melting and boiling points of nonmetals, in contrast to the comparatively high values for metals. Nonmetals, when solid, are usually either soft and powdery or brittle. Carbon and sulfur are examples. Nonmetals are thermal and electrical insulators rather than conductors and are dull rather than lustrous. Five of the nonmetals are gases at room temperature: oxygen and nitrogen, which constitute the bulk of the atmosphere; hydrogen; and the toxic halogens chlorine and fluorine.

The third class of elements, when grouped according to properties, coincides with Group 0 of our earlier classification scheme. These are the six noble gases, first identified at the close of the nineteenth century. For many years they were thought to be chemically inert, but since 1961 chemists have produced a number of noble gas compounds. In some classification systems, therefore, the noble gases are considered nonmetals.

It is useful at this point to reemphasize the difference between physical and chemical properties. Most of the foregoing has been a discussion of physical properties; we can describe chemical properties only in terms of interactions among the elements. For example, metals show little tendency to combine chemically with one another, but they readily form compounds with most nonmetals. Thus most metals react with oxygen to form oxides, and many combine with chlorine and sulfur to form chlorides and sulfides. On the other hand, nonmetals combine not only with metals but also among themselves. For example, when carbon, sulfur, or phosphorus burns, it unites with oxygen to form an oxide. Other combinations produce silicon carbide and bromine fluoride.

Just as there have been exceptions to several of our earlier useful generalizations, so there are fuzzy areas in our classification of the elements according to their properties. There is no difficulty with the noble gases, but some elements that look like metals behave chemically like nonmetals. A familiar example is aluminum; the perplexing problem of what to do with these elements is often solved by classifying them in a separate group called *metalloids*. Referring once more to Figure 5.18 we see that the metalloids, except beryllium, lie on a zigzag line between the metals and the nonmetals. All the metals are to the left; all the nonmetals, except hydrogen, are to the right. As you might expect, the elements farthest from the dividing line display the most extreme metallic and nonmetallic properties. Fluorine is the most nonmetallic element; francium, the most metallic. This trend of increasingly metallic character, from upper right to lower left, is one of the most important features of the periodic table. It is also useful to note that, though there are many exceptions (H, He, Be, B, Al, Sn, Pb, Bi), most atoms with one, two, or three electrons in the outer energy level are metals, whereas those with four or more are nonmetals.

Questions

8. Without referring to a periodic table, answer the questions i–v below for atoms with the following electron configurations:

   a. $1s^2, 2s^2, 2p^3$ b. $1s^2, 2s^1$ c. $1s^2, 2s^2, 2p^6, 3s^2$
   d. $1s^2, 2s^2, 2p^1$ e. $1s^2, 2s^2, 2p^6, 3s^2, 3p^6, 3d^3, 4s^2$ f. $1s^2, 2s^2, 2p^6$
   g. 2, 8, 2   h. 2, 8, 18, 8   i. 2, 8, 5
   j. 2, 8, 18, 4   k. 2, 8, 11, 2

   (i) In what period will it be found?
   (ii) Is it a representative element, transition element, or rare earth?
   (iii) In what group will it lie?
   (iv) Is it (probably) a metal or a nonmetal?
   (v) What are the first two quantum numbers of its differentiating electron?

9. What physical properties are characteristic of metals? Of nonmetals?

10. What chemical properties are characteristic of metals? Of nonmetals?

11. What are metalloids?

12. Determine the total number of elements classed as
   a. noble gases   b. metals   c. metalloids   d. nonmetals

5.7 Atomic Properties

We have now discussed, at least in general terms, some physical and chemical properties of the elements. In the last few pages we have paid special attention to how trends in such properties are related to the arrangement of the elements in the periodic table. In contrast to these properties of the bulk substances we now want to look at some properties of the atoms themselves. Among these *atomic properties* are several we have already encountered: atomic number, atomic mass, nuclear charge, and electron configuration. Another is the *atomic radius*. This is simply the size of the atom, but since the occupied orbitals in an atom are best represented as fuzzy electron clouds, size is a somewhat nebulous quality. Nonetheless it is often useful to approximate each atom as a sharply bounded sphere, and there are several logical approaches to estimating appropriate radii.

The most common method for estimating atomic radii requires that the distance between two chemically bonded nuclei be determined experimentally. For example, the distance between the two nuclei in the Cl_2 molecule can be measured by several techniques as 0.198 nm, which yields a value of 0.0990 nm for the radius of the Cl atom. This is called the *nonpolar covalent radius* of chlorine. (We shall discuss *nonpolar covalent bonds* in Chapter 6.) A similar measurement yields a nonpolar covalent radius for bromine of 0.114 nm. These values are confirmed by the internuclear distance in the ClBr molecule, which is 0.213 nm. Values for the nonpolar covalent radii are given in Figure 5.19, along with circles that represent the relative sizes of the atoms.

Just as we saw trends in physical and chemical properties in the periodic table, so now should we look for trends in atomic radii. From top to bottom in any group, there is a steady increase in size. This is not surprising, since an atom in the third period has electrons in three energy levels, whereas an atom in the second period has electrons in only two, and so on. However, the decrease in atomic size from left to right within a period is less easily explained. After all, we are adding electrons as we move to the right, but we are also increasing the nuclear charge while merely putting the differentiating electron in an energy level already occupied by one or more electrons. The increased nuclear charge exerts a greater attraction for these outer energy level electrons, and the result is a more compact atom.

In addition to nuclear charge and the outermost occupied energy level, there is a third factor that influences atomic radii. This is the *shielding effect*, which is apparent especially among the transition elements. From left to right in the sixth and seventh periods, the atomic sizes of the transition elements do not decrease as rapidly (in some cases they increase) as do those of representative elements. We explain this observation by arguing that as more *d* electrons are added to the inner energy level, they shield the outer energy level electrons from the attractive positive charge of the nucleus. Thus the increasing nuclear charge has only a small effect on the atomic radius. A still more complicated factor that becomes important in large atoms is the mutual repulsion of the many electrons circulating about the nucleus. With so many factors at work, it is no wonder that trends in atomic size are less obvious among the heavy elements. But where a trend can be observed, it is worth remembering.

A second important atomic property is the *ionization energy*. The process of *ionization* is the origin of the canal rays we discussed in Chapter 4, and we shall now describe the process in more detail. An *ion* is an atom or group of atoms that has acquired an electric charge, either by losing or gaining one or more electrons. Ions form, for example, in the upper atmosphere, where cosmic radiation has enough energy to eject electrons from uncharged atoms and molecules, thus the name for this part of the atmosphere—the *ionosphere*. Such ionization reactions can also be produced under controlled conditions in the laboratory. The ionization energy of an atom is the minimum energy necessary to remove the most loosely held electron from the gaseous atom in its ground state, thus producing a positive ion. After the first electron has been removed,

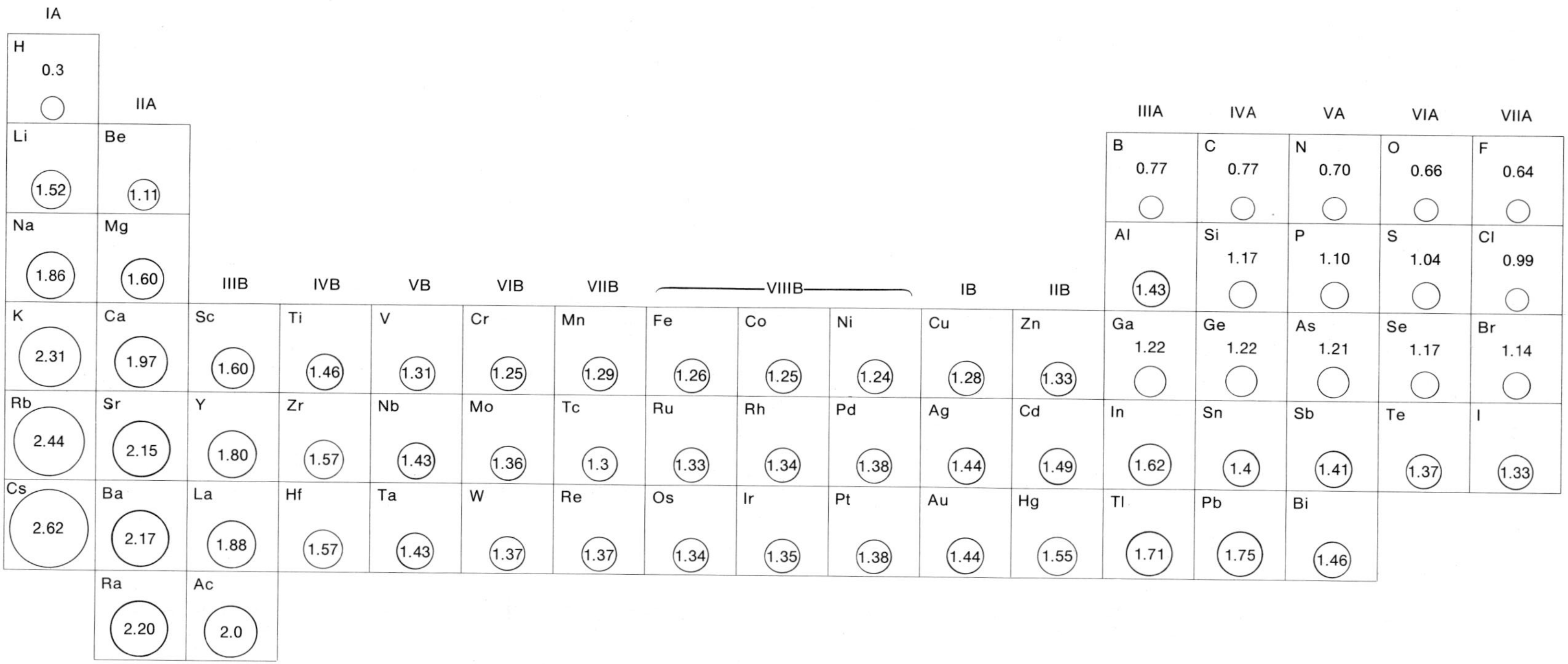

Figure 5.19 Covalent radii of the elements.

Table 5.8 The First Ionization Energies for the Elements of Group IA

Element	I.E. (kcal/g at. wt)
H	313
Li	124
Na	120
K	100
Rb	96
Cs	89

a second may also be removed, then a third, and so on. Thus an atom may also have a *second ionization energy*, a *third ionization energy*, and so forth. Table 5.8 gives the first ionization energies for the elements of Group IA.

The trend in ionization energies is consistent with our discussion of atomic sizes: It becomes easier to remove the outermost electron from an atom as one moves down a periodic group, since the electron is farther from the nucleus and thus less strongly held. It therefore takes 124 kcal of energy to remove the outermost electron from the second energy level of 6.02×10^{23} lithium atoms but only 100 kcal to remove the electrons from the fourth energy level of the same number of potassium atoms. The ionization energy for hydrogen is much higher than the others listed in Table 5.7, since there are no intervening electrons to shield the $1s$ electron from the attractive positive charge of the nucleus. This argument also explains the small atomic radius of the hydrogen atom.

Table 5.9 gives eight successive ionization energies for each of the elements of the third period. In every case the energies rise as additional electrons are removed; however, in each series there is a point at which the ionization energies

Table 5.9 Eight Successive Ionization Energies for the Elements of the Third Period (kcal/mol)

	1st	2nd	3rd	4th	5th	6th	7th	8th
Na	120	1088	1649	2275	3174	3956	4784	6072
Mg	175	345	1842	2951	3243	4278	5152	6118
Al	138	432	653	2760	3519	4370	5543	6555
Si	186	375	771	1037	3841	4715	5668	6969
P	242	453	695	1182	1495	5060	6049	7107
S	239	538	705	1088	1668	2024	6463	7567
Cl	299	547	918	1231	1559	2224	2722	8004
Ar	363	635	941	1375	1725	2100	2852	3289

increase dramatically. For example, we can remove the first three electrons from a mole of aluminum atoms with 138 kcal, 432 kcal, and 653 kcal, respectively. The fourth electron, however, requires 2760 kcal! This unique point for each atom is marked on Table 5.9 with a heavy black line. For sodium the change comes after one electron; for magnesium, after two; for aluminum, after three; and so on. Here we have direct experimental evidence that supports the theory of electron configurations. In every case, the more easily removed electrons to the left of the black line are those in the outermost energy level of the atom. These electrons are more loosely held than those (to the right of the line) which populate a lower energy level.

Figure 5.20 presents some of the same data in another way. The first ionization energies for the atoms of the second and third periods are plotted against atomic number. As we might expect, the ionization energies generally increase from left to right across a period as atomic radii decrease. Also first ionization energies for third period elements are lower than those for second period elements, since outer energy level electrons are less tightly bound in the larger third period atoms. Furthermore, two other features of the data corroborate our earlier statements about the stability of filled and half-filled sublevels. First, note that the energy required to remove an s electron from a filled sublevel (elements 4 and 12) is higher than that required to remove a lone p electron (elements 5 and 13). Second, it requires more energy to remove a p electron from the half-filled sublevel (elements 7 and 15) than to remove a p electron from a sublevel containing 4 electrons (elements 8 and 16).

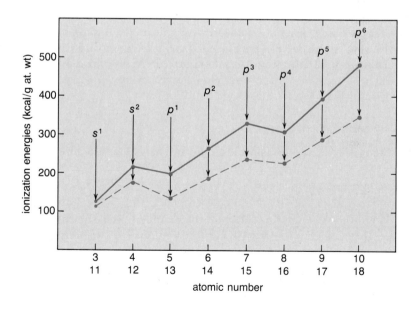

Figure 5.20 First ionization energies for second and third period elements: second period ———; third period ———.

5.8 Electron Formulas for Atoms

The chemical formulas we have discussed so far have been able to tell us only the relative numbers of atoms of each element that combine to form a compound. In Chapter 6, however, we shall begin to discuss chemical bonding, and we shall be interested in the arrangement of atoms in a molecule and the distribution of outer energy level electrons. Thus we need a formula that tells us something about the electronic structure of each atom or molecule. The **electron formula** for an atom does this simply by surrounding the symbol for the element by a number of dots equal to the number of electrons in the outermost energy level. From these, as we shall see in Chapter 6, we can assemble electron formulas, or *Lewis structures*, for molecules and ions. Electron formulas for the atoms of hydrogen and helium are

$$H\cdot \qquad He\colon$$

Electron formulas for the atoms of the second period are given below. Notice that dots are paired when the electrons they represent occupy the same orbital:

$$Li\cdot \quad Be\colon \quad \overset{}{\underset{\cdot}{B}}\colon \quad \cdot\overset{}{\underset{\cdot}{C}}\colon \quad \cdot\overset{\cdot\cdot}{\underset{\cdot}{N}}\colon \quad \cdot\overset{\cdot\cdot}{\underset{\cdot\cdot}{O}}\colon \quad \colon\overset{\cdot\cdot}{\underset{\cdot\cdot}{F}}\colon \quad \colon\overset{\cdot\cdot}{\underset{\cdot\cdot}{Ne}}\colon$$

Example 5.5 Show how the electron formula for Ge, germanium, can be determined either from its atomic number (32) or from its periodic table position.

Solution

a. To draw the electron formula for an element, we need to know only the number of electrons in the outer energy level. If we know the atomic number, we can immediately write the electron configuration in terms of two quantum numbers, which in turn yields the number of outer energy level electrons. Element 32 has the configuration

$$1s^2, 2s^2, 2p^6, 3s^2, 3p^6, 3d^{10}, 4s^2, 4p^2$$

b. There are four electrons in the outermost (fourth) energy level, so the electron formula is

$$\cdot\overset{}{\underset{\cdot}{Ge}}\colon$$

c. We can obtain the same information from the position of the element in the periodic table. Germanium is in Period 4, Group IVA. Since it is in an A group, it is a representative element; thus the group number (IV) is the number of electrons in the outer energy level

$$\cdot\overset{}{\underset{\cdot}{Ge}}\colon$$

d. Note that two electrons are paired; two are not. The paired electrons represent the two electrons in the single 4s orbital; each of the other two occupy one 4p orbital.

Questions and Problems

5.1 Electromagnetic Radiation

13. Calculate the frequency of blue-green light having a wavelength in vacuum of 5.00×10^{-7} m.

14. In Chapter 1 we defined the meter in terms of the wavelength of a certain kind of light. Calculate the frequency of this light.

15. Calculate the wavelength of yellow light, which has a frequency of 5.00×10^{14} Hz in vacuum.

16. What is meant by saying that matter, electricity, and radiant energy are all particulate in nature? What is the unit in each case?

17. Calculate the energy in joules per photon of a representative frequency in each of the following ranges of the electromagnetic spectrum:
  a. radio waves   b. infrared   c. green light
  d. ultraviolet   e. X rays   f. gamma rays

18. What are the wavelength and frequency of radiation whose quanta have energies of 1.3×10^{-12} J?

5.3 Electrons, Quantum Numbers, and Energy States

19. What is meant by saying that matter and radiant energy both have a "dual nature"?

20. What is meant by a "particle" of light?

21. From the limitations placed on quantum numbers, show that
  a. there are three orbitals in a p sublevel.
  b. there is a maximum of 14 electrons in the f sublevel.
  c. there is a maximum of 18 electrons in the third energy level.

22. Arrange the following sets of quantum numbers in order of increasing energy. If two sets represent electrons of identical energy, put them on the same level.
  a. $(2, 0, 0, +\tfrac{1}{2})$ b. $(3, 2, -1, +\tfrac{1}{2})$ c. $(1, 0, 0, -\tfrac{1}{2})$ d. $(3, 2, 0, +\tfrac{1}{2})$
  e. $(2, 1, -1, -\tfrac{1}{2})$ f. $(3, 1, 1, +\tfrac{1}{2})$ g. $(1, 0, 0, +\tfrac{1}{2})$ h. $(4, 3, 2, -\tfrac{1}{2})$

23. Identify the elements whose atoms contain, in the outermost energy level, only the electrons with the indicated quantum numbers.

Element A:	n	l	m	s	Element B:	n	l	m	s
	2	0	0	$+\tfrac{1}{2}$		3	0	0	$+\tfrac{1}{2}$
	2	0	0	$-\tfrac{1}{2}$		3	0	0	$-\tfrac{1}{2}$
	2	1	1	$+\tfrac{1}{2}$	Element C:	2	0	0	$+\tfrac{1}{2}$
	2	1	0	$+\tfrac{1}{2}$		2	0	0	$-\tfrac{1}{2}$
	2	1	-1	$+\tfrac{1}{2}$		2	1	1	$+\tfrac{1}{2}$

24. Which of the following sets of four numbers cannot be quantum numbers for an electron? Explain.
  a. $(2, 1, -1, +\tfrac{1}{2})$ b. $(5, 1, 1, +\tfrac{1}{2})$ c. $(2, 2, 0, -\tfrac{1}{2})$ d. $(3, 2, 1, +\tfrac{1}{2})$
  e. $(1, 0, 0, -\tfrac{1}{2})$ f. $(3, 2, -3, +\tfrac{1}{2})$ g. $(0, 1, 0, -\tfrac{1}{2})$ h. $(1, -1, 0, +\tfrac{1}{2})$

25. Give a possible set of four quantum numbers for the differentiating electron in each of the following atoms:

    a. calcium   b. selenium   c. cesium   d. krypton   e. iron

5.4 Periodic Classification of the Elements

26. Early chemists discovered a triad relationship among the elements. There were certain groups of three elements, with similar properties, such that the atomic weight of the second element was midway between the atomic weights of the first and third elements. Cl, Br, and I constitute such a triad. Find others.

27. Listed below are some properties of calcium (element 20) and barium (element 56). Predict the properties for the element strontium; then compare your predictions with values from the *Handbook of Chemistry and Physics*.

	Calcium	Barium
Atomic weight	40.08	137.34
Specific gravity	1.54	3.51
Melting point	842°C	725°C
Boiling point	1487°C	1140°C
Formula of oxide	CaO	BaO

5.5 Electron Configurations and the Periodic Table

28. Using the information on p. 119, write the electron configurations in terms of one quantum number for the elements between sodium and calcium.

29. Using the information on p. 119, write the electron configurations in terms of one and two quantum numbers for the elements between scandium and zinc.

30. Which of the following electron configurations cannot represent a ground state atom?

    a. 2, 8   b. 2, 8, 3   c. 2, 8, 18   d. 2, 2, 6   e. 2, 8, 18, 8   f. 2, 8, 8, 3

31. Which of the following electron configurations cannot represent a ground state atom?

    a. $1s^2, 2s^2, 2p^6, 3s^2, 3p^6, 4s^1$ b. $1s^2, 2s^2, 2p^2$
    c. $1s^2, 2s^2, 2p^2, 3s^2$ d. $1s^2, 2s^2, 2p^6, 2d^{10}, 3s^2$
    e. $1s^2$ f. $1s^2, 2s^2, 2p^6, 3s^2, 3p^6, 3d^3, 4s^2$
    g. $1s^2, 2s^2, 2p^6, 3s^2, 3p^6, 3d^4, 4s^2$

32. Which of the following electron configurations cannot represent a ground state atom?

    a. 1s [↑↓]

    b. 1s [↑↓] 2s [↑↓] 2p [↑↓][ ][ ]

    c. 1s [↑↓] 2s [↑↓] 2p [↑][↑][↑][↑]

    d. 1s [↑↓] 2s [↑↓] 2p [↑↓][↑↓][↑]

    e. 1s [↑↓] 2s [↑]

    f. 1s [↑↓] 2s [↑↓] 2p [↑↓][↑↓][↑↓] 3s [↑]

5.6 Categories of Elements and Periodic Table Position

33. Without reference to a periodic table, determine the group number for the elements with the following atomic numbers:
    a. 4   b. 20   c. 17   d. 30   e. 21
    f. 15 g. 28   h. 10   i. 3

34. Answer the questions i–vii for each of the following elements, one term of whose electron configurations is supplied.
    a. $3d^3$ b. $3p^6$ (none in $4s$)   c. $2p^4$ d. $2s^2$ (none in $2p$)
    e. $3p^3$ f. $3s^1$ g. $3p^2$
    (i) How many energy levels are there in the atom?
    (ii) How many electrons are there in the outer energy level?
    (iii) How many electrons in the atom have a second quantum number with a value of zero?
    (iv) What are the first and second quantum numbers of the differentiating electron?
    (v) In what period of the periodic table will the element be found?
    (vi) In what group of the periodic table will the element be found?
    (vii) What is the atomic number of the element?

35. List all the elements in the first four periods of the periodic table whose atoms have
    a. only one electron in the s orbital of the outermost energy level.
    b. only one electron in the p orbitals of the outermost energy level.
    c. only one electron in the d orbitals.

36.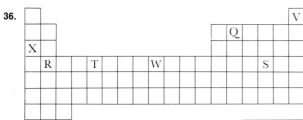

Consider the hypothetical elements in the preceding periodic table.
    a. Which are representative elements? Transition elements? Rare earth elements? Noble gases?
    b. How many occupied energy levels are there in atoms of elements X, T, V, Q, Y, and Z?
    c. How many electrons are there in the outer energy levels of the atoms of elements Q, R, S, T, V, W, and X?

37. For each of the elements listed below, identified only by period and group, answer questions i–iv that follow.
    a. Period 3, Group IIIA   b. Period 5, Group IIIB
    c. Period 2, Group 0   d. Period 4, Group VIA
    (i) Is the element a representative element? A transition element? A rare earth? A noble gas?
    (ii) What is the atomic number of the element?

(iii) How many electrons are there in the outermost energy level of the atom?
(iv) Is the differentiating electron in an *s*, *p*, *d*, or *f* sublevel?

38. Use the periodic table on the inside of the front cover to determine which elements in the first four periods have atoms with one unpaired electron, which have two unpaired electrons, which have three, and so on.

39. Locate two places in the periodic table where elements are not in order according to atomic weight.

40. What would be the atomic number of the first element in the eighth period of the table, if it existed?

41. Place the symbols for the hypothetical elements described below in their proper places in a diagram of the first four periods of the periodic table. If you place the symbols in the diagram *in the order described*, the results will spell out an important principle.

a. The atomic number of Wo is 4.
b. On has the largest ionization energy of any element in Group VIIA.
c. The electron configuration of A ends with $3s^2$.
d. Al is the second transition element (in order of atomic number) with a half-filled *d* sublevel.
e. T has only three electrons.
f. L has four electrons in the energy level for which $n = 2$.
g. Es has two electrons in a *d* sublevel.
h. Nt has seven filled sublevels and a single electron in the eighth sublevel.
i. W has 10 *p* electrons.
j. Om, Nu, and Ec are representative elements, each with three unpaired electrons. Their atomic weights are in the order Ec < Om < Nu.
k. Ev has six unpaired electrons.
l. Am and Sf are transition elements with three unpaired electrons. Sf is the heavier of the two.
m. Ue and Or are in the same period *and* in the same group. Or contains two more protons than Ue.
n. Ll contains five sublevels, all of which are completely filled.
o. Element At has the second smallest atom in Group IVA.
p. Rs has six electrons in the $4p$ sublevel.
q. Tr and M are nonmetals, each with two unpaired electrons. M is the heavier atom.
r. Q and Ua both have 10 electrons in the $3d$ sublevel, but Q has 1 electron in the $4s$ sublevel, whereas Ua has 2.
s. S has only two occupied energy levels, both of which are completely filled.
t. N contains no neutrons.
u. Th is a transition element.
v. O is a noble gas.
w. The electron formula for Ve is Ve:.
x. I and Be each have seven electrons in the outermost energy level. The Be atom contains 18 more protons than the atom of I.

y. Of and Ha each contain one electron in their outermost energy level, but Ha has more energy levels.

z. E, Ny, and Um are metalloids in the second, third, and fourth periods, respectively.

5.7 Atomic Properties

42. Plot the atomic radius against atomic number for each group of representative elements and for each period. Where does the largest relative change take place in each group?

43. Refer to Table 5.8 and answer the following questions:
  a. Why is the first ionization energy of Al *lower* than that of Si but the second ionization energy of Al *greater* than that of Si?
  b. Why is the first ionization energy of sulfur *lower* than that of phosphorus?
  c. Why is the second ionization energy of Mg greater than the first?

44. Where in the periodic table would you find the elements
  a. with the lowest ionization energies?
  b. with the highest ionization energies?
  c. with filled outermost energy levels?
  d. with half-filled sublevels?
  e. with eight electrons in their outermost energy levels?
  f. with two electrons in their outermost energy levels?
  g. with one electron in their outermost energy levels?
  h. with the most metallic character?
  i. with the least metallic character?
  j. with differentiating electrons in d orbitals?
  k. with differentiating electrons in p orbitals?
  l. with differentiating electrons in f orbitals?
  m. with differentiating electrons in s orbitals?
  n. whose atoms have electrons in five energy levels?
  o. with the smallest atomic diameters?

5.8 Electron Formulas for Atoms

45. Draw Lewis structures for atoms of each of the third period elements.

46. Draw a single electron formula for each of the following *groups* of atoms. Use X as the symbol for the element.
  a. fluorine, chlorine, bromine, iodine
  b. oxygen, sulfur, selenium, tellurium
  c. nitrogen and phosphorus
  d. neon, argon, krypton, xenon, radon
  e. lithium, sodium, potassium, rubidium, cesium
  f. beryllium, magnesium, calcium, strontium, barium
  g. aluminum

Discussion Starters

47. Compare and contrast the following terms.
  a. velocity (of light), frequency   b. wavelength, amplitude
  c. infrared, ultraviolet   d. electron configuration, electron formula

48. Explain or account for the following true statements:

  a. It requires less voltage to produce an electric spark between two electrodes if the electrodes are illuminated with ultraviolet radiation.
  b. Any orbital can accommodate only two electrons.
  c. The first three quantum numbers are sometimes said to describe the size, shape, and orientation of an orbital.
  d. Within each sublevel, one electron goes into each orbital before any orbital is filled with two electrons.
  e. Atoms of chromium and copper have only one electron in the $4s$ sublevel.
  f. The vanadium atom has three electrons in the $3d$ sublevel but is in periodic Group VB.
  g. There are 10 transition elements in a period but only eight B groups.
  h. Sodium and potassium are in the same group of the periodic table and hence have similar properties. However, they impart different colors to a flame, sodium gives a yellow color and potassium, a violet color.
  i. Atomic size decreases from left to right within a period of the periodic table.
  j. Atomic size increases with atomic number within a group of the periodic table.
  k. Ionization energies generally increase from left to right within a period of the periodic table, but the first ionization energy of aluminum is less than that of magnesium, and the first ionization energy of oxygen is less than that of nitrogen.
  l. Ionization energies decrease with increasing atomic number within a group of the periodic table.

Chemical Bonding—Electron Formulas

In Chapter 4 we looked at the subatomic particles that comprise all matter. We then put these elementary building blocks together and in Chapter 5 discussed the properties of atoms and the classification of the elements. In this chapter we shall take the next step, looking at how these elements combine to form stable compounds. And we shall see later that the same principles of chemical bonding are involved in the simplest diatomic gases as in the complex molecules of living tissue.

One of the first questions that comes to mind is why chemical bonds form at all. A full answer must await the discussion in Chapter 11, but we can state a general principle here. The formation of stable chemical bonds, like all other spontaneous physical and chemical processes, tends to minimize the potential energy of the system. Just as stones roll downhill, so do isolated atoms tend to form chemical bonds. The product contains less potential energy than the isolated atoms; hence energy is evolved when atoms combine.

Another question is why elements combine in just the ratio they do. Why is the formula for water H_2O rather than HO (as Dalton thought) or H_3O? Why is the formula of table salt NaCl and not Na_2Cl or $NaCl_2$? To answer these questions we must recall that noble gases are much less reactive than other elements. This suggests that the electron configurations of the noble gases are especially stable. Therefore, it is not surprising that when stable chemical bonds are formed, the electron configuration of each atom is usually changed—by losing, gaining, or sharing electrons—to that of one of the noble gases; that is, when atoms react, they tend to acquire a more stable electron configuration. Since this tendency is usually a matter of acquiring eight electrons in the outermost energy level, this generalization is often referred to as the *octet rule*. It is a useful generalization and can be used to predict correctly the formulas for water and table salt, but as we shall see later in this chapter, there are exceptions.

Let us now look in more detail at the nature of the chemical bonds themselves. We shall consider two types: *covalent* and *ionic* (or *electrovalent*). A **covalent bond** is formed when electron pairs are shared between the bonded atoms. An **ionic bond,** on the other hand, is formed when one or more electrons are transferred from one atom to another, producing charged particles called **ions.** The atom that loses electrons becomes positively charged and is called a **cation;** the atom that gains electrons becomes negatively charged and is called an **anion.** The electrostatic attraction between cations and anions then produces stable ionic bonds.

Before going on, we should emphasize that the notions of covalent and ionic bonds are merely convenient concepts. As we shall see in Section 6.3, few chemical bonds should be regarded as purely one or the other; most bonds are a blend of the two types. Nonetheless, it is useful to begin by talking about the features of ionic and covalent bonds as if they were always distinguishable.

Questions

1. Why do chemical bonds form?
2. Distinguish between the two main types of chemical bonds.
3. How are ions formed from neutral atoms?

6.1 The Ionic Bond

We shall consider the ionic bond by first recalling the electron formulas for the isolated atoms of elements of the first three periods of the periodic table:

IA	IIA	IIIA	IVA	VA	VIA	VIIA	0
H·							He:
Li·	Be:	B:	·C:	·N:	·O:	:F:	:Ne:
Na·	Mg:	Al:	·Si:	·P:	·S:	:Cl:	:Ar:

Since chemical bonding usually involves transferring or sharing outer energy level electrons, we may infer that the **valence,** or combining capacity, of atoms of the representative elements is governed by the number of *valence electrons*—the electrons in the outermost energy level (the *valence energy level*). As an example, consider the elements of Groups IA and VIIA.

Notice that the elements in Group IA, with the exception of hydrogen, can attain a noble gas electron configuration by gaining seven electrons or, more conveniently, by losing one. (In the case of hydrogen, the electron configuration of helium might be acquired by gaining one electron to form the hydride ion, H^-. This often occurs, but more commonly hydrogen, a nonmetal, shares electrons to form covalent bonds.) Conversely, the elements in Group VIIA can acquire the electron configuration of one of the noble gases by losing seven electrons or by gaining one. Since the highly charged ions that would result from the transfer of many electrons are rarely stable, ionic bonds usually involve the transfer of a small number of electrons. Hence, elements of Group IA (hydrogen excluded) form ionic bonds by losing one electron rather than by gaining seven, and the elements of Group VIIA form ionic bonds by gaining one electron. Thus, the formation of sodium fluoride might be represented as

$$Na\cdot + :\!\overset{..}{\underset{..}{F}}\!: \longrightarrow \left[Na^+ + :\!\overset{..}{\underset{..}{F}}\!:^-\right] \quad \text{or} \quad NaF$$

Notice that the sodium ion is represented only by the symbol and the charge. Since sodium has transferred its single valence electron, the valence energy level is empty and the ion has a $+1$ charge. The fluoride ion is represented by the symbol, the charge, and eight dots for the eight electrons now in its outermost energy level.

If we now consider compound formation between an element of Group IIA and an element of Group VIIA, we discover that two atoms from Group VIIA are required for each atom from Group IIA. For example, in the formation of magnesium fluoride, each fluorine atom can gain only one electron, but the magnesium atom must lose two:

$$\text{Mg:} + 2\,\text{:}\!\ddot{\underset{..}{\text{F}}}\!\text{:} \longrightarrow \left[\text{Mg}^{2+} + 2\,\text{:}\!\ddot{\underset{..}{\text{F}}}\!\text{:}^{-}\right] \quad \text{or} \quad \text{MgF}_2$$

If we continue to extend the principles of valence and the octet rule, we find that magnesium and nitrogen react in the ratio of three to two. Each magnesium atom loses two electrons, but each nitrogen atom gains three:

$$3\,\text{Mg:} + 2\,\cdot\!\ddot{\text{N}}\!\text{:} \longrightarrow \left[3\,\text{Mg}^{2+} + 2\,\text{:}\!\ddot{\underset{..}{\text{N}}}\!\text{:}^{3-}\right] \quad \text{or} \quad \text{Mg}_3\text{N}_2$$

Similarly,

$$\text{Mg:} + \cdot\!\ddot{\text{S}}\!\text{:} \longrightarrow \left[\text{Mg}^{2+} + \text{:}\!\ddot{\underset{..}{\text{S}}}\!\text{:}^{2-}\right] \quad \text{or} \quad \text{MgS}$$

$$3\,\text{Li}\cdot + \cdot\!\ddot{\text{N}}\!\text{:} \longrightarrow \left[3\,\text{Li}^{+} + \text{:}\!\ddot{\underset{..}{\text{N}}}\!\text{:}^{3-}\right] \quad \text{or} \quad \text{Li}_3\text{N}$$

In each of these five cases, the ions inside the brackets constitute the *formula unit* of an ionic compound. This is the smallest collection of the given ions that is electrically neutral. A mole of an ionic compound is 6.02×10^{23} formula units. In the case of the compound composed of lithium and nitrogen ions (lithium nitride), a mole contains 6.02×10^{23} [$3\,\text{Li}^+ + \text{:}\!\ddot{\underset{..}{\text{N}}}\!\text{:}^{3-}$] units or 6.02×10^{23} $\text{:}\!\ddot{\underset{..}{\text{N}}}\!\text{:}^{3-}$ ions and $(3)(6.02 \times 10^{23})$, or 1.81×10^{24} Li^+ ions. Once we know the formula unit of a compound, we can immediately write the chemical formula, as shown in the examples above.

In each of the compounds we have considered so far, all the constituent atoms acquired the electron configuration of one of the noble gases, either by gaining or losing electrons. We say that the resulting ions are *isoelectronic* with the noble gases. Thus the nitride ion ($\text{:}\!\ddot{\underset{..}{\text{N}}}\!\text{:}^{3-}$) is isoelectronic with neon, and the sulfide ion ($\text{:}\!\ddot{\underset{..}{\text{S}}}\!\text{:}^{2-}$) is isoelectronic with argon. We may generalize this observation by saying that monatomic anions usually acquire the electron configuration of one of the noble gases. The same is true of cations formed from the atoms of the metals of Groups IA, IIA, and IIIB, and aluminum. Whereas the negative ions attain stable electron configurations by adding electrons to the p orbitals in the outermost energy level, the situation is more complex with positive ions. Representative metals usually lose electrons only from the outermost energy level, but transition metals often lose electrons from both the outermost and the next-to-outermost energy levels. For example, elements of Groups IA and IIA lose electrons only from the outermost *s* orbitals, and aluminum loses two electrons from the *s* orbital and one electron from a *p* orbital of the outermost energy level. Metals from Group IIIB, however, lose two electrons from the outermost *s* orbital and one electron from a *d* orbital in the next-to-outermost energy level. As we shall see in Section 6.5, most ions formed from the transition metals beyond Group IIIB are not isoelectronic with noble gases.

As Figure 6.1 illustrates, ionic radii can differ markedly from the radii of the corresponding atoms. The ionic radii of metallic cations are smaller than the corresponding atomic radii, whereas the anions that are formed from nonmetals are larger than the corresponding neutral atoms. This observation is easy to justify

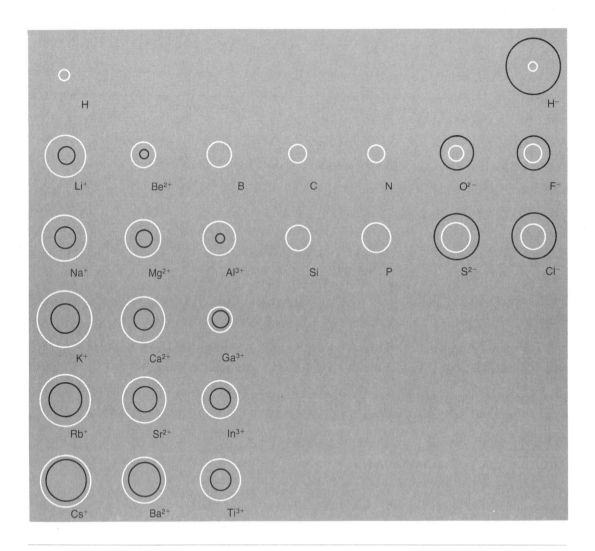

Figure 6.1 Relative atomic and ionic radii of some common elements (ionic radii indicated in color).

since metallic ions are formed by losing outer energy level electrons, whereas nonmetallic ions are formed by capturing outer energy level electrons. The process of anion formation, which involves no change in nuclear charge, should be contrasted with our earlier observation (Section 5.7) that atomic radii decrease as we move from left to right within a period, increasing both the number of valence electrons and the nuclear charge. Without this benefit of additional nuclear charge, added valence electrons merely increase electrostatic repulsion within the valence energy level, which in turn increases the ionic radius.

Questions

4. How many valence electrons are there in each atom of
   a. Na   b. C   c. Cl   d. B   e. Mg

5. Write the electron formulas for ions that are isoelectronic with argon and have the following charges: $+1, +2, -1, -2, -3$.

6. Write electron formulas for ionic compounds formed from the following pairs of elements:
   a. potassium and chlorine   b. radium and iodine   c. barium and sulfur
   d. aluminum and fluorine   e. calcium and phosphorus

7. Calculate the *total* number of ions in 1.00 mol of each of the following substances:
   a. NaCl   b. $CaCl_2$ c. Mg_3P_2 d. K_3N e. CaO   f. Al_2S_3

6.2 The Covalent Bond

Whereas ionic bonds involve the transfer of electrons between two dissimilar atoms (usually a metal and a nonmetal), covalent bonding results when electrons are shared between atoms of nonmetals. Covalent bonds are usually formed between similar or even identical atoms. As with ionic bonding, however, the ratio of atoms in a molecule can often be predicted from the number of valence electrons in each atom. Let us consider some typical covalently bound molecules.

Hydrogen atoms combine to form diatomic hydrogen molecules:

$$2\ H\cdot \longrightarrow H\colon H \quad \text{or} \quad H\text{—}H \quad \text{or} \quad H_2$$

Either a pair of dots or a dash between atoms represents a shared pair of electrons (a covalent bond). In this text, we shall generally use a pair of dots. The two electrons in the hydrogen molecule are equally attracted to the two nuclei and fill the first energy level of both atoms. Hence, *both* hydrogen atoms have electron configurations isoelectronic with helium. In forming compounds, hydrogen usually acquires this electron configuration.

As we noted when discussing ionic bonds, atoms of Group VIIA, which are called the *halogens*, have seven valence electrons. Like hydrogen, these atoms are one electron short of having a noble gas electron configuration. Therefore, we expect the elements in Group VIIA, like hydrogen, to exist in nature as diatomic molecules. The formation of these molecules from isolated atoms is illustrated by the following:

$$2 :\!\ddot{\underset{..}{F}}\!: \longrightarrow :\!\ddot{\underset{..}{F}}\!:\!\ddot{\underset{..}{F}}\!: \text{ or } F_2$$

$$2 :\!\dot{\underset{..}{I}}\!: \longrightarrow :\!\ddot{\underset{..}{I}}\!:\!\ddot{\underset{..}{I}}\!: \text{ or } I_2$$

We might also expect the halogens to form diatomic molecules with hydrogen. Indeed, hydrogen combines readily with each of the halogens. For example,

$$H\cdot + :\!\dot{\underset{..}{Cl}}\!: \longrightarrow H:\!\ddot{\underset{..}{Cl}}\!: \text{ or } HCl$$

$$H\cdot + :\!\dot{\underset{..}{Br}}\!: \longrightarrow H:\!\ddot{\underset{..}{Br}}\!: \text{ or } HBr$$

Interhalogen compounds also exist:

$$:\!\dot{\underset{..}{Br}}\!: + :\!\dot{\underset{..}{Cl}}\!: \longrightarrow :\!\ddot{\underset{..}{Br}}\!:\!\ddot{\underset{..}{Cl}}\!: \text{ or } BrCl$$

$$:\!\dot{\underset{..}{I}}\!: + :\!\dot{\underset{..}{Cl}}\!: \longrightarrow :\!\ddot{\underset{..}{I}}\!:\!\ddot{\underset{..}{Cl}}\!: \text{ or } ICl$$

Just as we used the octet rule to determine the ratio of ions in ionic compounds, we can use it to predict the formulas of covalent compounds. We must only remember that hydrogen needs two, not eight, electrons to complete its outer shell. Therefore, since the elements of Group VIA require two additional electrons to acquire a noble gas electron configuration, hydrogen and the halogens combine with atoms from Group VIA in a two-to-one ratio:

$$2 H\cdot + \cdot\!\dot{\underset{..}{S}}\!: \longrightarrow H:\!\ddot{\underset{..}{S}}\!:H \text{ or } H_2S$$

$$2 :\!\dot{\underset{..}{F}}\!: + \cdot\!\dot{\underset{..}{O}}\!: \longrightarrow :\!\ddot{\underset{..}{F}}\!:\!\ddot{\underset{..}{O}}\!:\!\ddot{\underset{..}{F}}\!: \text{ or } OF_2$$

Likewise, we can conclude that Group VIIA elements (or hydrogen) and Group VA elements usually combine in a three-to-one ratio:

$$3 :\!\dot{\underset{..}{Cl}}\!: + \cdot\!\dot{P}\!: \longrightarrow :\!\ddot{\underset{..}{Cl}}\!:\!\overset{:\ddot{\underset{..}{Cl}}:}{P}\!:\!\ddot{\underset{..}{Cl}}\!: \text{ or } PCl_3$$

$$3 H\cdot + \cdot\!\dot{N}\!: \longrightarrow H:\!\overset{H}{\underset{..}{N}}\!:H \text{ or } NH_3$$

Finally, the ratio between the halogens or hydrogen and atoms of Group IVA is four-to-one:

$$4 H\cdot + \cdot\!\dot{Si}\!: \longrightarrow H:\!\overset{H}{\underset{H}{Si}}\!:H \text{ or } SiH_4$$

$$4 :\!\dot{\underset{..}{Cl}}\!: + \cdot\!\dot{C}\!: \longrightarrow :\!\ddot{\underset{..}{Cl}}\!:\!\overset{:\ddot{\underset{..}{Cl}}:}{\underset{:\ddot{\underset{..}{Cl}}:}{C}}\!:\!\ddot{\underset{..}{Cl}}\!: \text{ or } CCl_4$$

Note that before four bonds can be formed with carbon or silicon, the electron configurations of these atoms must be changed from ·C: and Si: to ·C· and ·Si·. We shall discuss this phenomenon further in Chapter 7.

Question

8. Write electron formulas for compounds formed from the following pairs of elements.
   a. H and Cl   b. Si and Cl   c. O and F
   d. H and I   e. P and Br

6.3 Electronegativities and Bond Polarity

As we emphasized earlier, most chemical bonds should not be regarded as purely ionic or purely covalent. The ionic and covalent bonds represent two extreme types—complete transfer of electrons in the first case and equal sharing of electrons in the second. Between these extremes are more realistic intermediate cases where pairs of electrons are shared unequally. Bonds of this type are said to be *partially ionic*, or *polar covalent*.

The *electronegativity scale* helps the chemist to describe these chemical bonds. On this scale, each element is assigned an electronegativity value, which is indicative of an atom's attraction for a pair of bonding electrons. Electronegativities cannot be determined experimentally but are usually derived from a formula devised by Linus Pauling (b. 1901). Fluorine, the most electronegative element, is arbitrarily assigned an electronegativity value of four, with all the other elements falling in line below it according to Pauling's formula. For example, boron, with an electronegativity value of two, has about half the attraction for a shared pair of electrons that fluorine has. (The noble gases are exceptions; since they have little tendency to form chemical bonds, they are not assigned electronegativity values.) In general, the electronegativity of the elements increases from left to right in a period and decreases from top to bottom in a group. Therefore, the most electronegative elements are in the upper right-hand corner, and the most electropositive elements (those with low electronegativities and small ionization energies) are in the lower left-hand corner (Figure 6.2).

When a bond forms between unlike atoms, the element with the greater electronegativity more strongly attracts the bonding pair of electrons, thus gaining a partial negative charge while producing a corresponding partial positive charge on the less electronegative atom. This *dipole*—a pair of equal but opposite charges separated by a finite distance—characterizes both ionic and polar covalent bonds. The calculated value for the *dipole moment* of such a bond is the product of one of the equal charges and the distance between the charges.

We may now ask how one can predict whether a bond between two atoms will be ionic, polar covalent, or *nonpolar covalent* (where the electron pair is shared equally). As is frequently the case in chemistry, there are no hard and fast rules, but electronegativity differences are often useful clues. If the difference

IA																	0
1 H 2.1	IIA											IIIA	IVA	VA	VIA	VIIA	2 He —
3 Li 1.0	4 Be 1.5											5 B 2.0	6 C 2.5	7 N 3.0	8 O 3.5	9 F 4.0	10 Ne —
11 Na 0.9	12 Mg 1.2	IIIB	IVB	VB	VIB	VIIB	——	VIIIB	——	IB	IIB	13 Al 1.5	14 Si 1.8	15 P 2.1	16 S 2.5	17 Cl 3.0	18 Ar —
19 K 0.8	20 Ca 1.0	21 Sc 1.3	22 Ti 1.5	23 V 1.6	24 Cr 1.6	25 Mn 1.5	26 Fe 1.8	27 Co 1.8	28 Ni 1.8	29 Cu 1.9	30 Zn 1.6	31 Ga 1.6	32 Ge 1.8	33 As 2.0	34 Se 2.4	35 Br 2.8	36 Kr —
37 Rb 0.8	38 Sr 1.0	39 Y 1.2	40 Zr 1.4	41 Nb 1.6	42 Mo 1.8	43 Tc 1.9	44 Ru 2.2	45 Rh 2.2	46 Pd 2.2	47 Ag 1.9	48 Cd 1.7	49 In 1.7	50 Sn 1.8	51 Sb 1.9	52 Te 2.1	53 I 2.5	54 Xe —
55 Cs 0.7	56 Ba 0.9	57–71 — 1.1–1.2	72 Hf 1.3	73 Ta 1.5	74 W 1.7	75 Re 1.9	76 Os 2.2	77 Ir 2.2	78 Pt 2.2	79 Au 2.4	80 Hg 1.9	81 Tl 1.8	82 Pb 1.8	83 Bi 1.9	84 Po 2.0	85 At 2.2	86 Rn —
87 Fr 0.7	88 Ra 0.9	89 Ac 1.1															

Figure 6.2 Electronegativities of the elements.

between the electronegativities of bonding atoms is large, the bond is considered essentially ionic, and we write the electron formula as ionic. The bond between francium and fluorine, where the difference in electronegativities is 3.3, is the extreme case. For NaCl, the difference is only 2.1, but NaCl is still essentially an ionic compound.

The difference between the electronegativities of hydrogen and chlorine is still less (0.9). Here we cannot claim that the bond is ionic; yet neither are the electrons shared equally. The bond in hydrogen chloride, therefore, is polar covalent. Chlorine, with the greater electronegativity, acquires a partial negative charge and the hydrogen atom acquires a partial positive charge. Thus the polarity of the hydrogen-chlorine bond is designated as

$$\overset{\delta^+}{\text{H}}—\overset{\delta^-}{\text{Cl}}$$

where the Greek letter δ (delta) is used to signify a partial charge. An alternative method of designating a polar bond is to use an arrow with a "positive" tail, pointing in the direction of the partial negative charge:

$$\overset{\longrightarrow}{\text{H}—\text{Cl}}$$

Other polar bonds are represented in the same way:

$$\overset{\delta^+}{\text{Cl}}-\overset{\delta^-}{\text{F}} \quad \text{or} \quad \overset{\longrightarrow}{\text{Cl}-\text{F}}$$

$$\overset{\delta^+}{\text{H}}-\overset{\delta^-}{\text{N}} \quad \text{or} \quad \overset{\longrightarrow}{\text{H}-\text{N}}$$

$$\overset{\delta^+}{\text{C}}-\overset{\delta^-}{\text{O}} \quad \text{or} \quad \overset{\longrightarrow}{\text{C}-\text{O}}$$

$$\overset{\delta^+}{\text{C}}-\overset{\delta^-}{\text{Cl}} \quad \text{or} \quad \overset{\longrightarrow}{\text{C}-\text{Cl}}$$

It is necessary to mention here that although electronegativity is perhaps the most useful simple concept for predicting bond types, there are other factors that must be considered. Among these are atomic size and oxidation number (see Section 6.10). For example, when it is possible for two elements to form more than one compound, the different compounds can have different bond types. $SnCl_2$, a solid at room temperature, has essentially ionic bonds, whereas $SnCl_4$, a liquid at room temperature, has covalent bonds.

If a molecule contains polar bonds, the molecule itself may or may not be polar, depending on the spatial distribution of the polar bonds. Although the molecule is electrically neutral—it contains the same number of positive and negative charges—the distribution of partial charges, if asymmetrical, can produce a molecular dipole. Hydrogen chloride, like all diatomic molecules containing a polar bond, is polar; the symmetrical carbon tetrachloride (CCl_4) molecule is not, since the contributions of the four polar bonds cancel one another. We shall say more about molecular dipoles in Chapter 7.

A bond in which a pair of electrons is shared equally between two atoms is nonpolar. Nonpolar covalent bonds occur only when the bonded atoms have the same electronegativity. Thus the bonds in diatomic molecules such as H_2, Cl_2, and F_2 and in larger molecules such as P_4 and S_8 are nonpolar. All such molecules, containing only nonpolar bonds, are themselves nonpolar.

In conclusion, let us point out a trend within the periodic table that should be clear from what we have said about ionization energies and electronegativities. **Chemical activity**—the tendency of an element to take part in a chemical reaction—increases as one goes down within a group on the left side of the periodic table but decreases as one goes down within a group on the right side of the table. Thus francium is the *most* active metal of Group IA, and astatine is the *least* active nonmetal of Group VIIA (Figure 6.3).

Questions

9. Criticize the following statement:
It is impossible to state with assurance that a given bond is completely covalent or completely ionic.

10. Arrange the bonds in each of the following groups in order of increasing polarity (see Figure 6.2).
  a. H—F, H—Cl, H—I, H—Br
  b. Br—Cl, Cl—F, Br—I, Br—F
  c. I—Cl, I—Br, H—F, S—F, P—Cl

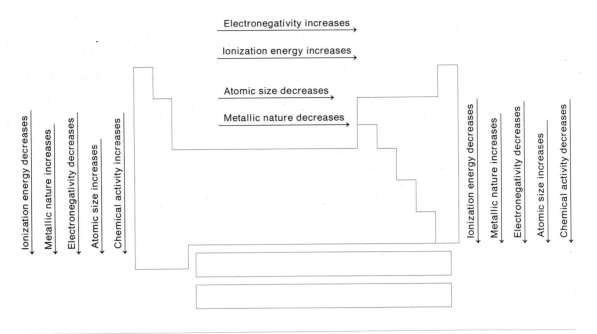

Figure 6.3 Trends of the periodic table.

11. In each pair of elements listed below, which has the greater
    a. ionization energy?   b. electronegativity?   c. atomic diameter?
    d. chemical activity?   e. metallic nature?
    (i) K or Rb   (ii) Cl or S   (iii) Cl or Br
    (iv) Na or Mg (v) C or Si

12. Molecules of both HCl and CCl_4 have polar bonds. Explain why HCl molecules are polar but CCl_4 molecules are not.

6.4 Multiple Covalent Bonding

In our discussion of covalent bonding so far, we have limited ourselves to molecules containing hydrogen or one of the halogens, each of which needs only one electron to complete a noble gas electron configuration. Now we want to consider more complicated molecules.

Nitrogen, like most elementary gases, is most stable as a diatomic molecule. However, when one attempts to write an electron formula for a nitrogen molecule using the method that worked in Section 6.2, trouble arises:

$$2 \cdot \ddot{N}: \longrightarrow \cdot \ddot{N}\!:\!\ddot{N}\cdot$$

There simply are not enough valence electrons to give each nitrogen atom the eight it needs to complete its outermost energy level. In such cases, the atoms are often bonded together by more than one covalent bond. Here, the two nitrogen atoms can acquire a noble gas electron configuration if we show them bound by three covalent bonds—a *triple bond*:

$$2 \cdot \ddot{N}: \longrightarrow :N:::N: \quad \text{or} \quad N_2$$

The triple bond produces a very stable diatomic nitrogen molecule. It is one of the most stable molecules known.

A slightly more complex case is the ethyne (commonly called acetylene) molecule, C_2H_2. Here again, it is impossible, using only single bonds, to redistribute the 10 valence electrons (4 from each of the two carbon atoms and 1 from each of the two hydrogen atoms) so that each atom has a noble gas electron configuration. However, if the two carbon atoms are linked with a triple bond, each atom is able to fill its valence energy level:

$$2\,H\cdot + 2\cdot\ddot{C}{:}\,(\cdot\dot{C}\cdot) \longrightarrow H{:}C{:::}C{:}H \quad \text{or} \quad C_2H_2$$

Ethene (C_2H_4) is another molecule that contains a multiple covalent bond. Also called ethylene, it is a by-product of gasoline production and is used to manufacture polyethylene, a common plastic. In this case, two pairs of bonding electrons are shared, producing a *double bond*:

$$4\,H\cdot + 2\cdot\ddot{C}{:}\,(\cdot\dot{C}\cdot) \longrightarrow H{:}\overset{H}{\ddot{C}}{::}\overset{H}{\ddot{C}}{:}H \quad \text{or} \quad C_2H_4$$

In carbon dioxide there are two double bonds:

$$:\ddot{O}::C::\ddot{O}:$$

Though not restricted to the second period, multiple bonding is less common among elements of other periods of the periodic table.

Questions

13. Multiple bonding is most characteristic of which elements?

14. Account for the relative chemical inertness of the nitrogen molecule.

6.5 Metal Ions That Deviate from Noble Gas Electron Configurations

Many of the heavier elements ($Z > 21$), especially transition metals, form stable ions that violate the octet rule. Consequently, they are not isoelectronic with the noble gases. This is not surprising, since most of these elements would have to gain or lose at least four electrons to achieve a noble gas electron configuration; such highly charged ions are rarely stable. We shall now consider three of these more complicated stable electron configurations.

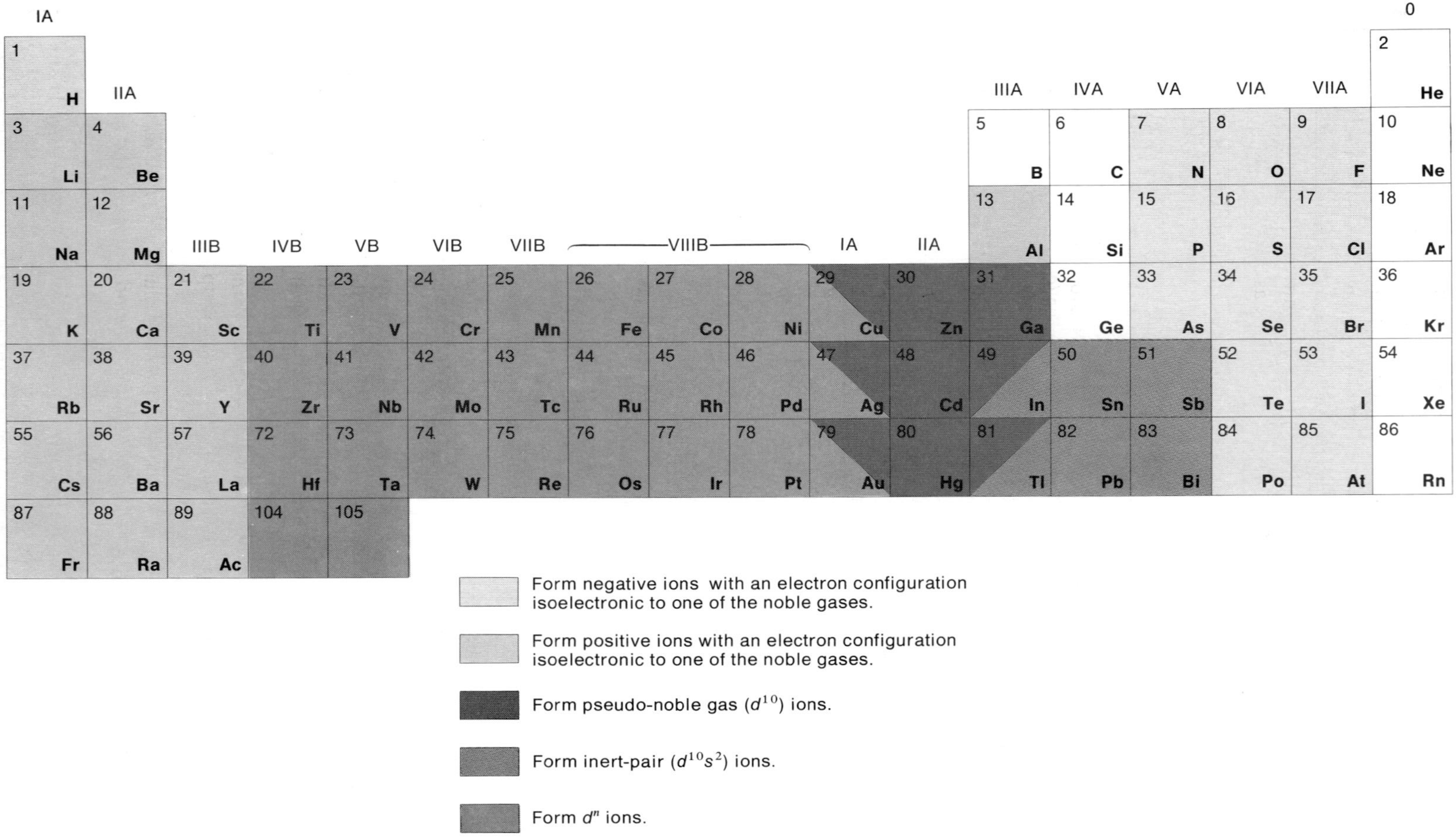

Figure 6.4 Types of monatomic ions.

First among these is an electron configuration with 18 electrons in the outermost energy level. Ions of this type are sometimes known as the *pseudo-noble gas ions*, or, since their outermost energy level configuration is $(n-1)s^2$, $(n-1)p^6$, $(n-1)d^{10}$ as the d^{10} ions. (n is used here to designate the outermost energy level of the neutral atom.) This configuration is characteristic of elements of periodic groups IB, IIB, and IIIA below aluminum (Figure 6.4). Thus we have Cu^+, Ag^+, Zn^{2+}, Cd^{2+}, Hg^{2+}, Ga^{3+}, In^{3+}, and Tl^{3+}, as shown in Table 6.1.

Table 6.1 Electron Configurations of d^{10} Ions

Element	Atomic Number	Electron Configuration of Isolated Atom (two outermost energy levels only)	Ion	Electron Configuration of Ion (outermost energy level only)
Cu	29	$3s^2, 3p^6, 3d^{10}, 4s^1$	Cu^+	$3s^2, 3p^6, 3d^{10}$
Ag	47	$4s^2, 4p^6, 4d^{10}, 5s^1$	Ag^+	$4s^2, 4p^6, 4d^{10}$
Au	79	$5s^2, 5p^6, 5d^{10}, 6s^1$	Au^+	$5s^2, 5p^6, 5d^{10}$
Zn	30	$3s^2, 3p^6, 3d^{10}, 4s^2$	Zn^{2+}	$3s^2, 3p^6, 3d^{10}$
Cd	48	$4s^2, 4p^6, 4d^{10}, 5s^2$	Cd^{2+}	$4s^2, 4p^6, 4d^{10}$
Hg	80	$5s^2, 5p^6, 5d^{10}, 6s^2$	Hg^{2+}	$5s^2, 5p^6, 5d^{10}$
Ga	31	$3s^2, 3p^6, 3d^{10}, 4s^2, 4p^1$	Ga^{3+}	$3s^2, 3p^6, 3d^{10}$
In	49	$4s^2, 4p^6, 4d^{10}, 5s^2, 5p^1$	In^{3+}	$4s^2, 4p^6, 4d^{10}$
Tl	81	$5s^2, 5p^6, 5d^{10}, 6s^2, 6p^1$	Tl^{3+}	$5s^2, 5p^6, 5d^{10}$

Another stable configuration is characterized by 2 electrons in the n level and 18 electrons in the $n-1$ level: $(n-1)s^2$, $(n-1)p^6$, $(n-1)d^{10}$, ns^2. Ions of this type are called the *inert-pair ions*, or the $d^{10}s^2$ ions. The metals of periodic groups IIIA, IVA, and VA below the fourth period form ions of this type (Figure 6.4). Hence we have In^+, Tl^+, Sn^{2+}, Pb^{2+}, Sb^{3+}, and Bi^{3+} (Table 6.2).

Notice that thallium (Tl) and indium (In) appear in both Table 6.1 and Table 6.2. Both of these elements can form stable d^{10} ions and stable $d^{10}s^2$ ions. Such multiple ionic valences are unusual among the representative elements; however, except among the elements of Groups IIIB and IIB, multiple valences are common among the transition elements. The Group IIIB elements exhibit primarily a +3 valence, and the elements of Group IIB have only a +2 valence. (For this generalization to be rigorously true, we must consider the common (but exceptional) Hg_2^{2+} ion as a mercury atom bound to a Hg^{2+} ion, rather than two Hg^+ ions.)

Table 6.2 Electron Configurations of $d^{10}s^2$ Ions

Element	Atomic Number	Electron Configuration of Isolated Atom (two outermost energy levels only)	Ion	Electron Configuration of Ion (outermost energy level only)
In	49	$4s^2, 4p^6, 4d^{10}, 5s^2, 5p^1$	In^+	$4s^2, 4p^6, 4d^{10}, 5s^2$
Tl	81	$5s^2, 5p^6, 5d^{10}, 6s^2, 6p^1$	Tl^+	$5s^2, 5p^6, 5d^{10}, 6s^2$
Sn	50	$4s^2, 4p^6, 4d^{10}, 5s^2, 5p^2$	Sn^{2+}	$4s^2, 4p^6, 4d^{10}, 5s^2$
Pb	82	$5s^2, 5p^6, 5d^{10}, 6s^2, 6p^2$	Pb^{2+}	$5s^2, 5p^6, 5d^{10}, 6s^2$
Sb	51	$4s^2, 4p^6, 4d^{10}, 5s^2, 5p^3$	Sb^{3+}	$4s^2, 4p^6, 4d^{10}, 5s^2$
Bi	83	$5s^2, 5p^6, 5d^{10}, 6s^2, 6p^3$	Bi^{3+}	$5s^2, 5p^6, 5d^{10}, 6s^2$

The final category of stable ions can be formed from the transition metals between periodic groups IIIB and IIB (Figure 6.4). These ions are designated d^n *ions* (where n ranges from 1 to 9) because they contain n electrons beyond a noble gas electron configuration. Included in this category are Fe^{2+}, Co^{2+}, Ni^{2+}, Cu^{2+}, Fe^{3+}, and Cr^{3+}.

The stability of the d^{10}, $d^{10}s^2$, and d^n ions is difficult to understand at first glance. After all, none of them can claim the stable electron configuration of a noble gas. Most of them can be explained, however, if we adopt three rules, each soundly rooted in atomic theory:

1. Atoms and ions gain a measure of stability by filling energy sublevels or by half-filling d energy sublevels.

2. The fourth ionization energy is so large that $+4$ ions are rarely stable.

3. The transition metal atoms always lose their outermost s electrons before they lose any d electrons.

We can now understand the stability of the d^{10} ions. To achieve the configuration of a noble gas, these ions would need to lose 10 additional electrons—quite an unfavorable prospect. Indeed, to lose any additional electrons would be to forfeit the stability of the filled d energy sublevel. Likewise, the $d^{10}s^2$ ions benefit from the filled s sublevel. Furthermore, we can see how In and Tl can form either stable d^{10} ions $(+3)$ or stable $d^{10}s^2$ ions $(+1)$, whereas Pb, for example, does not form d^{10} ions $(+4)$.

To look at some typical d^n ions, we shall consider the transition metals of the third period (Table 6.3). Notice first that the Sc^{3+} ion is not a d^n ion; instead, it has a noble gas electron configuration. Also, the Zn^{2+} and Cu^+ ions are d^{10} ions. But the other ions of these transition elements are all d^n ions. These ions are often more puzzling than the d^{10} and $d^{10}s^2$ ions; nonetheless, some

Table 6.3 Electron Configurations of Isolated Atoms and Some Ions of the First Series of the Transition Elements*

Element	Atom (3d, 4s)	Ionic Charge	Ions (3d, 4s)
Sc	3d: ↑ _ _ _ _ 4s: ↑↓	+3	3d: _ _ _ _ _ 4s: _
Ti	3d: ↑ ↑ _ _ _ 4s: ↑↓	+3	3d: ↑ _ _ _ _ 4s: _
V	3d: ↑ ↑ ↑ _ _ 4s: ↑↓	+3	3d: ↑ ↑ _ _ _ 4s: _
Cr	3d: ↑ ↑ ↑ ↑ ↑ 4s: ↑	+3	3d: ↑ ↑ ↑ _ _ 4s: _
		+2	3d: ↑ ↑ ↑ ↑ _ 4s: _
Mn	3d: ↑ ↑ ↑ ↑ ↑ 4s: ↑↓	+2	3d: ↑ ↑ ↑ ↑ ↑ 4s: _
Fe	3d: ↑↓ ↑ ↑ ↑ ↑ 4s: ↑↓	+3	3d: ↑ ↑ ↑ ↑ ↑ 4s: _
		+2	3d: ↑↓ ↑ ↑ ↑ ↑ 4s: _
Co	3d: ↑↓ ↑↓ ↑ ↑ ↑ 4s: ↑↓	+3	3d: ↑↓ ↑ ↑ ↑ ↑ 4s: _
		+2	3d: ↑↓ ↑↓ ↑ ↑ ↑ 4s: _
Ni	3d: ↑↓ ↑↓ ↑↓ ↑ ↑ 4s: ↑↓	+2	3d: ↑↓ ↑↓ ↑↓ ↑ ↑ 4s: _
Cu	3d: ↑↓ ↑↓ ↑↓ ↑↓ ↑↓ 4s: ↑	+2	3d: ↑↓ ↑↓ ↑↓ ↑↓ ↑ 4s: _
		+1	3d: ↑↓ ↑↓ ↑↓ ↑↓ ↑↓ 4s: _
Zn	3d: ↑↓ ↑↓ ↑↓ ↑↓ ↑↓ 4s: ↑↓	+2	3d: ↑↓ ↑↓ ↑↓ ↑↓ ↑↓ 4s: _

* The electron configuration through the 3p sublevel ($1s^2$, $2s^2$, $2p^6$, $3s^2$, $3p^6$), which is common to all of them, is not shown.

general observations are in order. First, in accordance with our general rules, no ions with a charge of $+4$ appear in Table 6.3. In addition, we can see the stabilizing effect of the half-filled d sublevel. Mn^{2+} is the only common ion of manganese, and Fe^{3+} is more stable than Fe^{2+}, which has an extra $3d$ electron.

A distinguishing feature of the d^n ions is that all have unpaired electrons. Since the spin of electrons causes them to act like small magnets, material composed of ions or atoms that contain unpaired electrons will be weakly attracted by a magnetic field. The field aligns the unpaired spins and magnetizes the substance. Most substances lose their magnetism when the field is removed and are said to be *paramagnetic*. A few *ferromagnetic* elements, however, such as iron, cobalt, and nickel, may retain their aligned spins and their magnetism after the magnetic field is withdrawn. Atoms or ions in which all electrons are paired show no such magnetic properties and are called *diamagnetic*. For reasons related to the orbital motion of the electrons, they are weakly repelled by magnetic fields. All the ions we have discussed except the d^n ions are diamagnetic.

Questions

15. In terms of electron configurations there are four major categories of ions. What are they?

16. Which elements in the transition groups form ions isoelectronic with one of the noble gases?

6.6 Molecules That Deviate from Noble Gas Electron Configurations

So far we have avoided discussion of compounds containing boron or beryllium since they do not behave as we might expect. Instead of forming ionic compounds like the metals below them in Groups IIA and IIIA, boron and beryllium form strongly covalent compounds, even with highly electronegative fluorine. Experiment has shown that BF_3 contains three equivalent covalent bonds and that BeF_2 contains two equivalent covalent bonds. Yet to write appropriate electron formulas, we must violate the octet rule, giving boron and beryllium fewer than eight electrons in their outermost energy levels:

$$\text{Be}: + 2\,:\!\ddot{\text{F}}: \longrightarrow :\!\ddot{\text{F}}\!:\!\text{Be}\!:\!\ddot{\text{F}}: \quad \text{or} \quad BeF_2$$

$$\text{B}: + 3\,:\!\ddot{\text{F}}: \longrightarrow :\!\ddot{\text{F}}\!:\!\text{B}\!:\!\ddot{\text{F}}: \quad \text{or} \quad BF_3$$
$$:\!\ddot{\text{F}}:$$

Notice in each case that the $2s$ electrons in the central atom are unpaired during bond formation. Thus Be: becomes ·Be·, and B: becomes ·B·.

There are also molecules in which the central atom has more than 8 electrons in its outermost energy level. For example, in PCl_5, phosphorus has 10 electrons in its valence energy level. Here again, s electrons must be unpaired during bond formation.

$$·\!\dot{\text{P}}: \longrightarrow ·\!\dot{\text{P}}·$$

$$·\!\dot{\text{P}}· + 5\,:\!\ddot{\text{Cl}}: \longrightarrow \begin{array}{c} :\!\ddot{\text{Cl}} \quad \ddot{\text{Cl}}: \\ \diagdown\diagup \\ P \\ \diagup\diagdown \\ :\!\ddot{\text{Cl}} \quad | \quad \ddot{\text{Cl}}: \\ :\!\ddot{\text{Cl}}: \end{array} \quad \text{or} \quad PCl_5$$

Other examples in which the central atom has 10 electrons include

$$:\!\ddot{\underset{\cdot}{Te}}\!\cdot \longrightarrow \cdot\!\overset{\cdot\cdot}{\underset{\cdot}{Te}}\!\cdot$$

$$\cdot\!\overset{\cdot\cdot}{\underset{\cdot}{Te}}\!\cdot + 4:\!\overset{\cdot\cdot}{\underset{\cdot\cdot}{Cl}}\!:\; \longrightarrow \quad \begin{array}{c} :\!\overset{\cdot\cdot}{\underset{\cdot\cdot}{Cl}} \quad \overset{\cdot\cdot}{\underset{\cdot\cdot}{Cl}}: \\ Te \\ :\!\overset{\cdot\cdot}{\underset{\cdot\cdot}{Cl}} \quad \overset{\cdot\cdot}{\underset{\cdot\cdot}{Cl}}: \end{array} \quad \text{or} \quad TeCl_4$$

$$:\!\overset{\cdot\cdot}{\underset{\cdot\cdot}{Cl}}\!:\; \longrightarrow\; :\!\overset{\cdot\cdot}{\underset{\cdot}{Cl}}\!:$$

$$:\!\overset{\cdot\cdot}{\underset{\cdot}{Cl}}\!:\; +\; 3:\!\overset{\cdot\cdot}{\underset{\cdot}{F}}\!:\; \longrightarrow\; \begin{array}{c} \cdot\!\overset{\cdot\cdot}{\underset{\cdot\cdot}{Cl}} \\ :\!F\; \big|\; F\!: \\ :\!F\!: \end{array} \quad \text{or} \quad ClF_3$$

$$:\!\overset{\cdot\cdot}{\underset{\cdot\cdot}{Xe}}\!:\; \longrightarrow\; :\!\overset{\cdot}{\underset{\cdot\cdot}{Xe}}\!:$$

$$:\!\overset{\cdot}{\underset{\cdot\cdot}{Xe}}\!:\; +\; 2:\!\overset{\cdot\cdot}{\underset{\cdot}{F}}\!:\; \longrightarrow\; \begin{array}{c} \cdot\!\overset{\cdot\cdot}{\underset{\cdot\cdot}{Xe}}\!\cdot \\ :\!F \quad F\!: \end{array} \quad \text{or} \quad XeF_2$$

The outermost energy level can also be expanded to 12. For example,

$$\cdot\!\overset{\cdot}{\underset{\cdot\cdot}{S}}\!:\; \longrightarrow\; :\!\overset{\cdot}{\underset{\cdot}{S}}\!:$$

$$\cdot\!\overset{\cdot}{\underset{\cdot\cdot}{S}}\!:\; +\; 6:\!\overset{\cdot\cdot}{\underset{\cdot}{F}}\!:\; \longrightarrow\; \begin{array}{c} :\!F\!: \\ :\!F \quad F\!: \\ S \\ :\!F \quad F\!: \\ :\!F\!: \end{array} \quad \text{or} \quad SF_6$$

$$:\!\overset{\cdot\cdot}{\underset{\cdot}{Br}}\!\cdot\; \longrightarrow\; :\!\overset{\cdot}{\underset{\cdot}{Br}}\!:$$

$$:\!\overset{\cdot}{\underset{\cdot}{Br}}\!:\; +\; 5:\!\overset{\cdot\cdot}{\underset{\cdot}{F}}\!:\; \longrightarrow\; \begin{array}{c} :\!F \quad F\!: \\ Br \\ :\!F \quad F\!: \\ :\!F\!: \end{array} \quad \text{or} \quad BrF_5$$

$$:\!\overset{\cdot\cdot}{\underset{\cdot\cdot}{Xe}}\!:\; \longrightarrow\; :\!\overset{\cdot}{\underset{\cdot}{Xe}}\!:$$

$$:\!\overset{\cdot}{\underset{\cdot}{Xe}}\!:\; +\; 4:\!\overset{\cdot\cdot}{\underset{\cdot}{F}}\!:\; \longrightarrow\; \begin{array}{c} :\!F \quad F\!: \\ Xe \\ :\!F \quad F\!: \end{array} \quad \text{or} \quad XeF_4$$

All the compounds described above, despite their violation of the octet rule, do exist. Some of them are used commercially. Sulfur hexafluoride (SF_6) is used as a gaseous insulator for electrical equipment. Chlorine trifluoride (ClF_3) and bromine pentafluoride (BrF_5) are used to oxidize rocket fuels (see Chapter 18). The compounds of xenon are especially significant. Before 1962, the elements of Group 0 were referred to as "inert" gases, but then XeF_4 was prepared. This was the first preparation of a stable binary compound of a noble gas.

Several things should be noted about the molecules with the expansions beyond eight electrons in the outermost energy level of the central atom. First, the expansion of the number of electrons in the outermost energy level beyond the normal eight requires the use of *d* orbitals. Since there are no *d* orbitals in the first or second energy levels, the central atoms in these compounds are limited to elements below the second period in the periodic table. The tendency of an atom to acquire more than eight electrons in the outermost energy level increases as one moves down within a group. Second, the element other than the one occupying the central position is often fluorine or another element consisting of small atoms, because spatial arrangements do not permit the close proximity of five or six large atoms about one central atom; that is, expansion beyond eight electrons is favored by a large central atom and small peripheral atoms.

We shall encounter several of these compounds again in Chapter 7 when we introduce the concept of *hybridization*. The unpairing of electrons and the expansion of outer energy levels will then be discussed in more detail. We shall also discuss deviations from the octet rule involving molecules with an odd number of valence electrons.

Questions

17. Which atoms are most likely to form molecules in which they have fewer than eight electrons in their outermost energy levels?

18. Which atoms are most likely to form molecules in which they have more than eight electrons in their outermost energy levels? Which never have more than eight?

19. Which atoms are most likely to be peripheral atoms in a molecule where the central atom has more than eight outer energy level electrons?

6.7 Resonance

Occasionally, more than one electron formula is necessary to represent a molecule accurately. Let us take SO_2 as an example. Sulfur dioxide might plausibly be represented as containing one single and one double bond:

$$\cdot \ddot{S} \cdot + 2 \cdot \ddot{O} \colon \longrightarrow \colon \ddot{O} \colon S \colon\colon \ddot{O} \colon \quad \text{or} \quad SO_2$$

Indeed, the octet rule is satisfied for all three atoms. However, experimental evidence indicates that the two sulfur-oxygen bonds in SO_2 are equivalent. Neither is a typical single or double bond; both have characteristics between the double and single bond. A convenient way to represent this situation is to write two electron formulas with a double-headed arrow between them:

$$\colon \ddot{O} \colon \ddot{S} \colon\colon \ddot{O} \colon \longleftrightarrow \colon \ddot{O} \colon\colon \ddot{S} \colon \ddot{O} \colon$$

By this convention, we indicate that the structure of the SO_2 molecule is something between the two hypothetical structures written. The SO_2 molecule is real, but the two structures represented above are not. Sulfur dioxide is *not* a mixture of two kinds of molecules, nor does each molecule oscillate between two structures. Hence the sulfur dioxide molecule is often said to be a *resonance hybrid* of two electron formulas. Note that the structures that contribute to a hybrid differ only in the location of electron pairs. There is no rearrangement of individual atoms. We shall come back to SO_2 and take a three-dimensional look at it in Chapter 7.

A sulfur atom can also combine with three atoms of oxygen:

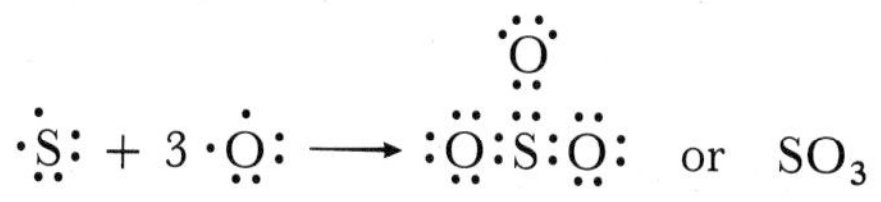

$$\cdot\ddot{S}: + 3 \cdot \ddot{O}: \longrightarrow :\ddot{O}:\ddot{S}:\ddot{O}: \quad \text{or} \quad SO_3$$

Again we have written a plausible electron formula but, as with SO_2, measurements of physical and chemical properties demand that all sulfur-oxygen bonds in this molecule be equivalent. Thus its structure is best represented as a resonance hybrid of three electron formulas, reflecting the three possible positions of the double bond:

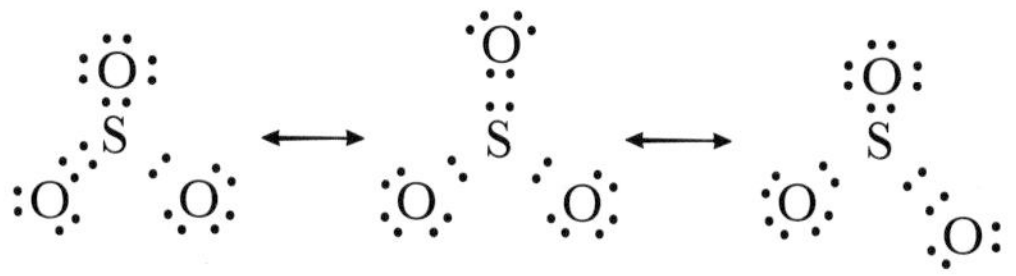

Resonance should also be anticipated in other cases where two or more plausible electron formulas can be written, each differing only in the positions of electrons. It must be emphasized that every atom must remain in the same position in all forms.

6.8 The Coordinate Covalent Bond

As we have seen, boron can form stable compounds with only six electrons in the outermost energy level. Often, however, these compounds will react with compounds containing an unshared electron pair, permitting boron to acquire eight electrons in its outermost energy level. For example, BF_3 reacts with NH_3:

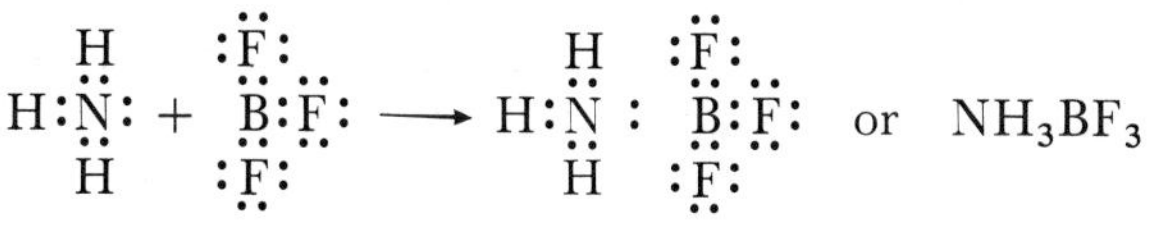

$$H:\ddot{N}:H + \ddot{B}:\ddot{F}: \longrightarrow H:\ddot{N}:\ddot{B}:\ddot{F}: \quad \text{or} \quad NH_3BF_3$$

Here the boron and nitrogen atoms are united by a *coordinate covalent* bond, a bond in which both of the shared electrons are contributed by one atom. Once the bond is formed though, it is no different from any other covalent bond.

This same concept of bonding is often used to account for *polyatomic ions*. An example is the common ammonium ion (NH_4^+). Its formation can be represented as a reaction between a hydrogen ion (H^+) and a molecule of ammonia (NH_3), where a bond is formed by sharing a pair of electrons contributed by the ammonia molecule:

$$H\!:\!\underset{H}{\overset{H}{N}}\!: + H^+ \longrightarrow H\!:\!\underset{H}{\overset{H}{N}}\!:\!H \quad \text{or} \quad NH_4^+$$

Note that all four hydrogen atoms in the ammonium ion are identical and that none can be identified as the one bonded to the nitrogen atom by the electrons contributed entirely by the ammonia molecule.

Unlike the ammonium ion, most polyatomic ions are negatively charged and contain one or more oxygen atoms. The reason is not hard to find. Oxygen has six electrons in its outer energy level and can readily act as an electron-pair acceptor in the formation of coordinate covalent bonds. The electron formulas of several common ions of this kind are listed in Table 6.4.

In writing the electron formula of a polyatomic ion, one must account carefully for the number of electrons in the valence energy level. Some will have been contributed by each atom in the ion, but those responsible for the ion's charge must have another source. We can account for these extra electrons

Table 6.4 Electron Formulas of Some Common Polyatomic Ions

Hydroxide	:Ö:H⁻	OH^-
Hypochlorite	:C̈l:Ö:⁻	ClO^-
Chlorite	:C̈l:Ö:⁻ :Ö:	ClO_2^-
Sulfate	:Ö:S̈:Ö:²⁻ :Ö:	SO_4^{2-}
Sulfite	:Ö:S̈:Ö:²⁻ :Ö:	SO_3^{2-}
Phosphate	:Ö:P̈:Ö:³⁻ :Ö:	PO_4^{3-}
Nitrate	:Ö: .N̈. .Ö: :Ö:⁻ ⟷ :Ö: .N̈. :Ö: :Ö:⁻ ⟷ :Ö .N̈. :Ö: :Ö:⁻	NO_3^-
Nitrite	N̈::Ö:⁻ :Ö: ⟷ N̈:Ö: .Ö.	NO_2^-
Carbonate	:Ö: .C̈. :Ö: :Ö:²⁻ ⟷ :Ö .C̈. :Ö: :Ö:²⁻ ⟷ :Ö: .C̈. :Ö :Ö:²⁻	CO_3^{2-}
Borate	:Ö: .B̈. :Ö: :Ö:³⁻	BO_3^{3-}

only by looking at the entire compound of which the ion is a part. Let us consider sodium phosphate as an example. The electron formula of the phosphate ion (PO_4^{3-}) shows 32 electrons. Five are contributed by the phosphorus atom, six by each oxygen, and three by atoms that do not appear in the formula of the ion. In this case, however, it is clear that the extra electrons are contributed by sodium atoms, which in turn become positively charged. The complete electron formula for this ionic compound is

$$3\,Na^+ + \begin{bmatrix} \ddot{\underset{..}{O}}\mathbin{:} \\ \ddot{\underset{..}{O}}\mathbin{:}\ddot{P}\mathbin{:}\ddot{\underset{..}{O}}\mathbin{:} \\ \ddot{\underset{..}{O}}\mathbin{:} \end{bmatrix}^{3-} \quad \text{or} \quad Na_3PO_4$$

Resonance is common among polyatomic ions as well as among molecules. Examples are the nitrite ion, the nitrate ion, and the carbonate ion. The last two of these are isoelectronic. Notice also that the borate ion is another example in which boron has only six electrons in the outermost energy level.

6.9 Summary of Generalizations for Deducing Electron Formulas

The electron formulas for most compounds can be deduced by following the general procedures listed below.

1. Write the electron formulas for the isolated atoms of each element in the compound.

Determine whether the compound is essentially ionic or covalent. Generally, a compound is ionic if it contains a metal or an ammonium ion. In deciding whether an element is a metal, consider its location in the periodic table and the number of electrons in the outermost energy level. Metals usually have three or fewer outer energy level electrons. Exceptions are boron and hydrogen, which are essentially nonmetallic, and sometimes beryllium. Elements with four or more electrons in the outermost energy level usually behave as nonmetals; however, tin and lead, with four electrons in the outer energy level, and bismuth, with five, commonly behave as metals.

3. If the compound is ionic, determine the charge on each ion and the necessary stoichiometry; then represent the compound as we did in Section 6.1.

4. For polyatomic ions and covalent compounds, determine the total number of electrons in the outermost energy levels of all the atoms. For molecules, simply total the valence electrons from each atom. For ions, account also for electrons that must be added or subtracted to give the indicated charge.

5. Arrange the atoms in molecules and polyatomic ions to provide a skeleton structure. Often there is only one reasonable arrangement. In other cases, however, you may need experimental information to arrange the atoms correctly. Note that identical atoms rarely bond to one another except in diatomic molecules, peroxides and superoxides (Chapter 20), and some extended molecules and ions (later chapters).

6. Join covalently bound atoms by placing a pair of dots between them.

7. Place dots around each symbol so that eight electrons occupy the outermost energy level for each atom (two for each hydrogen). If there are too few electrons to provide a noble gas electron configuration for each atom in this way, consider multiple bonding. Remember that boron and beryllium are exceptions. When there are more electrons than required to provide a noble gas electron configuration for each atom, the central atom probably has more than eight electrons in its outermost energy level (Section 6.6).

8. Write resonance structures for molecules and polyatomic ions if two or more plausible electron formulas differ only in the positions of electron pairs. The number of shared and unshared electron pairs must be the same in all the formulas.

Write electron formulas for the indicated compounds in Examples 6.1 to 6.7.

Example 6.1 The compound between rubidium and sulfur:

Solution

a. We must first determine the electron formulas of the isolated atoms of rubidium and sulfur. Rubidium is in Group IA and sulfur is in Group VIA. Since each is a representative element, the number of electrons in the outermost energy level is the same as the group number. Hence we have

$$\text{Rb}\cdot \quad \text{and} \quad \cdot\ddot{\underset{\cdot\cdot}{\text{S}}}:$$

b. Rubidium, like all the elements in Group IA, is a metal; hence the bond between rubidium and sulfur is ionic.

c. In ionic compounds metals lose electrons and nonmetals accept electrons. In this case electrons are transferred from the rubidium atoms to the sulfur atoms.

d. The number of electrons lost by the rubidium atoms must equal the number gained by the sulfur atoms. Since each rubidium atom must lose one electron and each sulfur atom must gain two, the ratio of rubidium atoms to sulfur atoms is two to one.

e. After the transfer of electrons, each rubidium ion will have a charge of $+1$, since each will have one more proton than electron. The sulfur ions (called *sulfide* ions) will have -2 charges since they will contain two more electrons than protons. Rubidium now has no electrons in its outermost energy level, and sulfur has eight. Thus we write the electron formula for the compound between rubidium and sulfur as

$$[2\ Rb^+ + :\!\ddot{\underset{..}{S}}\!:^{2-}]$$

Example 6.1 illustrates the application of the generalizations above in writing an electron formula for an ionic compound. Examples 6.2 and 6.3 show the application of these generalizations in writing formulas of covalent compounds.

Example 6.2 The compound between nitrogen and fluorine:

Solution

a. Again, we first write the electron formulas for the isolated atoms:

$$\cdot\!\ddot{N}\!:\quad \text{and} \quad :\!\ddot{\underset{..}{F}}\!:$$

b. Since both elements are nonmetals (four or more electrons in the outer energy level), we expect the compound to be covalent.

c. Since the fluorine atom needs to share one pair of electrons and the nitrogen atom needs to share three, the ratio of the fluorine atoms to the nitrogen atoms must be three to one. Thus the electron formula for the compound is

$$:\!\ddot{\underset{..}{F}}\!:\!\ddot{N}\!:\!\ddot{\underset{..}{F}}\!:$$
$$:\!\ddot{\underset{..}{F}}\!:$$

Example 6.3 $BeCl_2$

Solution

a. The electron formulas of the isolated atoms are

$$Be\!:\quad \text{and} \quad :\!\dot{\underset{..}{Cl}}\!:$$

b. Although its position in the periodic table and the number of electrons in its outer energy level suggest that beryllium is a metal, we have seen that some beryllium compounds are covalent. This is such a case. When in doubt with atoms like beryllium you should consult a reference book to be sure. The electron formula for $BeCl_2$ is

$$:\!\ddot{\underset{..}{Cl}}\!:\!Be\!:\!\ddot{\underset{..}{Cl}}\!:$$

Writing electron formulas for ternary (three-element) compounds involves generalizations 4, 5, 6, and 7. The application of these rules is illustrated in Examples 6.4 and 6.5.

Example 6.4 $NaClO_4$:

Solution

a. The electron formulas of the isolated atoms are

$$Na\cdot \quad :\overset{.}{\underset{..}{Cl}}: \quad \text{and} \quad \cdot\overset{.}{\underset{..}{O}}:$$

b. Any bond involving a Group IA element is ionic, but the four oxygen atoms and the chlorine atom are bonded by shared electrons (covalent bonds). In this case each sodium atom transfers one electron to the ClO_4 group, leaving a sodium ion with a $+1$ charge and giving the ClO_4 group a -1 charge.

c. There are now 32 outer energy level electrons in the ClO_4^- ion: 7 from the chlorine atom, 24 from the four oxygen atoms, and 1 from the sodium atom.

d. The electron formula of the compound is

$$\left[Na^+ \; + \; :\overset{..}{\underset{..}{O}}: \; \overset{\overset{..}{\underset{..}{O}}:}{\underset{\overset{..}{\underset{..}{O}}:}{Cl}} :\overset{..}{\underset{..}{O}}:^- \right]$$

This example is typical in that the oxygen atoms in most polyatomic ions are bonded to the central atom rather than to each other.

Example 6.5 $NaCN$:

Solution

a. The electron formulas for the isolated atoms are

$$Na\cdot \quad \cdot\overset{.}{\underset{.}{C}}: \quad \text{and} \quad \cdot\overset{.}{\underset{.}{N}}:$$

b. Since sodium is a typical metal, the compound is ionic. Both carbon and nitrogen are nonmetals, however; hence they must be covalently bonded. The compound thus consists of Na^+ ions and CN^- ions.

c. Each CN^- ion contains 10 outer energy level electrons: 4 from the carbon atom, 5 from the nitrogen atom, and 1 from the sodium atom.

d. To bond the carbon and nitrogen atoms with only 10 electrons, we must use a triple bond. Thus the electron formula for the compound is

$$[Na^+ \; + :C:::N:^-]$$

Electron formulas are often written for ions as well as for complete compounds. Example 6.6 is concerned with an ion which is best represented as a resonance hybrid.

Example 6.6 NO_2^-:

Solution

a. The electron formulas we need are

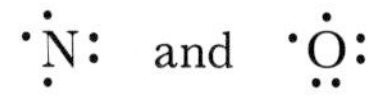

$\cdot \ddot{N}\colon$ and $\cdot \ddot{O}\colon$

b. Because nitrogen and oxygen are nonmetals, they are bonded covalently in NO_2^- ions.

c. Each NO_2^- ion contains 5 electrons from the nitrogen atom, 12 from the two oxygen atoms, and 1 transferred from an atom which does not appear in the formula of the ion—a total of 18. The electron formula for this ion is

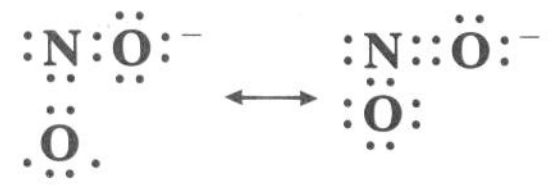

$\begin{matrix} :\ddot{N}\colon\ddot{O}\colon^- \\ \cdot\ddot{O}\cdot \end{matrix} \longleftrightarrow \begin{matrix} :N\colon\colon\ddot{O}\colon^- \\ \colon\ddot{O}\colon \end{matrix}$

We must indicate the resonance hybrids since we can write two equally valid electron formulas for the NO_2^- ion.

Example 6.7 illustrates the procedure for writing electron formulas where the number of electrons in the outermost energy level is greater than eight.

Example 6.7 Cs_2SiF_6:

Solution

a. The electron formulas for the atoms are

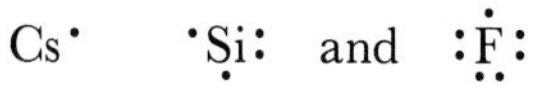

$Cs\cdot$ $\cdot \ddot{Si}\colon$ and $\colon\ddot{F}\colon$

b. Since cesium is a metal, the compound is ionic.

c. Silicon and fluorine are covalently bonded nonmetals; thus the compound is composed of Cs^+ ions and SiF_6^{2-} ions.

d. Each SiF_6^{2-} ion contains 48 electrons: 4 from the silicon atom, 7 from each fluorine atom, and 2 from the two cesium atoms.

e. The only way to bond six fluorine atoms and one silicon atom is to bond each fluorine atom to the silicon atom. (Fluorine atoms do not bond to each other except in F_2.) Since silicon is below the second period, it has d orbitals and can expand its outer energy level to more than 8 electrons. The electron formula for this compound is

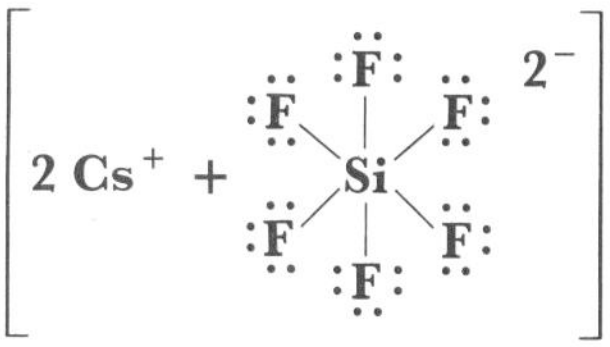

$$2\,Cs^+ + \left[\begin{matrix} & :\ddot{F}: & \\ :\ddot{F} & | & \ddot{F}: \\ & \diagdown Si \diagup & \\ :\ddot{F} & | & \ddot{F}: \\ & :\ddot{F}: & \end{matrix} \right]^{2-}$$

6.10 Oxidation Numbers

To provide a convenient method of electron bookkeeping, we now introduce the concept of *oxidation numbers* (or *oxidation states*). This notion is closely related to electron configurations and electronegativities. Oxidation numbers are assigned to each element of a compound on the basis of several generalizations:

1. Uncombined elements have an oxidation number of 0, regardless of the number of atoms in each structural unit. The oxidation state of the elements in each of the following is 0: H_2, Fe, S_8, O_3, He.

2. The oxidation number of a simple (that is, monatomic) ion is equal to the ionic charge.

3. In covalent compounds, each shared pair of electrons is assigned to the more electronegative element sharing the electrons. The atoms of the compound are then treated as simple ions.

4. The sum of the oxidation numbers of a neutral compound must equal 0.

5. The sum of the oxidation numbers in a polyatomic ion must equal the ionic charge. Examples 6.8 and 6.9 illustrate the application of these generalizations.

Example 6.8 Ascertain the oxidation numbers of each element in $SiCl_4$.

Solution

a. The electron formula of this compound is indicated below. Since chlorine is the most electronegative element, the shared electron pairs are assigned to chlorine as indicated.

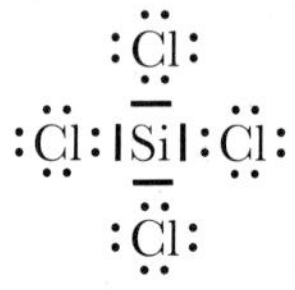

$$\begin{array}{c} :\!\ddot{\underline{Cl}}\!: \\ :\!\ddot{\underline{Cl}}\!:\!|Si|\!:\!\ddot{\underline{Cl}}\!: \\ :\!\ddot{\underline{Cl}}\!: \end{array}$$

b. The hypothetical assignment of electrons results in an oxidation number of -1 for chlorine (each chlorine atom gains one electron) and $+4$ for silicon (silicon loses four electrons).

c. The sum of the oxidation numbers is $+4 + 4(-1)$ or **0.**

Example 6.9 What is the oxidation number of each element in Na_2SeO_4?

Solution

a. This compound consists of Na^+ ions and SeO_4^{2-} ions.

b. The oxidation number of a simple (monatomic ion) is the same as the ionic charge. Thus, sodium has a $+1$ oxidation number.

c. The electron formula of SeO_4^{2-} ion is

$$\left[\begin{array}{c} ::\ddot{O}:: \\ ::\ddot{O}::|\underline{Se}|::\ddot{O}:: \\ ::\ddot{O}:: \end{array} \right]^{2-}$$

d. The electrons shared between Se and O are regarded as belonging to O; thus Se is $+6$ and each O is -2.

e. The sum of oxidation numbers in SeO_4^{2-} is $+6 + 4(-2)$ or -2, the same as the ionic charge.

Table 6.5 illustrates the application of these rules to additional binary compounds and polyatomic ions. As shown, oxidation numbers are conventionally

Table 6.5 Oxidation Numbers

Compound or Ion	Type of Bond	Ionic Charge if Ionic	Number of Bonds and Most Electronegative Element if Covalent	Oxidation Numbers	Sum of Oxidation Numbers
KCl	Ionic	K^+ Cl^-		$(+1)(-1)$ KCl	$+1 + (-1) = 0$
$CaCl_2$	Ionic	Ca^{2+} Cl^-		$(+2)(-1)$ $CaCl_2$	$+2 + 2(-1) = 0$
Na_3N	Ionic	Na^+ N^{3-}		$(+1)(-3)$ Na_3N	$3(1) + (-3) = 0$
HCl	Covalent		1(Cl)	$(+1)(-1)$ HCl	$+1 + (-1) = 0$
CCl_4	Covalent		4(Cl)	$(+4)(-1)$ CCl_4	$+4 + 4(-1) = 0$
SF_6	Covalent		6(F)	$(+6)(-1)$ SF_6	$+6 + 6(-1) = 0$
XeF_4	Covalent		4(F)	$(+4)(-1)$ XeF_4	$+4 + 4(-1) = 0$
CO_2	Covalent		4(O)	$(+4)(-2)$ CO_2	$+4 + 2(-2) = 0$
SO_3	Covalent		4(O)	$(+6)(-2)$ SO_3	$+6 + 3(-2) = 0$
SO_4^{2-}	Covalent		4(O)	$(+6)(-2)$ SO_4^{2-}	$+6 + 4(-2) = -2$
PO_4^{3-}	Covalent		4(O)	$(+5)(-2)$ PO_4^{3-}	$+5 + 4(-2) = -3$

written above each symbol in a formula. We shall return to the concept of oxidation numbers in Chapter 17, where we shall use it to write chemical equations.

Question

20. Indicate the oxidation numbers for each element in the following compounds or ions:
    a. HBr   b. $AlCl_3$ c. BaS   d. PCl_5 e. SO_2 f. P_4
    g. Ca_3P_2 h. $SnCl_4$ i. $SnCl_2$ j. BrF_5 k. XeF_4 l. ClF_3
    m. PO_4^{3-} n. NH_4^+ o. ClO_4^- p. ClO_3^- q. NO_3^- r. SO_3^{2-}

6.11 Introduction to Chemical Nomenclature

Each of the well over 2 million chemical compounds has a name as well as a formula. Some of these names have been determined by popular usage and give no clue to the composition of the compound. These are called *trivial* or *common* names and were adopted long before systematic nomenclature had been developed. Some trivial names are

water	H_2O
ammonia	NH_3
phosphine	PH_3
hydrazine	N_2H_4

If all compounds were named in this way, mastery of chemical nomenclature would of course be an impossible task. As we shall see in this section, however, there are systematic methods of naming common inorganic compounds. These systematic names, like formulas, specify the compositions of the compounds.

Positive Ions. Most positive ions are monatomic, and each is given the name of the metal from which it is derived. For example,

Na^+	sodium ion
Ba^{2+}	barium ion
Al^{3+}	aluminum ion

These names are unambiguous, since each of these metals forms only one ion. When a metal can form more than one ion, however, the oxidation number of the ion is enclosed in parentheses following the name of the metal. This is called the *Stock System*, after Alfred Stock (1876–1946), who first proposed it. Examples include

Fe^{2+}	iron(II) ion	Au^+	gold(I) ion
Fe^{3+}	iron(III) ion	Au^{3+}	gold(III) ion

A less desirable but still commonly used method for distinguishing among different ions of the same metal is to drop the ending (usually *um*) from the Latin or English name of the metal and then add the suffix *-ous* to indicate the smaller ionic charge or *-ic* to indicate the larger ionic charge (Table 6.6). The examples above then become

Fe^{2+}	ferrous ion	Au^+	aurous ion
Fe^{3+}	ferric ion	Au^{3+}	auric ion

Finally, the names of two common, positively charged polyatomic ions are

H_3O^+ hydronium ion

NH_4^+ ammonium ion

Table 6.6 Ions Using Latin Stem in Name

English Name	Latin Name	Latin Stem	Name of Ion with Higher Oxidation Number	Name of Ion with Lower Oxidation Number
Copper	Cuprum	Cupr	(+2) cupric	(+1) cuprous
Gold	Aurum	Aur	(+3) auric	(+1) aurous
Iron	Ferrum	Ferr	(+3) ferric	(+2) ferrous
Tin	Stannum	Stann	(+4) stannic	(+2) stannous
Mercury	Hydrargyrum	(not used)	(+2) mercuric	(+1) mercurous

Negative Ions. The names of monatomic negative ions combine the stem of the name with the suffix *-ide* (Table 6.7). A few polyatomic negative ions also have the *-ide* ending, for example, hydroxide (OH^-) and cyanide (CN^-).

Many polyatomic anions contain oxygen. The relative amounts of oxygen are indicated by a combination of prefixes and suffixes. The number of oxygen atoms in the anion increases as one goes down the following list:

    hypo ___ ite

        ___ ite

        ___ ate

    per ___ ate

Table 6.7 Names of Some Common Monatomic Negative Ions

Element Name	Stem	Name of Ion	Formula
Bromine	Brom-	Bromide	Br^-
Chlorine	Chlor-	Chloride	Cl^-
Fluorine	Fluor-	Fluoride	F^-
Hydrogen	Hydr-	Hydride	H^-
Iodine	Iod-	Iodide	I^-
Oxygen	Ox-	Oxide	O^{2-}
Phosphorus	Phosph-	Phosphide	P^{3-}
Sulfur	Sulf-	Sulfide	S^{2-}

For example, there are four negative ions containing chlorine and oxygen:

ClO^- hypochlorite ion

ClO_2^- chlorite ion

ClO_3^- chlorate ion

ClO_4^- perchlorate ion

Notice that none of the prefixes, suffixes, or combinations of prefixes and suffixes indicates any specific number of oxygen atoms. They indicate only the *relative number* of oxygen atoms in a series of compounds. If there are only two oxygen-containing anions in the series, only the suffixes *-ite* and *-ate* are used:

NO_2^- nitrite ion

NO_3^- nitrate ion

Some other common polyatomic negative ions are

$C_2H_3O_2^-$	acetate ion	MnO_4^-	permanganate ion
CO_3^{2-}	carbonate ion	CrO_4^{2-}	chromate ion
AsO_4^{3-}	arsenate ion	$Cr_2O_7^{2-}$	dichromate ion
$C_2O_4^{2-}$	oxalate ion	SO_3^{2-}	sulfite ion
PO_4^{3-}	phosphate ion	SO_4^{2-}	sulfate ion

The names of polyatomic anions containing hydrogen are formed by adding the word *hydrogen* to the name:

HCO_3^- hydrogen carbonate ion

HS^- hydrogen sulfide ion

HSO_3^- hydrogen sulfite ion

HPO_4^{2-} hydrogen phosphate ion

$H_2PO_4^-$ dihydrogen phosphate ion

For monohydrogen ions with a -1 charge, one may also use the prefix *bi-*. Thus HSO_3^- is also called the *bisulfite* ion, and HS^-, the *bisulfide* ion.

Ionic Compounds. Just as the positively charged ion always appears first in the formula, the cation is always named first. It is followed by the name of the anion. For example,

    NaBr   sodium bromide

    $FeSO_4$ iron(II) sulfate

    $NaHCO_3$ sodium hydrogen carbonate or sodium bicarbonate

Example 6.10 Assign names to the following compounds: (a) Na_2S, (b) $Ca(ClO_4)_2$, (c) $Mg_2(PO_4)_3$.

Solution

a. Since Na^+ is the sodium ion and S^{2-} is the sulfide ion, the compound is **sodium sulfide.**

b. Since Ca^{2+} is the calcium ion and ClO_4^- is the perchlorate ion, the compound is **calcium perchlorate.**

c. Since Mg^{2+} is the magnesium ion and PO_4^{3-} is the phosphate ion, the compound is **magnesium phosphate.**

In naming the compounds in Example 6.10, it was not necessary to indicate the oxidation numbers of the positive ions, since the ions of metals of Groups IA and IIA show only one oxidation number. In the following example, oxidation numbers must be indicated in the names. In order to do this one must remember the charges of common negative ions.

Example 6.11 Name the following compounds: (a) $CuCl_2$, (b) $Fe_2(SO_4)_3$.

Solution

a. Since the chloride ion is a -1 ion, copper must have a $+2$ oxidation number; hence, $CuCl_2$ is **copper(II) chloride (or cupric chloride).**

b. Since the sulfate ion is a -2 ion, iron must have a $+3$ oxidation number so that the $+6$ on the two iron ions balances the -6 on the three sulfate ions. Therefore, $Fe_2(SO_4)_3$ is **iron(III) sulfate (or ferric sulfate).**

Binary Compounds of Nonmetals. The elements are named in the order they appear in the formula. The name of the second element is modified with the suffix *-ide* (Table 6.7). Examples of pairs of nonmetals that form only one compound are

    HI   hydrogen iodide

    HBr   hydrogen bromide

    H_2S hydrogen sulfide

If two nonmetallic elements form more than one compound, Greek prefixes are used to designate the number of atoms of each element:

one	mono-	six	hexa-
two	di-	seven	hepta-
three	tri-	eight	octa-
four	tetra-	nine	ennea-
five	penta-	ten	deca-

The prefix *mono-* is often dropped. For example,

    SO_3 sulfur trioxide

    CCl_4 carbon tetrachloride

    PCl_3 phosphorus trichloride

    PCl_5 phosphorus pentachloride

    N_2O dinitrogen oxide

    N_2O_3 dinitrogen trioxide

    P_4S_{10} tetraphosphorus decasulfide

Acids. Aqueous solutions rich in hydronium ions are called **acids.** Acids often form when covalent, hydrogen-containing compounds dissolve in water. For example, hydrochloric acid is formed when hydrogen chloride dissolves in water:

$$HCl + H_2O \longrightarrow H_3O^+ + Cl^-$$

Because acids have different properties than the anhydrous compounds, they are given different names. If the acid is derived from a binary compound, we replace the word *hydrogen* with the prefix *hydro-*, substitute the suffix *-ic* for *-ide*, and add the word *acid*. The following are examples:

Formula	Name of Anhydrous Compound	Acid Name
HCl	Hydrogen chloride	Hydrochloric acid
HBr	Hydrogen bromide	Hydrobromic acid
H_2S	Hydrogen sulfide	Hydrosulfuric acid

For the names of ternary acids (acids derived from anhydrous compounds containing three elements), we drop the word *hydrogen*, change *-ite* to *-ous* or *-ate* to *-ic*, and add the word *acid*. The following are examples:

Formula	Name of Anhydrous Compound	Acid Name
H_2SO_4	Hydrogen sulfate	Sulfuric acid
H_2SO_3	Hydrogen sulfite	Sulfurous acid
H_3PO_4	Hydrogen phosphate	Phosphoric acid
H_3PO_3	Hydrogen phosphite	Phosphorous acid
$HC_2H_3O_2$	Hydrogen acetate	Acetic acid
H_2CO_3	Hydrogen carbonate	Carbonic acid
HNO_3	Hydrogen nitrate	Nitric acid
HNO_2	Hydrogen nitrite	Nitrous acid

Chemical names and formulas of some common compounds are listed in Table 6.8.

Table 6.8 Formulas, Chemical Names, and Use of Common Substances

Common Name	Formula	Chemical Name	Partial List of Uses
Baking soda	$NaHCO_3$	Sodium hydrogen carbonate	Ingredient of baking powder, antacid, mouthwash, fire extinguishers
Battery acid	H_2SO_4	Sulfuric acid	Storage battery; by far the most widely used industrial chemical
Engravers acid	HNO_3	Nitric acid	Manufacture of explosives and fertilizers
Lime	CaO	Calcium oxide	Steel industry, manufacture of glass, dietary supplement
Limestone	$CaCO_3$	Calcium carbonate	Building stone, metallurgy
Lye	NaOH	Sodium hydroxide	Manufacture of rayon, soap, and detergents
Muriatic acid	HCl	Hydrochloric acid	Chemical intermediate, food processing
Norway fertilizer	NH_4NO_3	Ammonium nitrate	Fertilizer and explosives
Potash	KOH	Potassium hydroxide	Electrolyte in fuel cells, chemical intermediate
Sal ammoniac	NH_4Cl	Ammonium chloride	Soldering flux, manufacture of dry cells
Saltpeter	KNO_3	Potassium nitrate	Gunpowder, fertilizers, rockets
Slaked lime	$Ca(OH)_2$	Calcium hydroxide	Mortars and plasters, cements
Soda ash	Na_2CO_3	Sodium carbonate	Paper manufacture, water treatment

Questions

21. What do the following suffixes signify?
    a. *-ide*   b. *-ous*   c. *-ic*

22. Which of the following compounds contains the most oxygen? The least oxygen?
    a. sodium chlorate   b. sodium chlorite   c. sodium perchlorate

23. What are the names of two frequently encountered positive polyatomic ions?

24. Distinguish between
    a. cuprous oxide and cupric oxide
    b. sulfuric acid and hydrosulfuric acid
    c. hydrogen chloride and hydrochloric acid

Questions and Problems

6.1 The Ionic Bond

25. What is the relation between each of the following and the tendency of an atom to lose or gain electrons when forming a bond?
    a. Location in periodic table
    b. Number of electrons in outermost energy level
    c. Electronegativity

6.2 The Covalent Bond

26. Given the hypothetical elements, V, W, X, Y, and Z, with atomic numbers 6, 8, 9, 11, and 12, respectively, write electron formulas for three binary ionic compounds and three binary covalent compounds that might be formed among them.

6.5 Metal Ions Which Deviate from Noble Gas Electron Configurations

27. Distinguish between the following:
    a. electronegativity and ionization energy
    b. polar covalent and nonpolar covalent
    c. d^{10} ions and d^n ions

28. Identify the following ions as isoelectronic with a noble gas, a d^{10} ion, a $d^{10}s^2$ ion, or a d^n ion.
    a. Al^{3+} b. Mn^{2+} c. In^+ d. N^{3-} e. Sn^{2+} f. Sc^{3+}
    g. Bi^{3+} h. O^{2-} i. Cd^{2+} j. Rb^+ k. Cu^+ l. Cu^{2+}
    m. Ga^{3+} n. Br^- o. Pb^{2+} p. Cr^{3+} q. Co^{2+} r. Cl^-

29. What are the electron configurations in terms of two quantum numbers for the following ions?
    a. Mg^{2+} b. Cu^{2+} c. Zn^{2+} d. Mn^{2+}
    e. Fe^{2+} f. Fe^{3+} g. Sn^{2+} h. Ga^+
    i. Ga^{3+} j. P^{3-} k. Cu^+ l. O^{2-}

30. Account for the properties in column A by matching each with the appropriate term in column B.

A	B
diamagnetism	unpaired electrons
combining capacity	paired electrons
paramagnetism	unshared electrons
	valence electrons

6.6 Molecules Which Deviate from Noble Gas Electron Configurations

31. Indicate which atoms
    a. are most likely to form covalent bonds.
    b. are most likely to transfer s electrons only.
    c. are most likely to transfer s and d electrons.
    d. are most likely to transfer p electrons only.
    e. are most likely to transfer s and p electrons.
    f. are most likely to form multiple bonds.
    g. are most likely to form paramagnetic ions.
    h. are most likely to expand the number of electrons in their outermost energy level beyond eight.
    i. form ions by losing only electrons from the outermost energy level.
    j. form ions by losing electrons from the outermost and the next-to-outermost energy levels.
    k. form ions by gaining electrons in the p sublevel of the outermost energy level.

32. Copy the diagram below, which represents the first four periods of the periodic table. Then fill in the table with the symbols for the hypothetical elements described below. If you place the symbols correctly, and *in the order described*, the result will spell out an important fact about chemical bonding.

   a. El is the most metallic element in this portion of the periodic table.
   b. Ma is the only element in Group IA that rarely transfers an electron.
   c. M is often the central atom of a molecule in which it has only six outer energy level electrons.
   d. The atoms of Ny will not transfer, gain, or lose electrons. Furthermore, they have no p orbitals.
   e. The At^+ ion is isoelectronic with the Ny atom.
   f. The Tr^{3+} ion is isoelectronic with a noble gas. Tr is a transition element.
   g. The He^{2+} ion has a half-filled d sublevel.
   h. Ct is the most electronegative element in the periodic table.
   i. To atoms have eight electrons in the outermost energy level. The outer level has no d orbitals.
   j. The A^{2-} ion is isoelectronic with To.

k. Both H and Ve are halogens. Ve is less active than H.
l. The St^{2+} ion is a d^{10} ion.
m. Each atom of Re and Er will form four covalent bonds. Er is more metallic than Re. Atoms of both have d orbitals in their outermost energy levels.
n. E forms E^{3-} ions. E atoms never have more than eight electrons in their outermost energy levels.
o. Qu^{2+} ions are isoelectronic with To atoms.
p. O sometimes forms molecules in which it has only four electrons in its outermost energy level.
q. I and En are in the same group. En is more metallic than I. I^{3+} ions are formed by the loss of one p and two s electrons.
r. Ec is the most metallic element in Group IIA in this portion of the table.
s. The Ou^{3+} ion has a half-filled d sublevel.
t. The Mo^+ ion is a d^{10} ion.
u. Ei and Gy are in the same group. Gy is less electronegative than Ei. The Gy^{3+} ion is a $d^{10}s^2$ ion.
v. The electron formula for an isolated atom of G is $\cdot \ddot{G} :$. Le is in the same group as G. Le is less chemically active than G.
w. Ac is a representative metal.
x. Sr is a nonmetal.
y. The Te^{3+} ion has four unpaired electrons. R is in the same group as Te.
z. T and L are noble gases. The L atom is larger than T.
aa. The Si^{3+} ion contains two unpaired electrons.
bb. Each isolated atom of Nt contains six unpaired electrons.
cc. On forms a d^n ion.

6.8 Bonds in Which Both Electrons Are Provided by One of the Bonded Atoms

33. Identify the following compounds as essentially covalent or essentially ionic:
    a. $KClO_4$ b. SO_2 c. $Pb(NO_3)_2$ d. $(NH_4)_2SO_4$
    e. H_2O f. H_2SO_3 g. XeF_4

34. Indicate what is wrong with the following proposed electron formulas.

    a. $Rb \colon H$

    b. $:\ddot{Br}: \ddot{B} :\ddot{Br}:$ with $:\ddot{Br}:$ above

    c. $:\ddot{O}:S:\ddot{O}:$

    d. Xe with four $\ddot{F}:$ around it

    e. $\left[P^{3+} + 3 :\ddot{Cl}:^- \right]$

    f. $\left[H^+ + :\ddot{O}: H^- \right]$ (for water)

    g. $\left[3\, Na^+ + :\ddot{P}:\ddot{O}:\ddot{O}:^{3-} \right]$ with $:\ddot{O}:$ above and below P

    h. $Mg:\ddot{Cl}:\ddot{O}:$ with $:\ddot{O}:$ above and below

    i. $Cl:\ddot{C}:Cl$ with Cl above and Cl below

180   Chapter 6

6.9 Summary of Generalizations for Deducing Electron Formulas

35. Write electron formulas for possible binary compounds formed from the following pairs of elements.
- a. potassium and chlorine
- b. barium and phosphorus
- c. silicon and chlorine
- d. selenium and barium
- e. phosphorus and bromine
- f. calcium and oxygen
- g. arsenic and hydrogen
- h. iodine and bromine
- i. cesium and hydrogen

36. Write electron formulas for the species listed below.
- a. $RbIO_4$
- b. OF_2
- c. CO
- d. $MgSO_3$
- e. Na_4SiO_4
- f. NH_4Cl
- g. $NaBrO_3$
- h. $KICl_4$
- i. CO_2
- j. $SiCl_4$
- k. KNO_2
- l. K_3AsO_3
- m. $Ca(ClO_3)_2$
- n. $Al_2(SO_4)_3$
- o. Na_2SiF_6
- p. BrF_5
- q. ClF_3
- r. PF_5
- s. O_3
- t. XeF_4
- u. $MgCO_3$
- v. $Ba_3(PO_4)_2$
- w. $NaBrF_4$
- x. HCN

37. Show that each molecule in column A is isoelectronic with at least one ion in column B.

A	B
$SiCl_4$	NO_2^-
N_2	CN^-
SiF_4	PO_4^{3-}
SF_6	SiF_6^{2-}
CH_4	SO_4^{2-}
XeF_4	NH_4^+
	ICl_4^-
	CO_3^{2-}
	H_3O^+

6.11 Introduction to Chemical Nomenclature

38. Write the formulas and assign the names of the chlorides, oxides, nitrides, hypochlorites, sulfates, hydroxides, and phosphates of
- a. potassium
- b. calcium
- c. aluminum

39. Name the following:
- a. $PbCrO_4$
- b. $Fe(ClO_3)_3$
- c. CaO
- d. Au_2O
- e. $Cr(OH)_3$
- f. I_2O_5
- g. CuS
- h. $HClO_4$
- i. $HgCl_2$
- j. Cr_2O_3
- k. $Mg_3(PO_4)_2$
- l. $Ca(HCO_3)_2$
- m. H_3PO_3
- n. HI
- o. FeS
- p. $AuCl_3$
- q. $SnCl_2$
- r. P_4O_{10}
- s. OF_2
- t. NH_4NO_3
- u. SF_6
- v. XeF_4
- w. BF_3
- x. CuO
- y. HNO_2
- z. $Zn(C_2H_3O_2)_2$

40. Write the formulas of
- a. cuprous oxide
- b. ferrous sulfate
- c. dinitrogen pentaoxide
- d. mercury(II) chloride
- e. sodium chlorate
- f. chlorous acid
- g. phosphine
- h. potassium hydroxide

i. sodium cyanide
j. tin(II) nitrate
k. calcium sulfite
l. copper(II) sulfate
m. nitrous acid
n. phosphoric acid
o. hydrogen sulfide
p. iron(II) chloride
q. silver oxide
r. barium acetate
s. sulfur trioxide
t. carbon tetrachloride
u. stannic chloride
v. hydrochloric acid
w. sodium phosphide
x. mercuric chloride
y. calcium carbonate
z. sodium hydrogen sulfate

Discussion Starters

41. Explain or account for the following true statements:
    a. There is a SF_4 but no OF_4.   b. There is a BrF_5 but no ClF_5.

Covalent Bonding—Molecular Structure

In Chapter 6 we discussed ionic and covalent bonding, and we used electron formulas to depict the bonds formed in stable molecules. However, there is much more to be learned about the nature of the chemical bond—especially the covalent bond that is the basis of most chemical compounds. For example, many molecules cannot be represented by electron formulas, and others exhibit properties inconsistent with their simple electron formulas. But the most important shortcoming of the electron formulas we have studied so far is that they can tell us little about the shapes of molecules. For this, and for the many useful conclusions that can be drawn from a clear understanding of molecular geometry, we must turn to more sophisticated ways of representing the covalent bond.

7.1 Molecular Shape and Polarity

As an example of important properties that depend on molecular geometry, we shall look first at polarity. We have already seen that polar bonds arise when a pair of electrons is shared by two atoms with different electronegativities. Thus diatomic molecules such as HCl, BrCl, and HF are polar. In a polyatomic molecule, however, polar bonds do not necessarily imply a polar molecule. In fact, it is often the case that the bonds are arranged in such a way that their polarities cancel. In such cases the molecules are nonpolar. For example, consider the hypothetical triatomic molecule BA_2, in which element A is more electronegative than element B. In this molecule each B—A bond will be polar, but whether the whole molecule is polar depends on its shape. If the three atoms are arranged in a straight line, with B in the center, as shown in Figure 7.1a, the partial charges that appear as a result of the two polar bonds simply cancel; that is, the atom B is both the center of positive charge and the center of negative charge, since the two partially charged A atoms are on opposite sides and equidistant from it. Since there is no separation of the centers of positive and negative charge, the molecule is nonpolar. If the BA_2 molecule is bent, however, as shown in Figure 7.1b, the centers of positive and negative charge do not coincide, and the molecule is polar.

It is useful here to define *bond angle*, since it will be important in describing more complex molecules. The **bond angle** is the angle between the lines which could be drawn connecting two atoms to a third atom. Thus $BeCl_2$, a linear

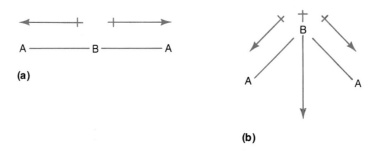

Figure 7.1 Relation between shape and polarity of a triatomic molecule. Both of the diagrams represent molecules containing polar bonds. However, only the bent molecule is polar.

molecule, has a bond angle of 180°. It is an example of the first case above and is nonpolar, although both Be—Cl bonds are polar. An example of the second case, in which the bond angle is less than 180°, is OF_2.

As you might expect, polar and nonpolar covalent substances behave differently when subjected to an electric field. Both differ strongly from ionic compounds. Ionic substances, which are nearly always solids at room temperature, are **strong electrolytes;** that is, they are good conductors of electricity when melted or when in solution. The current is carried by the movement of ions, not, as in metallic conductors, by the movement of electrons. Thus, ionic substances will not conduct a current in the solid state since the ions are not free to move. In contrast, most covalent substances, whether polar or nonpolar, will not conduct a current under any conditions since they contain neither ions nor free electrons; hence they are referred to as *nonelectrolytes*. However, some covalent substances can form ions in solution by reacting with water. If the reaction is complete, or nearly so, the resulting solution is a strong electrolyte; if only a fraction of the molecules react to form ions, the solution is called a **weak electrolyte.**

Although most polar covalent molecules will not conduct a current, they do respond to an electric field. They tend to orient themselves so that the portion of each molecule bearing the partial positive charge points toward the negative pole and the portion bearing the partial negative charge points toward the positive pole (see Figure 7.2). This tendency of a molecule to orient itself in an electric field depends on its *dipole moment*, which we first discussed in Chapter 6. Thus we say that a very polar molecule has a large dipole moment, whereas a nonpolar molecule has a dipole moment of zero.

In addition to providing clues to such physical properties as polarity, the geometry of a molecule is an important factor in determining its chemical, and ultimately its biological, properties. For example, starch and cellulose, both of which are large and complex molecules containing many glucose units bonded together, differ only in the geometry of the bond between the glucose units. Yet our bodies can digest starch, using it as a source of energy, whereas they cannot digest cellulose. Thus the shapes of molecules are important in understanding how life processes are carried on, how genetic characteristics are transmitted from

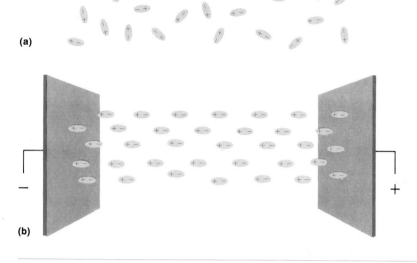

Figure 7.2 The effect of an electric field upon polar molecules. **(a)** Kinetic energy tends to keep the molecules in disorder. **(b)** The molecules tend to line up under the influence of the electric field.

one generation to another, and even how life began. Clearly, studying the shapes of molecules is not just a theoretical exercise.

7.2 Orbitals and Bonding

To write the electron formula for a molecule, we need to know only how many electrons are in the outer energy level of the constituent atoms. To predict the shape of the molecule, however, we need to ask which orbitals are occupied in the outer energy level. Since we shall begin our discussion with hydrogen, hydrogen fluoride, water, and ammonia, we shall first write the electron configurations of the elements we need in terms of four quantum numbers:

	$1s$	$2s$	$2p$
H	↑		
N	↑↓	↑↓	↑ ↑ ↑
O	↑↓	↑↓	↑↓ ↑ ↑
F	↑↓	↑↓	↑↓ ↑↓ ↑

Since we are going to be talking about the shapes of molecules, it is important to recall that the second quantum number is related to the shape of the electron orbital. The s orbitals are spherical, whereas the three p orbitals are dumbbell-shaped, each lying along a different axis (Figure 5.10).

How then shall we assemble atoms, each with its set of electron orbitals, to form covalently bound molecules? We shall take as our first principle that a covalent bond results when an electron pair occupies a region of space common to two atoms. Thus we can represent a bond by overlapping a pair of orbitals, one each from two atoms. Such an overlap should be seen as a region in which we are especially likely to find the shared pair of electrons. For example, we can depict the hydrogen molecule as two hydrogen atoms, brought close enough together that their two $1s$ orbitals (one from each atom) overlap. The shared electron pair is most likely to be found in the region of overlap (Figure 7.3). The HF molecule requires an overlap of the $1s$ orbital of the hydrogen atom with one of the $2p$ orbitals of the fluorine atom, as shown in Figure 7.4. Naturally, all diatomic molecules, such as H_2 and HF, must be linear.

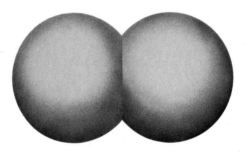

Figure 7.3 The hydrogen molecule.

A triatomic molecule, such as water, is a more complex case. Each oxygen atom has two unpaired electrons in two p orbitals. Since these p orbitals lie at 90° from one another, we would expect to find a bond angle of 90° in any molecule where oxygen is the central atom, as shown for water in Figure 7.4. The actual bond angle in water is about 104.5°. (Molecules like water, which have bond angles less than 180°, are called *bent* molecules.) We shall look at another approach to predicting the geometry of the water molecule on p. 189.

The nitrogen atom has three unpaired electrons in three orbitals. Therefore we expect the three nitrogen-hydrogen bonds in an ammonia molecule (NH_3) to be at right angles to each other, as shown in Figure 7.4. Again the actual bond angles are somewhat larger than predicted; about 107.3°. As with water, we shall take a closer look at ammonia on p. 189.

So far we have explained with some success the formation of some simple molecules in terms of orbital overlap. We can predict the number of bonds from the number of orbitals containing one electron. Furthermore, we have been able to make reasonable predictions about the geometry of the molecules; however,

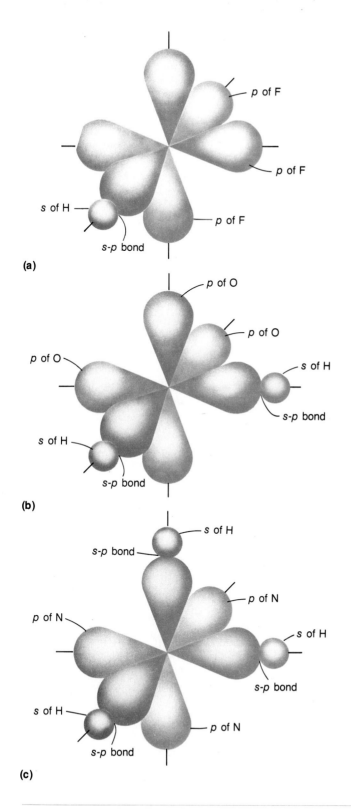

Figure 7.4 The expected models of the **(a)** hydrogen fluoride molecule, **(b)** water molecule, and **(c)** ammonia molecule. More realistic models of these molecules are shown on p. 189.

there have been notable deviations from our predictions of bond angles. We shall now look at ways of explaining these deviations.

7.3 Hybridization of Orbitals

If we approach the carbon-containing compounds with the same bonding concepts that worked for H_2, HF, H_2O, and NH_3, we encounter considerable difficulty. However, by introducing two new concepts, we not only solve the dilemma of the carbon compounds but also provide a powerful theory for explaining bonding in many more complex molecules and provide a more adequate explanation for the bond angles we observed in H_2O and NH_3.

The electron configuration for carbon is

$$\begin{array}{ccc} 1s & 2s & 2p \\ [\uparrow\downarrow] & [\uparrow\downarrow] & [\uparrow\,|\,\uparrow\,|\,\;] \end{array}$$

Since the isolated carbon atom has two unpaired electrons in orbitals at an angle of 90° from one another, we might predict the existence of a bent CH_2 molecule. However, the simplest stable molecule containing only hydrogen and carbon is methane, which has the formula CH_4. Its stability can be explained by its having a noble gas electron configuration, whereas CH_2 could not. Thus we must look for a mechanism by which carbon can form four bonds rather than the two that its electron configuration would lead us to expect. If we assume that one of the $2s$ electrons is moved to the empty $2p$ orbital, we have the required four unpaired electrons. This new electron configuration, obtained by *promotion* of a $2s$ electron, is an excited configuration (Figure 7.5). Of course, the promotion of an electron

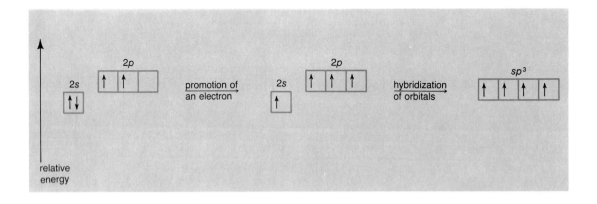

Figure 7.5 Changes in the electron configuration in the outermost energy level of the carbon atom in forming the CH_4 molecule.

requires energy, but the energy released by the formation of four stable bonds is more than enough to make promotion favorable.

Promotion thus explains the number of bonds in methane, but it does not explain the experimental evidence that the four bonds are equivalent. In fact, we would expect three equivalent bonds, at about 90° to each other, involving the $2p$ orbitals and a fourth nonequivalent bond involving the $2s$ orbital. The answer is to introduce a second new concept—**hybridization.** Orbitals in the same atom that have similar energies often combine to form a new set of equivalent orbitals, called *hybrid orbitals*. For example, in the carbon atom one $2s$ orbital and three $2p$ orbitals mix to form four new equivalent orbitals. Each of the new orbitals can be thought of as 75 percent p orbital and 25 percent s orbital, and each is directed to one of the corners of a tetrahedron (Figure 7.6). These orbitals are designated sp^3 orbitals. Note that the superscript refers to the number of p orbitals involved, *not* the number of electrons. Now our picture of methane agrees with experimental observations: The molecule has four equivalent C—H bonds, each 109.5° from the others.

We also begin to see a way of better explaining the shapes of the water and ammonia molecules. Although we can satisfactorily explain many of the physical and chemical properties of these compounds in terms of pure p orbitals, the experimentally determined bond angles suggest hybrid orbitals. The angles of 104.5° for water and 107.3° for ammonia are much closer to the 109.5° of the hybrid bonds than to the 90° of the pure p orbitals. For water we picture two of the hybrid orbitals of oxygen as bonding with the hydrogen $1s$ orbitals, and the other two as containing unshared electrons. Similarly, for ammonia, three of the hybrid orbitals of nitrogen form bonds with hydrogen atoms, and the fourth contains a pair of unshared electrons.

Regardless of whether we use pure p orbitals or hybrid orbitals to explain the structure of water and ammonia molecules, we describe their shapes as bent and *trigonal pyramidal*, respectively. The three hydrogen atoms of ammonia can be visualized in either case as being at the three corners of the base of a trigonal pyramid, with the nitrogen atom at the apex. Notice, however, that in such cases as H_2O and NH_3, we must distinguish between the *electron-pair geometry* and the *molecular shape*. If there are no unshared electron pairs in a molecule, the electron-pair geometry and molecular shape are the same. For methane, both are tetrahedral. But with water and ammonia we must be more careful. If we regard these molecules as having hybridized orbitals, the electron-pair geometry for both is tetrahedral, although we have described the molecular shape of the water molecule as bent and that of the ammonia molecule as trigonal pyramidal (Figure 7.6).

The tetrahedral arrangement of electron pairs in sp^3 orbitals is an example of the *valence shell electron-pair repulsion (VSEPR) theory*. This theory states that the mutual repulsion of valence electron pairs produces an arrangement in which the electron pairs are as far apart as possible. Four pairs of electrons will be farther from each other in the tetrahedron (109.5°) than in any other structure. Thus in ions and molecules analogous to the methane molecule, where the central atom is nonmetallic and has no unshared electron pairs associated with it, sp^3 hybridization is always involved and the electron-pair geometry and molecular

Number of Unshared Electron Pairs on Central Atom	Example	Shape	Two representations of the electron-pair geometry and shape. (a) The arrangement of sp^3 orbitals and their overlap with s orbitals of the hydrogen atoms to form s-sp^3 bonds. (b) A simple diagram of the geometries.	
			(a)	(b)
0	CH_4	tetrahedral		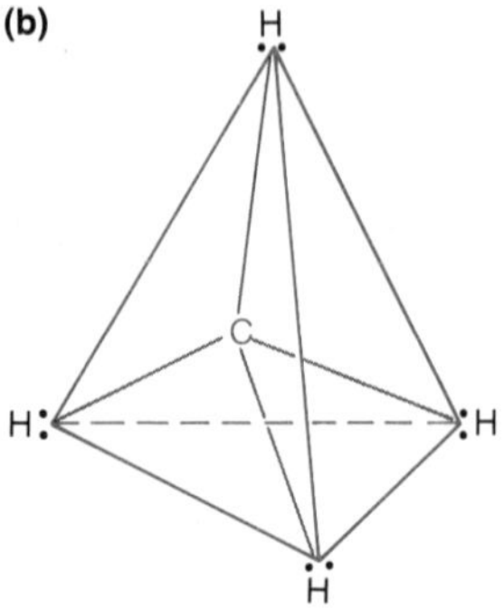
1	NH_3	trigonal pyramidal		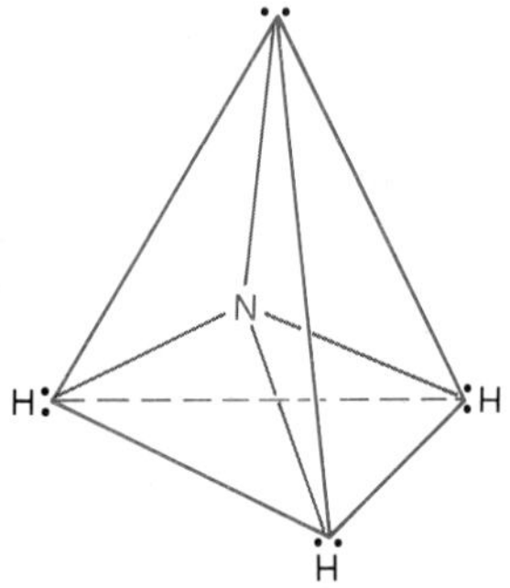
2	H_2O	bent		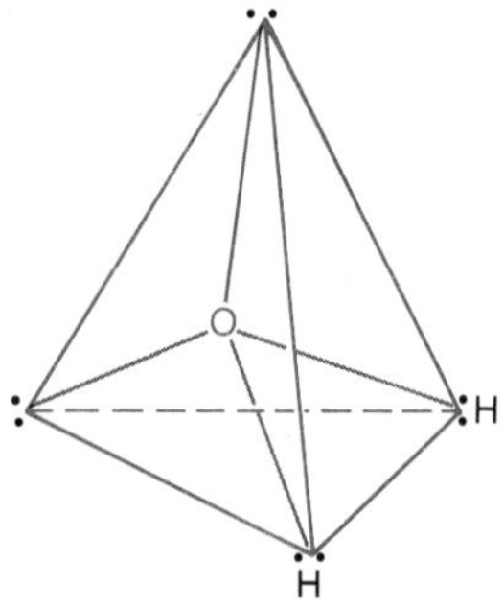

Figure 7.6 Shapes of entities (molecules and ions) characterized by four pairs of electrons associated with the central atom (sp^3 hybridization and tetrahedral electron-pair geometry).

geometry are always tetrahedral. Molecules and ions such as SiF_4, GeH_4, SO_4^{2-}, and NH_4^+ are sp^3-hybridized and have tetrahedral shapes.

Molecules with unshared electron pairs, however, such as water and ammonia, are often more difficult to characterize. For example, the bond angles in H_2S, H_2Se, H_2Te, PH_3, and SbH_3 are very close to 90°. In NF_3, PF_3, AsF_3, and SCl_2, on the other hand, bond angles are much closer to 109.5° than to 90°. Recall once more, however, that regardless of how we choose to describe electron-pair geometries in these cases, the molecular shapes remain the same. Molecules of OF_2, H_2S, and SCl_2, where the central atom is bonded to two atoms, are bent, and molecules or ions such as PH_3, PF_3, and SO_3^{2-}, where the central atom is bonded to three atoms, are trigonal pyramidal.

In the ions and molecules we have looked at so far, the central atoms have been able to acquire a full complement of eight electrons in their outer energy levels, two in each of their four sp^3 hybrid orbitals. Other types of hybrid orbitals can be formed, however, especially in cases where the central atoms do not achieve noble gas electron configurations or where multiple bonds are formed.

As an example of an atom that does not have a noble gas electron configuration, let us look first at the beryllium atom in BeF_2. The electron configuration of the unexcited atom, as shown in Figure 7.7, reveals a pair of outer energy level electrons. Since in this compound covalent bonds require unpaired electrons, however, a 2s electron must be promoted to a 2p orbital. Hybridization then produces two equivalent orbitals. Since they arise from one s and one p orbital, these are designated *sp* hybrid orbitals. According to the valence shell electron-pair repulsion theory, these orbitals will be on opposite sides of the beryllium atom; hence the electron-pair geometry of *sp* orbitals is linear. Since all the outer energy level electrons in beryllium are in *sp* orbitals and since none are unshared, the electron-pair geometry of beryllium is linear, as is the shape of the beryllium fluoride molecule (Figure 7.8).

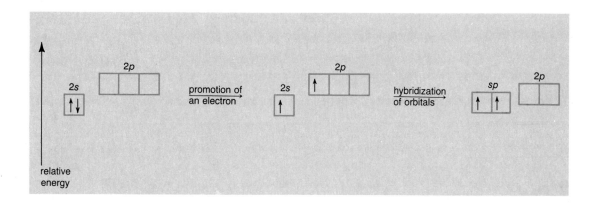

Figure 7.7 Changes in the electron configuration in the outermost energy level of the beryllium atom in forming the BeF_2 molecule.

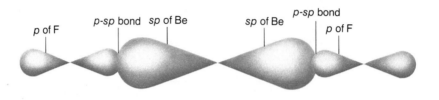

Figure 7.8 The BeF$_2$ molecule.

Another compound, this one a dangerously poisonous one, having sp bonding is dimethyl mercury, $(CH_3)_2Hg$:

$$\begin{array}{ccc} H & & H \\ \ddot{} & & \ddot{} \\ H\!:\!\!\ddot{C}\!:\!Hg\!:\!\ddot{C}\!:\!H \\ \ddot{} & & \ddot{} \\ H & & H \end{array}$$

This compound has been found in polluted water, and high levels of it have been detected in fish. It is synthesized by microscopic organisms from less toxic elemental mercury, which can arise from either natural or industrial wastes. Dimethyl mercury is highly toxic, causing birth defects as well as damage to the brain, liver, and kidneys.

There are two types of bonds in dimethyl mercury. Each bond between carbon and hydrogen results from the overlap of the s orbital of the hydrogen atom and an sp^3 orbital of the carbon atom. Each Hg—C bond, however, involves the overlap of the remaining sp^3 orbital of the carbon atom and an sp orbital on the mercury atom. Therefore each carbon atom has a tetrahedral arrangement of bonds around it, whereas the mercury atom has a linear arrangement around it.

Boron compounds such as boron trifluoride are examples of a third type of hybridization. Boron is also another example of an atom that does not achieve a complete outer energy level of electrons. Its ground state electron configuration is

$$\begin{array}{ccc} 1s & 2s & 2p \\ \boxed{\uparrow\downarrow} & \boxed{\uparrow\downarrow} & \boxed{\uparrow\,\,\,\,\,\,\,} \end{array}$$

To form three covalent bonds, boron can promote one $2s$ electron to a $2p$ orbital. This excited state of boron is shown in Figure 7.9. As we might expect, however, the one s orbital and two p orbitals are mixed to form three equivalent sp^2 hybrid orbitals. The hybrid orbitals will be as far apart as possible; thus sp^2 orbitals are oriented in a plane 120° apart. Since the orbitals can be thought of as directed to the corners of a planar equilateral triangle, we call the electron-pair geometry produced by sp^2 hybridization *trigonal planar*. Since the boron trifluoride molecule contains only bonds formed from sp^2 orbitals and since boron contains no unshared electrons, the molecular shape is also trigonal planar (Figure 7.10).

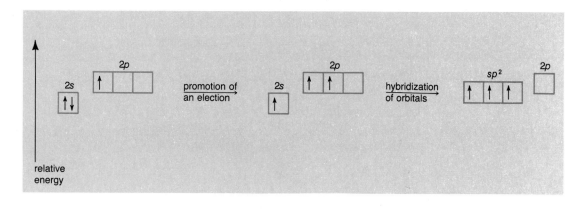

Figure 7.9 Changes in the electron configuration in the outermost energy level of the boron atom in forming the BF₃ molecule.

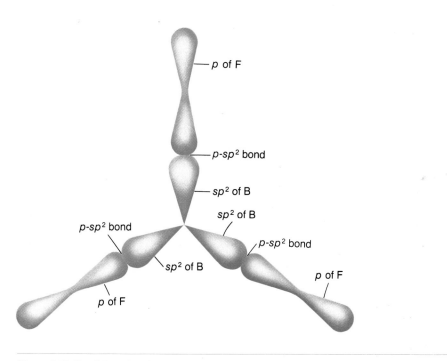

Figure 7.10 The BF₃ molecule.

In contrast to beryllium and boron, which often deviate from noble gas electron configurations by having fewer than eight outer energy level electrons, some atoms expand beyond eight. An example is phosphorus in PCl₅, which forms five identical hybrid orbitals by using a 3d orbital. Its unexcited electron

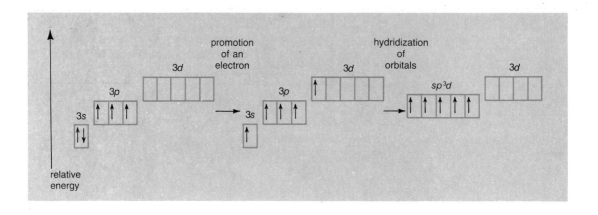

Figure 7.11 Changes in the electron configuration in the outermost energy level of the phosphorus atom in forming the PCl$_5$ molecule.

configuration is

If one 3s electron is promoted to a 3d orbital, the result is the configuration shown in Figure 7.11. Notice that by promoting the 3s electron to the 3d orbital, we have bypassed the 4s orbital. (Recall that in the normal filling order, 4s precedes 3d.) We probably would not have predicted this promotion, but the stability of the hybrid orbitals that result more than make up for the extra cost in energy.

We now have five unpaired electrons available for bonding. As we have learned to expect, these electrons occupy not one s orbital, three p orbitals, and one d orbital but five sp^3d orbitals of equivalent energy. The maximum distance between electron pairs is provided for if they are directed to the corners of a *trigonal bipyramid* (Figure 7.12).

Until now we have had little trouble deducing electron-pair geometries and molecular shapes from the geometries of hybrid orbitals, regardless of whether the hybrid orbitals were used for bonding or contained unshared electrons. For example, we learned that sp^3 orbitals have tetrahedral geometry. Therefore, the electron-pair geometry and molecular shape of CH$_4$ are obviously tetrahedral. The electron-pair geometry of NH$_3$ is also tetrahedral and, regardless of which orbital contains the unshared electron pair, the shape of the molecule will be trigonal pyramidal. Regardless of where we put the two pairs of unshared electrons in water, the molecule will be bent. In the trigonal bipyramid, however, we must be more careful where we put unshared electron pairs. Although the sp^3d orbitals have identical energies, they are not geometrically equivalent. Look again at Figure 7.12 and notice that the apex positions

Number of Unshared Electron Pairs on Central Atom	Example	Shape	Two representations of electron-pair geometry and shape. (a) The arrangement of sp^3d orbitals and their overlap with p orbitals of the peripheral atoms to form p-sp^3d bonds. (b) A simple diagram of the geometries.
0	PCl_5	trigonal bipyramidal	(a) and (b) 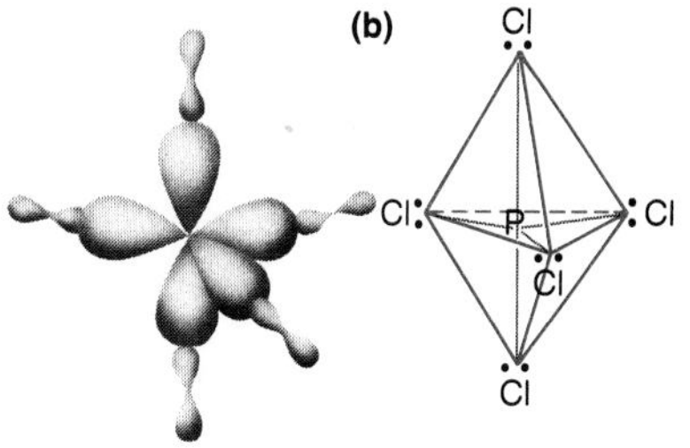 diagrams
1	$TeCl_4$	distorted tetrahedron	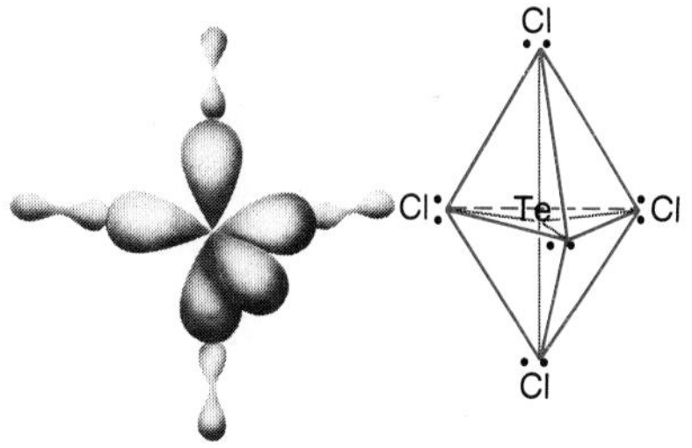 (a) and (b) diagrams
2	ClF_3	T shaped	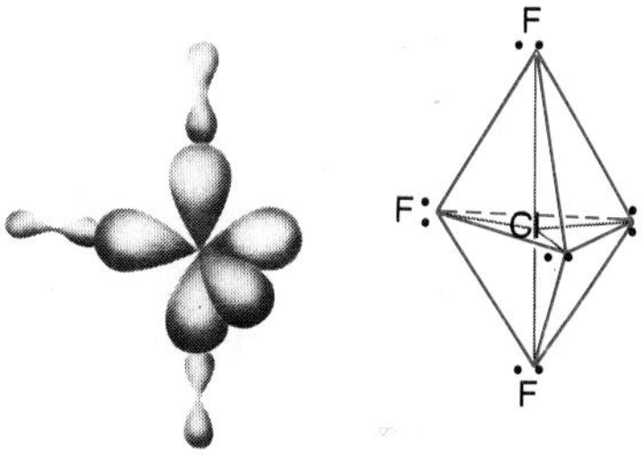 (a) and (b) diagrams
3	XeF_2	linear	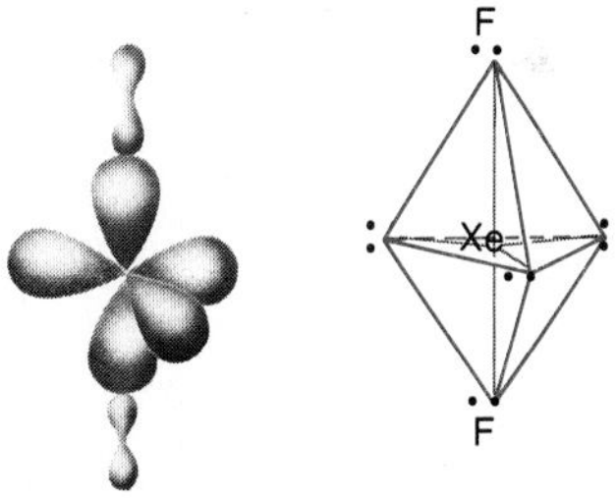 (a) and (b) diagrams

Figure 7.12 Shapes of entities (molecules and ions) characterized by five pairs of electrons associated with the central atom (sp^3d hybridization and trigonal bipyramidal electron-pair geometry).

are different from the three corners of the plane. Therefore, since unshared electron pairs need more space than bonding pairs (that explains the slight deviations from tetrahedral electron-pair geometries in NH_3 and H_2O), they tend to favor the "roomier" corner positions in the trigonal bipyramid. Thus, depending on the number of unshared pairs, we find the following molecular geometries:

Example	Number of Unshared Pairs	Molecular Shape (geometry)
$TeCl_4$	1	Distorted tetrahedron
ClF_3	2	T-shaped
XeF_2	3	Linear

The SF_6 molecule requires six covalent bonds, which are formed simply by extending the hybridization concepts we have already covered. The electron configuration of sulfur is

If we promote one paired $3p$ electron and one paired $3s$ electron to $3d$ orbitals (again passing over the $4s$ orbital), the result is the electron configuration shown in Figure 7.13. Again, the six orbitals containing electrons are hybridized, this time producing six equivalent sp^3d^2 orbitals directed to corners of an octahedron (Figure 7.14).

The iodine in IF_5 is surrounded by six pairs of electrons; thus it too has sp^3d^2 hybridization and an octahedral electron-pair geometry. However, it

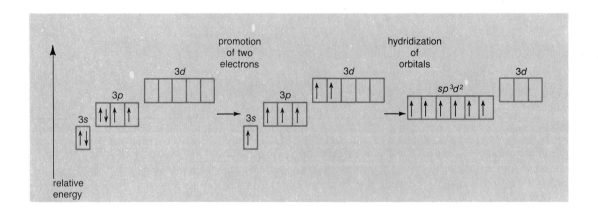

Figure 7.13 Changes in the electron configuration in the outermost energy level of the sulfur atom in forming the SF_6 molecule.

Number of Unshared Electron Pairs on Central Atom	Example	Shape	Two representations of electron-pair geometry and shape. (a) The arrangement of sp^3d^2 orbitals and their overlap with p orbitals of the peripheral atoms to form p-sp^3d^2 bonds. (b) A simple diagram of the geometries.
0	SF_6	octahedral	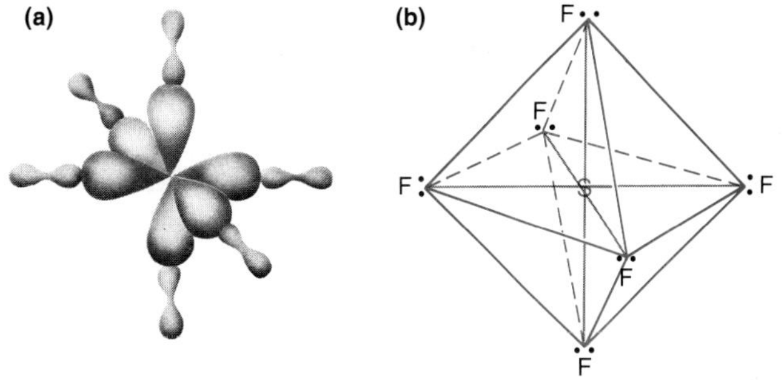
1	IF_5	square pyramidal	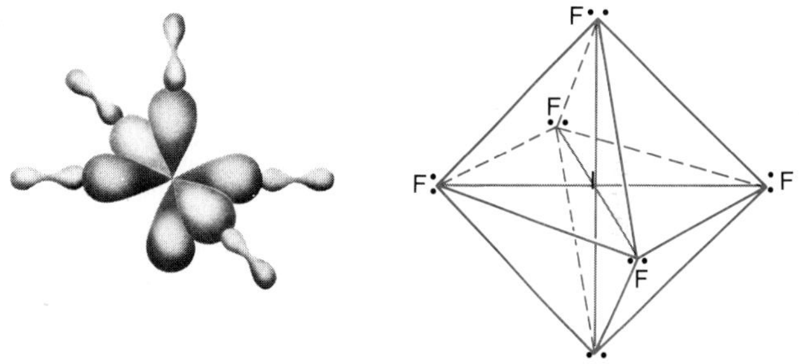
2	XeF_4	square planar	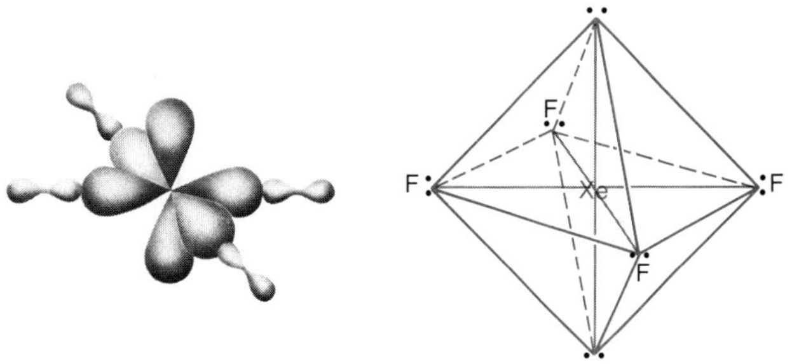

Figure 7.14 Shapes of entities (molecules and ions) characterized by six pairs of electrons associated with the central atom (sp^3d^2 hybridization and octahedral electron-pair geometry).

differs from the SF_6 molecule in that one of the electron pairs is unshared. Since all corners of the octahedron are geometrically equivalent, it does not matter where we put the unshared pair. The molecular geometry of IF_5 is square pyramidal (Figure 7.14).

The XeF_4 molecule is another molecule with sp^3d^2 hybridization and an octahedral electron-pair geometry, but the xenon in this molecule has two

Covalent Bonding—Molecular Structure

unshared electron pairs. Here we must put the unshared pairs as far apart as possible—at opposite corners of the octahedron—hence the molecular geometry is square planar (Figure 7.14).

Questions

1. Distinguish between electron-pair geometry and molecular shape.

2. Complete the following table, showing the relations among types of hybrid bonding, electron-pair geometry, and molecular shape. (Assume that all outer energy level electrons occupy hybrid orbitals.)

Hybrid Bonding	Number of Unshared Pairs of Electrons	Electron-Pair Geometry	Molecular Shape
sp	0		
	1		
sp^2	0		
	1		
	2		
sp^3	0		
	1		
	2		
	3		
sp^3d	0		
	1		
	2		
	3		
sp^3d^2	0		
	1		
	2		

7.4 Geometry of Multiple Bonds

We now want to extend our discussion of hybridization to molecules containing multiple bonds. Most of the principles we used in Section 7.3 will remain valid, but there will be one important difference: In a molecule with multiple bonds, not all the electrons in the outer energy level participate in the hybridization—some remain in pure orbitals. Let us begin by listing several experimental observations about ethene, the simplest carbon-hydrogen compound containing a double bond:

1. The strength of the double bond is about 1.75 times that of a single bond.
2. All atoms in the ethene molecule are in the same plane.
3. There is no rotation about the double bond.

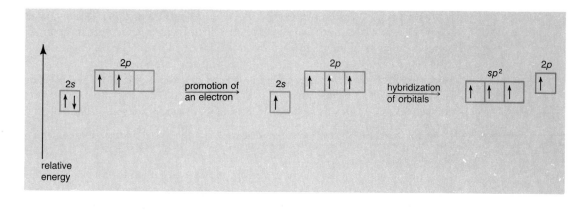

Figure 7.15 Changes in the electron configuration in the outermost energy level of the carbon atom in the forming of the C_2H_4 molecule.

We have learned that sp^2 orbitals (such as those in the BF_3 molecule) have a planar trigonal geometry. This suggests that the planar C_2H_4 molecule might involve sp^2 hybridization. If we think of one s orbital and two p orbitals of each carbon hybridizing to form three sp^2 orbitals, we can account for the planar structure of the ethene molecule (Figure 7.15). Two of the sp^2 orbitals form bonds to two hydrogen atoms, and the third forms one of the bonds between the carbon atoms. However, this leaves the fourth electron in the outermost energy level of each carbon atom in a pure p orbital (Figure 7.16). These two p orbitals are at right angles to the plane of the sp^2 bonds and their overlap forms the second bond between the carbon atoms.

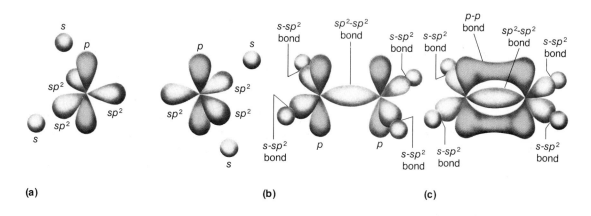

Figure 7.16 Formation of the C_2H_4 molecule. **(a)** Orbitals involved in bonding on two carbon atoms and four hydrogen atoms. **(b)** Formation of the five sigma bonds in the C_2H_4 molecule. **(c)** Formation of the pi bond between the two carbon atoms in the C_2H_4 molecule.

This second carbon-to-carbon bond is different from any we have seen so far. It involves the edges of *p* orbitals, and it forms not in line between the bonded atoms but above and below that line (see Figure 7.16c). This bond is called a *pi bond*, whereas the more conventional carbon-carbon bond is an example of a *sigma bond*. All single bonds and one bond of a double bond must be sigma bonds. They always lie in line between the bonded atoms, and they may involve a variety of orbitals. In ethene the carbon-carbon sigma bond is an sp^2-sp^2 bond, and the carbon-hydrogen bonds are s-sp^2 bonds. Pi bonds most frequently involve *p* orbitals, but in complex atoms higher orbitals may be involved.

Let us now try to justify a couple of the experimental observations that we noted above: the restricted rotation about the C—C bond and the planar structure of the molecule.

In a sigma bond (see Figure 7.16b), the extent of orbital overlap does not change when the bonded atoms are rotated around the axis of the bond. Therefore, if there are no other constraints, free rotation is usually allowed about a single bond. However, rotation about a double bond reduces the amount of overlap in the pi bond or breaks it completely. Since this requires energy, rotation about a double bond is severely restricted.

Since the pi bond in ethene is parallel to the companion sigma bond, it does not affect the geometry established by the sigma framework. In fact, we may generally deduce electron-pair geometries and molecular shapes from the orientations of sigma bonds and unshared electron pairs only. Thus both the molecular geometry and the electron-pair geometry of ethene are planar, despite the *p* electrons sticking out of the plane.

In the carbon dioxide molecule, there are two double bonds ($\ddot{\text{O}}::\text{C}::\ddot{\text{O}}$). In this case we must reserve two of the *p* electrons of the excited carbon atom for forming pi bonds; thus we have one *s* and one *p* orbital left to form hybrid bonds (Figure 7.17). We saw previously that *sp* bonding is linear; hence we would expect the molecular geometry of CO_2 to be linear also.

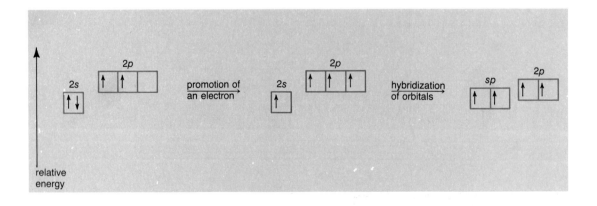

Figure 7.17 Changes in the electron configuration in the outermost energy level of the carbon atom in forming the CO_2 or C_2H_2 molecule.

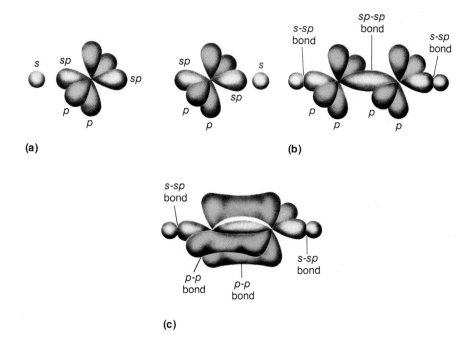

Figure 7.18 Formation of the C_2H_2 molecule. **(a)** Orbitals involved in bonding two carbon atoms and two hydrogen atoms. **(b)** Formation of the three sigma bonds in the C_2H_2 molecule. **(c)** Formation of the two pi bonds between the two carbon atoms in the C_2H_2 molecule.

By approaching the acetylene (C_2H_2) molecule as we have ethene and carbon dioxide, we find that the carbon-carbon triple bond consists of one sigma bond and two pi bonds. The sigma bond involves one *sp* hybrid orbital from each carbon atom, and the two pairs of pure *p* orbitals form the pi bonds. Each C—H bond involves the second *sp* orbital from carbon and the 1*s* orbital from hydrogen. Figure 7.18 illustrates these bonds in the acetylene molecule.

Questions

3. How many pi bonds are there in a
   a. single bond?   b. double bond?   c. triple bond?
4. Compare pi and sigma bonds with regard to
   a. overlap of orbitals.   b. strength.   c. rotation about the bond.

7.5 Resonance and Delocalized Electrons

In Chapter 6 we discussed resonance hybrids, arguing that molecules such as SO_2 and SO_3 could not be adequately represented by single electron formulas.

We can now obtain a better picture of these molecules by taking a three-dimensional look at their bonding orbitals and by using what we know about hybridization and pi bonds. In SO$_2$ we know that the sulfur atom forms two bonds and retains an unshared pair of electrons, so it is reasonable to suggest the hybridization of three outer orbitals as shown below, where one sp^2 orbital contains the unshared electrons.

The other two sp^2 orbitals overlap with oxygen orbitals to form the sigma framework shown in Figure 7.19. That leaves the unhybridized p orbital, containing two electrons, at right angles to the plane of the sigma framework. This orbital overlaps with p orbitals from each of the two oxygen atoms, forming a *pi cloud* above and below the sigma framework and parallel to it (Figure 7.19). There are four electrons in this cloud—two from sulfur and one from each oxygen—but they do not remain identified with the atoms that contributed

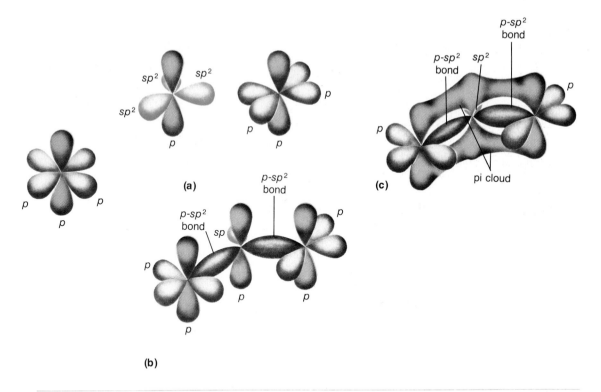

Figure 7.19 Formation of the SO$_2$ molecule. **(a)** Orbitals available for bonding one sulfur atom and two oxygen atoms. **(b)** Formation of the two sigma bonds in the SO$_2$ molecule. **(c)** Formation of the pi cloud spanning the sulfur atom and two oxygen atoms above and below the sigma framework.

them. Since they are shared by more than two bonded atoms, they are called *delocalized electrons*. Since the pi cloud is always parallel to the sigma framework, these delocalized electrons do not affect the electron-pair geometry around the sulfur atom, which is planar trigonal. The molecular shape of SO_2 is bent.

A second example in which delocalized electrons play a part is SO_3. The hybridization for this molecule is more difficult to work out than for SO_2, but the result is similar. Three sp^2 orbitals on the sulfur atom form sigma bonds with p orbitals from three oxygen atoms, and a single p orbital on the sulfur atom becomes part of a delocalized pi cloud above and below the sigma framework. In order to understand the SO_3 molecule better, Figure 7.20 presents its formation in a stepwise fashion. This is a theoretical presentation and one should not think of the actual formation of this molecule as taking place by steps. There are six delocalized electrons in the pi cloud, and they are shared by all four atoms in the molecule. As in SO_2, the electron-pair geometry around sulfur is planar trigonal, and since there are no unshared electron pairs in the central atom of SO_3, the molecule is also planar trigonal.

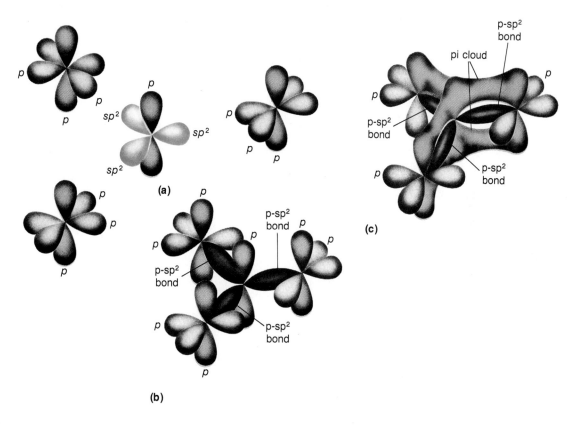

Figure 7.20 Formation of the SO_3 molecule. **(a)** Orbitals available for bonding one sulfur atom and three oxygen atoms. **(b)** Formation of three sigma bonds in the SO_3 molecule. **(c)** Formation of the pi cloud spanning the sulfur atom and three oxygen atoms above and below the sigma framework.

Question

5. What type of hybrid orbital would you expect to find in nitrogen in each of the following?
    a. NH_3 b. NO_2^- c. NO_3^- d. CN^- e. NH_4^+

7.6 Summary of Rules for Predicting Molecular Geometries

In general, we can predict the electron-pair and molecular geometries of a molecule or ion by adopting the following procedure:

1. Write the electron formula.

2. Add the number of electron pairs in the sigma bonds and the number of unshared electron pairs assigned to the central atom. (Do not include electron pairs in pi bonds.)

3. Determine the type of hybridization of the central atom by distributing these electrons as follows: one pair to an s orbital, *up to three* pairs to p orbitals, and the remaining pairs to d orbitals. The resulting hybridization determines the electron-pair geometry as follows:

sp	linear
sp^2	trigonal
sp^3	tetrahedral
sp^3d	trigonal bipyramidal
sp^3d^2	octahedral

4. Finally, count the number of unshared electron pairs, and determine the molecular shape from the table below.

Electron-Pair Geometry	Number of Unshared Electron Pairs	Molecular Geometry
Trigonal	0	Trigonal
	1	Bent
Tetrahedral	0	Tetrahedral
	1	Trigonal pyramidal
	2	Bent
	3	Linear
Trigonal bipyramidal	0	Trigonal bipyramidal
	1	Distorted tetrahedron
	2	T-shaped
	3	Linear
Octahedral	0	Octahedral
	1	Square pyramidal
	2	Square planar

Using the concepts in Chapters 5 and 6, you should now be able to predict the electron formulas as well as the shapes of most simple molecules and ions. From the shape of the molecule it is possible to ascertain whether the molecule is polar or nonpolar. This is illustrated in Examples 7.1 and 7.2.

Example 7.1 Ascertain whether the Cl_2O molecule is polar or nonpolar.

Solution

a. Using the generalizations listed on p. 165, draw the electron formula.

$$:\!\ddot{\underset{..}{O}}\!:\!\ddot{\underset{..}{Cl}}\!:$$
$$:\!\ddot{\underset{..}{Cl}}\!:$$

b. From the electronegatives of chlorine and oxygen (p. 152), determine whether the bonds are polar or nonpolar. Since oxygen is more electronegative than chlorine, chlorine-oxygen bonds are polar.

c. Using the generalizations listed on p. 204, determine the shape of the molecule. From rules 2 and 3 we conclude the molecule is characterized by sp^3 hybridization and tetrahedral electron-pair geometry. From rule 4 we conclude the shape of the molecule is bent.

d. From the shape of the molecule, determine whether the polar bonds cancel each other or are additive.

Since the molecule is bent, the bonds are additive and the molecule is **polar.**

Example 7.2 Determine whether the molecule CSe_2 is polar or nonpolar.

Solution

a. Using the generalizations listed on p. 165, draw the electron formula.

$$:\!\ddot{Se}\!::\!C\!::\!\ddot{Se}\!:$$

b. From the electronegativities of selenium and carbon (p. 152), determine whether the bonds are polar or nonpolar. Since carbon is slightly more electronegative than selenium, the selenium-carbon bonds are polar.

c. Using the generalizations listed on p. 204, determine the shape of the molecule. From rules 2 and 3 we conclude that the molecule is characterized by sp hybridization and linear electron-pair geometry. From rule 4 we understand the molecular shape is also linear.

Table 7.1 Predicting Electron-Pair Geometry and Shape

Formula	Electron Formula	Type of Hybridization	Electron Pair Geometry	Number of Unshared Pairs in Central Atom	Shape
$BeCl_2$	:Cl:Be:Cl:	sp	Linear	0	Linear
ClO_4^-	:O:Cl:O: with :O:⁻ above and :O: below	sp^3	Tetrahedral	0	Tetrahedral
CN^-	:C:::N:⁻	sp	Linear	1	Linear
NO_2^-	:N:O:⁻ ⟷ :N::O:⁻	sp^2	Planar trigonal	1	Bent
SiF_6^{2-}	(octahedral F around Si)²⁻	sp^3d^2	Octahedral	0	Octahedral
ICl_2^-	:Cl:I:Cl:⁻	sp^3d	Trigonal bipyramidal	3	Linear

d. From the shape of the molecule, determine whether the polar bonds cancel each other or are additive.

$$Se \longrightarrow C \longleftarrow Se$$

Since the two sigma bonds are arranged linearly, there is no separation of the centers of positive and negative charge and the molecule is **nonpolar.**

Table 7.1 provides additional illustrations of the generalizations in this chapter. Some ions and molecules do not exist as discrete structural units; instead, they are parts of much larger extended structures. These structures, which often form by repetition of smaller subunits, follow the same rules for bonding as simpler molecules, but because of their complexity and diversity we shall defer discussion of them until later (especially Chapters 9, 22, and 23). In Chapter 9 we shall also come back to look at the special concepts involved in metallic bonding.

Sidelight
Molecular Orbitals

So far we have taken two approaches to understanding bond formation in simple molecules. First we looked at bonds simply as the overlap of pure atomic orbitals. In many simple cases, this concept is adequate, but our failure to predict correctly the bond angles in H_2O and NH_3 suggested that bond formation is often more complicated. Indeed when we encountered methane, with its four identical bonds, our simple approach failed badly. We then introduced the ideas of electron promotion and hybridization and found that we could account not only for the bond angles in water and ammonia and our observations about methane but also for the molecular geometries of a host of other molecules. For some molecules, however, chief among them the paramagnetic molecules, neither the simple atomic orbital approach nor the hybridized atomic orbital approach is sufficient.

We first mentioned paramagnetism in Chapter 6 during our discussion of the d^n ions of transition elements. Recall that ions and, hence, molecules with unpaired electrons respond strongly to a magnetic field and are called *paramagnetic*. It is these unpaired electrons that are difficult to account for using atomic orbital approaches to bonding. Before considering the solution to this difficulty, however, we should emphasize that it is not only among compounds of the transition elements that paramagnetism can be found. Such simple molecules as NO, NO_2, and ClO_2 are also paramagnetic. Each contains an odd number of electrons (hence one of them must be unpaired); thus they are called *odd molecules*. Another familiar paramagnetic molecule, this one with two unpaired electrons, is O_2.

To account for the oxygen molecule, we must look at bonding in an entirely new way. In both atomic orbital approaches we thought of stable bonds as formed by the overlap of either pure or hybridized orbitals. Each atomic orbital remained essentially unchanged; it merely overlapped with an atomic orbital from another atom. In this third approach, however, we shall think of atomic orbitals from different atoms as merging to form new *molecular orbitals*. In so doing, the participating atomic orbitals lose their identity, and the molecular orbitals are associated with the molecule as a whole.

In principle, theoretical chemists should be able to calculate the shapes and relative energies of molecular orbitals from the equations that describe atomic orbitals. In practice, however, precise calculations are exceedingly complex for all but the simplest molecules. Consequently we shall limit ourselves to diatomic molecules, where the shapes and energies of molecular orbitals can be safely deduced from what we know of atomic orbitals.

As you might expect, the simplest molecular orbitals are those formed by the merger of the simplest atomic orbitals. Figure 7.21 therefore depicts how two 1s orbitals combine to produce two molecular orbitals. However, even this simple example illustrates several important points. First, the number of molecular orbitals is always equal to the number of atomic orbitals combined to produce them. Second, just as with atomic orbitals, stable bonds demand

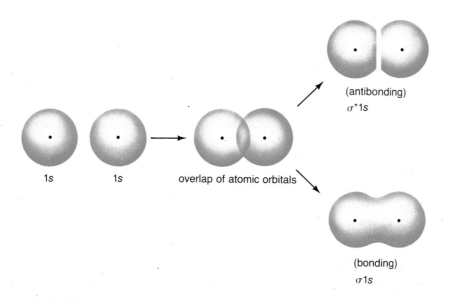

Figure 7.21 The 1s molecular orbitals.

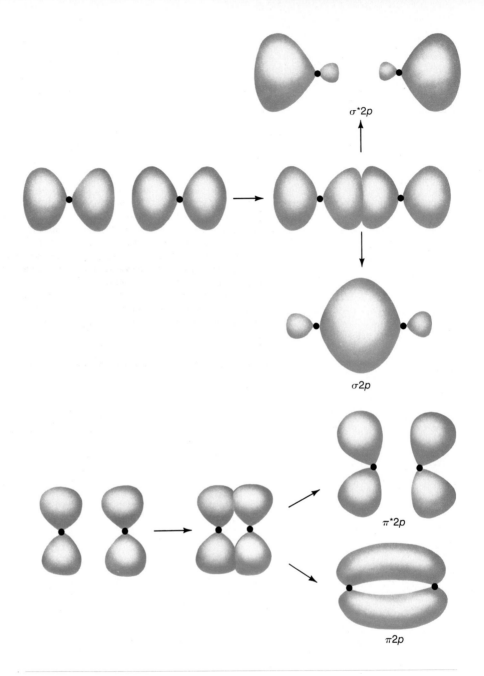

Figure 7.22 The 2p molecular orbitals.

molecular orbitals with an enhanced electron density between the nuclei. Thus the lower energy orbital in Figure 7.21 is called a *bonding orbital*, whereas the higher energy orbital is called an *antibonding orbital*. Finally, again like atomic orbitals, each molecular orbital has a capacity of two paired electrons. And of course electrons tend to fill the lower energy bonding orbitals before occupying the antibonding orbitals.

The nomenclature of these molecular orbitals deserves mention. The Greek letter sigma (σ) is used to designate orbitals that are symmetrical about a line connecting the two nuclei; that is, if σ *orbitals* are rotated about the internuclear axis, they will not change in appearance. Each molecular orbital is also identified by the atomic orbitals that gave rise to it, in this case 1s atomic orbitals. Finally, an asterisk (*) designates an antibonding orbital. Thus Figure 7.21 illustrates a $\sigma 1s$ orbital and a $\sigma^* 1s$ orbital.

Molecular orbitals formed from two 2s atomic orbitals are completely analogous to $\sigma 1s$ and $\sigma^* 1s$. They are designated $\sigma 2s$ and $\sigma^* 2s$, and both lie at higher energies than $\sigma^* 1s$. Again, the antibonding orbital ($\sigma^* 2s$) lies at a higher energy than the bonding orbital ($\sigma 2s$). Thus in terms of increasing energy, the order of the four molecular orbitals we have discussed so far is $\sigma 1s$, $\sigma^* 1s$, $\sigma 2s$, and $\sigma^* 2s$.

When we look at molecular orbitals formed from p atomic orbitals, we find a complication that did not arise when we combined s orbitals: p orbitals can combine end to end, producing σ orbitals, or they can combine side by side, producing *pi (π) orbitals*. Therefore, if we combine six 2p orbitals, three from each of two atoms, we shall produce six molecular orbitals: two σ orbitals (one bonding, one antibonding) and four π orbitals (two identical bonding orbitals, two identical antibonding orbitals). The two σ orbitals and two of the four π orbitals are illustrated in Figure 7.22. Notice that the designations are consistent with those we have already discussed. Note also that each bonding orbital links the two nuclei with regions of high electron density, whereas in the antibonding orbitals the electrons are more likely to be found well away from the internuclear axis. Finally, it is useful to notice the similarity between the $\pi 2p$ orbital and the pi cloud formed by the overlap of atomic orbitals. This suggests that in many cases the molecular orbital and atomic orbital approaches will produce similar pictures of covalent bonding, but the two concepts must not be confused.

Let's now look at the energy levels of the six 2p molecular orbitals. We have learned that the atomic 2p orbitals all have the same energy (we say that they are *degenerate*). However, this is obviously not the case with the molecular 2p orbitals. For one thing, the antibonding orbitals must lie at a higher energy level than the bonding orbitals. Furthermore, there is a clear difference between the sigma and pi orbitals. The pair of $\pi 2p$ and the pair of $\pi^* 2p$ orbitals are identical, however, so their energies are degenerate. Thus when we add the 2p orbitals to the 1s and 2s orbitals, we find an energy level diagram like that in Figure 7.23. For some molecules the $\sigma 2p$ orbital lies below the $\pi 2p$ orbitals; in others, it lies above the $\pi 2p$ orbitals. We have shown this uncertainty with the dashed lines in Figure 7.23. (The reason for this variation in energy level is complex, but it involves the mixing of molecular orbitals in some molecules which of course affects their energies and the internuclear distances in the molecules formed.)

We now have an energy level diagram of 10 discrete orbitals, which we can begin to fill with electrons much the way we filled atomic orbitals. The most important difference, however, is that here the diagram represents the orbitals of a diatomic molecule, not the orbitals of isolated atoms. Nonetheless, in arriving at a molecular electron configuration, we follow the same rules we learned when filling atomic orbitals, namely:

1. Electrons fill the orbitals with the lowest energy first.

2. Each orbital can accommodate two paired electrons.

3. When orbitals are degenerate, each receives one electron before any receives two.

Using these rules we can produce the molecular electron configurations shown in Figure 7.24. The figure also includes the total number of electrons in the molecule, the net number of bonding electrons, and the bond order, which can be thought of as the number of bonds between the

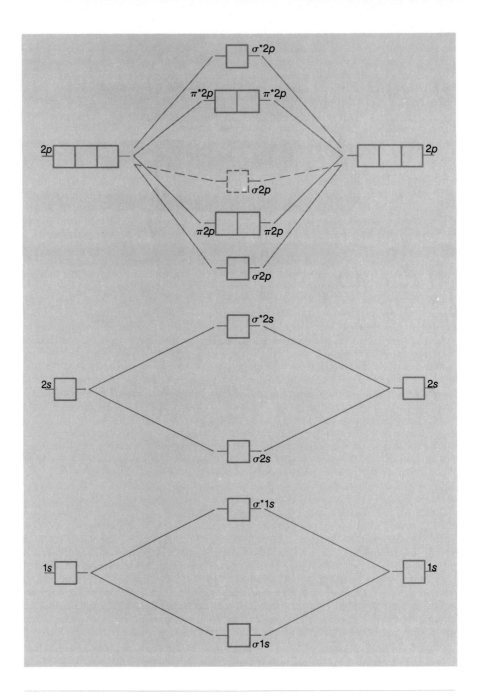

Figure 7.23 Energy levels for molecular orbitals.

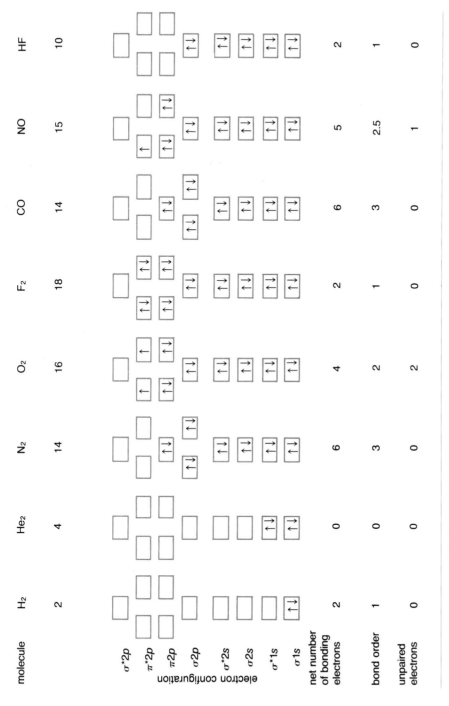

Figure 7.24 Molecular orbital electron configurations for some diatomic molecules.

pair of atoms. The number of electrons in each molecule is simply the sum of the number of electrons in each atom. The net number of bonding electrons is the difference between the number in bonding orbitals and the number in antibonding orbitals. The bond order is the net number of bonding electrons divided by two.

The electron configurations in Figure 7.24 agree remarkably well with experimental observations. For example, the He_2 molecule, with an equal number of bonding and antibonding electrons, does not exist. On the other hand, the figure correctly predicts the stability of the N_2 molecule and that it has one σ and two π bonds, and it agrees with our earlier conclusion that the molecule contains a triple bond. Most impressively, however, it accounts for the two unpaired electrons in O_2 and the weaker paramagnetism of NO. It even shows why NO^+ is a very stable ion and why the superoxide ion, O_2^-, and the peroxide ion, O_2^{2-}, are observed.

Questions

i. For each of the following species, write the molecular orbital electron configuration. Assume for a, d, and f that $\sigma 2p$ is filled before $\pi 2p$. Which of these could exist? Calculate the bond order of each.

    a. CN^- b. He_2 c. HHe^+ d. Ne_2
    e. Li_2 f. NO^+ g. B_2 (paramagnetic)

ii. Which of the following nitrogen oxide molecules are properly termed "odd molecules"?

N_2O NO N_2O_3 NO_2 N_2O_4 N_2O_5

Questions and Problems

7.2 Orbitals and Bonding

6. What postulate of geometry is the basis of the following statement: "Of course any diatomic molecule must be linear"?

7.3 Hybridization of Orbitals

7. Explain why the hybrid orbital theory is necessary to account for the structure of the CH_4 molecule.

8. What experimental evidence regarding the water molecule is explained better by hybrid orbitals than by pure p orbitals?

9. What is the basic postulate of the valence shell electron-pair repulsion theory?

10. The sp^3 orbitals are equivalent, and we can think of an unshared pair of electrons as residing in any one of the four orbitals. The five sp^3d orbitals are not geometrically equivalent, however, and when there is an unshared pair of electrons, it always occupies one of the orbitals directed at the corners of the plane rather than one of the orbitals directed at the apexes of the trigonal bipyramid. Explain.

11. What is meant by hybridization of orbitals?

12. Indicate the distribution of valence electrons around the central atom of the following molecules before and after bond formation.
    a. CH_4 b. BeF_2 c. BF_3 d. PCl_5
    e. SF_4 f. SF_6 g. XeF_4 h. XeF_2

7.4 Geometry of Multiple Bonds

13. Give the formula of a molecule or ion which contains each of the following:
    a. no pi bonds   b. one pi bond   c. two pi bonds   d. double bond
    e. triple bond   f. sp orbitals   g. sp^2 orbitals   h. sp^3 orbitals

14. Distinguish between each member of the following pairs of terms:
    a. sigma bonds and pi bonds   b. polar bonds and polar molecules

15. Explain the absence of rotation about a double bond.

7.5 Resonance and Delocalized Electrons

16. Why is the following electron formula for sulfur dioxide incorrect?

$:\ddot{O}::\ddot{S}:$
    $:\ddot{O}:$

17. How are the orbitals in the outer energy level of the sulfur atom likely to be hybridized in each of the following?
    a. SF_6 b. SF_4 c. SO_2 d. SO_3 e. SO_4^{2-} f. SO_3^{2-}

18. How would you expect the orbitals in the outer energy level of the carbon atom to be hybridized in each of the following?
    a. CH_4 b. CO_2 c. C_2H_4 d. C_2H_2 e. CO_3^{2-}

7.6 Summary of Principles for Predicting Molecular Geometries

19. Determine the type of hybridization, electron-pair geometry, and molecular shape for molecules formed from the following pairs of elements.
    a. germanium and chlorine   b. arsenic and bromine
    c. phosphorus and hydrogen

20. Determine the type of hybridization, electron-pair geometry, and shape for the species listed below.
    a. IO_4^- b. OF_2 c. CO   d. SO_3^{2-} e. SiO_4^{4-}
    f. NH_4^+ g. BrO_3^- h. ICl_4^- i. CO_2 j. $SiCl_4$
    k. NO_2^- l. AsO_4^{3-} m. ClO_3^- n. SO_4^{2-} o. SiF_6^{2-}
    p. BrF_5 q. ClF_3 r. PF_5 s. O_3 t. XeF_4
    u. CO_3^{2-} v. PO_4^{3-} w. BrF_4^- x. HCN   y. SiF_4

21. Indicate whether the following molecules are polar or nonpolar.
    a. CBr_4 b. XeF_2 c. CO   d. HCN   e. NH_3
    f. SO_2 g. BF_3 h. Cl_2O i. H_2S j. SF_6

Discussion Starter

22. Explain how a molecule may contain polar bonds and yet be a nonpolar molecule.

Gases

In the first seven chapters of this book we have discussed the structure of atoms, we have seen how that structure influences the properties of the elements, and we have arranged the elements in a periodic table according to their electron configurations. Finally we outlined the principles that dictate how atoms combine to form stable compounds. In short we have laid the foundation for the task of investigating and explaining how matter behaves in our world. To continue, however, we must change our point of view somewhat. Instead of looking closely at the properties of individual atoms, which are the building blocks for all matter, we shall now take a step back to ask what **physical states** matter takes, how these states differ, and what properties each possesses.

Our first observation may seem a prosaic one but it is a necessary one, namely, that we find matter in three forms: solid, liquid, and gas. A practical description of any substance must therefore describe which of these physical states we shall find it in under ordinary conditions or, more specifically, at room temperature and pressure. Under these conditions, we know that metals such as iron or copper are solid, gasoline and water are liquid, and the air around us is a gas. We also know, however, that we can find water as a solid (ice) or a gas (steam). Indeed, all substances undergo these same changes of state, given the appropriate conditions. Thus as we study each physical state, we may assume that we are studying phenomena common to all elements and compounds, even though we can only rarely expect to encounter solid gasoline or gaseous iron.

We shall begin in earnest with an explanation for why we have chosen gases as a starting point. They are after all the least visible and hence the least familiar of all substances. Nevertheless it is best to begin here since gases are far less complicated than liquids and solids. The interactions among the atoms and molecules in the more *condensed* states can be complex; in gases the atoms and molecules are farther apart and the interactions are easier to understand. Thus we begin with gases because they will provide a firm basis on which to build our study of liquids and solids.

But how do we study the properties of a substance that scarcely seems to be a substance? How do we measure its volume if it promptly fills any container we put it in? And how do we weigh a substance, indeed how do we describe it at all, if we can't see it, feel it, or smell it? To answer these questions, let us take a historical approach. In so doing we find that much was known about the properties of solids and liquids long before serious work commenced on the gases. After all, the solids and liquids can be manipulated and experimented upon, all the time displaying visible changes. In the seventeenth century, the attention of some finally turned to the study of air.

8.1 Air, the Atmosphere, and Atmospheric Pressure

Seventeenth-century scientists were concerned with the problem of pumping water from a well. No matter how constructed, no pump was able to raise water more than approximately 10 meters. Galileo Galilei (1564–1642), Italian scientist and philosopher, began experiments which were continued after his death by a student of his, Evangelista Torricelli (1608–1647). The action of the pumps had been explained with the saying, "Nature abhors a vacuum." It was claimed that when the air in the pipe was removed by the pump, the water was then "drawn up" into the vacuum created. But why didn't the water continue to rise in the evacuated pipe regardless of its length? Torricelli provided the answer when he proposed that "we live at the bottom of an ocean of air," explaining that the water in the pipe was *pushed up* by *atmospheric pressure* rather than *drawn up* by the vacuum. This atmospheric pressure was simply the force (per unit area) exerted on the surface of the water in the well by the weight of the gaseous atmosphere above it. Torricelli went on to reason that since mercury is 13.6 times as dense as water, atmospheric pressure would support a column of mercury only 1/13.6 as high as water. He therefore constructed the first *barometer* by filling a glass tube (closed at one end) with mercury and then inverting it in a container of mercury. As predicted, the level of the mercury fell to 0.76 m (Figure 8.1). At that height the pressure of the column of mercury on the mercury in the open container was exactly balanced by the pressure of the air. A few years later, at the suggestion of Blaise Pascal (1623–1662), the French mathematician and philosopher, a similar barometer was carried up a mountain in France to show that atmospheric pressure decreases with altitude. As a result, Pascal was able to affirm that it is the weight of our atmosphere that supports columns of water or mercury in evacuated tubes. A dramatic demonstration of the force of the atmosphere was made about 1654 by Otto Von Guericke (1602–1686), who was the inventor of the vacuum pump. Using his pump to evacuate the volume between two copper hemispheres, he was able to put on a crowd-pleasing show in Magdeburg, Germany (Figure 8.2). He hitched a team of horses to each hemisphere and astonished the crowd by showing that the horses were unable to separate the hemispheres. When he opened a valve that readmitted air to the evacuated volume, the hemispheres fell apart of their own weight.

Today scientists use several different units to measure pressure. But as we have seen, pressure is force per unit area, so let's look at units of force before considering units of pressure. By definition, force is mass times acceleration:

$$F = ma$$

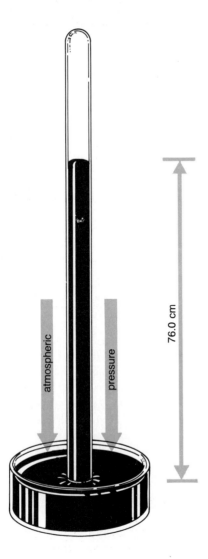

Figure 8.1 A barometer.

The SI unit of force is the *newton* (**N**), which is the force of a 1-kilogram mass being accelerated at 1 meter per second per second:

$$N = \frac{1 \text{ kg m}}{s^2}$$

Therefore, the SI unit of pressure is the *pascal* (**Pa**), which is the pressure due to 1 newton of force acting on an area of 1 square meter:

$$Pa = \frac{N}{m^2}$$

Figure 8.2 Von Guericke's hemispheres.

A more familiar unit, found for example on air-pressure gauges for automobile tires, is the pound per square inch (psi). Still another unit is the most familiar one to scientists: Since the mercury barometer is such a convenient device for measuring pressure, it is often natural to express pressure in millimeters or centimeters of mercury. In honor of Torricelli, we call a millimeter of mercury a *torr*. Meterologists also measure pressure with a mercury barometer, but they usually report its height in inches: "The pressure today is 29.34 and rising." A final unit of pressure is the *atmosphere*, which is especially convenient when we deal with extremely high pressures. One atmosphere is equal to 760 mm of Hg, a value that has been adopted as *standard pressure*. Table 8.1 summarizes several

Table 8.1 Units of Pressure

Name	Abbreviation	Standard Pressure
Pascals	Pa	1.013×10^5 Pa
Newtons per square meter	N/m^2	1.013×10^5 N/m^2
Pounds per square inch	psi	14.7 psi
Millimeters of mercury	mm of Hg	760 mm of Hg
Inches of mercury	in. of Hg	29.9 in. of Hg
Torrs	torr	760 torr
Atmospheres	atm	1.00 atm

units of pressure and gives for each the value of standard pressure. The most important units to remember are the pascal, the torr, the atmosphere, and millimeters of mercury.

Question

1. Consider the following pairs of barometers, identical in all respects except the one given. In which of each pair will the liquid level be higher?
    a. Tube A is in a partially evacuated air chamber; tube B is outside the chamber.
    b. Tube A is on a mountaintop; tube B is at sea level.
    c. Tube A is a mercury barometer; tube B is filled with a less dense fluid.
    d. The cross-sectional area of tube A is twice that of tube B.

8.2 Qualitative Descriptions of Gases

Before we look in some detail at the laws that govern the behavior of gases, we should review some of their more familiar qualitative properties. For example, all but a few gases are colorless. The familiar gases of the atmosphere—nitrogen, oxygen, carbon dioxide, and water vapor—as well as hydrogen and carbon monoxide, are all invisible. Most that we think we can see are not gases at all. When we "see" steam, we really see a suspension of droplets of liquid water. Smoke is not visible because of the gases it might contain but because of the solids or liquids carried along by the gases. Soot (carbon), ash (oxides), or droplets of water thus produce clouds of black, gray, or white smoke. The few gases that are colored are seldom seen outside the laboratory. Chlorine, for example, is a yellow-green gas and nitrogen dioxide is reddish brown.

When we turn to another of our senses the story is different: Many gases have characteristic odors. In fact, all odors are due to gases since it is only as vapor that substances reach our olfactory apparatus. The odor that we ascribe to a liquid or solid is instead the odor of the gas rising from it. For example, the smell of the household ammonia solution is actually the odor of the gas, NH_3. Another characteristic smell, around the chemistry lab as well as near leaking sewers, is the "rotten eggs" odor of H_2S. However, many common gases have no odor: Oxygen, nitrogen, hydrogen, carbon monoxide, carbon dioxide, and water vapor are odorless as well as colorless.

Many common gases have other physiological effects as well. Ammonia not only has a characteristic odor but also is a powerful lachrymator, stimulating the tear ducts. Other gases irritate the mucous membranes of the nose and throat, causing a choking sensation. An example is the industrial effluent, sulfur dioxide. Some of these qualitative characteristics of common gases are summarized in Table 8.2.

Since none of these physiological properties is common to all gases, however, they do not help us distinguish gases from solids and liquids. To do this we

Table 8.2 Colors and Odors of Some Common Gases

Name	Formula	Color	Odor	Physiological Effect
Elements				
Hydrogen	H$_2$	Colorless	Odorless	
Nitrogen	N$_2$	Colorless	Odorless	
Oxygen	O$_2$	Colorless	Odorless	Supports life
Fluorine	F$_2$	Pale greenish yellow	Pungent, irritating	Poisonous
Chlorine	Cl$_2$	Yellow-green	Pungent, Irritating	Poisonous
Noble gases	(He, Ne, Ar, Kr, Xe, Rn)	Colorless	Odorless	
Compounds				
Carbon monoxide	CO	Colorless	Odorless	Poisonous
Carbon dioxide	CO$_2$	Colorless	Odorless	
Nitrous oxide	N$_2$O	Colorless	Mild, sweet	Anesthetic
Nitrogen dioxide	NO$_2$	Reddish brown	Pungent, sweet	Poisonous
Propane	C$_3$H$_8$	Colorless	Almost odorless	
Ammonia	NH$_3$	Colorless	Sharp	Lachrymator
Hydrogen sulfide	H$_2$S	Colorless	Rotten eggs	Poisonous
Sulfur dioxide	SO$_2$	Colorless	Sharp	Irritant
Hydrogen chloride	HCl	Colorless	Sharp	Irritant
Hydrogen cyanide	HCN	Colorless	Bitter almonds	Poisonous

can turn to density. The densities of gases, under ordinary conditions, range from less than 1×10^{-4} g/cm^3 to almost 1×10^{-2} g/cm^3. For solids and liquids, the range is between 0.3 and 21 g/cm^3. Thus at room temperature and pressure, we should be able to distinguish the densest gas from the least dense liquid.

Some qualitative properties of gases are also characteristic. Gases *diffuse*; that is, they spread out to occupy completely any container. If we place an open container of ammonia in one corner of a room, we soon notice the gas throughout the room. Gases also expand when heated at constant pressure and contract when cooled—hot air ballons make use of this principle. Solids and liquids also show this property of *thermal expansion* and *contraction* but to such a slight extent that we may infer an entirely different mechanism. Finally, gases can be *compressed* if we apply pressure. Liquids and solids are practically incompressible.

Thus we can describe gases as being less dense than other states of matter and as having properties of diffusion, compressibility, and thermal expansion and contraction. The explanations for some of these qualitative observations will become clear as we now turn to the quantitative properties of gases.

Question

2. Although odor is not a useful property for distinguishing one physical state from another, it sometimes helps us distinguish one substance from another. What gases can you recognize by odor?

8.3 The Gas Laws

The mathematical statements that summarize the relationships among the volume, temperature, and pressure of a gas are called *gas laws*. Although these laws have now been given a theoretical basis, they were first derived from simple observations of the behavior of gases. The first, formulated by Robert Boyle (1627–1691) and hence called **Boyle's law,** describes the relationship between the volume and the pressure of a gas.

Boyle took a curved glass tube, closed at only one end, and poured mercury into the open end until it stood at the same level in each arm of the tube, as shown in Figure 8.3a. Since the two levels were the same, he assumed that the

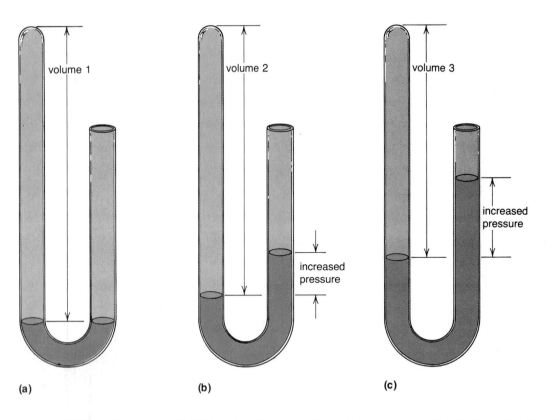

Figure 8.3 Boyle's pressure-volume experiment.

gas (air) in the closed portion of the tube was at the same pressure as the air in the open arm of the tube, that is, at atmospheric pressure. With the confined gas as the "sample" for his experiment, Boyle then subjected it to changes in pressure. He poured more mercury into the tube and observed the changes in the two levels (Figure 8.3b). He was able to measure the reduced volume of the sample, and he could calculate the pressure on the sample as the difference between the mercury levels plus the atmospheric pressure. After a series of measurements, Boyle's data looked much like those in Table 8.3.

Table 8.3 Pressure-Volume Data for a Confined Gas (temperature constant)

Volume of Gas (ml)	Difference in Mercury Levels (cm)	Total Pressure (cm of Hg)	V × P (ml × cm of Hg)
50.00	0	76.00	3.800×10^3
43.93	9.95	85.95	3.776×10^3
39.45	20.15	96.15	3.793×10^3
35.54	30.48	106.48	3.784×10^3
32.77	39.81	115.81	3.795×10^3

A graph of this sample data appears in Figure 8.4. As the pressure increases, the volume of the gas decreases. The relationship can be represented algebraically by either of the following statements:

$$V = k \left(\frac{1}{P}\right)$$ *Volume is inversely proportional to pressure.* (k is a proportionality constant.)

or

$$PV = k$$ *The product of the volume and the pressure is constant.* ($P_1 V_1 = P_2 V_2 = P_3 V_3 =$ etc.)

The last column of Table 8.3 convincingly demonstrates this relationship. The product of V and P varies only slightly over the range of pressures measured, and the variation is within experimental error. If we do a similar experiment at much higher pressures, however, the variation is more pronounced. Thus it is only a theoretically perfect gas, which we call an *ideal gas*, that maintains a constant PV product under all conditions. Nonetheless, under ordinary conditions, deviations from *ideal gas behavior* are small, and Boyle's law is an accurate if not rigorous generalization: *The volume of a confined gas at a constant temperature is inversely proportional to the pressure.*

Examples 8.1 and 8.2 illustrate some applications of Boyle's law.

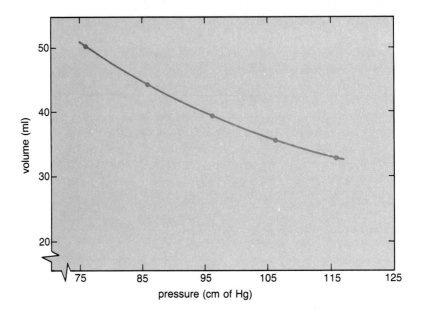

Figure 8.4 The relationship between the volume and the pressure of a confined gas at constant temperature (Boyle's law).

Example 8.1 A sample of gas occupies 10.0 liters at 740 mm of Hg. If the temperature is constant, what will its volume be at 770 mm of Hg?

Solution

a. Since according to Boyle's law $P_1 V_1 = P_2 V_2$, we can easily calculate the new volume V_2 if the other three quantities are given:

$$V_2 = \left(\frac{P_1}{P_2}\right) V_1$$

However, in this first problem, we prefer to take a more "reasoned" approach.

b. Note first the *kind* of change we expect in the volume. Will it increase or decrease? Since volume and pressure are inversely proportional, an *increase* in pressure will produce a *decrease* in volume.

c. The pressure increase is represented by the fraction

$$\frac{770 \text{ mm of Hg}}{740 \text{ mm of Hg}}$$

Thus the volume decrease is represented by the fraction

$$\frac{740 \text{ mm of Hg}}{770 \text{ mm of Hg}}$$

d. The new volume will therefore be

$$(10.0 \text{ liters}) \left(\frac{740 \text{ mm of Hg}}{770 \text{ mm of Hg}} \right) = 9.61 \text{ liters}$$

Example 8.2 Meterological balloons that are sent to high altitudes are only partially inflated when released since they will expand at the lower pressures of the upper atmosphere. If the volume of a balloon at 1.00 atm is 1000 m³, at what pressure will its volume be 32,000 m³? Assume that the temperature is constant.

Solution

a. The volume will increase by a factor of 32:

$$\frac{32,000 \text{ m}^3}{1,000 \text{ m}^3}$$

Since volume and pressure are inversely proportional, we expect the pressure to decrease by the same factor.

b. Thus when the volume of the balloon is 32,000 m³, the atmospheric pressure will be

$$(1.00 \text{ atm}) \left(\frac{1,000 \text{ m}^3}{32,000 \text{ m}^3} \right) = 0.0312 \text{ atm}$$

Boyle discovered the relationship between the pressure and volume of a gas in the seventeenth century. It was not until late in the eighteenth century, however, that two French physicists independently discovered the part that temperature plays in the gas laws. Both Jacques Charles (1746–1823) and Joseph Gay-Lussac (1778–1850) discovered the relationship between volume and temperature that we now know as **Charles' law.**

We have already observed that a gas confined at constant pressure expands when it is heated. To quantify this observation, Charles and Gay-Lussac made the kinds of measurements summarized in Table 8.4. The volumes of 10-g samples

Table 8.4 Volume-Temperature Data for Several Gas Samples at 1.00 atm

Temperature (°C)	Volume of 10.0-g of O_2 (liters)	Volume of 10.0-g of CO_2 (liters)	Volume of 10.0-g of N_2 (liters)
60	8.54	6.21	9.76
30	7.77	5.65	8.88
0	7.00	5.09	8.00
−30	6.23	4.53	7.12
−60	5.46	3.97	6.24

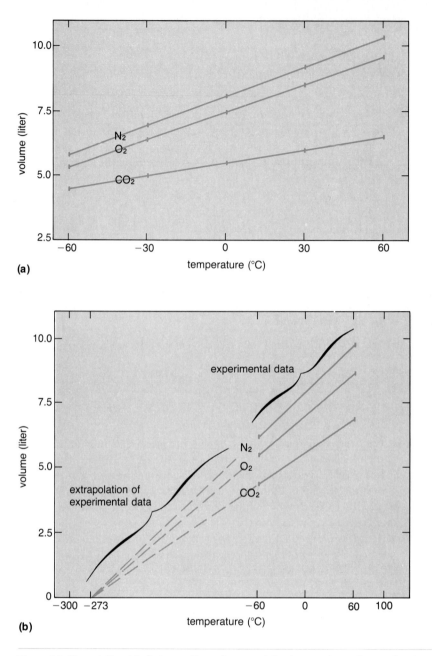

Figure 8.5 The relationship between the volume and the temperature of a confined gas at constant pressure **(a)**. In **(b)** the graph has been extended to 0 volume.

of several gases are tabulated for a range of temperatures, with the pressure constant at 1.00 atm. A plot of these data appears in Figure 8.5. We can now draw several conclusions:

1. As we have already seen, when temperature increases, volume increases.

2. For each gas the relationship between temperature and volume is linear.

3. If all three lines are extended to zero volume, they each intersect the temperature axis at approximately $-273°C$.

We may therefore make the following statement of the relationship between Celsius temperature and volume:

$$V = k(°C + 273) \quad \text{(pressure constant)}$$

Again, k is a constant. This relationship requires that as the temperature approaches $-273°C$, the volume of gas approaches zero. It should therefore not surprise us to find that real gases deviate from this ideal behavior at low temperatures. Indeed, all known substances liquefy well above $-273°C$. Nonetheless, we can extend our definition of an ideal gas by insisting that it obey this relationship under all conditions. Furthermore, all real gases obey it closely under ordinary conditions. The temperature of $-273°C$ (or more precisely $-273.16°C$) becomes the firm basis of the Kelvin temperature scale. Zero kelvins, which we call *absolute zero*, is the temperature at which a substance has lost all thermal energy. For more complicated theoretical reasons, this temperature is in principle unattainable. Using the Kelvin scale, we can now state Charles' law as a simple proportion:

$$V = kT \quad \text{\textit{The volume of a sample of gas at constant pressure is directly proportional to the temperature in kelvins.}}$$

or

$$\frac{V}{T} = k \quad \text{\textit{The quotient } V/T \textit{ for a sample of gas at constant pressure is constant.}}$$

Standard pressure has already been designated as 1.00 atm. *Standard temperature* is designated as $0.0°C$ or 273 K. Together, 273 K and 1.00 atm are known as *standard conditions* of temperature and pressure and often abbreviated STP.

Example 8.3 A balloon contains 12.0 m³ of helium at 6:00 A.M. when the temperature is 20°C. If atmospheric pressure remains unchanged, what will the volume of the balloon be at noon when the temperature has risen to 28°C?

Solution

a. Express both temperatures on the Kelvin scale:

$$20°C + 273 = 293 \text{ K}$$

$$28°C + 273 = 301 \text{ K}$$

b. Since volume and temperature in kelvins are directly proportional, an increase in temperature must be accompanied by a proportional increase in volume. Thus both the temperature increase and the volume increase are represented by the ratio

$$\frac{301 \text{ K}}{293 \text{ K}}$$

c. The new volume will be

$$(12.0 \text{ m}^3)\left(\frac{301 \text{ K}}{293 \text{ K}}\right) = 12.3 \text{ m}^3$$

Finally, we can determine experimentally the relationship between the pressure and temperature of a gas at constant volume. We find *the pressure and the Kelvin temperature of a gas at constant volume are directly proportional* ($P = kT$). We can summarize our findings in one algebraic statement:

a. Pressure and volume are inversely proportional.

$$PV = k_1$$

b. Pressure and Kelvin temperature are directly proportional.

$$\frac{P}{T} = k_2$$

c. Volume and Kelvin temperature are directly proportional.

$$\frac{V}{T} = k_3$$

or

$$\frac{PV}{T} = k$$

Example 8.4 If a sample of gas has a volume of 2.34 ml at $-60°C$ and 2.00 atm, what will its volume be at STP?

Solution

a. Volume and Kelvin temperature are directly proportional; therefore, an increase in volume must accompany the increase in temperature from $-60°C$ to $0.0°C$. Thus we can represent the increase in volume by the ratio

$$\frac{273 \text{ K}}{213 \text{ K}}$$

b. Volume and pressure are inversely proportional. A decrease in pressure will thus tend to increase the volume. Since the pressure decreases by the ratio

$$\frac{1.00 \text{ atm}}{2.00 \text{ atm}}$$

the volume must increase by the ratio

$$\frac{2.00 \text{ atm}}{1.00 \text{ atm}}$$

c. We now find the new volume by multiplying the original volume by the ratio that accounts for the temperature change and by the ratio that accounts for the pressure change:

$$(2.34 \text{ ml}) \underbrace{\left(\frac{273 \text{ K}}{213 \text{ K}}\right)}_{\text{temperature effect}} \underbrace{\left(\frac{2.00 \text{ atm}}{1.00 \text{ atm}}\right)}_{\text{pressure effect}} = 6.00 \text{ ml}$$

Questions

3. Calculate the new volume for 1.00 liter of a gas at constant temperature if
   a. the pressure is increased from 1.00 atm to 2.00 atm.
   b. the pressure is reduced from 1.00 atm to 0.333 atm.
   c. the pressure is increased from 760 mm of Hg to 1140 mm of Hg.
4. Calculate the resulting volume at constant pressure if
   a. the temperature of a 10.0-liter sample of gas increases from $0.0°C$ to $(1.00 \times 10^2)°C$.
   b. the temperature of a 10.0-liter sample of gas drops from 373 K to 273 K.
   c. the temperature of a 10.0-liter sample of gas changes from $427°C$ to (3.00×10^2) K.

8.4 Combining Volumes and the General Gas Law Equation

To extend our understanding of the behavior of gases, we must now turn to their chemical properties. Again, some of the pioneering work was done by Joseph Gay-Lussac, who studied the *combining volumes of gases*. Gay-Lussac noticed that when he passed a spark through a mixture of oxygen and hydrogen, causing them to combine explosively, they always combined in a ratio of two volumes of hydrogen to one volume of oxygen. In another experiment he found that two volumes of carbon monoxide always combined with one volume of oxygen to form two volumes of carbon dioxide. He determined that ammonia combined in a 1:1 volume ratio with either of the gases HCl or BF_3. Although other scientists of the time accused him of rounding off his values to produce the observed whole-number ratios, he summarized his observations in the **law of combining volumes:** *Volumes of gases react with each other in simple whole-number ratios.*

As we shall see, the implications of this law are profound. However, many scientists, including John Dalton, rejected it. However, within 3 years following the publication of Gay-Lussac's work, an Italian physicist, Amedeo Avogadro (1776–1856), proposed an explanation. In exchange for his insight, the scientific world refused to accept his theory for almost 50 years.

The first of Avogadro's assumptions, now called **Avogadro's hypothesis,** was that *equal volumes of gases, under identical conditions of temperature and pressure, contain the same number of reactive particles.* His second assumption was that the *elemental gases are diatomic.* (The noble gases were still unknown.) If equal volumes of ammonia and hydrogen chloride react to form a salt and if hydrogen reacts with fluorine liter for liter, it seems reasonable that there must be as many particles of ammonia as of hydrogen chloride in each liter and as many of hydrogen as of fluorine. To explain that one volume of hydrogen reacts with one volume of fluorine to produce *two* volumes of hydrogen fluoride, it is necessary to assume that the reacting particles of the gases must divide into two parts. Thus we now describe *molecules* of most of the elemental gases as consisting of two atoms: O_2, N_2, F_2, Cl_2, and H_2.

Since we have already discussed the stoichiometry of chemical reactions, we know that a mole of any substance contains the same number of atoms or molecules. Thus, Avogadro's hypothesis implies that a mole of any gas should occupy the same volume. This volume, known as the **molar volume,** is equal to 22.4 liters at STP. Another experimental quantity that we should recall from Chapter 3 is the number of molecules in a mole of any substance. Called *Avogadro's number*, its value is 6.02×10^{23}. To summarize these quantities and relationships, one mole of O_2 weighs 32.0 g, contains 6.02×10^{23} molecules, and occupies a volume of 22.4 liters at STP.

We can now include the number of moles in our equation describing the behavior of an ideal gas. Since the volume of a gas (as well as of a liquid or solid) at constant temperature and pressure is directly proportional to the number of moles (or molecules), the equation becomes

$$\frac{PV}{T} = kn$$

where n stands for the number of moles. The value of the proportionality constant in this equation can be determined by substituting the values for temperature and pressure at which 1.00 mol of a gas occupies a volume of 22.4 liters, namely, 273 K and 1.00 atm.

$$k = \frac{PV}{Tn} = \frac{(1.00 \text{ atm})(22.4 \text{ liters})}{(273 \text{ K})(1.00 \text{ mol})}$$

$$k = 0.0821 \frac{(\text{liter})(\text{atm})}{(\text{mol})(\text{K})}$$

This constant is generally represented by the letter R and called the *gas law constant*. The equation $PV = nRT$ is called the *general gas law equation* or the *ideal gas law equation*.

Example 8.5 A large balloon is to be inflated with helium at constant temperature and pressure. If 52.0 g of helium inflate the balloon to a volume of 2.90×10^2 liters, how much nitrogen must be added to inflate it to a volume of 1.00×10^3 liters?

Solution

a. The fraction that represents the volume change is

$$\frac{1.00 \times 10^3 \text{ liters}}{2.90 \times 10^2 \text{ liters}}$$

Since volume and quantity are directly proportional, this fraction also represents the quantity change.

b. The number of moles of gas originally in the balloon was

$$(52.0 \text{ g}) \left(\frac{1.00 \text{ mol}}{4.00 \text{ g}} \right) = 13.0 \text{ mol}$$

c. The quantity of gas necessary to inflate the balloon to a volume of 1.00×10^3 liters is

$$(13.0 \text{ mol}) \left(\frac{1.00 \times 10^3 \text{ liters}}{2.90 \times 10^2 \text{ liters}} \right) = 44.8 \text{ mol}$$

d. The quantity of nitrogen to be added is

$$\begin{array}{r} 44.8 \text{ mol gas} \\ - 13.0 \text{ mol He} \\ \hline 31.8 \text{ mol gas} \end{array}$$

e. The weight of nitrogen to be added is

$$(31.8 \text{ mol}) \left(\frac{28.0 \text{ g}}{1.00 \text{ mol}} \right) = 8.90 \times 10^2 \text{ g}$$

Example 8.6 A sample of a gas with a mass of 55.7 g occupies 14.3 liters at 20°C and 740 mm of Hg. Calculate the molecular weight of the gas.

Solution

a. Since the molecular weight of a substance is numerically the same as the weight in grams of one mole, we must first determine the number of moles in this sample.

b. At standard conditions, 22.4 liters of the gas would contain 1.00 mol. We may now calculate the effect on this quantity as we impose the conditions specified in the problem.

c. The quantity will change in proportion to the volume:

$$\frac{14.3 \text{ liters}}{22.4 \text{ liters}}$$

d. The quantity will change inversely with the temperature:

$$\text{temperature increase}: \frac{293 \text{ K}}{273 \text{ K}}$$

$$\text{quantity decrease}: \frac{273 \text{ K}}{293 \text{ K}}$$

e. Finally, the quantity will change in proportion to the pressure:

$$\frac{740 \text{ mm of Hg}}{760 \text{ mm of Hg}}$$

f. Thus, the number of moles in the sample is

$$(1.00 \text{ mol}) \left(\frac{273 \text{ K}}{293 \text{ K}}\right) \left(\frac{740 \text{ mm of Hg}}{760 \text{ mm of Hg}}\right) \left(\frac{14.3 \text{ liters}}{22.4 \text{ liters}}\right) = 0.579 \text{ mol}$$

g. If the sample weighs 55.7 g and contains 0.579 mol, the weight of one mole is

$$\frac{55.7 \text{ g}}{0.579 \text{ mol}} = 96.2 \text{ g/mol}$$

Question

5. In each of the following cases, tell whether the volume of the gas will increase, decrease, or remain the same. If it is impossible to predict, say so.
    a. Pressure increases, temperature and quantity decrease.
    b. Pressure and temperature both increase, and quantity decreases.
    c. Pressure, temperature, and quantity increase.
    d. Pressure increases, and temperature and quantity remain constant.
    e. Temperature decreases, and pressure and quantity remain constant.

f. Quantity increases, and pressure and temperature remain constant.
g. Pressure and temperature increase, and quantity remains constant.
h. Pressure increases, temperature decreases, and quantity remains constant.

8.5 Stoichiometry and Gas Volumes

Avogadro's hypothesis, embodied in the ideal gas law equation, tells us that we can study the stoichiometry of reactions in the gas phase in terms of the volumes of the reactants and products. For example, we can read the equation

$$N_2(g) + 3 H_2(g) \longrightarrow 2 NH_3(g)$$

several ways. We can say that one mole (or one molecule) of nitrogen reacts with three moles (or three molecules) of hydrogen to produce two moles (or two molecules) of ammonia. But since molar ratios are also volume ratios, we can also say that, at constant temperature and pressure, 1 liter of nitrogen reacts with 3 liters of hydrogen to produce 2 liters of ammonia.

Example 8.7 At STP what volume of oxygen is required for the complete combustion of 35.6 liters of methane (CH_4)?

Solution
a. The equation for the complete combustion of methane is

$$CH_4(g) + 2 O_2(g) \longrightarrow CO_2(g) + 2 H_2O(l)$$

b. Thus each liter of CH_4 reacts with 2 liters of O_2. The conversion factor we need is

$$\left(\frac{2 \text{ liters } O_2}{1 \text{ liter } CH_4}\right)$$

c. For the complete combustion of 35.6 liters of methane, we need

$$(35.6 \text{ liters } CH_4) \left(\frac{2 \text{ liters } O_2}{1 \text{ liter } CH_4}\right) = 71.2 \text{ liters } O_2$$

8.6 Dalton's Law of Partial Pressures

The relationship between the quantity and pressure of a gas allows us to draw some important conclusions about the properties of one gas in a mixture of several. We know that 1.00 mol of O_2 in a 22.4-liter container at 0.0°C exerts

Gases

a pressure of 1.00 atm and that a second mole will increase the pressure to 2.00 atm. In fact, we shall measure the same total pressure regardless of whether the second mole is O_2 or some other gas, as long as no chemical reaction occurs between the two gases. This observation leads to John Dalton's **law of partial pressures:** *In a mixture of gases, each gas exerts the same pressure it would exert if it alone occupied the container.* We call this pressure the *partial pressure*. For example, in Figure 8.6a, two 1.00-liter vessels are separated by a partition. Container A is filled with a gas at 1.00 atm, and container B is filled with a different gas at 2.00 atm. We shall assume that they are at the same temperature. In Figure 8.6b, the partition has been removed. The two gases mix and fill the entire container, but let's look at them separately. Gas *A* expands to twice its original volume; therefore its partial pressure is reduced by one-half to 0.500 atm. The same thing happens to gas *B*: It now occupies a 2.00-liter container; thus its partial pressure is 1.00 atm. The total pressure P_T in the system is simply the sum of the partial pressures, or 1.50 atm. In general, for a mixture of unreactive gases;

$$P_T = P_A + P_B + P_C + \cdots$$

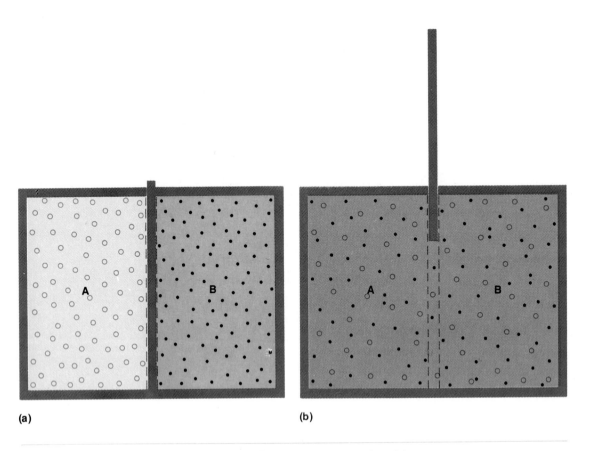

Figure 8.6 Schematic illustration for Dalton's law of partial pressures.

where P_T is the total pressure and P_A, P_B, etc., are the partial pressures of the components.

Example 8.8 Consider two containers similar to those in Figure 8.6, one with a volume of 0.200 liter and the other with a volume of 0.300 liter. The first container is filled with gas A at a pressure of 740 mm of Hg; the second is filled with gas B at 700 mm of Hg. If the connecting valve is opened, and if no interaction between the gases takes place, what will be the final partial pressures of the two gases and what will be the final pressure in the system?

Solution

a. Gas A will increase in volume from 0.200 liter to 0.500 liter. We represent this volume change with the fraction

$$\frac{0.500 \text{ liter}}{0.200 \text{ liter}}$$

Since pressure is inversely proportional to volume, we represent the drop in pressure by

$$\frac{0.200 \text{ liter}}{0.500 \text{ liter}}$$

b. The new pressure of gas A is

$$(740 \text{ mm Hg}) \left(\frac{0.200 \text{ liter}}{0.500 \text{ liter}} \right) = 296 \text{ mm of Hg}$$

c. Following the same reasoning for gas B, we find its final partial pressure to be

$$(700 \text{ mm of Hg}) \left(\frac{0.300 \text{ liter}}{0.500 \text{ liter}} \right) = 420 \text{ mm of Hg}$$

The final pressure, at any point in the system, is the sum of the two partial pressures:

$$296 \text{ mm of Hg} + 420 \text{ mm of Hg} = 716 \text{ mm of Hg}$$

8.7 The Kinetic Molecular Theory

This chapter so far has dealt with gas laws that were derived originally from empirical observations. These "laws" are perfectly true only for the hypothetical ideal gas, although real gases under ordinary conditions do not deviate from them significantly. To reinforce our faith in these laws we shall now turn to a general theory of gas behavior known as the *kinetic molecular theory*. This theory, which is widely regarded as offering the best explanation for the behavior of ideal gases, can be summarized as follows:

A. *All gases are composed of molecules.* (In the case of the noble gases, the molecule is the lone atom.)
  1. The volume occupied by these molecules is negligible compared to the total volume occupied by the gas.
  2. Attractions between molecules in the gas phase are negligible.
  3. The molecules of a gas are in constant, rapid, random motion, traveling in straight lines until they encounter other molecules or the walls of the container.
  4. All collisions are perfectly elastic.

B. *The velocities of the molecules of a gas depend on the temperature, and at the same temperature the average kinetic energies of the molecules of all gases are the same.*

The first set of statements says that gases may be considered to be free of most of the complications of liquids and solids: The molecules are far apart, they are in rapid motion, and they interact only by colliding elastically. The constant, rapid, random, straight-line motion of the molecules explains the property of all gases of diffusing to fill any container. The relatively large distances between molecules in the gas phase explains the compressibility of gases. By assuming that differences in sizes of molecules and variations in attractions between molecules are negligible in the gas phase, the theory explains the *general* applicability of the gas laws. It is not that the molecules of different gases are not different, but because they are so far apart, the differences do not appreciably affect the behavior of gases (at ordinary conditions) in response to changes in volume, temperature, pressure, and quantity. An elastic collision is one in which there is *no net loss of energy*. If energy were lost during collisions—for example, to the environment outside the container—the gas might eventually collapse to the bottom of the container.

When energy in the form of heat is added to a sample of a gas, it results on a molecular level in an increase in the velocity of the molecules. If the gas is confined in a rigid container, this means that the number of collisions per second between gas molecules and the walls of the container will increase. We sense or measure this as an increase in pressure. On the other hand, if the sample is confined in a container at constant pressure, the increase in the rate of collisions with the walls of the container will result in an increase in volume.

To understand the second part of statement B completely, more must be said about kinetic energy. As pointed out in Chapter 1, kinetic energy is the energy of motion. It depends only on the mass and the velocity of the object in motion. It is easy to see how both of these factors play a part in the energy of an object: A slowly moving bowling ball can do as much work (say, on a window pane) as a much faster golf ball. In mathematical terms,

$$ke = \tfrac{1}{2}mv^2$$

where m is mass and v is velocity. Statement B, therefore, says that at a given temperature a heavy molecule will be traveling more slowly than a light one. In fact, their masses and velocities must be such that their kinetic energies are equal.

Question

6. What do we mean by
   a. an elastic collision?
   b. an ideal gas?
   c. a partial pressure?

8.8 Molecular Velocities

We would now like to see how molecular velocities (and thus diffusion rates) depend on the masses of the molecules. From the definition of kinetic energy, we know that if the kinetic energies of two molecules are equal, the lighter molecule must be moving faster. We can work out how much faster as follows. Let the subscripts denote the average values for the molecules in gas a and gas b.

$$ke_a = \tfrac{1}{2} m_a v_a^2 \quad \text{and} \quad ke_b = \tfrac{1}{2} m_b v_b^2$$

and since

$$ke_a = ke_b$$

then

$$\tfrac{1}{2} m_a v_a^2 = \tfrac{1}{2} m_b v_b^2$$

or

$$\frac{m_a}{m_b} = \frac{v_b^2}{v_a^2}$$

Therefore, under identical conditions, the molecular masses of two gas samples will be inversely proportional to the squares of their velocities. In the more common form below, this relationship is known as **Graham's law:**

$$\frac{\sqrt{m_a}}{\sqrt{m_b}} = \frac{v_b}{v_a}$$

Under the same conditions, a mole of gas a will contain exactly as many molecules as a mole of gas b. Thus the ratio of the molecular weights of the two pure gases, as well as the ratio of their densities, are the same as the ratio of their molecular masses. Therefore,

$$\frac{\sqrt{mw_a}}{\sqrt{mw_b}} = \frac{v_b}{v_a} \quad \text{and} \quad \frac{\sqrt{d_a}}{\sqrt{d_b}} = \frac{v_b}{v_a}$$

where mw is molecular weight and d is density.

Example 8.9 Compare the relative molecular velocities of the two gases, SO_2 and CH_4.

Solution

a. The molecular weights of the two gases are

$$mw_{SO_2} = 64$$
$$mw_{CH_4} = 16$$

b. The molecular velocities are inversely proportional to the square roots of the molecular weights:

$$\frac{v_{CH_4}}{v_{SO_2}} = \frac{\sqrt{mw_{SO_2}}}{\sqrt{mw_{CH_4}}} = \frac{\sqrt{64}}{\sqrt{16}} = 2$$

Thus under the same conditions, the average molecular velocity of methane is twice that of sulfur dioxide.

Although we now know something about the "average" behavior of a gas sample, not much has been said as yet about the individual molecules. In fact, their velocities span a range on either side of the average value. The distribution of molecular velocities in a typical sample of gas might look like that in Figure 8.7. (More will be said about distributions like this in Chapter 12.) Under ordinary conditions each molecule in such a sample will travel only about 1×10^{-5} cm between collisions, either with other molecules or with the walls of the container. However, each molecule suffers about 1×10^{10} such collisions every second; thus we sometimes hear that the hydrogen molecule travels about "a mile per second."

Question

7. Arrange these gases in descending order of rate of diffusion:

$CO_2 \quad CO \quad SO_2 \quad CH_4 \quad Cl_2 \quad H_2 \quad Ar \quad He$

(Assume the same temperature.)

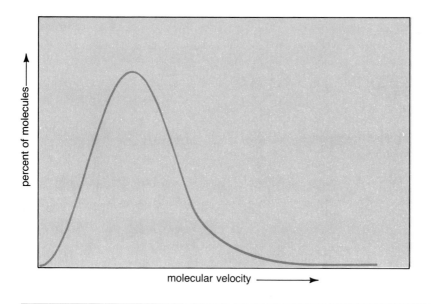

Figure 8.7 Distribution of molecular velocities in a gas sample.

8.9 Deviations from the Gas Laws

The kinetic molecular theory satisfactorily explains the relationships among the measurable properties (volume, pressure, and temperature) of a gas in terms of molecular masses and velocities. The theory allows us to regard each molecule as a dimensionless point in space that behaves in a simple way when it collides either with another molecule or with the walls of the container. Thus if we heat a gas while holding its volume constant, the velocities of the molecules will increase, there will be more frequent and more energetic collisions with the walls of the container, and hence we shall measure an increase in pressure. We see the same result if we reduce the volume of the gas or increase its quantity—more collisions and thus a higher pressure. There are, however, conditions under which we must discard the assumptions of the kinetic molecular theory. Under these conditions gases deviate significantly from ideal behavior.

At high pressures, we can no longer disregard the dimensions of a molecule compared to the average intermolecular distance. Therefore we must now think of the volume of a sample as made up partly of incompressible molecules; hence a further pressure increase will not reduce the volume as much as Boyle's law predicts. At low temperatures, another assumption breaks down. Since the molecules are moving more slowly, intermolecular attractions become significant. The lower the temperature, the greater the relative effect of the mutual attraction; hence as we lower the temperature, the volume decreases more than predicted by the ideal gas laws.

Question

8. The physical behavior of gases deviates from the laws of Boyle and Charles at extremely low temperatures and high pressures. How would their behavior be affected by the opposite extremes, namely, high temperature and low pressure?

Questions and Problems

8.1 Air, the Atmosphere, and Atmospheric Pressure

9. A weather report gives the barometric pressure as 29.34 in. Convert this value to atmospheres.

10. Using the values for density given below, calculate the reading on a water barometer when the atmospheric pressure is 1.00 atm.

water: 1.00 g/ml

mercury: 13.6 g/ml

8.2 Qualitative Descriptions of Gases

11. The densities of a typical gas, liquid, and solid might be 7.00×10^{-4} g/ml, 7.00×10^{-1} g/ml, and 7.00 g/ml, respectively. What volume would 1.00 g of each of these occupy?

8.3 The Gas Laws

12. For each of the pressures below, calculate the new volume for a gas that occupies 1.35×10^3 ft^3 at 745 mm of Hg. Assume constant temperature.
   a. 824 mm of Hg
   b. 632 mm of Hg
   c. 2.34 atm
   d. 0.468 atm

13. For each of the temperatures below, calculate the new volume for a gas that occupies 27.0 ml at 23°C. Assume constant pressure.
   a. (1.00×10^2)°C
   b. 0.0°C
   c. −43°C

14. A sample of a gas at 0.0°C is originally at a pressure of 14.7 lb/in^2. Assuming constant volume, calculate the new temperature for the pressures below.
   a. 12.3 lb/in.^{2}
   b. 15.4 lb/in.^{2}

15. What must the new pressure on a sample of gas be if its volume of 1.25×10^2 liters at 1.00 atm is reduced by 10.0 percent, at constant temperature?

16. Calculate the volume that 243 ml of a gas at 744 mm of Hg and 23°C will occupy at 874 mm of Hg and −15°C.

8.4 Combining Volumes and the General Gas Law Equation

17. Tell whether the following changes will increase, decrease, or have no effect upon the temperature of a gas. If it is impossible to predict, say so.
   a. Quantity increases, and volume and pressure remain constant.
   b. Pressure and volume both decrease, and quantity remains constant.
   c. Pressure increases, and volume decreases and quantity remains constant.
   d. Pressure decreases and volume increases, and quantity remains constant.
   e. Pressure increases, and volume and quantity decrease.
   f. Pressure and volume both increase, and quantity decreases.
   g. Pressure, volume, and quantity all increase.

18. Predict the effect of the following changes on the pressure of a gas. If it is impossible to predict, say so.
   a. Temperature increases, and quantity and volume remain constant.
   b. Quantity decreases, and temperature and volume remain constant.
   c. Volume increases, and temperature and quantity remain constant.
   d. Temperature and volume both increase, and quantity remains constant.
   e. Temperature and quantity both decrease, and volume remains constant.
   f. Quantity increases, temperature decreases, and volume remains constant.
   g. Quantity decreases, temperature increases, and volume remains constant.
   h. Temperature increases, and volume and quantity decrease.
   i. Quantity increases, and temperature and volume decrease.
   j. Volume decreases, temperature increases, and quantity remains constant.

19. Calculate the resulting pressure at constant volume if
   a. the temperature of a sample of gas having an initial pressure of 1.00 atm is raised from 20.0°C to (1.00×10^2)°C.
   b. the number of moles of gas in a container having an initial pressure of 744 mm of Hg is changed from 2.34 mol to 5.83 mol.
   c. the temperature of a sample of gas at 2.20 atm is reduced from 455 K to 325 K and the number of molecules in the sample is changed from 1.34×10^{23} to 2.47×10^{24}.

20. One mole of a gas at 1.00 atm undergoes the following changes. Determine the new pressure assuming that all conditions not mentioned remain constant.
   a. Kelvin temperature is doubled.
   b. An additional mole of gas is added and Kelvin temperature is tripled.
   c. Kelvin temperature is reduced to half the original, and one-third of the gas molecules are allowed to escape.
   d. Volume is tripled, and Kelvin temperature is halved.
   e. Two moles of gas are added, the Kelvin temperature is reduced to one-fourth of the original, and the volume is tripled.

8.5 Stoichiometry and Gas Volumes

21. Calculate the volume (at STP) of the carbon dioxide produced by the complete combustion of 5.00 lb of charcoal.

$$C + O_2 \longrightarrow CO_2$$

22. A spark is passed through a mixture containing 7.50 g of hydrogen and 42.4 g of oxygen. Calculate the volume (at STP) of water produced.

23. What volume of NO_2 at 23°C and 1.50 atm will be produced by the complete combustion of 23.8 liters (at STP) of ammonia?

$$4 NH_3 + 7 O_2 \longrightarrow 4 NO_2 + 6 H_2O$$

8.6 Dalton's Law of Partial Pressures

24. Container S holds 3.00 liters of helium at 745 mm of Hg, and container T holds 7.00 liters of argon at 635 mm of Hg. If a partition between the containers is removed and the gases allowed to mix, what will be the resulting partial pressures and the total pressure?

25. A gas container with a volume of exactly 22.4 liters is filled with H_2 and He at 0.0°C. If the partial pressure of the H_2 is 2.50 atm and the He is 760 mm of Hg, how many moles of each are present?

8.7 The Kinetic Molecular Theory

26. A sample of N_2 weighs 38.0 g.
  a. How many moles are in the sample?
  b. At STP what volume would the sample occupy?
  c. How many molecules does it contain?

27. Compare 1.00 mol of H_2S, C_2H_6, Ar, and O_2 with respect to
  a. weight   b. number of molecules   c. number of atoms

28. At STP 1.00 liter of CO
  a. contains how many moles?
  b. contains how many molecules?
  c. weighs how much?

29. Calculate the total weight of the gases in a mixture containing 1.45×10^{23} molecules of nitrogen, 3.47 mol of helium, 0.234 g of argon, and 17.5 liters of O_2 at STP.

30. Which of the following contains the greatest number of molecules?
  a. 1.00 liter of SO_2 at STP   b. 4.35×10^{-2} mol of O_2 c. 14.0 g of C_2H_6

31. Which of the following would occupy the greatest volume at STP?
  a. 12.0 g of argon
  b. 0.333 mol of ammonia
  c. 1.94×10^{23} molecules of oxygen

32. Which of the following samples of gas would weigh most?
  a. 355 ml of O_2 at 27°C and 740 mm of Hg
  b. 2.34×10^{-2} mol of neon
  c. 1.04×10^{22} molecules of ethane (C_2H_6)

33. The atmosphere contains 75.6 percent (by weight) nitrogen, 23.1 percent oxygen, 1.28 percent argon, 0.048 percent carbon dioxide, and traces of other gases. What are the relative numbers of moles of each gas in any sample of air?

*__34.__ On a day when the barometric pressure is 754 mm of Hg, what would be the partial pressures of the gases in Question 33?

*__35.__ What is the average density of air at STP? (See Question 33.)

36. Atmospheric pressure is frequently given as 14.7 lb/in.^{2}. Calculate the number of molecules in the imaginary cylinder of air above each square inch of the earth's surface. Use the value for the average density of air calculated in Question 35.

37. Which of two equal volumes of pure SO_2 and pure CO_2 (at STP) would have the greater
   a. number of molecules?
   b. total weight?
   c. average molecular velocity?
   d. number of moles?
   e. molecular mass?
   f. average molecular kinetic energy?

8.8 Molecular Velocities

38. If the velocity of the hydrogen molecule is approximately 1 mps at ordinary temperatures, estimate the velocity of the radon molecule.

39. Gas A has a molecular velocity one-fourth that of He under the same conditions. What is its molecular weight?

40. If the velocity of hydrogen is 1.00 mps at 20°C, what will be its velocity at $-156°C$?

*__41.__ Ammonia and hydrogen chloride gas are released simultaneously into opposite ends of a glass tube 1.00×10^3 cm in length. Both ends of the tube are immediately stoppered. The two gases will diffuse through the length of the tube. When they contact each other, they will react to form NH_4Cl, which will appear as a ring of white solid on the inside of the tube. Where will this ring appear?

42. Calculate the densities of the following gases at STP:
   a. CO_2 b. HBr c. C_2H_6

43. A certain gas has a density of 1.43 g/liter at STP. What will its density be if the Kelvin temperature is doubled (at constant pressure)?

44. Calculate the molecular weight of a gas if 15.0 g occupies 9.74 liters at 20.0°C and 742 mm of Hg.

45. The density of an unknown gas at 27°C and 2.24 atm is 2.73 g/liter. Calculate the molecular weight of this gas.

46. What is the weight in grams of one molecule of HBr?

47. One molecule of an unknown gas is found to weigh 7.31×10^{-23} g. Calculate the molecular weight.

48. How many molecules of air are there at STP in an empty room that measures 15.0 ft × 20.0 ft × 18.0 ft?

49. What pressure (in atmospheres) would be exerted by each of the following in a 1.00-liter container at 0.0°C?
   a. 32.0 g of O_2 b. 6.02×10^{22} molecules of H_2
   c. 0.235 mol of CO_2

50. What volume (in liters) would be occupied at 760.0 mm of Hg and 0.0°C by each of the following?
   a. 6.02×10^{24} molecules of HBr   b. 3.02 mol of HCl
   c. 2.45 g of HF

51. What temperature change will be required to maintain a constant pressure in a rigid flask containing 24.4 g of oxygen at 0.0°C if 25.0 percent of the gas is allowed to escape?

52. A cathode-ray tube having a volume of 1.25×10^2 ml is evacuated to a pressure of 3.00×10^{-8} mm of Hg. If the temperature is 20.0°C, how many molecules of gas remain in the tube?

*53. A weather balloon is filled with helium. At ground level, where the temperature is 25°C and the pressure is 755 mm of Hg, it has a diameter of 4.34 ft. When the balloon reaches an altitude where the temperature is −15°C and the atmospheric pressure is 4.30×10^2 mm of Hg, what will its diameter be?

54. The velocity of the hydrogen molecule at ordinary temperatures is approximately 1 mps; however, when a container of hydrogen is opened, it takes several minutes for the hydrogen to diffuse to the far corners of the room. Explain.

55. Gases are shipped in high-pressure containers to conserve space. A tank containing oxygen has a capacity of 2.20×10^2 ft^3 and is shipped at a pressure of 2.20×10^2 lb/in.^{2} and 25°C.
   a. What volume would the oxygen in the tank occupy at STP?
   b. What weight of oxygen does it contain?

56. What volume of oxygen at STP will be produced by heating 135 g KClO$_3$?

$$2\ KClO_3 \longrightarrow 2\ KCl + 3\ O_2$$

57. A 2.20-g sample of a compound containing only carbon and hydrogen was burned to produce 3.36 liters (at STP) of CO$_2$ and 3.60 g of liquid water. If the molecular weight of the compound is 44.0, what is its molecular formula?

58. A mixture of 3.24 g of CO$_2$ and 4.68 g of O$_2$ exerts a pressure of 22.0 lb/in.^{2}. Calculate the partial pressure of each gas.

*59. The following table includes a number of variables that affect the behavior of gases. Each horizontal row represents the changes imposed on a sample of gas. Fill in the blanks or explain why it is impossible to do so. The following abbreviations are used:

- ke   kinetic energy of molecule
- v   velocity of molecule
- m   mass of molecule
- i   increase
- d   decrease
- c   constant
- P   pressure
- T   temperature
- Q   number of moles or molecules
- V   volume

ke	m	v	Q	P	V	T
c		i	c	i		
i	c		c	d		
c	c	c	i		i	c
				i	i	d
	i			c	c	c
	i		c	i		c
			i		c	
		c	c		c	

*60. At what temperature would a liter of nitrogen at 1.00 atm weigh exactly 1.20 g?

61. A liter of gas weighs 1.10 g at 25°C and 7.40 × 10² mm of Hg. What would a liter of this gas weigh at STP?

62. 3.20 g of an unknown gas has a volume of 1.34 liters at STP. What would the density of this gas be at 127°C and 3.00 atm?

Discussion Starters

63. Why are there no gases at 0.0 K?
64. Why do pressures increase in automobile tires on a hot day?
65. Why do gases not follow Boyle's and Charles' laws exactly?

Liquids and Solids—The Condensed Phases

We were able to describe the empirical properties of a gas by considering only its temperature, pressure, and volume. These in turn were adequately explained under most conditions if we thought of the gas as a collection of independent point-masses, colliding elastically, and having kinetic energies dependent only on the temperature. In liquids and solids, which we call the *condensed phases*, pressure and volume will cease to be important variables, but the interactions among molecules will present new complications. Perhaps the best way to introduce these complications and to begin our discussion of the condensed phases is to consider the condensation of gases to liquids and then to look more closely at the transitions among the three physical states.

9.1 Condensation

We based our understanding of the behavior of gases on the central assumption that the gas molecules are far enough apart that we can neglect both the attractions among the molecules and the sizes of the molecules. This is a reasonable assumption at low pressure and high temperature. Even at ordinary temperatures and pressures, the deviations of such gases as oxygen, nitrogen, and hydrogen from the predictions based on this assumption are insignificant. If, however, we greatly increase the pressure and decrease the temperature, intermolecular distances decrease, kinetic energies diminish, and the behavior of the gas deviates significantly from Charles' and Boyle's laws. Indeed, if we continue to increase the pressure and decrease the temperature, the substance no longer behaves as a gas, and we say that it has *condensed*: It has become a liquid. Since we know that no gases exist at 0 K, it follows that we can condense all gases simply by decreasing the temperature, regardless of the pressure. However, it is *not* true that we can condense all gases simply by increasing the pressure. In fact, all substances have a **critical temperature,** above which they are always gases, regardless of the pressure. Above the critical temperature, attractive forces among the molecules are never sufficient to overcome their kinetic energies; hence they cannot exist in a condensed phase. The pressure necessary to liquefy a gas at its critical temperature is called the **critical pressure.** Some critical temperatures and pressures are listed in Table 9.1. Note that since

Table 9.1 Some Critical Temperatures and Critical Pressures

Substance	Critical Temperature (°C)	Critical Pressure (atm)
Argon	−122	48
Carbon dioxide	31.1	73
Chlorine	144.0	77.3
Helium	−267.9	2.336
Hydrogen	−239.9	13
Nitrogen	−149.9	34
Oxygen	−118.5	50.5
Sulfur dioxide	157.2	77.7
Water	374.0	217.7

intermolecular attraction is stronger among polar molecules than nonpolar ones, the nonpolar substances, such as hydrogen and nitrogen, have much lower critical temperatures. Note also that in general the heavier the molecule of a nonpolar substance, the more readily it liquefies. We can rationalize this observation by arguing that larger molecules have stronger intermolecular attractive forces (p. 258); hence they condense more readily. Finally, note the unusually high critical temperature of water. The explanation is an intermolecular force that we shall describe when we look more closely at all these molecular interactions in Section 9.8.

Questions

1. Is it possible to liquefy nitrogen at room temperature? Is it possible to liquefy chlorine at room temperature? Explain your answers.

2. Would you expect F_2 to have a higher or lower critical temperature than Cl_2? Explain.

9.2 Evaporation

Evaporation is a common phenomenon, but at first glance it is not obvious why a volume of liquid well below its boiling point should slowly disperse into the gas phase. The reason again lies in the conflict between molecular motion and intermolecular attraction. In a liquid the motion of most molecules is restricted by the forces that hold them to each other. But there is a range of molecular velocities, much as we saw in the gas phase. If we plot the number of liquid molecules against their velocities, we find a curve such as that in Figure 9.1. When a fast-moving molecule reaches the surface of the liquid, it overcomes intermolecular attractions and escapes into the atmosphere.

Since it is the more energetic molecules that escape during evaporation, the liquid loses a disproportionate amount of kinetic energy, and its temperature

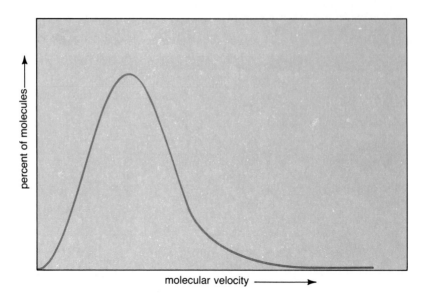

Figure 9.1 Distribution of molecular velocities in a liquid sample.

drops. (This is why the evaporation of perspiration provides cool relief on hot days.) If the evaporation takes place from an open noninsulated container, however, heat from the surroundings will maintain the temperature of the liquid, as well as the supply of high-energy molecules. Thus evaporation will continue until the liquid disappears.

From this explanation of evaporation, we can see that, at a constant temperature, heat is required to vaporize a quantity of liquid. The name we give to the amount of heat required to convert 1.00 mol of liquid to 1.00 mol of vapor at a specified temperature is the **molar heat of vaporization.** Heats of vaporization for several substances, recorded as customary at the normal boiling point (p. 253), are shown in Table 9.2. The same trends we observed in the critical temperatures are evident here, namely, that polar substances have

Table 9.2 Heats of Vaporization

Substance	Boiling Point (°C)	Heat of Vaporization (kcal/mol)
Argon	−186	1.56
Chlorine	−34.1	2.44
Helium	−269	0.020
Hydrogen	−253	0.108
Nitrogen	−196	0.67
Oxygen	−183	0.81
Sulfur dioxide	−10	6.07
Water	100	9.71

higher heats of vaporization than do nonpolar substances and that among nonpolar substances heats of vaporization generally increase with molecular weight.

Questions

3. Why does evaporation have a cooling effect?

4. Which member of the following pairs of compounds would you expect to have the higher molar heat of vaporization? Why?
    a. SO_2 or CO_2 b. $BeCl_2$ or OF_2 c. HCl or CH_4
    d. CCl_4 or CH_4 e. Cl_2 or HBr

9.3 Liquid-Vapor Equilibrium

If we want to prevent the molecules in a liquid-vapor system from escaping into the atmosphere, we can close the vessel that contains them. By restricting the vapor molecules to the space above the liquid, we ensure that they will once again strike the surface of the liquid. Each time this happens, the intermolecular attractions between the gaseous and liquid phase molecules may be great enough to overcome the kinetic energy of the gaseous molecule, in which case the gaseous molecule will return to the liquid state.

    When we first close the container, the number of molecules leaving the liquid state will exceed the number returning to it. Soon, however, after the number of molecules in the vapor state has increased, molecules will be returning to the liquid state (condensing) at the same rate they are escaping from it (evaporating). From this point on, if no heat is lost or gained through the walls of the container, the temperature will remain constant and the amounts of vapor and liquid in the container will not change. This phenomenon is an example of a **dynamic physical equilibrium:** a condition in which two opposing physical processes take place at exactly the same rate. At equilibrium the number of vapor molecules does not change; however, vapor molecules are constantly returning to the liquid state and liquid molecules are constantly entering the vapor state (Figure 9.2b).

    We call the gas pressure exerted by the vapor molecules, in equilibrium with the liquid, the *vapor pressure*. If we now raise the temperature, more molecules will have sufficient energy to escape from the surface of the liquid; thus for a time molecules escape more rapidly than they are recaptured. However, if we maintain this higher temperature, not increasing it further, equilibrium will be reestablished. The number of molecules in the vapor state will now be greater; therefore the vapor pressure will be higher (Figure 9.2c). Table 9.3 shows this trend in the vapor pressure of water.

    In addition to temperature changes, we might look at other stresses that we could apply to a liquid-vapor equilibrium. For example, if we remove some vapor from the closed container, more liquid will evaporate until the equilibrium is reestablished. Similarly, if we increase the pressure above the surface of the liquid, the vapor molecules will condense until the equilibrium is established

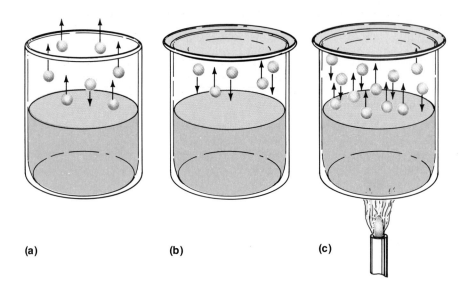

Figure 9.2 Vapor pressure. **(a)** Vapor molecules escape from an open container; hence rate of evaporation exceeds rate of condensation and equilibrium is not possible. **(b)** In a closed container vapor molecules cannot escape and equilibrium is reached (rate of evaporation and rate of condensation are equal). The pressure exerted by the vapor molecules above the surface of the liquid is the vapor pressure. **(c)** At a higher temperature more molecules are in the vapor state when equilibrium is attained; hence the vapor pressure is greater.

Table 9.3 Vapor Pressure of Water (in mm of Hg) at Various Temperatures (Celsius)

Temperature	Pressure	Temperature	Pressure
0	4.6	25	23.8
5	6.5	26	25.2
10	9.2	27	26.7
15	12.8	28	28.4
16	13.6	29	30.0
17	14.5	30	31.8
18	15.5	40	55.3
19	16.5	50	92.5
20	17.5	60	149.4
21	18.7	70	233.7
22	19.8	90	525.8
23	21.1	100	760.0
24	22.4		

again. These reactions to stress of a system in equilibrium are governed by a generalization known as **Le Chatelier's principle.** It states that *if a stress is applied to a system at equilibrium, the equilibrium will shift in the direction that relieves the stress applied.* This principle will be developed more fully in Section 13.2.

We can conveniently measure the vapor pressure of a liquid by inserting a small amount of it above the surface of the mercury in a barometer tube as shown in Figure 9.3. The vapor pressure of the liquid therefore counteracts the

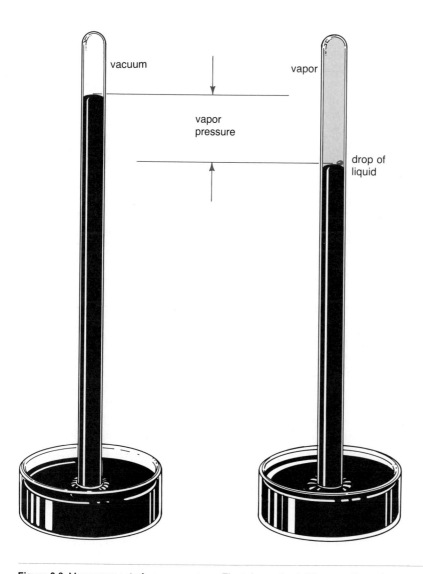

Figure 9.3 Measurement of vapor pressure. The tube on the left is an ordinary barometer. The tube on the right contains a small amount of liquid in equilibrium with its vapor above the mercury. The difference in height of the mercury column is the vapor pressure (usually expressed in millimeters of Hg).

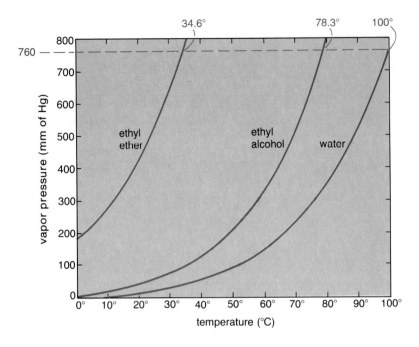

Figure 9.4 Vapor pressures of three common substances at various temperatures. The intersection of each curve with the dashed line (760 mm) represents the normal boiling point for that liquid.

atmospheric pressure and pushes the mercury in the tube down. We can then compare the reading of this barometer with the height of mercury in a standard barometer. If we measure several liquids this way, we soon confirm an obvious fact: Different liquids have different vapor pressures at the same temperature. Mercury, for example, must have negligible vapor pressure at room temperature for the barometer to measure atmospheric pressure accurately. The vapor pressures for three other liquids, water, ethyl alcohol, and ether, are plotted against temperature in Figure 9.4. Again, we see the effect of intermolecular forces: Water, with strong intermolecular attractive forces, has a low vapor pressure, whereas ether, with only weak forces holding it in the liquid state, has a high vapor pressure.

Questions

5. How can vapor pressure be measured?

6. If liquid A has a vapor pressure of 30.9 mm of Hg at 20°C and liquid B has a vapor pressure of 78.0 mm of Hg at the same temperature, what can we say about the relative strengths of the intermolecular forces in liquids A and B?

7. Why does vapor pressure increase with temperature?

9.4 Collecting Gas Over Water

Practical applications of some of the foregoing principles come up frequently in the laboratory. For example, we often collect water-insoluble gases by water displacement, using an apparatus as shown in Figure 9.5. In such experiments we must remember that the receiving bottle will contain a mixture of two gases—the reaction product and the water vapor. When the water level inside the receiving bottle is the same as that in the outer container, the pressures are of course also equal (see Figure 9.6). We must therefore determine the partial pressure of the reaction product by subtracting the vapor pressure of the water from atmospheric pressure.

Example 9.1 The volume of a gas collected over water is 8.00×10^2 ml when the atmospheric pressure is 748 mm of Hg and the temperature is 22°C. What volume will the dry gas occupy at STP?

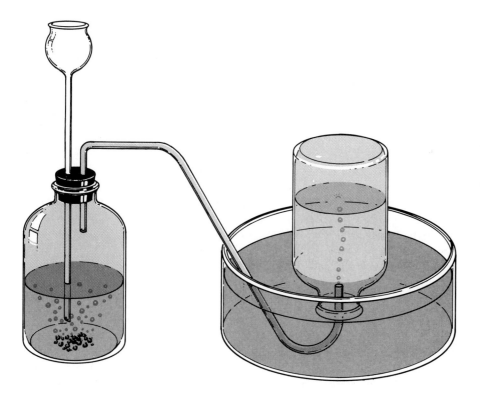

Figure 9.5 Collecting gas over water.

Liquids and Solids—The Condensed Phases

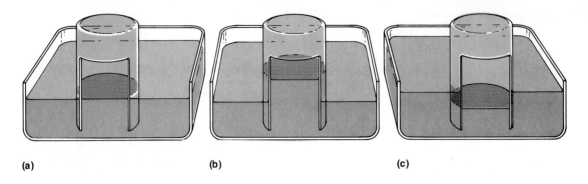

Figure 9.6 Relation between water level and pressure inside gas receiver. **(a)** Pressure gas = atmospheric pressure. **(b)** Pressure of gas < atmospheric pressure. **(c)** Pressure of gas > atmospheric pressure.

Solution

a. We assume that the water level inside the container has been adjusted (Figure 9.6) so that the pressure inside is equal to atmospheric pressure: 748 mm of Hg. This pressure is then the sum of the partial pressures of the water vapor and of the gas collected. We find the vapor pressure of water at 22°C in Table 9.3.

$$\begin{matrix} \text{total pressure of gas} \\ \text{mixture in container} \end{matrix} - \begin{matrix} \text{partial pressure} \\ \text{of water vapor} \end{matrix} = \begin{matrix} \text{partial pressure} \\ \text{of dry gas} \end{matrix}$$

$$748 \text{ mm of Hg} - 19.8 \text{ mm of Hg} = 728 \text{ mm of Hg}$$

b. From this point, we can solve the problem as we did in Chapter 8. The volume decreases in the same proportion that the pressure increases:

$$(8.00 \times 10^2 \text{ ml}) \left(\frac{728 \text{ mm of Hg}}{760 \text{ mm of Hg}} \right)$$

c. The volume decreases in the same proportion that the Kelvin temperature decreases.

$$(8.00 \times 10^2 \text{ ml}) \left(\frac{728 \cancel{\text{ mm of Hg}}}{760 \cancel{\text{ mm of Hg}}} \right) \left(\frac{273 \text{ K}}{295 \text{ K}} \right) = 709 \text{ ml}$$

Question

8. If oxygen is collected over water, the total pressure inside the container is the sum of the partial pressures of what gases?

9.5 Boiling Points

Generally, for a molecule to escape from the liquid state, it must *reach the surface of the liquid* with enough kinetic energy to overcome intermolecular attractions. As the temperature increases, the kinetic energies of the molecules increase and a greater number of molecules escape at the surface of the liquid; thus the vapor pressure increases. At a characteristic temperature for each liquid, however, vaporization begins to take place throughout the body of the liquid. At this temperature, which we call the **boiling point,** *the vapor pressure of the liquid is equal to the atmospheric pressure.* Thus the molecules have enough kinetic energy to enter the vapor phase *before reaching the liquid surface.* If at this point we increase the rate at which we add heat, we shall not increase the temperature but merely promote a more rapid formation of bubbles.

Since a liquid boils when its vapor pressure just exceeds the pressure above the liquid, it follows that the boiling point varies with atmospheric pressure. For example, water boils at 100°C when the atmospheric pressure is 760 mm of Hg, but if the atmospheric pressure is less than 760 mm of Hg, the water will boil below 100°C. Thus we can boil water at any temperature below 100°C by reducing the external pressure to the vapor pressure at that temperature (see Table 9.3). Similarly, of course, if the external pressure is greater than 760 mm of Hg, water boils above 100°C. For example, if the external pressure is 4.70 atm, water boils at 150°C. The boiling points listed in reference books assume an atmospheric pressure of 760 mm of Hg and are called the *normal boiling points.*

We may now turn briefly to a practical point. Since cooking involves chemical and physical changes that occur at high temperatures, food naturally cooks more rapidly at higher temperatures. Since atmospheric pressure varies with elevation, boiling points will be lower in the mountains than along the Gulf Coast. Thus a cook in Denver will find it easier to *boil* water to cook an egg than a cook in New Orleans, but the mile-high egg will be *cooked* more slowly.

Questions

9. At temperatures below the boiling point, what requirements must be met for a molecule to escape from the liquid to the vapor state? What requirement must be met at the boiling point?

10. Is it possible to boil water at 40.0°C? Explain.

11. What will the boiling point of water be on a mountain peak where the atmospheric pressure is 526 mm of Hg? (See Table 9.3.)

9.6 The Solid-Liquid Equilibrium

Having discussed the liquid-vapor system at length, we shall now find that we can draw many parallel conclusions about solid-liquid equilibria. For example, when we add heat to a mixture of ice and water at 0°C, the temperature does not increase. Instead, as we observed at the boiling point of a liquid, the heat is used to change the physical state of the substance. The heat energy frees some of the molecules from the fairly fixed positions characteristic of the *solid state* by overcoming intermolecular forces, thus allowing the molecules to move with the greater freedom we recognize in liquids. We call the amount of heat required to melt 1.00 mol of solid at constant temperature the *molar heat of fusion*. Like the heat of vaporization, it varies widely from substance to substance (Table 9.4). It is also worth noting that the heat of fusion of a substance is always smaller than its heat of vaporization. It takes more energy to overcome completely the intermolecular forces in a liquid than merely to disrupt the forces that give solids their rigidity. For much the same reason most substances expand only slightly when melted: Mobile molecules in the liquid state are only slightly farther apart than more rigidly bound molecules in the solid state. (We shall look later at the exceptional case of water, which expands when it solidifies—the practical consequences of which the antifreeze in your car is designed to avoid.)

Table 9.4 Heats of Fusion

Substance	Melting Point (°C)	Heat of Fusion (kcal/mol)
Argon	−190	0.281
Chlorine	−101	1.54
Hydrogen	−259	0.0280
Nitrogen	−210	0.172
Oxygen	−219	0.106
Water	0.0	1.44
Sulfur dioxide	−72.7	2.06

We can extend our analogy with the liquid-vapor system by looking at a liquid at its *freezing point*. In an insulated container there will be a dynamic equilibrium between molecules in the solid state and those in the liquid state. If we add or remove small quantities of heat, we simply increase or reduce the proportion of molecules in the liquid state as the equilibrium responds to the stress. At 1.00 atm the temperature of the equilibrium state is called the *normal freezing point*. Since there is little volume change during freezing, freezing points vary only slightly with atmospheric pressure.

Questions

12. Why is the molar heat of fusion of a substance smaller than its molar heat of vaporization?

13. Why does a change in pressure have a greater effect on the boiling point than on the freezing point?

9.7 Phase Diagrams

We might now look for a convenient way to depict the relationship between the pressure and temperature of a substance and its physical states. We can do this graphically with a *phase diagram*. A typical example appears in Figure 9.7. The three areas in the diagram represent the three phases of the substance; thus we can predict its physical state for any temperature and pressure. At points on the lines that separate two physical states, the two phases are in equilibrium. Therefore all points on the line *BC* are boiling points, and all points on *BD* are freezing points. As we have seen, the temperature of the boiling point increases dramatically with increasing pressure, whereas the freezing point is relatively insensitive to pressure changes. The freezing point does, however, increase slightly with increasing pressure. This means that higher pressures favor the solid state in this example, which suggests that the solid is denser than the liquid. We shall say more about this conclusion when we study the phase diagram for water.

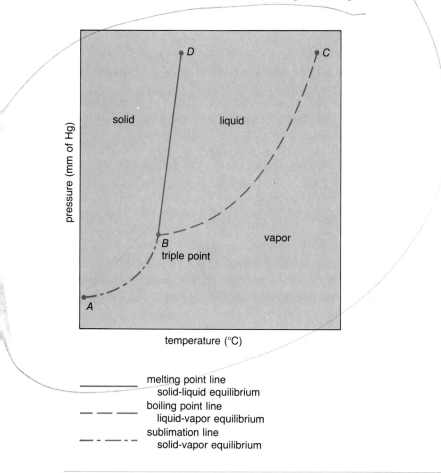

Figure 9.7 Phase diagram for a liquid which contracts on freezing.

Liquids and Solids—The Condensed Phases

Line *AB*, which separates the solid from the vapor in Figure 9.7, represents an equilibrium that we have not yet studied. It suggests that under certain conditions we can establish an equilibrium between solid and vapor. A familiar example is "dry ice" (solid carbon dioxide), which cannot exist as a liquid below 5.1 atm. Instead of melting, it *sublimes*—passes directly into the vapor state. Any point on *AB* is therefore a *sublimation* point. Since there is a large volume change during sublimation, as during vaporization, sublimation temperatures increase dramatically with increasing pressure, much as boiling points do. Since the vapor pressures of liquids increase with temperature, we expect the same of the vapor pressures of solids. We can demonstrate this easily enough by heating iodine crystals and noting the increase in the color of the vapor or by noticing the odor of mothballs (naphthalene) as they are heated.

Point *B* in Figure 9.7 is the unique point at which all three phases coexist in equilibrium. We call this the *triple point*. With this final concept, we are now prepared to look in some detail at a phase diagram for water—a pure substance with which we each have had experience in all its three phases.

Figure 9.8 is a schematic representation of the phase diagram for water. Perhaps its most interesting characteristic is the slope of line $B'D'$. This line indicates that the freezing point decreases slightly as the pressure rises. Thus the higher pressure must favor the liquid phase. This means that in this unusual case the liquid is the more condensed phase, and we expect water to expand upon freezing. (As we saw in the more typical case of Figure 9.7, the solid is usually denser than the liquid.) To see how slight is the effect of pressure on the freezing point, however, note that the triple point for water is 0.01°C and 4.58 mm of Hg. The freezing point decreases only to 0.00°C when we raise the pressure to 760 mm of Hg. The unusual relationship between pressure and the freezing point of water appears in a number of places, from cracked engine blocks in winter to the skating rink where the pressure of the skate blade ensures a layer of lubricating water beneath the skate.

Line $B'C'$ is the vapor pressure curve for liquid water. Notice that it does not extend beyond the critical point (374°C, 218 atm), above which there is not a well-defined transition between liquid and vapor states. Line $A'B'$ is the vapor pressure curve for ice, which theoretically extends to zero pressure and zero kelvins. This curve says that below 0.0°C (and at ordinary pressures) there can be no liquid water but that if the partial pressure of water vapor is low enough, ice will "disappear" by sublimation. Since this process is in effect a combination of melting and vaporization, we find that at constant temperature the *heat of sublimation* is the sum of the heats of fusion and vaporization:

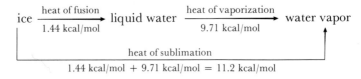

Each of these processes, sublimation, fusion, and vaporization, are **endothermic** (require the addition of heat). Conversely, condensation processes are **exothermic** (release heat).

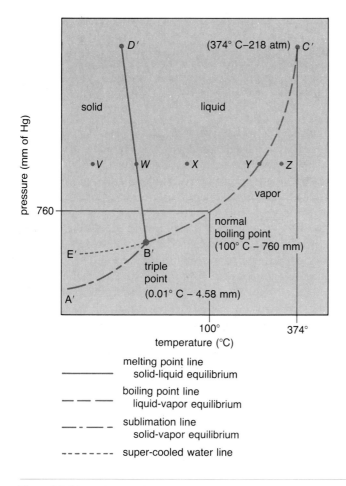

Figure 9.8 Phase diagram (not drawn to scale) for water.

Under certain conditions, we can cool liquid water at atmospheric pressure below its normal freezing point. We call this *metastable* state *supercooled water*. Its vapor pressure curve appears as the dashed line $B'E'$ in Figure 9.8.

Phase diagrams permit one to ascertain what phases are present at any combinations of temperature and pressure. In Figure 9.8 at point V, only ice is present. At point W, ice and liquid water are at equilibrium. At point X, only liquid water is present; at point V, liquid water and vapor are at equilibrium. Finally, at point Z, only vapor is present.

Questions

14. How do boiling points change when the atmospheric pressure increases? How do most freezing points change? How does the freezing point of water change?

15. Why does ice disappear even in cold (below 0°C) dry weather?

16. Most phase diagrams consist of three areas, three lines, and a single point where the lines meet. What do each of these represent?

9.8 Intermolecular Forces

We have repeatedly invoked intermolecular forces as being involved in changes of physical state. We shall now try to make sense of these forces by looking at their origins. First we shall divide them into three types, *London dispersion forces* (sometimes called van der Waals forces), *dipole attractions*, and *hydrogen bonding*, and then consider each in turn.

The weakest and most prevalent of these forces are the London dispersion forces; they are present in all substances and are the sole source of attraction between nonpolar molecules. We can best explain the nature of these forces by referring to Figure 9.9. As one molecule approaches another, its nucleus is repelled by the positive charge of the other nucleus but is attracted by its electrons. Also, the electron clouds of the approaching molecules repel each other. These electrostatic interactions thus produce transient unsymmetrical electron distributions in the two molecules. Therefore, even a collection of nonpolar molecules will at any instant have induced dipole moments that produce weak attractive forces between the molecules. Of course electrons are always in motion, as are the molecules (even in the solid phase), so the polarizations constantly change directions. Nonetheless, the net effect of these transitory induced dipoles is a mutual attraction.

Since the London dispersion forces are generally weak, they are significant only in condensed phases where intermolecular distances are small. The strength of the force varies from substance to substance, depending strongly on the size and shape of the molecules. For example, the greater the molecular weight, the farther the outermost electrons are from the nucleus and hence the more readily

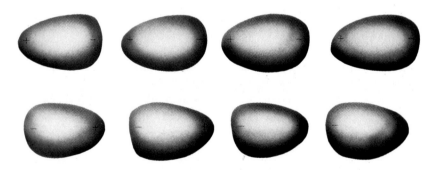

Figure 9.9 The London dispersion forces are caused by temporary unsymmetrical distribution of electric charges. The positive and negative areas of the molecules are constantly shifting as molecules change positions.

the molecule can be polarized. Therefore London dispersion forces tend to increase as molecular weights increase. Table 9.5 is a good illustration of this since heats of vaporization and boiling points depend heavily on intermolecular forces in the liquid state. Since all the substances in Table 9.5 are nonpolar, the weak London force is the only attractive intermolecular force; thus all but Br_2 and I_2 are gases under normal conditions.

If the weak, transient dipole of the London force produces a measurable intermolecular attraction, we would expect permanent dipole moments to be even more important sources of intermolecular forces. We can check this conclusion by comparing the properties of polar and nonpolar molecules having similar molecular weights. For example, HCl (a polar molecule with molecular weight 36.5) boils at $-84.9°C$ and has a heat of vaporization of 3.865 kcal/mol. By contrast, F_2 (molecular weight 38.0) boils at $-188°C$ and has a heat of vaporization of 0.782 kcal/mol. Similarly, SO_2 (molecular weight 64) boils at $-10°C$ and has a heat of vaporization of 6.07 kcal/mol, whereas Cl_2 (molecular weight 70.9) boils at $-34.1°C$ and has a heat of vaporization of 2.439 kcal/mol. Thus we may say that in polar molecules the electrostatic attraction between centers of unlike charge promotes the stability of condensed phases.

The third type of intermolecular force is by far the strongest: Hydrogen bonds are often one-tenth as strong as typical ionic or covalent bonds. When a hydrogen atom is bonded to a small, strongly electronegative atom, such as fluorine, oxygen, or nitrogen, the difference in electronegativities produces a strongly polarized covalent bond. The hydrogen atom thus becomes little more than a bare proton, which then seeks to share an unshared electron pair of a nearby molecule. The resulting intermolecular hydrogen bond greatly stabilizes the liquid phase of molecules like hydrogen fluoride, water, and

Table 9.5 Boiling Points and Heats of Vaporization of Some Elements

Substance	Molecular Weight	Normal Boiling Point (°C)	Heat of Vaporization (kcal/mol)
Rare Gases			
He	4.00	−269	0.020
Ne	20.2	−246	0.422
Ar	39.9	−186	1.558
Kr	83.8	−153	2.158
Xe	131	−108	3.021
Rn	222	−62	3.92
Halogens			
F_2	38.0	−188	0.782
Cl_2	70.9	−34.1	2.439
Br_2	160	58.2	3.59
I_2	254	186	4.99

ammonia. We therefore call these liquids *associated liquids*. We can depict them as shown here, where solid lines represent covalent bonds and dotted lines represent hydrogen bonds:

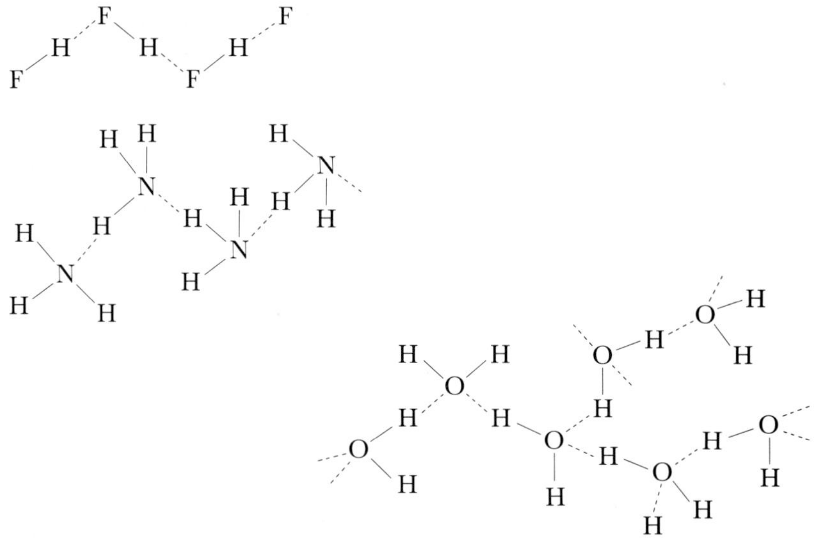

Since fluorine is so strongly electronegative, the hydrogen bonds in hydrogen fluoride are especially strong. In fact, some chemists consider the hydrogen bonds in hydrogen fluoride strong enough to be thought of as chemical bonds rather than merely intermolecular forces. At room temperature, for example, hydrogen fluoride behaves in some ways like a molecule whose average molecular weight is 120, suggesting a formula of H_6F_6. At 32°C, however, it behaves like H_2F_2; at higher temperatures, like HF. Since this temperature dependence is characteristic of hydrogen bonding but not normal chemical bonding, we shall stay with the formula HF.

Since each hydrogen atom and each unshared electron pair can participate in only one hydrogen bond, both hydrogen fluoride (with one hydrogen atom in each molecule) and ammonia (with one pair of unshared electrons in each molecule) can associate only in chainlike structures such as those shown above. Water molecules, on the other hand, can interlink in complex associations since each molecule has two unshared electron pairs and two hydrogen atoms. In water each oxygen atom is surrounded by a tetrahedral array of four hydrogen atoms—two bonded by covalent bonds and two by longer, weaker hydrogen bonds. Of course the hydrogen bonds are not permanent, and the tetrahedral structure of water constantly changes as the hydrogen bonds break and reform.

Figure 9.10 dramatically illustrates the effect of hydrogen bonding on the stability of the liquid phase. In compounds of hydrogen and the elements of Group IVA, the boiling point rises as the molecular weight increases. This we expect since London dispersion forces are the only intermolecular attractions present in these liquids, and we know that London forces increase with molecular weight. In compounds of hydrogen and the elements of Groups VA, VIA, and

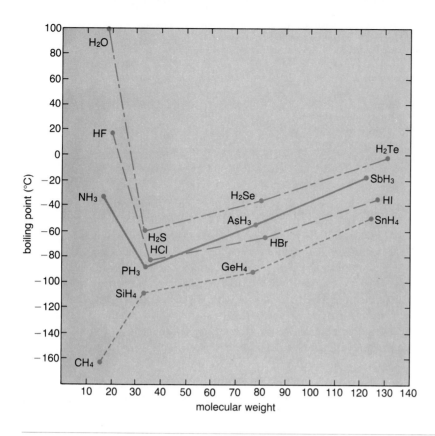

Figure 9.10 Boiling points of the binary-hydrogen compounds of the elements of Groups IVA, VA, VIA, and VIIA, showing the abnormally high boiling points of ammonia, water, and hydrogen fluoride (atmospheric pressure = 760 mm of Hg).

VIIA, however, the lightest molecules are way out of line. We attribute these unusually high boiling points to hydrogen bonding. The behavior of water is especially remarkable: Its boiling point is much higher than that of HF, although the hydrogen bonds in HF are stronger. To explain this we must simply recall that each water molecule can form two hydrogen bonds with neighboring molecules, whereas hydrogen fluoride can form only one per molecule.

We can also invoke hydrogen bonding in water to explain its unusually high heat of fusion (1.44 kcal/mol), specific heat [1.00 cal/(g)(°C)], and heat of vaporization (9.71 kcal/mol). Since the hydrogen bonding is so pervasive in the condensed phases of water, it takes extra energy to dissociate the molecules when we melt ice or heat or boil water. To emphasize this point we can compare the specific heat of water with that of three other liquids, each illustrating the effect of a different intermolecular attractive force (Table 9.6). The specific heat of ethanol ($C_2H_5O_H$) is about one-half that of water; ethanol exhibits hydrogen bonding, but each molecule can form only one such bond. Chloroform ($CHCl_3$) is a polar molecule, stabilized in the condensed phases only by dipole attractions;

Table 9.6 Specific Heats of Some Liquids

Substance	Formula	Specific Heat [cal/(g) (°C)]
Water	H_2O	1.00
Ethanol (ethyl alcohol)	C_2H_5OH	0.535
Chloroform	$CHCl_3$	0.232
Carbon tetrachloride	CCl_4	0.198

hence its specific heat is lower than that of ethanol but higher than that of carbon tetrachloride, which is nonpolar. The conspicuously high specific heat of water manifests itself in many ways, one example being the moderating influence lakes and oceans often have on the temperatures of adjoining land areas. Since the temperature of the water fluctuates slowly, it can serve as both a source of cool breezes on warm days and a reservoir of heat during cold nights.

Finally, hydrogen bonding plays a part, though a more complicated one, in two other unusual properties of water: First, liquid water reaches its maximum density at 3.98°C; second, as we have already noticed, water expands when it freezes. As we cool water, molecular motion decreases; thus the average distance between molecules diminishes. This effect alone, therefore, would cause the density to increase as the temperature drops. However, the decrease in temperature also promotes hydrogen bonding, and hydrogen-bonded aggregates do not pack as tightly as unassociated molecules. This means that this second trend counteracts the first. The result—one that we could not have predicted accurately—is that the first trend predominates above 3.98°C and the second trend below 3.98°C. At the freezing point, hydrogen bonding binds water into a solid aggregate, and the density drops dramatically.

Example 9.2 How many calories are required to change 2.00×10^2 g of ice at $-10.0°C$ to steam at $110.0°C$? Assume a specific heat of 0.500 cal/(g)(°C) for ice and steam.

Solution

a. We must consider five steps separately, as shown below and then add the heat required for each step to find the total for the whole process.

$$\text{ice at } -10.0°C \xrightarrow{b} \text{ice at } 0°C \xrightarrow{c} \text{water at } 0°C \xrightarrow{d} \text{water at } 100°C$$
$$\text{steam at } 110°C \xleftarrow{f} \text{steam at } 100°C \xleftarrow{e}$$

b. To raise the temperature of the ice from $-10°C$ to $0°C$, that is, $10°C$:

$$(2.00 \times 10^2 \text{ g}) \left[\frac{0.500 \text{ cal}}{(1.00 \text{ g})(1.00°C)} \right] (10.0°C) = 1.00 \times 10^3 \text{ cal}$$

c. To melt the ice (see Table 9.4 for heat of fusion):

$$(2.00 \times 10^2 \text{ g}) \left(\frac{1.44 \text{ kcal}}{1.00 \text{ mol}}\right)\left(\frac{1000 \text{ cal}}{1 \text{ kcal}}\right)\left(\frac{1.00 \text{ mol}}{18.0 \text{ g}}\right) = 1.60 \times 10^4 \text{ cal}$$

d. To heat the liquid water from 0°C to 100°C:

$$(2.00 \times 10^2 \text{ g}) \left[\frac{1.00 \text{ cal}}{(1.00 \text{ g})(1.00°C)}\right](100°C) = 2.00 \times 10^4 \text{ cal}$$

e. To vaporize the water (see Table 9.2 for heat of vaporization):

$$(2.00 \times 10^2 \text{ g}) \left(\frac{9.71 \text{ kcal}}{1.00 \text{ mol}}\right)\left(\frac{1000 \text{ cal}}{1 \text{ kcal}}\right)\left(\frac{1.00 \text{ mol}}{18.0 \text{ g}}\right) = 1.08 \times 10^5 \text{ cal}$$

f. To heat the steam from 100°C to 110°C, that is, 10°C:

$$(2.00 \times 10^2 \text{ g}) \left[\frac{0.500 \text{ cal}}{(1.00 \text{ g})(1.00°C)}\right](10.0°C) = 1.00 \times 10^3 \text{ cal}$$

g. Thus the total for the process is the sum of parts b through f.

$$(1.00 \times 10^3 \text{ cal}) + (16.0 \times 10^3 \text{ cal}) + (20.0 \times 10^3 \text{ cal})$$
$$+ (108 \times 10^3 \text{ cal}) + (1.00 \times 10^3 \text{ cal}) = 146 \times 10^3 \text{ cal} = \mathbf{1.46 \times 10^5 \text{ cal}}$$

We shall now add an important footnote to our discussion of hydrogen bonding in small molecules like HF, H_2O, and NH_3. Many larger organic and biological molecules also contain groups that promote intermolecular associations. For example, all protein molecules contain

$$\text{C=Ö} \quad \text{and} \quad \underset{H}{\overset{\ddot{N}}{|}}\text{C}$$

groups. Therefore hydrogen bonds like the following, either between different molecules or between two groups in the same large molecule, are important influences on the structure and properties of proteins.

$$\text{C=Ö} \cdots \text{H} - \underset{|}{\overset{\ddot{N}}{}} - \text{C}$$

Finally, we want to look at a very different manifestation of intermolecular forces: *surface tension*. In the body of a liquid, every molecule is subject to intermolecular attractions in all directions (Figure 9.11). At the surface, however, molecules are attracted laterally and inward but not upward. Thus the surface

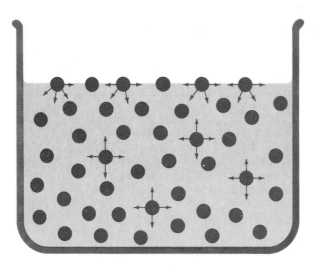

Figure 9.11 Schematic representation of intermolecular forces in a liquid.

tends to contract; the molecules crowd close together in the denser "skin" at the surface. This is the reason that liquids form spherical drops when they fall freely. The liquid simply assumes the shape with the least surface area in response to the unbalanced intermolecular forces at its surface. As we might expect, water has a very high surface tension, again because of hydrogen bonding. Thus some insects are able to walk on the surface of the water, and with care one can even make a needle float although it is many times denser than water.

Questions

17. What do we mean by intermolecular forces? Arrange the following in order of increasing strength:
    a. dipole attractions   b. London dispersion forces   c. hydrogen bonds

18. From each of the following pairs, select the substance that you think will have the higher molar heat of vaporization. Justify your choice.
    a. HF or NH_3 b. H_2S or H_2O c. GeH_4 or H_2Se d. CH_4 or SnH_4

$$\text{e. } H:\overset{H}{\underset{H}{\overset{..}{C}}}:\overset{..}{\underset{..}{O}}:\overset{H}{\underset{H}{\overset{..}{C}}}:H \quad \text{or} \quad H:\overset{H}{\underset{H}{\overset{..}{C}}}:\overset{H}{\underset{H}{\overset{..}{C}}}:\overset{..}{\underset{..}{O}}:H$$

9.9 The Kinetic Theory of Liquids

Liquids have several important properties that distinguish them from solids and gases. Like solids they are only slightly compressible, but they have no characteristic shape; they diffuse more slowly than gases but more rapidly than solids; and they vaporize more rapidly than solids. However, accounting for these properties as well as the others we have discussed in this chapter is not simple. In fact, it is much easier to give an adequate account of both gases and solids. In one we can assume complete disorder; in the other, nearly complete order. Liquids lie in the uncomfortable region between the extremes. Nonetheless, we can make a few useful observations.

One of the first important observations about the nature of the liquid phase was made in 1828 by Robert Brown (1773–1858) when he noticed that microscopic grains of pollen suspended in water are in constant random motion. He showed that this motion (which we now call *Brownian movement*) was not due to external forces such as convection or stirring, but he was not able to explain its source. Sir William Ramsey (1852–1916) provided the explanation much later when he proposed that the zigzag paths of the suspended particles were the results of collisions with water molecules in constant motion. Thus we must think of liquids as composed of molecules free to move, though subject to more constraints than in the gas phase. The most significant of these constraints is that the volume of most liquids is only slightly greater than that of an equal amount of solid. Clearly the motion of the molecules is restricted. But we can explain both the free motion of the molecules and the high density by thinking of liquids as full of "holes," each about the size of a molecule. These holes account for only about 3 percent of the volume of the liquid (hence the liquid is nearly as dense as the solid), but they also provide a mechanism for molecular motion and for the fluidity of the liquid. Molecules shift into adjacent vacant holes, hence creating holes in different places. Thus we can think of the mobility of the liquid phase as due to the mobility of these holes. In the same way, these shifting holes provide an explanation for diffusion in liquids: The diffusing substances spread by simply moving from vacant hole to vacant hole.

Question

19. Compare gases and liquids with regard to
    a. distance between molecules.
    b. freedom of molecular motion.
    c. rate of diffusion.
    d. expansion with increase in temperature.

9.10 Properties of Solids

As we suggested in Section 9.9, we shall find it easier to provide a theoretical explanation for the structure of solids than for liquids. Nonetheless, solids display an impressive variety of physical properties. Metals are shiny and malleable, are good electrical conductors, and have high melting points. Ionic solids likewise have high melting points but are hard and brittle and do not conduct electricity unless melted. Still other solids, like sulfur and phosphorus, are soft, dull, and powdery, are nonconductors, and have low melting points. And of course some substances exist as solids only at lower temperatures or higher pressures than we often encounter. Nonetheless, all substances can be solidified under the right conditions, and this suggests a first characteristic property of the solid state: *All solid pure substances that melt without decomposition have a well-defined melting point.* Tables 9.7 and 9.8 suggest the wide range of these melting points.

Table 9.7 Melting Points of Some Elements

Substance	(°C)
He	−272.2
N_2	−209.86
Cl_2	−100.98
P_4 (white)	44.1
S_8 (rhombic)	112.8
Al	660
Fe	1535
C (diamond)	>3550

Table 9.8 Melting Points of Some Compounds

Substance	(°C)
CO	−199
CS_2	−110.8
CO_2	−56.6
CCl_4	−23.0
$C_{10}H_8$	80.5
$FeCl_3$	306
NaCl	801
Al_2O_3	2045
MgO	3600

The second important property of all solid pure substances is that they have a regular, repeating structural pattern called a *crystal lattice*. We often see this regularity in large crystalline solids, but others have crystal structures too small to see. In all true solids, however, we can completely describe the structure of the substance by describing a simple *unit cell*, which is the smallest repeated portion of the crystal lattice. For example, the simplest three-dimensional crystalline

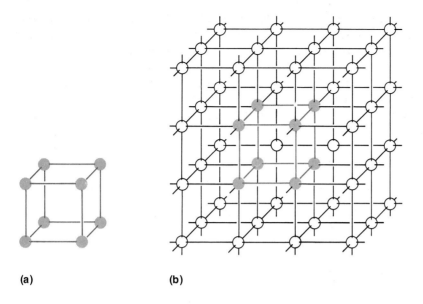

Figure 9.12 The simple cubic unit cell **(a)** and lattice **(b)**.

pattern or lattice is shown in Figure 9.12a. In this *simple cubic* lattice, an identical atom or molecule occupies each corner of a cube. As Figure 9.12b shows, this unit cell can be reproduced in three dimensions to *generate* the whole crystal.

The simple cubic lattice is not the most efficient lattice, however, with respect to the close packing of the units which occupy the positions. Other cubic lattices have as unit cells the *body-centered cube* and the *face-centered cube*. A comparison of the packing efficiency of these three lattices may serve to make their respective geometries more understandable.

Before examining the cubic lattice in detail, it might be a good idea to review some of the geometry of the cube. The cube is a *regular* solid (Figure 9.13a). This means that length, width, and height are equal. Thus the volume (V) of a cube is found by cubing the length of an edge (E).

volume = length × width × height

$$V = E \times E \times E = E^3$$

Any face of a cube is a square (Figure 9.13b). Thus the length of a face diagonal (F) has a relationship to the length of an edge, (E), which may be found through the application of the Pythagorean theorem.

$$F^2 = E^2 + E^2 = 2E^2$$
$$F = E\sqrt{2}$$

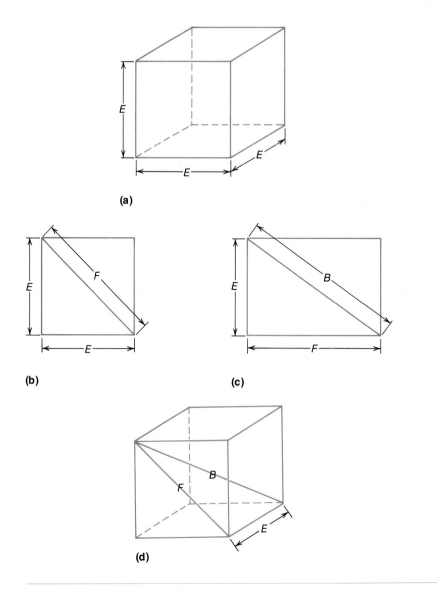

Figure 9.13 The geometry of the cube.

The body diagonal of a cube (Figure 9.13c) is the diagonal (B) of a rectangle with the cube edge (E) as one of its sides and the face diagonal (F) as another. The relationships among the body diagonal (B), the face diagonal (F), and the edge (E) of a cube may be found by another application of the Pythagorean theorem.

$$B^2 = E^2 + F^2$$

Substitute $F = \sqrt{2}E$.

$$B^2 = E^2 + (\sqrt{2}E)^2$$
$$B^2 = 3E^2$$
$$B = \sqrt{3}E$$
$$B^2 = E^2 + F^2$$

Substitute $E = F/\sqrt{2}$.

$$B^2 = \left[\frac{F}{\sqrt{2}}\right]^2 + F^2$$
$$B = \sqrt{\frac{3}{2}}F = \frac{\sqrt{6}}{2}F$$

Consider a simple cubic crystal lattice and its unit cell. If each position is occupied by a sphere and the packing is such that all spheres are in contact, each sphere is surrounded by six others. The *coordination number* of this lattice is 6. Coordination number is the number of "nearest neighbors" for any sphere (Figure 9.14). The portion of the unit cell which is occupied can be estimated from Figure 9.15. *Only one-eighth of each sphere* is within a given cubic unit cell. Another way of saying the same thing is that each corner, and thus each sphere, is common to eight unit cells.

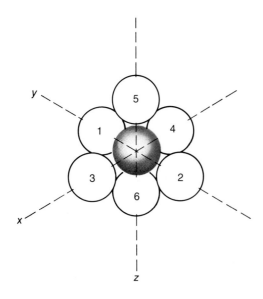

Figure 9.14 The coordination number of each sphere in a simple cubic lattice is 6.

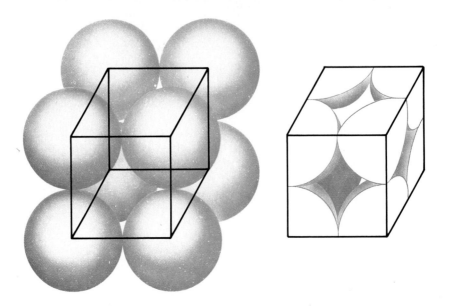

Figure 9.15 Models of the simple cubic unit cell.

Example 9.3 Using r as the radius of one of the theoretical spheres, calculate the percent of the volume of the simple cubic unit cell which is occupied.

Solution

a. The spheres are assumed to be in contact with each other, so the edge length (E) of the unit cell is $2r$. Then the volume of the unit cell is

$$V = E^3$$
$$V = (2r)^3 = 8r^3$$

b. There are eight eighths of spheres within the cell, or the equivalent of one whole sphere. The volume of space occupied is the volume of one sphere of radius (r).

$$V = \tfrac{4}{3}\pi r^3$$

c. Calculate the percent of the volume of the simple cubic unit cell which is occupied.

$$\text{percent occupied} = \left(\frac{\text{volume of spheres}}{\text{volume of cell}}\right)(100)$$

$$= \left(\frac{\tfrac{4}{3}\pi r^3}{8r^3}\right)(100)$$

$$= 52.37\%$$

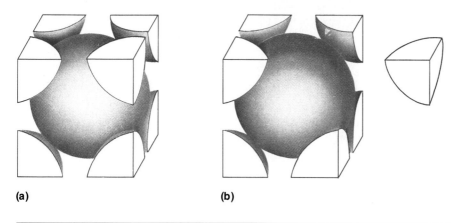

Figure 9.16 Models of the unit cell of the body-centered cubic lattice. **(a)** Unit cell. **(b)** Unit cell with one corner sphere removed.

The closest packing of spheres possible with the body-centered cubic arrangement brings the spheres into contact only along the body diagonal (Figure 9.16). This body diagonal is the dimension most readily available in terms of r for this cell. Each body-centered sphere is in contact with eight corner spheres, and each corner sphere is in contact with eight body-centered spheres in other unit cells. The coordination number for the body-centered cubic lattice is 8.

Example 9.4 Calculate the percent of the volume of a body-centered cube which is occupied.

Solution

a. The length of the body diagonal is $4r$.

b. The relationship between the body diagonal (B) of a cube and the edge (E) (derived through application of the Pythagorean theorem), is

$$B = \sqrt{3}E$$

c. If the body diagonal is $4r$, then the edge is found by substitution.

$$4r = \sqrt{3}E$$

$$E = \frac{4r}{\sqrt{3}}$$

d. The volume of the cube is

$$V = E^3$$

$$V = \left[\frac{4r}{\sqrt{3}}\right]^3$$

e. Although the body-centered cube is a larger structure (relative to the radius of our theoretical sphere) than the simple cube, it contains more spheres. In addition to one-eighth of each of the eight corner spheres, it contains one sphere wholly within the unit cell. The occupied space within the unit cell is equivalent to two spheres.

$$2\left(\tfrac{4}{3}\pi r^3\right)$$

f. Calculate the percent of the volume of the body-centered cube which is occupied.

$$\text{percent occupied} = \left(\frac{\text{volume of spheres}}{\text{volume of cell}}\right)(100)$$

$$= \frac{2\left(\tfrac{4}{3}\pi r^3\right)}{(4r/\sqrt{3})^3}(100) = 68.02\%$$

The face-centered cube is a more efficient arrangement than either of the other two which have been examined. In this unit cell (Figure 9.17) spheres

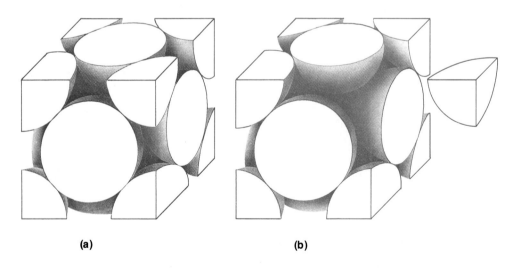

(a)   (b)

Figure 9.17 Models of the unit cell of the face-centered cubic lattice. **(a)** Unit cell. **(b)** Unit cell with one corner sphere removed.

are in contact only along the diagonal of a face. The coordination number of a sphere in a face-centered lattice is 12.

Example 9.5 Calculate the percent of the volume of a face-centered cube which is occupied.

Solution

a. The geometric relationship between the length of a face diagonal (F) and the edge length (E) is the relationship between the diagonal of a square and its side, namely,

$$F = E\sqrt{2}$$

b. The length of the face diagonal is $4r$. Calculate the edge length.

$$4r = E\sqrt{2}$$

$$E = \frac{4r}{\sqrt{2}}$$

c. Calculate the volume of the cube.

$$V = E^3$$

$$V = \left[\frac{4r}{\sqrt{2}}\right]^3$$

d. The face-centered unit cell contains one-eighth of each of the eight corner spheres, but in addition each one of the six faces contains a sphere squarely in the middle of it, bisected by the boundary plane of the unit cell. Thus there are six half-spheres and eight eighth-spheres, or a total of four spheres within the cell.

e. The volume of occupied space within the cell is

$$4(\tfrac{4}{3}\pi r^3)$$

f. Calculate the percent of the face-centered unit cell which is occupied.

$$\text{percent occupied} = \left(\frac{\text{volume of spheres}}{\text{volume of cell}}\right)(100) = \frac{4(\tfrac{4}{3}\pi r^3)}{(4r/\sqrt{2})^3}(100)$$

$$= 74.04\%$$

We have repeatedly explained the existence of certain forms of matter in terms of stability. Although it is merely a different expression of the same thing, we shall explain the existence of the various crystal lattices in terms of the efficiency of their packing. The third of the three forms considered, the face-centered

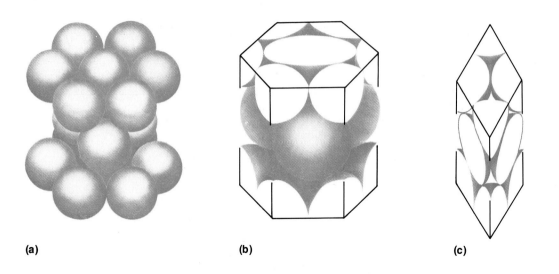

Figure 9.18 Hexagonal lattice **(a)** and **(b)**; unit cell **(c)**.

cubic lattice, is the most efficient of the three and is, in fact, one of the most efficient packing arrangements which exist. This lattice and the *hexagonal lattice* (Figure 9.18) are called, specifically, *closest-packed crystal lattices*. Where bonding forces between units (spheres) are nondirectional and equivalent for all units, the nature of the lattices is determined by purely geometric considerations. These two arrangements (hexagonal and face-centered cubic) place the maximum number of units (spheres) as close together as is permitted by mutual repulsions.

The cubic and hexagonal systems are only two of the seven *crystal systems*. The various crystal systems are most simply differentiated in terms of the relative lengths of the axes of the crystal and the sizes of the angles between them. The cubic system has three axes of equal length and three angles of 90°. The hexagonal system has two equal axes, two 90° angles, and one 120° angle. The seven systems are diagrammed in Figure 9.19, and the relationships among lengths of axes and sizes of angles between axes are given in Figure 9.20.

Question

20. Consider one unit cell in a simple cubic lattice. How many other unit cells share a face, an edge, or a corner with the given unit cell?

9.11 Molecular Solids

Having dispensed with the characteristics common to all solids, we shall now look in turn at each of four types of solids: *molecular solids, ionic solids, metallic solids,*

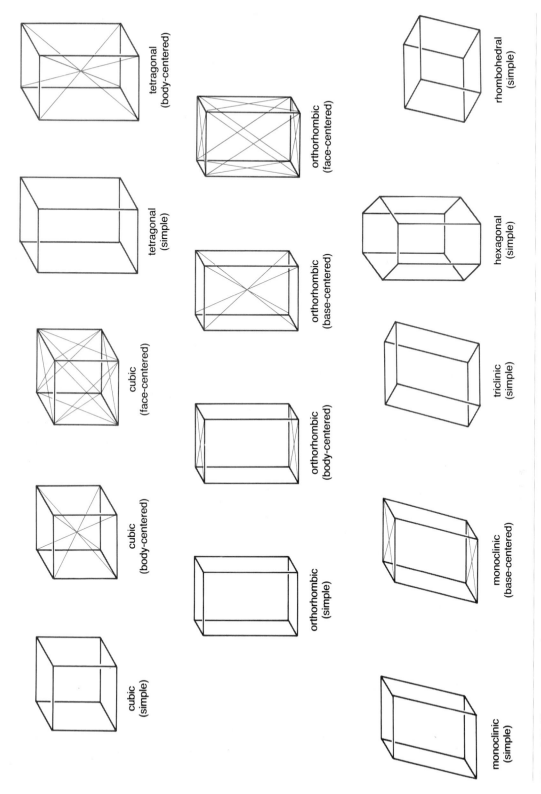

Figure 9.19 The seven crystal systems.

Liquids and Solids—The Condensed Phases

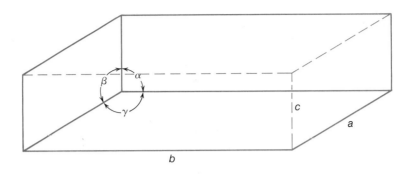

cubic	$a = b = c$	$\alpha = \beta = \gamma = 90°$
tetragonal	$a = b \neq c$	$\alpha = \beta = \gamma = 90°$
orthorhombic	$a \neq b \neq c$	$\alpha = \beta = \gamma = 90°$
monoclinic	$a \neq b \neq c$	$\alpha = \gamma = 90°; \beta \neq 90°$
triclinic	$a \neq b \neq c$	$\alpha \neq \beta \neq \gamma \neq 90°$
hexagonal	$a = b \neq c$	$\alpha = \beta = 90°; \gamma = 120°$
rhombohedral	$a = b = c$	$\alpha = \beta = \gamma \neq 90°$

Figure 9.20 Relationships among the lengths of the axes *a*, *b*, and *c* and the angles between the axes α, β, and γ.

and *network solids*. Each will be distinguished by the particles that occupy the lattice points and by the forces that hold them together.

Nonmetallic solids in which each lattice point is occupied by a molecule are called *molecular solids*. The family of molecular solids is therefore large and diverse: The molecules can range from the monatomic noble gases to large enzymes and proteins. Likewise, the forces that hold the crystals together range from the weak London dispersion forces to hydrogen bonds. The hardness and the melting points of the solids vary accordingly.

The simplest of all crystals are the solid noble gases. Each lattice point is occupied by an identical atom, bound only by the weakest intermolecular force, the London dispersion force. Therefore solid noble gases have low melting points and are soft and easily crushed. Other nonpolar, nonmetallic substances (such as N_2, O_2, F_2, Cl_2, Br_2, I_2, CO_2, and CCl_4) have similar properties. When studying liquids we saw that the strength of the London dispersion force increased with molecular weight; therefore, we might expect melting points to reflect this trend in nonpolar molecular solids. Table 9.9 bears out our expectations—for example, among the noble gases and the elements of Group VIIA. We would see a similar trend among the compounds CH_4, CF_4, CCl_4, CBr_4, and CI_4.

Many of the nonmetals that are solids at room temperature have more complex molecules than the monatomic and diatomic molecules we have seen in most gases. A crystal of elemental boron, for example, has 12 atoms at each point of a closest-packed lattice. The 12 atoms in turn form a geometric figure known as an *icosahedron* (Figure 9.21). In solid sulfur the basic structural unit is

Table 9.9 Melting Points of Some Molecular Solids

		Noble Gases	
		Molecular Weight	Melting Point (°C)
	He	4.0026	−272.2
	Ne	20.183	−248.67
	Ar	39.948	−189.2
	Kr	83.80	−156.6
	Xe	131.30	−119.9
	Rn	222.00	−71.0

	Nonpolar Elements			Nonpolar Compounds	
	Molecular Weight	Melting Point (°C)		Molecular Weight	Melting Point (°C)
He	4.0026	−272.2	CH_4	16.04	−182.48
H_2	2.0159	−259.14	C_2H_6	30.07	−183.3
N_2	28.0134	−209.86	SiH_4	32.12	−185
O_2	31.9988	−218.4	CF_4	87.99	−184
F_2	18.998	−219.62	SiF_4	104.08	−90.2
Cl_2	70.96	−100.98	CCl_4	153.82	−22.99
Br_2	159.81	−7.2	$SiCl_4$	169.90	−70
I_2	253.809	113.5			

			Polar Compounds				
	Molecular Weight	Melting Point (°C)		Molecular Weight	Melting Point (°C)		Molecular Weight	Melting Point (°C)
*HF	20.01	−83.1	*H_2O	18.015	0.00	*NH_3	17.03	−77.7
HCl	36.46	−114.8	H_2S	34.08	−85.5	PH_3	34.00	−133
HBr	80.92	−88.5	H_2Se	80.98	−60.4	AsH_3	77.95	−116.3
HI	127.91	−50.8	H_2Te	129.62	−49	SbH_3	124.77	−88

*Extensively hydrogen-bonded.

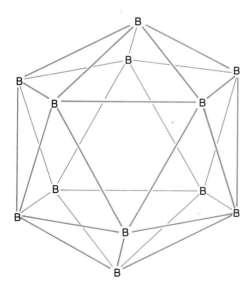

Figure 9.21 The boron (B_{12}) icosahedron.

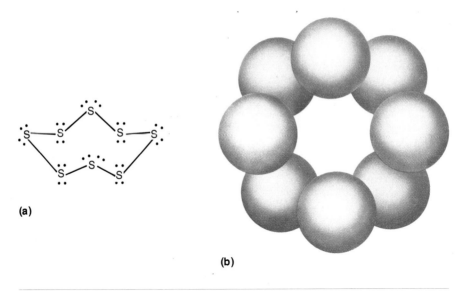

Figure 9.22 The S_8 molecule. **(a)** Electron formula. **(b)** Model.

a nonplanar eight-membered ring of sulfur atoms (Figure 9.22) that occupies the lattice points in either a *rhombic* (short for *orthorhombic*) or a *monoclinic* lattice, depending on the temperature. We call these different forms of sulfur *allotropes*. They differ only in the way the S_8 molecules are arranged. Phosphorus is another example of an allotropic element, but here the difference is the arrangement of atoms. White phosphorus is a white solid in which the structural unit is the P_4 molecule, four atoms bonded together in a tetrahedral unit (Figure 9.23). These units then occupy positions in a cubic lattice. Other allotropes of phosphorus

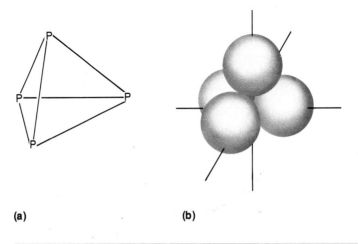

Figure 9.23 The P_4 molecule. **(a)** Electron formula. **(b)** Model.

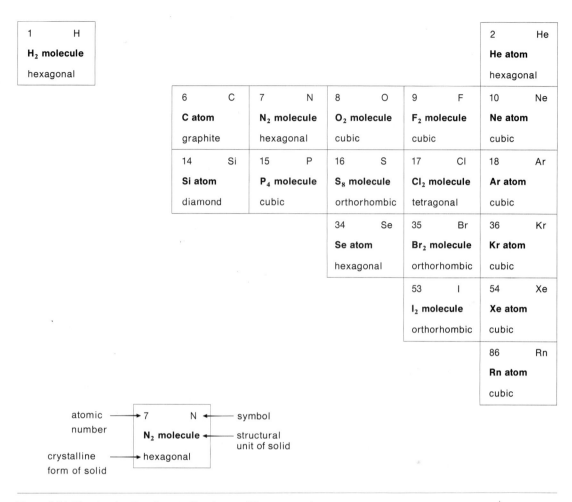

Figure 9.24 Structural unit and crystalline forms of the nonmetals.

contain extended sheets or chains of phosphorus atoms. Selenium is similar to sulfur, but it forms hexagonal and monoclinic allotropes. Figure 9.24 summarizes the properties of these and other nonmetallic, elemental solids. Of these, only carbon, silicon, and some of the allotropes of phosphorus are not molecular solids.

Many nonpolar compounds behave much as the nonmetallic elements. Others, however, have areas of localized charge or nonspherical shapes and hence arrange themselves in special ways within the lattice. Carbon dioxide, for example, orients itself as shown in Figure 9.25.

Crystals that benefit from dipole attractions and hydrogen bonds are in most ways similar to those bound only by London dispersion forces. They merely tend to be harder and more brittle (the stronger the force, the more rigidly it binds the molecule in the crystal) and to have higher melting points. An extreme example, which can hardly be considered a molecular solid at all, is the extensively hydrogen-bonded ice crystal. We shall look more closely at ice in Section 9.16.

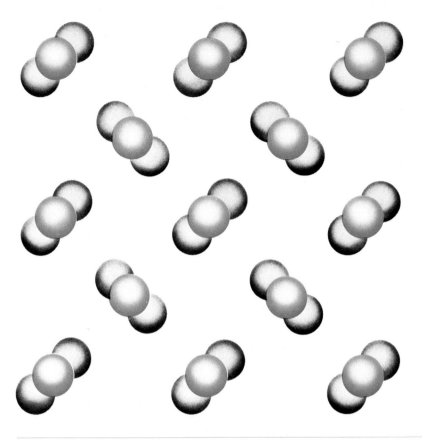

Figure 9.25 Arrangement of CO_2 molecules in one plane of the solid CO_2 crystal: Oxygen atoms are oriented toward carbon atoms in adjacent molecules.

Question

21. Explain the differences in melting points for the following pairs of substances.
    a. Cl_2, Br_2 b. Cl_2, CCl_4 c. Cl_2, HBr d. Cl_2, F_2

9.12 Metallic Solids

In many ways the structure of metallic solids is similar to that of solids we have already considered. However, we must invoke an entirely different binding force to explain the properties that distinguish metals: thermal and electrical conductivity, luster, malleability, and high melting points.

Since we have defined electric current as a flow of electrons, we can begin by assuming that in a metallic crystal there are electrons free to move about when we apply an electrical potential. This means that neither covalent bonding nor

IA	IIA	IIIB	IVB	VB	VIB	VIIB	VIIIB			IB	IIB	IIIA	IVA	VA	VIA	VIIA	0
3 Li 179																	
11 Na 97.8	12 Mg 651											13 Al 660					
19 K 63.7	20 Ca 842–8	21 Sc 1539	22 Ti 1675	23 V 1890	24 Cr 1890	25 Mn 1244	26 Fe 1535	27 Co 1495	28 Ni 1453	29 Cu 1083	30 Zn 419	31 Ga 29.8	32 Ge 937				
37 Rb 38.9	38 Sr 769	39 Y 1495	40 Zr 1852	41 Nb 2468	42 Mo 2610	43 Tc 2140	44 Ru 2250	45 Rh 1966	46 Pd 1552	47 Ag 961	48 Cd 321	49 In 157	50 Sn 232	51 Sb 631			
55 Cs 28.5	56 Ba 725	57 La 920	72 Hf 2150	73 Ta 2996	74 W 3410	75 Re 3180	76 Os 3000	77 Ir 2410	78 Pt 1769	79 Au 1063	80 Hg −38.9	81 Tl 304	82 Pb 328	83 Bi 271	84 Po 254		
87 Fr (27)	88 Ra 700	89 Ac 1050															

58 Ce 795	59 Pr 3127	60 Nd 1024	61 Pm (1027)	62 Sm 1072	63 Eu 826	64 Gd 1312	65 Tb 1356	66 Dy 1407	67 Ho 1461	68 Er 1497	69 Tm 1545	70 Yb 824	71 Lu 1652
90 Th 1750	91 Pa (1230)	92 U 1132	93 Np 640	94 Pu 640	95 Am 850	96 Cm	97 Bk	98 Cf	99 Es	100 Fm	101 Md	102 No	103 Lw

Figure 9.26 Melting points of the metals (°C).

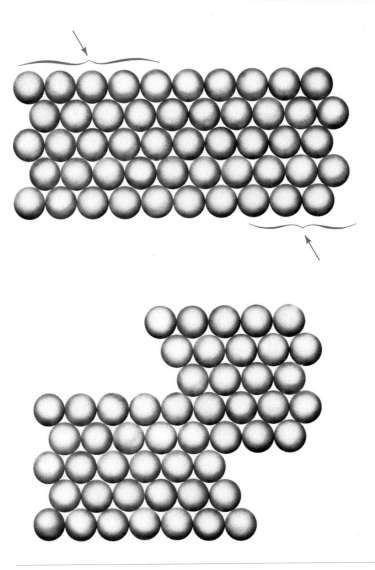

Figure 9.27 Slippage planes in a metal crystal.

ionic bonding is involved in metallic solids—neither would account for free-moving electrons. Nonetheless, a binding force much stronger than the London dispersion forces is necessary to explain the high melting points of metals. The force must be nondirectional (unlike hydrogen bonding, for example) since most metals are malleable, not rigid and brittle.

We can best account for all these properties by visualizing the metallic crystal as a lattice of positive metal ions with their outer energy level electrons stripped away, or *delocalized*, to form a "sea of electrons." In some ways this delocalization is similar to that we described in SO_2 and SO_3 (Section 7.5), where the electrons were delocalized over three or four atoms. In metals, however, the electrons are free to move throughout the crystal. With this concept we can easily explain the flow of free electrons necessary for a metal to conduct electric current. Furthermore, the attraction between the positive ions and the sea of electrons adequately accounts for the strong bonding in metallic crystals.

The attraction of the positive metal ions for the delocalized electrons is the bonding force in the metallic crystal. It might be expected, then, that those metals with more outer energy level electrons per atom would have higher melting points. The values given in Figure 9.26 for IA and IIA metals illustrate this. For other metals there are other factors to be considered, such as size of the ion, nuclear charge, and type of crystal lattice.

Figure 9.27 shows what happens when planes in a metallic crystal slip past one another, as they might when the metal is hammered into a sheet. Because all lattice points are occupied by identical ions and because the forces binding the crystal together are strong but nondirectional, this slippage does not disrupt the integrity of the crystal. Thus metals are malleable. In polar molecular solids, on the other hand, the strong directional forces will not permit slippage; in nonpolar molecular solids, the intermolecular forces are so weak that the crystal crumbles.

Finally, the delocalized electrons account for the luster of metals. These electrons, not restricted to orbitals of specific energies, absorb light of any frequency and then reemit it, which we see as reflection, or shine.

Questions

22. What are some properties that distinguish metals from other solids?

23. Which properties of metals rule out the covalent bond as the explanation for bonding?

24. Which properties of metals rule out the ionic bond as the explanation for bonding?

25. Which properties of metals rule out London dispersion forces as the explanation of the bonding?

9.13 Ionic Solids

Ionic solids are fundamentally different from both metals and molecular solids in that they do not have identical units at all lattice points. Each crystal is composed of positive and negative ions, each positive ion surrounded by negative ions and each negative ion surrounded by positive ions. Figure 9.28 shows the crystal lattice positions of the sodium chloride crystal. The Na^+ ions, considered by themselves, form a face-centered cubic crystal, as do the chloride ions, also considered by themselves. The whole crystal can be looked at as two interpenetrating lattices. The unit cell, however, is as shown. Whether the corners are occupied by the anion or the cation is irrelevant. The unit cell in either case generates the whole lattice if moved along in all directions. Assuming the ions to be in contact along an edge, the edge length is equal to two sodium ion radii

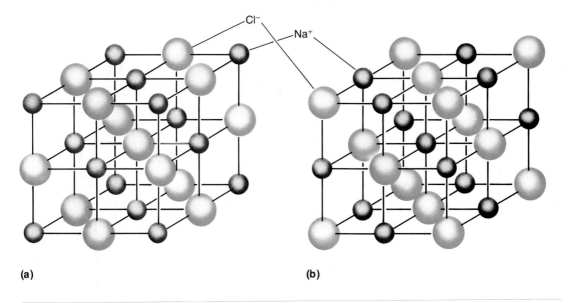

Figure 9.28 Crystal lattice positions in the sodium chloride crystal.

plus two chloride ion radii (thus equal to the sodium ion diameter plus the chloride ion diameter). The coordination number of an ion in an ionic crystal is the number of nearest ions *of opposite charge*. The coordination number of either ion in NaCl is 6.

The electrostatic force which maintains the ionic lattice is very strong; thus these crystals have high melting points, but the force is completely nondirectional in nature so that it places no restriction upon the lattice type. Because the positions are occupied by different units, one plane cannot easily be slipped over another, for this would bring mutually repulsive units into close proximity (Figure 9.29). As a result, these crystals are hard and brittle, as compared with the malleable metal crystal or the soft molecular crystals.

We owe our understanding of crystal structure and our knowledge of spacings and dimensions in specific crystals to the development of the process of X-ray diffraction. X rays are electromagnetic radiation with wavelengths in the range of 10^{-10} to 10^{-12} m. Like all electromagnetic radiation, X rays exhibit the phenomena of *interference* and *diffraction*. As in the earlier discussion of wave characteristics, it is helpful to picture the analogous water waves.

What happens when advancing waves strike a wall—but a wall in which there is a small opening? Most of the waves are dissipated against the wall, but at the opening the waves cause a disturbance which results in a new set of waves, going out in all directions (Figure 9.30). The wavelength of the new waves is the same as that of the original waves. Light displays this diffraction effect.

Interference effects may also be more easily understood through a consideration of water waves. When two rocks are dropped into a pool at separate

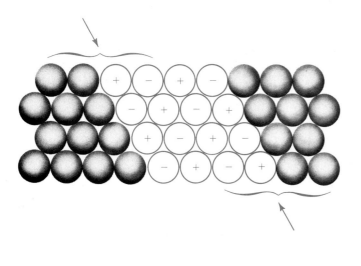

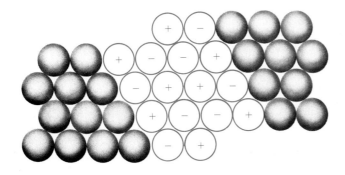

Figure 9.29 Planes in an ionic solid cannot move into new positions relative to each other by slipping, because of mutual repulsions.

positions, interference occurs at the spot where the advancing wave fronts from one disturbance encounter those from the other. To simplify our example, suppose that the waves are identical as to velocity, frequency, and amplitude. Then two extreme situations may occur. If the trough of one wave reaches a spot at the same time that the crest of the other wave arrives, the waves will nullify each other—cancel each other out. On the other hand, if the two crests arrive simultaneously, the resulting waves will be doubled in amplitude. In the second case we say that the waves were *in phase* with each other and that the result was *constructive interference*. In the first case—and in all cases where the two waves are *out of phase*—*destructive interference* is the result.

In X-ray diffraction, a thin beam of X rays of a single wavelength is allowed to fall upon a crystal. The scattering of the rays which pass through

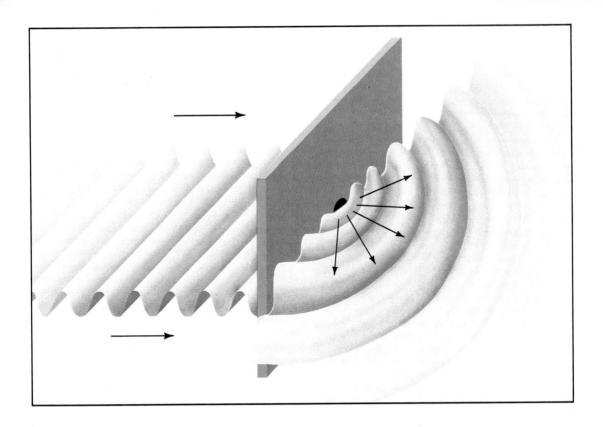

Figure 9.30 Waves striking a wall which contains a small opening.

the crystal is detected on a photographic plate placed behind the crystal (Figure 9.31).

The X rays are reflected from the planes of the crystal. Where the reflections from a series of planes are in phase, a strong beam will be seen as a spot on the photographic plate. If reflected rays are out of phase, of course, destructive interference will result and no spot will be observed. Figure 9.32 gives some idea of how the distance between planes of a crystal together with the wavelength of the incident X-ray beam determine whether reinforcement or destruction will result.

As X-ray 1 reaches point A, X-ray 2 reaches point B. Ray 2 travels farther than ray 1 by the distances BC and CD. When ray 2 reaches point D, in order for the two rays to be in phase with each other, the wave of 2 must be at exactly the same stage of its development as it was at point B. In other words, the distance BCD must be some whole-number multiple of the wavelength. This distance can be related through trigonometry to d, the distance between planes.

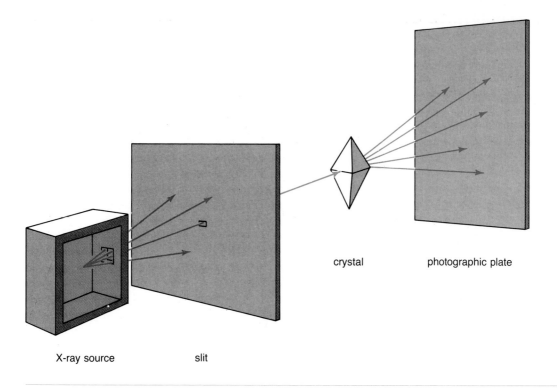

Figure 9.31 X-ray diffraction of a crystal (schematic).

If

$$BCD = n\lambda$$
$$BC = \frac{n\lambda}{2}$$

Then

$$\sin \theta = \frac{BC}{d} = \frac{n\lambda}{2d}$$

The angle of incidence (the angle between the ray of light and the reflecting plane) can be measured, the wavelength is a known value, and setting $n = 1, 2, 3, \ldots$ permits calculation of d. The equation above, in its usual form

$$n\lambda = 2d \sin \theta$$

is called the *Bragg equation*, for William Henry Bragg and William Lawrence

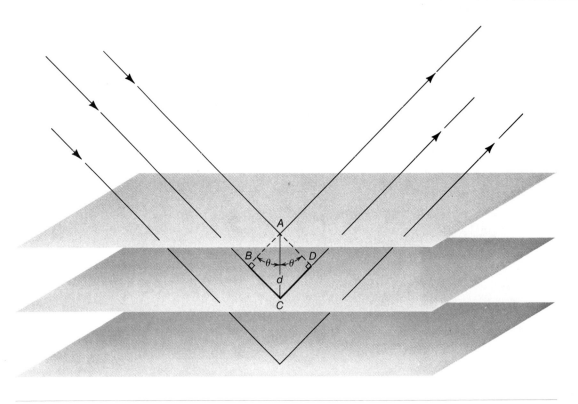

Figure 9.32 Distance between planes of a crystal and reflection of light rays.

Bragg, father and son, who derived it in 1913. Figure 9.33 shows a typical pattern of spots which might be obtained by X-ray diffraction.

In addition to spacing distances in crystals, ionic radii are also deduced from diffraction data. Figure 9.34 gives a set of values for some of the more common ions. The radius of an ion is in no way like the radius of a sphere used to represent an ion in a model or illustration. An ionic radius is the "effective radius" of the ion and depends to some extent on the nearness of and the charge of neighboring ions. Nevertheless, a correlation between these ionic radii and lattice type can be found.

In binary compounds of the *AB* type (one-atom-to-one-atom ratio) a comparison of the *radius ratio* (cation to anion) with the coordination number leads to some interesting conclusions. For a radius ratio greater than 0.732, a coordination number of 8 is most likely. For radius ratio values between 0.732 and 0.414, a coordination number of 6 is most likely; between 0.414 and 0.225, tetrahedral coordination occurs; between 0.225 and 0.155, trigonal coordination occurs. For radius ratios smaller than 0.155, only linear relationships are possible. We can explain these correlations in a way which should make them seem logical. Suppose a cation is surrounded by anions only a little larger than itself in a body-centered cubic arrangement. The eight anions are equidistant

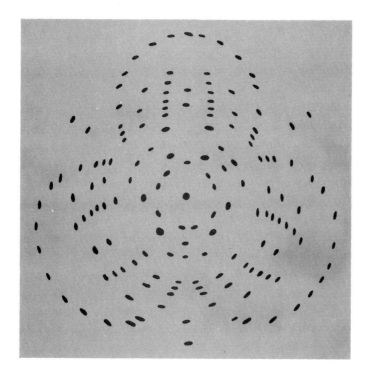

Figure 9.33 A typical X-ray diffraction pattern.

from each other and sufficiently far apart so that their mutual repulsions are not so great as their attractions for the cation. However, suppose we consider other cations in place of the first, of smaller and smaller radius. The anions, in drawing close to the cation, will also draw closer and closer to each other until a point is reached where their mutual repulsions prevent the maintenance of eight anions around the cation. Theoretically, this happens when the radius ratio reaches 0.732. From that point on, as we continue to consider smaller and smaller cations, only six anions can be clustered around each cation. Eventually, however, a place will be reached where a stable configuration is no longer possible with a coordination number of 6, and the lattice will change to one with tetrahedral relationships—and so on.

In compounds of the AB_2 or A_2B type, the coordination number of the ion present to the greater extent is one-half that of the coordination number of the other ion. Further complications are introduced into ionic structure by the presence of polyatomic ions such as the very common nitrate, sulfate, or carbonate ions.

There is no molecule of an ionic substance because no group of ions is separate and distinct from any other group of ions in the lattice. For this reason, formulas for ionic compounds are empirical formulas which give only the ratio in which atoms of the various elements are present in the compound.

IA	IIA	IIIB	IVB	VB	VIB	VIIB	VIIIB			IB	IIB	IIIA	IVA	VA	VIA	VIIA	0
Li^+ 0.060																	
Na^+ 0.097	Mg^{2+} 0.065											Al^{3+} 0.050		N^{3-} 0.171	O^{2-} 0.140	F^- 0.136	
K^+ 0.133	Ca^{2+} 0.099	Sc^{3+} 0.081	Ti^{2+} 0.090	V^{3+} 0.074	Cr^{3+} 0.069	Mn^{2+} 0.080	Fe^{3+} 0.064 Fe^{2+} 0.076	Co^{3+} 0.063 Co^{2+} 0.074	Ni^{2+} 0.072	Cu^{2+} 0.069 Cu^+ 0.096	Zn^{2+} 0.074	Ga^{3+} 0.062		P^{3-} 0.212	S^{2-} 0.184	Cl^- 0.181	
Rb^+ 0.148	Sr^{2+} 0.113	Y^{3+} 0.093								Ag^+ 0.126	Cd^{2+} 0.097	In^{3+} 0.081	Sn^{2+} 0.112		Se^{2-} 0.198	Br^- 0.195	
Cs^+ 0.169	Ba^{2+} 0.135	La^{3+} 0.115								Au^+ 0.137	Hg^{2+} 0.110	Tl^{3+} 0.095	Pb^{2+} 0.120	Bi^{3+} 0.120		I^- 0.216	
Fr^+ 0.176	Ra^{2+} 0.140																

Figure 9.34 Radii of some simple ions [Radii are given in nanometers (1 nm = 1×10^{-9} m)].

Example 9.6 Using the ionic radii values from Figure 9.34, calculate the density of NaCl.

Solution

a. The edge length of the unit cell converted to centimeters for the density calculation is

$$2(9.7 \times 10^{-11} \text{ m})\left(\frac{10^2 \text{ cm}}{1 \text{ m}}\right) + 2(1.81 \times 10^{-10} \text{ m})\left(\frac{10^2 \text{ cm}}{1 \text{ m}}\right)$$
$$= 5.56 \times 10^{-8} \text{ cm}$$

b. The volume of the unit cell is

$$(5.56 \times 10^{-8} \text{ cm})^3 = 1.72 \times 10^{-22} \text{ cm}^3$$

c. Eight sodium corner ions (Figure 9.28a) contribute eight eighths or one ion to the unit cell. Six sodium face-centered ions contribute one-half ion each, or three ions. The total is four sodium ions per unit cell.

d. Twelve chloride ions located on the edges of the cube contribute one-fourth ion each or three chloride ions, and together with the one body-centered chloride ion they give a total of four chloride ions per unit cell.

e. Four sodium ions weigh

$$(4 \text{ ions})\left(\frac{1.00 \text{ mol}}{6.02 \times 10^{23} \text{ ions}}\right)\left(\frac{23.0 \text{ g}}{1.00 \text{ mol}}\right) = 1.53 \times 10^{-22} \text{ g}$$

f. Four chloride ions weigh

$$(4 \text{ ions})\left(\frac{1.00 \text{ mol}}{6.02 \times 10^{23} \text{ ions}}\right)\left(\frac{35.4 \text{ g}}{1.00 \text{ mol}}\right) = 2.35 \times 10^{-22} \text{ g}$$

g. The weight of the unit cell is

$$(1.53 \times 10^{-22} \text{ g}) + (2.35 \times 10^{-22} \text{ g}) = 3.88 \times 10^{-22} \text{ g}$$

h. The density of the unit cell and thus of crystalline sodium chloride is

$$\frac{3.88 \times 10^{-22} \text{ g}}{1.72 \times 10^{-22} \text{ cm}^3} = 2.26 \text{ g/cm}^3$$

Questions

26. What properties of an ionic crystal show that it contains no delocalized electrons?

27. Explain the difference between constructive and destructive interference.

9.14 Network Solids

Unlike molecular solids, where discrete molecules are bound together by intermolecular forces, neither ionic solids nor metallic solids contain discrete molecular units. In NaCl it is impossible to assign a Cl^- ion to a specific Na^+ ion. Likewise, in a metal we cannot identify a single uncharged atom since all share the sea of electrons. Thus the formulas for metals and for ionic compounds are only empirical formulas. The same is true for network solids, which contain atoms of one or more elements, covalently bonded to one another in a large array. Again, we cannot isolate a discrete molecule.

Diamond is a perfect network solid. It contains carbon atoms tetrahedrally bonded to one another with strong covalent bonds (Figure 9.35). The strength of these bonds gives diamond its high melting point and since the bonds are directional, they produce a hard, rigid crystal rather than a malleable one. The crystalline regularity of the diamond is also responsible for its renowned beauty.

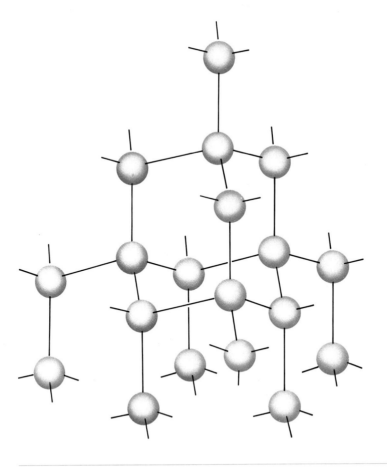

Figure 9.35 Structure of diamond.

Several other network solids, such as silicon carbide (SiC), silicon dioxide (quartz, SiO_2), and boron nitride (BN) also have the tetrahedral diamond structure and thus are hard, high-melting crystals.

Although silicon and phosphorus as well as carbon can form network solids, most network solids are compounds. Among these compounds are many combinations of oxygen and other nonmetals. In most, each nonmetal atom forms four tetrahedral bonds with oxygen atoms and each oxygen atom forms one or two bonds with nonmetal atoms. For example, the nonmetals of the third period and below form discrete polyatomic ions with oxygen, such as ClO_4^-, SO_4^{2-}, PO_4^{3-}, SiO_4^{4-}, SeO_4^{2-}, and IO_4^-. If two of these tetrahedra share an oxygen atom, they form more complex units such as chlorine heptaoxide, Cl_2O_7, and the pyrosulfate and pyrophosphate ions, $S_2O_7^{2-}$ and $P_2O_7^{4-}$:

$$\begin{array}{ccc}
:\ddot{O}:\quad:\ddot{O}: & :\ddot{O}:\quad:\ddot{O}:^{2-} & :\ddot{O}:\quad:\ddot{O}:^{4-} \\
:\ddot{O}:\ddot{Cl}:\ddot{O}:\ddot{Cl}:\ddot{O}: & :\ddot{O}:\ddot{S}:\ddot{O}:\ddot{S}:\ddot{O}: & :\ddot{O}:\ddot{P}:\ddot{O}:\ddot{P}:\ddot{O}: \\
:\ddot{O}:\quad:\ddot{O}: & :\ddot{O}:\quad:\ddot{O}: & :\ddot{O}:\quad:\ddot{O}:
\end{array}$$

In beryl [$Be_3Al_2(Si_6O_{18})$], the cyclic anion shares six oxygen atoms (Figure 9.36). It should not surprise us, therefore, to find that when we fit these units

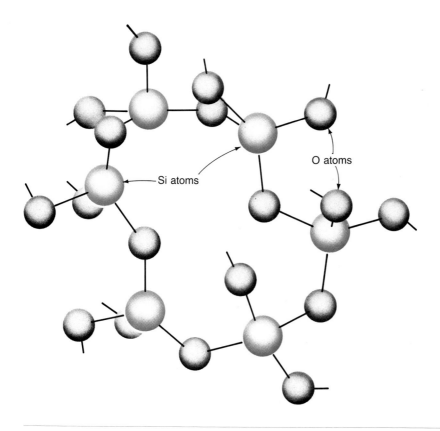

Figure 9.36 Anion ($Si_6O_{18}^{12-}$) of beryl, $Be_3Al_2(Si_6O_{18})$.

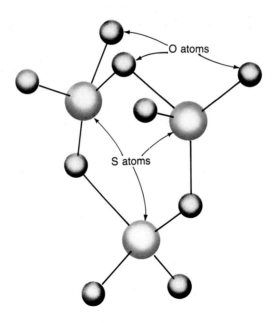
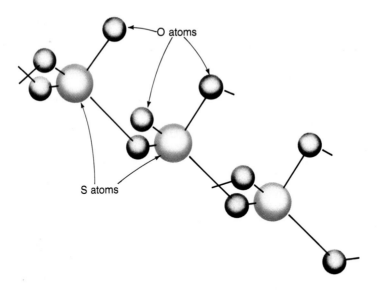

Figure 9.37 The two molecular forms of SO$_3$.

into a lattice even more extensive sharing often occurs. It becomes impossible to identify a discrete molecular or ionic unit; hence we have network solids rather than molecular or ionic solids.

The tendency of these oxides to share oxygen atoms extensively depends on the electronegativity of the central nonmetal atom. For example, the electronegativity of chlorine (3.0) is close to that of oxygen (3.5) so that bonding electrons are approximately equally shared between the atoms. Since oxygen does not have an excess of bonding electrons, it has little tendency to form additional covalent bonds, and Cl_2O_7 is the largest chlorine-oxygen unit we find. But as the electronegativity of the central atom decreases, the oxygen atom receives a larger share of the bonding electrons; hence we find more examples of extended bonding. Sulfur (electronegativity 2.5), in addition to forming discrete pyrosulfate ions, forms two types of sulfur trioxide crystals, one consisting of discrete ring molecules and the other of endless chains (Figure 9.37).

Phosphorus (electronegativity 2.1) forms a variety of compounds with oxygen in which one, two, three, or four oxygen atoms in each tetrahedron are shared with other phosphorus atoms. The *metaphosphoric acids*, for example, have the empirical formula $(HPO_3)_n$, where n may have the values 3, 4, 5, ... Some of these acids are ring molecules and thus form molecular solids. Others, however, where n is very large, consist of long, branched-chain molecules and are sticky, viscous liquids that form glasslike solids (see pseudosolids, p. 301).

The electronegativity of silicon is only 1.8; thus it is not surprising that, among the compounds and ions of the nonmetals, silicon and oxygen form the greatest variety of network solids. The importance of these silicates cannot be overemphasized. Minerals such as flint, onyx, granite, quartz, agate, talc, mica, and asbestos; precious and semiprecious stones such as emerald, amethyst, opal, and jade; and many manufactured products such as cement, brick, glass, and various ceramics are all silicates.

If two oxygen atoms in each (SiO_4) tetrahedron are shared between silicon atoms, the results are chainlike network solids: the single-chain *pyroxenes* and the double-chain *amphiboles* (Figure 9.38). Asbestos is an example of an amphibole. If each tetrahedral unit shares three oxygen atoms, the silicates form sheets, such as talc and mica (Figure 9.39). Both the fiberlike and sheetlike silicates are anions, and the fibers and sheets are held together by ionic attractions for metal cations dispersed among them.

Finally, all four oxygen atoms in each tetrahedral unit can be shared, producing a three-dimensional lattice. Quartz (Figure 9.40) and the feldspars are examples. Table 9.10 summarizes the variety of oxygen-containing compounds and ions formed with chlorine, sulfur, phosphorus, and silicon.

Before leaving the network solids, we shall note that not all involve tetrahedral bonding (sp^3 hybridization). For example, the boron-oxygen compounds are sp^2 hybrids, with each boron atom forming three bonds with oxygen atoms and each oxygen atom forming one or two bonds with boron atoms. (Recall that boron compounds are often exceptions to the octet rule.) As with silicates, there are discrete borate units, BO_3^{3-}, $B_2O_5^{4-}$, and $B_6O_{12}^{6-}$ (a cyclic anion),

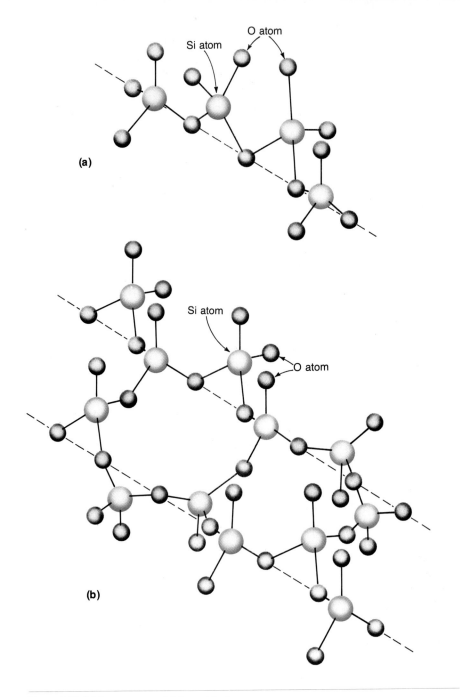

Figure 9.38 (a) Pyroxenes and (b) amphiboles.

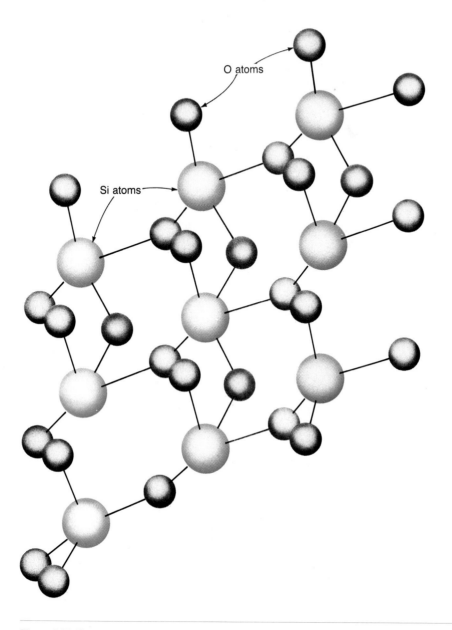

Figure 9.39 Sheetlike silicates.

as well as chainlike and sheetlike borates. Boric acid also has a sheetlike structure because of extensive hydrogen bonding between H_3BO_3 units. This hydrogen bonding makes boric acid one of the few inorganic acids that is solid at room temperature.

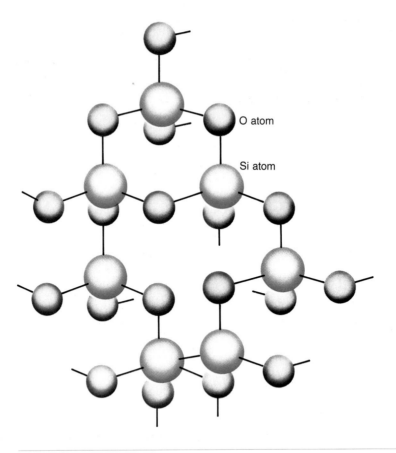

Figure 9.40 Quartz (SiO$_2$).

Table 9.11 tabulates the most important characteristics of molecular, metallic, ionic, and network solids and gives examples in each class.

9.15 Graphite

As we might expect, there are exceptions to our rigid classification of solids. Graphite, the form in which we usually find carbon, is a common example. In contrast to diamond, where each carbon atom forms four sp^3-hybridized sigma bonds, each carbon atom in graphite forms three sp^2-hybridized sigma bonds. Graphite, therefore, consists of two-dimensional sheets of carbon atoms as shown in Figure 9.41. The bonding force between the sheets of atoms is more complicated. For many years it was thought that the only forces involved were London dispersion forces, whose weakness explained the lubricating properties of graphite: The sheets could thus slip easily past one another. During World War II, however, graphite was found to lose its lubricating property in aircraft

Table 9.10 Compounds and Ions of Oxygen and Third-Period Nonmetals

Element, Electronegativity		Single Tetrahedron (discrete ion)	One Oxygen Atom Shared Between Tetrahedra	Two Oxygen Atoms Shared Between Tetrahedra	Three Oxygen Atoms Shared Between Tetrahedra	Four Oxygen Atoms Shared Between Tetrahedra
				Rings / Chains	Sheets	Three-dimensional networks
Cl	3.0	ClO_4^-	Cl_2O_7			
S	2.5	SO_4^{2-}	$S_2O_7^{2-}$	$(SO_3)_3$ / $(SO_3)_n$		
P	2.1	PO_4^{3-}	$P_2O_7^{4-}$	$Na_5P_3O_{10}$ (sodium tripolyphosphate) / $(HPO_3)_n$ Metaphosphoric acids		
Si	1.8	SiO_4^{4-}	$Si_2O_7^{6-}$	$(SiO_3)_3^{6-}$ $(SiO_3)_6^{6-}$ Emerald / Pyroxenes and amphiboles Asbestos	Mica Talc Clays	$(SiO_2)_n$ Quartz Feldspars Zeolites

Table 9.11 Characteristics of Solids

Class		Units Occupying Lattice Points	Lattice Forces	Examples
Molecular	Nonpolar	Nonpolar molecules	London forces	Ne, CO_2, Cl_2
	Polar	Polar molecules	London forces, dipole attractions	HBr, CH_3Cl, H_2S
	H-bonded	Polar molecules containing at least one unshared pair of electrons and one H atom on a small, highly electronegative atom such as F, O, or N.	London forces, dipole attractions, H-bond	HF, H_2O, NH_3
Ionic		Alternating positive and negative ions	Electrostatic attraction of positive ions for negative ions	NaCl, $Ca(NO_3)_2$, $Zn(OH)_2$
Metallic		Positive ions	Attraction of positive ions for delocalized electrons	K, Fe, Cu
Network		Atoms	Covalent bonds	C_n (diamond) SiO_2, $(HPO_3)_n$

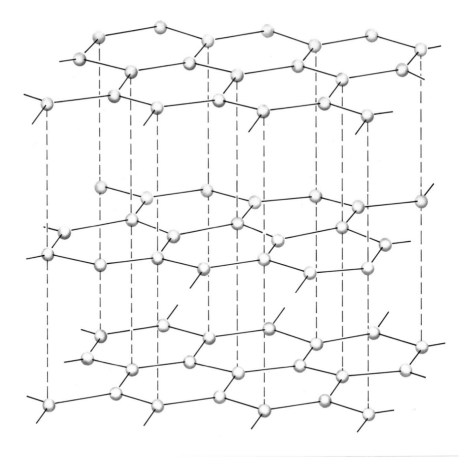

Figure 9.41 Structure of graphite.

at high altitudes. We now know that this was a result of the low pressure. Under normal conditions gas molecules are trapped between the layers of graphite, weakening their attraction for one another; at lower pressures the gas is drawn off, and the London forces between the layers destroy the lubricating property.

So far we might regard graphite as a simple network solid, in spite of its interesting properties. However, it also shares some properties with metals. Each carbon atom in graphite has a p orbital and an electron not involved in sigma bonding. These electrons become delocalized throughout a single layer because of the overlapping of the p orbitals, forming what we might think of as a vast pattern of pi bonds. This sea of electrons therefore conducts electricity parallel to the carbon planes, but graphite remains an insulator if a potential is applied perpendicular to the planes.

Question

28. In what way does graphite resemble
    a. a metal?   b. a molecular solid?   c. a network solid?

9.16 Ice

Another substance that does not fit neatly into any of our categories is ordinary ice. The structural unit of water is a polar molecule, but the extent of the hydrogen bonding in ice makes it a poor example of a molecular solid. These bonds are so extensive that in many ways we can compare the crystalline structure of ice to that of a three-dimensional network solid like diamond. Each oxygen atom forms four bonds: two covalent bonds and two hydrogen bonds (Figure 9.42). Each hydrogen atom in turn forms two bonds: one covalent bond and one hydrogen bond. Of course the hydrogen bonds are not so strong as the covalent bonds in a true network solid, but the ice crystal is much harder and has a higher melting point than we would expect for a simple polar molecular solid.

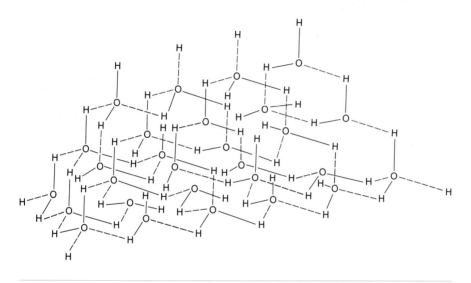

Figure 9.42 Structure of ice.

9.17 Pseudosolids

Finally we shall take a brief look at some substances that can only loosely be called solids. Glasses, tar, and many plastics are more accurately described as *pseudosolids* or *amorphous solids*. These materials meet neither of the two criteria for true solids: They do not have a melting point; instead they soften as heat is applied, becoming less and less viscous. They do not have a regular, crystalline structure. Usually the structural unit in these materials is a large, complex molecule. They can be crystallized into true solids, but the temperature must be lowered very slowly to allow the large units to move into lattice positions. If they are cooled more rapidly, the molecules are trapped in a random, liquidlike arrangement, but with strong intermolecular forces holding them rigidly in place. Thus an ordinary windowpane is merely a supercooled liquid. In fact, if we look closely at a very old windowpane, we shall find the glass thicker at the bottom of the pane than at the top, and we can see ripples where the glass has flowed under the influence of gravity.

Questions and Problems

9.2 Evaporation

29. Calculate the number of calories absorbed when
  a. 1.00 kg of oxygen vaporizes.   b. 1.00 kg of helium vaporizes.

9.4 Collecting Gas Over Water

30. If we collect hydrogen over water at 28°C and 7.40×10^2 mm of Hg pressure, what fraction of the gas is water vapor?

31. What is the molecular weight of a gas if 2.83 g occupy 8.00×10^2 ml over water at 25°C and 7.50×10^2 mm of Hg?

32. 4.50×10^2 ml of an unknown gas, collected over water at 23°C and 747 mm of Hg, weighs 1.39 g. What is its molecular weight? What is its molecular formula if it is composed of 92.3 percent carbon and 7.7 percent hydrogen?

*33. How many milliliters of water (assume a density of 1.00 g/ml) would you obtain from the condensation of 1.00×10^2 liters of steam at 2.00 atm of pressure and (5.00×10^2)°C?

9.6 The Solid-Liquid Equilibrium

34. Using the figures tabulated in Table 9.4, calculate the number of calories liberated when
  a. 1.00 kg of hydrogen at -259°C freezes.
  b. 1.00 kg of chlorine at -101°C freezes.

35. How many calories of heat are required to melt 1.00×10^3 kg of SO_2? (See Table 9.4.)

36. What weight of ice could be melted by the heat liberated when 3.00×10^2 g of steam condenses at (1.00×10^2)°C?

37. What weight of water could be vaporized at (7.00×10^1)°C by the heat liberated when 1.00 kg of water at 0°C freezes?

*38. What will be the final temperature when 4.00 kg of water at 20.0°C and 4.00×10^2 g of ice are mixed? Assume that no heat is gained or lost from the system.

9.7 Phase Diagrams

39. Distinguish between the following terms:
  a. boiling point and normal boiling point
  b. heat of vaporization and heat of fusion
  c. heat of fusion and heat of sublimation
  d. total pressure and vapor pressure

9.8 Intermolecular Forces

*40. Justify the existence of the ionic compound KHF_2.

41. For each of the following pairs of substances, indicate which you think will have the lower boiling point. Give the reason for your choice.
  a. CH_4 and NH_3 b. CH_4 and SiH_4 c. GeH_4 and AsH_3

42. Which of the following would you expect to exhibit hydrogen bonding? Indicate your reasons.
  a. CH_3OH b. H_2Se c. CH_3NH_2 d. CH_3F e. CH_3OCH_3
  f. C_8H_{18} g. HBr h. $(CH_3)_3\text{-}CH$ i. SiH_4

43. Does water contract or expand in going from
  a. 12°C to 3.98°C?   b. 20°C to 60°C?
  c. 3.98°C to 0°C?   d. 3.0°C to 2.0°C?

44. List the unusual characteristics of water that distinguish it from most other liquids.

9.10 Properties of Solids

45. Discuss ways in which the spheres used in a model of a crystal differ from the atoms or ions they represent.

46. Distinguish between the following terms:
  a. face diagonal and body diagonal   b. crystal system and crystal lattice
  c. lattice and unit cell

47. What volume will one mole of neon atoms occupy if the radius of the atom is 1.39×10^{-10} m and the crystal is face-centered cubic?

48. Use the value from Question 47 to calculate the density of solid neon.

9.12 Metallic Solids

49. In what ways does the metallic crystal resemble
  a. the ionic crystal?   b. network solids?   c. molecular solids?

50. The gold crystal is face-centered cubic, with the edge length of the unit cell being 4.08×10^{-10} m. What is the radius of the gold ion?

51. Iron crystallizes in a body-centered cubic structure with an edge length for the unit cell of 2.91×10^{-10} m. What is the radius of the iron ion?

*52. Copper crystallizes in a face-centered cubic crystal. The density of copper is 8.92 g/cm³. What is the length of an edge of the copper unit cell? What is the radius of the copper ion?

9.13 Ionic Solids

53. Use the form of the unit cell in Figure 9.28b, with the chloride ions at the corners, and verify that the density is the same as that calculated from Figure 9.28a in Example 9.6.

54. Use Figure 9.34 to calculate the radius ratio for each of the following substances. Then predict the probable coordination number of each ion.
    a. KBr   b. NaI   c. MgS   d. MgO
    e. CaS   f. MgSe   g. SrO   h. RbCl

55. An ionic crystal consists of a face-centered cubic arrangement in which each face-center position is occupied by an ion of element W and each corner position, by an ion of element Z. Give the empirical formula for the compound.

56. An ionic crystal consists of a body-centered cubic arrangement in which each body-center position is occupied by an ion of element X and each corner position, by an ion of element Y. Give the empirical formula for the compound.

*57. X rays of wavelength 1.5374×10^{-10} m were reflected from a crystal with a measured angle of $16°10'$. Determine the minimum spacing of crystal planes from this data.

9.14 Network Solids

58. Explain the reasons for the differences in the following pairs of melting points:
    a. Cl_2: $-101°C$ and Br_2: $-7.2°C$
    b. HCl: -115 and Cl_2: -101
    c. Cl_2: -101 and   HBr: -88.5
    d. SiO_2: 1610   and CO_2: -56.6
    e. K: 63.7   and   Ca: 842

59. Draw electron formulas for the following discrete units.
    a. $(SO_3)_3$ b. H_3BO_3 c. $H_4B_2O_5$ d. the pyrosilicate ion

60. Would you expect to find more extensive network solids involving bromine-to-oxygen bonds than chlorine-to-oxygen bonds? Explain.

9.15 Graphite

61. Give examples in which electrons are delocalized over
    a. three atoms.   b. four atoms.
    c. an entire layer of a network solid.   d. an entire crystal.

62. What would you expect the bond angles to be in the following solids?
    a. diamond   b. boric acid   c. graphite   d. quartz

63. Give the coordination number of
    a. carbon in diamond.
    b. carbon in graphite.
    c. silicon in quartz.
    d. oxygen in quartz.
    e. silicon in silicon carbide.
    f. aluminum in a face-centered cubic structure.
    g. calcium in a body-centered cubic structure.

64. Explain these properties of graphite in terms of its structure:
   a. electrical conductivity   b. lubricating ability   c. high melting point

9.17 Pseudosolids

65. Prepare a chart or outline that provides the following information for each class of solids (nonpolar molecular solids, polar molecular solids, hydrogen-bonded molecular solids, metallic solids, ionic solids, and network solids).
   a. nature of units occupying lattice sites
   b. nature of bonding forces
   c. electrical conductivity
   d. hardness
   e. range of melting points
   f. several examples

66. Which properties of a solid are most directly affected by each of the following?
   a. whether the bonding forces are directional or nondirectional
   b. strength of the bonding forces
   c. having identical or nonidentical units at lattice points

67. Identify the dominant forces in the crystal lattice of each of the following substances:
   a. calcium carbonate   b. graphite   c. sulfur   d. ice
   e. argon   f. iron   g. aluminum sulfate   h. quartz

68. Classify each of the following as molecular crystals, metallic crystals, ionic crystals, network solids, or pseudosolids:
   a. aluminum   b. table salt   c. glass   d. ice
   e. dry ice   f. copper sulfate   g. gold   h. quartz

69. In which of the five classes given in Question 68 do we find substances with the following properties?
   a. high electrical conductivity   b. brittleness
   c. malleability   d. high melting points

Discussion Starters

70. Explain or account for the following:
   a. Water becomes cooler when some of it evaporates.
   b. Different liquids have different vapor pressures at the same temperature.
   c. Water has an unusually high boiling point.
   d. A lake freezes from the top rather than from the bottom.
   e. Water expands when it freezes.
   f. Some insects can walk on water even though their densities are greater than that of water.
   g. Chlorine has a much higher boiling point than hydrogen.
   h. Water has a much higher boiling point than hydrogen sulfide.
   i. On very cold nights farmers often put tubs of water in storage areas for fruits and vegetables.
   j. Large lakes have a moderating influence on the temperatures of adjoining areas.
   k. Water has an unusually high heat of vaporization.
   l. Water is densest at $3.98°C$.
   m. An increase in pressure lowers the melting point of ice.
   n. A meal can be cooked faster in a pressure cooker than in an open container.

o. Ice disappears on a cold (below 0°C), dry, winter day.
p. Farmers irrigate their crops when there is a danger of frost.
q. "Dry ice" is better than ordinary ice for preserving ice cream.
r. Liquids, not gases, are used in hydraulic brakes.
s. Wet clothes dry when hung outdoors even when the temperature is below freezing.
t. A freely falling drop of water assumes a nearly spherical shape.
u. Heat is necessary to vaporize a liquid at constant temperature.
v. A motionless skater will find that the blades of his skates sink slowly into the ice.
w. A car will skid more readily on ice when the temperature is just below the freezing point.
x. The heat of vaporization of a substance is always several times greater than its heat of fusion.
y. Water can be made to boil at 60°C.
z. The vapor pressure of a liquid is independent of the amount present.
aa. When we write the boiling point of a liquid, we must specify the atmospheric pressure; when we write the freezing point, we do not.
bb. Water that seeps into the tiny crevices of a rock help break it down when the temperature reaches freezing.
cc. One should be more concerned about the flammability of diethyl ether

$$\begin{array}{ccccc} H & H & & H & H \\ \vdots & \vdots & \ddots & \vdots & \vdots \\ H:C:&C:&\ddot{O}:&C:&C:H \\ \vdots & \vdots & \ddots & \vdots & \vdots \\ H & H & & H & H \end{array}$$

than that of butyl alcohol

$$\begin{array}{ccccc} H & H & H & H & \\ \vdots & \vdots & \vdots & \vdots & \ddots \\ H:C:&C:&C:&C:&\ddot{O}:H \\ \vdots & \vdots & \vdots & \vdots & \\ H & H & H & H & \end{array}$$

dd. Graphite is a nonmetal but conducts electricity.
ee. The atomic weight of aluminum (27.0) is only one-third that of rubidium (85.5), but the density of aluminum is 2.70 g/cm^3, whereas that of rubidium is only 1.53 g/cm^3.
ff. Glass is rigid but is not a true solid.
gg. The discrete SO_3 molecule contains delocalized electrons but this compound does not conduct electricity.

Solutions

10

We can begin our discussion of solutions by seeking a useful way of distinguishing between homogeneous and heterogeneous mixtures. As we suggested in Chapter 1, however, homogeneity is likely to be a matter of definition. How small must the solute particles be, and how well dispersed, before we can call a mixture a solution? Thus we find another situation in chemistry where rigid distinctions—in this case between solutions and heterogeneous mixtures—are only convenient approximations. Nonetheless, we shall rely on a classification according to the size of the solute particles dispersed in the solvent. Solute and solvent were defined in Chapter 3.

If the diameter of each particle of solute, whether a single molecule or an aggregate of molecules, is less than 10^{-7} cm, we shall regard the mixture as a **true solution.** If the diameter of the particles is between 10^{-7} and 10^{-4} cm, we still have a homogeneous mixture, but instead of a solution we shall call it a **colloidal suspension.** Colloidal particles are usually large enough to be seen with an electron microscope. If the dispersed particles are still larger, we shall consider the suspension a **heterogeneous mixture.** Usually such coarse suspensions will separate spontaneously, just as particles of silt will settle to the bottom of a still, muddy lake.

We shall not discuss heterogeneous mixtures at length in this chapter, but lest they be dismissed as unimportant, we shall say a word more about them. Nonhomogeneous mixtures are, in fact, far more common in our daily lives than colloidal suspensions or true solutions. Many paints, lacquers, varnishes, and ceramics are nonhomogeneous suspensions, as are most raw ores and fuels. The properties and behaviors of these materials are of great industrial and economic importance. But heterogeneous mixtures are even more complex than homogeneous ones, and we shall find that solutions are challenging enough for the present.

Question

1. Explain why the term *homogeneous mixture* is applicable to solutions.

10.1 Types of Solutions and Colloids

We usually think of solutions as being liquids, but mixtures of gases are often even more ideal examples—just consider the air around us. Furthermore there is no reason not to expect solid solutions as well. In fact, since solutes and solvents can exist in any of the three physical states, we might expect nine kinds of solutions and nine kinds of colloidal suspensions. Tables 10.1 and 10.2 list nine types of solutions but only eight types of colloids. Any gas-in-gas mixture *must* be regarded as a true solution, since both solute and solvent particles will always have diameters less than 10^{-7} cm. Thus there are no gas-in-gas colloidal suspensions.

Of the 17 types of solutions and colloidal suspensions, liquid solutions are the most common and most interesting. Accordingly we shall devote the rest of this chapter to a more detailed look at solutions in which the solvent is a liquid.

Table 10.1 Solutions

Type	Example
Gas-in-gas	Any mixture of gases; air
Gas-in-liquid	Oxygen in water; CO_2 in soft drinks
Gas-in-solid (rare)	Hydrogen in palladium metal
Liquid-in-gas	Fog
Liquid-in-liquid	Methanol in water; oil in gasoline
Liquid-in-solid	Mercury in silver (dental amalgam)
Solid-in-gas	Smoke
Solid-in-liquid	Sugar in water; table salt in water
Solid-in-solid	Some (but not all) alloys

Table 10.2 Colloids

Type		Example
Liquid-in-gas	Aerosols	Fog
Solid-in-gas		Smoke
Liquid-in-liquid	Emulsions	Milk
Solid-in-liquid	Sols	Milk of magnesia
	Gels	Gelatin, jellies
Gas-in-liquid	Foams	Whipped cream
Gas-in-solid		Polyurethane foam, Ivory soap
Liquid-in-solid	Solid emulsions	Butter, cheese
Solid-in-solid	Solid sols	Some alloys

Questions

2. Why is there no gas-in-gas colloid?

3. Define colloid in terms of particle size.

10.2 Gas-in-Liquid Solutions

When a solution has been produced, three things will have occurred:

1. Particles of the solute will have been separated from each other.
2. Particles of the solvent will have been separated from each other.
3. Particles of the solute and of the solvent will have become intimately mixed.

Energy is required in processes 1 and 2 because the attraction that these particles have for each other must be overcome. However, energy is released in process 3 since the mixing, if it occurs at all, occurs because of the attraction of solvent particles for particles of the solute.

In a pure liquid, for example, intermolecular forces hold molecules close together while allowing them to flow over and past each other. When a gas dissolves in the liquid, these forces must be overcome as the gas molecules disperse among the liquid molecules. A gas will dissolve appreciably in a liquid only if the attraction between solute and solvent molecules is as great as the attraction between solvent molecules alone. The solubility of the gas is thus a measure of the interaction between solute and solvent. The nonpolar molecules of gases such as oxygen and nitrogen have very little attraction for water molecules. They are unable to overcome the strong hydrogen bonding in water; thus these gases dissolve only slightly. The very polar molecules of other gases such as ammonia and hydrogen chloride interact more strongly with the water molecules; thus their solubilities are greater.

The effect of these interactions appears in what we measure as the **heat of solution,** represented by the symbol, ΔH. If a strong attraction between solute and solvent favors the formation of a solution, heat will be evolved (an *exothermic* process) and ΔH is designated as negative. If the molecular interactions do not favor the solution, ΔH is designated as positive and heat must be added to dissolve the solute (an *endothermic* process). Since most heats of solution for gaseous solutes in liquid solvents are negative, let's look more closely at the exothermic case. Once we have dissolved as much gas as possible in the solvent, we can say that the system is at equilibrium. We shall have more to say later about saturated solutions as systems at equilibrium, but meanwhile we shall write this process as follows:

$$\text{solute}(g) + \text{solvent}(l) \rightleftharpoons \text{solution}(l) + \text{heat}$$

We can now use Le Chatelier's principle (Chapter 9) to conclude that raising the temperature (adding heat to the system) will cause the equilibrium to shift to the left. In general, therefore, an increase in temperature will decrease the solubility of a solute with a negative heat of solution. Since most gases fit this description in liquid solvents, we expect their solubilities to decrease as the temperature rises.

Le Chatelier's principle also explains the effect of pressure on the solubility of gases: As the pressure increases, the system tends in the direction that reduces its volume. Thus gas molecules tend to dissolve since they occupy less volume in solution. For gases that do not react extensively with the solvent, we can make a precise statement of this tendency: *The weight of gas that will dissolve in a given amount of liquid at constant temperature is directly proportional to the partial pressure of the gas.* This relationship is known as **Henry's law.**

There are many everyday examples of the effects of temperature and pressure on the solubility of gases in liquids. For example, soft drinks contain carbon dioxide gas that has been added under pressure. When we open a bottle, the pressure is reduced and the gas begins to escape from the liquid, causing the familiar "fizz" (see Figure 10.1). An increase in temperature also decreases the solubility of the CO_2, so carbonated beverages are refrigerated before they

(a)   (b)

Figure 10.1 (a) The reduction of the pressure on a carbonated soft drink by the removal of the cap causes it to "fizz" or to foam over as the CO_2 escapes. **(b)** Ice cubes in a carbonated beverage keep the temperature low and the liquid retains the dissolved CO_2.

are opened. A less familiar but more important example of a gas-in-liquid solution is the very small amount of oxygen dissolved in our lakes and streams. This oxygen is essential for aquatic life. This is one of the reasons why industrial waste, which may raise the temperature of the water in a lake or stream and reduce the amount of dissolved oxygen, is such a serious threat to aquatic life.

Example 10.1 At room temperature 1.00×10^2 g of water will dissolve 0.00470 g of O_2, or 58.7 g of NH_3. Calculate the ratio of molecules of gas to molecules of water in each case.

Solution

a. Determine the number of molecules in 1.00×10^2 g of water:

$$(1.00 \times 10^2 \text{ g}) \left(\frac{1.00 \text{ mol}}{18.0 \text{ g}}\right) \left(\frac{6.02 \times 10^{23} \text{ molecules}}{1.00 \text{ mol}}\right)$$

$$= 3.34 \times 10^{24} \text{ molecules}$$

b. Determine the number of molecules in 4.70×10^{-3} g of O_2:

$$(4.70 \times 10^{-3} \text{ g}) \left(\frac{1.00 \text{ mol}}{32.0 \text{ g}}\right) \left(\frac{6.02 \times 10^{23} \text{ molecules}}{1.00 \text{ mol}}\right)$$

$$= 8.84 \times 10^{19} \text{ molecules}$$

c. Determine the approximate ratio of molecules of O_2 to molecules of water:

$$\frac{8.84 \times 10^{19} \text{ molecules } O_2}{3.34 \times 10^{24} \text{ molecules } H_2O} = \frac{2.65 \times 10^{-5}}{1} \text{ or approximately } \frac{3}{100,000}$$

d. Determine the number of molecules in 58.7 g of NH_3:

$$(58.7 \text{ g}) \left(\frac{1.00 \text{ mol}}{17.0 \text{ g}}\right) \left(\frac{6.02 \times 10^{23} \text{ molecules}}{1.00 \text{ mol}}\right) = 2.08 \times 10^{24} \text{ molecules}$$

e. Determine the approximate ratio of molecules of NH_3 to molecules of water:

$$\frac{2.08 \times 10^{24} \text{ molecules } NH_3}{3.35 \times 10^{24} \text{ molecules } H_2O} \approx \frac{3}{5}$$

In a saturated solution of oxygen in water there are approximately 3 molecules of oxygen for every 100,000 molecules of water, whereas in a saturated ammonia solution there are approximately 3 molecules of ammonia for every 5 molecules of water! We've already alluded to the reasons for this tremendous difference. Oxygen molecules are nonpolar and are attracted to water molecules only by weak London dispersion forces. The very polar ammonia molecules, on the other hand, can form strong hydrogen bonds with the water molecules.

Questions

4. What factors determine the solubility of a gas in a liquid?

5. Which would you expect to be more soluble in water, Cl_2 or H_2S?

6. What steps can be taken to preserve the carbonation of a drink?

10.3 Liquid-in-Liquid Solutions

Some liquids are infinitely soluble in one another. Methanol and water, for example, can be mixed in any proportion to form a solution; thus we say they are completely *miscible*. Similarly, motor oil and gasoline are completely miscible. But water and oil are so *immiscible* that they appear to form no solution whatsoever. We can summarize the reasons for these and other relationships between solute and solvent by quoting a convenient phrase: "like dissolves like." We can see what this means by looking at the chart below.

Solvent	Solute	
Nonpolar	Nonpolar	Attractions among solute molecules and among solvent molecules are both weak, so that although attractions between particles of solute and particles of solvent are also weak, random mixing does take place.
Nonpolar	Polar	The forces of attraction among the solute particles are so much greater than attractions between solute particles and solvent particles that there is little tendency for them to separate and dissolve.
Polar	Nonpolar	The situation parallels that of the nonpolar solvent with polar solute. Forces among the solvent molecules impede any intrusion by solute particles.
Polar	Polar	This situation is similar to that of the nonpolar solvent and nonpolar solute. Forces between particles of the different substances are similar to those between particles of the same substance; random mixing takes place freely.

These observations can be related to the processes outlined on p. 309. As shown in Figure 10.2, the heat of solution is the sum of three terms; the energy required to separate solvent molecules to make room for solute molecules, $\Delta H_1 > 0$; the energy required to separate solute molecules from one another, $\Delta H_2 > 0$; and the energy released as solute molecules are attracted to solvent molecules, $\Delta H_3 < 0$. Only if process 3 releases more energy than is consumed by processes 1 and 2 will the overall process be exothermic. This is seldom the case for solids

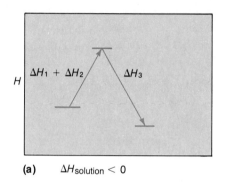 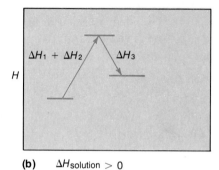

(a) $\Delta H_{solution} < 0$ (b) $\Delta H_{solution} > 0$

Figure 10.2 Relationships which determine the nature of the heat of solution (liquid-in-liquid).
ΔH_1 = energy used to separate solvent molecules.
ΔH_2 = energy used to separate solute molecules.
ΔH_3 = energy released by attraction of solute molecules to solvent molecules.
(a) When $|\Delta H_1| + |\Delta H_2| < |\Delta H_3|$, the heat of solution is negative.
(b) When $|\Delta H_1| + |\Delta H_2| > |\Delta H_3|$, the heat of solution is positive.

dissolving in liquids but not uncommon for a mixture of two liquids. Therefore, although most liquids (like gases) have negative heats of solution, there are many exceptions to the rule. When we do encounter a positive heat of solution for a solute-solvent pair, we expect heat to be absorbed from the surroundings as mixing occurs. For this process at equilibrium, we write

$$\text{solute}(l) + \text{solvent}(l) + \text{heat} \rightleftharpoons \text{solution}(l) \qquad \Delta H > 0$$

If we add heat to this system, we expect the equilibrium to shift to the right; therefore, in general terms, for a solute with a positive heat of solution the solubility increases as the temperature increases. Pressure, on the other hand, has little effect on liquid-in-liquid equilibria since there is little volume difference between the pure liquids and the solution.

Question

7. What factors determine the solubility of a liquid in a liquid?

10.4 Solid-in-Liquid Solutions

We can look at solutions of solids in liquids much the same way that we look at liquid-in-liquid solutions. Again the heat of solution is a reflection of three processes: the separation of the particles of the solid solute, for which ΔH_1 (the *lattice energy*) is always positive; the separation of the solvent molecules, $\Delta H_2 > 0$; and the association of the solute particles with the solvent molecules $\Delta H_3 < 0$. Figure 10.3 illustrates these processes for KCl and water. The final step in the solution process, the association of the solvent molecules with the solute, is known

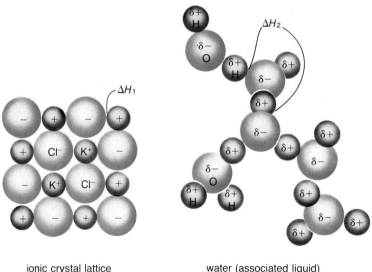
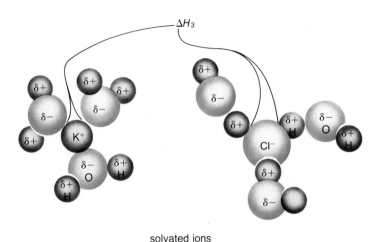

Figure 10.3 The dissolving of an ionic crystal in water.
ΔH_1 = lattice energy.
ΔH_2 = energy of intermolecular attraction in water.
ΔH_3 = energy of attractions between water molecules and solute ions.
For exothermic processes: $|\Delta H_3| > |\Delta H_1| + |\Delta H_2|$

as *solvation* (or, in aqueous solution, *hydration*), and we call the *net* energy change for the solvent ($\Delta H_2 + \Delta H_3$), the *solvation energy*.

As we saw before, the solution process will be exothermic if the final step, represented by ΔH_3, releases more energy than required by the first and second steps (ΔH_1 and ΔH_2). This is often not the case for solid solutes, especially ionic solutes dissolving in water. These processes are endothermic, drawing heat from

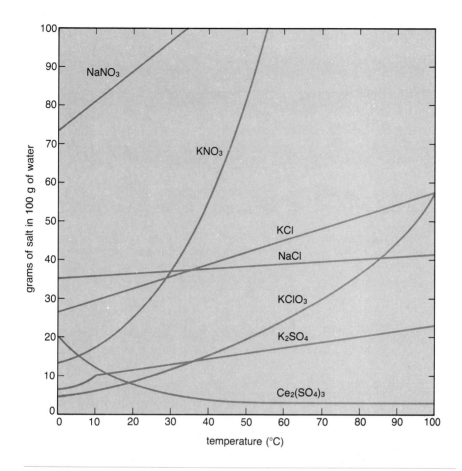

Figure 10.4 Solubility of various solids in water.

the surroundings. Thus as we see in Figure 10.4 the solubility of many ionic salts in water increases as the temperature increases.

Example 10.2 Calculate the ratio of ions to water molecules in a solution containing 37.0 g of KCl/1.00×10^2 ml of water.

Solution

a. Determine the number of ions in 37.0 g of KCl.

$$(37.0 \text{ g})\left(\frac{1.00 \text{ mol}}{74.6 \text{ g}}\right)\left(\frac{6.02 \times 10^{23} \text{ formula units}}{1.00 \text{ mol}}\right)\left(\frac{2 \text{ ions}}{1 \text{ formula unit}}\right)$$
$$= 5.97 \times 10^{23} \text{ ions}$$

(Note that we use the term *formula unit* for KCl, an ionic compound.)

b. The density of water is 1.00 g/ml; thus 1.00×10^2 ml of water weighs 1.00×10^2 g. Determine the number of molecules in 1.00×10^2 g of water:

$$(1.00 \times 10^2 \text{ g}) \left(\frac{1.00 \text{ mol}}{18.0 \text{ g}} \right) \left(\frac{6.02 \times 10^{23} \text{ molecules}}{1.00 \text{ mol}} \right)$$

$$= 33.4 \times 10^{23} \text{ molecules}$$

c. Express the ratio of ions to molecules of water:

$$\frac{5.97 \times 10^{23} \text{ ions}}{33.4 \times 10^{23} \text{ molecules H}_2\text{O}} \approx \frac{3 \text{ ions}}{17 \text{ molecules H}_2\text{O}}$$

The solubility of KCl is similar, on a particle basis, to the solubility of ammonia in water.

10.5 Concentrations

Now that we have discussed solutions in general terms, we must think about how to express their compositions quantitatively. We saw in Chapter 3 that one way to do this is to give the *molarity* of the solution: the number of moles of solute per liter of solution. A similar expression of concentration is *formality* (F): the number of gram formula weights per liter of solution. There are still other ways. We can state the ratio of the weight or number of moles of solute to the weight or volume of solvent. Or we can give the *mole fraction* of one of the constituents: the ratio of the number of moles of that constituent to the total number of moles. Below is a summary of several of these ways of expressing concentration, followed by several problems illustrating them.

Concentration

1. Grams of solute per liter of solution.
2. Moles of solute per liter of solution (molarity, M).
3. Grams of solute per kilogram of solvent.
4. Moles of solute per kilogram of solvent (molality, m).
5. Weight percent of solute.
6. Mole fraction of solute (χ, Greek *chi*).

Example 10.3 The contents of a bottle of concentrated HCl are described on the label as 37.0 percent by weight, specific gravity = 1.20. (The solvent is water.) Express the concentration as grams of solute per liter of solution.

Solution

a. The label tells us that 1.00×10^2 g of this solution contains 37.0 g of pure HCl, thus 63.0 g of water. From this information we can write a host of conversion factors:

$$\frac{37.0 \text{ g HCl}}{1.00 \times 10^2 \text{ g solution}} \quad \text{or} \quad \frac{1.00 \times 10^2 \text{ g solution}}{37.0 \text{ g HCl}} \quad \text{or}$$

$$\frac{63.0 \text{ g H}_2\text{O}}{1.00 \times 10^2 \text{ g solution}} \quad \text{or} \quad \frac{1.00 \times 10^2 \text{ g solution}}{63.0 \text{ g H}_2\text{O}} \quad \text{or}$$

$$\frac{37.0 \text{ g HCl}}{63.0 \text{ g H}_2\text{O}} \quad \text{or} \quad \frac{63.0 \text{ g H}_2\text{O}}{37.0 \text{ g HCl}}$$

b. The specific gravity (Section 1.7) of a substance is the ratio between its density and the density of water at the same temperature. The density of water is 1.00 g/ml, and thus the density of the HCl solution is

$$\frac{1.20 \text{ g solution}}{1.00 \text{ ml solution}}$$

c. The problem asks for the number of grams of solute per liter of solution. Using the density of the solution and the appropriate conversion factor from among those in step a, we can now calculate the answer:

$$\left(\frac{1.20 \text{ g solution}}{1.00 \text{ ml solution}}\right)\left(\frac{10^3 \text{ ml solution}}{1.00 \text{ liter solution}}\right)\left(\frac{37.0 \text{ g HCl}}{1.00 \times 10^2 \text{ g solution}}\right)$$

$$= \textbf{444 g HCl/1.00 liter solution}$$

Example 10.4 Express the concentration of the solution in Example 10.3 as molarity.

Solution

a. Molarity is the number of moles of solute per liter of solution:

$$\frac{\text{moles solute}}{1.00 \text{ liter solution}}$$

b. The molecular weight of HCl ($35.4 + 1.00 = 36.4$) provides the necessary conversion factor to change grams of HCl per liter of solution, as obtained in Example 10.3, to moles of HCl per liter of solution:

$$\left(\frac{4.44 \times 10^2 \text{ g HCl}}{1.00 \text{ liter solution}}\right)\left(\frac{1.00 \text{ mol HCl}}{36.4 \text{ g HCl}}\right) = \frac{12.2 \text{ mol HCl}}{1.00 \text{ liter solution}} \quad \text{or} \quad \textbf{12.2 } M$$

Example 10.5 Express the concentration of the solution in Example 10.3 as grams of HCl per kilogram of H_2O.

Solution

a. From Example 10.3, 1.00 liter of solution weighs

$$(1.00 \text{ liter solution}) \left(\frac{10^3 \text{ ml solution}}{1 \text{ liter solution}} \right) \left(\frac{1.20 \text{ g solution}}{1.00 \text{ ml solution}} \right) = 1.20 \times 10^3 \text{ g},$$

and contains 4.44×10^2 g of HCl.

b. Therefore the weight of water in 1.00 liter of solution is

$$\begin{array}{r} 1200 \text{ g of solution} \\ -\ 444 \text{ g HCl} \\ \hline 760 \text{ g } H_2O \end{array} \quad \text{or} \quad 7.6 \times 10^2 \text{ g of water}$$

(Note that only two digits in this result are significant because the last zero in 1200 was not given in the data.)

c. Hence 4.44×10^2 g of HCl is dissolved in 7.6×10^2 g of water.

d. Determine the number of grams of HCl per kilogram of water:

$$\left(\frac{4.44 \times 10^2 \text{ g HCl}}{7.6 \times 10^2 \text{ g } H_2O} \right) \left(\frac{10^3 \text{ g } H_2O}{1 \text{ kg } H_2O} \right) = \left(\frac{5.8 \times 10^2 \text{ g HCl}}{1.0 \text{ kg } H_2O} \right)$$

Note that although we chose to do this problem from the results of Example 10.3, we could also have done it directly from one of the conversion factors in part a of Example 10.3:

$$\frac{37.0 \text{ g HCl}}{63.0 \text{ g } H_2O}$$

Example 10.6 Express the concentration of the solution of Example 10.3 as molality.

Solution

a. Molality is the number of moles of solute per kilogram of solvent:

$$\frac{\text{moles solute}}{1.00 \text{ kg solvent}}$$

b. The molecular weight of HCl again provides the needed conversion factor:

$$\left(\frac{5.8 \times 10^2 \text{ g HCl}}{1.0 \text{ kg H}_2\text{O}}\right)\left(\frac{1.00 \text{ mol HCl}}{36.4 \text{ g HCl}}\right) = \frac{16 \text{ mol HCl}}{1.00 \text{ kg H}_2\text{O}} \quad \text{or} \quad 16 \text{ m}$$

Example 10.7 Express the concentration of the solution of Example 10.3 in terms of the mole fraction of the solute.

Solution

a. From Example 10.5 we know that 1.00 liter of solution contains 7.6×10^2 g of water. The molecular weight of water ($16.0 + 2.02 = 18.0$) provides the conversion factor we need to determine the number of moles of water in a liter of solution:

$$\left(\frac{7.6 \times 10^2 \text{ g H}_2\text{O}}{1.00 \text{ liter solution}}\right)\left(\frac{1.00 \text{ mol H}_2\text{O}}{18.0 \text{ g H}_2\text{O}}\right) = \frac{42 \text{ mol H}_2\text{O}}{1.00 \text{ liter solution}}$$

b. In Example 10.4 we calculated that there were 12.2 mol HCl in 1.00 liter of solution; thus the total number of moles of H_2O and of HCl in 1.00 liter of solution is

$$42 \text{ mol} + 12.2 \text{ mol} = 54 \text{ mol}$$

c. The mole fraction is the ratio of moles of solute to total moles:

$$\text{mole fraction HCl} = \chi \text{ HCl} = \frac{12.2 \text{ mol HCl}}{54 \text{ mol (total)}} = 0.22$$

To summarize, we can express the concentration of this one solution of HCl in water in any of the following ways:

Weight percent:	36.0%
Grams of solute per liter of solution:	4.44×10^2 g/liter of solution
Molarity:	12.2 M
Grams of solute per kg of solvent:	5.8×10^2 g/kg solvent
Molality:	16 m
Mole fraction of solute (χ):	0.22

The following problems are practical examples of the kinds of calculations chemists often find necessary in the laboratory.

Example 10.8 If a reaction requires 0.173 mol of HNO_3, what volume of a 0.200 M solution must be used?

Solution

a. Since we know that 1.00 liter of solution contains 0.200 mol HNO_3, we can calculate the answer in a single step:

$$(0.173 \text{ mol } \cancel{HNO_3}) \left(\frac{1.00 \text{ liter solution}}{0.200 \text{ mol } \cancel{HNO_3}} \right) = 0.865 \text{ liter solution}$$

Example 10.9 If 27.0 ml of a 3.45 M solution is diluted to 150.0 ml, what is the concentration of the diluted solution?

Solution

a. To express the concentration of the diluted solution, we need know only its volume (which is given) and the number of moles of solute it contains.

b. Calculate the number of moles of solute in the sample:

$$(27.0 \text{ \cancel{ml cone. solution}}) \left(\frac{1 \text{ \cancel{liter}}}{10^3 \text{ \cancel{ml}}} \right) \left(\frac{3.45 \text{ mol}}{1.00 \text{ \cancel{liter cone. solution}}} \right)$$

$$= 0.0932 \text{ mol}$$

c. Divide the number of moles in the dilute solution by the volume of the solution to obtain its molarity:

$$\left(\frac{0.0932 \text{ mol}}{150 \text{ \cancel{ml} solution}} \right) \left(\frac{10^3 \text{ \cancel{ml}}}{1 \text{ liter}} \right) = \mathbf{0.621 \ M}$$

Example 10.10 What volume of 2.86 M NaOH solution must be used to prepare 0.500 liter of 0.646 M solution?

Solution

a. Calculate the number of moles of NaOH needed. (Designate the more dilute solution as A and the more concentrated solution as B.)

$$(0.500 \text{ liter A}) \left(\frac{0.646 \text{ mol NaOH}}{1.00 \text{ liter A}} \right)$$

b. Calculate the volume of concentrated solution necessary to provide this amount of NaOH.

$$(0.500 \text{ \cancel{liter A}}) \left(\frac{0.646 \text{ \cancel{mol NaOH}}}{1.00 \text{ \cancel{liter A}}} \right) \left(\frac{1.00 \text{ liter B}}{2.86 \text{ \cancel{mol NaOH}}} \right) = 0.113 \text{ liter B}$$

Example 10.11 What volume of 0.0445 M HCl can be prepared from 2.26 liters of 1.50 M HCl?

Solution

a. Calculate the number of moles of HCl available. (Designate the more dilute solution as A and the more concentrated solution as B.)

$$(2.26 \text{ liters B}) \left(\frac{1.50 \text{ mol HCl}}{1.00 \text{ liter B}} \right)$$

b. Calculate the volume of 0.0445 M solution which can be prepared from this many moles of HCl.

$$(2.26 \cancel{\text{liters B}}) \left(\frac{1.50 \cancel{\text{mol HCl}}}{1.00 \cancel{\text{liter B}}} \right) \left(\frac{1.00 \text{ liter A}}{0.0445 \cancel{\text{mol HCl}}} \right) = \mathbf{76.2 \text{ liters A}}$$

Example 10.12 Calcium chloride and phosphoric acid react in solution according to the equation below. What volume of 0.200 M H_3PO_4 solution will be required to react completely with 0.165 liter of 0.680 M $CaCl_2$?

Solution

a. $3 \text{ CaCl}_2(aq) + 2 \text{ H}_3\text{PO}_4(aq) \longrightarrow \text{Ca}_3(\text{PO}_4)_2(s) + 6 \text{ HCl}(aq)$

b. To solve this problem we must first find the number of moles of $CaCl_2$ involved, then use the coefficients from the chemical equation to determine how many moles of H_3PO_4 are required to react with this amount, and finally calculate the volume of 0.200 M solution required.

c. Calculate the number of moles of $CaCl_2$:

$$[0.165 \text{ liter CaCl}_2(aq)] \left(\frac{0.680 \text{ mol CaCl}_2}{1.00 \text{ liter CaCl}_2(aq)} \right)$$

d. Calculate the number of moles of H_3PO_4 required:

$$[0.165 \text{ liter CaCl}_2(aq)] \left(\frac{0.680 \text{ mol CaCl}_2}{1.00 \text{ liter CaCl}_2(aq)} \right) \left(\frac{2 \text{ mol H}_3\text{PO}_4}{3 \text{ mol CaCl}_2} \right)$$

e. Calculate the volume of 0.200 M H_3PO_4 required:

$$[0.165 \cancel{\text{liter CaCl}_2(aq)}] \left(\frac{0.680 \cancel{\text{mol CaCl}_2}}{1.00 \cancel{\text{liter CaCl}_2(aq)}} \right) \left(\frac{2 \cancel{\text{mol H}_3\text{PO}_4}}{3 \cancel{\text{mol CaCl}_2}} \right)$$

$$\times \left(\frac{1.00 \text{ liter H}_3\text{PO}_4(aq)}{0.200 \cancel{\text{mol H}_3\text{PO}_4}} \right) = \mathbf{0.374 \text{ liter H}_3\text{PO}_4(aq)}$$

Questions

8. A solution of sugar in water is prepared by adding 10.0 g of sugar to 90.0 g of water. What is the concentration of this solution, expressed as the percent of sugar by weight?

9. Would you expect the specific gravity of the solution in Question 8 to be greater or less than 1.00? Why?

10. A solution of glucose is prepared by dissolving 1.00 mol of glucose in 1.00 kg of water. Express the concentration as mole fraction of glucose.

11. What is the molality of the solution in Question 10?

12. Would you expect the molality of 1.0 M sucrose solution to be greater than, less than, or the same as the molarity? Why?

10.6 Saturation

In common usage, to say that something is **saturated** means that it contains "all it can hold." For chemists, the term has much the same meaning, but as we shall see, there's much more to be said about the phenomenon of saturation in solutions. In Section 10.2 we said that when we write an equation expressing a dynamic equilibrium between solute and solvent on the one side and solution on the other, we are describing a saturated solution. As an example, consider what happens when we add an excess of sugar to a beaker of water. As we stir the mixture, the sugar will dissolve rapidly at first but will soon seem to stop, leaving undissolved sugar on the bottom of the beaker. The system is now at equilibrium. But from what we know of equilibria we can infer that the sugar has not stopped dissolving, instead it is dissolving at the same rate that it is crystallizing out of the solution and being deposited on the bottom. There is no *net* change because the two opposite processes are occurring at equal rates. This solution is said to be *saturated*. It contains the maximum amount of solute that can be dissolved in the given amount of solvent, under the given conditions. Even an "insoluble" salt in contact with solvent produces just such an equilibrium. The term *insoluble* merely indicates that the saturated solution of this salt has an extremely *low* concentration. The solubility of a substance is always understood to refer to a saturated solution.

Notice also that we have described a saturated solution as being in contact with undissolved solute. If not, there can be no assurance that the solution is not *unsaturated* or *supersaturated*. An unsaturated solution, of course, is one in which we might increase the concentration simply by adding more solute. The phenomenon of supersaturation is more interesting. Consider an example. Crystalline sodium acetate ($NaC_2H_3O_2$) has a positive heat of solution; thus if we raise the temperature, the solubility will increase. Let us suppose that we have prepared a saturated solution of $NaC_2H_3O_2$ at room temperature but that we have left no solid $NaC_2H_3O_2$ in contact with the solution. If we heat the solution to 35°C, it of course becomes unsaturated: We can add more solute and it will dissolve completely. Let us then saturate the solution at the higher temperature, again leaving no solute in contact with the solution. Now the solution can be cooled carefully to room temperature with no crystalline $NaC_2H_3O_2$ appearing. The solution is now supersaturated. It is in a metastable state, like a carefully built house of cards. To disturb the supersaturated solution, we need only to drop in a "seed crystal" of sodium acetate or to shake the container. Crystallization will begin immediately and will proceed until the rate of crystallization is equaled by the rate of dissolution—thus we are left once more with a saturated solution.

10.7 Physical Properties of Solutions

In many cases we can discuss the physical properties of solutions in terms of the properties of the pure solute and pure solvent. The most important of these physical properties is vapor pressure. When we dissolve a *nonvolatile* solute (one whose vapor pressure is essentially zero at room temperature) in a liquid solvent, the vapor pressure of the solution is lower than that of the pure solvent. We can calculate the drop in vapor pressure (ΔP) knowing only the mole fraction of the solute (χ_{solute}) and the vapor pressure of the pure solvent ($P_{solvent}$):

$$\Delta P = (\chi_{solute})(P_{solvent})$$

This is a statement of **Raoult's law** and is the basis of all *colligative properties* of solutions. *Colligative properties depend only on the relative numbers of solute particles and solvent molecules in the solution* and do *not* depend on the specific identity of each of these particles. As we shall see in the following discussion, the boiling point, freezing point, and osmotic pressure of solutions are colligative properties.

Before going on with our discussion, however, we should describe two assumptions that we shall be making. First, we shall be talking only about *ideal solutions*—those that obey Raoult's law. Some solutions are not so well behaved but we shall reserve discussion of these interesting cases until later. Second, we shall consider only solutions of nonvolatile solids in liquids. Thus we need to consider only the vapor pressure of the solvent. With these assumptions in hand we can now look at an illustration of the behavior described by Raoult's law and then seek an explanation.

In Chapter 9 we described what happens in a closed system of water and the air above it. In a short time the air above the water becomes saturated with water vapor, and a dynamic equilibrium is established in which evaporation and condensation take place at the same rate. Let us now modify this system as shown in Figure 10.5. In Figure 10.5a, the two beakers contain equal volumes of liquid; one contains pure water and the other, a solution of sugar in water. In Figure 10.5b, we see what happens after a few hours. The water level has dropped, and the volume of the sugar solution has increased. The explanation is not difficult if we understand Raoult's law and the nature of dynamic equilibria. Raoult's law demands that the equilibrium vapor pressure for the sugar

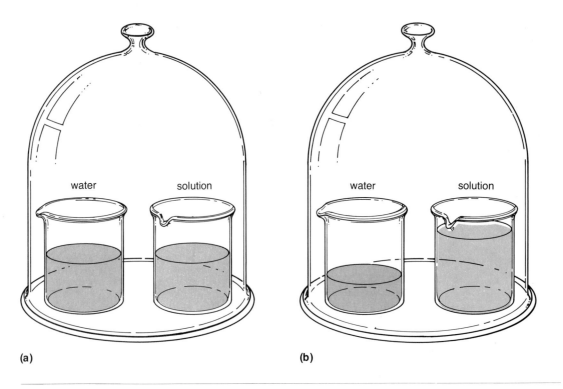

Figure 10.5 The effect of dissolved sugar on the vapor pressure of water.

solution be lower than for the water; thus the system in the bell jar cannot be at equilibrium. Instead, the vapor pressure is somewhere between the two equilibrium values. As a result, condensation occurs faster than evaporation at the surface of the sugar solution, and evaporation takes place faster than condensation at the surface of the water.

We can explain the lowering of vapor pressure in a solution by considering the mechanism of evaporation. The probability that a solvent molecule will escape into the vapor phase depends on the number of molecules of solvent at the surface of the solution. If some of the solvent molecules are crowded out by dissolved solute molecules, the rate of vaporization—thus the vapor pressure—diminishes and we find that *the vapor pressure of a solution of a nonvolatile solute is equal to the mole fraction of the solvent times the vapor pressure of the pure solvent:*

$$P_{\text{solution}} = (\chi_{\text{solvent}})(P_{\text{solvent}})$$

This is merely a restatement of Raoult's law.

Example 10.13 The vapor pressure of water at 70.0°C is 234 mm of Hg. If 725 g of sucrose (molecular weight = 342) is dissolved in 1.00 liter of water at 70.0°C, what will be the vapor pressure of the solution? Assume that the density of the water is 1.00 g/ml.

Solution

a. To calculate the vapor pressure of the solution using Raoult's law, we need the mole fraction of the solvent. This requires that we find the number of moles of solvent and the total number of moles in the solution. First, determine the number of moles of sucrose:

$$(725 \text{ g sucrose}) \left(\frac{1.00 \text{ mol sucrose}}{342 \text{ g sucrose}} \right) = 2.12 \text{ mol sucrose}$$

b. Determine the number of moles of water:

$$(1.00 \text{ liter } H_2O) \left(\frac{10^3 \text{ ml}}{1 \text{ liter}} \right) \left(\frac{1.00 \text{ g}}{1.00 \text{ ml}} \right) \left(\frac{1.00 \text{ mol}}{18.0 \text{ g}} \right) = 55.6 \text{ mol } H_2O$$

c. Thus the total number of moles is

$$55.6 \text{ mol } H_2O + 2.12 \text{ mol sucrose} = 57.7 \text{ mol (total)}$$

d. Determine the mole fraction of the solvent:

$$\frac{55.6 \text{ mol } H_2O}{57.7 \text{ mol (total)}} = 0.964 = \chi_{\text{solvent}}$$

e. From Raoult's law the vapor pressure of the solution is

$$(0.964)(234 \text{ mm of Hg}) = \mathbf{226 \text{ mm of Hg}}$$

Since the boiling point of a liquid is merely the temperature at which its vapor pressure equals the atmospheric pressure and since we have seen that the vapor pressure of a solution varies in a regular way with the concentration of a nonvolatile solute, we must expect the boiling point to do likewise. At a given temperature the vapor pressure drops in a regular way as we add solute; thus the boiling point must rise. The extent of the rise depends on the solvent and is characterized by a *boiling point elevation constant*, abbreviated $K_{\text{b.p.}}$, for each solvent. The values for this constant, which is equal to the boiling point elevation caused by each 1.00 mol of solute in 1.00 kg of solvent, vary widely for different solvents: for water, $K_{\text{b.p.}} = 0.52°C$; for benzene, $K_{\text{b.p.}} = 2.53°C$; and for carbon tetrachloride, $K_{\text{b.p.}} = 5.03°C$. From the definitions of $K_{\text{b.p.}}$ and molality, therefore, we conclude that *the elevation of the boiling point of a solvent is equal to the boiling point elevation constant for that solvent times the molality of the solution*:

$$\Delta \text{b.p.} = (K_{\text{b.p.}})(m)$$

Example 10.14 What will be the boiling point of a solution prepared by dissolving 25.0 g of sucrose in 1.00×10^2 g of water?

Solution

a. We must first determine the molality of the solution. Calculate the number of moles of sucrose per gram of water:

$$\left(\frac{25.0 \text{ g sucrose}}{1.00 \times 10^2 \text{ g H}_2\text{O}}\right)\left(\frac{1.00 \text{ mol sucrose}}{342 \text{ g sucrose}}\right)$$

b. Determine the molality of the solution (number of moles of sucrose per kilogram of water):

$$\left(\frac{25.0 \text{ g sucrose}}{1.00 \times 10^2 \text{ g H}_2\text{O}}\right)\left(\frac{1.00 \text{ mol sucrose}}{342 \text{ g sucrose}}\right)\left(\frac{1000 \text{ g H}_2\text{O}}{1 \text{ kg H}_2\text{O}}\right)$$

$$= 0.731 \frac{\text{mol sucrose}}{\text{kg H}_2\text{O}}$$

c. Determine the boiling point elevation of the solution:

$$\Delta\text{b.p.} = (0.731 \text{ m})(0.52°\text{C}) = 0.38°\text{C}$$

d. Add the boiling point elevation to the boiling point of the solvent:

$$100.00°\text{C} + 0.38°\text{C} = \mathbf{100.38°\text{C}}$$

Just as molecules of dissolved solute interfere with the vaporization of solvent at the boiling point, so they interfere with the crystallization of solvent molecules at the freezing point. Thus we must lower the temperature still more to freeze the solvent out of solution. This drop in the freezing point can be predicted from the *freezing point depression constant*, abbreviated $K_{f.p.}$, which is characteristic for each solvent. The freezing point depression constant, which measures the drop in the freezing point of 1.00 kg of solvent for each 1.00 mol of solute, is 1.86°C for water, 5.12°C for benzene, and 31.8°C for carbon tetrachloride. *The lowering of the freezing point of a solvent is equal to the freezing point depression constant for that solvent times the molality of the solution:*

$$\Delta\text{f.p.} = (K_{f.p.})(m)$$

Example 10.15 At what temperature will a 0.701 M aqueous solution of sucrose (molecular weight = 342) freeze? The density of the solution is 1.09 g/ml.

Solution

a. The depression of the freezing point depends on the molality of the solution; thus we must convert the concentration from molarity to molality. 1.00 liter of the solution weighs

$$(1.00 \text{ liter solution}) \left(\frac{1000 \text{ ml solution}}{1 \text{ liter solution}} \right) \left(\frac{1.09 \text{ g solution}}{1.00 \text{ ml solution}} \right)$$
$$= 1.09 \times 10^3 \text{ g solution}$$

b. The weight of sucrose in this liter is

$$\left(\frac{0.701 \text{ mol sucrose}}{1.00 \text{ liter solution}} \right) \left(\frac{342 \text{ g sucrose}}{1.00 \text{ mol sucrose}} \right) = 240 \text{ g sucrose}$$

c. The weight of water in the liter of solution is therefore

$$1090 \text{ g solution} - 240 \text{ g sucrose} = 8.5 \times 10^2 \text{ g H}_2\text{O}$$

(Note that the last zero in the result is not significant.)

d. The molality of the solution is

$$\left(\frac{0.701 \text{ mol sucrose}}{0.85 \text{ kg H}_2\text{O}} \right) = 0.82 \text{ m}$$

e. The freezing point depression of this solution is

$$\Delta \text{f.p.} = (1.86°\text{C})(0.82 \text{ m}) = 1.5°\text{C}$$

f. Subtract the freezing point depression from the freezing point of the solvent to obtain the freezing point of the solution:

$$0.00°\text{C} - 1.5°\text{C} = -1.5°\text{C}$$

In Figure 10.6 the phase diagram for a dilute aqueous solution of a nonvolatile solute illustrates the changes that the freezing point, the boiling point, and the vapor pressure of a solvent undergo when solute is added. Notice, however, that the line describing the equilibrium between solid and vapor has not changed since we expect to find no solute in either the gas phase or in the solid water that freezes out of solution.

Freezing point depression is a property often used to determine the molecular weights of unknown solutes. If we know $K_{\text{f.p.}}$ for the solvent, the depression of the freezing point gives the molality. Thus if we know the weight of the solute, we can easily calculate its molecular weight.

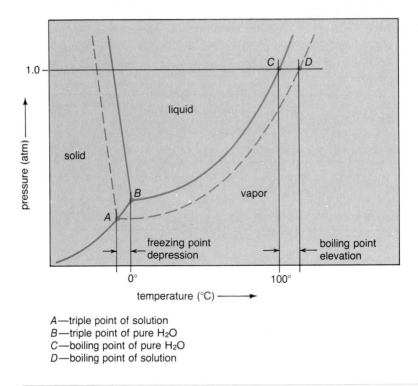

A—triple point of solution
B—triple point of pure H₂O
C—boiling point of pure H₂O
D—boiling point of solution

Figure 10.6 Phase diagrams for pure liquid (solvent): ——— and for solution: ———.

Example 10.16 Calculate the molecular weight of a solute if 1.86 g, dissolved in 25.0 g of benzene, produces a solution that freezes at 2.65°C. (The freezing point of pure benzene is 5.50°C, and its freezing point depression constant is 5.12°C.)

Solution

a. Determine the depression of the freezing point:

$$5.50°C - 2.65°C = 2.85°C$$

Sidelight
Fresh Water from Salt Water

The world was jolted into an awareness of the fuel crisis in the 1970s. Projections predicting the depletion of fossil fuel resources have now become familiar. Perhaps we shall profit from this experience and the experiences of drought-stricken Californians in 1976–1977 and face up to another impending shortage. Demand for usable fresh water is already outstripping supplies in many areas of the world, including the American Southwest. There remains of course one large reservoir of water that is practically untapped: the oceans. Ninety-seven percent of the earth's water contains between 18,000 and 35,000 ppm (parts per million, or grams of solute per 1 million grams of solution) salt. We need only devise a reasonable way to remove this impurity.

Desalination methods are invariably based upon the physical properties of aqueous solutions. One well-known method is *distillation*. If we boil salt water (at the elevated temperature required by the molality of the solution), the vapor is almost entirely pure H_2O since the dissolved salts are nonvolatile. When this steam is recondensed, it contains none of the solute. Several large saline stills have been developed in the Southwest to exploit just this property. However, the high energy requirement for this process remains a major disadvantage except where demand is high. A second method of desalination is *crystallization*. If we freeze an aqueous salt solution (at the lower temperature required by the molality of the solution), the ice that forms is pure H_2O. In commercial processes, the ice crystals can then be separated mechanically from the rest of the solution, rinsed, and then melted to yield pure water.

An ecological problem that attends both of these desalination processes is getting rid of the salt after it has been removed from the water. Desalination plants producing 100,000 gallons of usable water every day, even from brackish water (containing only a few thousand parts per million salt) must dispose of a ton or two of salt each day. One "wild idea" for producing fresh water—one with few ecological consequences—is to tow Antarctic icebergs to the southwestern California coast. Since icebergs are sometimes more than a square mile in area and several hundred feet thick, they could provide a tremendous reservoir of pure water. The disadvantage in this case (ignoring for the moment the problem of insulating the iceberg for the trip) is the cost of the energy required to move an iceberg several thousand miles.

b. Calculate the molality:

$$m = \frac{\Delta \text{b.p.}}{K_{\text{b.p.}}} = \frac{2.85°C}{5.12°C} = 0.557\ m$$

c. From the data, calculate the concentration of solute in grams per kilogram of benzene:

$$\left(\frac{1.86\ \text{g solute}}{25.0\ \text{g benzene}}\right)\left(\frac{1000\ \text{g benzene}}{1\ \text{kg benzene}}\right) = 74.4\ \frac{\text{g solute}}{\text{kg benzene}}$$

d. Use this concentration expression together with the molality to calculate the gram molecular weight of the solute.

$$\left(\frac{74.4\ \text{g solute}}{1.00\ \text{kg solvent}}\right)\left(\frac{1.00\ \text{kg solvent}}{0.557\ \text{mol solute}}\right) = 134\ \text{g/mol}\quad \textbf{(solute)}$$

As we have said, colligative properties depend solely on the number of solute particles in solution and not on the kind of particles. Therefore we must be especially careful with solutions of electrolytes. For example, a solution of potassium chloride contains potassium ions *and* chloride ions. Thus a 1.00-molal solution of KCl contains 12.04×10^{23} rather than 6.02×10^{23} particles dissolved in each 1.00 kg of solvent; that is, the solution is 1.00 m with respect to K^+ ions but also 1.00 m with respect to Cl^- ions. Thus we expect the freezing point of a 1.00 m aqueous solution of potassium chloride to be $(2.00 \times -1.86)°C$ or $-3.70°C$. When we measure the freezing point, however, it is $-3.25°C$. We explain this deviation from the expected value in terms of interionic attraction.

Table 10.3 Observed and Expected Freezing Points of Aqueous Potassium Chloride Solutions

Molality	Observed Freezing Point (°C)	Expected Freezing Point (°C)	Percent Deviation
2.0	−6.44	−7.44	−13.4
1.0	−3.25	−3.72	−12.6
0.5	−1.66	−1.86	−10.8
0.2	−0.679	−0.744	−8.8
0.1	−0.345	−0.372	−7.3
0.05	−0.175	−0.186	−5.9
0.02	−0.0713	−0.0744	−4.2
0.01	−0.0361	−0.0372	−3.0
0.005	−0.0182	−0.0186	−2.2
0.001	−0.00366	−0.00372	−1.6

Although we usually regard the KCl as completely ionized, the ions are not entirely independent of each other. The attraction between positive and negative ions produces *ion clusters*. These clusters diminish the effects of the individual ions; thus the *effective* number of particles in a 1.00 m solution of potassium chloride is slightly less than 12.04×10^{23}.

As the concentration of an ionic solution decreases, the distance between ions increases; thus we would expect a smaller proportion of the ions to remain in ion clusters. Therefore we would predict that our calculated freezing point depressions more accurately reflect the measured freezing points as we dilute the solution. As shown in Table 10.3, our expectations are borne out by the experiment.

For salts containing more than two ions per formula unit, of course, we must extend our observations about KCl. A 1.00 m aqueous solution of barium chloride ($BaCl_2$), for example, should freeze near $-5.58°C$ [$(3.00 \times -1.86)°C$]. The experimental value is $-5.20°C$.

Questions

13. If an aqueous solution of a nonvolatile nonelectrolyte boils at 100.26°C, at what temperature should it freeze?

14. What is the molality of the solution in Question 13?

10.8 Osmosis

Osmosis is a vital process, upon which both plant and animal life depend. The concepts of dynamic equilibria used to understand vapor pressure lowering in solutions can be applied to explain this phenomenon.

In Figure 10.7 two compartments are separated by a membrane porous enough to allow free passage to water but impermeable to large solute molecules. Such *semipermeable membranes* are not hypothetical partitions as they might seem: Cellophane, some kinds of paper, and many animal and plant membranes have this property. We now put water into one compartment and a solution of sugar in the other. We can see the parallel between this experiment and the one depicted in Figure 10.5. There we observed the difference in vapor pressures; here we shall see a difference in rates of diffusion. The water molecules pass freely through the membrane in both directions, but they pass less frequently from B to A since, because of the dissolved solute, there are fewer of them in the vicinity of the membrane. The result is apparent after a time, as the illustration shows: If the semipermeable membrane is rigid, the level of the sugar solution will rise. To prevent this happening we would have to apply an external pressure to the sugar solution, thus equalizing the diffusion rates in both directions. This pressure is the *osmotic pressure* of the solution, a colligative property.

This osmotic pressure, for dilute solutions at a constant temperature is directly proportional to the concentration of the solution:

$$P_{os} = k\left(\frac{n}{V}\right)$$

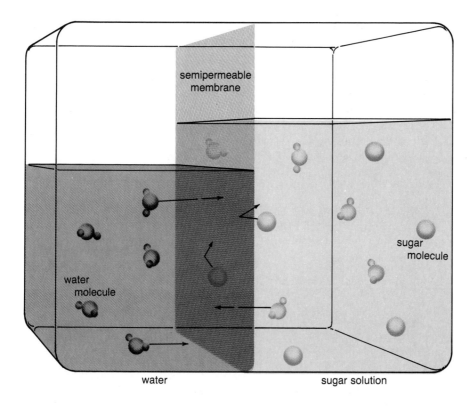

Figure 10.7 Osmosis.

where n is the number of moles of solute and V is the total volume of the solution. Therefore, if $\bar{V}$ is the *molar volume* (the volume of solution containing 1.00 mole of solute),

$$P_{os}(\bar{V}) = k$$

The osmotic pressure is also directly proportional to the absolute temperature:

$$P_{os} = k'T$$

We can now combine these two expressions (note their similarity to the laws of Boyle and Charles) into the *van't Hoff equation*, which bears a striking resemblance to the ideal gas equation:

$$P_{os}(\bar{V}) = RT$$

in which R is again the ideal gas constant, with a value of $0.08206 \frac{\text{(liter) (atm)}}{\text{(mol) (K)}}$.

Like other colligative properties, osomtic pressure is not affected by molecular properties, such as the molecular weight of the solute. Instead, it depends only on the number of particles in solution. The parallel between the van't Hoff equation and the ideal gas equation also tells us that the osmotic pressure is equal to the pressure that one mole of the solute (in the gas phase) would exert if confined to a volume $\bar{V}$ at temperature T.

Example 10.17 An aqueous solution contains 3.5 g of a certain protein in each 1.00×10^2 ml of solution at 20.0°C. If the molecular weight of the protein is 6.8×10^4, what osmotic pressure will this solution have?

Solution

a. Calculate the number of moles of protein in 3.5 g:

$$(3.5 \text{ g}) \left(\frac{1.0 \text{ mol}}{6.8 \times 10^4 \text{ g}} \right) = 5.1 \times 10^{-5} \text{ mol}$$

b. Calculate the osmotic pressure. Note that we can solve the problem as if it were a simple ideal gas law problem.

$$(760 \text{ mm Hg}) \left(\frac{5.1 \times 10^{-5} \text{ mol}}{1.00 \text{ mol}} \right) \left(\frac{293 \text{ K}}{273 \text{ K}} \right) \left(\frac{22.4 \text{ liter}}{0.100 \text{ liter}} \right) = 9.3 \text{ mm of Hg}$$

Figure 10.8 shows one simple way to measure osmotic pressure. The beaker contains pure water. The thistle tube inverted in it is sealed with a semipermeable membrane and contains the solution whose osmotic pressure we are measuring.

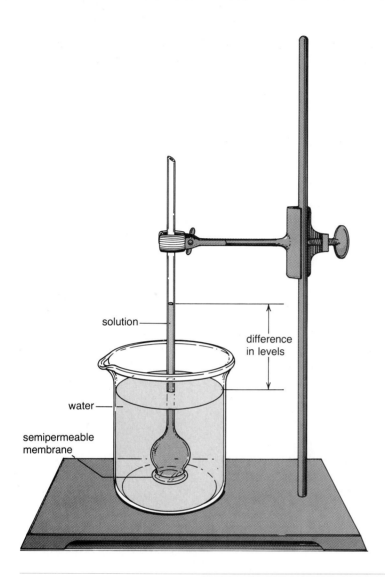

Figure 10.8 Measuring osmotic pressure.

The two liquid levels must initially be the same, but then water will diffuse *into* the solution in the thistle tube faster than it will diffuse *out*. At equilibrium, when the two diffusion rates are the same, the difference in levels gives the osmotic pressure (100 mm of H_2O = 7.35 mm of Hg = 9.67×10^{-3} atm).

Example 10.18 If the osmotic pressure of the solution in Example 10.17 were to be measured by the device illustrated in Figure 10.8, what would be the height of the water in the tube?

Solution To solve this problem we must calculate the height of a water column that corresponds to a mercury column 9.3 mm high. Since the density of mercury is 13.6 times that of water, the water column must be 13.6 times as high as the mercury column, or

$$\left(\frac{13.6 \text{ mm H}_2\text{O}}{1.00 \text{ mm Hg}}\right)(9.3 \text{ mm Hg}) = 126 \text{ mm H}_2\text{O}$$

The next example demonstrates why it is easier to determine the molecular weight of a large molecule by measuring the osmotic pressure than by measuring the freezing point depression.

Example 10.19 What would be the freezing point change of the aqueous solution in Example 10.17? The specific gravity of the solution is 1.020.

Solution

a. Determine the weight of 1.000 liter of solution:

$$(1.000 \text{ liter})\left(\frac{1000 \text{ ml}}{1 \text{ liter}}\right)\left(\frac{1.020 \text{ g}}{1.00 \text{ ml}}\right) = 1020 \text{ g}$$

b. Determine the weight of water contained in 1.000 liter of solution:

$$1020 \text{ g} - 35 \text{ g} = 985 \text{ g H}_2\text{O}$$

c. Express the concentration as molality:

$$\left(\frac{5.1 \times 10^{-5} \text{ mol}}{985 \text{ g H}_2\text{O}}\right)\left(\frac{1.00 \times 10^3 \text{ g H}_2\text{O}}{1 \text{ kg H}_2\text{O}}\right) = 5.2 \times 10^{-5} \frac{\text{mol}}{\text{kg H}_2\text{O}}$$

$$= 5.2 \times 10^{-5} \text{ m}$$

d. Thus the freezing point would be lowered by

$$(5.2 \times 10^{-5})(1.86°\text{C}) = (9.7 \times 10^{-5})°\text{C}, \left(\text{approximately } \frac{1}{10,000} °\text{C}\right)$$

It is clearly easier to measure a column of water 13 cm high than a freezing point only one ten-thousandth of a degree less than that of water.

Osmosis and osmotic pressures are of critical importance to most living systems. The roots of plants absorb moisture by osmosis, and in animals digestive and excretory functions depend on osmosis. Plant-cell walls, the walls of intestines, and kidney tissues all contain semipermeable membranes.

Two solutions of the same pressure separated by a semipermeable membrane are said to be *isotonic* when no net flow occurs in either direction. When solutions are added to the blood or to body tissues, utmost care must be taken that they are isotonic with the fluids of the body. A *hypertonic* solution, one with a higher osmotic pressure, will cause fluid to flow from the body tissues and can cause cells to collapse. *Hypotonic* solutions, ones with lower osmotic pressures, have the opposite effect: Fluid flows into the cells causing them to swell and burst. The sting we feel when water is applied to the tender tissues of a burn is caused by just this phenomenon. An isotonic solution would be painless. A most dramatic use of hypotonic solutions is the ancient practice of driving dry wooden pegs into rocks and then soaking them in water. The tissues of the wood absorb water, swell irresistibly, and split the rocks.

Questions and Problems

10.1 Types of Solutions and Colloids

15. Which of the following familiar materials are homogeneous?
  a. the air we breathe
  b. the water we drink
  c. paper
  d. the wood in a pencil
  e. the lead in a pencil
  f. the glass in a window
  g. a key
  h. cement
  i. steel girders
  j. milk
  k. copper wire
  l. ice
  m. aluminum foil

10.2 Gas-in-Liquid Solutions

16. In each pair below, which would you expect to be more soluble in benzene? (Benzene is a nonpolar liquid.)
  a. CH_4 HF b. CO C_2H_6 c. NO_2 C_3H_8

17. The solubility of nitrogen is 1.90 mg/1.00×10^2 ml of water at 20°C and 1.00 atm pressure. What is the solubility of nitrogen in water at 5.00 atm and 20°C?

10.3 Liquid-in-Liquid Solutions

18. Calculate the ratio of molecules of solute to molecules of solvent in a 50.0 percent (by weight) solution of glycerol ($C_3H_8O_3$) in water.

19. What aqueous solution of ethanol (C_2H_5OH), expressed in percent by weight, contains one molecule of ethanol for every five molecules of water?

10.4 Solid-in-Liquid Solutions

20. Differentiate between solvation energy and lattice energy.

21. The solubility of lithium sulfate (in grams of salt per 100 grams of water) is 43.6 at 0°C and 35.0 at 100°C. The solubility of potassium sulfate is 7.35 at 0°C and 24.1 at 100°C. Explain these values in terms of hydration and lattice energies.

22. When NH_4NO_3 is added to water, the temperature of the water drops markedly. However, when $LiNO_3$ is added to water, the temperature increases. Explain these observations in terms of lattice energy and hydration energy.

10.5 Concentrations

23. The four gases below are arranged in order of decreasing solubility in water, with solubilities expressed in grams of solute per 100 g of water. Calculate the mole fraction of solute in each case, and then rearrange the gases in order of decreasing solubility, with concentrations expressed in this way.

Cl_2	0.729
H_2S	0.385
O_2	0.00434
CO	0.00284

24. Concentrated aqueous ammonia, as obtained commercially, is a 29.2 percent (by weight) aqueous solution with a specific gravity of 0.900. Express this concentration in terms of
   a. grams of NH_3 per liter of solution.
   b. molarity.
   c. grams of NH_3 per kilogram of water.
   d. molality.
   e. liters of NH_3 (at STP) per kilogram of water.
   f. mole fraction of NH_3.

25. The information listed below is generally provided on the labels of concentrated acids supplied to laboratories. Express the concentrations of each acid in the five other ways we have used in this chapter.

Acid	Specific Gravity	Percent by Weight
Hydroiodic acid (HI)	1.50	47
Hydrobromic acid (HBr)	1.50	49
Hydrochloric acid (HCl)	1.18	37
Hydrofluoric acid (HF)	1.17	48

26. What volume of $NH_3(g)$ at STP is necessary to prepare 2.00 liters of 0.100 M aqueous solution?

27. What volume of methanol is necessary to prepare 3.25 liters of a 10.0 percent (by volume) aqueous solution?

28. What weight of sucrose ($C_{12}H_{22}O_{11}$) is necessary to prepare 1.24 liters of a 17.3 percent (by weight) aqueous solution? (Specific gravity = 1.25.)

29. What weight of NaCl is necessary to prepare 2.00×10^2 ml of 3.27 M aqueous solution?

30. For each case below, express the concentration (as molarity) that results when the given solution is diluted to 1.00 liter.
   a. 255 ml of 3.47 M HCl.
   b. 1.36×10^{-2} liter of 0.175 M NaOH.

c. 3.81×10^2 g of 32.4 percent (by weight) $NaNO_3$ solution.
d. 5.61×10^{-1} liter of 13.4 percent (by volume) ethanol (C_2H_5OH) solution. (Specific gravity of ethanol = 0.789.)

31. 55.0 ml of 2.46 M K_2SO_4 is mixed with 75.0 ml of 3.17 M Na_2SO_4. Assuming that no reaction takes place among the ions, calculate the concentration of each ion in the final solution.

32. Calculate the percent by weight, the molality, and the mole fraction of solute for each case:
  a. 60.0 g of methanol (CH_3OH) in 145 g of water.
  b. 1.23 g of naphthalene ($C_{10}H_8$) in 25.6 g of benzene (C_6H_6).
  c. 354 g of phosphoric acid (H_3PO_4) in 826 g of water.
  d. 45.0 g of benzene (C_6H_6) in 125 g of acetone (C_3H_6O).

33. Complete the following table for aqueous solutions of the substance in the first column.

Solute	Grams Solute per Liter of Solution	Molarity	Molality	Specific Gravity	Percent (by wt)	Mole Fraction (solute)
H_2SO_4				1.85	96.0	
Sucrose $C_{12}H_{22}O_{11}$	585		2.73			
Ethanol C_2H_5OH		9.21				0.249
KI	414			1.30		

34. What volume of water must be added to change the concentration of 1.00 liter of a 6.00 M NaOH solution to 3.00 M? to 1.00 M? to 0.100 M?

35. To what volume must 0.376 liter of a 1.28 M solution of H_3PO_4 be diluted to reduce the concentration to 0.100 M?

36. A reaction mixture requires the addition of 1.22 g of $AgNO_3$. How many milliliters of a 10.0 percent solution (specific gravity = 1.09) of $AgNO_3$ must be used?

37. The concentration of glucose ($C_6H_{12}O_6$) in human blood is approximately 100 mg/100 ml of blood. What is its molarity?

38. For each case below, the amount of solute under A is used to prepare the amount of (aqueous) solution under B. Calculate the molarity.

	A	B
a.	217 g $KClO_3$	0.350 liter
b.	104 g $NiCl_2$	2.46 liters
c.	4.34×10^2 liters (STP) NH_3	1.00 liter
d.	2.46×10^{-1} liter (STP) CO_2	1.53 liters

10.6 Saturation

39. Calculate the approximate ratio of molecules or ions of solute to molecules of water for each of the following saturated solutions. Concentrations are in amount of solute per 1.00×10^2 g of water.
   a. Argon (Ar), gas: 5.6 cc at STP
   b. Carbon tetrachloride (CCl_4), liquid: 0.0970 g
   c. Silver chloride (AgCl), solid: 8.9×10^{-5} g
   d. Sodium chloride (NaCl), solid: 35.7 g

40. A saturated solution of KNO_3 at 20°C is 2.77 M, but at 60°C a saturated solution is 7.17 M. As 1.00 liter of the 7.17 M solution cools to 20°C, what weight of KNO_3 will crystallize, if supersaturation is avoided?

10.7 Physical Properties of Solutions

41. A solution containing 15.6 g of an unknown nonelectrolyte in 25.0 g of water boils at 100.94°C. What is the molality of the solution? What is the molecular weight of the solute?

42. Water at the top of a mountain boils at 98.0°C because of the reduced atmospheric pressure. How much salt (NaCl) would have to be added to a kilogram of water to bring its boiling point back to 100.0°C? (Assume no interaction between Na^+ ion and Cl^- ion.)

43. Naphthalene melts at 80.60°C and has a molal freezing point depression constant of 6.85°C. If 3.20 g of sulfur, dissolved in 25.0 g of naphthalene, produces a solution that melts at 77.17°C, what is the correct chemical formula for the elemental sulfur?

44. Calculate the freezing point of a 24.2 percent solution of ethanol (C_2H_5OH) in water.

45. A glycerol-water mixture freezes at -15.5°C. Express the concentration of the solution as percent by weight of glycerol ($C_3H_8O_3$).

46. Camphor melts at 179.0°C (1.00 atm). A 0.24 m solution of a nonelectrolyte in camphor melts at 169.9°C. Calculate the molal freezing point depression constant for camphor.

*__47.__ Ethylene glycol ($C_2H_6O_2$) is common antifreeze. How many quarts of antifreeze must be added to 3 gal of water to protect a car radiator against freezing for temperatures down to -20°F? The specific gravity of ethylene glycol is 1.11.

48. At what temperature would the solution in Problem 47 boil?

49. a. Predict how much the vapor pressure of 1.00 liter of water at 100.0°C will drop if 42.4 g of LiCl is added.
   b. If the experiment described above is done, the vapor pressure is found to drop by 25.5 mm of Hg. Give a reason for the deviation from the predicted value.

50. Arrange the following in order of their increasing effect on the vapor pressure of 1.00 liter of water:
   a. 164 g of $Ca(NO_3)_2$
   b. 37.3 g of KCl
   c. 513 g of sucrose ($C_{12}H_{22}O_{11}$)
   d. 649 g of glycerol ($C_3H_8O_3$)

10.8 Osmosis

51. A glucose solution prepared for intravenous injection is 0.615 molal and has a specific gravity of 1.20. What is the osmotic pressure of blood at body temperature (37.0°C)?

52. An aqueous solution contains 4.3 g of an unidentified macromolecule in 1.00×10^2 ml of solution. The osmotic pressure of the solution is 11.7 mm of Hg at 20.0°C. Calculate the approximate molecular weight of the dissolved substance.

53. If a bag made of a semipermeable membrane is filled with a 0.6 m aqueous solution of glucose and immersed in a 0.1 m aqueous glucose solution, what will happen?

Discussion Starters

54. If all other factors are equal, which member of each of the following pairs is more soluble? Explain your answer.
 a. A solute with a high lattice energy or one with a low one?
 b. A solute with a high hydration energy or one with a low one?
 c. A solute that liberates energy when it dissolves or one that absorbs energy?

55. Criticize the following statements.
 a. Dissolving NaCl is a physical process.
 b. Silver iodide is an insoluble compound.
 c. Pure air is a solution.

56. If a solid evolves heat when it dissolves, its solubility decreases with temperature; if it absorbs heat when it dissolves, its solubility increases with increasing temperature. Explain.

57. A 1.00 m solution of sodium hydroxide in water would be expected to freeze at $-3.72°C$. However, the actual freezing point is $-3.44°C$. Explain.

11 Thermodynamics

Although our concern shall continue to be the structure of matter and the physical and chemical changes it can undergo, this chapter will have a very different flavor from those that have preceded it. Here we shall be calculating the heat, work, and energy involved during physical changes and chemical reactions, and we shall introduce the science that provides the theoretical foundations for these calculations. In the end we'll find that all the processes we are interested in proceed according to a few precepts that form the core of this science that we call *thermodynamics*.

The basic concept in thermodynamics is energy, which as we saw in Chapter 1 is the capacity to do work. Energy can take many forms, some of which we've already encountered: We first mentioned heat in Chapter 1, and it has come up repeatedly since; electrical energy was involved in our discussion of subatomic particles in Chapter 5, as was electromagnetic radiation (radiant energy); and we mentioned kinetic energy in our discussion of the theory of gases. But energy alone does not tell the whole story. Only by deriving other concepts from the basic one of energy can we begin to do what we would like, namely, to predict the conditions under which a process occurs spontaneously and to calculate how much energy it will absorb or release. As an illustration of the job before us, as well as the inadequacy of our intuition to deal with it, let's look at an example.

Our experience suggests that spontaneous processes give off energy: We are familiar with things that fall down, cool off, or unwind. We never expect a spring to absorb energy and thereby rewind itself or a kettle of water to absorb enough heat from the air to boil. Nonetheless, spontaneous endothermic reactions do occur. An everyday example is the ice cube that absorbs heat when it melts on exposure to air at room temperature. An even more interesting example is the contrast between the soluble salts calcium chloride and potassium chloride. When we dissolve calcium chloride, the resulting solution is warmer than the water we began with. With potassium chloride the solution is cooler; yet both processes are spontaneous. This just goes to show that the job of thermodynamics is not always an easy one.

Question

1. Classify the following changes as exothermic or endothermic:
   a. melting of an ice cube   b. burning a match
   c. evaporation of water

11.1 Heat and Work

In 1789 when Antoine Lavoisier (see p. 34) published a list of the 33 known elements he included heat, or "caloric," among them. It is easy to see why he and his contemporaries might make this mistake, for heat does seem to be something that flows from an object of higher temperature to one of lower temperature. However, one contemporary of Lavoisier, Benjamin Thomson, Count Rumford (1753–1814), did not subscribe to the common belief. While director of munitions manufacture in Bavaria, he noticed that great quantities of heat were given off when brass cannons were bored. Water around the cannons boiled during the process. Thomson subsequently performed experiments that demonstrated a direct relationship between the heat produced and the mechanical work done. James Joule (1818–1889), a Manchester brewer, extended Thomson's work by making more careful measurements of this relationship. He measured the mechanical work necessary to raise the temperature of a known quantity of water, and his results were remarkably close to the modern value of 4.184×10^3 J/(°C) (kg). Of course Joule was an Englishman, so he expressed his results in foot-pounds of work, pounds of water, and Fahrenheit degrees, but his name

Table 11.1 Units

$$\text{velocity} = \frac{\text{distance}}{\text{time}}$$

$$\text{acceleration} = \frac{\text{distance}}{(\text{time})^2}$$

$$\text{momentum} = \frac{(\text{mass})(\text{distance})}{(\text{time})}$$

$$\text{force} = \frac{(\text{mass})(\text{distance})}{(\text{time})^2}$$

$$\text{energy} = \frac{(\text{mass})(\text{distance})^2}{(\text{time})^2}$$

CGS System	SI Units
Force: $1 \text{ dyne} = \dfrac{(1 \text{ g})(1 \text{ cm})}{(1 \text{ s})^2}$	$1 \text{ newton} = \dfrac{(1 \text{ kg})(1 \text{ m})}{(1 \text{ s})^2}$
Energy: $1 \text{ erg} = \dfrac{(1 \text{ g})(1 \text{ cm})^2}{(1 \text{ s})^2}$	$1 \text{ joule} = \dfrac{(1 \text{ kg})(1 \text{ m})^2}{(1 \text{ s})^2}$

$1 \text{ newton} = 1 \times 10^5 \text{ dynes}$
$1 \text{ joule} = 1 \times 10^7 \text{ ergs}$
$1 \text{ calorie} = 4.184 \text{ joules}$

is now given to the metric unit for work (1.00 joule = 0.738 ft-lb = $\dfrac{1.00 \text{ kg m}^2}{1.00 \text{ s}^2}$ = 1.00 Nm). Table 11.1 provides a review of important units. Joule's first papers on the *mechanical equivalent of heat* were rejected outright by the British Royal Society, but within 10 years his ideas had won wide support and "caloric" was no longer considered an element.

Now that work is playing an important role in our discussion of thermodynamics, we should look more carefully at how we measure it. We shall define **mechanical work** as follows:

mechanical work = force × distance through which the force operates

A special kind of mechanical work, which will be especially important in chemical thermodynamics, is **pressure-volume work**, or $P \, \Delta V$ work. As an example of $P \, \Delta V$ work, consider a cylinder of gas with a weightless piston, as shown in Figure 11.1. In Figure 11.1a the gas pressure exactly balances the atmospheric pressure. However, if we heat the gas, it will expand, and the piston will move to the position shown in Figure 11.1b. Clearly work has been done: The force of the expanding gas has moved the piston.

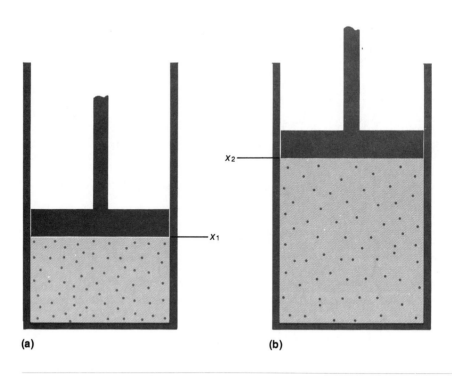

Figure 11.1 The volume of a gas under constant pressure is increased by an increase in temperature.

Let us now show that we can calculate the amount of work done by considering only the pressure of the gas and the change in volume (ΔV):

work = force × distance through which the force acts

work = $F(X_2 - X_1) = F \Delta X$

But since pressure is defined as force per unit area,

$$P = \frac{F}{A}$$

$$F = PA$$

Thus,

work = $PA \Delta X$

or, since the area of the piston (A) times the distance it moves (ΔX) is equal to the change in volume (ΔV),

work = $P \Delta V$

This work done by the expanding gas represents a loss in internal energy by the gas, and so is designated as *negative* work:

$$w = -P \Delta V$$

(In the next section we look more closely at the sign conventions used in expressing heat and work in relation to a system and to its surroundings.) To confirm that $P \Delta V$ does indeed represent work, we can substitute the appropriate units.

$$P \Delta V = \left(\frac{\text{newton}}{\text{meter}^2}\right)(\text{meter}^3) = (\text{newton})(\text{meter}) = \text{joule}$$

Example 11.1 Calculate the work in joules done at 1.00 atm by a gas in expanding from 1.00 liter to 1.10 liters. (The discussion of pressure units in Chapter 8 may be helpful.) For work done by the gas, $P \Delta V < 0$.

a. To express $P \Delta V$ work in joules, we must express pressure in newtons per square meter and volume in cubic meters.

b. $(1.00 \text{ atm}) \left(\frac{1.013 \times 10^5 \text{ N/m}^2}{1.00 \text{ atm}}\right) = 1.01 \times 10^5 \text{ N/m}^2$

c. $(0.10 \text{ liter}) \left(\frac{10^3 \text{ cm}^3}{1 \text{ liter}}\right) \left(\frac{1 \text{ m}}{10^2 \text{ cm}}\right)^3 = 1.0 \times 10^{-4} \text{ m}^3$

d. $P\Delta V = (1.01 \times 10^5 \text{ N/m}^2)(1.0 \times 10^{-4} \text{ m}^3) = 1.0 \times 10^1 \text{ Nm}$

e. $P\Delta V = -w; w = -\mathbf{1.0 \times 10^1 \text{ J}}$

Question

2. Can you give an example where heat flows from one object to another but does *not* raise the temperature of the second object?

11.2   Thermodynamic Systems

We now know that in thermodynamics we're going to be dealing in some way with heat, work, and energy. To go further we must define precisely the *systems* we shall be dealing with. Each system that undergoes a physical or chemical change can in principle be separated from its *surroundings* by distinct boundaries. Together each system and its surroundings constitute the *universe* of the event (or reaction). We can define the *state* of each system by specifying enough independent variables that another experimenter can reproduce the system exactly. For instance, we can describe a system consisting of one mole of an ideal gas by specifying only two of the three variables: temperature, volume, and pressure. The third is then determined by the ideal gas law equation if the gas behaves ideally. In this example the variables P, T, and V are called **state functions**. The values of state functions depend only on the state of a system and not in any way on the history of the system. The temperature of a system in no way depends on whether it was *heated* to that temperature or *cooled* to that temperature. Another example of a state function is energy. Two important quantities that are *not* state functions are work and heat. To illustrate, consider as a system a quantity of water at 20°C. We may raise the temperature of this system 1°C by transferring a certain quantity of heat to it. One might then attempt to make some judgement as to the quantity of heat contained by the system. But the same effect might be produced by setting into motion a paddle wheel or other mechanical device within the system. As Thomson discovered with his cannon, the temperature of a system may be elevated by transferring a quantity of work to it. This, then, would be the "same" system attained by heating, but it certainly would *not* contain the same quantity of heat. It then becomes obvious that the "quantity of heat" or "quantity of work" contained by a system does depend on the path by which the system attained a given state. Thus heat and work are not state functions.

If a system does not exchange matter with its surrounding, we call it a *closed* system; if it exchanges neither energy nor matter, it is an *isolated* system. To close a system, we need only to seal the container that encloses it. To isolate a system, however, we must also insulate it to ensure that it neither gains nor loses heat and must ensure that no work is done on or by it.

As we have already emphasized, the most important quantity in thermodynamics is energy. Therefore, for each system, we shall be interested in its *total internal energy*, which we give the symbol U. This is the sum of the kinetic and potential energies of all the atoms, ions, and molecules in the system. Since we cannot directly measure the energy of a system, however, an even more important quantity is ΔU, the *change in the internal energy* of a system during a chemical or physical change. For most of the systems we'll be interested in, we can describe changes in internal energy in terms of heat and work. (Remember that we can relate heat and work by using the mechanical equivalent of heat, 4.184 J = 1.000 cal.) We shall let q represent the *heat absorbed by the system* from its surroundings and w represent the *work done on the system*. *Heat evolved from a system* will be designated by $-q$, and *work done by the system* will be designated by $-w$. We can now write for the change in internal energy of a system

$$\Delta U = q + w$$

Since energy is a state function, the change in energy of a system depends only on its initial and final states. However, both q and w depend on the path the system follows to get from its initial to its final state. We shall say more about this in the following sections.

11.3 The First Law of Thermodynamics

The **first law of thermodynamics** parallels the *law of the conservation of mass*: It states that *energy is conserved*. We can state it concisely as follows:

$$\Delta U_{system} = -\Delta U_{surroundings}$$

or

$$\Delta U_{system} + \Delta U_{surroundings} = 0$$

These statements say that any energy that leaves a system will appear in its surroundings, and vice versa. Thus there can be no net change of energy in the universe. (We shall find, later, that mass and energy can be converted from one to the other; but for all reactions except certain *nuclear reactions* (Chapter 25), mass changes that accompany energy changes are too small to be measured.)

When we measure energy changes in the laboratory, two especially convenient procedures are available. We can carry out the reaction in a container open to the atmosphere, thus ensuring that the pressure remains constant, or we can carry it out in a closed, rigid container, in which case the volume must remain constant. An example of the constant-volume method is the use of a simple *calorimeter*, such as the one shown in Figure 11.2. A calorimeter is a device that permits an experimenter to measure the absorption or evolution of heat during a reaction. In the example shown in the figure, a U-shaped reaction container is immersed in a known quantity of water in an insulated vessel. When the U-tube

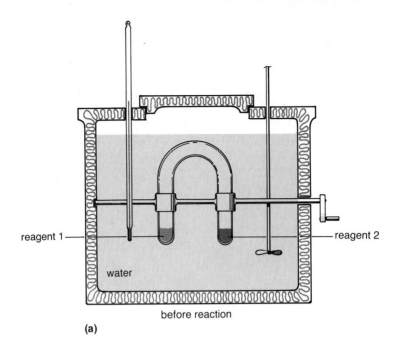
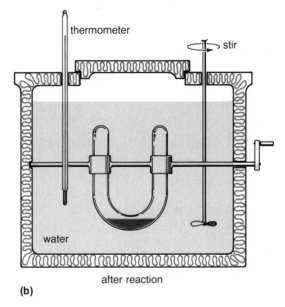

Figure 11.2 A calorimeter.

is inverted, the reactants mix, and the water temperature rises or drops, depending on whether heat is given off or absorbed. Some constant-volume calorimeters are heavily constructed to withstand large pressure changes and hence are called *bomb calorimeters*. Other calorimeters are constructed as constant-pressure calorimeters. The symbol q_v is used for heat evolved or absorbed at constant volume.

Example 11.2 The temperature of 1.00 kg of water in a calorimeter rises 11.6°C when 1.48 g of carbon burns. Calculate the heat given off (q_v) for each mole of carbon. Assume no heat loss to the calorimeter.

Solution

a. Since the specific heat of water is 1.00 cal/(g)(°C), the heat liberated is

$$\left[\frac{1.00 \text{ cal}}{(1.00 \text{ g})(1.00°\text{C})}\right](11.6°\text{C})(1000 \text{ g})$$

b. The heat given off for each gram of carbon is

$$\left[\frac{1.00 \text{ cal}}{(1.00 \text{ g})(1.00°\text{C})}\right](11.6°\text{C})(1000 \text{ g})\left(\frac{1}{1.48 \text{ g C}}\right)$$

Therefore the heat liberated for each mole of carbon is

$$\left[\frac{1.00 \text{ cal}}{(1.00 \text{ g})(1.00°\text{C})}\right](11.6°\text{C})(1000 \text{ g})\left(\frac{1}{1.48 \text{ g C}}\right)\left(\frac{12.0 \text{ g C}}{1.00 \text{ mol}}\right)$$

$$= \textbf{94,000 cal/mol or 94.0 kcal/mol}$$

Using an isolated constant-volume calorimeter is especially useful since it ensures that no pressure-volume work is done. Therefore, for most reactions (where pressure-volume work is the only kind of work we need to worry about), we can measure ΔU simply by measuring q_v:

$$\Delta U = q_v + w$$
$$w = 0$$
$$\Delta U = q_v$$

However, it is often more convenient to carry out reactions in vessels open to the atmosphere. In this case, the pressure remains constant but work is often done against the pressure of the atmosphere as the volume of the system changes. An increase in volume ($\Delta V > 0$) represents work done on the surroundings. Thus

$$w = -P\Delta V$$

To measure the energy change, therefore, we must measure the work done as well as q_p, the heat absorbed:

$$\Delta U = q_p + w$$

or

$$\Delta U = q_p - P \Delta V$$

We should emphasize that this equation is valid only if (1) the external pressure remains constant, (2) the internal pressure is the same as the external pressure when the reaction is complete, and (3) only $P \Delta V$ work is done.

Here it is useful to define a new thermodynamic quantity, **enthalpy**. Enthalpy (H) is a state function and is defined by the equation

$$H = U + PV$$

Thus at constant pressure

$$\Delta H = \Delta U + P \Delta V$$

or since

$$\Delta U = q_p - P \Delta V$$
$$\Delta H = q_p$$

It is because of this relation that H was defined and is used as a basis for heat calculations and tabulations in chemistry. It is a general result that

$$\Delta U = q_v$$
$$\Delta H = q_p$$

This is true for any process, as ΔU and ΔH are state functions. Thus if you have ΔH and ΔU for a given reaction, you immediately know how much heat would be evolved if the reaction were run at constant pressure or at constant volume.

The relation between the two is best shown as

$$U = H - PV$$
$$\Delta U = \Delta H - \Delta(PV)$$

which reduces to

At constant pressure: $\quad \Delta U = \Delta H - P \Delta V$

At constant volume: $\quad \Delta U = \Delta H - V \Delta P$

or, if the gas is ideal so that $PV = nRT$,

$$\Delta U = \Delta H - \Delta(nRT)$$

In reactions where gases are not present, the $P\Delta V$ work is very small compared to the total energy change, so we can safely estimate that the change in internal energy is approximately equal to the change in enthalpy. When only gases are involved, however, $P\Delta V$ will depend on Δn, the change in the number of moles; therefore, we can calculate ΔU exactly.

Example 11.3 If we burn 1.00 mol of propane gas (C_3H_8) at constant pressure, the reaction evolves 526.0 kcal/mol at 25.0°C and 1.00 atm ($\Delta H < 0$). This represents a change in enthalpy. Calculate ΔU, which is equal to q_v.

Solution

a. The equation for the reaction is

$$C_3H_8(g) + 5\,O_2(g) \longrightarrow 3\,CO_2(g) + 4\,H_2O(l)$$

Therefore, 6 mol of gaseous reactants form 3 mol of gaseous products. If the pressure remains constant, the volume must decrease, and work will be done by the surroundings *on the system*. (Note that we can ignore the insignificant volume of the liquid water produced during the reaction.)

b. If we let V_a and n_a represent the volume and number of moles of gaseous reactants and V_b and n_b represent the volume and number of moles of gaseous products, then

$$PV_a = n_a RT \quad \text{and} \quad PV_b = n_b RT$$

Or, subtracting the first equation from the second,

$$P(V_b - V_a) = (n_b - n_a)RT$$
$$P\Delta V = \Delta nRT$$

where, as usual, the symbol Δ stands for the change in a quantity. Substituting for Δn, R, and T,

$$P\Delta V = (-3.00 \text{ mol})\left[\frac{1.99 \text{ cal}}{(K)(\text{mol})}\right](298 \text{ K})$$
$$= -1.78 \times 10^3 \text{ cal} \quad \text{or} \quad -1.78 \text{ kcal}$$

As a check, note that $P\Delta V < 0$ (reduction in volume) implies $w > 0$, which means work was done *on* the system.

c. $\Delta U = \Delta H - P\,\Delta V$

$\Delta U = (-526.0 \text{ kcal}) - (-1.78 \text{ kcal})$

$\Delta U = -524.2 \text{ kcal}$

If we had used ΔH to estimate ΔU, our error would have been very small:

$$\frac{1.8 \text{ kcal}}{524 \text{ kcal}} \times 100\% = 0.34\%$$

Example 11.4 Show that the following two values for the gas law constant R are equal:

$$0.08206\,\frac{(\text{liter})(\text{atm})}{(\text{K})(\text{mol})} \quad \text{and} \quad 1.987\,\frac{\text{cal}}{(\text{K})(\text{mol})}$$

$(1.000 \text{ atm} = 1.013 \times 10^1 \text{ N/cm}^2)$

Solution

a. We need only show that 0.08206 liter-atm $= 1.987$ cal. We can do this by converting both to joules, since both are in units of energy (work).

b. $(1.987 \text{ cal})\left(\dfrac{4.184 \text{ J}}{1.000 \text{ cal}}\right) = 8.314 \text{ J}$

c. $(0.08206 \text{ liter-atm})\left(\dfrac{10^3 \text{ cm}^3}{1 \text{ liter}}\right)\left(\dfrac{1.013 \times 10^1 \text{ N/cm}^2}{1.000 \text{ atm}}\right) = 8.313 \times 10^2 \text{ N cm}$

$= (8.313 \times 10^2 \text{ N cm})\left(\dfrac{1 \text{ m}}{10^2 \text{ cm}}\right) = 8.313 \text{ N m} = 8.313 \text{ J}$

Question

3. Complete and balance the following equations; then determine whether ΔU or ΔH for the reaction at 25°C will have the greater absolute value. (All reactions are highly exothermic, and all reactants and products except water are gases.)
   a. $C_3H_8 + O_2$ b. $CH_4 + O_2$ c. $C_2H_2 + O_2$

11.4 Thermochemical Equations

Thermochemical equations are much like the equations we studied during our discussion of stoichiometry. In addition to providing information about mass relationships, however, a thermochemical equation provides thermodynamic information. For example,

$$2\ C_2H_6(g) + 7\ O_2(g) \longrightarrow 4\ CO_2(g) + 6\ H_2O(l) \qquad \Delta H° = -746\ \text{kcal}$$

Here $\Delta H°$ is the change in enthalpy for the reaction (see below for explanation of superscript in $\Delta H°$). This quantity can be treated as an additional term in the equation: When 2 mol of ethane and 7 mol of oxygen react, the products are 4 mol of CO_2 gas, 6 mol of liquid water, and 746 kcal of heat. (Note the sign of $\Delta H°$. If $\Delta H < 0$, then $q_p < 0$, and the reaction is exothermic.)

Since ΔH depends on the physical states of the reactants and products, as well as the temperature and pressure, we must carefully identify the conditions under which ΔH was measured or calculated. This is the origin of the concept of **standard states.** The standard state of a substance is the most stable form in which it occurs at 1 atm and the given temperature. Most tables supply ΔH values at 25°C. The superscript (°) on a symbol indicates that the value is for a reaction at 1 atm where the reactants and products are in their standard states at a specified temperature. For example, $\Delta H°$ for the reaction given in the first paragraph of this section represents the heat evolved at constant pressure when 2 mol of ethane and 7 mol of oxygen are completely converted to 4 mol of CO_2 gas and 6 mol of H_2O *liquid* water. (The water is normally liquid at 25°C; thus combustion reactions and ΔH values traditionally refer to the formation of liquid water.)

Sometimes it is inconvenient to measure ΔH experimentally. Furthermore, we cannot calculate ΔH for a reaction from the absolute enthalpies of the reactants and products, since we can't measure the *absolute* enthalpies any more than we can measure absolute internal energies (p. 345). However, *relative* enthalpies have been measured for many substances, and a vast amount of data has been tabulated in standard reference works such as the *Handbook of Chemistry and Physics*. One of the most useful quantities available is the *standard enthalpy of formation* normally tabulated at 25°C which we denote with the symbol $\Delta H°_f$. (Table 11.1 lists values for $\Delta H°_f$ for several common substances.) The standard enthalpy of formation of a substance is defined as the difference between the molar enthalpy of the substance in its standard state and the total enthalpy of its constituent elements in their standard states. In other words, $\Delta H°_f$ for all elements in their standard states has been arbitrarily designated as zero. The *standard enthalpy of combustion* is also available for many organic compounds (see Table 11.3). It is defined as the heat evolved when 1.00 mol of the organic substance is burned to $H_2O(l)$ and $CO_2(g)$ in an excess of oxygen at 1.00 atm.

Example 11.5 How much methane must be burned to $H_2O(l)$ and $CO_2(g)$ to provide enough energy to prepare 15.0 g of NO from its constituent elements? Assume that the changes in temperature do not affect the enthalpies.

Solution

a. We find in Tables 11.2 and 11.3 that the enthalpy of formation of NO is 21.6 kcal/mol and that the enthalpy of combustion of CH_4 is -213 kcal/mol.

b. Since these enthalpies are molar quantities, we must next determine the number of moles of NO to be produced.

$$(15.0 \text{ g}) \left(\frac{1.00 \text{ mol NO}}{30.0 \text{ g}} \right)$$

Table 11.2 Standard Enthalpies of Formation at 25°C and 1 atm

Substance	Physical State	Standard Enthalpy of Formation (kcal/mol)
C (graphite)	Solid	0
C (diamond)	Solid	0.453
CCl_4	Liquid	-32.4
CO	Gas	-26.4
CO_2	Gas	-94.0
CH_4	Gas	-17.9
C_2H_6	Gas	-20.2
Cl_2	Gas	0
HBr	Gas	-8.66
HCl	Gas	-22.1
F_2	Gas	0
HF	Gas	-64.8
H_2	Gas	0
H_2O	Gas	-57.8
H_2O	Liquid	-68.3
I_2	Solid	0
I_2	Gas	14.9
HI	Gas	6.20
N_2	Gas	0
NO	Gas	21.6
NO_2	Gas	7.93
N_2O	Gas	19.49
NH_3	Gas	-11.0
O_2	Gas	0
O_3	Gas	34.1
P_4	Solid	0
P_4	Gas	14.1
PH_3	Gas	1.3
P_4O_{10}	Solid	-713
PCl_5	Gas	-95.4

Table 11.3 Standard Enthalpies of Combustion* at 25°C and 1 atm

Substance	Physical State	Standard Enthalpy of Combustion (kcal/mol)
C (graphite)	Solid	−94.0
C (diamond)	Solid	−94.5
CO	Gas	−67.6
CH_4	Gas	−213
C_2H_6	Gas	−373
H_2	Gas	−68.3
C_2H_4	Gas	−337
C_2H_2	Gas	−311
C_6H_6	Gas	−781
C_3H_8	Gas	−526

* Products of combustion for hydrocarbons are $CO_2(g)$ and $H_2O(l)$.

c. Therefore, the energy required to produce 15.0 g of NO from nitrogen and oxygen is

$$(15.0 \text{ g}) \left(\frac{1.00 \text{ mol NO}}{30.0 \text{ g}} \right) \left(\frac{21.6 \text{ kcal}}{1.00 \text{ mol NO}} \right) = 10.8 \text{ kcal}$$

d. The number of moles of CH_4 that must be burned to produce this energy is

$$(10.8 \text{ kcal}) \left(\frac{1.00 \text{ mol } CH_4}{213 \text{ kcal}} \right)$$

e. Therefore, the number of grams of methane that must be burned is

$$(10.8 \text{ kcal}) \left(\frac{1.00 \text{ mol } CH_4}{213 \text{ kcal}} \right) \left(\frac{16.0 \text{ g}}{1.00 \text{ mol } CH_4} \right) = \mathbf{0.811 \text{ g}}$$

The values for the enthalpies tabulated in Tables 11.2 and 11.3 do more than merely provide information about the formation and combustion of numerous substances at 1 atm. They also allow us to calculate enthalpy changes for reactions that have never been carried out in the laboratory. For any reaction at 25°C and 1 atm, the enthalpy change is equal to the sum of the enthalpies of formation for the products (each multiplied by its coefficient from the balanced equation), minus the sum of the enthalpies of formation for the reactants (each multiplied by its coefficient from the equation):

$$\Delta H°(\text{reaction}) = \Delta H°_f(\text{products}) - \Delta H°_f(\text{reactants})$$

This is an application of **Hess's law:** *When a reaction can be regarded as the sum of two or more reactions, the enthalpy change of the total reaction is the sum of the enthalpy changes for the component reactions.*

Hess's law is valid because enthalpy is a state function. The enthalpy of a compound is independent of how we obtain it. For example, the value for the standard enthalpy of formation of CO_2 is the same whether we get the CO_2 by burning carbon in an excess of oxygen:

$$C(s) + O_2(g) \longrightarrow CO_2(g)$$

or by a two-step process that first produces CO as an intermediate:

$$2\,C(s) + O_2(g) \longrightarrow 2\,CO(g)$$
$$2\,CO(g) + O_2(g) \longrightarrow 2\,CO_2(g)$$

The enthalpy change for the first process is given in Table 11.2 as -94.0 kcal/mol. We can check this value as follows, using only the values for the enthalpy of formation for CO (Table 11.2) and the enthalpy of combustion for CO (Table 11.3):

$$2\,C(s) + O_2(g) \longrightarrow 2\,CO(g)$$

$$\Delta H_f^\circ\ CO(g) = -26.4\ \text{kcal/mol}$$

$$\Delta H_{\text{reaction}}^\circ = \left(\frac{-26.4\ \text{kcal}}{1.00\ \text{mol}}\right)(2.00\ \text{mol}) = -52.8\ \text{kcal}$$

$$2\,CO(g) + O_2(g) \longrightarrow 2\,CO_2(g)$$

$$\Delta H_{\text{combustion}}^\circ\ CO(g) = -67.6\ \text{kcal/mol}$$

$$\Delta H_{\text{reaction}}^\circ = \left(\frac{-67.6\ \text{kcal}}{1.00\ \text{mol}}\right)(2.00\ \text{mol}) = -135.2\ \text{kcal}$$

$$\begin{array}{lll}
2\,C(s) + O_2(g) \longrightarrow 2\,CO(g) & \Delta H_{\text{reaction}}^\circ = -52.8\ \text{kcal} \\
2\,CO(g) + O_2(g) \longrightarrow 2\,CO_2(g) & \Delta H_{\text{reaction}}^\circ = -135.2\ \text{kcal} \\
\hline
2\,C(s) + 2\,O_2(g) \longrightarrow 2\,CO_2(g) & \Delta H_{\text{reaction}}^\circ = -188.0\ \text{kcal}
\end{array}$$

For one mole of CO_2,

$$\Delta H_{\text{reaction}}^\circ = -94.0\ \text{kcal} = \Delta H_f^\circ\ CO_2(g)$$

Example 11.6 Using Hess's law, calculate the enthalpy change for the following reaction at 25°C and 1 atm.

$$CH_4(g) + 4\,Cl_2(g) \longrightarrow CCl_4(l) + 4\,HCl(g)$$

Solution

a. From Table 11.2, we must first find the enthalpies of formation for each of the reactants and products:

$$CH_4(g): \Delta H_f^\circ = -17.9 \text{ kcal}$$
$$Cl_2(g): = 0$$
$$CCl_4(g): = -32.4 \text{ kcal}$$
$$HCl(g): = -22.1 \text{ kcal}$$

b. Then, from Hess's law,

$$\Delta H_{\text{reaction}}^\circ = [\Delta H_f^\circ \; CCl_4(l) + 4\,\Delta H_f^\circ \; HCl(g)]$$
$$- [\Delta H_f^\circ \; CH_4(g) + 4\,\Delta H_f^\circ \; Cl_2(g)]$$

c. Substitute and solve:

$$\Delta H_{\text{reaction}}^\circ = \left[(1 \text{ mol})\left(\frac{-32.4 \text{ kcal}}{1.00 \text{ mol}}\right) + (4 \text{ mol})\left(\frac{-22.1 \text{ kcal}}{1.00 \text{ mol}}\right)\right]$$
$$- \left[(1 \text{ mol})\left(\frac{-17.9 \text{ kcal}}{1.00 \text{ mol}}\right) + (4 \text{ mol})\left(\frac{0 \text{ kcal}}{1.00 \text{ mol}}\right)\right]$$
$$= -120.8 \text{ kcal} + 17.9 \text{ kcal}$$

$$\Delta H_{\text{reaction}}^\circ = -102.9 \text{ kcal}$$

Example 11.7 Use heats of combustion to calculate the standard enthalpy of formation (ΔH_f°) at 25°C for ethane.

Solution

a. Write the equation for the combustion of ethane.

$$2\,C_2H_6(g) + 7\,O_2(g) \longrightarrow 4\,CO_2(g) + 6\,H_2O(l)$$

b. The enthalpy change for each mole of C_2H_6 undergoing this reaction is the standard enthalpy of combustion tabulated in Table 11.3:

$$\Delta H_{\text{reaction}}^\circ = \Delta H_{\text{combustion}}^\circ \; C_2H_6(g) = -373 \text{ kcal/mol}$$

c. And from Table 11.2,

$$\Delta H_f^\circ \; O_2(g) = 0$$
$$\Delta H_f^\circ \; CO_2(g) = -94.0 \text{ kcal/mol}$$
$$\Delta H_f^\circ \; H_2O(l) = -68.3 \text{ kcal/mol}$$

d. Applying Hess's law,

$$\Delta H^\circ_{reaction} = [4\,\Delta H^\circ_f\,CO_2(g) + 6\,\Delta H^\circ_f\,H_2O(l)]$$
$$- [2\,\Delta H^\circ_f\,C_2H_6(g) + 7\,\Delta H^\circ_f\,O_2(g)]$$

e. Letting X stand for the standard enthalpy of formation of ethane, we can now substitute and solve:

$$\left(\frac{-373\,\text{kcal}}{1.00\,\cancel{\text{mol}\,C_2H_6}}\right)(2\,\cancel{\text{mol}\,C_2H_6}) =$$

$$\left[(4\,\cancel{\text{mol}})\left(\frac{-94.0\,\text{kcal}}{1.00\,\cancel{\text{mol}}}\right) + (6\,\cancel{\text{mol}})\left(\frac{-68.3\,\text{kcal}}{1.00\,\cancel{\text{mol}}}\right)\right]$$

$$-\left[(2\,\cancel{\text{mol}})\left(\frac{X\,\text{kcal}}{1.00\,\cancel{\text{mol}}}\right) + (7\,\cancel{\text{mol}})\left(\frac{0\,\text{kcal}}{1.00\,\cancel{\text{mol}}}\right)\right]$$

$$-746\,\text{kcal} = -785.8\,\text{kcal} - 2X\,\text{kcal}$$

$$2X\,\text{kcal} = -40\,\text{kcal}$$

$$X = \frac{-20\,\text{kcal}}{1.00\,\text{mol}\,C_2H_6(g)}$$

$$\Delta H^\circ_f\,C_2H_6(g) = -20\,\text{kcal/mol}$$

Question

4. Use the thermochemical equation below to answer the questions that follow.

$$2\,C_2H_6(g) + 7\,O_2(g) \longrightarrow 4\,CO_2(g) + 6\,H_2O(l) \qquad \Delta H^\circ = -746\,\text{kcal}$$

a. What is the molar heat of combustion at 25°C of ethane?
b. How much heat is produced when 1.00 g of ethane burns?

11.5 Bond Enthalpies

The dissociation of a diatomic gas into its component atoms is an especially important reaction. As we shall soon see, it provides a convenient index of *bond strengths*. We call the ΔH for such a reaction the *dissociation enthalpy* or the *enthalpy of atomization* (sometimes given as heats of formation of gaseous atoms). For example, it takes 104 kcal to dissociate 1.00 mol of $H_2(g)$ into $H(g)$ atoms and 37 kcal to dissociate 1.00 mol of $F_2(g)$ into $F(g)$ atoms (see Table 11.4). As we shall see in Example 11.9, atomization enthalpies are available for more complex substances than diatomic molecules, for example, graphite.

Using these dissociation enthalpies to calculate bond strengths is another application of Hess's law. For example, to calculate the enthalpy of the H—Cl bond, we can use the dissociation enthalpy of $H_2(g)$ and $Cl_2(g)$. The complete

Table 11.4 Dissociation Enthalpies (1 atm, 25°C)

$H_2(g)$	104 kcal/mol
$N_2(g)$	226
$O_2(g)$	118
$F_2(g)$	37
$Cl_2(g)$	58
$Br_2(g)$	46
$I_2(g)$	36
$CO(g)$	257
$CO_2(g)$	384*

* Note that this value is for two bonds per molecule.

reaction for the formation of $HCl(g)$ from the gaseous elements is

$$H_2(g) + Cl_2(g) \longrightarrow 2\ HCl(g)$$

We can represent it as the sum of two processes:

Bond breaking:
$$H_2(g) \longrightarrow 2\ H(g)$$
$$Cl_2(g) \longrightarrow 2\ Cl(g)$$

Bond formation: $\quad 2\ H(g) + 2\ Cl(g) \longrightarrow 2\ HCl(g)$

The energy for the complete reaction is the standard enthalpy of formation for $HCl(g)$, times the number of moles formed:

$$\Delta H_f^\circ = -22.1\ \text{kcal/mol}$$
$$\Delta H_f^\circ \times (2.00\ \text{mol}) = -44.2\ \text{kcal}$$

The enthalpies for the bond-breaking step are the dissociation enthalpies of the elements:

$$H_2(g) \longrightarrow 2\ H(g) \qquad \Delta H^\circ = 104\ \text{kcal/mol}$$
$$Cl_2(g) \longrightarrow 2\ Cl(g) \qquad \Delta H^\circ = 58\ \text{kcal/mol}$$

One should always remember that bonds are reasonably stable entities—energy is required to break them and energy is given off when they are formed.

Using the symbol D° with the appropriate subscript for dissociation enthalpy at 1 atm, 25°C, we can now complete the calculation as follows:

$$\Delta H_f^\circ\ HCl(g) \times (2.00\ \text{mol}) = D_{H_2}^\circ + D_{Cl_2}^\circ + \Delta H \quad (2\ \text{mol H—Cl bonds})$$
$$\Delta H(2\ \text{mol of H—Cl bonds}) = -44.2\ \text{kcal} - (104\ \text{kcal} + 58\ \text{kcal})$$
$$= -206\ \text{kcal}$$
$$\Delta H\ \text{for 1 mol of H—Cl bonds} = -103\ \text{kcal}$$

The minus sign tells us that 103 kcal are evolved when 1 mol of H—Cl bonds is formed. We can verify the calculated bond enthalpy for the H—Cl bond by looking up the experimentally measured dissociation enthalpy for HCl(g), which is 103 kcal/mol.

Example 11.8 Calculate the bond enthalpy of the HF molecule.

Solution

a. Reaction:

$$\underbrace{H_2(g) + F_2(g) \longrightarrow 2\,H(g) + 2\,F(g)}_{\text{bond breaking}} \underbrace{\longrightarrow 2\,HF(g)}_{\text{bond formation}}$$

b. Enthalpy values:

$$D^\circ_{H_2} = 104 \text{ kcal/mol}$$
$$D^\circ_{F_2} = 37 \text{ kcal/mol}$$
$$\Delta H^\circ_f \text{ HF}(g) = -64.8 \text{ kcal/mol}$$

c. Apply Hess's law:

$$\Delta H^\circ_f(2 \text{ mol}) \text{ HF}(g) = D^\circ_{H_2} + D^\circ_{F_2} + \Delta H(2 \text{ mol H—F bonds})$$
$$\Delta H(2 \text{ mol H—F bonds}) = -130 \text{ kcal} - (104 \text{ kcal} + 37 \text{ kcal})$$
$$= -271 \text{ kcal}$$

d. ΔH for formation of 1.00 mol H—F bonds = **−136 kcal**

Dissociation enthalpies are especially useful in estimating the bond enthalpies in polyatomic molecules since we cannot measure them directly. These estimates require some approximations, but the results are usually accepted as very close to the true values. The major approximation is the assumption that the strength of a given bond is constant regardless of its molecular environment. If this is true, we can say that the bond enthalpy for a given bond in a polyatomic molecule is the *average energy necessary to break a mole of bonds of this type in gaseous molecules*. Note that we have specified the *gas* state. As an illustration, consider the following example.

Example 11.9 Calculate the bond enthalpy for the C—H bond from the enthalpy of formation of methane, the atomization enthalpy of graphite (172 kcal/mol), and the dissociation enthalpy of $H_2(g)$.

Solution

a. A reaction by which the C—H bond strength might be evaluated is

$$\underbrace{C(s) + 2\,H_2(g) \longrightarrow C(g) + 4\,H(g)}_{\text{bond breaking}} \underbrace{\longrightarrow CH_4(g)}_{\text{bond formation}}$$

b. Enthalpy values:

$$D^\circ_{H_2} = 104 \text{ kcal/mol}$$
$$\text{atomization enthalpy } C(s) = 172 \text{ kcal/mol}$$
$$\Delta H^\circ_f\, CH_4(g) = -17.9 \text{ kcal/mol}$$

c. Apply Hess's law:

$$\Delta H^\circ_f\, CH_4(g) = D^\circ_{H_2}(2 \text{ mol}) + \text{at. energy } C(s) + \Delta H(4 \text{ mol C—H bonds})$$
$$\Delta H(4 \text{ mol C—H bonds}) = -17.9 \text{ kcal} - (208 \text{ kcal} + 172 \text{ kcal})$$
$$= -398 \text{ kcal}$$

d. Divide by 4 to obtain the average enthalpy of the C—H bond.

$$\frac{-396 \text{ kcal}}{4} = -99.5 \text{ kcal/mol}$$

(This answer is valid under the rules of significant figures but is something of an overestimate of the validity of the result because of the average nature of the quantity.)

11.6 Entropy

From what we have said so far, as well as from our own experience, we know that energy can be converted from one form to another. For example, in hydroelectric power stations potential energy is converted to electrical energy, and in our automobiles chemical energy becomes kinetic energy. From the *first law of thermodynamics* we know that during these processes energy can be neither created nor destroyed. However, the law fails to tell us one important thing: Why will one process occur, whereas its reverse will not? Sometimes our intuition is sufficient. We know very well that stones roll downhill rather than up (though in light of the first law the two events are equally likely), but our intuition is not likely to be such a faithful guide in predicting whether chemical reactions proceed or not. For this we shall need the more powerful concept of *entropy*.

Before defining entropy, however, let's introduce the new concepts of *reversible* and *irreversible* processes by describing an ideal experiment, which in turn will serve as a springboard for our discussion of spontaneity and entropy. Consider a cylinder filled with an ideal gas as shown in Figure 11.3. Suppose we now *instantaneously* withdraw the piston so that the volume of the cylinder doubles. We can now look at the statement of the first law

$$\Delta U = q + w$$

and ask what happens as the gas spontaneously expands into the new volume. For an ideal gas, since there are no interactions among the molecules, the internal energy depends only on the temperature and is independent of pressure and volume. Therefore, if we maintain the system at constant temperature, $\Delta U = 0$ and $q = -w$. Furthermore, since the gas did not expand against an external pressure, $P \Delta V = 0$; hence no work was done and no heat was exchanged with the surroundings.

Now let us recompress the gas to its original state and once again look at the thermodynamic changes. Again we shall carry out the process *isothermally* (at constant temperature), but we shall do it *reversibly*. By this we mean that we shall do it so slowly that the pressure we apply to the piston is always only

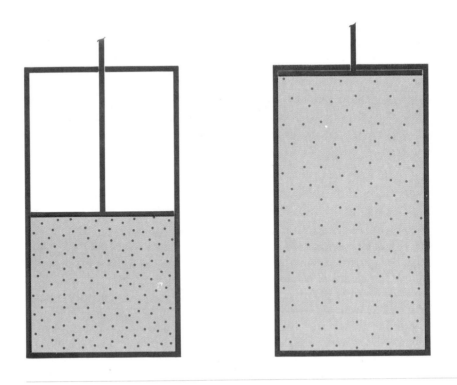

Figure 11.3 The diffusion of a gas.

infinitesimally greater than the pressure in the cylinder. Since we have not changed the temperature of the gas during this compression, we can again say that $\Delta U = 0$. However, we have done work on the system, and since $q = -w$, heat must flow from the system into the surroundings. Therefore, if we look at the complete cyclic process (expansion and recompression), we can draw some interesting conclusions. We find the gas in its original state (pressure, temperature, and volume); yet heat has flowed from it into the surroundings. To see the importance of this heat flow, let's look at another cyclic process.

Again consider the cylinder in Figure 11.3. This time, however, we shall withdraw the piston reversibly and then recompress the gas reversibly. During the first half of this cyclic process, work is done by the system; during the second half, an equal amount of work is done on the system. Likewise, during the expansion heat flows into the system, whereas during the compression an equal amount of heat flows out. Thus during this completely reversible cyclic process there has been no net heat flow; whereas in the first cyclic process, which included a spontaneous, irreversible step, there was a net flow of heat. To clarify the important difference between these two cyclic processes, we shall define a new state function S, which we call **entropy**. We shall define the change in the entropy of a system, during a reversible, isothermal process, as follows:

$$\Delta S = \frac{q_{\text{rev}}}{T} \quad \text{(valid only for a constant } T\text{)}$$

where T is the temperature in kelvins and q_{rev} is the heat absorbed by the system during the reversible change. Now let us go back to our cycles (Figure 11.4) for the last time, first to the reversible cycle: During the reversible expansion the system does work and absorbs heat. The entropy of the system increases, but the entropy of the surroundings decreases by an equal amount. During the reversible compression, the surroundings do work on the system and heat flows from the system. The entropy of the system decreases and the entropy of the surroundings increases by the same amount. In fact, we can generalize this observation: *For any reversible process the entropy of the universe remains constant.*

Now, consider the cycle that included the spontaneous expansion. During the spontaneous irreversible expansion there was no heat flow, but that tells us nothing about ΔS since we defined ΔS only for reversible processes. Instead, note that the state of the system following the irreversible expansion is the same as after the reversible expansion; that is, it has the same temperature, pressure, volume, energy, etc. Therefore, since we have demanded that entropy be a function only of the state of the system, the entropy of the system has increased just as it did during the reversible expansion. However, unlike the reversible case the surroundings have not changed at all while the gas expanded spontaneously. For the surroundings, $\Delta S = 0$. Thus the total entropy change for the universe has been positive. We complete the discussion by stating one of the most important laws in science: *For any spontaneous (irreversible) process, the entropy of the universe increases.* This is one statement of the **second law of thermodynamics.** There are many other statements equivalent to this which are of interest both practically and theoretically. One is that absolute zero (total lack

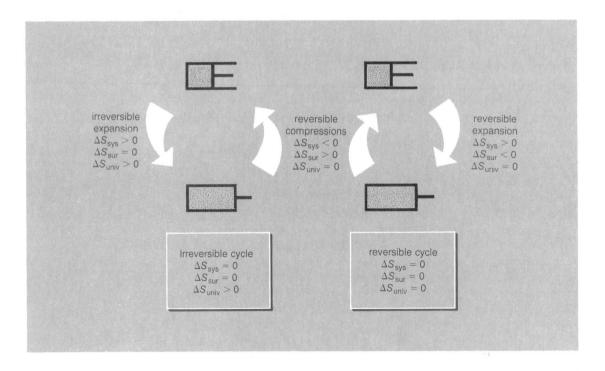

Figure 11.4 Entropy changes during expansion and compression of a gas.

of motion) is unattainable. Another, of major interest in these days of concern over energy use, is that it is impossible to convert heat into work with 100 percent efficiency. Thus burning hydrocarbon fuels to produce heat which is used to produce steam to run turbines to generate electricity must always involve a waste of a certain percent of the stored energy of the fuel (normally at least 35 percent) as heat to be exhausted. The interest in fuel cells (Chapter 18) is partially because they involve direct conversion from chemical energy to electrical energy without intermediate conversion into heat. Thus the fuel cell can in principle be operated to produce more energy in useful form from fuel.

But what is this entropy that we've gone to such pains to define? Can we relate it to some physical property of matter? It turns out that we can in many cases visualize entropy and for some processes predict the direction in which entropy will change. We can do this by thinking of entropy as a measure of disorder. As the gas in our cylinder expanded isothermally, its entropy increased: The molecules of a diffused gas are more disorderly than those in a compressed gas. This tendency toward randomness is one of the fundamental driving forces for changes of matter. The second law of thermodynamics tells us that all spontaneous changes tend to reduce the order of the universe.

There are local exceptions of course. For example, the human body continuously converts oxygen, water, and a variety of nutrients into a complex and highly ordered system. But as it does, the surroundings suffer large positive entropy changes. When we step back and look at the entire universe, the entropy is ever increasing.

To make this intuitive notion of randomness more exact, let's consider a simpler "system": 10 ordinary pennies. If we throw this fistful of pennies on the table, we shall get one of many random arrangements of heads and tails. There are in fact 2^{10} ways the pennies might fall; thus we can say that the system of 10 pennies has a large number of essentially equivalent variations. If, however, we constrain each coin to be a head, there is only one possible arrangement. Likewise, the molecules in a gas at high temperature can have a huge variety of positions and velocities, whereas in a crystal the possibilities are much more restricted. Therefore, we shall define a *thermodynamic probability*, Ω, that is proportional to the number of possible variations in a system. This is a measure of the randomness of a thermodynamic system. With this, we can show entropy to be proportional to the natural logarithm of Ω:

$$S = k \ln \Omega$$

where k is a constant.

Since the logarithm of 1 is zero, we must find a system with a thermodynamic probability of 1, that is, with only one possible configuration, if we want to define the zero point of entropy. We have encountered this ideal system before: It is a perfect crystal at 0 K. Thus the **third law of thermodynamics** follows: *The entropy of any pure perfect crystal at absolute zero is zero.* (We should add that it is impossible in principle to reach absolute zero; nonetheless, it serves as a basis for all absolute entropy measurements.) As the temperature increases, the entropy increases. The higher the temperature, the more freedom each atom or molecule enjoys and thus the more configurations available to the system. Furthermore, the entropy changes dramatically (even at constant temperature) when the system changes physical state. It is not hard to see that a gas is less ordered than a liquid, which in turn is less ordered than a solid.

Though we shall not prove it here, it can be shown that entropy changes for a pure substance are related to its specific heat. Therefore, it is possible, by measuring the specific heat and accounting for phase changes, to calculate the absolute entropy of any substance at any temperature. During the last several decades researchers have made such calculations and have produced extensive compilations of absolute entropies, which we give the symbol $S°$. Values for several compounds and elements appear in Table 11.5. Note that $S°$ values are given in calories rather than kilocalories and that since they are absolute values, they do not equal zero for elements in their standard states (unlike $\Delta H_f°$). We can now use these absolute entropies to calculate the entropy change for any chemical reaction, much as we calculated enthalpy changes from heats of formation. Since entropy too is a state function, this is merely an extension of Hess's law.

Table 11.5 Absolute Entropies [cal/(K)(mol)] at 25°C and 1 atm

Substance	Physical State	Absolute Entropy
C (graphite)	Solid	1.37
C (diamond)	Solid	0.568
CCl_4	Liquid	51.7
CO	Gas	47.2
CO_2	Gas	51.1
CH_4	Gas	44.5
C_2H_6	Gas	54.8
Cl_2	Gas	53.3
HCl	Gas	44.6
F_2	Gas	48.4
HF	Gas	41.5
H_2	Gas	31.2
H_2O	Gas	45.1
H_2O	Liquid	16.7
I_2	Solid	27.7
I_2	Gas	62.3
HI	Gas	49.3
N_2	Gas	45.8
NO	Gas	50.3
NO_2	Gas	57.4
N_2O	Gas	52.58
NH_3	Gas	46.0
O_2	Gas	49.0
O_3	Gas	57.1
F_2O	Gas	59.0
P_4	Solid	39.3
P_4	Gas	66.9
P_4O_{10}	Solid	54.7
PH_3	Gas	50.2

Example 11.10 What is $\Delta S°$ for the reaction in which 1.00 mol of $CO_2(g)$ is produced from graphite and oxygen at 25°C?

Solution

a. We obtain absolute entropies for the reactants and products from Table 11.5:

$$S° \text{ C} = 1.37 \text{ cal}/(K)(\text{mol})$$
$$S° \text{ O}_2 = 49.0 \text{ cal}/(K)(\text{mol})$$
$$S° \text{ CO}_2 = 51.1 \text{ cal}/(K)(\text{mol})$$

b. From the equation for the reaction

$$C(s) + O_2(g) \longrightarrow CO_2(g)$$

we write the expression for $\Delta S°$ according to Hess's law:

$$\Delta S° = [S° \text{ CO}_2(g)] - [S° \text{ C}(s) + S° \text{ O}_2(g)]$$

$$\Delta S° = \left[(1 \text{ mol})\left(\frac{51.1 \text{ cal}}{1.00 \text{ (mol)(K)}}\right)\right]$$
$$- \left[(1 \text{ mol})\left(\frac{1.37 \text{ cal}}{1.00 \text{ (mol)(K)}}\right) + (1 \text{ mol})\left(\frac{49.0 \text{ cal}}{1.00 \text{ (mol)(K)}}\right)\right]$$

$$= 0.7 \text{ cal/(mol)(K)}$$

Questions

5. In which situation is entropy greater?
  a. A set of dominoes at the start or at the end of a game.
  b. The ingredients of a cake before or after the batter is prepared for the oven.
  c. The pieces of a jigsaw puzzle before or after the puzzle is assembled.
  d. Ice or steam.

6. Which system has a higher entropy, 1 mol N_2 and 3 mol H_2 or 2 mol NH_3?

11.7 Free Energy

We now know that a spontaneous change is one in which the entropy of the universe increases. Only ideal, reversible processes occur with no net entropy change. It is important to emphasize that this criterion for spontaneity is not the same as requiring that the entropy of the system increase. For example, we might easily calculate a positive ΔS for a system for some chemical reaction, but whether the reaction would actually occur in an open flask also depends on ΔS for the surroundings—a more elusive quantity. Therefore, to be able to predict whether a process will occur spontaneously by measuring only changes in the system, we must introduce a final thermodynamic quantity—**free energy.**

Free energy, G, is sometimes called the Gibbs free energy, for Willard Gibbs (1839–1903), an American physicist who first enunciated the principles of spontaneity. Like enthalpy and entropy, free energy is a state function. At constant temperature and pressure, the change in free energy for a physical or chemical change is given by the following equation:

$$\Delta G = \Delta H - T \Delta S$$

At 1 atm pressure,

$$\Delta G° = \Delta H° - T \Delta S°$$

where T is the Kelvin temperature. We see, therefore, that both the entropy and enthalpy changes in the system play a part in determining whether a reaction will proceed spontaneously or not. From what we have said about the importance of entropy, this may seem curious until we realize that ΔH for the system is closely related to ΔS for the *surroundings*. Therefore, for a given reaction,

if $\Delta G > 0$, the reaction is not spontaneous

if $\Delta G < 0$, the reaction is spontaneous

if $\Delta G = 0$, the reaction is at equilibrium

This tells us that exothermic reactions (where ΔH is negative) tend to be spontaneous, as do reactions where the entropy increases. However, both of these conditions need not be present. For example, an endothermic reaction draws heat from the surroundings and thus tends to reduce its entropy, but if ΔS for the system is sufficiently positive, even this endothermic reaction will proceed spontaneously.

We can now explain the origin of the term "free" energy. If we take total energy changes due to bond breaking and formation, ΔH, and subtract from it the energy which must be used to produce expansion of gases and any necessary order, $T \Delta S$, we are left with the excess energy from the reaction, G, which can be used—to run a motor, light a light bulb, or otherwise perform useful work. To be spontaneous, a set of reactants must contain more than *just enough* energy to produce products. This excess or free energy is G.

As an example of a spontaneous reaction where the entropy of the system *decreases*, consider the following example.

Example 11.11 Show that the reaction between sodium and chlorine to form NaCl is spontaneous at 25°C and 1 atm. $\Delta H°$ for the reaction is -98.2 kcal/mol NaCl and $\Delta S°$ is -21.6 cal/(K) (mol NaCl).

Solution

a. We calculate $\Delta G°$ by substituting in the expression:

$$\Delta G° = \Delta H° - T \Delta S°$$
$$= -98,200 \text{ cal/mol} - (298 \text{ K}) [(-21.6 \text{ cal}/(K)(\text{mol}))]$$
$$= -91,800 \text{ cal/mol NaCl}$$

b. Since $\Delta G°$ is negative, the reaction is spontaneous. Here the effect of a negative $\Delta S°$ is overwhelmed by the quantity of heat evolved by the reaction.

Since the temperature is a factor in the expression for $\Delta G°$, we might expect some reactions to occur only when the temperature is favorable. We can regard

$\Delta S°$ and $\Delta H°$ as independent of temperature (because they do not change very rapidly as long as there is no phase change.)

Example 11.12 Show that the reaction $H_2 + I_2 \rightarrow 2\ HI$ does not occur spontaneously at 25°C and 1 atm but does at 200°C and 1 atm. $\Delta H°$ is 6200 cal/mol of HI and $\Delta S°$ is 19.7 cal/(K)(mol HI), and neither vary appreciably with temperature.

Solution

a. First we calculate $\Delta G°$ at 25°C:

$$\begin{aligned}\Delta G° &= \Delta H° - T\,\Delta S° \\ &= 6200\ \text{cal/mol} - (298\ \text{K})[19.7\ \text{cal/(K)(mol)}] \\ &= 329\ \text{cal/mol}\end{aligned}$$

b. Then we calculate $\Delta G°$ at 200°C:

$$\begin{aligned}\Delta G° &= \Delta H° - T\,\Delta S° \\ &= 6200\ \text{cal/mol} - (473\ \text{K})[19.7\ \text{cal/(K)(mol)}] \\ &= -3118\ \text{cal/mol}\end{aligned}$$

This example illustrates the importance of temperature when $\Delta H°$ and $\Delta S°$ have the same sign. At low temperatures, $\Delta H°$ is the controlling factor; whereas, at higher temperatures, $T\,\Delta S°$ prevails.

This example also allows us an opportunity to touch on two other topics related to free energy—subjects that we shall discuss at length in later chapters. First, even though free energy provides a criterion for spontaneity, it says nothing about how fast a spontaneous reaction will occur—will it proceed explosively or so slowly that we won't even be aware of it? We shall take up this question when we study *kinetics* in Chapter 12. Second, when we say that H_2 and I_2 react spontaneously at 200°C to form HI, we do not mean that *all* the hydrogen and iodine molecules react to produce hydrogen iodide but merely that the forward reaction, $H_2 + I_2 \rightarrow 2\ HI$, is *more favorable* than the reverse reaction, $2\ HI \rightarrow H_2 + I_2$. We'll say more about this topic in Chapter 13. This, incidentally, is a good place to illustrate what happens when $\Delta G = 0$: The forward and reverse reactions are equally likely; the result is that we see no spontaneous change in the system, and we say that it is at equilibrium.

Example 11.13 Calculate the change in free energy when 1 mol of ice melts at 1 atm at the three following temperatures: 10.0°C, 0.0°C, and −10.0°C. Use 1.436 kcal/mol for the heat of fusion for water and 5.257 cal/(mol)(K) for the change in entropy for the melting process.

Solution

a. At 10.0°C

$$\Delta G = 1.436 \text{ kcal} - (283.0 \text{ K}) \left(\frac{5.257 \times 10^{-3} \text{ kcal}}{1.00 \text{ K}} \right)$$

$$= 1.436 \text{ kcal} - 1.488 \text{ kcal}$$
$$= -0.052 \text{ kcal}$$

At 10.0°C, ΔG is negative and the ice melts spontaneously.

b. At 0.0°C,

$$\Delta G = 1.436 \text{ kcal} - (273.0 \text{ K}) \left(\frac{5.257 \times 10^{-3} \text{ kcal}}{1.00 \text{ K}} \right)$$

$$= 1.436 \text{ kcal} - 1.435 \text{ kcal}$$
$$= 0.001 \text{ kcal}$$

At 0.0°C, $\Delta G = 0$; hence it is equally favorable for ice to melt and for water to freeze: The liquid-solid system is at equilibrium.

c. At $-10.0°$C,

$$\Delta G = 1.436 \text{ kcal} - (263.0 \text{ K}) \left(\frac{5.257 \times 10^{-3} \text{ kcal}}{1.00 \text{ K}} \right)$$

$$= 1.436 \text{ kcal} - 1.383 \text{ kcal}$$
$$= 0.053 \text{ kcal}$$

At $-10.0°$C, ΔG is positive and ice does not melt spontaneously. In fact, the reverse process is spontaneous; hence water will freeze at $-10.0°$C.

Free energies, like enthalpies, cannot be measured directly. However, changes in free energy (ΔG) can. Therefore, we can define a standard free energy of formation ($\Delta G_f°$) exactly analogous to the enthalpy of formation. Likewise, we declare that the standard free energy of formation for the most stable form of each element at a specified temperature (usually 25°C) and 1 atm is zero. Table 11.6 lists the values of $\Delta G_f°$ for a number of compounds and elements.

Example 11.14 Determine whether the following reaction between $NH_3(g)$ and $O_2(g)$ is spontaneous at 25°C.

$$2\,NH_3(g) + 2\,O_2(g) \longrightarrow N_2O(g) + 3\,H_2O(l)$$

Table 11.6 Standard Free Energies of Formation at 25°C

Substance	Physical State	Standard Free Energy of Formation (kcal/mol)
$AgCl$	Solid	−26.22
$BaCO_3$	Solid	−272.2
$Ba(NO_3)_2$	Solid	−190.0
BaO	Solid	−126.3
Br_2	Liquid	0.00
HBr	Gas	−12.72
C (graphite)	Solid	0.00
C (diamond)	Solid	0.69
CCl_4	Liquid	−16.4
CH_4	Gas	−12.14
C_2H_2	Gas	50.00
C_2H_6	Gas	−7.86
CH_3OH	Liquid	−39.73
C_2H_5OH	Liquid	−41.77
CO_2	Gas	−94.26
CS_2	Liquid	15.2
$CaCO_3$	Solid	−269.78
CaO	Solid	−144.4
$Ca(OH)_2$	Solid	−214.3
Cl_2	Gas	0.0
HCl	Gas	−22.77
$CuCO_3$	Solid	−123.8
CuO	Solid	−30.4
F_2	Gas	0.0
HF	Gas	−66.08
H_2	Gas	0.0
H_2O	Gas	−54.64
H_2O	Liquid	−56.69
I_2	Gas	4.63
I_2	Solid	0.00
HI	Gas	0.31
$MgCO_3$	Solid	−246.0
MgO	Solid	−136.13
N_2	Gas	0.00
N_2O	Gas	24.76
NO	Gas	20.72
NO_2	Gas	12.39
NH_3	Gas	−3.976
NH_4Cl	Solid	−48.73
O_2	Gas	0.00
O_3	Gas	39.06
P_4	Gas	5.82
PH_3	Gas	2.21
PCl_3	Gas	−68.42
PCl_5	Gas	−77.59
SO_2	Gas	−71.79
SO_3	Liquid	−104.67
H_2S	Gas	−7.892

Solution

a. $\Delta G°_{reaction} = [\Delta G°_f\, N_2O(g) + 3\,\Delta G°_f\, H_2O(l)] - [2\,\Delta G°_f\, NH_3(g) + 2\,\Delta G°_f\, O_2(g)]$

b. Substitute values from Table 11.6:

$$\Delta G°_{reaction} = \left[(1.00\text{ mol})\left(\frac{24.76\text{ kcal}}{1.00\text{ mol}}\right) + (3.00\text{ mol})\left(\frac{-56.99\text{ kcal}}{1.00\text{ mol}}\right)\right.$$
$$\left. - \left[(2.00\text{ mol})\left(\frac{-3.976\text{ kcal}}{1.00\text{ mol}}\right) + (2.00\text{ mol})(0)\right]\right.$$
$$= -145.31 + 7.952$$
$$= -137.36\text{ kcal}$$

c. $\Delta G° < 0$; therefore the reaction as written is spontaneous at 25°C.

We conclude this chapter with the explanation for those spontaneous *and* endothermic dissolving processes encountered in Chapter 10. The ionic salt KCl, for example, dissolves in water, although heat is absorbed from the surroundings ($\Delta H > 0$) in the process. ΔG for this spontaneous process must therefore be negative. Thus from the definition of free energy ($\Delta G = \Delta H - T\Delta S$) it follows that ΔS is positive and sufficiently large to offset the positive ΔH. Furthermore, as we have already argued from Le Chatelier's principle, the process becomes more favorable as the temperature rises, since the $T\Delta S$ term becomes even more dominant. Incidentally, it's not hard to see why ΔS increases so dramatically: In going from a well-ordered crystal to a solution, the randomness of the solute is likely to increase significantly. Lest we generalize too hastily from this example, however, a very different case should be considered. When CuSO$_4$ dissolves in water, the entropy *decreases*. The solvating water molecules become so well-ordered about each solute ion that they offset the entropy increase in the solute itself. In this example the solution process is exothermic ($\Delta H < 0$), however, and ΔG is again negative.

Questions and Problems

11.4 Thermochemical Equations

7. Using Hess's law, calculate the enthalpy change for each of the following reactions at 25°C and 1 atm. (Note that these are merely reaction statements, not equations.)
   a. $P_4(s) + O_2(g) \longrightarrow P_4O_{10}(s)$
   b. $F_2(g) + HCl(g) \longrightarrow HF(g) + Cl_2(g)$
   c. $NH_3(g) + O_2(g) \longrightarrow NO_2(g) + H_2O(l)$

8. Use heat of combustion data to calculate the standard enthalpy of formation (kilocalories per mole) at 25°C for
   a. C_2H_4 b. C_2H_2 c. C_3H_8

11.5 Bond Enthalpies

9. Calculate the bond enthalpy of the HBr and HI molecules and compare them with HF and HCl.

10. Calculate the enthalpy of the H—O bond in water.

11. Calculate and compare the C—O bond enthalpies in CO and CO_2.

11.6 Entropy

12. In which situation is entropy greater?
   a. Milk immediately after milking or after standing for several hours.
   b. Billiard balls racked up before the game or after being struck by the cue ball.
   c. Dynamite before or after explosion.
   d. During or shortly after a chemistry lecture.
   e. Carbon dioxide before or after the sublimation of "dry ice."
   f. The contents of a flask before or after the following reaction takes place:

$$H_2(g) + Cl_2(g) \longrightarrow 2\ HCl(g)$$

   g. A closed vessel before or after the burning of methane

$$CH_4(g) + 2\ O_2(g) \longrightarrow CO_2(g) + 2\ H_2O(l)$$

13. For each of the following changes, tell whether you expect the enthalpy and entropy of the system to increase or decrease.
   a. the boiling of water
   b. the freezing of water
   c. the burning of coal
   d. the leaking of air from a tire

14. Compare entropy and enthalpy for the solid, liquid, and gaseous states. Arrange the following processes in order of increasing ΔS: boiling, sublimation, melting, increasing the temperature 10°C in the same physical state. Restrict your consideration to the substance undergoing the change, not the surroundings.

15. A certain reaction, A $\longrightarrow$ B, is both spontaneous and endothermic. What can you say about the structures of A and B?

11.7 Free Energy

16. Balance the following equations (using smallest whole-number coefficients possible) and calculate ΔH, ΔS, and ΔG for each reaction. Refer to Tables 11.2 and 11.4 for absolute entropies and enthalpies of formation. Assume all reactants and products to be in the physical states indicated in the tables. Furthermore, assume that all reactions take place at 25°C.
   a. $N_2 + O_2 \longrightarrow NO_2$
   b. $O_2 \longrightarrow O_3$
   c. $H_2O + F_2 \longrightarrow HF + O_2$
   d. $HCl + O_2 \longrightarrow H_2O + Cl_2$
   e. $PH_3 + O_2 \longrightarrow P_4O_{10} + H_2O$
   f. $NH_3 + O_2 \longrightarrow NO_2 + H_2O$

17. Classify the reactions in Question 16 as exothermic or endothermic.

18. Classify the reactions in Question 16 as spontaneous or nonspontaneous.

19. The melting point of chlorine is $-101°C$, and the heat of fusion is 1.53 kcal/mol. Calculate the change in entropy for the freezing of chlorine and compare it to the value calculated for water.

20. The melting point for magnesium metal is 650°C, and the heat of fusion is 2.2 kcal/mol. Calculate the change in entropy for the melting of magnesium.

21. Below are several pairs of values for changes in enthalpy and entropy for a reaction. Determine which reactions are definitely spontaneous and which might be spontaneous.
 a. $\Delta H > 0, \Delta S > 0$ b. $\Delta H > 0, \Delta S < 0$
 c. $\Delta H < 0, \Delta S > 0$ d. $\Delta H < 0, \Delta S < 0$

22. What is the change in free energy when 1 mol of water vaporizes at 100°C and 1 atm? (Heat of vaporization of water = 9.71 kcal/mol.)

Chemical Kinetics 12

12.1 Rates of Reactions

So far we have looked at chemical reactions from two important standpoints. First we studied stoichiometry by expressing reactions as chemical equations. This allowed us to quantify the amounts of reactants and products in a chemical process. Second, by adding an enthalpy term to each equation we produced thermochemical equations. Using these equations, along with additional thermodynamic data, we calculated free energy changes and thus established a criterion for spontaneity. But until now we have only hinted at a final important topic: the **rates of reactions.** If we looked at things naively, we might think that spontaneous reactions should be fast reactions and that the more thermodynamically favorable a process, the more likely it is to occur. But this is emphatically not the case. In Chapter 11 we found that an exothermic process is not necessarily spontaneous. We must now add that *a spontaneous reaction is not necessarily a fast reaction.* Hydrogen and oxygen combine spontaneously and explosively when ignited; iron rusts spontaneously but at quite a different rate. Likewise, carbon disulfide, a common liquid solvent, has a free energy of formation of 21 kcal/mol. This tells us that CS_2 must eventually decompose into elemental carbon and sulfur; yet we can store it indefinitely in an ordinary container with little evidence of decomposition.

A number of things affect the speed of a reaction, some of which are common to our everyday experience. If the reaction involves a solid, pulverizing that solid will undoubtedly speed up the reaction. Wood shavings and sawdust burn much more rapidly than logs. We can also usually count on an increase in the temperature to increase the speed of a reaction. Or we can increase the amount of one reactant available, just as we increase the supply of oxygen when we blow on a glowing ember. In this chapter we shall take up these and other aspects of **chemical kinetics,** which is the study of factors that affect the rates of chemical reactions. Ultimately, the goal of chemical kinetics is to understand how and why chemical reactions take place and how they may be controlled. The application of chemical kinetics to industrial processes is often economically important, and its use in understanding and controlling the metabolic processes of the human body is of even greater interest. Unfortunately these important processes are also extremely complex. We shall therefore look at simpler, if less interesting, examples at first.

Although the rate of a reaction in the laboratory may be obvious if the reaction produces a gas or if colored reactants produce colorless products, rate measurements are frequently very complicated experimental procedures. Nevertheless the concept of reaction rate is quite simple: *The rate of a reaction is the change in the concentration of a reactant or a product per unit of time.* For the reaction A + B ⟶ C, we can express the rate as $-\Delta[A]/\Delta t$, $-\Delta[B]/\Delta t$, or $\Delta[C]/\Delta t$, where the brackets denote the concentrations of the enclosed species in moles per liter. Of course the sign of $\Delta[C]/\Delta t$ is opposite that of the other two expressions since as the equation is written, [C] will increase as [A] and [B] decrease. We shall see reaction rates expressed in various units, but the most common is moles per liter per second. For gas phase reactions, pressure changes are sometimes a convenient measure of reaction rates.

The most suitable reactions for simple rate measurements are those that are free of the complication of reverse reactions. For example, from a practical standpoint, the precipitation of ionic solids from solution proceeds in only one direction, but such reactions proceed so rapidly that their rates are difficult to measure. Another class of reactions free from reverse reactions are those that produce a gas, which then escapes. For example,

$$2\ H_2O_2(l) \longrightarrow 2\ H_2O(l) + O_2(g)$$

Since oxygen is relatively insoluble in water, it will escape from an open container as soon as it is produced; hence the reaction will continue until all the hydrogen peroxide decomposes. (In this chapter we shall consider only reactions of this sort. In the next chapter, however, we shall include the reverse reaction in our studies and see that it leads directly to an understanding of chemical equilibrium.)

Example 12.1 By measuring the volume of oxygen evolved at constant temperature during the decomposition of hydrogen peroxide, a student obtained the data in Table 12.1. Using this data, graph the total volume of oxygen evolved as a function of time.

Table 12.1 Student Data

Elapsed Time (min)	Total Volume of O_2 Evolved (ml)
1.00	2.43
2.00	4.78
3.00	7.01
4.00	9.07
5.00	10.90
6.00	12.43
7.00	13.58
8.00	14.27
9.00	14.41
10.00	14.41

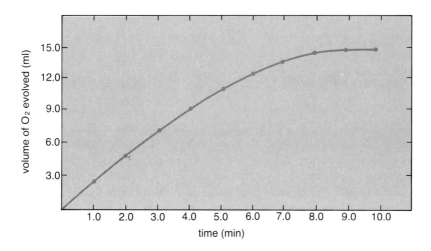

Figure 12.1 Graph of student data.

Solution The result should look like Figure 12.1, from which we see that oxygen is evolved rapidly at first and then less rapidly as the hydrogen peroxide is used up. After 9.00 min, essentially all the H_2O_2 has decomposed. The graph also makes apparent another difficulty in measuring rates of reactions: *The rate is constantly changing.* In this example, it is constantly decreasing. In other cases the reaction rate increases constantly. With the rate constantly changing, how shall we choose to express it? A real study of rates requires the application of calculus, but some approaches are suggested in what follows.

Example 12.2 Use the data of Example 12.1 to determine the average rate of the reaction during the first 2.00 min, 4.00 min, 6.00 min, 8.00 min, and 10.00 min. Notice that we continue to express rate as volume of O_2 evolved per minute. We could easily convert it to moles per liter per minute.

Solution

a. The average rate for a period is found by dividing the volume of oxygen evolved during that period by the length of the period.

b. Average rate for

$$2.00 \text{ min} = \frac{4.78 \text{ ml}}{2.00 \text{ min}} = \mathbf{2.39 \text{ ml/min}}$$

$$4.00 \text{ min} = \frac{9.07 \text{ ml}}{4.00 \text{ min}} = \mathbf{2.27 \text{ ml/min}}$$

$$6.00 \text{ min} = \frac{12.43 \text{ ml}}{6.00 \text{ min}} = \mathbf{2.07 \text{ ml/min}}$$

Chemical Kinetics 375

$$8.00 \text{ min} = \frac{14.27 \text{ ml}}{8.00 \text{ min}} = 1.78 \text{ ml/min}$$

$$10.00 \text{ min} = \frac{14.41 \text{ ml}}{10.00 \text{ min}} = 1.44 \text{ ml/min}$$

The average reaction rate for a specific time interval is often easy to calculate but is seldom useful to the kineticist. A better way to express the speed of a reaction is the *instantaneous rate of reaction*, which in this case is the change in total volume of oxygen per unit of time, $\Delta V/\Delta t$, where the unit of time is made infinitesimally small. Graphically the instantaneous rate at any instant is the *slope of the curve* at that point. Two examples are illustrated in Figure 12.2. Note also that beyond 9.00 min the slope of the curve becomes zero, indicating that the reaction has ceased. As in this example, the instantaneous reaction rate is often easiest to evaluate near the beginning of the reaction since although the rate is more rapid during this period, it is *changing more slowly*. But wherever it is evaluated, a rate is pertinent only for the time at which it is measured.

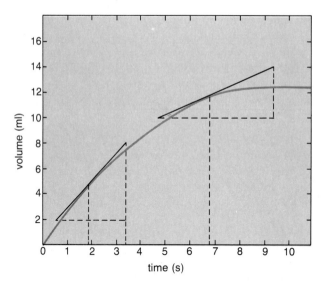

Figure 12.2 The instantaneous rate at any instant is the slope of the rate curve at that point.

We might now ask what factors influence the rate of a chemical reaction. The answer to this question will, in fact, occupy us for the rest of the chapter. We shall look first at the effects of the physical condition of the reactants, the temperature, and the concentrations of the reactants. At the end of the chapter we shall look at a fourth factor: *catalysis*.

Questions

1. List several units in which the rate of a reaction might be measured.

2. Explain how a substance whose decomposition is thermodynamically favorable might not seem to decompose at all.

3. Why are precipitations of ionic compounds less suitable than some other reactions for rate measurements?

4. What is the physical meaning of the numerical value for the slope of a rate curve?

12.2 Factors Affecting Rates: Chemical and Physical Condition of Reactants

The chemical identity of a substance determines *what* reactions it will undergo and, of course, determines whether it will react faster with substance B or with substance C. But what we are more interested in now are the factors that affect the rate of a given reaction. These factors are usually physical rather than chemical.

When we talk about the physical condition of a substance, we are referring both to its physical state (solid, liquid, gas) and to macroscopic properties such as particle size. A reaction between two substances will proceed more rapidly if the substances are liquids than if they are solids and more rapidly still if they are gases. For much the same reasons, small solid particles react more readily than large ones. In flour mills and coal mines, the danger of fire and explosion is greatest when the combustible coal or flour is present as dust, well mixed with air.

12.3 Factors Affecting Rates: Temperature

A rise in temperature will always increase the rate of an elementary (one-step) chemical reaction. (Reactions that appear to slow down as the temperature increases are invariably made up of two or more component reactions, each of which responds differently to the temperature change.) In fact, for many room-temperature reactions—enough to provide us a useful rule of thumb—a 10°C increase in temperature will double the rate of the reaction. There is much to be learned from this surprising observation, so let's look at it closely. To simplify matters, we shall consider a gas phase reaction.

For a reaction to occur between two molecules, they must come to within some specific distance of each other. Although it is something of an oversimplification, we call such an encounter a *collision* between molecules. At STP, about 10^{32} such collisions occur in a liter of gas each second. That means that each molecule undergoes several *billion* collisions every second. This in turn tells us that by no means does a reaction occur each time two molecules collide. If it did, we would expect all gas phase reactions to occur explosively and, in general, they do not. Many take place at slow, controllable rates.

In Chapter 8 we calculated the energy of a molecule of a gas from its mass and velocity:

$$ke = \tfrac{1}{2}mv^2$$

We observed that within any sample of a gas there is a wide distribution of velocities and thus a similar distribution of energies (Figure 12.3) but that the average kinetic energy of a gas depends only on the temperature. Since we also derived a relationship between temperature and velocity, we should now be able to calculate the effect on average kinetic energy of a 10°C rise in temperature.

Example 12.3 By what percent are the average molecular velocity and the average kinetic energy of a gas increased when its temperature is increased from 27.0°C to 37.0°C?

Solution

a. The average velocity is proportional to the square root of the Kelvin temperature. Therefore

$$\frac{V_2}{V_1} = \frac{\sqrt{T_2}}{\sqrt{T_1}} = \frac{\sqrt{310.0}}{\sqrt{300.0}} = 1.016$$

The average velocity increases by **1.6 percent**.

b. The average kinetic energy is proportional to the square of the velocity; hence it is directly proportional to the Kelvin temperature:

$$\frac{KE_2}{KE_1} = \frac{T_2}{T_1} = \frac{310.0}{300.0} = 1.033$$

Thus the average kinetic energy increases by **3.3 percent**.

Example 12.3 clearly demonstrates that the increase in reaction rate for a ten-degree rise in temperature does not depend directly on the small increase in average kinetic energy. Instead we must consider the distribution of energies, much as we used this idea to explain evaporation from a liquid (p. 245). Only a fraction of the molecules had enough energy to escape into the gas phase. Here we shall argue that only a small percentage of the gas molecules, those in the high-energy tail of the distribution, have enough energy to react when they collide. Note that this is consistent with our observation that most collisions do not produce a reaction. In Figure 12.3a the shaded area under the curve represents this small fraction of reactive molcules. In Figure 12.3b we see the

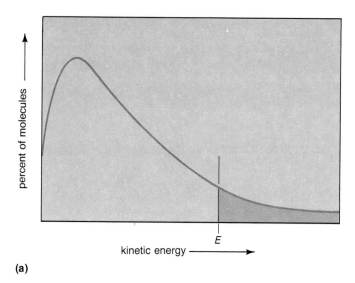

Figure 12.3 A 10° increase in temperature doubles the *fraction* of molecules with energy high enough to react upon collision ($>E$).

effect of temperature. The shape of the curve changes only slightly as the temperature rises, but the *fraction* of the molecules with enough energy to react when they collide doubles and so does the rate of the reaction. We should admit that not all collisions between high-energy molecules produce a reaction: In some cases, the molecules must also be properly oriented with respect to one another.

Chemical Kinetics 379

Questions

5. What do we mean when we say that a collision between molecules is a necessary but not a sufficient criterion for reaction?

6. What two criteria must be met for a collision between molecules to result in a reaction?

12.4 Factors Affecting Rates: Concentrations

In studying the effect of reactant concentration on reaction rates we shall not be able to take quite as theoretical an approach as we took with temperature. Nonetheless, we can write a *general rate-law equation*, which will serve as the basis for our discussion. For a reaction in which two substances, A and B, yield a third, C, we can write

$$\text{rate} = k[A]^m[B]^n$$

The values of m and n in this expression, which *must* be obtained from experimental measurements and not from the coefficients in the chemical equation, determine the *order* of the reaction. We say that the reaction is of order m in A, order n in B, and order $m + n$ overall. The proportionality constant, k, also called the *rate constant*, must likewise be evaluated by experiment.

To make these ideas clearer, we can look at an example of a reaction having the form $A + B \rightarrow C$. Let us assume that experiment yields the following data, where the subscript i denotes initial concentrations:

$[A]_i$ (mol/liter)	$[B]_i$ (mol/liter)	Initial Rate = $\Delta[C]/\Delta t$ (mol/liter) s
0.011	0.035	3.9×10^{-5}
0.011	0.070	7.8×10^{-5}
0.022	0.035	15.6×10^{-5}

These data show that for our example the reaction rate doubles when we double the initial concentration of B; thus we conclude that the rate is directly proportional to [B]:

$$\text{rate} \propto [B]$$

When we double the concentration of A, on the other hand, the reaction rate quadruples. Thus

$$\text{rate} \propto [A]^2$$

We can now combine our findings to produce a rate-law equation:

$$\text{rate} = k[A]^2[B]$$

This of course is simply a restatement of the general rate-law equation with the values of m and n filled in for this particular example. (If we liked, we could have evaluated k as well.) This reaction is therefore second order in A, first order in B, and third order overall. We should add that in some cases a reaction might be zero order in a reactant, which tells us that the reaction rate is independent of the concentration of that reactant. We may encounter fractional orders as well.

If a reaction is first order overall, its rate law will be of the form

$$\text{rate} = k[A]$$

We recognize this as the equation of a straight line ($y = ax$). Thus if we plot rate against the concentration of A, we expect a straight line whose slope gives us the rate constant k. For reaction orders higher than one, evaluation of the rate constant and the exact order can be more difficult, but in general both can be determined from appropriate experimental data.

Example 12.4 The reaction of potassium bromate with potassium bromide in acidic aqueous solution takes place according to the following equation:

$$BrO_3^-(aq) + 5\,Br^-(aq) + 6\,H^+(aq) \longrightarrow 3\,Br_2(aq) + 3\,H_2O(l)$$

In a laboratory experiment to investigate the kinetics of this reaction, the reagent phenol (C_6H_5OH) was added to the reaction mixture. Phenol reacted with the Br_2 as it was formed, thus eliminating the complication of a reverse reaction. The time required for 0.002 mol/liter of bromate ion to be consumed was measured. Four trials were run, so that the concentration of each reactant could be doubled in turn while the others were held constant. Table 12.2 shows

Table 12.2 Time Required for 0.002 mol/liter of Bromate Ion To Be Consumed at 25°C*

	Initial [BrO_3^-]	Initial [Br^-]	Initial [H^+]	Time (s)
1.	0.0333	0.0667	0.100	42.6
2.	0.0333	0.133	0.100	21.5
3.	0.0667	0.0667	0.100	21.1
4.	0.0333	0.0667	0.200	10.7

* J. R. Clark, "The Kinetics of the Bromate-Bromide Reaction," *J. Chem. Ed.*, **47**, 775–778 (Nov. 1970).

the results of this experiment. Determine the rate law and the order for this reaction; then evaluate the rate constant.

Solution

a. When the Br^- ion concentration was doubled (Trial 2 compared to Trial 1), the reaction time was cut in half (from 42.6 s to 21.5 s). In other words, the rate doubled. Thus the rate is proportional to the first power of the concentration of Br^-. We find the same relationship for the BrO_3^- ion (Trial 3 compared to Trial 1).

b. When the H^+ ion concentration was doubled (Trial 4 compared to Trial 1), the reaction time decreased from 42.6 s to 10.7 s: The rate increased fourfold. Thus the rate is proportional to the second power of the concentration of H^+.

c. The rate law for the reaction is

$$\text{rate} = k[Br^-][BrO_3^-][H^+]^2$$

d. The reaction is first order in both Br^- and BrO_3^-, second order in H^+, and fourth order overall.

e. To evaluate the rate constant, we must now substitute the concentration and rate values from any of the four trials into the rate law. Units for the rate constant will, in general, be specified by the rate and concentration units and by the order of the reaction.

From Trial 1,

$$\text{rate} = \frac{0.00200 \text{ mol/liter}}{42.6 \text{ s}} = 4.69 \times 10^{-5} \frac{\text{mol/liter}}{\text{s}}$$

$$4.69 \times 10^{-5} = k(6.67 \times 10^{-2})(3.33 \times 10^{-2})(1.00 \times 10^{-1})^2$$

$$k = \frac{4.69 \times 10^{-5}}{2.22 \times 10^{-5}} = \mathbf{2.11}$$

Here the units turn out to be $\text{liter}^3/\text{mol}^3 \text{ s}$.

The examples we have used so far have been very elementary ones. It would be optimistic to assume that all reactions can be accurately represented with such simple statements. For example, the homogeneous gas phase reaction between hydrogen and bromine would seem relatively simple. Its stoichiometry certainly is:

$$H_2(g) + Br_2(g) \longrightarrow 2 \text{ HBr}(g)$$

But the rate law for this reaction is complicated.

$$\text{rate} = \frac{k_1[H_2][Br_2]^{1/2}}{1 + k_2 \frac{[HBr]}{[Br_2]}}$$

The rate depends not only on the concentrations of H_2 and Br_2 but also on the concentration of the product, HBr. Furthermore the expression includes two *different* rate constants with different temperature dependencies. As we shall see later, such complexity suggests that the reaction involves more than one simple step, despite the uncomplicated stoichiometry.

Many ingenious and precise methods are available for measuring the rates of reactions. For the decomposition of hydrogen peroxide

$$2\ H_2O_2(l) \longrightarrow 2\ H_2O(l) + O_2(g)$$

or the decomposition of dinitrogen pentoxide

$$2\ N_2O_5(g) \longrightarrow 4\ NO_2(g) + O_2(g)$$

the kineticist might measure the volume of oxygen evolved during the reaction using an arrangement like that illustrated in Figure 12.4. To separate the two gaseous products in the latter reaction, we might carry it out in CCl_4: NO_2 is quite soluble in CCl_4, but O_2 is not. If there are ionic species involved in the

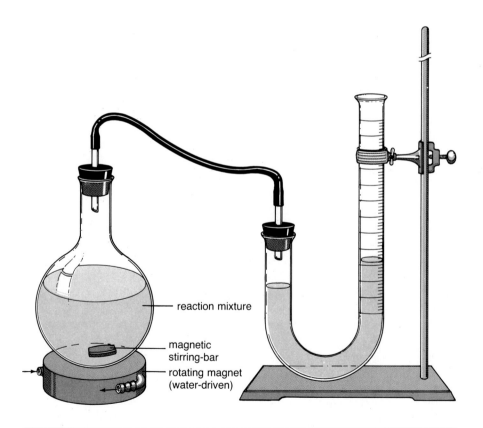

Figure 12.4 Experimental setup for measuring the rate of evolution of a gas from a reaction.

reaction, the conductance of the solution can sometimes be used to measure changes in concentration. If visible color changes occur during the reaction, colorimetric methods can be used. Finally, kineticists often depend on spectroscopic methods, where for example changes in absorbance at a characteristic wavelength can be monitored with a spectrophotometer during the reaction.

Sometimes there is no convenient way to observe the progress of the reaction as it occurs. In such cases, it may be necessary to remove small samples from the reaction mixture as the reaction continues. The chemist must then *quench* the reaction in the small sample, perhaps by cooling it quickly or diluting it.

A technique used to analyze the dependence of rate on concentration is to ensure that all reactants except one are present in great excess. Since the concentrations of these reactants can be regarded as essentially constant during the reaction, the effects of the other reactant can be accurately gauged.

Example 12.5 When the number of moles of gas on the left-hand side of a chemical equation is different from the number of moles of gas on the right-hand side, the rate of the reaction can be monitored by measuring the pressure. The data below were obtained for the reaction between nitric oxide and chlorine:

$$2\ NO(g) + Cl_2(g) \longrightarrow 2\ NOCl(g)$$

Determine the order of the reaction and evaluate the rate constant. Assume that the rate law has a simple form.

Solution

	P_{NO}(atm)	P_{Cl_2}(atm)	Rate = $\Delta P/\Delta t$
Trial 1	0.0223	0.0223	5.1×10^{-3}
Trial 2	0.0446	0.0446	4.0×10^{-2}
Trial 3	0.0223	0.0446	1.0×10^{-2}

a. If we compare Trial 3 to Trial 1, it is apparent that doubling the pressure and thus the concentration of Cl_2 doubles the rate of the reaction.

b. Comparing Trial 2 with Trial 3 shows that doubling the concentration of NO quadruples the rate of the reaction.

c. The rate law for the reaction is therefore

$$\text{rate} = k[NO]^2[Cl_2]$$

d. The reaction is first order with respect to Cl_2, second order with respect to NO, and thus a third-order reaction.

e. Finally, we can evaluate the rate constant by substituting the data from any of the three trials into the rate equation. Here, the units of the rate constant are $\text{atm}^{-2}\ \text{s}^{-1}$.

$$\text{Trial 1} \qquad\qquad \text{Trial 2}$$
$$5.1 \times 10^{-3} = k(0.0223)^2(0.0223) \qquad 4.0 \times 10^{-2} = k(0.0446)^2(0.0446)$$
$$k = 4.6 \times 10^2 \qquad\qquad k = 4.5 \times 10^2$$

$$\text{Trial 3}$$
$$1.0 \times 10^{-2} = k(0.0223)^2(0.0446)$$
$$k = 4.5 \times 10^2$$

The **half-life** of a reaction is the time required for the reaction to go halfway to completion. First-order reactions are unique in that the half-life for a given reaction is independent of the initial concentration of reactant. This provides an easy method for recognizing a first-order reaction. A good example is the decomposition of N_2O_5 in carbon tetrachloride at 45°C. Table 12.3 shows some typical data. If we plot the concentration of N_2O_5 as a function of time, as shown in Figure 12.5a, we are in a position to evaluate the order of the reaction using our concept of half-life. The initial concentration of 2.33 mol/liter decreases to 1.17 mol/liter (one-half the original value) in 18.6 min. This concentration in turn decreases to 0.58 in 18.4 min. Likewise, a concentration of 2.00 mol/liter diminishes to 1.00 mol/liter in 18.5 min. Since these half-lives are equal, we can conclude that the reaction is first order. If we want to corroborate this finding and evaluate the rate constant, we can use yet another technique of the kineticist. In Table 12.3, Δc is the change in concentration of N_2O_5 and Δt is the time interval between two adjacent data points. The rate given in the table is $\Delta c/\Delta t$ and thus is an average rate for the time interval given; nonetheless, it is a good approximation for the instantaneous rate. If we now calculate the quotient, rate/$[N_2O_5]$, for each time interval, we find that it remains constant (Figure 12.5b), thus confirming that the reaction is first order. The value of rate/$[N_2O_5]$ is the rate constant. Radioactive decay processes (Chapter 25) are first-order reactions and thus radioactive isotopes are characterized by the lengths of their half-lives.

Table 12.3 The Decomposition of N_2O_5 in Carbon Tetrachloride at 45°C

Time (min)	$[N_2O_5]$	Δc	Δt	Rate = $\Delta c/\Delta t$	k = rate/$[N_2O_5]$
0	2.33				
		0.25	3.1	0.081	0.039
3.1	2.08				
		0.17	2.2	0.077	0.040
5.3	1.91				
		0.24	3.5	0.069	0.041
8.8	1.67				
		0.31	5.7	0.054	0.040
14.5	1.36				
		0.25	5.5	0.046	0.041
20.0	1.11				
		0.39	13.3	0.029	0.040
33.3	0.72				
		0.17	7.5	0.023	0.042
40.8	0.55				
		0.21	15.0	0.014	0.041
55.8	0.34				

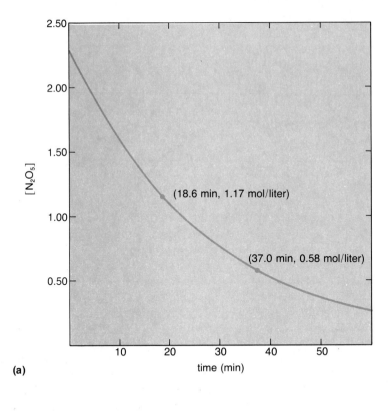

(a)

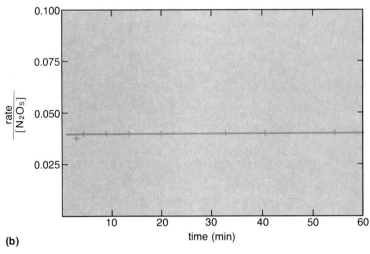

(b)

Figure 12.5 (a) Graph of the decomposition of N_2O_5 in CCl_4 at 45°C. (b) Graph of rate/$[N_2O_5]$ versus time for the data of Table 12.3.

Questions

7. What is a rate law?

8. For the reaction

$$H_2O_2(aq) + 2\ H^+(aq) + 2\ I^-(aq) \longrightarrow I_2(aq) + 2\ H_2O(l)$$

the rate statement is

$$\text{rate} = k[H_2O_2][I^-]$$

a. What is the order of this reaction?
b. What order is this reaction with respect to $[H^+]$?

12.5 Reaction Mechanisms

Developing methods for measuring reaction rates and for investigating how those rates depend on temperature and reactant concentrations is only part of the kineticist's work. The other part, the theoretical half, is using this experimental information to describe *mechanisms* for chemical reactions. A **mechanism** is a series of individual steps that, taken together, account for both the stoichiometry and the rate law observed for the overall reaction. We call these individual one-step reactions *elementary reactions*, and each *must be a simple molecular event*.

For each of these elementary reactions we can write a simple rate equation. We do this using the **principle of mass action,** which states that *the rate at which reactant molecules are converted to product molecules is proportional to the concentration of each substance participating in that elementary reaction*. (As we apply this principle in the examples below, note that we always write the equation so that all coefficients are equal to one. For example, instead of writing $2\ A \longrightarrow B$, we write $A + A \longrightarrow B$.) The number of molecules participating in an elementary reaction identifies the *molecularity* of the reaction: *unimolecular, bimolecular,* or *termolecular*.

The rate for a unimolecular elementary reaction is proportional to the concentration of the one reactant.

Equation: $\quad A_2 \longrightarrow A + A$
Proportionality statement: $\quad \text{rate} \propto [A_2]$
Rate law: $\quad \text{rate} = k[A_2]$

A bimolecular reaction might be of the type

$$A + B \longrightarrow AB$$

For this the proportionality statement would be

$$\text{rate} \propto [A][B]$$

and the rate law

$$\text{rate} = k[A][B]$$

Another type of bimolecular reaction is

$$A + A \longrightarrow A_2$$

Proportionality statement: $\text{rate} \propto [A][A]$ or $\text{rate} \propto [A]^2$

Rate law: $\text{rate} = k[A]^2$

A termolecular reaction might be any of these types.

Equation: $A + B + B \longrightarrow AB_2$
Proportionality statement: $\text{rate} \propto [A][B]^2$
Rate law: $\text{rate} = k[A][B]^2$

Equation: $A + B + C \longrightarrow ABC$
Proportionality statement: $\text{rate} \propto [A][B][C]$
Rate law: $\text{rate} = k[A][B][C]$

Equation: $A + A + A \longrightarrow A_3$
Proportionality statement: $\text{rate} \propto [A]^3$
Rate law: $\text{rate} = k[A]^3$

Let us reemphasize that *the principle of mass action holds only for elementary (one-step) reactions.*

Using what we know about elementary reactions, we can now look at more complex chemical reactions—the kind we actually encounter in the laboratory. All reactions, of course, are not complicated. Some second-order reactions, first order in each of two reactants, are themselves elementary reactions. All that is necessary for reaction in such cases is that a molecular collision occur in which (1) the sum of the energies of the two molecules exceeds some minimum value and (2) the orientation of the molecules is such that the reaction can occur. If we increase the concentration of either reactant, we increase the number of high-energy molecules and the chances for appropriately oriented collisions. Therefore the reaction rate rises proportionately. Often, however, the experimental rate laws cannot be explained with such a simple mechanism. Recall, for example, the reaction between H_2 and Br_2 in Section 12.4. In these more complicated (and more common) reactions, the overall reaction is the sum of a series of elementary reactions. Sometimes one of these reactions will be significantly slower than any of the others and will therefore determine the rate and the order of the reaction. We call this step the *rate-determining step*.

An example of the effect of a rate-determining step, and of the difficulties faced by chemists who propose reaction mechanisms, is the reaction between H_2 and I_2: $H_2(g) + I_2(g) \longrightarrow 2\,HI(g)$. The simple second-order rate law (rate = $k[H_2][I_2]$) was taken for years to show that the reaction involved a simple collision between H_2 and I_2. Recent evidence, however, suggests a two-step mechanism:

(1) $I_2 \rightleftharpoons 2\,I$

(2) $2\,I + H_2 \longrightarrow 2\,HI$

Step 2 is the rate-determining step, for which we can write

$$\text{rate} = k[I]^2[H_2]$$

It can be shown experimentally that in the reaction mixture $[I]^2 \propto [I_2]$, so the rate law for step 2 reduces to a simple second-order law, which in turn becomes the rate law for the overall reaction: rate = $k[H_2][I_2]$.

This example also illustrates another rule by which the kineticist lives: A proposed mechanism is never proved, but it can be disproved by new evidence.

Question

9. What is the principle of mass action and what are the restrictions upon its application?

12.6 Theory of the Activated Complex

To look more deeply into the matter of reaction mechanisms, let us consider an elementary reaction that takes place between molecules of substances A_2 and B_2 to form substance AB. The equation for this reaction is

$$A_2 + B_2 \longrightarrow 2\,AB$$

and the second-order rate law is

$$\text{rate} = k[A_2][B_2]$$

The mechanism postulated for this reaction demands the formation of a high-energy complex, A_2B_2, which is produced during a collision between a molecule of A_2 and a molecule of B_2, provided they possess enough energy and are appropriately oriented. This *activated complex* or *transition state* can then decay in either of two ways—back to A_2 and B_2 or forward to two molecules of AB. We can depict this reaction with an energy diagram such as that shown in Figure 12.6. If we follow the progress of the reaction in this diagram, we see that the kinetic energy of molecules A_2 and B_2 is converted first into the high potential energy of the activated complex and then back into the kinetic energy of the products. Figure 12.6 illustrates an exothermic reaction since the energy level of the products is lower than that of the reactants, but we might just as well have used an endothermic reaction as an example, as shown in Figure 12.7. The change in enthalpy for the reaction has no bearing on its rate; the reaction rate depends only on the rate at which the activated complex is formed. This in turn depends on the energy difference between reactants and the activated complex. We call this difference the *activation energy*. The easier it is for the reactants to form activated complexes, the faster the reaction. For example, as the temperature increases, the number of molecules with enough kinetic energy to climb the activation-energy barrier also increases; thus the reaction rate increases. The activated complex is at a condition of very high potential energy, with some bonds partially broken and some partially made. It is unobservable because of its very short life, but we can sometimes make reasonable guesses about its structure.

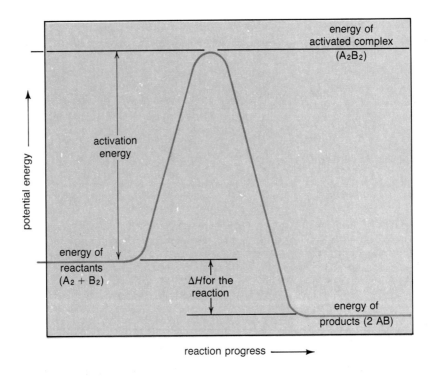

Figure 12.6 Energy diagram for an exothermic reaction taking place through an activated complex.

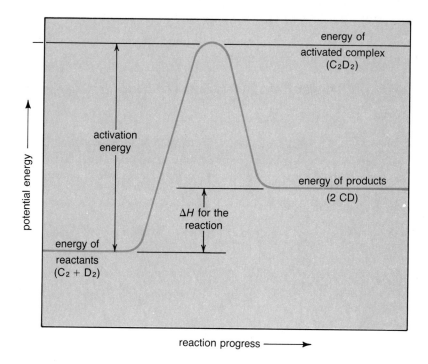

Figure 12.7 Energy diagram for an endothermic reaction taking place through an activated complex.

Sidelight
The Arrhenius Equation

The relationship between temperature and reaction rates is not a simple one. To describe how the rate constant k depends on the Kelvin temperature T, we must resort to the empirically based *Arrhenius equation*:

$$k = Ae^{-E_a/RT}$$

In this equation, A is a constant, E_a is the activation energy, and R is the gas constant. Since the temperature appears in the denominator of a negative exponent, an increase in T produces an increase in k. The exponential relationship between the rate constant and the temperature also agrees with our observation that small changes in temperature have large effects on the rates. If we rewrite the equation

$$\ln k = \ln A - \frac{E_a}{RT}$$

and then plot the natural logarithm of k versus $1/T$, we see that the result will be a straight line with slope $-E_a/R$. Thus we can evaluate the activation energy for a reaction if we know the rate constant at several temperatures.

This concept of the activated complex is not merely an academic exercise; it is commonly invoked to explain the kinetics of chemical reactions. An example is the reaction between the interhalogen compound ICl and hydrogen:

Equation: $2\,ICl(g) + H_2(g) \longrightarrow I_2(g) + 2\,HCl(g)$

Rate law: $\text{rate} = k[H_2][ICl]$

Since the exponents in the rate law do not correspond to the coefficients in the chemical equation, the reaction must comprise more than one elementary step. One mechanism proposed for the reaction consists of a slow, rate-determining, bimolecular first step:

$$H_2 + ICl \longrightarrow \underset{(H_2ICl)}{\text{activated complex}} \longrightarrow HI + HCl$$

followed by a fast reaction between HI and a second molecule of ICl:

$$ICl + HI \longrightarrow HCl + I_2$$

The overall rate is thereby the rate of the first elementary reaction. Of course the mechanism must also be consistent with the observed stoichiometry. We find that it is by adding the two elementary steps:

$$\begin{array}{l} H_2 + ICl \longrightarrow \cancel{HI} + HCl \\ ICl + \cancel{HI} \longrightarrow HCl + I_2 \\ \hline H_2 + 2\,ICl \longrightarrow 2\,HCl + I_2 \end{array}$$

Another reaction in which an activated complex forms in a rate-determining step is that between the hydrogen peroxide and iodide ions in acidic solution. The stoichiometry shows that hydrogen ions are involved in the reaction; yet they do not appear in the rate law:

Equation: $H_2O_2(aq) + 2\,H^+(aq) + 2\,I^-(aq) \longrightarrow 2\,H_2O(l) + I_2(aq)$

Rate law: $\text{rate} = k[H_2O_2][I^-]$

Apparently the rate-determining step does not involve H^+. One suggestion is that there is a slow, bimolecular first step:

$$H_2O_2 + I^- \longrightarrow \underset{(H_2O_2I^-)}{\text{activated complex}} \longrightarrow OH^- + HOI$$

followed by two faster steps:

$$H^+ + OH^- \longrightarrow H_2O$$
$$HOI + H^+ + I^- \longrightarrow I_2 + H_2O$$

The rate therefore depends only on the concentrations of H_2O_2 and I^-, and the mechanism is consistent with the observed stoichiometry:

$$\begin{aligned} H_2O_2 + I^- &\longrightarrow \cancel{OH^-} + \cancel{HOI} \\ H^+ + \cancel{OH^-} &\longrightarrow H_2O \\ \underline{\cancel{HOI} + H^+ + I^- \longrightarrow I_2 + H_2O} \\ H_2O_2 + 2\,H^+ + 2\,I^- &\longrightarrow I_2 + 2\,H_2O \end{aligned}$$

Questions

10. Can you think of some examples of exothermic reactions with large activation-energy barriers?

11. Can you think of some examples of exothermic reactions with zero activation-energy barriers?

12.7 Chain Reactions

Not all reactions will produce a graph like the one shown in Figure 12.2, where the reaction slows down steadily as it proceeds. The rates of some **chain reactions**, for instance, increase constantly until the reactants are consumed. The reaction between chlorine and methane to produce methyl chloride and hydrogen chloride is an example of such a chain reaction. Light is necessary for this reaction to proceed at room temperature, but it will take place in the dark if the temperature is above 250°C. The mechanism that best accounts for the observations about this reaction can be represented as follows:

(1) $\;:\!\ddot{\underset{\cdot\cdot}{Cl}}\!:\!\ddot{\underset{\cdot\cdot}{Cl}}\!: \xrightarrow{\text{heat or light}} 2\,:\!\ddot{\underset{\cdot\cdot}{Cl}}\!\cdot$

(2) $\;:\!\ddot{\underset{\cdot\cdot}{Cl}}\!\cdot\; +\; H\!:\!\underset{H}{\overset{H}{C}}\!:\!H \;\longrightarrow\; H\!:\!\ddot{\underset{\cdot\cdot}{Cl}}\!:\; +\; H\!:\!\underset{H}{\overset{H}{C}}\!\cdot$

(3) $\;H\!:\!\underset{H}{\overset{H}{C}}\!\cdot\; +\; :\!\ddot{\underset{\cdot\cdot}{Cl}}\!:\!\ddot{\underset{\cdot\cdot}{Cl}}\!: \;\longrightarrow\; H\!:\!\underset{H}{\overset{H}{C}}\!:\!\ddot{\underset{\cdot\cdot}{Cl}}\!:\; +\; :\!\ddot{\underset{\cdot\cdot}{Cl}}\!\cdot$

The first step breaks the chlorine-chlorine bond and demands either light or high temperatures. The result is a pair of highly reactive chlorine atoms, each with an unpaired valence electron. Since the concentration of this reactive species remains small, two chlorine atoms are not likely to meet. Each is more likely to encounter a chlorine molecule or a methane molecule.

When a chlorine atom reacts with a chlorine molecule, no net change results:

$$\left(:\!\ddot{\underset{\cdot\cdot}{Cl}}\!\cdot\; +\; :\!\ddot{\underset{\cdot\cdot}{Cl}}\!:\!\ddot{\underset{\cdot\cdot}{Cl}}\!:\; \longrightarrow\; :\!\ddot{\underset{\cdot\cdot}{Cl}}\!:\!\ddot{\underset{\cdot\cdot}{Cl}}\!:\; +\; :\!\ddot{\underset{\cdot\cdot}{Cl}}\!\cdot\right)$$

If, however, the chlorine atom collides with a methane molecule, the chlorine extracts a hydrogen atom and produces a hydrogen chloride molecule (step 2). The remnant of the methane molecule

$$\left(\text{H}:\overset{\text{H}}{\underset{\text{H}}{\text{C}}}\cdot \right)$$

is an unstable group; hence it will seek to form a new bond and once again regain its stable electron configuration. It is unlikely that it will strike another $CH_3\cdot$ group or a second chlorine atom since as we have said above, these reactive species remain rare throughout the reaction. Furthermore if it strikes a methane molecule and extracts a hydrogen atom from it, no net change will result:

$$\left(\text{H}:\overset{\text{H}}{\underset{\text{H}}{\text{C}}}\cdot + \text{H}:\overset{\text{H}}{\underset{\text{H}}{\text{C}}}:\text{H} \longrightarrow \text{H}:\overset{\text{H}}{\underset{\text{H}}{\text{C}}}:\text{H} + \text{H}:\overset{\text{H}}{\underset{\text{H}}{\text{C}}}\cdot \right)$$

However, if the $CH_3\cdot$ group strikes a chlorine molecule and extracts a chlorine atom, a molecule of methyl chloride will be produced along with another reactive chlorine atom (step 3). We can see therefore that once the reaction is initiated, every elementary reaction that is likely to occur in a mixture of Cl_2 and CH_4 produces exactly as many reactive species as it consumes. The reaction, in other words, is self-propagating.

Sometimes the steps in a chain reaction produce *more* reactive species than they consume. Such reactions, which we call *branched-chain reactions*, will therefore proceed at an ever-increasing rate, even if we remove the source of energy necessary to start the reaction. A familiar example is the reaction between H_2 and O_2 to produce water, which proceeds explosively after a single spark. We can represent its complex mechanism as follows:

(1) $H_2 \longrightarrow H\cdot + H\cdot$

(2) $H\cdot + O_2 \longrightarrow \cdot\ddot{\text{O}}:H + :\ddot{\text{O}}\cdot$

(3) $:\ddot{\text{O}}\cdot + H_2 \longrightarrow \cdot\ddot{\text{O}}:H + H\cdot$

(4) $\cdot\ddot{\text{O}}:H$ (from step 2) $+ H_2 \longrightarrow H_2O + H\cdot$

(5) $\cdot\ddot{\text{O}}:H$ (from step 3) $+ H_2 \longrightarrow H_2O + H\cdot$

(6) $H\cdot + \cdot\ddot{\text{O}}:H \longrightarrow H_2O$

(7) $H\cdot + H\cdot \longrightarrow H_2$

Note that each hydrogen atom produced in step 1 can, through steps 2, 3, 4, and 5, produce not only two molecules of water but also *three* additional reactive hydrogen atoms. Steps 6 and 7 involve the recombination of two reactive species, whose concentrations remain relatively low; hence they occur much less frequently than the propagation steps.

Question

12. What is the difference between a simple chain reaction and a branched-chain reaction?

12.8 Factors Affecting Rates: Catalysts

We shall now conclude this chapter with a discussion of the fourth factor that may influence the rate of a reaction: the presence of a catalyst. A **catalyst** is a substance that influences the speed of a reaction but is unchanged after the reaction is terminated. Catalysts do not initiate reactions nor do they affect the thermodynamic favorability of reactions. They merely increase the reaction rates. Furthermore many catalysts are highly specific; that is, each catalyst will affect the speed of only a single reaction rather than reactions in general.

We can best understand catalysts as substances that lower the activation energy of a reaction. Nonetheless, in spite of a great deal of research on catalysts, there are very few chemical reactions, if any, for which the exact mechanism of catalysis is understood.

One possible theory of catalytic action is that a distinct intermediate compound (not merely an activated complex) is formed, so that the reaction takes place in two steps instead of one. We can illustrate this mechanism and contrast it to an uncatalyzed reaction with the following equations:

$$\left.\begin{array}{l} A + C \longrightarrow AC \\ AC + B \longrightarrow AB + C \end{array}\right\} \text{two rapid reactions involving a catalyst } C$$

$$A + B \longrightarrow AB \qquad \text{a single slow reaction without a catalyst}$$

Here we must assume that the time required for the first two reactions together is less than that required for the uncatalyzed reaction. For this to be true we must depict these hypothetical reactions on an energy diagram as shown in Figure 12.8. The activation energy for the catalyzed reaction is less than the activation energy for the uncatalyzed reaction. This type of catalyst is often called a *carrier catalyst*. *Enzymes* (Chapter 23) provide this kind of catalysis for many biological reactions.

Another catalytic mechanism is called *contact catalysis*. An example is the effect of platinum on the reaction between hydrogen and oxygen. These two gases do not ordinarily react appreciably when mixed at room temperature. However, if a small amount of finely divided platinum is added to the mixture, the reaction proceeds rapidly, though not by the mechanism discussed in Section 12.7. The mechanism here involves *adsorption*, a process by which the gas molecules are held to the surface of the metal by weak physical or chemical forces. Once some of the gas molecules are adsorbed, any of several factors may increase the speed of reaction. The bond of the adsorbed molecule might be weakened or stretched, making the molecule more reactive, or, once adsorbed, the molecule might be better positioned for a reactive collision.

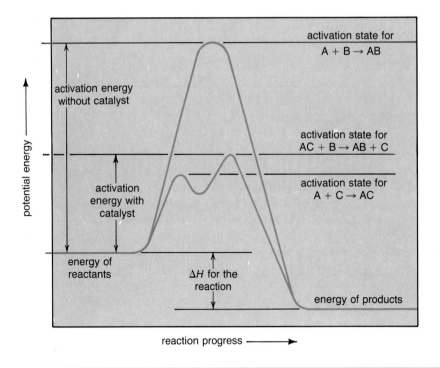

Figure 12.8 Effect of a catalyst.

Questions

13. What is the difference between a carrier catalyst and a contact catalyst?

14. What is the effect of a catalyst on the ΔH of a reaction?

Questions and Problems

12.1 Rates of Reactions

15. List several things that might be done to alter the speed of a reaction.

16. What is meant by saying that a reaction is thermodynamically but not kinetically favorable?

17. The concentration of H_2O_2, whose decomposition rate was to be measured, was initially 3.4×10^{-3} mol/liter. It was 9.2×10^{-4} mol/liter 24 min later. Express the average reaction rate for this time interval in two ways.

18. The rate of decomposition of H_2O_2 was determined by collecting the gas evolved in a buret. After 3.3 min, 10.4 ml of gas had been collected; after 6.9 min, 17.3 ml of gas

had been collected. Calculate the average rate for the first 3.3 min and for the first 6.9 min.

19. The rate of decomposition of azomethane (CH_3NNCH_3) was studied. The rate was measured by monitoring the pressure of azomethane in millimeters of Hg. Plot the data given below; then draw the slope at $t = 20$ min. What instantaneous rate does this give?

t (min)	0	15	30	48	75
pressure (mm)	36.2	30.0	24.9	19.3	13.1

12.3 Factors Affecting Rates: Temperature

20. By what factors are the average molecular velocity and the average kinetic energy of a gas increased when its temperature is increased from 23°C to 53°C? By what factor would this temperature increase be expected to increase a typical reaction rate?

21. If the rate of a reaction is 0.020 (mol/liter)/s at 25°C, what would you expect the rate to be at 35°C? At 65°C?

12.4 Factors Affecting Rates: Concentration

22. The hypothetical reaction, $A + B \rightarrow C$, has a rate law:

$$\text{rate} = k[A]^x[B]^y$$

If the rate doubles when the concentration of A is doubled and B is held constant, but the rate increases fourfold when the concentration of B is doubled and A is held constant, what are the values of x and y?

23. The rate of the reaction of hydrogen peroxide depends on the concentration of iodide ion present. The rate of decomposition was measured for various concentrations of H_2O_2 and of KI (at constant temperature and pressure). The data appear below. Determine the order in each substance, write the rate law, and evaluate the rate constant.

Rate (ml/min)	[H_2O_2]	[KI]
0.90	0.15	0.033
1.7	0.30	0.033
1.9	0.15	0.066

24. Give the order in each reactant and the overall order for the reaction

$$A + B + C \rightarrow D + E$$

which obeys the following rate law:

$$\text{rate} = k[A][B]^2$$

25. Describe a graphical method for using experimental data to determine whether a reaction is first order or not.

26. The half-life for the decomposition of N_2O_5 at 30°C is 10.0 min. What fraction of a given sample of N_2O_5 will decompose in 30.0 min?

27. After 40.0 min at 500°C, only 12.5 percent of a sample remains. What is the half-life for the first-order decomposition of the sample?

*28. If the original sample in Question 26 weighed 243 g, how much would remain after 2.00 hr?

29. The following experimental data were obtained for the reaction

$$2\,A + 3\,B \longrightarrow C + 2\,D$$

[A]	[B]	Rate = $\Delta[C]/\Delta t$ [(mol/liter)/s]
0.0510	0.0420	2.3×10^{-2}
0.102	0.0420	9.2×10^{-2}
0.0510	0.0840	4.6×10^{-2}

Assuming that the rate law has a simple form, determine the reaction order in each reactant and then evaluate the rate constant.

30. The following experimental data have been determined for the reaction

$$NH_4^+(aq) + NO_2^-(aq) \longrightarrow N_2(aq) + 2\,H_2O(l) \quad \text{(acid solution)}$$

$[NH_4^+]$	$[NO_2^-]$	Rate = $\Delta[N_2]/\Delta t$ [(mol/liter)/s]
0.0092	0.098	34.9×10^{-8}
0.0092	0.049	16.6×10^{-8}
0.0488	0.196	315×10^{-8}
0.0249	0.196	156×10^{-8}

Determine the rate law for the reaction, and evaluate the rate constant.

31. For the reaction $A + 2\,B \longrightarrow C$, the following experimental data were obtained. Derive a rate law for the reaction, and evaluate the rate constant.

[A]	[B]	Rate = $\Delta[C]/\Delta t$ [(mol/liter)/s]
0.40	0.32	0.012
0.80	0.32	0.024
0.80	0.64	0.096

32. For the reaction

$$2\,NO(g) + 2\,H_2(g) \longrightarrow N_2(g) + 2\,H_2O(g)$$

at 1100°C, the following data have been obtained:

[NO]	[H$_2$]	Rate = Δ[NO]/Δt [(mol/liter)/s]
5.0 × 10^{-3}	2.0 × 10^{-3}	2.4 × 10^{-5}
15 × 10^{-3}	2.0 × 10^{-3}	2.2 × 10^{-4}
15 × 10^{-3}	4.0 × 10^{-3}	4.4 × 10^{-4}

Derive a rate law for the reaction, and obtain a value for the rate constant.

12.7 **Chain Reactions**

33. What is a chain reaction? A branched-chain reaction? How can a chain reaction be controlled?

Discussion Starters

34. Compare and contrast:

instantaneous rate	average rate
homogeneous reaction	heterogeneous reaction
rate law	rate constant
activated complex	activation energy
order of a reaction	molecularity of a reaction

13 Molecular Equilibria

Until now, in our discussions of stoichiometry, thermodynamics, and kinetics, we have assumed that chemical reactions go to completion—that is, that they continue until one or more of the reactants are exhausted. Only a few times have we hinted that this may not be strictly true. In this chapter, when we look more closely at our assumption, we shall find that *in a closed, isolated system no reaction goes to completion*. Instead, the initial concentrations of the reactants increase or decrease for a time and then remain constant; subsequently there is no net change in the concentrations of reactants or products. At this point the system is in a state of dynamic equilibrium—much like the equilibrium established in a closed container between a liquid and its vapor, between vaporization and condensation. Here the equilibrium is between two *opposing reactions*. We shall look at such a system in detail in the following section.

In many cases, of course, the equilibrium concentration of one of the reactants will be very small, say 10^{-15} mol/liter. For *practical* purposes, we can say that such a reaction has gone to completion. Furthermore, if we do not carefully isolate a reaction in order to maintain equilibrium, it may in fact go to completion. A piece of charcoal in an open container will burn completely since both heat and the gaseous products are allowed to escape. In a closed, isolated system, however, all chemical reactions produce only an equilibrium between reactants and products.

Our study of chemical equilibria will tie together several topics we have already covered, especially thermodynamics and kinetics. In this chapter, therefore, we shall use what we know about these two subjects to approach chemical equilibria. In addition, the concept of equilibrium will come up again when we introduce electrochemistry in Chapter 18.

13.1 The Equilibrium Constant

To take a more careful look at chemical equilibrium, let us consider the reaction

$$H_2(g) + I_2(g) \rightleftharpoons 2\,HI(g)$$

We shall begin by placing measured quantities of hydrogen and iodine in a reaction vessel at a temperature of 764 K. If we then monitor the progress of the reaction, we find that at first the concentrations of the reactants decrease quite rapidly. However, the rate of decrease diminishes until the concentrations of reactants become constant. During the same time interval, of course, the concentration of hydrogen iodide rises, rapidly at first and then more slowly, from zero to a constant value (Figure 13.1). When these constant values have been established (at the time t in Figure 13.1), all *net* chemical change ceases. Equilibrium has been established. Yet reactions continue to occur: Hydrogen and iodine combine to form hydrogen iodide ($H_2 + I_2 \rightarrow 2\ HI$), and hydrogen iodide decomposes into hydrogen and iodine ($2\ HI \rightarrow H_2 + I_2$). The two reactions merely occur at exactly the same rate.

At first, when the concentrations of H_2 and I_2 were relatively high, the *forward reaction* predominated. We therefore observed a rapid net increase in the concentration of hydrogen iodide. As the amounts of hydrogen and iodine available for combination decreased, however, so did the rate of the forward reaction. At the same time the amount of HI available for decomposition increased and so did the rate of the *reverse reaction*. Eventually, the rate of decomposition equaled the rate of formation; hence we observed no further net change. (Notice that we indicated this reversibility with a pair of arrows when we wrote the chemical equation.)

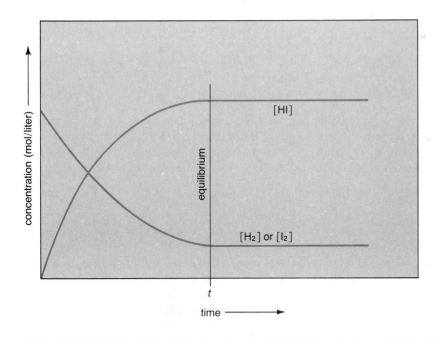

Figure 13.1 Approach to equilibrium for the reaction

$$H_2(g) + I_2(g) \rightleftarrows 2\ HI(g)$$

Initially the reaction mixture consists of only H_2 and I_2. The initial concentration of HI is zero. The curves show the changing concentrations of the components of the system with time. When the two curves become straight lines, equilibrium has been established.

Table 13.1 summarizes the data from several experiments, each of which began with different concentrations of H_2, I_2, and HI. Depending on the initial concentrations the equilibrium concentrations of each component vary widely. The last column of the table, however, tells us something very important. If we divide the square of the equilibrium concentration of HI by the equilibrium concentrations of H_2 and I_2, we find a number that is constant for all seven trials. The small deviations can be attributed to experimental errors.

Table 13.1 Experimental Data for the Equilibrium: $H_2(g) + I_2(g) \rightleftharpoons 2\,HI(g)$ at 764 K*

Trial	Initial Concentrations (moles/liter)			Equilibrium Concentrations (moles/liter)			$\dfrac{[HI]^2}{[H_2][I_2]}$
	H_2	I_2	HI	H_2	I_2	HI	
1	0.0109	0.0115	0	0.00228	0.00284	0.0172	45.7
2	0.0106	0.0129	0	0.00172	0.00406	0.0178	45.4
3	0.0106	0.0103	0	0.00219	0.00287	0.0169	45.4
4	0.0110	0.0107	0	0.00265	0.00231	0.0166	45.0
5	0	0	0.0228	0.00260	0.00260	0.0176	45.8
6	0	0	0.0166	0.00190	0.00190	0.0128	45.4
7	0	0	0.0154	0.00198	0.00198	0.0134	45.0

* A. H. Taylor and R. H. Cerst, *J. Amer. Chem. Soc.*, **63**, 1377 (1941).

To look for a similar pattern in another reaction, let us turn to the reaction between nitrogen and hydrogen. Though this is a rather simple reaction, it is considerably more significant than that between hydrogen and iodine. In 1898 Sir William Crookes predicted dire things for civilization: He contended that the world's arable land could not keep up with the demand for food. His prophecy was based on the observation that plants consume nitrogen compounds from the soil but that they are unable to use the abundant free (elemental) nitrogen from the atmosphere. Unless the soil is replenished with combined nitrogen, it thus becomes unfit for crops in a few years. Crookes maintained that natural sources of combined nitrogen were not sufficient and that unless chemists developed a process for *nitrogen fixation*—by which nitrogen compounds could be economically made from elementary nitrogen—civilization would come to an end. Crookes was not a crackpot; he was one of the most respected scientists of his time. His concerns were real.

Scientists met the challenge, and a number of commercial methods of nitrogen fixation have been developed. The most successful of these was developed by Fritz Haber (1868–1934) less than a decade after Crookes expressed his fears. We can represent this process by the deceptively simple equation

$$N_2(g) + 3\,H_2(g) \rightleftharpoons 2\,NH_3(g)$$

In practice this process demands a good deal of complicated chemistry, and we shall discuss it again later in the chapter. For the moment, however, we can consider the experimental data listed in Table 13.2. This table presents the equilibrium concentrations of N_2, H_2, and NH_3 following several experiments conducted at 673 K. Again we find a pattern in the data. If the concentration of ammonia is squared and then divided by the concentration of nitrogen times the cube of the concentration of hydrogen, the result is constant.

Table 13.2 Experimental Data for the Equilibrium: $N_2(g) + 3\,H_2(g) \rightleftharpoons 2\,NH_3(g)$ at 673 K

Equilibrium Concentration (moles/liter)			$\dfrac{[NH_3]^2}{[N_2][H_2]^3}$
N_2	H_2	NH_3	
0.0416	0.130	0.00679	0.504
0.122	0.365	0.0548	0.506
0.192	0.576	0.137	0.512
0.373	1.02	0.451	0.514

If we were to continue with other reactions, we would find similar relationships. For each, an **equilibrium constant** relates the equilibrium concentrations of reactants and products. In each case, we can calculate the constant, given one set of equilibrium concentrations and a balanced chemical equation. To illustrate, let the following general equation represent a gas phase reaction in which w moles of A react with x moles of B to form y moles of C and z moles of D:

$$wA + xB \rightleftharpoons yC + zD$$

The equilibrium constant for this reaction is then given by the expression

$$\frac{[C]^y[D]^z}{[A]^w[B]^x} = K$$

In general, *the equilibrium constant is the product of the concentrations of the products, each raised to the power of its respective coefficient in the equation, divided by the product of the concentrations of the reactants, each raised to the power of its respective coefficient in the equation.*

Before we go on, two observations are in order. First, the equilibrium constant must of course be independent of initial concentrations, but it does depend on temperature. We shall say more about this temperature dependence later. Second, we could just as easily have defined the equilibrium constant with product concentrations in the denominator and reactant concentrations in the numerator. Convention alone dictates that products go on top and reactants, on the bottom. Likewise by convention we shall specify concentrations in moles per liter, unless we specifically state otherwise. Our equilibrium constants are therefore *concentration equilibrium constants*.

Equilibrium constants for different reactions range from exceedingly small to very large numbers. A large value of K means that at equilibrium the concentrations of the products are large and the concentrations of the reactants are small. The larger the value, the more nearly complete the reaction. Conversely, a small value of K indicates that at equilibrium the concentrations of the products are small and the concentrations of the reactants are large. The smaller the value, the less product present at equilibrium.

Questions

1. Write the expression for K for each of the following gas phase equilibria.
   a. $2\ SO_2 + O_2 \rightleftharpoons 2\ SO_3$
   b. $N_2O_4 \rightleftharpoons 2\ NO_2$
   c. $CO + 2\ H_2 \rightleftharpoons CH_3OH$
   d. $2\ O_3 \rightleftharpoons 3\ O_2$
   e. $2\ NO \rightleftharpoons N_2 + O_2$
   f. $COCl_2 \rightleftharpoons CO + Cl_2$
   g. $2\ NO_2 + 7\ H_2 \rightleftharpoons 2\ NH_3 + 4\ H_2O$
   h. $4\ H_2 + CS_2 \rightleftharpoons CH_4 + 2\ H_2S$

2. What would you conclude about the relative equilibrium concentrations of reactants and products for the reaction $A + B \rightleftharpoons C + D$ if the
   a. value of K is very large?
   b. value of K is equal to 1.00?
   c. value of K is very small?

13.2 Le Chatelier's Principle and Chemical Equilibrium

All systems tend to reach equilibrium spontaneously (though sometimes slowly) and to remain in that condition unless some external stress is applied. If a stress is applied, the system shifts toward either reactants or products, and new equilibrium concentrations are established. How the system responds to the stress is governed by Le Chatelier's principle, which we first encountered in Chapter 9: *If a stress is applied to a system at equilibrium, the equilibrium will shift to relieve the stress applied.*

One way to shift an equilibrium is to add one of the components of the equilibrium mixture. For example, if hydrogen is added to the following equilibrium system,

$$H_2(g) + I_2(g) \rightleftharpoons 2\ HI(g)$$

the stress is the excess hydrogen. The rate of the reaction to the right will therefore increase to reduce the hydrogen concentration. When the new equilibrium is established, we shall find the hydrogen concentration larger than before the addition but smaller than immediately after the addition. Of course the equilibrium shift also decreases the concentration of I_2 and increases the concentration of HI. As we add more and more H_2, the reaction will shift further to the right and consume more and more I_2. It is important to emphasize, however, that no component of an equilibrium mixture can be *completely* consumed.

A second method of applying stress, thus shifting an equilibrium, is to remove one of the components of the system. If some iodine molecules are removed, the reaction must shift to the left to make up for the reduced I_2 concentration. When equilibrium is reestablished, the concentration of iodine will be smaller than before the removal but larger than immediately after the removal. The hydrogen concentration will be larger than it was originally; the hydrogen iodide concentration, smaller.

A third very important way to apply stress is to change the temperature. It has been established that the equilibrium constant K is a function of temperature; thus equilibrium concentrations would be expected to change as the temperature changes. As we saw in Chapter 10, an *increase* in temperature favors endothermic reactions (that is, it increases the ratio of products to reactants); whereas, a *decrease* in the temperature favors exothermic reactions. At the risk of repeating the arguments in Chapter 10, we can see this more clearly if we think of the energy provided by an increase in temperature as a reactant in the chemical equation. In an endothermic reaction energy is consumed to form the products:

$$\text{energy} + H_2(g) + I_2(g) \rightleftharpoons 2\,HI(g)$$

Thus an increase in the temperature provides excess energy and the reaction shifts to the right. In an exothermic reaction energy is a product:

$$N_2(g) + 2\,H_2(g) \rightleftharpoons 2\,NH_3(g) + \text{energy}$$

If energy is added to this mixture, the equilibrium will shift to the left; thus it is a decrease in temperature that favors the product. Note that when the temperature of a reaction is changed, the numerical value of the equilibrium constant changes; when the concentrations of reactants and products are changed, the equilibrium constant does *not* change.

Finally, we can shift some equilibria by altering the pressure. However, this can occur only with a reaction in which the pressure changes as the reaction proceeds. Changes in pressure will have little effect on reactions in which all components of the equilibrium mixture are solids or liquids. In many gas phase equilibria, though, a change in pressure can be regarded as a stress on the system. For example, consider an equilibrium mixture of the gases H_2, N_2, and NH_3 in a closed container:

$$3\,H_2(g) + N_2(g) \rightleftharpoons 2\,NH_3(g)$$

If the pressure on the reaction container is increased (at constant temperature), the equilibrium will shift in the direction that tends to reduce the total pressure.

Since there are four moles of gas on the left side of the equation for every two on the right, we can expect the equilibrium to shift to the right: The relative amount of NH_3 will increase. To look at things slightly differently, we can say that the increase in pressure increases concentrations. The reaction thus shifts in the direction that reduces the total concentration of gas (in moles per liter).

Whichever way we choose to look at the reaction between hydrogen and iodine ($H_2 + I_2 \rightleftharpoons 2\,HI$), we can see that pressure has no effect on the equilibrium. Two moles of gas react to form two moles of gas; thus pressure does not favor one side or the other.

Table 13.3 summarizes the effects of stress on chemical equilibria.

Table 13.3 Summary of the Effects of Changes on a System at Equilibrium

Change	Equilibrium	K
Concentration of one of the reactants is increased.	Shifts to the right	No effect
Concentration of one of the reactants is decreased.	Shifts to the left	No effect
Concentration of one of the products is increased.	Shifts to the left	No effect
Concentration of one of the products is decreased.	Shifts to the right	No effect
Pressure is increased (by a decrease in volume).	Shifts in the direction that produces the smaller number of gas molecules. If the number of gas molecules is the same on both sides of the equation, the equilibrium will not be affected.	No effect
Pressure is decreased (by an increase in volume).	Shifts in the direction that produces the larger number of gas molecules. If the number of gas molecules is the same on both sides of the equation, the equilibrium will not be affected.	No effect
Temperature is increased.	Exothermic reactions shift to the left; endothermic reactions shift to the right.	Increases for endothermic reactions; decreases for exothermic reactions.
Temperature is decreased.	Exothermic reactions shift to the right; endothermic reactions shift to the left.	Decreases for endothermic reactions; increases for exothermic reactions.

Questions

3. What do we mean by a shift in equilibrium? Use an example to explain.

4. What are three ways of shifting a chemical equilibrium?

5. How is the equilibrium $N_2(g) + 3\,H_2(g) \rightleftharpoons 2\,NH_3(g)$ affected by the following changes? The reaction to the right is exothermic.
- a. Pressure is decreased.
- b. Nitrogen is added.
- c. Ammonia is added.
- d. Hydrogen is removed.
- e. Ammonia is removed.
- f. Temperature is increased.

13.3 Quantitative Aspects of Chemical Equilibrium

To better understand Le Chatelier's principle, we shall apply what we've discussed to several specific chemical problems. In each, assume that all reactants and products are gases.

Example 13.1 An equilibrium mixture of the reaction

$$CO_2(g) + 2\,SO_3(g) \rightleftharpoons CS_2(g) + 4\,O_2(g)$$

was found upon analysis to contain the components of this system in the following concentrations at 100°C.

CO_2: 0.500 mol/liter

SO_3: 5.00 mol/liter

CS_2: 6.00 mol/liter

O_2: 1.00 mol/liter

Calculate the concentration equilibrium constant for the reaction.

Solution

a. The expression for the concentration equilibrium constant is

$$\frac{[CS_2][O_2]^4}{[CO_2][SO_3]^2} = K$$

b. Substitute the appropriate equilibrium concentrations:

$$\frac{(6.00)(1.00)^4}{(0.500)(5.00)^2} = \mathbf{0.480\ mol^2/liter^2}$$

The unit (moles² per liter²) for K in Example 13.1 is not common to all equilibrium constants; it should be obvious that units for equilibrium constants will vary as the expressions vary. For this reason, they are often omitted completely, but the following conventions must be followed carefully:

1. Write the balanced chemical equation from which the expression for K is derived with the smallest possible whole-number coefficients.

2. Adhere strictly to the definition of K given on p. 403.

3. Express all concentrations in moles per liter.

Example 13.2 Using the equilibrium constant calculated in Example 13.1, what must the concentration of SO_3 be if the other components of an equilibrium mixture have the following concentrations?

CS_2: 8.00 mol/liter

O_2: 2.00 mol/liter

CO_2: 12.0 mol/liter

Solution

a. The expression for the equilibrium constant has only a single unknown quantity—the concentration of SO_3:

$$\frac{[CS_2][O_2]^4}{[CO_2][SO_3]^2} = 0.480$$

$$\frac{(8.00)(2.00)^4}{(12.0)[SO_3]^2} = 0.480$$

$$[SO_3]^2 = \frac{(8.00)(2.00)^4}{(12.0)(0.480)}$$

$$[SO_3] = \sqrt{\frac{(8.00)(16.0)}{(12.0)(0.480)}} = \mathbf{4.71\ mol/liter}$$

Example 13.3 The sulfur dioxide in the atmosphere comes from the burning of fossil fuels (coal and petroleum), primarily by electric utilities, in private homes and in industrial plants. A second pollutant, nitrogen dioxide, is produced in internal combustion engines. It is possible that these two gases are the sources of the highly corrosive droplets of H_2SO_4 found in the air of industrial cities:

$$SO_2(g) + NO_2(g) \rightleftharpoons NO(g) + SO_3(g)$$

$$SO_3(g) + H_2O(g) \longrightarrow H_2SO_4(l)$$

What is the equilibrium constant for the first of these two reactions if, after we place 4.00 mol of SO_2 and 3.00 mol of NO_2 in a 1.00-liter reaction vessel, we find 2.00 mol of SO_3 in the equilibrium mixture?

Solution

a. The expression for K is

$$\frac{[SO_3][NO]}{[SO_2][NO_2]} = K$$

b. From the equation we see that one mole of NO is formed for every mole of SO_3 formed. Since we started with no NO or SO_3 present, the equilibrium mixture must contain equal amounts of SO_3 and NO. Therefore, $[NO] = 2.00$ mol/liter.

c. From the equation we also see that for every mole of SO_3 formed, one mole of NO_2 and one mole of SO_2 have reacted; thus at equilibrium the original concentrations of both SO_2 and NO_2 will have been reduced by 2.00 mol/liter:

$$[SO_2] = \text{(original concentration)} - \text{(amount reacting)}$$
$$= (4.00 - 2.00) = 2.00 \text{ mol/liter}$$

$$[NO_2] = \text{(original concentration)} - \text{(amount reacting)}$$
$$= (3.00 - 2.00) = 1.00 \text{ mol/liter}$$

d. We now have the values necessary to calculate the equilibrium constant:

$$K = \frac{(2.00)(2.00)}{(2.00)(1.00)} = \mathbf{2.00}$$

If we know the equilibrium constant of a system, we can predict the behavior of a nonequilibrium mixture by calculating the *reaction quotient* Q. To do this we simply substitute the nonequilibrium concentrations into the expression for the equilibrium constant. If the result is smaller than the equilibrium constant ($Q < K$), the reaction to the right will predominate until equilibrium is reached; if $Q > K$, the reaction to the left will predominate. If $Q = K$, the reaction is, of course, already at equilibrium. Q, like K, is generally given without units.

Example 13.4 For each of the following reaction mixtures, A, B, and C, determine whether the system is at equilibrium and, if not, which way the reaction will go to reach equilibrium.

Reaction: $2 CO(g) + O_2(g) \rightleftharpoons 2 CO_2(g)$, $K = 0.250$

	A	B	C
$[CO]$:	2.00 mol/liter	2.00 mol/liter	1.00 mol/liter
$[O_2]$:	2.00 mol/liter	3.00 mol/liter	1.00 mol/liter
$[CO_2]$:	2.00 mol/liter	1.00 mol/liter	0.500 mol/liter

Solution

a. Evaluate Q for each reaction mixture:

$$\text{A} \qquad \qquad \text{B} \qquad \qquad \text{C}$$

$$\frac{(2.00)^2}{(2.00)^2(2.00)} = 0.500 \qquad \frac{(1.00)^2}{(2.00)^2(3.00)} = 0.0833 \qquad \frac{(0.500)^2}{(1.00)^2(1.00)} = 0.250$$

b. In A, $Q > K$, so the reaction to the left will predominate.
In B, $Q < K$, so the reaction to the right will predominate.
In C, $Q = K$, so this is an equilibrium mixture.

Example 13.5 Large quantities of hydrogen are required for the Haber synthesis of ammonia. Part of this hydrogen is produced by an industrial process called the *water gas reaction*. In this process steam is passed over hot carbon:

$$C(s) + H_2O(g) \qquad CO(g) + H_2(g)$$

The CO formed during this reaction can then be heated with more steam, producing more hydrogen.

$$CO(g) + H_2O(g) \rightleftharpoons CO_2(g) + H_2(g)$$

What will be the concentration of each component of this second system if 3.00 mol of CO and 2.00 mol of H_2O are placed in a 1.00-liter reaction vessel and allowed to reach equilibrium? The value of K is 2.00.

Solution

a. Let x be the number of moles of CO that have been consumed when equilibrium is attained. Then x mol of H_2O have also reacted, producing x mol of CO_2 and H_2; thus at equilibrium

$$[CO] = (3.00 - x) \text{ mol/liter}$$
$$[H_2O] = (2.00 - x) \text{ mol/liter}$$
$$[CO_2] = x \text{ mol/liter}$$
$$[H_2] = x \text{ mol/liter}$$

b. We can now substitute these equilibrium concentrations into the expression for the equilibrium constant K:

$$\frac{[CO_2][H_2]}{[CO][H_2O]} = K = \frac{x^2}{(3.00 - x)(2.00 - x)} = 2.00$$

c. This is a second-degree equation, which we can solve using the quadratic formula.

$$x = \frac{-b \pm \sqrt{b^2 - 4ac}}{2a}$$

d. To use this formula, however, we must first rewrite our equation in standard form: $ax^2 + bx + c = 0$.

$$x^2 = (3.00 - x)(2.00 - x)(2.00)$$
$$x^2 = (6.00 - 5.00x + x^2)(2.00)$$
$$x^2 = 12.0 - 10.0x + 2.00x^2$$
$$x^2 - 10.0x + 12.0 = 0$$

Thus $a = 1$, $b = -10.0$, and $c = 12.0$.

e. Substitute into the quadratic formula:

$$x = \frac{10 \pm \sqrt{100 - 48}}{2}$$

$$x = 1.40 \quad \text{or} \quad 8.60$$

f. This equation, like all quadratic equations, has two roots, that is, two solutions for x. Whenever we use the equation to solve physical problems, however, we can count on one root being a physical impossibility. In this case, if x were 8.60, the equilibrium concentrations of H_2O and CO would be negative numbers—obviously an absurd result.

g. Thus the concentrations at equilibrium are

$$[CO] = 3.00 - x = 3.00 - 1.40 = \mathbf{1.60 \text{ mol/liter}}$$
$$[H_2O] = 2.00 - x = 2.00 - 1.40 = \mathbf{0.60 \text{ mol/liter}}$$
$$[CO_2] = x = \mathbf{1.40 \text{ mol/liter}}$$
$$[H_2] = x = \mathbf{1.40 \text{ mol/liter}}$$

h. We can check these values by substituting them in the expression for the equilibrium constant:

$$\frac{(1.40)(1.40)}{(0.60)(1.60)} = 2.0$$

Example 13.6 Both CO and CO_2 are emitted in automobile exhausts. Both are colorless and odorless gases. Carbon monoxide is a highly poisonous gas, which deprives the body of oxygen by reacting with hemoglobin. In terms of weight, it is our most common air pollutant. In the presence of oxygen and at high temperatures, the following equilibrium exists:

$$2\ CO(g) + O_2(g) \rightleftharpoons 2\ CO_2(g)$$

Calculate the equilibrium constant for this reaction if the equilibrium concentrations of CO, O_2, and CO_2 are 2.00, 4.00, and 2.00 mol/liter, respectively. How many moles per liter of CO must be added to increase the equilibrium concentration of CO_2 to 3.00 mol/liter?

Solution

a. To evaluate K, substitute the equilibrium concentrations in the appropriate expression:

$$K = \frac{[CO_2]^2}{[CO]^2[O_2]} = \frac{(2.00)^2}{(2.00)^2(4.00)} = 0.250$$

b. The addition of CO is a stress upon the equilibrium. The equilibrium will therefore shift to relieve that stress—in this case to the right.

c. Let x be the number of moles per liter of CO that are added. Therefore, as soon as the CO is added, its concentration is $(2.00 + x)$ mol/liter. But at equilibrium the concentration of CO_2 must increase by 1 mol/liter, so according to the equation the concentration of CO must decrease by the same amount; hence the new equilibrium is

$$[CO] = (2.00 + x - 1.00) \quad \text{or} \quad (1.00 + x)$$

d. For every mole of CO that reacts, 0.500 mol of O_2 must also react; hence when the new equilibrium is established, $[O_2] = 3.50$.

e. We can now substitute values in the expression for the equilibrium constant and solve for x:

$$0.250 = \frac{(3.00)^2}{(1.00 + x)^2(3.50)}$$

$$0.875x^2 + 1.75x - 8.125 = 0$$

$$x = \frac{-1.75 \pm \sqrt{31.50}}{2(0.875)}$$

$$= \frac{-1.75 \pm 5.61}{1.75}$$

$$= \mathbf{2.20\ mol} \quad (\text{or } -4.21 \text{ mol, which is absurd})$$

$$[CO] = 3.20\ \text{mol/liter}$$
$$[CO_2] = 3.00\ \text{mol/liter}$$
$$[O_2] = 3.50\ \text{mol/liter}$$

f. To confirm our solution, we again use the expression for K:

$$\frac{(3.00)^2}{(3.20)^2(3.50)} = 0.250$$

Question

6. Why are the units for K usually omitted?

13.4 K_p, the Equilibrium Constant in Terms of Partial Pressures

We can rearrange the general gas law equation, $PV = nRT$, to give an expression for the concentration of a gas:

$$\frac{n}{V} = \frac{P}{RT}$$

From this we see that n/V, the concentration in moles per liter, is proportional to pressure (temperature constant). Thus for systems in which all reactants are gases we can write an equilibrium constant, K_p, in terms of partial pressures. The general expression for K_p, for the gas phase reaction $wA + xB \rightleftharpoons yC + zD$, is

$$\frac{(P_C)^y(P_D)^z}{(P_A)^w(P_B)^x} = K_p$$

where each P is the partial pressure *in atmospheres* of the gas designated by the subscript. The units for K_p, as for K, are usually omitted since they vary according to the coefficients of the balanced chemical equation. In general, the values for K and K_p will differ for the same reaction; only if the balanced equation has the same number of moles on both sides will the two equilibrium constants be equal.

13.5 Free Energy and Equilibrium

We are now prepared to link some of the most important ideas in chemistry. In this section we shall tie together the concepts of spontaneity, which arose in Chapter 11 from our study of thermodynamics, and equilibrium. In Section 13.6, we shall make clearer the link between equilibrium and the kinetics of elementary reactions. We can begin by reemphasizing that in a closed system no chemical reaction goes to completion. Thus when we say that a reaction occurs spontaneously, that it has a negative free energy change, we can only mean that there is a greater tendency for the reaction to occur as written than in the opposite direction.

For a gas phase system, we can relate the standard free energy change for a reaction to the equilibrium constant using the following equation:

$$\Delta G° = -2.303RT \log K_p$$

We shall not derive this relationship, but it has its origin in thermodynamic principles we have already studied. It confirms what we expected. The more negative the value of $\Delta G°$, the larger the value of K_p, thus the more product we find at equilibrium. When $\Delta G°$ is positive, on the other hand, $\log K_p$ must be negative, and K_p must be a fraction. The larger the value of $\Delta G°$, the more reactant we shall find at equilibrium.

We can rewrite this equation in a form convenient for calculating equilibrium constants at 25°C by rearranging it and substituting $T = 298$ K and $R = 1.987 \times 10^{-3}$ kcal/(mol)(K):

$$\log K_p = \frac{-\Delta G°}{1.364}$$

We should also emphasize that both of these equations are strictly valid only for ideal gases. The equations are fairly accurate for real gases under ordinary conditions, but they are usually inadequate for reactions in solution. Especially when dealing with ionic solutions, we cannot relate free energy changes to equilibrium constants calculated from the concentrations of reactants and products. Instead of concentrations, we would have to use *activities*, which are concentrations multiplied by a correction factor. (Activities can be thought of as *effective concentrations*.) Rather than broach the complicated subject of activities, we shall look only at gas phase reactions when relating free energy to equilibrium constants.

Example 13.7 Use standard free energy data (Table 11.5) to calculate K_p at 25°C for the reaction

$$2\ NO_2(g) \rightleftharpoons N_2O_4(g)$$

Solution

a. $\Delta G°$ for the reaction is

$$\Delta G°_f\ (N_2O_4) - 2\ \Delta G°_f\ (NO_2)$$

$$= (1\ \text{mol})\left(\frac{23.49\ \text{kcal}}{1.00\ \text{mol}}\right) - (2\ \text{mol})\left(\frac{12.39\ \text{kcal}}{1.00\ \text{mol}}\right) = -1.29\ \text{kcal}$$

b. From this we can calculate K_p:

$$\log K_p = \frac{-\Delta G°}{1.364}$$

$$\log K_p = \frac{-(-1.29)}{1.364} = 0.946$$

$$K_p = 8.83$$

c. Note that if we had written the reaction $N_2O_4 \rightleftharpoons 2\,NO_2$, $\Delta G°$ would equal 1.29 kcal, $\log K_p$ would equal -0.946, and K_p would equal 0.113, which is the reciprocal of the value calculated in step b.

In Chapter 11 great importance was attached to the fact that spontaneous reactions proceed with a decrease in free energy. We have now amplified that discovery by saying that spontaneous reactions do not go to completion but only to equilibrium. At equilibrium the favorable spontaneous reaction is counterbalanced by the reverse reaction. From these two observations we might therefore deduce that *at equilibrium the free energy of the system is at a minimum.* All systems move toward equilibrium with a decrease in free energy.

A useful way to apply this fundamental observation to gas phase reactions is the following equation, which involves the reaction quotient Q:

$$\Delta G = \Delta G° + 2.303 RT \log Q$$

When $Q < K$, $\Delta G < 0$, and the net reaction will proceed as written (from left to right) toward equilibrium. When $Q > K$, $\Delta G > 0$, and the reaction from right to left must dominate as the system moves toward equilibrium. When $Q = K$, $\Delta G = 0$, and the system is at equilibrium.

Example 13.8 NO_2 and N_2O_4 are placed in a reaction vessel at 25°C. The partial pressure of NO_2 is 0.500 atm, and that of N_2O_4 is 1.93 atm. Calculate Q, $\Delta G°$, and ΔG, then predict the direction of the reaction $NO_2 \rightleftharpoons N_2O_4$.

Solution

a. We calculate Q by substituting the initial partial pressures into the expression for the equilibrium constant:

$$2\,NO_2 \rightleftharpoons N_2O_4$$

$$K_p = \frac{(P_{N_2O_4})}{(P_{NO_2})^2} \quad \text{(at equilibrium)}$$

$$Q = \frac{(P_{N_2O_4})}{(P_{NO_2})^2} \quad \text{(initially)}$$

$$Q = \frac{1.93}{(0.500)^2} = 7.72$$

b. Calculate $\Delta G°$, as in Example 13.7:

$$\Delta G° = \Delta G°_f\, N_2O_4(g) - 2\,\Delta G°_f\, NO_2(g)$$

$$= (1\text{ mol}) \left(\frac{23.49 \text{ kcal}}{1.00 \text{ mol}}\right) - (2 \text{ mol}) \left(\frac{12.39 \text{ kcal}}{1.00 \text{ mol}}\right)$$

$$\Delta G° = -1.29 \text{ kcal for each mole of product}$$

c. Calculate ΔG:

$$\Delta G = \Delta G° + 2.303 RT \log Q$$

$$= \left(\frac{-1.29 \text{ kcal}}{1.00 \text{ mol}}\right) + (2.303) \left(\frac{1.987 \times 10^{-3} \text{ kcal}}{(1.00 \text{ K})(1.00 \text{ mol})}\right)(298 \text{ K})(\log 7.72)$$

$$= \frac{-1.29 \text{ kcal}}{1.00 \text{ mol}} + \frac{1.21 \text{ kcal}}{1.00 \text{ mol}}$$

$$= -0.08 \text{ kcal/mol}$$

d. Since $\Delta G < 0$, the initial concentrations are such that the reaction to the right (2 $NO_2 \rightarrow N_2O_4$) will predominate as the system approaches equilibrium. When the system reaches equilibrium, the partial pressure of N_2O_4 will have increased, and the partial pressure of NO_2 will have decreased.

13.6 Reaction Rates and Equilibrium

We now take the second step in tying together thermodynamics, kinetics, and the notion of equilibrium. Let us consider an elementary, one-step reaction

$$aA + bB \rightleftharpoons cC + dD$$

where the concentration equilibrium constant is

$$K = \frac{[C]^c[D]^d}{[A]^a[B]^b}$$

Since this is an elementary reaction, we can also write expressions for the rates of the forward and reverse reactions:

$$\text{rate}_{\text{forward}} = k_f[A]^a[B]^b$$

$$\text{rate}_{\text{reverse}} = k_r[C]^c[D]^d$$

At equilibrium these rates are equal; therefore

$$k_f[A]^a[B]^b = k_r[C]^c[D]^d$$

$$\frac{k_f}{k_r} = \frac{[C]^c[D]^d}{[A]^a[B]^b} = k$$

Thus for an elementary reaction the equilibrium constant is directly related to the rate constants for the opposing reactions. For example, when the rate constant for the forward reaction is greater than for the reverse reaction, $k > 1$. Thus we can also conclude that the free energy change for the reaction as written is negative.

To this discussion we must now add a disclaimer: Whereas the equilibrium constant for a reaction can always be written directly from the balanced chemical equation, we cannot always count on reaction-rate expressions being so simply related to the coefficients of the equations. As we saw in Chapter 12, some rate expressions are quite complex, thereby demonstrating that the reaction in fact comprises more than one elementary step.

We have discussed four important aspects of chemical reactions: *stoichiometry*, *thermodynamics*, *kinetics*, and *dynamic equilibrium*. We now know that these concepts are not entirely independent of one another; yet it is often necessary to consider each in turn when studying a reaction closely, for example, when selecting the most economical conditions for a commercial process. The chemistry of ammonia production is a good example, so we shall look at it in detail in Section 13.7.

13.7 The Ammonia Synthesis

More than three-fourths of the atmosphere is nitrogen. However, because of its strong triple bond, this essential nutrient is practically inert under ordinary conditions. As we have observed before, it must be *fixed* in reactive compounds— ammonium salts, nitrates, and nitrites—before it becomes useful for plants.

Example 13.9 (Stoichiometry) Calculate the weights of nitrogen and hydrogen that must react to produce 5 million tons of ammonia (total U.S. production for several months). What volume in cubic feet would this quantity of nitrogen occupy at STP?

Solution

a. The chemical equation is

$$N_2 + 3\,H_2 \rightleftharpoons 2\,NH_3$$

b. Regardless of the weight units we are using, the following conversion factors are valid:

$$\frac{2(17.0 \text{ weight units NH}_3)}{3(2.02 \text{ weight units H}_2)}$$

and

$$\frac{2(17.0 \text{ weight units NH}_3)}{1(28.0 \text{ weight units N}_2)}$$

c. 5×10^6 tons of ammonia will thus require

$$(5 \times 10^6 \text{ tons NH}_3)\left(\frac{1(28.0 \text{ tons N}_2)}{2(17.0 \text{ tons NH}_3)}\right) = 4 \times 10^6 \text{ tons N}_2$$

$$(5 \times 10^6 \text{ tons NH}_3)\left(\frac{3(2.02 \text{ tons H}_2)}{2(17.0 \text{ tons NH}_3)}\right) = 9 \times 10^5 \text{ tons H}_2$$

d. The volume of 4×10^6 tons N_2 at STP is

$$(4 \times 10^6 \text{ tons})\left(\frac{2 \times 10^3 \text{ lb}}{1 \text{ ton}}\right)\left(\frac{1 \text{ kg}}{2.2 \text{ lb}}\right)\left(\frac{10^3 \text{ g}}{1 \text{ kg}}\right)\left(\frac{1 \text{ mol}}{28 \text{ g N}_2}\right)\left(\frac{22.4 \text{ liters}}{1 \text{ mol}}\right)$$

$$= 3 \times 10^{12} \text{ liters}$$

e. Converted to cubic feet (STP), this is

$$(3 \times 10^{12} \text{ liters})\left(\frac{1 \text{ m}^3}{10^3 \text{ liters}}\right)\left(\frac{39.37 \text{ in.}}{1 \text{ m}}\right)^3\left(\frac{1 \text{ ft}}{12 \text{ in.}}\right)^3$$

$$= 100 \text{ billion ft}^3 \text{ N}_2 \quad \text{(at STP)}$$

Example 13.10 (Thermodynamics) The standard free energy of formation for NH_3 is negative (-3.98 kcal/mol). This means that the synthesis is spontaneous at 25°C and 1 atm. In Chapter 11, however, we found that the free energy of formation is a function of temperature. Use the enthalpies of formation and the absolute entropies from Tables 11.4 and 11.5, assume that they do not change much with temperature, and determine the approximate temperature at which the decomposition of NH_3 becomes spontaneous.

Solution

a. The change in entropy for the reaction is

$$\Delta S° = 2S° \text{ NH}_3(g) - [S° \text{ N}_2(g) + 3S° \text{ H}_2(g)]$$

$$\Delta S° = (2 \text{ mol})\left(\frac{46.01 \text{ cal}}{1.00 \text{ mol}}\right) - \left[(1 \text{ mol})\left(\frac{45.77 \text{ cal}}{1.00 \text{ mol}}\right) + (3 \text{ mol})\left(\frac{31.21 \text{ cal}}{1.00 \text{ mol}}\right)\right]$$

$$= -47.38 \text{ cal}$$

b. The change in enthalpy for the reaction is

$$\Delta H° = 2 \Delta H° \, NH_3(g) - [\Delta H° \, N_2(g) + 3 \Delta H° \, H_2(g)]$$

$$\Delta H° = (2 \text{ mol})\left(\frac{-11.04 \text{ kcal}}{1.00 \text{ mol}}\right)$$

$$- \left[(1 \text{ mol})\left(\frac{0 \text{ kcal}}{1.00 \text{ mol}}\right) + (3 \text{ mol})\left(\frac{0 \text{ kcal}}{1.00 \text{ mol}}\right)\right]$$

$$= -22.08 \text{ kcal}$$

c. Let x be the temperature at which ΔG is zero. This is the temperature at which the decomposition becomes spontaneous.

$$\Delta G = \Delta H - T \Delta S$$

$$0 = -22{,}080 \text{ cal} - x(-47.37 \text{ cal})$$

$$x = 466 \text{ K} \quad \text{or} \quad 193°C$$

We can therefore conclude that above 200°C the synthesis of ammonia is thermodynamically unfavorable.

Example 13.11 (Equilibrium) Use the data in Table 11.5 to calculate the equilibrium constant (K_p) at 25°C for the synthesis of ammonia.

Solution

a. We first calculate the free energy change for the reaction:

H_2: $\quad \Delta G_f° = 0$

N_2: $\quad \Delta G_f° = 0$

NH_3: $\quad \Delta G_f° = -3976$ cal/mol

$$\Delta G° = 2 \Delta G_f° \, (NH_3) - [\Delta G_f° \, (N_2) + 3 \Delta G_f° \, (H_2)]$$

$$\Delta G° = \left[(2 \text{ mol})\left(\frac{-3976 \text{ cal}}{1.00 \text{ mol}}\right)\right] - [(0) + 3(0)] = -7952 \text{ cal}$$

b. The relationship between K_p and $\Delta G°$ is

$$\Delta G° = -2.303 RT \log K_p$$

c. Substitute and solve for K_p:

$$\log K_p = \frac{-7952}{(-2.303)(1.99)(298)}$$

$$= 5.82$$

$$K_p = 6.6 \times 10^5 = \frac{(P_{NH_3})^2}{(P_{N_2})(P_{H_2})^3}$$

The large value for K_p tells us that at equilibrium the product (NH_3) is heavily favored at 25°C and 1 atm.

Example 13.12 (Equilibrium) How will the equilibrium concentrations be affected by changes in temperature and pressure?

Solution

a. In the reaction as written, four gas molecules react to form two gas molecules; thus an increase in pressure will cause the equilibrium to shift to the right. This means that at high pressures the proportion of ammonia in an equilibrium mixture will be greater than at lower pressures.

b. The synthesis is exothermic; as we have already seen in Example 13.10, an increase in temperature will thus favor the reactants (H_2 and N_2).

So far we have seen no reason why we could not produce large quantities of NH_3 simply by mixing huge amounts of H_2 and N_2 at high pressure and relatively low temperatures. Indeed, commercial processes do use pressures near 1000 atm to increase the yield of ammonia. But we have not yet said anything about kinetics. Unfortunately hydrogen and nitrogen react so slowly that despite our thermodynamic calculations the synthesis of ammonia is impractical at ordinary temperatures.

We must therefore compromise. If the temperature is raised to 500°C, the reaction rate increases, making the process practical, even though lower temperatures are much more favorable thermodynamically. It is after all far more profitable to obtain a 50 percent yield every hour than to obtain a 98 percent yield once a year! Finally, we should remember that there is another important way to influence reaction rates: catalysts. For the industrial production of ammonia, catalysts containing iron and iron oxides increase the reaction rate without affecting the equilibrium concentrations.

Questions and Problems

13.1 The Equilibrium Constant

7. What are the relative sizes of the equilibrium constants for the reaction $A + B \rightleftharpoons C + D$ if
   a. scarcely any product is formed?
   b. nearly all the reactant has been exhausted?
   c. product and reactant concentrations are nearly the same?

8. Which one of the following hypothetical gas phase reactions will be most nearly complete and which one will be least complete?
   a. $3R + J \rightleftharpoons 2L$ $K = 10^{14}$
   b. $Z + L \rightleftharpoons 2N$ $K = 10^{-7}$

c. $4 O + P \rightleftharpoons 3 T \qquad K = 2.3$
d. $5 X + Y \rightleftharpoons 4 W \qquad K = 10^{-18}$

9. A 1.00-liter reaction vessel in which the gas phase reaction $SO_2 + NO_2 \rightleftharpoons SO_3 + NO$ has reached equilibrium was found to contain the following:

   4.00 mol of SO_2
   1.00 mol of NO_2
   3.00 mol of SO_3
   2.00 mol of NO

Calculate K for this reaction.

10. A 2.00-liter reaction vessel in which the reaction $2 SO_2 + O_2 \rightleftharpoons 2 SO_3$ had reached equilibrium was found to contain the following:

    3.00 mol of SO_3
    102 g of O_2
    1.43×10^{24} molecules of SO_2

Calculate K for this reaction.

11. The following system is at equilibrium in a 500-ml flask:

$$N_2(g) + 3 H_2(g) \rightleftharpoons 2 NH_3(g)$$

Upon analysis, the flask is found to contain 0.300 mol of N_2, 0.600 mol of H_2, and 0.500 mol of NH_3. Calculate the equilibrium constant.

13.2 Le Chatelier's Principle and Chemical Equilibrium

12. What will be the effect upon the equilibrium

$$2 NH_3 + 7 O_2 \rightleftharpoons 6 H_2O + 4 NO_2 + \text{heat}$$

of the following? (All components of the system are gases.)
   a. Removal of O_2.
   b. Addition of NO_2.
   c. Addition of NH_3.
   d. Removal of H_2O.
   e. Increase in pressure.
   f. Decrease in temperature.

13. Indicate how the following equilibria would be affected by
    a. an increase in temperature.
    b. a decrease in pressure.
    c. an increase in the concentration of the underlined component.
   (All components are gases.)
   (i) $2 SO_2 + \underline{O_2} \rightleftharpoons 2 SO_3$ $\qquad \Delta H < 0$
   (ii) $N_2O_4 \rightleftharpoons 2 \underline{NO_2}$ $\qquad \Delta H > 0$
   (iii) $CO + 2 \underline{H_2} \rightleftharpoons CH_3OH$ $\qquad \Delta H < 0$
   (iv) $2 O_3 \rightleftharpoons 3 O_2$ $\qquad \Delta H < 0$
   (v) $2 NO \rightleftharpoons N_2 + O_2$ $\qquad \Delta H < 0$
   (vi) $COCl_2 \rightleftharpoons CO + \underline{Cl_2}$ $\qquad \Delta H < 0$
   (vii) $2 NO_2 + \underline{7 H_2} \rightleftharpoons 2 NH_3 + 4 H_2O$ $\qquad \Delta H < 0$
   (viii) $\underline{4 H_2} + CS_2 \rightleftharpoons CH_4 + 2 H_2S$ $\qquad \Delta H > 0$

14. List five ways one might shift the following gas phase equilibrium and predict the direction of the shift:

$$CO + H_2O \rightleftharpoons CO_2 + H_2 + \text{heat}$$

Did you include a change in pressure as one of these methods? Why or why not?

13.3 Quantitative Aspects of Chemical Equilibrium

15. 1.00 mol of H_2 and 1.00 mol of I_2 are allowed to reach equilibrium in a 1.00-liter reaction flask. At equilibrium the flask contains 0.600 mol of HI. What is the equilibrium constant for the reaction $H_2(g) + I_2(g) \rightleftharpoons 2\,HI(g)$?

16. a. Calculate the equilibrium constant for the reaction

$$2\,SO_2(g) + O_2(g) \rightleftharpoons 2\,SO_3(g)$$

if at equilibrium the concentrations of the components of the system are as follows:

$[SO_2] = 3.00$ mol/liter

$[O_2] = 2.00$ mol/liter

$[SO_3] = 4.00$ mol/liter

b. What will be the concentration of O_2 in a reaction vessel if the concentrations of SO_2 and SO_3 at equilibrium are 4.00 and 2.00 mol/liter, respectively? Assume the same temperature and pressure as in part a.

17. At a certain temperature the equilibrium constant for the gas phase reaction below was found to be 4.00.

$$2\,NO + Br_2 \rightleftharpoons 2\,NOBr$$

At equilibrium the concentrations of NOBr and Br_2 were found to be 3.00 and 1.00 mol/liter, respectively. What was the concentration of NO?

18. Consider the equation

$$2\,NO(g) \rightleftharpoons N_2(g) + O_2(g)$$

When 2.00 mol of NO are placed in a 1.00-liter reaction vessel and allowed to reach equilibrium at a certain temperature, 0.300 mol of O_2 is found in the reaction mixture. Calculate K at this temperature.

19. a. If the concentrations of A, B, C, and D are 1.00, 3.00, 2.00, and 6.00 mol/liter, respectively, at equilibrium, what is the equilibrium constant for the gas phase reaction $A + B \rightleftharpoons C + D$?

b. What is the concentration of A in moles per liter if the concentrations of B, C, and D are 5.00, 2.00, and 4.00 mol/liter, respectively, at equilibrium? Assume the same temperature and pressure as in part a.

c. If 1.00 mol of A and 4.00 mol of B are placed in a 1.00-liter reaction vessel and allowed to reach equilibrium, what will be the concentrations of A, B, C, and D? Assume the same temperature and pressure as in part a.

20. Equilibrium concentrations for the reaction

$$CO(g) + H_2O(g) \rightleftharpoons CO_2(g) + H_2(g)$$

were found to be

$[H_2] = 1.00$ mol/liter

$[CO_2] = 3.00$ mol/liter

$[CO] = 5.00$ mol/liter

$[H_2O] = 4.00$ mol/liter

If 2.00 mol/liter of H_2 is added, what will the concentrations be when equilibrium is reestablished?

21. Consider the following system:

$$SO_2(g) + NO_2(g) \rightleftharpoons SO_3(g) + NO(g)$$

At equilibrium the concentrations are

$[SO_2] = 3.00$ mol/liter

$[NO_2] = 2.00$ mol/liter

$[SO_3] = 4.00$ mol/liter

$[NO] = 3.00$ mol/liter

If 1.00 mol/liter of SO_3 is removed, what will be the concentration of each component when equilibrium is reestablished?

22. a. Calculate the equilibrium constant for the reversible reaction $CO(g) + Cl_2(g) \rightleftharpoons COCl_2(g)$ if at equilibrium the concentrations of CO, Cl_2, and $COCl_2$ are 3.00, 4.00, and 6.00 mol/liter, respectively.
b. How much $COCl_2$ must be removed to reduce the concentration of Cl_2 to 3.00 mol/liter?

23. The concentrations for the components of the system $CO(g) + Cl_2(g) \rightleftharpoons COCl_2(g)$ are as follows:

$[COCl_2] = 16.00$ mol/liter

$[CO] = 4.00$ mol/liter

$[Cl_2] = 2.00$ mol/liter

If 1.00 mol/liter of CO is removed from the reaction vessel, what will be the concentration of each component after equilibrium is reestablished?

24. At high temperatures H_2S dissociates into H_2 and S_2:

$$2\,H_2S(g) \rightleftharpoons 2\,H_2(g) + S_2(g)$$

At equilibrium the concentrations are

$$[H_2] = 3.00 \text{ mol/liter}$$
$$[S_2] = 1.00 \text{ mol/liter}$$
$$[H_2S] = 2.00 \text{ mol/liter}$$

How many moles per liter of S_2 must be added to increase $[H_2S]$ to 2.50 mol/liter?

25. The concentrations of the components of the system

$$SO_2(g) + NO_2(g) \rightleftharpoons SO_3(g) + NO(g)$$

at equilibrium are as follows:

$$[NO] = 2.00 \text{ mol/liter}$$
$$[SO_3] = 3.00 \text{ mol/liter}$$
$$[SO_2] = 0.500 \text{ mol/liter}$$
$$[NO_2] = 6.00 \text{ mol/liter}$$

How many moles per liter of SO_2 must be added to increase the concentration of NO to 4.00 mol/liter?

26. The concentrations at equilibrium for the components of the reaction $2\,SO_2(g) + O_2(g) \rightleftharpoons 2\,SO_3(g)$ are 4.00, 1.00, and 6.00 mol/liter, respectively. How many moles per liter of oxygen must be added to increase the concentration of SO_3 to 7.00 mol/liter?

27. At an elevated temperature, PCl_5 dissociates into PCl_3 and Cl_2. If 2.00 mol of PCl_5 are placed in a 1.00-liter vessel, what will be the concentration of Cl_2 at equilibrium? All components of the system are gases. $K = 0.0643$.

28. If 1.00 mol of HBr in a 1.00-liter reaction vessel is found to be 40.0 percent dissociated at a certain elevated temperature, what is the equilibrium constant for the reaction $2\,HBr \rightleftharpoons H_2 + Br_2$? All components of the system are gases.

29. Phosphorus pentachloride dissociates according to the following equation: $PCl_5(g) \rightleftharpoons PCl_3(g) + Cl_2(g)$. If 1.00 mol of PCl_5 in a 1.00-liter reaction vessel is 30.0 percent dissociated at equilibrium, what is the equilibrium constant for the reaction?

30. 1.00 mol of PCl_3 and 2.00 mol of PBr_3 are placed in a 1.00-liter reaction vessel and allowed to reach equilibrium: $PCl_3(g) + PBr_3(g) \rightleftharpoons PCl_2Br(g) + PClBr_2(g)$. If the value of K is 2.00, what is the concentration of each component of the system at equilibrium?

31. 2.00 mol of CO and 2.00 mol of H_2O are placed in a 1.00-liter container and allowed to reach equilibrium: $CO(g) + H_2O(g) \rightleftharpoons CO_2(g) + H_2(g)$. If the value for K is 4.00, what will be the concentration of each component at equilibrium?

32. The equilibrium constant for the hypothetical reaction $2\,A(g) \rightleftharpoons B(g) + C(g)$ is 3.00. What will be the concentration of each component of the system if 3.00 mol of A is placed in a 4.00-liter reaction vessel and allowed to reach equilibrium?

33. Under certain conditions, the equilibrium constant for the Haber process is 1.3. Determine whether the concentration of NH_3 will increase or decrease as the reaction mixtures described below approach equilibrium.

    a. $[N_2] = 0.560$ mol/liter
       $[H_2] = 0.760$ mol/liter
       $[NH_3] = 0.590$ mol/liter

    b. $[N_2] = 0.00249$ mol/liter
       $[H_2] = 0.645$ mol/liter
       $[NH_3] = 0.00123$ mol/liter

34. For the reaction of $W(g) + X(g) \rightleftharpoons Y(g) + Z(g)$, the equilibrium constant is 3.00. If 2.00 mol of W, 4.00 mol of X, 6.00 mol of Y, and 1.00 mol of Z are placed in a 1.00-liter reaction vessel, what will be the equilibrium concentration for each of the components of the system?

13.4 K_p, the Equilibrium Constant in Terms of Partial Pressures

35. At 0.0°C and 1.00 atm, an equilibrium mixture of the reaction

$$N_2O_4(g) \rightleftharpoons 2\, NO_2(g)$$

is found to have the following partial pressures:

$$P_{N_2O_4} = 0.80 \text{ atm}$$

$$P_{NO_2} = 0.20 \text{ atm}$$

    a. Calculate K_p for this equilibrium.
    b. Convert the partial pressures to moles per liter and calculate K.

13.5   Free Energy and Equilibrium

36. From free energy data (Table 11.5), calculate K_p at 25°C for each of the following gas phase equilibria:

    a. $PCl_5(g) \rightleftharpoons PCl_3(g) + Cl_2(g)$
    b. $2\, HF(g) \rightleftharpoons H_2(g) + F_2(g)$
    c. $2\, NH_3(g) \rightleftharpoons N_2(g) + 3\, H_2(g)$

14 Acids and Bases

The last few chapters have dealt with some of the more theoretical aspects of elementary chemistry: kinetics, thermodynamics, and the concept of dynamic equilibria. It is now time to turn to chemistry as we find it in the laboratory and in industry, to reactions between acids and bases, to electrochemical reactions, and to a description of some of the elements. All the while, of course, we shall be carrying with us the concepts we have learned in the first half of the book; they will be invaluable for explaining and tying together the diverse phenomena we are about to describe. Let us begin with acids and bases, where an understanding of the electronic structure of the atoms, chemical bonding, and equilibria will be essential.

So far we have discussed acids and bases in only a general way. We now want to discuss these classes of compounds in more detail, since concepts underlying the classification of acids and bases are crucial ones in many fields of chemistry. We shall find that there are several ways to define acids and bases but that the definitions are complementary and not contradictory. The first definitions we shall look at are expansions of the short definitions we gave in Chapter 6, but to make sense of these we must first discuss an important property of water that we have not yet considered.

14.1 Autoionization of Water

We have assumed that water is a covalent compound—most liquids are—thus we would not expect it to conduct an electrical current. Only if we dissolved an ionic solute in water would we expect the solution to be a conductor. However, if we use a sensitive galvanometer (an instrument for measuring the intensity of an electrical current), we can show that a current can be made to flow even in pure water. This demonstrates that some of the water molecules themselves must ionize according to the following equation:

$$H_2O \rightleftharpoons H^+ + OH^-$$

Since the H^+ ion is only a bare proton unscreened by electrons, it is a very unstable species: We would not expect to find lone protons in water or in aqueous solutions. Thus we often describe each proton in aqueous solutions as *hydrated*. We now rewrite the ionization of water as

$$H_2O + H_2O \rightleftharpoons H_3O^+ + OH^-$$

This equation represents the process as the transfer of a proton from one molecule of water to another, unquestionably a more likely picture of what actually happens. We have learned that each hydrogen atom in a water molecule can be tenuously attached by a hydrogen bond to the oxygen atom of another water molecule; it is not hard to imagine that occasionally a proton leaves one water molecule and goes to another, thus forming an H_3O^+ ion, which is called a *hydronium ion*. This is not a common event, however; at any instant only about 1 out of 550 million water molecules, or 1.8×10^{-7} percent, are ionized. This means that the concentrations of H_3O^+ and OH^- are only 10^{-7} mol/liter at 25°C in pure water.

Although representing the autoionization of water as producing H_3O^+ and OH^- ions is sufficient for most purposes, it does have its shortcomings. It seems more likely that most protons are linked to more than one water molecule. Thus we should represent the hydrated protons as $H_5O_2^+$, $H_7O_3^+$, $H_9O_4^+$, etc. Furthermore, each OH^- ion is likely to be hydrated by surrounding unionized water molecules. Because of these complications we shall often find the ionization of water represented merely as

$$H_2O(l) \rightleftharpoons H^+(aq) + OH^-(aq)$$

where (aq) signifies a hydrated ion.

There are advantages and disadvantages to each of these methods of representing the ionization of water. In this text we shall use all three methods, depending on which seems most appropriate. For example, we shall use the hydronium ion (H_3O^+) when we wish to show the role of water in removing a proton from a neutral molecule:

$$HX + H_2O \rightleftharpoons H_3O^+ + X^-$$

where X stands for a nonmetal atom or a group of nonmetal atoms.

Question

1. How many H^+ and OH^- ions are present at any instant in 1.00 liter of pure water at 25°C?

14.2 The Arrhenius Concept

When Arrhenius originally proposed his theory of acids and bases in 1884, the distinction between ionic and covalent compounds was not completely clear. Modern understanding of the nature of matter has thus required considerable modification of the original Arrhenius concept. The modified Arrhenius theory defines acids and bases in terms of the water equilibrium; hence this concept is applicable only to aqueous solutions. *Those substances that upset the water equilibrium by releasing hydrogen ions are known as* **acids,** *and those that upset the water equilibrium by releasing hydroxide ions are known as* **bases.**

If we add an acid to water, the excess H⁺ ions apply a stress to the system at equilibrium

$$H_2O(l) \rightleftharpoons H^+(aq) + OH^-(aq)$$

thus shifting it to the left. Some of the hydrogen ions combine with hydroxide ions to form water. When equilibrium is restored, the concentration of hydrogen ions is greater than it is in pure water, and the concentration of hydroxide ions is less than it is in pure water. Thus in an acid solution the hydrogen ion concentration is (1) greater than 10^{-7} mol/liter and (2) greater than the hydroxide ion concentration in the same solution. An acidic solution also has a hydroxide ion concentration (1) smaller than 10^{-7} mol/liter and (2) smaller than the hydrogen ion concentration in the same solution.

Similarly, when we add a base to water, the equilibrium shifts to the left, this time in response to the excess hydroxide ions. When equilibrium is restored, the hydroxide ion concentration is greater than it is in pure water. Therefore a basic solution is characterized by a hydroxide ion concentration (1) greater than 10^{-7} mol/liter and (2) greater than the hydrogen ion concentration in the same solution. Conversely, the hydrogen ion concentration in a basic solution is (1) smaller than 10^{-7} mol/liter and (2) smaller than the hydroxide ion concentration in the same solution.

If a solute does not affect the water equilibrium, the hydrogen ion and hydroxide ion concentrations remain equal. We thus refer to the resulting solution as *neutral*.

Questions

2. Classify the solutions described below as acidic, basic, or neutral.
   a. Hydrogen ion concentration is 10^{-8} mol/liter.
   b. Hydrogen ion concentration is 10^{-12} mol/liter.
   c. Hydrogen ion concentration is greater than the hydrogen ion concentration in pure water.
   d. Hydrogen ion concentration is equal to the hydroxide ion concentration.
   e. Hydrogen ion concentration is 10^{-2} mol/liter.
   f. Hydroxide ion concentration is 10^{-6} mol/liter.
   g. Hydrogen ion concentration is greater than the hydroxide ion concentration.
   h. Hydroxide ion concentration is less than the hydroxide ion concentration in pure water.
   i. Hydroxide ion concentration is less than the hydrogen ion concentration in pure water.
   j. Hydrogen ion concentration is 10^{-7} mol/liter.

14.3 Characterization and Classification of Arrhenius Bases

The hydroxides of the metals in Group IA are among the most common Arrhenius bases. These compounds are both ionic and very soluble, and since they dissociate completely in solution, we call them *strong bases*. Each of them produces a hydroxide ion concentration essentially equal to the concentration of the solute itself; that is, a 1.00 M solution of NaOH is effectively 1.00 M in hydroxide ions and 1.00 M in sodium ions. These basic solutions can be prepared by dissolving either the hydroxide or the oxide of the appropriate Group IA metal. If the hydroxide is used, the compound simply dissociates in solution. The oxide, however, must first react with water to produce hydroxide ions:

$$Na_2O(s) + H_2O(l) \longrightarrow 2\ Na^+(aq) + 2\ OH^-(aq)$$

The hydroxides of most of the metals of Group IIA also dissociate completely in water and thus can be regarded as strong bases. All of them, however, are only slightly soluble. The hydroxides of the transition elements by contrast do not dissociate completely and are *weak bases*. Most metallic oxides, all of which are often called *basic oxides*, are likewise only weak bases. Only the oxides of Group IA metals can be counted on to react with water to produce very concentrated basic solutions.

Another common weak base is ammonia. It contains no hydroxide ions itself; but in solution it places a stress upon the water ionization equilibrium by reacting with water to produce hydroxide ions:

$$NH_3(aq) + H_2O(l) \rightleftharpoons NH_4^+(aq) + OH^-(aq)$$

The solution therefore becomes basic. Ammonia is highly soluble in water, and it is only because the equilibrium above lies well to the left that we must consider it merely a weak base.

In this section we have described Arrhenius bases in terms of *solubility* and *degree of ionization*. It is important that you distinguish between these two terms. If a hydroxide is classified as insoluble or slightly soluble, it is not possible to have a solution of this compound with a high OH^- ion concentration, even if it is classified as a strong base. For example, a saturated solution of $Ca(OH)_2$, a strong base, has a low OH^- ion concentration because it is only slightly soluble. Aqueous solutions of some highly soluble bases such as NH_3 cannot have a high OH^- ion concentration because they ionize only slightly. Hydroxides of Group IA metals can form solutions with a very high OH^- ion concentration because they are both very soluble and dissociate completely in water. Finally, the hydroxides of the transition metals are both slightly soluble and weak bases.

Questions

3. Distinguish between weak bases and strong bases. Give an example of each.

4. What metal oxides would you dissolve in water to form a strongly basic solution?

5. What gas would you dissolve in water to form a weakly basic solution?

14.4 Characterizations and Classification of Arrhenius Acids

Several of the most common acids are binary (two-element) hydrogen compounds. Examples of these binary acids are

$$HCl(g) \longrightarrow H^+(aq) + Cl^-(aq)$$
$$HBr(g) \longrightarrow H^+(aq) + Br^-(aq)$$
$$HI(g) \longrightarrow H^+(aq) + I^-(aq)$$

All of these compounds are highly soluble in water and ionize completely; thus these hydrogen halides are *strong acids*. On the other hand, hydrogen fluoride is highly soluble in water, but it ionizes weakly:

$$HF(g) \rightleftharpoons H^+(aq) + F^-(aq)$$

Hydrofluoric acid is therefore only a *weak acid*.

Nitric acid is an example of a *ternary acid*—one that contains three elements. It ionizes completely in water and hence is a strong acid:

$$HNO_3(l) \longrightarrow H^+(aq) + NO_3^-(aq)$$

Nitric acid, like many other ternary acids, is formed by a reaction between a nonmetallic oxide and water. In this case the nonmetallic oxide is dinitrogen pentoxide (N_2O_5):

$$N_2O_5(s) + H_2O(l) \longrightarrow 2\,H^+(aq) + 2\,NO_3^-(aq)$$

Nitrous acid, HNO_2, is an example of a weak, ternary acid. Here again (as with HF) we have a highly soluble acid that is a weak acid because it does not ionize completely in solution.

Incidentally, nonmetallic oxides can be serious pollutants since they need only to react with water to produce corrosive acids. Example 13.3 presented one example. A second illustration begins with the metallic sulfides produced in some mining areas. These sulfides are converted to SO_2 and SO_3 by microorganisms, and these compounds in turn dissolve in water to form H_2SO_3 and H_2SO_4, respectively. Water draining from mining areas can therefore be highly acidic.

The nonmetal oxides, both soluble and insoluble, are called *acidic oxides*. Those oxides which can be formed by the removal of water from acid are also called *acid anhydrides*.

In addition to classifying acids as strong or weak, we can classify them by the number of ionizable hydrogen atoms in each molecule. Most of the examples discussed so far have been *monoprotic acids*; those with more than one ionizable hydrogen atom in each molecule are called *polyprotic acids*. Sulfuric acid is a common example of the latter. Since sulfuric acid has two hydrogen atoms which ionize, it is specifically referred to as a diprotic acid.

$$H_2SO_4(aq) \longrightarrow H^+(aq) + HSO_4^-(aq)$$
$$HSO_4^-(aq) \rightleftharpoons H^+(aq) + SO_4^{2-}(aq)$$

Sulfuric acid is unusual in that the first ionization is complete (in solutions with concentrations below 6 M), whereas the second hydrogen is only partially ionized. We can therefore think of sulfuric acid as a strong acid and HSO_4^- as a weak acid. Most other polyprotic acids are weak acids in all stages of ionization. Phosphoric acid is an example. Phosphoric acid is called a *triprotic acid* because all three hydrogen atoms are ionizable.

$$H_3PO_4(aq) \rightleftharpoons H_2PO_4^-(aq) + H^+(aq)$$
$$H_2PO_4^-(aq) \rightleftharpoons HPO_4^{2-}(aq) + H^+(aq)$$
$$HPO_4^{2-}(aq) \rightleftharpoons PO_4^{3-}(aq) + H^+(aq)$$

For all polyprotic acids we can count on the first ionization taking place to a greater extent than the second and the second to a much greater extent than the third.

Table 14.1 lists the most common Arrhenius acids, each classified as weak or strong and as monoprotic, diprotic, or triprotic.

Table 14.1 Common Acids

Strong	Weak Monoprotic
HCl	HF
HBr	HNO_2
HI	HClO
$HClO_4$	HCN
$HClO_3$	$HC_2H_3O_2$
HNO_3	
H_2SO_4 (first stage only)	
	Weak Diprotic
	H_2S
	H_2SO_3
	H_2CO_3
	Weak Tripotic
	H_3PO_4

Question

6. Write the formula of an acid illustrating each of the following:
   a. binary acid
   b. ternary acid
   c. polyprotic acid
   d. monoprotic acid
   e. weak acid
   f. strong acid

14.5 Some Typical Reactions of Arrhenius Acids and Bases

The most characteristic reaction of Arrhenius acids is the *neutralization* of bases. (Of course we could just as easily think of the base neutralizing the acid.) **Neutralization** may be defined as the reaction between hydrogen ions from the acid and hydroxide ions from the base to form water. The compound that remains in solution after neutralization thus consists of the positive ion from the base and the negative ion from the acid and is called a **salt**. The reaction between hydrochloric acid and sodium hydroxide is a typical example of a neutralization reaction between a strong acid and a strong base:

$$HCl + NaOH \longrightarrow NaCl + H_2O$$

This equation adequately represents the reaction, but it does have certain deficiencies. It does not indicate that the reaction involves ionic species: We can therefore represent the reaction better as

$$(H^+ + Cl^-) + (Na^+ + OH^-) \longrightarrow (Na^+ + Cl^-) + H_2O$$

This equation explicitly tells us that the compounds HCl (in solution), NaOH, and NaCl consist of ions that move more or less independently of one another. The parentheses, however, show which ions were originally bound together before they were dissolved. This format for writing the formulas of compounds that ionize or dissociate in solution is called the *open form*. If a compound is insoluble or if it consists mainly of un-ionized molecules, as water does, we still write it in the *closed form*.

The first of the two equations above, in which we made no distinction between ionized and un-ionized compounds, is called an *overall equation*. The second equation is an *ionic equation*.

We can write still a third type of equation (not only for neutralization reactions), in which we cancel identical species that appear on both sides of the ionic equation. The result is a *net ionic equation*. Thus for the reaction between NaOH and HCl, the net ionic equation is

$$H^+(aq) + OH^-(aq) \longrightarrow H_2O(l)$$

We can illustrate the reaction between a weak acid and a strong base with the following equations.

Overall: $HC_2H_3O_2 + NaOH \longrightarrow NaC_2H_3O_2 + H_2O$

Ionic: $HC_2H_3O_2 + (Na^+ + OH^-) \longrightarrow (Na^+ + C_2H_3O_2^-) + H_2O$

Net ionic: $HC_2H_3O_2(aq) + OH^-(aq) \longrightarrow C_2H_3O_2^-(aq) + H_2O(l)$

We have written acetic acid ($HC_2H_3O_2$) in the closed form, since it ionizes only slightly in solution.

For the neutralization of a weak base by a strong acid:

Overall: $HCl + NH_3 \longrightarrow NH_4Cl$

Ionic: $(H^+ + Cl^-) + NH_3 \longrightarrow (NH_4^+ + Cl^-)$

Net ionic: $H^+(aq) + NH_3(aq) \longrightarrow NH_4^+(aq)$

Finally, the reaction between a weak base and a weak acid:

Overall: $NH_3 + HC_2H_3O_2 \longrightarrow NH_4C_2H_3O_2$

Ionic and net ionic: $NH_3 + HC_2H_3O_2 \longrightarrow (NH_4^+ + C_2H_3O_2^-)$

In this case, since neither the weak acid nor the weak base produces many ions in solution, the ionic and net ionic equations are identical.

We have seen that only a few metallic oxides are soluble in water, thus producing basic solutions. Many, however, will dissolve in acidic solution, for example,

Overall: $MgO + 2\,HCl \longrightarrow MgCl_2 + H_2O$

Ionic: $MgO + 2\,(H^+ + Cl^-) \longrightarrow (Mg^{2+} + 2\,Cl^-) + H_2O$

Net ionic: $MgO(s) + 2\,H^+(aq) \longrightarrow Mg^{2+}(aq) + H_2O(l)$

Also typical of many Arrhenius acids is the reaction with some metals to produce salts and hydrogen. Hydrochloric acid reacts with zinc crystals to form zinc chloride and hydrogen:

Overall: $Zn + 2\,HCl \longrightarrow ZnCl_2 + H_2$

Ionic: $Zn + 2\,(H^+ + Cl^-) \longrightarrow (Zn^{2+} + 2\,Cl^-) + H_2$

Net ionic: $Zn(s) + 2\,H^+(aq) \longrightarrow Zn^{2+}(aq) + H_2(g)$

Note that the net ionic equation, in which the negative ion does not appear, will be the same whenever a strong acid reacts with zinc to produce hydrogen. If the reacting acid is a weak one, the equations will look like the following example.

Overall: $Mg + 2\,HC_2H_3O_2 \longrightarrow Mg(C_2H_3O_2)_2 + H_2$

Ionic and net ionic: $Mg + 2\,HC_2H_3O_2 \longrightarrow (Mg^{2+} + 2\,C_2H_3O_2^-) + H_2$

Again the ionic and net ionic equations are identical. (We shall look more closely at the reactions between metals and acids in Chapter 17.)

Acids also react with salts. Most such reactions fall into one of two classes: reactions of strong acids with salts of weak acids and reactions of nonvolatile acids with salts of volatile acids. An example of the first type is the reaction of HNO_3 with insoluble Ag_3PO_4:

Overall: $\quad Ag_3PO_4 + 3\,HNO_3 \longrightarrow 3\,AgNO_3 + H_3PO_4$

Ionic: $\quad Ag_3PO_4 + 3\,(H^+ + NO_3^-) \longrightarrow 3\,(Ag^+ + NO_3^-) + H_3PO_4$

Net ionic: $\quad Ag_3PO_4(s) + 3\,H^+(aq) \longrightarrow 3\,Ag^+(aq) + H_3PO_4(aq)$

Since Ag_3PO_4 is insoluble in water, we write it throughout in the closed form. Like many insoluble salts of weak acids, it dissolves in strong acids only as it reacts. We can explain this in terms of Le Chatelier's principle. In the presence of a strong acid, a stress is applied to the system at equilibrium

$$Ag_3PO_4(s) \rightleftharpoons 3\,Ag^+(aq) + PO_4^{3-}(aq)$$

because the phosphate ion is the anion of a weak acid and thus has a great affinity for protons:

$$PO_4^{3-}(aq) + H^+(aq) \rightleftharpoons HPO_4^{2-}(aq)$$

Phosphate ions are thus removed from the system, and the first equilibrium shifts to the right.

In some reactions between strong acids and the salts of weak acids, the weak acid that is produced decomposes into water and the acid anhydride, for example,

Overall: $\quad 2\,HCl + CaCO_3 \longrightarrow CaCl_2 + CO_2 + H_2O$

Ionic: $\quad 2\,(H^+ + Cl^-) + CaCO_3 \longrightarrow (Ca^{2+} + 2\,Cl^-) + CO_2 + H_2O$

Net ionic: $\quad 2\,H^+(aq) + CaCO_3(s) \longrightarrow Ca^{2+}(aq) + CO_2(g) + H_2O(l)$

It is important to observe that insoluble salts of strong acids, such as AgCl, are insoluble even in strong acids because the anions of strong acids have no appreciable affinity for protons. Consequently, there is no reaction available to shift the solution equilibrium of the insoluble salt.

To understand the reactions between nonvolatile acids and the salts of volatile acids, we must first describe what we mean by *volatile* and *nonvolatile*, at least in this context. Table 14.2 lists several common concentrated acids. The first four acids in this list can be thought of as volatile in the sense that they are unstable to heat and evolve gases at relatively low temperatures. Phosphoric acid and sulfuric acid are also unstable to heat and evolve gases, but these acids require much higher temperatures; hence they are spoken of as nonvolatile acids.

The following overall equations illustrate how a volatile acid can be prepared from a nonvolatile acid and the appropriate salt.

$$H_2SO_4(l) + NaCl(s) \longrightarrow NaHSO_4(s) + HCl(g)$$

$$H_2SO_4(l) + NaNO_3(s) \longrightarrow NaHSO_4(s) + HNO_3(g)$$

Table 14.2 Composition of Common Concentrated Acids

Acid	Molarity	Specific Gravity	Weight Percent
Hydroiodic	5.5	1.5	47
Hydrobromic	9.0	1.5	49
Hydrochloric	12.0	1.18	37
Nitric	15.8	1.42	70
Phosphoric	14.7	1.7	85
Sulfuric	18.0	1.84	96

Since concentrated sulfuric acid contains little water, only a little of the water-soluble gas (HCl or HNO_3) dissolves—the rest escapes as a gas. To prepare HBr and HI, phosphoric acid must be used, since sulfuric acid reacts with Br^- and I^- ions to produce free bromine and iodine.

Questions

7. What type of solvent might dissolve a water-insoluble metallic oxide?

8. Write net ionic equations for the reactions in aqueous solution between the following compounds:
   a. HI and KOH   b. NaOH and $HC_2H_3O_2$ c. HNO_3 and NH_3
   d. HF and NaOH   e. $HClO_4$ and Mg   f. HNO_2 and Mg
   g. HCl and MnS (insoluble in water)
   h. HNO_3 and Ag_3AsO_4 (insoluble in water)
   i. HCl and $BaCO_3$ (insoluble in water)

14.6 Hydrolysis

In a broad sense, the term **hydrolysis** refers to any reaction where water is a reactant. Usually, however, we apply this term only to those reactions in which a positive or negative ion reacts with water to change the hydrogen and hydroxide ion concentrations naturally present in the water. For example, when we add acetate ions (or other anions that form weak acids) to water, the solution becomes more basic:

$$C_2H_3O_2^-(aq) + H_2O(l) \rightleftharpoons HC_2H_3O_2(aq) + OH^-(aq)$$

For the purpose of later discussions, it is important to emphasize that this equilibrium lies well to the left, but enough hydroxide is generated to produce a distinctly basic solution. Likewise, we can produce an acidic solution by adding to water a cation that produces a weak base in solution:

$$NH_4^+(aq) + H_2O(l) \rightleftharpoons H_3O^+ + NH_3(aq)$$

The difficulty with this discussion, of course, is that it is impossible to add only anions or only cations to a solution. The acetate ion in the first equation must be paired with a cation in some salt, and the ammonium ion must be accompanied by an anion. If, however, we choose, say, $NaC_2H_3O_2$ and NH_4Cl, the Na^+ and Cl^- ions will not affect the equations we have written. They are ions of a strong base and a strong acid, respectively, and they will remain in solution, having no effect on the water equilibrium.

Other examples abound of the hydrolysis of the salts of weak acids and weak bases. Recall, for example, that the hydroxides of the metals outside of Groups IA and IIA are weak (and often insoluble) bases. If $CuCl_2$ is dissolved in water, the solution becomes acidic because of the following reaction:

$$Cu^{2+}(aq) + H_2O(l) \rightleftharpoons Cu(OH)^+(aq) + H^+(aq)$$

Since no copper(II) hydroxide is precipitated, the hydrolysis is shown as a one-step reaction.

We've seen in the examples above that the anions of strong acids and cations of strong bases do not affect the water equilibrium. Thus we draw the following conclusions: The solution of any salt formed by reacting a strong base and a weak acid is basic, the solution of any salt formed by reacting a weak base and a strong acid is acidic, and a solution of any salt from the reaction of a strong base and a strong acid is neutral. A solution of a salt of a weak base and a weak acid may be acidic, basic, or neutral, depending on the relative attraction the positive ions have for the hydroxide ion and the negative ions for the hydrogen ions.

Since the hydrogen ions of polyprotic acids are ionized one at a time, we might expect that the hydrolysis of anions from such acids would likewise occur in steps. Thus we have

$$S^{2-}(aq) + H_2O(l) \rightleftharpoons HS^-(aq) + OH^-(aq)$$
$$HS^-(aq) + H_2O(l) \rightleftharpoons H_2S(aq) + OH^-(aq)$$

and

$$PO_4^{3-}(aq) + H_2O(l) \rightleftharpoons HPO_4^{2-}(aq) + OH^-(aq)$$
$$HPO_4^{2-}(aq) + H_2O(l) \rightleftharpoons H_2PO_4^-(aq) + OH^-(aq)$$
$$H_2PO_4^-(aq) + H_2O(l) \rightleftharpoons H_3PO_4(aq) + OH^-(aq)$$

We shall often encounter such salts as $NaHSO_4$, $NaHCO_3$, NaH_2PO_4, Na_2HPO_4, $NaHS$, and $NaHSO_3$, which are formed by the incomplete neutralizations of acids. Such salts are called *acid salts* because they contain ionizable hydrogen atoms. However, acid salts do not always produce solutions with a hydrogen ion concentration greater than 10^{-7} mol/liter. If the salt is more easily hydrolyzed than ionized, it will form a basic solution. For example, $NaHCO_3$, Na_2HPO_4, and $NaHS$ form basic solutions; $NaHSO_4$, NaH_2PO_4, and $NaHSO_3$ form acidic solutions.

Questions

9. Write equations showing the hydrolysis of the following ions:
   a. NO_2^- b. NH_4^+ c. HCO_3^- d. Fe^{3+} e. CN^-

10. If an aqueous solution of the salt Na_2Te is basic, what can be deduced about the acid H_2Te?

11. If the salt $NaHC_4H_4O_6$ gives an acidic aqueous solution, what can you say about the processes of ionization and hydrolysis of the $HC_4H_4O_6^-$ ion?

14.7 The Brønsted-Lowry Acid-Base Concept

In 1923 Johannes Brønsted (1879–1947), a Danish chemist, and T. M. Lowry (1874–1936), an English chemist, independently proposed a definition of acids and bases that expanded considerably on that of Arrhenius. According to this new idea, *an acid is any molecule or ion that gives up a proton, and a base is any ion or molecule that accepts a proton*. Thus even the autoionization of water is an acid-base reaction:

$$H_2O + H_2O \rightleftharpoons H_3O^+ + OH^-$$
acid base acid base

In this equation, one water molecule acts as an acid; the other, as a base. Like a number of other molecules and ions, water has an ionizable hydrogen atom, which enables it to function as an acid, as well as an unshared pair of electrons, which permits it to accept a proton and thereby to function as a base.

According to the Brønsted-Lowry concept, every reaction between an acid and a base produces another acid-base pair, which is conjugate to the original pair. In the example above, the H_3O^+ ion is the conjugate acid (proton donor) of the "basic" H_2O molecule, and the OH^- ion is the conjugate base (proton acceptor) of the "acidic" H_2O molecule.

The Brønsted-Lowry definition allows us to think of every reaction in which a proton is transferred as an acid-base reaction. For example, when an Arrhenius acid dissolves in water, the water molecule serves as a base:

$$HCl + H_2O \longrightarrow H_3O^+ + Cl^-$$
acid base acid base

$$HC_2H_3O_2 + H_2O \rightleftharpoons H_3O^+ + C_2H_3O_2^-$$
acid base acid base

Or when the Arrhenius base NH_3 is dissolved in water, the water behaves as an acid:

$$H_2O + NH_3 \rightleftharpoons NH_4^+ + OH^-$$
acid base acid base

We can also represent the hydrolysis of metallic ions as acid-base reactions. For example, if we dissolve $FeCl_3$ in water, the Fe^{3+} ion becomes hydrated with six H_2O molecules. This hydrated ion in turn reacts as a Brønsted-Lowry acid:

$$\underset{\text{acid}}{Fe(H_2O)_6{}^{3+}} + \underset{\text{base}}{H_2O} \rightleftharpoons \underset{\text{acid}}{H_3O^+} + \underset{\text{base}}{Fe(OH)(H_2O)_5{}^{2+}}$$

We can explain the acidity of copper(II) chloride solution the same way (note that in contrast to the equation on p. 436, we have shown the hydrating water molecules here—a more accurate representation):

$$\underset{\text{acid}}{Cu(H_2O)_4{}^{2+}} + \underset{\text{base}}{H_2O} \rightleftharpoons \underset{\text{acid}}{H_3O^+} + \underset{\text{base}}{Cu(OH)(H_2O)_3{}^+}$$

These examples are illustrations of the behaviors of many metallic ions in aqueous solutions. Most +2 ions, except those in Group IIA, and nearly all +3 metal ions are hydrated in solution. More will be said about this phenomenon in Chapter 24.

Whereas the Arrhenius definition of acids and bases limited our attention to aqueous systems, the Brønsted-Lowry concept demands only that a proton be transferred. When ammonia and HCl *gases* are mixed, a dense, white cloud of the ionic solid NH_4Cl is formed. In this reaction, we think of the $NH_4{}^+$ ion as a conjugate acid and the Cl^- ion as a conjugate base:

$$\underset{\text{acid}}{HCl(g)} + \underset{\text{base}}{NH_3(g)} \rightleftharpoons \underset{\text{acid} \quad \text{base}}{NH_4Cl(s)}$$

Question

12. Label the two Brønsted-Lowry acids and the two Brønsted-Lowry bases in each of the following reactions:
   a. $HSO_4{}^- + H_2O \rightleftharpoons H_3O^+ + SO_4{}^{2-}$
   b. $HCO_3{}^- + H_2O \rightleftharpoons H_2CO_3 + OH^-$
   c. $HPO_4{}^{2-} + H_2O \rightleftharpoons H_2PO_4{}^- + OH^-$
   d. $CH_3NH_3{}^+ + H_2O \rightleftharpoons CH_3NH_2 + H_3O^+$

14.8 Relative Strengths of Brønsted-Lowry Acids and Bases

In any Brønsted-Lowry acid-base reaction, there is a competition between two acids to donate protons and a competition between two bases to accept protons. We can therefore determine the relative strengths of the two acids and of the two bases by looking at where the equilibrium lies. In an aqueous solution of hydrogen chloride, the reaction to the right:

$$\mathrm{HCl} + \mathrm{H_2O} \longrightarrow \mathrm{H_3O^+} + \mathrm{Cl^-}$$
$$\text{acid} \quad \text{base} \quad\quad \text{acid} \quad\; \text{base}$$

goes essentially to completion. Since water molecules have a much greater affinity for protons than do chloride ions, we can conclude that the water molecule is a stronger base than the chloride anion. Similarly the HCl molecule is a stronger acid than the hydronium ion, since it donates a proton more readily. By contrast, acetic acid ionizes only slightly in aqueous solution:

$$\mathrm{HC_2H_3O_2} + \mathrm{H_2O} \rightleftharpoons \mathrm{H_3O^+} + \mathrm{C_2H_3O_2^-}$$

Since this equilibrium lies well to the left, the hydronium ion must be a stronger acid than the acetic acid molecule; the acetate ion ($\mathrm{C_2H_3O_2^-}$), a stronger base than the water molecule. The importance of these two examples is that they illustrate how further comparisons would allow us to list acids and bases according to their relative strengths. Such a list appears in Table 14.3.

One important note should be added to this list of acids and bases. All the acids stronger than hydronium ion are completely ionized in water; thus we cannot compare their strengths in aqueous solution. (Because water eliminates the differences among these acids, it is said to have a *leveling effect* on them.) To compare these acids, therefore, we must use a solvent that is a weaker proton acceptor (that is, a weaker base) than the water molecule. One such solvent, in which these strong acids do not completely ionize, is methanol ($\mathrm{CH_3OH}$).

Table 14.3 bears out one of our earlier suppositions and allows us to draw some further conclusions. First, note that as we said earlier the first proton of a polyprotic acid ionizes more readily than the second; the second, more readily than the third. Thus $\mathrm{H_2SO_4}$ is a stronger acid than $\mathrm{HSO_4^-}$, and $\mathrm{H_3PO_4}$ is a stronger acid than $\mathrm{H_2PO_4^-}$, which in turn is a stronger acid than $\mathrm{HPO_4^{2-}}$.

Among the binary acids, we can observe several trends if we consider the position in the periodic table of the second element in the acid. First, the strengths of the acids increase from left to right within a period. For example, in the second period the strengths of the acids increase in the following order:

$$\mathrm{CH_4} < \mathrm{NH_3} < \mathrm{H_2O} < \mathrm{HF}$$

The strengths of the acids also increase from top to bottom within a periodic group. In Group VIIA the order is

$$\mathrm{HF} < \mathrm{HCl} < \mathrm{HBr} < \mathrm{HI}$$

And in Group VIA the order is

$$\mathrm{H_2O} < \mathrm{H_2S} < \mathrm{H_2Se} < \mathrm{H_2Te}$$

Table 14.3 Relative Strengths of Brønsted-Lowry Acids and Bases

	Acid	Base	
strong acids ↑ increasing acid strength	$HClO_4$	ClO_4^-	**weak bases**
	HI	I^-	
	HBr	Br^-	
	HCl	Cl^-	
	HNO_3	NO_3^-	
	H_2SO_4	HSO_4^-	
	$HClO_3$	ClO_3^-	
	H_3O^+	H_2O	
	H_2SO_3	HSO_3^-	
	HSO_4^-	SO_4^{2-}	
	$HClO_2$	ClO_2^-	
	H_3PO_4	$H_2PO_4^-$	
	$Fe(H_2O)_6^{3+}$	$Fe(OH)(H_2O)_5^{2+}$	
	H_3AsO_4	$H_2AsO_4^-$	
	H_2Te	HTe^-	
	HF	F^-	increasing base strength
	H_2Se	HSe^-	
	$HC_2H_3O_2$	$C_2H_3O_2^-$	
	$Al(H_2O)_6^{3+}$	$Al(OH)(H_2O)_5^{2+}$	
	H_2CO_3	HCO_3^-	
	H_2S	HS^-	
	$H_2PO_4^-$	HPO_4^{2-}	
	HSO_3^-	SO_3^{2-}	
weak acids	$HClO$	ClO^-	**strong bases**
	$HBrO$	BrO^-	
	NH_4^+	NH_3	
	HCO_3^-	CO_3^{2-}	
	HIO	IO^-	
	HS^-	S^{2-}	
	HPO_4^{2-}	PO_4^{3-}	
	H_2O	OH^-	
	NH_3	NH_2^-	
	CH_4	CH_3^-	↓

These trends would have been difficult to predict in advance, but having observed them we can now explain them in terms of two contradictory factors: electronegativity and atomic radius. As the electronegativity of the atom bonded to hydrogen increases, the shared pair of electrons will be displaced more strongly toward it and away from the hydrogen atom. Thus it becomes easier to ionize a proton. Within each period electronegativity is the dominant factor in determining the acidity of binary compounds. Within each group, however, atomic radius is a more important influence than electronegativity. The larger the atom bound to hydrogen, the more diffuse the valence electrons and hence the weaker the bond. Hydrogen iodide thus loses a proton more readily than hydrogen fluoride, despite the fact that fluorine is the more electronegative element.

We can also see a trend among the ternary acids having the general formula

where X is a nonmetal atom or a group of nonmetal atoms. The greater the electronegativity of X, the more the electrons will be displaced toward it and away from the proton. Thus we can expect the acidity of ternary acids to increase as the electronegativity of X increases. For example, the acidity of the hypohalous acids increases in the following order:

$$HOI < HOBr < HOCl$$

For acids that differ only in the number of oxygen atoms per molecule, we can simply extend our observations about ternary acids. The greater the number of oxygen atoms, the more strongly electrons are drawn from the hydrogen-oxygen bond and thus the stronger the acid. For example,

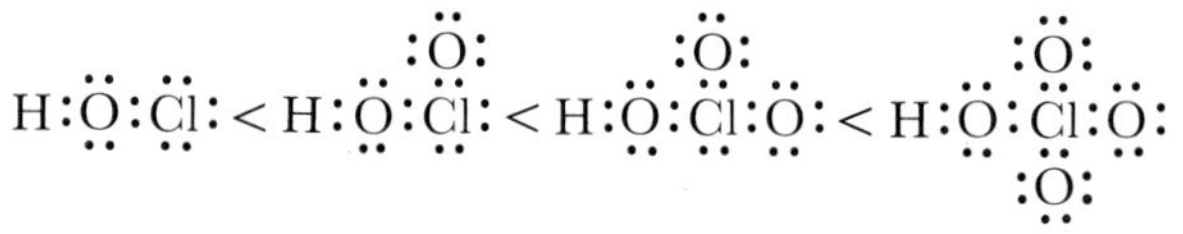

Questions

13. Select the stronger acid from each of the following pairs. Give your reasons.

a. H_2SO_4 and H_2SO_3 b. H_2O and HF
c. PH_3 and AsH_3 d. H_3AsO_4 and H_3PO_4
e. H_2Te and H_2Se f. HCl and H_2S
g. H_2SO_3 and HSO_3^- h. $H_2PO_4^-$ and HPO_4^{2-}
i. H_3BO_3 and H_2CO_3 j. $HClO_3$ and $HBrO_3$

14. Both $HClO_4$ and HNO_3 are completely ionized in water, but we say that $HClO_4$ is a stronger acid than HNO_3. Explain.

14.9 The Lewis Acid-Base Concept

In 1923, Gilbert N. Lewis (1875–1946) proposed a definition of acids and bases based upon the electron pair. According to this definition, *an acid is an electron-pair acceptor and a base is an electron-pair donor.* Any reaction in which a bond is formed using a pair of electrons from one of the atoms becomes a neutralization reaction. The Lewis definition expanded the class of acids in the same way that the Brønsted-Lowry definition enlarged our notion of bases. The Arrhenius definition of bases centered around the OH^- ion. Almost any molecule or ion with an unshared pair of electrons will accept a proton in the presence of a strong acid and hence can be thought of as a Brønsted-Lowry base. However, the definition of an acid remained pretty much tied to the H^+ ion under the Brønsted-Lowry definition. With the Lewis definition, acids are no longer tied to the H^+ but include any electron-pair acceptor.

Though the definitions differ, there is in fact little difference between a Lewis base and a Brønsted-Lowry base. Furthermore, since protons are electron acceptors, all Brønsted-Lowry acids (proton donors) ultimately serve as Lewis acids:

$$H^+ + :\overset{\overset{H}{..}}{\underset{\underset{H}{..}}{N}}:H \longrightarrow H:\overset{\overset{H}{..}}{\underset{\underset{H}{..}}{N}}:H \;\; {}^+$$

$$H^+ + :\overset{..}{\underset{..}{O}}:H^- \longrightarrow H:\overset{..}{\underset{..}{O}}:H$$

The importance of the Lewis concept is that several other types of ions and molecules can now be thought of as acids. Examples include compounds like BF_3 that contain an incomplete octet of electrons; BF_3 reacts with NH_3 as follows:

$$:\overset{\overset{:\ddot{F}:}{}}{\underset{\underset{:\ddot{F}:}{}}{\ddot{F}:B}} + :\overset{\overset{H}{..}}{\underset{\underset{H}{..}}{N}}:H \longrightarrow :\overset{\overset{:\ddot{F}:}{}}{\underset{\underset{:\ddot{F}:}{}}{\ddot{F}:B}}{-}\overset{\overset{H}{..}}{\underset{\underset{H}{..}}{N}}:H$$

where the line represents a coordinate covalent bond between nitrogen and boron. Metal ions serve as additional examples of Lewis acids:

$$Cu^{2+} + 4\, H:\overset{\overset{H}{..}}{\underset{\underset{H}{..}}{N}}: \longrightarrow \left[\begin{array}{c} H:\overset{\overset{H}{..}}{\underset{\underset{..}{..}}{N}}:H \\ H:\overset{\overset{H}{..}}{N}{-}Cu{-}\overset{\overset{H}{..}}{N}:H \\ H:\overset{\overset{..}{..}}{\underset{\underset{H}{..}}{N}}:H \end{array} \right]^{2+}$$

or

$$Cu^{2+} + 4\, H:\overset{\overset{..}{..}}{\underset{\underset{H}{..}}{O}}: \longrightarrow \left[\begin{array}{c} :\overset{..}{\underset{H}{O}}:H \\ H:\overset{..}{O}{-}Cu{-}\overset{..}{O}:H \\ H:\overset{..}{\underset{\underset{H}{..}}{O}}: \end{array} \right]^{2+}$$

We shall say more about these reactions in Chapter 24.

Molecules and ions containing central atoms that expand beyond the usual eight outer energy level electrons also function as Lewis acids:

$$SnCl_4 + 2\,Cl^- \longrightarrow SnCl_6^{2-}$$

$$I_2 + I^- \longrightarrow I_3^-$$

$$PCl_5 + Cl^- \longrightarrow PCl_6^-$$

Sometimes these compounds in which the central atom has d orbitals available for bonding are involved in more complicated reactions. An example is the hydrolysis of some metal halides. Tin(IV) chloride reacts vigorously with water to produce $Sn(OH)_4$. The mechanism is thought to involve four consecutive Lewis acid-base reactions, beginning with

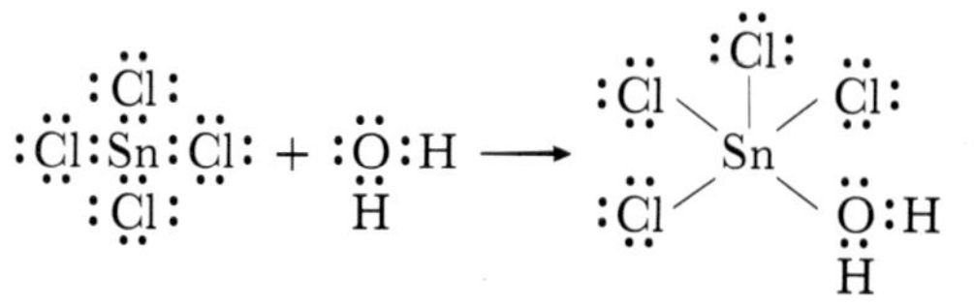

This unstable intermediate subsequently loses a H^+ ion and a Cl^- ion, forming

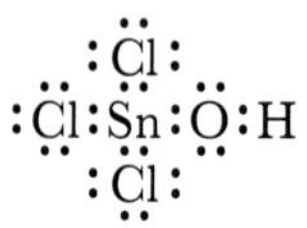

This process is repeated three times, resulting in $Sn(OH)_4$.

A final class of Lewis acids are compounds containing double bonds. Many of these compounds can accept a pair of electrons from a Lewis base, simultaneously shifting a second pair within the molecule. An illustration will make this process clearer:

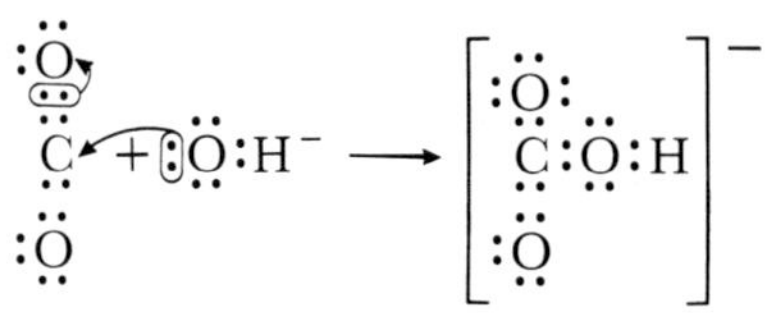

The hydroxide ion attacks the less electronegative of the double-bonded elements, causing an electron pair to shift to the more electronegative oxygen atom. Incidentally, this reaction is used in space to remove carbon dioxide from the gases exhaled by astronauts. Lithium hydroxide contains the Lewis base. A similar Lewis acid-base reaction is the formation of H_2SO_4 from SO_3 and H_2O. Here a proton as well as a pair of electrons must shift during the reaction:

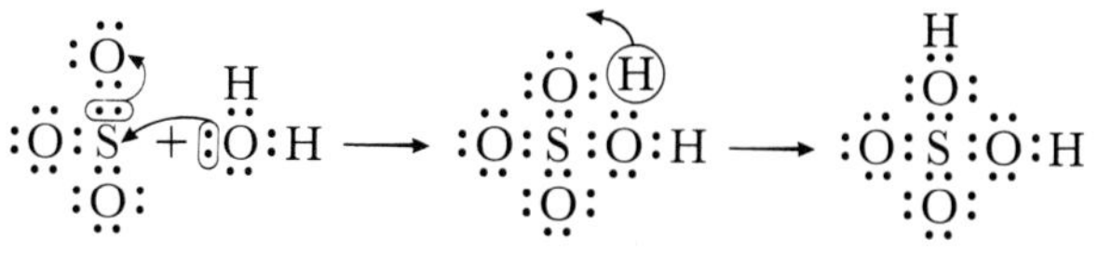

We shall find other examples of Lewis acid-base reactions in Chapter 24.

In conclusion we should emphasize that by no means does the Lewis concept replace that of either Arrhenius or Brønsted and Lowry. Each theory has its merits. Generally, in fact, when we speak simply of an *acid*, we mean an Arrhenius or Brønsted-Lowry acid. An electron-pair acceptor like BF_3 is specified as a *Lewis acid*.

Questions

15. Distinguish between the following terms:
    a. acid and base (Arrhenius)
    b. acid and base (Brønsted-Lowry)
    c. acid and base (Lewis)

16. Write the formulas for three compounds that are acids according to the Lewis definition but not the Brønsted-Lowry definition.

17. Write the formulas for three compounds or ions that are bases according to Brønsted and Lowry but not Arrhenius.

18. Write the formulas for three substances that can be either acids or bases according to the Brønsted-Lowry definition.

19. Is water an acid or base as defined by Lewis? Explain.

14.10 Acid-Base Trends in the Periodic Table

In previous chapters we have identified several trends in the periodic table. For example, we found the most metallic elements, those having the lowest electronegativities and ionization energies, at the lower left of the table. Conversely, the least metallic elements are at the upper right. In this chapter we have hinted at a relationship between the metallic character of an element and the acidity of its compounds: We have seen that metallic oxides tend to be basic, whereas nonmetallic oxides are acidic. We might, therefore, expect to find trends in the periodic table that are relevant to our study of acids and bases. We shall investigate these trends by asking whether the oxides of the elements we are discussing are basic or acidic oxides.

We might begin by looking at trends within each period. Unfortunately, the elements of the second period do not suit our purposes ideally since their properties are somewhat irregular; properties in the rest of the periods show more regular trends. The oxides of the third period elements, for instance, demonstrate a basic-to-acidic progression as we go from left to right in the period (Table 14.4). Sodium oxide dissolves readily in water to produce a strongly basic solution of sodium hydroxide. Magnesium forms an oxide that is much less soluble than sodium oxide, but it is basic enough to dissolve in any acid solution. At the other end of the period, chlorine, sulfur, phosphorus, and silicon are all nonmetals. The first three form oxides (Cl_2O_7, SO_3, and P_4O_{10}) that are acid anhydrides of decreasing strength—$HClO_4$ is one of the strongest

Table 14.4 Acid-Base Properties of the Oxides of Third Period Elements

Formula of Oxide	Classification	Solubility	Formula of Acid or Base
Na_2O	Basic	Soluble in water	NaOH
MgO	Basic	Insoluble in water, soluble in acid	$Mg(OH)_2$
Al_2O_3	Amphoteric	Insoluble in water, soluble in strong acids and strong bases	$Al(OH)_3$
SiO_2	Acidic	Insoluble in water, soluble in strong base	H_4SiO_4
P_4O_{10}	Acidic	Soluble in water	H_3PO_4
SO_3	Acidic	Soluble in water	H_2SO_4
Cl_2O_7	Acidic	Soluble in water	$HClO_4$

of all acids, H_2SO_4 is somewhat less strong, and H_3PO_4 is a weak acid. The oxide of the fourth (SiO_2) is very insoluble in water but is acid enough to dissolve in a strong base.

Of the third period elements, only aluminum remains. What is the nature of its oxide, standing as it does between the acidic and basic oxides? Aluminum oxide is insoluble in water but will dissolve in either strong acid or strong base. Since aluminum itself is a metalloid (p. 133), this intermediate behavior of its oxide should not surprise us. Furthermore if we carefully add a base to an acidic solution of aluminum ion,

$$Al^{3+}(aq) + 3\ OH^-(aq) \longrightarrow Al(OH)_3(s)$$

we precipitate a white, gelatinous solid that can be thought of as either an acid or a base. Aluminum hydroxide [$Al(OH)_3$] is an example of an *amphoteric hydroxide*—one that will dissolve in (react with) either acids or bases:

$$Al(OH)_3(s) + 3\ H^+(aq) \longrightarrow Al^{3+}(aq) + 3\ H_2O(l)$$
$$Al(OH)_3(s) + OH^-(aq) \longrightarrow Al(OH)_4^-(aq)$$

Compounds of fourth and fifth period elements show trends very similar to those of the third period. In the fifth period, we find two more elements with amphoteric hydroxides—tin and antimony.

There are also acid-base trends in the periodic groups. The most conspicuous example is Group VA (Table 14.5). The oxides of the nonmetals nitrogen, phosphorus, and arsenic, N_2O_5, P_4O_{10}, As_2O_5 (empirical formula, structure unknown) are acid anhydrides of decreasing strength—HNO_3 is a

Table 14.5 Acid-Base Properties of the Oxides of Group VA Elements

Formula of Oxide	Classification	Solubility	Formula of Acid or Base
N_2O_5	Acidic	Soluble in water	HNO_3
P_4O_{10}	Acidic	Soluble in water	H_3PO_4
As_2O_5	Acidic	Soluble in water	H_3AsO_4
Sb_2O_5	Amphoteric	Slightly soluble in water, soluble in acid or base	$HSb(OH)_6$
Bi_2O_3	Basic	Insoluble in water, soluble in base	$Bi(OH)_3$

strong acid; H_3PO_4, a weak acid; and H_3AsO_4, a very weak acid. Bismuth, on the other hand, which is a metal and is the last element in the group, forms a slightly basic oxide, Bi_2O_3. It is insoluble in water but will dissolve in acids. Finally, the oxide of antimony dissolves in acid or base since its hydroxide is amphoteric.

There is a third acid-base trend among the oxides of the elements, though this one is not a trend in the periodic table. Among the elements that exhibit two or more oxidation numbers, the oxides with more oxygen atoms tend to be more acidic. We shall find this behavior among the nonmetals and the transition metals and among the metals and metalloids of Groups IIIA, IVA, and VA. Among the nonmetals, it is quickly apparent: SO_3 is the anhydride of a stronger acid (H_2SO_4) than is $SO_2(H_2SO_3)$, and N_2O_5 is the anhydride of a stronger acid (HNO_3) than is $N_2O_3(HNO_2)$. The same relationship holds among the oxides of phosphorus, arsenic, and selenium, as well. Bismuth, which we have already considered, is an example of this same trend among the metals. Bi_2O_3 is the anhydride of a base [$Bi(OH)_3$], whereas bismuth(V) oxide is acidic enough to dissolve in concentrated sodium hydroxide, forming $NaBiO_3$.

We find more examples among the transition elements, for example, manganese. Manganese(II) oxide (MnO) is the anhydride of basic $Mn(OH)_2$. Manganese(IV) oxide (MnO_2), however, is so weakly basic that it is insoluble in many strong acids. Manganese(VI) oxide (MnO_3) is the anhydride of the unstable acid H_2MnO_4. Finally, manganese(VII) oxide (Mn_2O_7) reacts with water to form the purple permanganate ion of the strong acid $HMnO_4$. The same trend holds for chromium: CrO is the anhydride of basic $Cr(OH)_2$, Cr_2O_3 is the anhydride of the amphoteric hydroxide $Cr(OH)_3$, and CrO_3 reacts with water to form a strongly acidic mixture of chromic acid (H_2CrO_4) and dichromic acid ($H_2Cr_2O_7$).

Questions

20. Which of the oxides of iron, FeO or Fe_2O_3, would you expect to be more soluble in acid? Why?

21. If an oxide XO, which is insoluble in water, dissolves in both aqueous HCl and aqueous NaOH, what can you say about the metallic nature of X?

14.11 Equivalents and Normality

In all the stoichiometric problems we have solved so far, the first step has been to write a balanced equation. However, it is possible to solve stoichiometric problems without this step by employing the concept of *equivalents*. Instead of balancing an equation, we calculate the **equivalent weight** of each reactant and product, such that one equivalent of each reactant reacts to produce one equivalent of each product. A close look at this statement makes it clear that the equivalent weight of a compound depends not only on its formula but also on the reaction involved. For acid-base reactions, where this concept is especially useful, *the equivalent weight of an acid is that weight which furnishes one mole of hydrogen ions to the reaction, and the equivalent weight of a base is that weight which furnishes one mole of hydroxide ions to the reaction*: Thus for all monoprotic acids the equivalent weight in an acid-base reaction is the same as the formula weight, but each polyprotic acid has several possible equivalent weights. For example, in a reaction with sodium hydroxide, the equivalent weight of phosphoric acid will depend on how far the reaction goes:

$$H_3PO_4 + NaOH \longrightarrow NaH_2PO_4 + H_2O$$

$$H_3PO_4 + 2\,NaOH \longrightarrow Na_2HPO_4 + 2\,H_2O$$

$$H_3PO_4 + 3\,NaOH \longrightarrow Na_3PO_4 + 3\,H_2O$$

In the first reaction 1 mol of H_3PO_4 furnishes 1 mol of hydrogen ions; the equivalent weight of H_3PO_4 is identical to the formula weight. In the second reaction 0.5 mol of H_3PO_4 furnishes 1 mol of hydrogen ions; hence the equivalent weight is one-half the formula weight. Finally, in the third reaction the equivalent weight of H_3PO_4 is one-third the formula weight.

The idea of equivalents is so useful that concentrations of solutions are often expressed in terms of **normality** (N), which is defined as *the number of equivalents of solute per liter of solution*. The normality may of course vary from reaction to reaction since the equivalent weight of a substance may vary. By convention, however, the normality of an acid (or a base) is usually given as the molarity times the number of ionizable H^+ ions (or OH^- ions) in each formula unit.

Using normalities is often a helpful shortcut in acid-base calculations; however, we should bear in mind the disadvantages. First, despite our convention, the normality of, say, an acid solution will sometimes differ in different reactions. This ambiguity never arises when concentrations are expressed in molarity. Second, even in problems where we need not cope with the first difficulty, we must at least know which H^+ ions and OH^- ions are ionizable.

Acetic acid ($HC_2H_3O_2$) has only one ionizable proton—that is why we do not write it $H_4C_2O_2$—and phosphorous acid (H_3PO_3), despite the way its formula is written, has only two.

Example 14.1 How many equivalents are there in 78.0 g of H_2SO_4? Assume that both hydrogen atoms react.

Solution

a. The number of moles of H_2SO_4 in 78.0 g is

$$(78.0 \text{ g } H_2SO_4) \left(\frac{1 \text{ mol } H_2SO_4}{98.1 \text{ g } H_2SO_4} \right)$$

b. Because there are two moles of H^+ ions available from each mole of H_2SO_4, the number of equivalents is twice the number of moles:

$$(78.0 \text{ g } H_2SO_4) \left(\frac{1 \text{ mol } H_2SO_4}{98.1 \text{ g } H_2SO_4} \right) \left(\frac{2 \text{ equivalents } H_2SO_4}{1 \text{ mol } H_2SO_4} \right)$$

$$= 1.59 \text{ equivalents } H_2SO_4$$

Example 14.2 What is the normality of a solution that contains 17.2 g of $H_4P_2O_7$ in 425 ml of solution? Assume that all four hydrogen atoms are ionized.

Solution

a. For the solution described, the number of grams per liter is

$$\left(\frac{17.2 \text{ g } H_4P_2O_7}{425 \text{ ml}} \right) \left(\frac{1000 \text{ ml}}{1 \text{ liter}} \right)$$

b. From the molecular weight of $H_4P_2O_7$, we can calculate the number of moles per liter,

$$\left(\frac{17.2 \text{ g } H_4P_2O_7}{425 \text{ ml}} \right) \left(\frac{1000 \text{ ml}}{1 \text{ liter}} \right) \left(\frac{1.00 \text{ mol } H_4P_2O_7}{178 \text{ g } H_4P_2O_7} \right)$$

c. Since there are four moles of H^+ ions available from each mole of $H_4P_2O_7$, the number of equivalents per liter (normality) is four times the number of moles per liter (molarity):

$$\left(\frac{17.2 \text{ g } H_4P_2O_7}{425 \text{ ml}} \right) \left(\frac{1000 \text{ ml}}{1 \text{ liter}} \right) \left(\frac{1.00 \text{ mol } H_4P_2O_7}{178 \text{ g } H_4P_2O_7} \right)$$

$$\left(\frac{4 \text{ equivalents } H_4P_2O_7}{1 \text{ mol } H_4P_2O_7} \right) = \frac{0.909 \text{ equivalent } H_4P_2O_7}{1 \text{ liter}} = 0.909 \, N$$

Example 14.3 What is the equivalent weight of an acid if 0.381 g is neutralized by 32.1 ml of 0.0989 N base?

Solution

a. Determine the number of equivalents in 32.1 ml of 0.0989 N base:

$$(32.1 \text{ ml}) \left(\frac{1 \text{ liter}}{1000 \text{ ml}}\right) \left(\frac{0.0989 \text{ equivalent}}{1.00 \text{ liter}}\right) = 3.17 \times 10^{-3} \text{ equivalent}$$

b. Since by definition the number of equivalents of acid must equal the number of equivalents of base, there are 3.17×10^{-3} equivalent in 0.381 g of acid. The number of grams per equivalent is the equivalent weight; hence

$$\frac{0.381 \text{ g}}{3.17 \times 10^{-3} \text{ equivalent}} = \textbf{120 g/equivalent}$$

Example 14.4 If the acid in Example 14.3 was completely neutralized (all ionizable hydrogen atoms reacted) and if the formula weight of the acid is 480, how many ionizable hydrogen atoms are there in each molecule?

Solution

a. There are 480 g/mol and 120 g/equivalent; hence the number of equivalents per mole is

$$\left(\frac{480 \text{ g}}{1.00 \text{ mol}}\right)\left(\frac{1.00 \text{ equivalent}}{120 \text{ g}}\right) = 4 \text{ equivalents/mol}$$

b. In an acid-base reaction the number of equivalents per mole of acid is the same as the number of ionizable hydrogen atoms per molecule that react. Since all ionizable hydrogen atoms reacted in this problem, there must be **four ionizable hydrogen atoms in each molecule.**

Example 14.5 What weight of $Al(OH)_3$ will react with 314 ml of a 0.600 N solution of HCl?

Solution

a. Determine the number of equivalents of HCl reacting:

$$(314 \text{ ml}) \left(\frac{1 \text{ liter}}{1000 \text{ ml}}\right) \left(\frac{0.600 \text{ equivalent}}{1.00 \text{ liter}}\right) = 0.188 \text{ equivalent}$$

This must also be the number of equivalents of $Al(OH)_3$ reacting.

b. Determine the equivalent weight of $Al(OH)_3$:

$$\left(\frac{78.0 \text{ g}}{1.00 \text{ mol}}\right)\left(\frac{1 \text{ mol}}{3 \text{ equivalents}}\right)$$

c. The weight of 0.188 equivalent of $Al(OH)_3$ is therefore

$$(0.188 \text{ equivalent})\left(\frac{78.0 \text{ g}}{1.00 \text{ mol}}\right)\left(\frac{1 \text{ mol}}{3 \text{ equivalents}}\right) = 4.90 \text{ g}$$

Questions

22. How many equivalents of NaOH are required to react with 1.32 equivalents of H_2SO_4?

23. How many equivalents are there in 10.0 g of $Ca(OH)_2$? Assume a reaction where the $Ca(OH)_2$ is completely neutralized.

24. How many grams of H_2SO_4 are required to neutralize completely 131 g of $Al(OH)_3$?

25. How many grams of $CaCO_3$ will react with 48.0 ml of 3.0 N H_2SO_4?

26. What volume of 0.60 N HCl is required to neutralize 30.0 ml of 0.35 N NaOH?

Questions and Problems

14.2 The Arrhenius Concept

27. Give the modern Arrhenius definition of an acid and of a base.

28. What is the concentration of OH^- ions in a neutral solution?

14.3 Characterization and Classification of Arrhenius Bases

29. In moderately dilute aqueous solutions of ammonia, about 2 percent of the ammonia exists as NH_4^+ ion. What is the concentration of OH^- ions in a 0.010 M $NH_3(aq)$ solution?

14.4 Characterization and Classification of Arrhenius Acids

30. Name the following compounds. (You may want to review the nomenclature section in Chapter 6.)
    a. $HCl(g)$
    b. $HCl(aq)$
    c. $HClO(aq)$
    d. $HClO_2(aq)$
    e. $HClO_3(aq)$
    f. $HClO_4(aq)$
    g. $HNO_3(aq)$
    h. $HNO_2(aq)$
    i. $NH_3(g)$
    j. $NH_3(aq)$
    k. $Fe(OH)_2$
    l. $Fe(OH)_3$
    m. Cu_2O
    n. CuO
    o. $H_2S(g)$
    p. $H_2S(aq)$
    q. $H_2SO_4(aq)$
    r. $H_2SO_3(aq)$
    s. $HC_2H_3O_2(aq)$
    t. P_4O_{10}
    u. P_4O_6

14.5 Some Typical Reactions of Arrhenius Acids

31. Describe how you would prepare
  a. HNO_3 from $NaNO_3$.
  b. CO_2 from $CaCO_3$.
  c. HCl from NaCl.
  d. $CaSO_4$ from $Ca(OH)_2$.
  e. Na_2HPO_4 from H_3PO_4.

32. Given below are three reasons why solids dissolve, followed by a list of observations about the solubilities of several compounds. Select the most important reason each compound dissolves as described.

Reasons: (i) Neutralization and formation of molecular water.
(ii) The formation of a soluble molecular weak acid.
(iii) The formation of an unstable weak acid followed by decomposition and evolution of a gas.

  a. $Fe(OH)_3$ is insoluble in water but dissolves in HCl.
  b. $Ba_3(PO_4)_2$ and $BaSO_4$ are both insoluble in water. $Ba_3(PO_4)_2$ will dissolve in HCl, but $BaSO_4$ will not.
  c. CuO is insoluble in water but dissolves in HCl.
  d. Limestone ($CaCO_3$) is insoluble in water but dissolves in acid.

14.6 Hydrolysis

33. Classify aqueous solutions of each of the following as acidic, basic, or neutral. Use the Arrhenius definition of acids and bases.
  a. 1 M HCl
  b. 1 M NaCl
  c. 1 M $HC_2H_3O_2$
  d. 1 M H_2SO_4
  e. 1 M NaOH
  f. 1 M KNO_3
  g. 1 M $NaHSO_4$
  h. 1 M $NaC_2H_3O_2$
  i. 1 M $NaHCO_3$
  j. 1 M NH_3
  k. 1 M NH_4Cl
  l. H_2O
  m. 1 M NaH_2PO_4
  n. 1 M NaCN
  o. 1 M $NaClO_4$

34. How would you distinguish between members of the following pairs using acid-base reactions?
  a. NaCl and Na_2CO_3
  b. AgCl and Ag_3PO_4
  c. NH_4Cl and NaCl
  d. $NaHSO_4$ and Na_2SO_4

14.7 The Brønsted-Lowry Acid-Base Concept

35. Give an example of a reaction in which the water molecule reacts as a Brønsted base. Give one for the water molecule acting as a Brønsted acid.

36. List some other ions or molecules that can act either as a Brønsted acid or a Brønsted base.

37. Identify all the Brønsted acids and bases in the following reactions:
  a. $2\,HF + Na_2CO_3 \longrightarrow H_2CO_3 + 2\,NaF$
  b. $HBr + H_2O \longrightarrow H_3O^+ + Br^-$
  c. $NH_3 + NH_3 \rightleftarrows NH_4^+ + NH_2^-$ (in liquid ammonia)
  d. $S^{2-} + H_2O \rightleftarrows HS^- + OH^-$

e. $NH_3 + H_2O \rightleftharpoons NH_4^+ + OH^-$

f. $Al(H_2O)_6^{3+} + H_2O \rightleftharpoons Al(H_2O)_5OH^{2+} + H_3O^+$

g. $CO_3^{2-} + NH_4^+ \rightleftharpoons NH_3 + HCO_3^-$

38. Explain the observation that an aqueous solution of sodium chloride is neutral but an aqueous solution of sodium acetate is basic.

14.9 The Lewis Acid-Base Concept

39. Define neutralization in terms of each of the three acid-base theories: Arrhenius, Brønsted-Lowry, and Lewis.

14.11 Equivalents and Normality

*40. If 3.64×10^{-3} equivalent of NaOH is needed to neutralize 0.158 g of H_3XO_3, what is the atomic weight of X?

41. How many equivalents of H_3PO_4 are needed to neutralize 46.1 g of $Mg(OH)_2$?

42. If 0.68 g of an unknown base gives 0.028 mol of OH^-, what is the equivalent weight of the base?

43. If 0.0132 equivalent of HCl completely neutralizes 0.486 g of $M(OH)_2$, what is the atomic weight of M?

44. How many equivalents of acid are there in each of the following? Assume that all ionizable hydrogen atoms react.
   a. 3.60 liters of 4.00 N HCl.
   b. 5.00×10^2 ml of 2.00 M H_2SO_4.
   c. 1.00×10^2 ml of 3.00 M H_3PO_4.
   d. 2.50 liters of a solution that contains 2.00 mol of HCl and 3.00 mol of H_3PO_4.
   e. The quantity of acid that neutralizes 2.19 equivalents of base.

Discussion Starters

45. Some acid salts produce basic solutions when dissolved in water. Explain.

46. It is possible for a salt to undergo considerable hydrolysis and yet to produce a neutral solution when dissolved in water. Explain.

47. Criticize the following statements.
   a. The formula for the hydronium ion is H_3O^+.
   b. Acids always contain hydrogen.

48. Explain or account for the following.
   a. Water, a covalent compound, will conduct an electric current.
   b. Aluminum oxide will dissolve in either a strong base or a strong acid.
   c. Hydrogen chloride can be prepared by treating sodium chloride with concentrated sulfuric acid, but H_2SO_4 cannot be prepared by treating sodium sulfate with hydrochloric acid.
   d. An aqueous solution of NaCN is basic, but an aqueous solution of NaCl is neutral.
   e. An aqueous solution of NH_4Cl is acidic, but an aqueous solution of KCl is neutral.
   f. An aqueous solution of $NaHSO_4$ is acidic, but an aqueous solution of $NaHCO_3$ is basic.

g. Dilute sulfuric acid is often referred to as a strong acid; yet the $[H^+]$ is not equal to twice the molarity of the acid.
h. The acetate ion is a base.
i. The ammonium ion is an acid.
j. Water is both an acid and a base.
k. The $Cu(H_2O)_4^{2+}$ ion is an acid.

15

Acid-Base Equilibria

Wed - first part
Fri second part

In Chapter 13 we first applied the concepts of chemical equilibrium to problems involving only gases, arguing that gas phase systems presented the fewest complications. Now, however, we shall apply the same principles to acid-base reactions, usually in aqueous solution. A few new ideas will be necessary, but in general we'll simply be able to extend what we learned in Chapter 13.

15.1 The Water Equilibrium

The autoionization of water ($H_2O \rightleftharpoons H^+ + OH^-$), like any other equilibrium, has an equilibrium constant, which we can express in the usual way:

$$\frac{[H^+][OH^-]}{[H_2O]} = K$$

In this expression the square brackets, [ ], have the same meaning as they had in Chapter 13—concentration in moles per liter. To evaluate this expression, chemists have determined the hydrogen and hydroxide ion concentrations in pure water using electrical conductivity measurements. At 25°C the concentration of each is 1.0×10^{-7} mol/liter. To evaluate the concentration of water, $[H_2O]$, we need only its density at 25°C: 0.997 g/ml. The number of moles of water in 1.00 liter is thus 997 g/18 g, or 55. At equilibrium, therefore, we have

$$[H^+] = 1.0 \times 10^{-7} \text{ mol/liter}$$
$$[OH^-] = 1.0 \times 10^{-7} \text{ mol/liter}$$
$$[H_2O] = 55 - 1.0 \times 10^{-7}, \text{ or } 55 \text{ mol/liter}$$

Substituting into the expression for the equilibrium constant:

$$\frac{(1.0 \times 10^{-7})(1.0 \times 10^{-7})}{(55)} = K = 1.8 \times 10^{-16}$$

We could easily make use of this constant in the examples that will follow, but we can conveniently reduce it to an even simpler form. In any dilute aqueous solution, the concentration of un-ionized water will remain essentially the same: 55 mol/liter. Thus we can combine the equilibrium constant K with the constant concentration of water and write

$$[H^+][OH^-] = K[H_2O] = (1.8 \times 10^{-16})55$$

or

$$K_w = [H^+][OH^-] = 1.0 \times 10^{-14}$$

where K_w is the *ion-product constant of water*. Its value varies with temperature, but at 25°C, *in any aqueous solution, the product of the concentrations of hydrogen ions and hydroxide ions is always* 1.0×10^{-14}.

Question

1. Why is there no denominator in the expression for K_w?

15.2 Strong Acids and Strong Bases

Aqueous solutions of acids contain hydrogen ions from two sources: the ionization of water and the ionization of the acid. If the acid is a strong one, however, and if its concentration is greater than 10^{-4} M, we can neglect the contribution of the ionized water. For example, in a 0.10 M solution of hydrochloric acid, the acid ionizes completely, contributing 0.10 mol of hydrogen ions per liter of solution. This high concentration of hydrogen ions will shift the water autoionization equilibrium to the left and reduce both the concentration of hydroxide ions and the concentration of hydrogen ions from the autoionization of water. Therefore, the total concentration of hydrogen ions in this solution is 0.10 mol/liter plus something less than 1.0×10^{-7} mol/liter or, for all practical purposes, 0.10 M.

Since we now know the hydrogen ion concentration for this example, we can calculate the concentration of hydroxide ions using the ion-product constant of water:

$$K_w = [H^+][OH^-]$$

$$[OH^-] = \frac{K_w}{[H^+]} = \frac{1.0 \times 10^{-14}}{1.0 \times 10^{-1}} = 1.0 \times 10^{-13} \ M$$

What we have said about strong Arrhenius acids applies equally to solutions of strong Arrhenius bases. We must simply reverse the roles of hydrogen and hydroxide ions: In a solution of a strong base, we may ignore the contribution of water ionization to the concentration of hydroxide ions. If the base has one ionizable hydroxide ion, then the concentration of hydroxide ions equals the concentration of the base, and the concentration of hydrogen ions is $K_w/[OH^-]$.

For both acids and bases, it is important to emphasize that neither the hydrogen ion nor the hydroxide ion concentration is ever reduced to zero in an aqueous solution.

Example 15.1 What are $[H^+]$ and $[OH^-]$ in a 0.010 M solution of $HClO_4$?

Solution

a. Since perchloric acid is a strong acid (see Table 14.3), it ionizes completely. Therefore, the concentration of hydrogen ions from the acid is equal to the concentration of the acid itself. Compared to this, the concentration of hydrogen ions from the ionization of water is negligible. Therefore

$$[H^+] = 0.010\ M$$

b. Since

$$[H^+][OH^-] = K_w$$

$$[OH^-] = \frac{K_w}{[H^+]}$$

$$[OH^-] = \frac{1.0 \times 10^{-14}}{1.0 \times 10^{-2}} = 1.0 \times 10^{-12}\ M$$

Example 15.2 What concentration of NaOH would have a $[H^+]$ of 2.0×10^{-12}?

Solution

a. Since sodium hydroxide is a strong base, the concentration of NaOH is the same as the concentration of the hydroxide ion.

b. $[OH^-] = \dfrac{K_w}{[H^+]} = \dfrac{1.0 \times 10^{-14}}{2.0 \times 10^{-12}} = 5.0 \times 10^{-3}\ M$

Thus a 0.0050 M solution of NaOH would have a $[H^+]$ of $2.0 \times 10^{-12}\ M$.

Question

2. Why is neither the $[H^+]$ nor the $[OH^-]$ ever zero in an aqueous solution?

15.3 pH

Since solutions of acids or bases can be characterized by the concentrations of hydrogen and hydroxide ions alone, chemists have devised a shorthand way of expressing each. For any aqueous solution, the **pH** is defined as the negative logarithm (to the base 10) of the hydrogen ion concentration:

$$pH = -\log [H^+]$$

In pure water, therefore, where $[H^+] = 10^{-7}$ mol/liter, the pH is 7. If the hydrogen ion concentration is 10^{-12} mol/liter, the pH is 12. If the H^+ ion concentration is 10^{-3}, the pH is 3. Thus solutions with a pH less than 7 are acidic and those with a pH greater than 7 are basic.

Though we shall encounter it less often than pH, we shall also occasionally see the symbol **pOH**, which is the negative logarithm of the hydroxide ion concentration:

$$pOH = -\log [OH^-]$$

In neutral solutions, therefore, the pOH is 7. Basic solutions have a pOH less than 7, and acidic solutions have a pOH greater than 7. Since at 25°C,

$$[H^+][OH^-] = 1.0 \times 10^{-14},$$
$$-\log [H^+] + (-\log [OH^-]) = -\log 1.0 \times 10^{-14}$$

and, therefore, pH + pOH = 14.0 (commonly expressed in three significant figures).

Example 15.3 A sample of urine has a $[H^+]$ of 4.0×10^{-6} mol/liter. Determine its pH.

Solution

a. By definition,

$$\begin{aligned}
pH &= -\log (4.0 \times 10^{-6}) \\
&= -(\log 4.0 + \log 10^{-6}) \\
&= -(0.6 - 6) \\
&= -(-5.4) = \mathbf{5.4}
\end{aligned}$$

Example 15.4 If tears have a pH of 7.4, what is their $[H^+]$, pOH, and $[OH^-]$?

Solution

a. $\text{pH} = -\log[\text{H}^+] = 7.4$
   $\log[\text{H}^+] = -7.4 = -8 + 0.6$
   $[\text{H}^+] = 10^{-8} \times \text{antilog } 0.6$
   $[\text{H}^+] = \mathbf{4 \times 10^{-8}\ M}$

b. Since $\text{pH} + \text{pOH} = 14.0$,

   $\text{pOH} = 14.0 - \text{pH}$
   $= 14.0 - 7.4 = \mathbf{6.6}$

c. There are now two equally convenient ways to calculate $[\text{OH}^-]$. Since

   $[\text{OH}^-] = \dfrac{K_w}{[\text{H}^+]}$

   $[\text{OH}^-] = \dfrac{1.0 \times 10^{-14}}{4 \times 10^{-8}} = \mathbf{2 \times 10^{-7}\ M}$

or since

   $\text{pOH} = 6.6$
   $\log[\text{OH}^-] = -6.6 = -7 + 0.4$
   $[\text{OH}^-] = 10^{-7} \times \text{antilog } 0.4$
   $[\text{OH}^-] = \mathbf{2 \times 10^{-7}\ M}$

Question

3. Classify the following as characteristics of acidic, basic, or neutral solutions:
   a. pH = pOH   b. pH = 8   c. pH = 9
   d. pH = 3   e. pOH = 1   f. pOH = 11
   g. pH = 7   h. pOH = 13   i. pOH = 7

15.4 Weak Acids and Weak Bases

What we have said so far about the relation between hydrogen and hydroxide ion concentrations and about the calculation of pH and pOH applies equally to solutions of strong and weak acids and strong and weak bases. However, in solutions of weak acids, which are only partially ionized, we cannot assume that $[\text{H}^+]$ is the same as the concentration of the acid; nor in solutions of weak bases can we assume that $[\text{OH}^-]$ is the same as the concentration of the base.

We can represent the ionization of acetic acid, a typical weak acid, as follows:

$$HOAc + H_2O \rightleftharpoons H_3O^+ + OAc^-$$

(For simplicity, we have written the symbol OAc^- to represent the acetate ion $C_2H_3O_2^-$.) Following the principles of chemical equilibrium, we can then write an expression for an equilibrium constant for this reversible reaction:

$$K = \frac{[H_3O^+][OAc^-]}{[HOAc][H_2O]}$$

As we did when deriving the ion-product constant of water, however, we can simplify this expression further. Since in dilute solutions the concentration of water is essentially constant, we can multiply both sides by $[H_2O]$:

$$\frac{[H_3O^+][OAc^-]}{[HOAc]} = K[H_2O] = K_a$$

where $K_a = K[H_2O]$. Another way to derive this new constant is simply to omit the water of hydration when writing the equation for the ionization of acetic acid:

$$HOAc \rightleftharpoons H^+ + OAc^-$$

Now we can immediately write

$$K_a = \frac{[H^+][OAc^-]}{[HOAc]}$$

This expression for K_a is the same as the first, except that we have replaced H_3O^+ with H^+. In this chapter we shall use the second form.

This new equilibrium constant, K_a, is a measure of the ionization of weak acids. Later we shall derive analogous constants for the ionization of weak bases (K_b) as well as for the ionization of other weak electrolytes. Together these constants are called *ionization constants*. Like other equilibrium constants, they are usually determined experimentally, often from measurements of electrical conductance in solution.

Example 15.5 More acetic acid is produced commercially than any other organic acid. Its major uses are in the production of vinegar (which is at least 5 percent acetic acid) and in the manufacture of polymers such as vinyl acetate (p. 703). Concentrated acetic acid (99.5 percent) is called *glacial acetic acid*, since it can be frozen to an icelike solid; its freezing point is 16.7°C. If 0.020 M acetic acid is 3.0 percent ionized at 25°C, what is the value for K_a?

Solution

a. At the hypothetical moment after the acetic acid has been added to the water but before it ionizes, the concentrations of the components of the system are
[HOAc] = 0.020 M
[H$^+$] = 1.0 × 10^{-7} M
[OAc$^-$] = 0 M

$$[0.020] \quad [1.0 \times 10^{-7}] \quad [0]$$
$$\text{HOAc} \rightleftharpoons \text{H}^+ + \text{OAc}^-$$

b. After equilibrium has been established, 3.0 percent of acetic acid has been ionized. Therefore the concentration of ionized acetic acid is

$$\frac{(3.0)(0.020\ M)}{100} = 0.00060\ M \quad \text{or} \quad 6.0 \times 10^{-4}\ M$$

c. For every molecule of acetic acid that ionizes, a hydrogen ion and an acetate ion are produced; hence, at equilibrium the [H$^+$] from acetic acid is equal to [OAc$^-$]. Furthermore, the hydrogen ions from the acid repress the ionization of water and [H$^+$] from water is less than 1.0 × 10^{-7} M. Thus we can regard the total [H$^+$] in solution as 6.0 × 10^{-4} M.

$$[\text{H}^+] = [\text{OAc}^-] = 6.0 \times 10^{-4}\ M$$

d. The concentration of un-ionized acid at equilibrium is

$$0.020\ M - 0.00060\ M = 0.020\ M$$

e. We now have all the concentrations necessary to calculate K_a.

$$\frac{[\text{H}^+][\text{OAc}^-]}{[\text{HOAc}]} = K_a = \frac{(6.0 \times 10^{-4})^2}{0.020} = 1.8 \times 10^{-5}$$

Once an ionization constant has been determined, it can be used to calculate ionic concentrations, as the examples below illustrate. In general we shall understand ionization constants to apply to solutions at 25°C and to be accurate only to two significant figures.

Example 15.6 What are [H$^+$], [OAc$^-$], [OH$^-$], and the percent ionization of 0.010 M acetic acid? $K_a = 1.8 \times 10^{-5}$.

Solution

a. In each liter of solution some of the acetic acid (let's say x moles) ionizes to produce x moles of H$^+$ and x moles of OAc$^-$. The hydrogen ions thus produced repress the water equilibrium, so that we can again assume that the [H$^+$] from water is negligibly small. At equilibrium, therefore,

$$[H^+] = x$$
$$[OAc^-] = x$$
$$[HOAc] = 0.010 - x$$

The concentrations before any ionization has taken place and after equilibrium has been established are summarized below:

Before: 0.010   ~0   0
          HOAc $\rightleftharpoons$ H$^+$ + OAc$^-$
After: 0.010 − x   x   x

b. We can now substitute in the expression for K_a and solve the resulting quadratic equation for x:

$$\frac{[H^+][OAc^-]}{[HOAc]} = \frac{x^2}{0.010 - x} = 1.8 \times 10^{-5}$$

$$x^2 + 1.8 \times 10^{-5}x - 1.8 \times 10^{-7} = 0$$

Using the quadratic formula

$$x = \frac{-b \pm \sqrt{b^2 - 4ac}}{2a}$$

$$x = 4.2 \times 10^{-4} \, M$$

or

$$x = -4.4 \times 10^{-4} \, M$$

c. Since a negative concentration is impossible, both the hydrogen ion concentration and the acetate ion concentration are **4.2×10^{-4} M.**

d. Solving such quadratic equations can be a cumbersome and time-consuming task. Fortunately, we can often avoid this difficult chore. Since acetic acid is a weak acid, we could have predicted that x would be small compared to 0.010. Therefore, the denominator in the expression for K_a, that is, 0.010 − x, is practically equal to 0.010. Making this approximation, our calculation would have looked like this:

$$\frac{x^2}{0.010} = 1.8 \times 10^{-5}$$

$$x^2 = 1.8 \times 10^{-7}$$

$$x = 4.2 \times 10^{-4} \, M$$

e. The answer calculated by using the approximation agrees with the exact solution, thus demonstrating the validity of the approximation. In many similar problems involving weak electrolytes, we shall be making similar approximations. It is essential, however, to notice that a small number can be ignored only when it is *added to* or *subtracted from* a much larger one.

f. We can now determine $[OH^-]$ by dividing $[H^+]$ into K_w:

$$[OH^-] = \frac{1.0 \times 10^{-14}}{4.2 \times 10^{-4}} = 2.4 \times 10^{-11} \, M$$

g. We calculate the percent ionization by dividing the concentration of ionized acid at equilibrium (x) by the total acid concentration and then multiplying by 100:

$$\frac{4.2 \times 10^{-4}}{1.0 \times 10^{-2}} \times 100 = \mathbf{4.2\%}$$

Examples 15.5 and 15.6 indicate that 0.020 M and 0.010 M acetic acid solutions are 3.0 and 4.2 percent ionized, respectively. This illustrates a general phenomenon: The percent ionization of a weak electrolyte increases with dilution. (The value of the ionization constant, on the other hand, remains constant regardless of the concentration.) The approximation we made in Example 15.6 also becomes less valid as the concentration of the electrolyte decreases, since as the extent of ionization increases, the value of x becomes *relatively* larger. To illustrate, a 1.00×10^{-3} M acetic acid solution is 14 percent ionized (we would calculate 13 percent if we dropped the x in the denominator); a 1.00×10^{-4} M solution is 52 percent ionized (we would calculate only 42 percent using the approximation).

Frequently, before solving a problem, you will be unable to judge whether or not it is valid to drop the x from the denominator. In such a case, solve the problem using the approximation and then compare the value of x to the concentration of the un-ionized acid. If subtracting x from this larger number changes its value, the approximation was not justified, and the problem must be reworked using the quadratic formula.

Example 15.7 Formic acid (HCOOH) gets its name from the Latin word *formica*, meaning ant—it was originally produced in 1670 from a species of ants. Calculate the pH of a 0.0010 M solution of formic acid. The ionization constant for the reaction

$$HCOOH \rightleftharpoons HCOO^- + H^+$$

has the value 1.8×10^{-4}.

Solution

a. As in Example 15.6, we can think of x mol (per liter) of acid ionizing to produce x mol H^+ and x mol $HCOO^-$. Again we can ignore $[H^+]$ from water. Then at equilibrium

$$[H^+] = x$$

$$[HCOO^-] = x$$

$$[HCOOH] = 0.0010 - x$$

The concentrations before any ionization has taken place and after equilibrium has been established are summarized below:

	Before:	0.0010	~0	0
		$HCOOH \rightleftharpoons$	H^+ +	$HCOO^-$
	After:	$0.0010 - x$	x	x

b. Substituting in the expression for K_a, we have

$$\frac{[H^+][HCOO^-]}{[HCOOH]} = 1.8 \times 10^{-4}$$

$$\frac{x^2}{0.0010 - x} = 1.8 \times 10^{-4}$$

If we drop x from the denominator and solve, $x = 4.2 \times 10^{-4}$.

c. Applying the test for the validity of the approximation above, we subtract the calculated value of x (4.2×10^{-4}) from the concentration of the un-ionized acid (1.0×10^{-3}):

$$\begin{aligned} &0.0010 \\ -\,&0.00042 \\ \hline &0.0006 \end{aligned}$$

The value of x is *not* negligible compared to 0.0010; hence the approximation was not justified.

d. Therefore we must solve the quadratic equation in part b exactly. When we do this, we find $x = 3.5 \times 10^{-4}$. Thus $[H^+] = [HCOO^-] = 3.5 \times 10^{-4}$ M.

e. By definition,

$$pH = -\log [H^+]$$

$$pH = -\log (3.5 \times 10^{-4})$$

$$= 4 - \log 3.5 = \mathbf{3.46}$$

So far in this section we've dealt only with weak acids, but exactly the same principles are applicable to weak bases. As we have seen, ammonia forms a basic solution in water as follows:

$$NH_3 + H_2O \rightleftharpoons NH_4^+ + OH^-$$

The equilibrium constant for this reaction is

$$\frac{[NH_4^+][OH^-]}{[NH_3][H_2O]} = K$$

But again we can simplify things by multiplying both sides by the constant concentration of water:

$$\frac{[NH_4^+][OH^-]}{[NH_3]} = K[H_2O] = K_b$$

K_b is the ionization constant for aqueous ammonia, and it has the value 1.8×10^{-5}, coincidentally the same as the K_a for acetic acid. A list of other ionization constants appears in Table 15.1.

Example 15.8 What is the pH of 0.050 M aqueous ammonia?

Table 15.1 Ionization Constants at 25°C

Acids	
Acetic acid ($HC_2H_3O_2$)	1.8×10^{-5}
Carbonic acid (H_2CO_3)	$K_{a_1} = 4.3 \times 10^{-7}$
	$K_{a_2} = 5.6 \times 10^{-11}$
Formic acid (HCOOH)	1.8×10^{-4}
Hydrocyanic acid (aqueous HCN)	4.9×10^{-10}
Hydrosulfuric acid (aqueous H_2S)	$K_{a_1} = 9.0 \times 10^{-8}$
	$K_{a_2} = 1.2 \times 10^{-15}$
Nitrous acid (HNO_2)	4.6×10^{-4}
Phosphoric acid (H_3PO_4)	$K_{a_1} = 7.5 \times 10^{-3}$
	$K_{a_2} = 6.2 \times 10^{-8}$
	$K_{a_3} = 2.2 \times 10^{-13}$
Sulfuric acid (H_2SO_4) (K_{a_1} is too large to measure)	$K_{a_2} = 1.3 \times 10^{-2}$
Sulfurous acid (H_2SO_3)	$K_{a_1} = 1.5 \times 10^{-2}$
	$K_{a_2} = 1.0 \times 10^{-7}$
Base	
Ammonia (NH_3)	1.8×10^{-5}

Solution

a. Let x = the number of moles per liter of NH_3 that react with water to produce NH_4^+ and OH^- ions.

b. For every molecule of aqueous ammonia that reacts, one NH_4^+ ion and one OH^- ion are produced ($NH_3 + H_2O \rightleftharpoons NH_4^+ + OH^-$). Thus $x = [NH_4^+] = [OH^-]$. (Much as a weak acid represses the ionization of water by introducing a relatively large number of hydrogen ions, so does the reaction of ammonia with water dominate the autoionization of water. Thus we can ignore $[OH^-]$ from ionized water.) At equilibrium,

$$[OH^-] = [NH_4^+] = x$$
$$[NH_3] = 0.050 - x$$

c. We can now substitute these values in the expression for K_b and solve for x. Remember to solve the equation the easy way first—by dropping x from the denominator.

$$\frac{[NH_4^+][OH^-]}{[NH_3]} = K_b = 1.8 \times 10^{-5}$$

$$\frac{x^2}{(0.050 - x)} = 1.8 \times 10^{-5}$$

$$\frac{x^2}{0.050} = 1.8 \times 10^{-5}$$

$$x = 9.5 \times 10^{-4}$$

Since 9.5×10^{-4} does not affect the value of 0.050 upon subtraction with the correct use of significant figures, the dropping of x in the denominator is justified.

d. We now have $[OH^-] = [NH_4^+] = 9.5 \times 10^{-4}\ M$. By definition,

$$pOH = -\log[OH^-]$$
$$pOH = -\log(9.5 \times 10^{-4})$$
$$= 4 - \log 9.5$$
$$= 4 - 0.98 = 3.02$$

e. Since $pH + pOH = 14.0$,

$$pH = 14.0 - pOH$$
$$pH = 14.0 - 3.02 = \mathbf{11.0}$$

Questions

4. Indicate whether each of the following will increase, decrease, or remain the same as a 0.50 M solution of acetic acid is diluted to twice the volume.
   a. $[H^+]$
   b. pH
   c. $[OH^-]$
   d. pOH
   e. $[HOAc]$
   f. $[OAc^-]$
   g. percent ionization

5. If you are solving for $[H^+]$, is dropping the x from the denominator in the algebraic expression for K_a likely to be more valid for acid A or acid B in each of the following?
   a. Acid A ionizes more completely than acid B.
   b. The concentrations of the acids are the same, but the value of K_a for acid A is larger.
   c. The values for K_a are equal, but the concentration of acid A is greater.

15.5 Common-Ion Effect in Acid-Base Reactions

We can now calculate the ionic concentrations in solutions of weak acids and bases (as well as strong acids and bases, of course). However, our calculations can be upset if the solution also contains other ionic compounds. For example, in an aqueous solution of ammonia, the equilibrium

$$NH_3 + H_2O \rightleftharpoons NH_4^+ + OH^-$$

is disturbed if the solution contains another source of NH_4^+, say NH_4Cl. According to the principle of Le Chatelier the equilibrium will shift to the left and the concentration of OH^- will decrease. This is the **common-ion effect,** so named because NH_4^+ is common to both NH_4Cl and aqueous NH_3. Of course the solution also contains Cl^- ions, but aside from negligible interactions with other ions it has no effect on the equilibrium we're interested in.

The ionization of weak acids is suppressed in the same way by common ions. For example, the concentration of H^+ in dilute acetic acid decreases as sodium acetate is added.

Example 15.9 The familiar white deposit on glassware and windows in chemistry laboratories is ammonium chloride. It is a product of the reaction between the gases NH_3 and HCl, both common in the laboratory. In addition, NH_4Cl does have practical uses: One is its use as a flux in soldering. When it is heated to high temperature, it decomposes into HCl and NH_3. The HCl then reacts with the corrosion products (such as metal oxides) often found on the surface of metals, forming metal chlorides. These compounds, in turn, melt or vaporize at the high temperatures, leaving a clean surface to be soldered. What is $[OH^-]$ in a solution that is 0.050 M in ammonia and 0.030 M in NH_4Cl?

Solution

a. Let x = the number of moles per liter of aqueous NH_3 that have ionized when equilibrium is established ($NH_3 + H_2O \rightleftharpoons NH_4^+ + OH^-$). Then at equilibrium

$$[OH^-] = x$$
$$[NH_4^+] = 0.030 + x$$
$$[NH_3] = 0.050 - x$$

b. As in previous examples, we are ignoring the second source of hydroxide ions—the autoionization of water. The new feature of this problem is that $[NH_4^+]$ also has a second source—the completely ionized ammonium chloride.

c. We now substitute these concentrations in the expression for K_b. Again we have a quadratic equation, but again simplifications are possible. In Example 15.8 we were able to neglect x, compared to 0.050. Here, since NH_4Cl represses the ionization of ammonia, x will be still smaller; hence we can neglect it when added to 0.030 and also when subtracted from 0.050.

$$\frac{[NH_4^+][OH^-]}{[NH_3]} = K_b = \frac{(0.030 + x)x}{(0.050 - x)} = 1.8 \times 10^{-5}$$

$$\frac{0.030\, x}{0.050} = 1.8 \times 10^{-5}$$

$$x = [OH^-] = 3.0 \times 10^{-5} M$$

As we predicted, x is much smaller than 0.030 and 0.050; hence our approximation was justified.

Examples 15.8 and 15.9 illustrate very well the effect of a common ion. In a 0.050 M solution of NH_3 (Example 15.8) the hydroxide ion concentration was 9.5×10^{-4} mol/liter; when the solution was made 0.030 M in NH_4Cl (Example 15.9), the hydroxide ion concentration dropped thirtyfold.

Example 15.10 What will be the pH of the resulting solution if 0.50 liter of 2.0 M NaOH and 0.80 liter of 2.5 M HOAc are mixed?

Solution

a. We can first determine the number of moles of each solute from the molarity and volume of the two solutions before mixing:

$$\left(\frac{2.0 \text{ mol NaOH}}{1.0 \text{ liter}}\right)(0.50 \text{ liter}) = 1.0 \text{ mol NaOH}$$

$$\left(\frac{2.5 \text{ mol HOAc}}{1.0 \text{ liter}}\right)(0.80 \text{ liter}) = 2.0 \text{ mol HOAc}$$

Thus the solution is a mixture of 2.0 mol of acetic acid and 1.0 mol of sodium hydroxide. We know that acids and bases react to form salts, so 1.0 mol of the HOAc will react with 1.0 mol of NaOH in the following neutralization reaction, which goes essentially to completion:

$$\text{NaOH} + \text{HOAc} \longrightarrow \text{NaOAc} + \text{H}_2\text{O}$$
_{ionic   molec   ionic   molec}

Thus we can think of the final 1.30 liters of solution as containing 1.0 mol of acetic acid (which is slightly ionized) and 1.0 mol of sodium acetate (which is completely ionic).

b. Calculate the molarities of the NaOAc and HOAc:

$$\frac{1.0 \text{ mol}}{1.3 \text{ liters}} = 0.77 \ M$$

c. Let x = the number of moles per liter of HOAc ionized at equilibrium; then

$$[\text{H}^+] = x$$
$$[\text{OAc}^-] = 0.77 + x$$
$$[\text{HOAc}] = 0.77 - x$$

Again we have ignored the ionization of water as a source of hydrogen ions. The concentration of acetate ions is the sum of the concentrations of the completely ionized NaOAc and the slightly ionized HOAc. And the concentration of acetic acid is 0.77 minus the concentration of the molecules that are ionized.

d. We can now substitute these values into the expression for K_a, again dropping the x when it is added to or subtracted from a large number:

$$\frac{[\text{H}^+][\text{OAc}^-]}{[\text{HOAc}]} = K_a = \frac{x(0.77 + x)}{(0.77 - x)} = 1.8 \times 10^{-5}$$

$$\frac{x(0.77)}{0.77} = 1.8 \times 10^{-5}$$

$$x = [\text{H}^+] = 1.8 \times 10^{-5}$$

Notice that once more our approximation was valid, since 1.8×10^{-5} is insignificant compared to 0.77.

e. By definition,

$$\text{pH} = -\log [\text{H}^+]$$
$$\text{pH} = -\log (1.8 \times 10^{-5})$$
$$= 5 - \log 1.8$$
$$= 5 - 0.26 = \mathbf{4.74}$$

It is useful to look at the equation in d above, after we have made our customary approximation. This equation can be rearranged to give, in general,

$$\frac{[H^+]}{K_a} = \frac{[\text{weak acid}]}{[\text{anion of weak acid}]}$$

When the un-ionized acid and its anion have the same concentrations, as in Example 15.10, the hydrogen ion concentration is thus equal to the ionization constant. Likewise, for weak bases,

$$\frac{[OH^-]}{K_b} = \frac{[\text{weak base}]}{[\text{cation of weak base}]}$$

In conclusion we should add that the common ion in a solution of a weak base or a weak acid can be the hydroxide ion or hydrogen ion, respectively. Examples of that sort can be approached just as we've approached those in this section.

Questions

6. How is the pH of 0.20 M NH_3 affected by addition of the following?
   a. NH_4NO_3 b. NaOH   c. NH_4Cl d. H_2O e. HCl

7. How many moles of $NaC_2H_3O_2$ must be added to 0.25 liter of 0.30 M $HC_2H_3O_2$ to produce a solution where the $[H^+] = 1.8 \times 10^{-5}$ M?

15.6 Buffers

In many fields of chemistry, especially biochemistry, it is important to control the pH of reagents. To do this it is often necessary to use **buffers,** which are solutions to which acids and bases can be added without causing a correspondingly large change in the pH. (It is helpful to recall that we change the pH of a liter of pure water by two pH units merely by adding 10^{-5} mol of hydrogen or hydroxide ions.) To serve its purpose, therefore, a buffer solution must contain a substance that will neutralize any hydrogen ions added, as well as one that will neutralize any hydroxide ions added. It might contain either a weak acid and a corresponding salt or a weak base and a corresponding salt. For example, a solution of acetic acid and sodium acetate will serve as a buffer. If we add a strong acid, the acetate ions will react with most of the added hydrogen ions, merely increasing the concentration of acetic acid, and the pH will drop only slightly. If we add hydroxide ions, they will be neutralized by the acetic acid. The remaining acetic acid restores the concentration of hydrogen ions; hence the pH does not rise appreciably.

The buffer solutions most resistant to pH change contain equal concentrations of the weak acid (or base) and the salt. As we saw in Section 15.5, such solutions have hydrogen ion (or hydroxide ion) concentrations equal to their ionization constants. Thus different buffers can be selected for different pH ranges.

Example 15.11 What is the change in pH when 0.100 mol of sodium hydroxide is added to 1.00 liter of a solution containing 1.00 mol of sodium acetate and 1.00 mol of acetic acid?

Solution

a. Before the NaOH is added, [OAc$^-$] and [HOAc] are nearly equal. Thus,

$$\frac{[H^+][OAc^-]}{[HOAc]} = 1.8 \times 10^{-5}$$

$$[H^+] = 1.8 \times 10^{-5} \frac{[HOAc]}{[OAc^-]} = 1.8 \times 10^{-5} \left(\frac{1.00}{1.00}\right)$$

$$= 1.8 \times 10^{-5} \, M$$

$$pH = -\log(1.8 \times 10^{-5}) = 4.74$$

b. The 0.100 mol of sodium hydroxide reacts with 0.100 mol of acetic acid to form 0.100 mol of sodium acetate:

$$NaOH + HOAc \longrightarrow NaOAc + H_2O$$

Hence, after the NaOH is added, the solution is 0.90 M in acetic acid and 1.10 M in sodium acetate. If we can again assume that the [H$^+$] from water is negligible compared to that from acetic acid and that the concentration of acetic acid that ionizes is small compared to the concentration that does not, then

$$[H^+] = 1.8 \times 10^{-5} \frac{[HOAc]}{[OAc^-]}$$

$$[H^+] = (1.8 \times 10^{-5}) \left(\frac{0.90}{1.10}\right)$$

$$= 1.5 \times 10^{-5} \, M$$

$$pH = -\log(1.5 \times 10^{-5}) = 4.82$$

c. The pH change is

$$4.82 - 4.74 = \mathbf{0.08}$$

Example 15.12 What is the change in pH when 0.100 mol of hydrochloric acid is added to 1.00 liter of a solution containing 1.00 mol of sodium acetate and 1.00 mol of acetic acid?

Solution

a. This example is analogous to Example 15.11, so we can make parallel assumptions about the ionization of acetic acid and the contribution to [OH$^-$] from the ionization of water. Since the 0.100 mol of HCl reacts with an equal amount of sodium acetate, we then have a solution 1.10 M in acetic acid and 0.90 M in sodium acetate.

$$[H^+] = 1.8 \times 10^{-5} \frac{[HOAc]}{[OAc^-]}$$

$$= (1.8 \times 10^{-5}) \left(\frac{1.10}{0.90}\right)$$

$$= 2.2 \times 10^{-5} \, M$$

$$pH = -\log(2.2 \times 10^{-5}) = 4.66$$

b. The change in pH is

$$4.66 - 4.74 = -0.08$$

Examples 15.11 and 15.12 illustrate the effectiveness of a buffer solution. Recall that 0.100 mol of NaOH (or HCl) added to a liter of pure water would change the pH from 7.0 to 13.0 (or 1.0). Furthermore, acetic acid/acetate buffers can be used at pHs above or below 4.74 simply by adjusting the ratio of acid concentration to salt concentration. Of course there is a limit to their usefulness. For example, to prepare an acetic acid/acetate buffer at pH 6.0, we might make a solution 0.056 M in acetic acid and 1.0 M in sodium acetate. However, if we added 0.10 mol of NaOH to a liter of this buffer, the base would react with all the acetic acid, leaving a high concentration of OH$^-$ ions in solution. The effective pH range of a buffer is usually limited to about one pH unit above and below the pH where it is most effective. Thus the acetic acid/acetate buffer is useful between pH 3.74 and pH 5.74.

Buffers play an essential role in many biological systems, including the fluids of the human body. For example, they keep the pH of the blood very close to 7.4. Among the buffers in the blood are amino acids, which contain both an acid group (—COOH) and a basic group (—NH$_2$); dihydrogen phosphate ions (H$_2$PO$_4^-$) and hydrogen phosphate ions (HPO$_4^{2-}$), and carbonic acid (H$_2$CO$_3$) and bicarbonate ions (HCO$_3^-$). The last of these three systems is the most important.

When the blood becomes more acidic than normal, a condition called *acidosis*, bicarbonate ions react with the free hydrogen ions:

$$H^+(aq) + HCO_3^-(aq) \rightleftharpoons H_2CO_3(aq)$$

The excess carbonic acid then decomposes

$$H_2CO_3(aq) \longrightarrow H_2O(l) + CO_2(g)$$

and the CO_2 is expelled by the lungs. A water molecule thus replaces each hydrogen ion neutralized. The supply of HCO_3^- ions is replenished by the kidneys. When the blood becomes too basic (*alkalosis*), the hydroxide ions are neutralized by the carbonic acid:

$$OH^-(aq) + H_2CO_3(aq) \longrightarrow H_2O(l) + HCO_3^-(aq)$$

Another well-buffered biological fluid is the gastric juice in the stomach. Here the pH is maintained at a surprisingly low 1.6–1.8 but, as most of us know, excess acidity can be a problem. And of course there are many patent medicines containing sodium bicarbonate, magnesium hydroxide, and aluminum hydroxide that offer relief.

The pH of saliva is kept between 6.5 and 7.5; spinal fluid, between 7.3 and 7.5; and freshly excreted urine, about 6. During severe cases of acidosis, the kidneys provide high concentrations of HCO_3^- to the blood, and the pH of the urine can drop as low as 4.

Questions

8. Give three examples of buffer solutions.

9. How is the pH of blood maintained at 7.4?

15.7 Polyprotic Acids

When we turn from monoprotic to polyprotic acids, we must worry about several ionization equilibria, each with its own ionization constant, for each acid. As an example, we'll first consider hydrosulfuric acid (aqueous H_2S).

Natural sources of atmospheric H_2S include volcanoes and other sites of geothermal activity. A more prosaic source is decaying animal and plant material: The foul odor of rotten eggs is due largely to hydrogen sulfide; hence the name "rotten-egg gas." Petroleum refining processes also produce H_2S; thus refineries can contribute significantly to air pollution. In the laboratory, hydrogen sulfide is a common reagent, but it must be used with care since it is highly toxic. If inhaled it can cause headaches, nausea, and death. Because of its odor hydrogen sulfide can usually be detected at exceedingly low concentrations, but continued exposure causes a dulled olfactory sensitivity to it; hence toxic concentrations can build up unheeded.

Both of the hydrogens in hydrosulfuric acid ionize weakly. First, H_2S ionizes to produce hydrogen ions and hydrogen sulfide ions:

$$H_2S(aq) \rightleftharpoons HS^-(aq) + H^+(aq)$$

In the second step the hydrogen sulfide ions ionize to produce sulfide ions and more hydrogen ions:

$$HS^-(aq) \rightleftharpoons H^+(aq) + S^{2-}(aq)$$

Each of these equilibria has an ionization constant:

$$\frac{[H^+][HS^-]}{[H_2S]} = K_{a_1} = 9.0 \times 10^{-8}$$

and

$$\frac{[H^+][S^{2-}]}{[HS^-]} = K_{a_2} = 1.2 \times 10^{-15}$$

Thus the first hydrogen ionizes much more easily than the second. This is typical of weak polyprotic acids. In a solution of hydrogen sulfide, the hydrogen ion concentration depends almost entirely on the ionization of H_2S, and it is only negligibly larger than the hydrogen sulfide concentration. In general, for any polyprotic acid, if K_{a_1} (the first ionization constant) is 10^3 times larger than K_{a_2} (the second ionization constant), we can ignore the amount of H^+ ion produced by the second stage of the ionization when compared to the amount produced by the first stage.

Example 15.13 What is the hydrogen ion concentration in a solution saturated with hydrogen sulfide at room temperature? At room temperature the solubility of hydrogen sulfide is 0.10 mol/liter.

Solution

a. Hydrogen ions come from three sources: the ionization of H_2S, the ionization of HS^-, and the ionization of H_2O:

$$H_2S \rightleftharpoons H^+ + HS^-$$
$$HS^- \rightleftharpoons H^+ + S^{2-}$$
$$H_2O \rightleftharpoons H^+ + OH^-$$

The H^+ concentration from the last two sources, however, is negligibly small compared to $[H^+]$ from the first.

b. Let x equal the concentration of H_2S that has ionized at equilibrium. Then

$$[H_2S] = 0.10 - x$$
$$[H^+] = x$$
$$[HS^-] = x$$

c. Since only the ionization of H_2S is important, we need to write only an expression for K_{a_1}. Furthermore, since only a small fraction of the H_2S ionizes, x is very small compared to 0.10.

$$\frac{[H^+][HS^-]}{[H_2S]} = K_{a_1} = 9.0 \times 10^{-8}$$

$$\frac{x^2}{0.10 - x} = 9.0 \times 10^{-8}$$

$$x = [H^+] = 9.5 \times 10^{-5} \, M$$

In any system at equilibrium, all equilibria must be satisfied simultaneously. In a solution of H_2S there is only one $[H^+]$ and only one $[HS^-]$. Since we have already calculated values for these quantities, we can now substitute them into the expression for K_{a_2}, solving for $[S^{2-}]$. This is a point worth emphasizing: Once we accurately derive a value for the concentration of a species in an equilibrium system, we must use that value wherever that species appears. In the case of hydrogen sulfide, $[H^+] = [HS^-]$; hence $[S^{2-}]$ is equal to the second ionization constant:

$$\frac{[\cancel{H^+}][S^{2+}]}{[\cancel{HS^-}]} = K_{a_2} = 1.2 \times 10^{-15}$$

In an aqueous solution formed by dissolving only hydrogen sulfide, the sulfide ion concentration is always very nearly equal to the second ionization constant, regardless of the concentration of hydrogen sulfide. In fact, *the concentration of the divalent negative ion in any solution of a weak diprotic acid is about equal to the second ionization constant.*

In some cases it is useful to combine the expressions for K_{a_1} and K_{a_2} rather than work with them separately. For H_2S we have

$$\left(\frac{[H^+][\cancel{HS^-}]}{[H_2S]}\right)\left(\frac{[H^+][S^{2-}]}{[\cancel{HS^-}]}\right) = K_{a_1} K_{a_2} = \frac{[H^+]^2[S^{2-}]}{[H_2S]}$$

$$= (9.0 \times 10^{-8})(1.2 \times 10^{-15})$$

$$= 1.1 \times 10^{-22}$$

Expressions like this must be used with caution however. A hasty inspection might suggest that the $[H^+]$ is twice the $[S^{2-}]$. Of course this is not true; the ionization of H_2S proceeds in steps, and the first step ($H_2S \rightleftharpoons H^+ + HS^-$) takes place to a much greater extent than the second ($HS^- \rightleftharpoons H^+ + S^{2-}$). Thus the $[H^+]$ is invariably much larger than twice $[S^{2-}]$. Nonetheless, this combined expression is especially useful in cases where H^+ is a common ion.

Example 15.14 What hydrogen ion concentration is required to reduce the sulfide ion concentration to $2.0 \times 10^{-18} \, M$ in a saturated solution of hydrogen sulfide?

Solution

a. A saturated solution of H_2S is 0.10 M. Since a negligible fraction of H_2S ionizes, $[H_2S] = 0.10\ M$. Thus at equilibrium

$$[H^+] = x$$
$$[S^{2-}] = 2.0 \times 10^{-18}\ M$$
$$[H_2S] = 0.10\ M$$

b. We can now simply use the expression for the combined ionization constant and solve for x.

$$\frac{[H^+]^2[S^{2-}]}{[H_2S]} = K_{a_1}K_{a_2}$$

$$\frac{x^2(2.0 \times 10^{-18})}{0.10} = 1.1 \times 10^{-22}$$

$$x = [H^+] = 2.3 \times 10^{-3}\ M$$

Although most polyprotic acids are weak acids at each stage of ionization, sulfuric acid is not. At concentrations below about 1 M, the first ionization is complete, and no ionization constant can be written. The second hydrogen, however, does not ionize completely. Therefore, we can represent the second step as

$$HSO_4^-(aq) \rightleftharpoons H^+(aq) + SO_4^{2-}(aq)$$

The expression for the ionization constant is

$$\frac{[H^+][SO_4^{2-}]}{[HSO_4^-]} = K_{a_2} = 1.3 \times 10^{-2}$$

Example 15.15 Calculate the concentration of each ion in 0.10 M sulfuric acid.

Solution

a. The first stage of ionization of H_2SO_4 is complete

$$(H_2SO_4 \longrightarrow H^+ + HSO_4^-),$$

producing 0.10 M HSO_4^- and 0.10 M H^+. The second ionization is weaker, producing sulfate ions and more hydrogen ions ($HSO_4^- \rightleftharpoons H^+ + SO_4^{2-}$).

b. Let x equal the equilibrium concentration of SO_4^{2-}. Then at equilibrium

$$[SO_4^{2-}] = x$$
$$[HSO_4^-] = 0.10 - x$$
$$[H^+] = 0.10 + x$$

c. We can now substitute these values in the expression for K_{a_2} and solve for x:

$$\frac{[H^+][SO_4^{2-}]}{[HSO_4^-]} = K_{a_2} = \frac{(0.10 + x)x}{(0.10 - x)} = 1.3 \times 10^{-2}$$

If we assume that x is small compared to 0.10, we calculate that x is 1.3×10^{-2} M, which in turn shows that our assumption was invalid; hence we must solve a quadratic. The result is $x = 0.011$ M; thus

$$[H^+] = 0.11 \, M$$
$$[SO_4^{2-}] = 0.011 \, M$$
$$[HSO_4^-] = 0.09 \, M$$
$$[OH^-] = \frac{1.0 \times 10^{-14}}{0.11} = 9.1 \times 10^{-14} \, M$$

Questions

10. Write the equations for the three equilibria in an aqueous solution of H_2S. Which of these is the main source of hydrogen ions?

11. In general, what are the relative magnitudes of K_{a_1} and K_{a_2} for diprotic acids?

15.8 Hydrolysis

In Section 14.6 we saw that the hydroxide ion concentration in an aqueous solution increases during the hydrolysis of the anion of a weak acid. For example, we can represent the hydrolysis of the acetate ion as

$$OAc^-(aq) + H_2O(l) \rightleftharpoons HOAc(aq) + OH^-(aq)$$

The expression for the equilibrium constant is therefore

$$\frac{[HOAc][OH^-]}{[OAc^-][H_2O]} = K$$

As we have done before, however, we can simplify this expression by regarding [H$_2$O] as a constant. Thus we can write

$$\frac{[\text{HOAc}][\text{OH}^-]}{[\text{OAc}^-]} = K[\text{H}_2\text{O}]$$

This product, $K[\text{H}_2\text{O}]$, is called the *hydrolysis constant* and is represented by K_h.

It is possible to determine the numerical value of K_h experimentally, but it can also be calculated from the values we have already tabulated. To do this we multiply the top and the bottom of the expression for K_h by [H$^+$]:

$$\frac{[\text{HOAc}][\text{OH}^-][\text{H}^+]}{[\text{H}^+][\text{OAc}^-]} = K_h$$

Recalling the definition of the ion-product constant for water,

$$K_w = [\text{OH}^-][\text{H}^+]$$

and the ionization constant for acetic acid,

$$K_a = \frac{[\text{H}^+][\text{OAc}^-]}{[\text{HOAc}]}$$

we can now write K_h as the quotient of familiar terms:

$$K_h = \frac{K_w}{K_a}$$

The hydrolysis constants for other salts of weak acids have the same form, and the values depend only on the values of K_a.

Example 15.16 Sodium acetate is often added to foods as a preservative and buffering agent. What is the pH of 0.20 M sodium acetate?

Solution

a. For every acetate ion that hydrolyzes, one OH$^-$ ion and one HOAc molecule are formed:

$$\text{OAc}^- + \text{H}_2\text{O} \rightleftharpoons \text{HOAc} + \text{OH}^-$$

(Note that the Na$^+$ ion does not react and hence can be omitted from the equation.) Thus [OH$^-$] and [HOAc] are equal to the number of moles (per liter) of OAc$^-$ that are hydrolyzed.

Acid-Base Equilibria

b. Let x = the number of moles (per liter) of OAc^- that have hydrolyzed at equilibrium. If we ignore the negligible concentration of OH^- that is produced by the ionization of water, we thus have

$$[HOAc] = [OH^-] = x$$
$$[OAc^-] = 0.20 - x$$

c. Substitute these values in the expression for K_h, which in turn is equal to K_w/K_a, and solve for x:

$$\frac{[HOAc][OH^-]}{[OAc^-]} = K_h = \frac{1.0 \times 10^{-14}}{1.8 \times 10^{-5}}$$

$$\frac{x^2}{0.20 - x} = 5.6 \times 10^{-10}$$

$$\frac{x^2}{0.20} = 5.6 \times 10^{-10}$$

$$x = [OH^-] = 1.0 \times 10^{-5} \, M$$

Since x is indeed much smaller than 0.20, we were justified in dropping it from the denominator.

d. By definition,

$$pOH = -\log[OH^-]$$

therefore

$$pOH = -\log(1.0 \times 10^{-5}) = 5.0$$

e. Since

$$pH + pOH = 14.0,$$
$$pH = 14.0 - pOH$$
$$pH = 14.0 - 5.0 = \mathbf{9.0}$$

Example 15.17 If the pH of 0.10 M NaA solution is 10.3, what is the ionization constant for the hypothetical acid HA?

Solution

a. Since the solution has a pH greater than 7, it is alkaline. This means that NaA is the salt of a weak acid, hydrolyzing to form OH^- and un-ionized HA:

$$A^-(aq) + H_2O(l) \rightleftharpoons HA(aq) + OH^-(aq)$$

Recognizing that the concentration of OH⁻ and the concentration of HA are both equal to the number of moles (per liter) of A⁻ that are hydrolyzed, we can proceed as follows:

$$pH + pOH = 14.0$$
$$pOH = 14.0 - pH$$
$$= 14.0 - 10.3$$
$$= 3.7 = -\log[OH^-]$$
$$\log[OH^-] = -3.7 = -4 + 0.3$$
$$[OH^-] = 2 \times 10^{-4}\ M$$
$$[HA] = 2 \times 10^{-4}\ M$$
$$[A^-] = 0.10 - 2 \times 10^{-4} = 0.10\ M$$

b. Substitute the values for [HA], [OH⁻], and [A⁻] into the expression for K_h:

$$\frac{[HA][OH^-]}{[A^-]} = K_h = \frac{K_w}{K_a}$$

$$\frac{(2 \times 10^{-4})^2}{0.10} = \frac{1.0 \times 10^{-14}}{K_a}$$

$$K_a = 2 \times 10^{-8}$$

As we might expect, it is easy to extend the ideas above to the hydrolysis of salts of weak bases. For example, we can represent the hydrolysis of an ammonium ion as

$$NH_4^+(aq) \rightleftharpoons NH_3(aq) + H^+(aq)$$

[We could just as well have written $NH_4^+(aq) + H_2O(l) \rightleftharpoons NH_3(aq) + H_3O^+(aq)$]. The expression for the hydrolysis constant is

$$\frac{[NH_3][H^+]}{[NH_4^+]} = K_h$$

To relate the value of K_h for the ammonium ion to the more familiar K_w and K_b, we can multiply the top and the bottom of this expression by [OH⁻]:

$$\frac{[NH_3][H^+][OH^-]}{[NH_4^+][OH^-]} = K_h = \frac{[NH_3]}{[NH_4^+][OH^-]} \times [H^+][OH^-] = \frac{K_w}{K_b}$$

Example 15.18 We have already alluded to the use of ammonia as a fertilizer: About 75 percent of the ammonia produced in the United States is made for this purpose. Increasingly, anhydrous ammonia is being drilled directly into the ground. This works well if the ground is acidic, since the ammonia is quickly neutralized, producing ammonium salts. However, if the soil is alkaline, much of the ammonia vaporizes and is lost before it forms stable salts. Thus for alkaline soils, ammonium salts of strong acids, for example NH_4NO_3, serve as better fertilizers. Calculate the percent of ammonium nitrate that is hydrolyzed in a 0.030 M solution.

Solution

a. One H^+ ion and one NH_3 molecule are formed for every NH_4^+ ion hydrolyzed. Since the concentration of H^+ ions from the ionization of water is negligibly small, we can ignore it.

b. Let $x =$ the number of moles (per liter) of ammonium ion that are hydrolyzed; then at equilibrium

$$[H^+] = x$$
$$[NH_3] = x$$
$$[NH_4^+] = 0.030 - x$$

c. Substitute these values into the expression for K_h, which can be evaluated from K_w and K_b.

$$\frac{[H^+][NH_3]}{[NH_4^+]} = K_h = \frac{K_w}{K_b}$$

$$\frac{x^2}{(0.030 - x)} = \frac{1.0 \times 10^{-14}}{1.8 \times 10^{-5}} = 5.6 \times 10^{-10}$$

$$x = [\text{hydrolyzed } NH_4^+] = 4.1 \times 10^{-6} \, M$$

Again, we were justified in dropping x from the denominator of the expression for K_h.

d. The percent of the total $[NH_4^+]$ that is hydrolyzed is therefore

$$\frac{4.1 \times 10^{-6}}{0.030} \times 100\% = \mathbf{0.014\%}$$

Question

12. Why are there no tabulations of hydrolysis constants in this or other texts?

15.9 Indicators

There are several experimental methods for determining the pH of a solution. Some are very precise but usually involve instruments or lengthy analytical procedures. A much simpler method is the use of **indicators.** This method is not so precise as some, but it is accurate enough for many purposes.

Acid-base indicators are complex organic compounds with colors that are sensitive to the concentration of hydrogen ions. Many are weak acids (HIn), which ionize according to the equation, $HIn \rightleftharpoons H^+ + In^-$. (This equation is actually an oversimplification, since more complex changes—for example, bond rearrangements—usually accompany the ionization of indicators. Nonetheless, we can use this example as the basis of our discussion.) The un-ionized acid (HIn) has one color and the anion (In$^-$) has another. If we place such an indicator in a solution with a high hydrogen ion concentration, the equilibrium will shift to the left and the color of the un-ionized acid will predominate. On the other hand, if the solution has a low hydrogen ion concentration, the color of the ionized form predominates. The pH at which the color changes is characteristic of the indicator HIn.

Figure 15.1 lists a few of the many indicators available to the chemist. Most on this list are common only in the laboratory, but others are much less exotic. For example, red cabbage often changes colors when it is prepared for

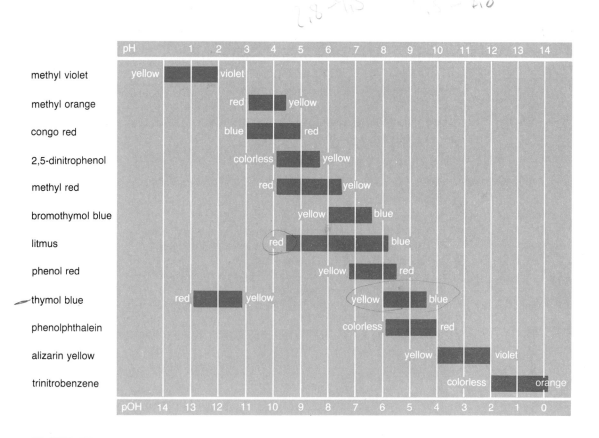

Figure 15.1 pH ranges in which some common indicators change colors.

the table. It is merely responding as an indicator to changes in pH. Figure 15.1 also shows the pH ranges in which these common indicators change color. Of course, the color change is never sharp, since there must always be a range in which the concentrations of HIn and In⁻ are comparable. For example, in a solution containing congo red, if [HIn]/[In⁻] is 10 or more, the eye sees only the blue HIn; if [HIn]/[In⁻] is 0.1 or less, the eye sees red. Between these points the eye can discern shades of violet or purple. To calculate the pH range through which the color changes from blue to red, we need the ionization constant for congo red ($K_a = 1 \times 10^{-4}$).

$$\text{HIn} \rightleftharpoons \text{H}^+ + \text{In}^-$$

$$\frac{[\text{H}^+][\text{In}^-]}{[\text{HIn}]} = K_a = 1 \times 10^{-4}$$

$$[\text{H}^+] = 1 \times 10^{-4} \frac{[\text{HIn}]}{[\text{In}^-]}$$

If the [HIn]/[In⁻] ratio is 10, [H⁺] = 10^{-3}, and the pH is 3. If the [HIn]/[In⁻] ratio is 0.1, [H⁺] = 10^{-5}, and the pH is 5. Thus congo red is blue when the pH is 3 or less and red when the pH is 5 or greater. Most indicators change color over a similar pH range. This calculation, however, suggests only the limits of a single indicator, used alone. By wisely using a combination of indicators, we can determine pH values much more accurately. For example, if a solution turns violet when we add a drop of methyl violet, we know the pH is greater than 2.0. If another sample of the same solution turns red when we add methyl orange, the pH must be less than 3.0. With other indicators the range could be narrowed still further.

Question

13. Using Figure 15.1, determine the approximate pH values of the following solutions. Indicate the uncertainty in each answer.
   a. A solution that turns methyl violet violet and methyl red red.
   b. A solution that turns thymol blue yellow and litmus red.
   c. A solution that turns phenophthalein red and trinitrobenzene colorless.
   d. A solution that turns methyl violet violet and congo red blue.
   e. A solution that turns thymol blue yellow and methyl red red.
   f. A solution that turns 2,5-dinitrophenol yellow and bromothymol blue yellow.

15.10 Titrations

Much of what we have learned in this chapter finds practical application in an analytical procedure called **titration.** We can often determine the concentration of a substance in a sample of known volume but unknown concentration by

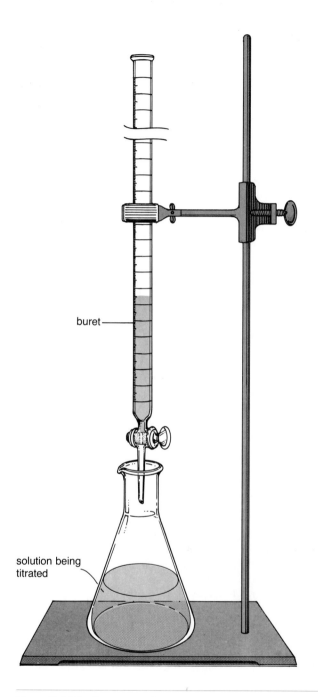

Figure 15.2 Titration.

Acid-Base Equilibria

adding a stoichiometrically equivalent amount of a second reagent, which we call the *standard* or *titrant*. If we accurately measure the amount of titrant added and have a precise way of recognizing the *equivalence point* (the point at which a stoichiometrically equivalent amount of titrant has been delivered), we can easily calculate the unknown concentration. The first requirement is met by using a *buret*, which is a graduated glass tube with a stopcock at the bottom (Figure 15.2); the second, by adding the appropriate indicator to the sample.

A titration is conducted simply by carefully adding the liquid titrant to the sample solution until the indicator changes color. This point is called the *end point*. Strictly speaking, the end point and the equivalence point are not *exactly* the same. (After all, some titrant is needed to react with the indicator.) However, the difference between the two is negligible if the correct indicator is chosen. To visualize the pH changes during a titration and to understand the role of the indicator, we can represent a typical titration by plotting the pH of the sample against the volume of titrant added. We call such a graph a *titration curve*.

The simplest acid-base titration curve represents the titration of a strong acid with a strong base. Figure 15.3 is the curve for the titration of 50.0 ml of

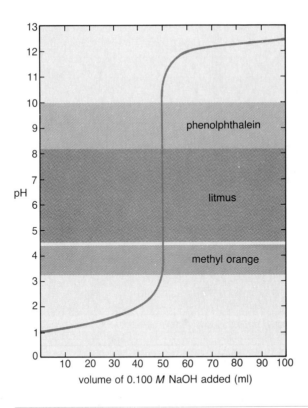

Figure 15.3 Titration of 50.0 ml of 0.10 *M* HCl with 0.10 *M* NaOH (strong acid with strong base).

0.10 M HCl with 0.10 M NaOH. The most important characteristic feature of this curve is the dramatic change in pH near the equivalence point. Adding only 0.2 ml of titrant causes the pH to rise six units. Titration curves also simplify the selection of a suitable indicator. We might appropriately choose any indicator that changes color within the pH range of the straight portion of the curve. In this case, that means any indicator that changes color between pH 3 and pH 10. All but three of the indicators listed in Figure 15.1 would serve. The pH ranges of methyl orange, litmus, and phenophthalein are shown on the titration curve.

If we use the same 0.10 M solution of NaOH to titrate a weak acid rather than a strong one, the titration curve has a very different look. Figure 15.4 shows the curve for the titration of 50.0 ml of 0.10 M acetic acid. Since acetic acid does not ionize completely, the pH at the beginning of the titration is higher than it was with HCl. Furthermore, as titrant is added, the pH rises more rapidly toward the equivalence point. We again reach the equivalence point after adding 50.0 ml of NaOH, but the pH is much higher and the linear portion of the curve near the equivalence point is smaller. Thus we must be more careful

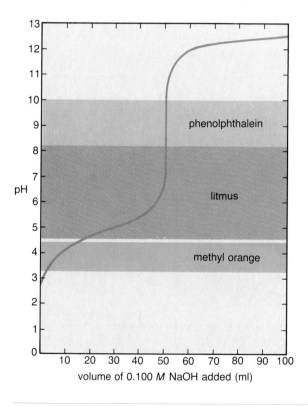

Figure 15.4 Titration of 50.0 ml of 0.10 M acetic acid with 0.10 M NaOH (weak acid with strong base).

when choosing an indicator. As shown in the figure, neither methyl orange nor litmus will do; both would change colors slowly as titrant is added, and methyl orange would change long before the equivalence point is reached. Phenolphthalein, on the other hand, is still a suitable choice, as would be phenol red or thymol blue.

The titration curves in Figures 15.3 and 15.4 can be used to emphasize two of the observations we have already made about neutralization and hydrolysis. A mixture of stoichiometrically equal amounts of HCl and NaOH produces a neutral (pH 7.0) solution of NaCl. When we mix NaOH and acetic acid, however, the result is an aqueous solution of NaOAc, which as we've seen, is not neutral. Hydrolysis of the acetate ions produces a basic solution. Paradoxically then all neutralization titrations do not produce neutral solutions.

Figures 15.5 and 15.6 are curves for two other titrations. The first, for the titration of 0.10 M NH$_3$ with 0.10 M HCl, shows the general features of a titration of a weak base with a strong acid. Methyl orange would be a suitable indicator. The second figure illustrates the titration of a weak acid, 0.10 M HOAc, with a weak base, 0.10 M NH$_3$. None of the indicators we have considered would be useful in this last case.

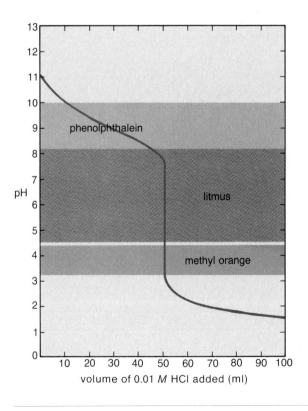

Figure 15.5 Titration of 50.0 ml of 0.10 M NH$_3$ with 0.10 M HCl (weak base with strong acid).

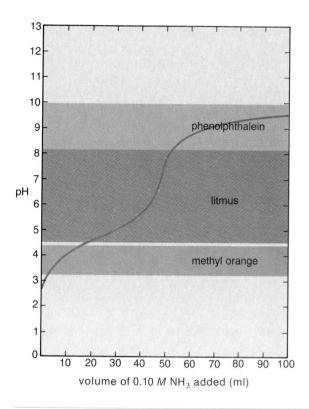

Figure 15.6 Titration of 50.0 ml of 0.10 M acetic acid with 0.10 M NH$_3$ (weak acid with weak base).

Question

14. Classify the solutions at the equivalence points in the following titrations as acidic, basic, or neutral:
  a. HBr and NH$_3$
  b. HC$_2$H$_3$O$_2$ and NaOH
  c. HI and KOH
  d. HClO$_4$ and NaOH
  e. HNO$_3$ and KOH

Questions and Problems

15.1 The Water Equilibrium

15. Calculate the ratio of H$^+$ ions to H$_2$O molecules in pure water at 25°C.

15.2 Strong Acids and Bases

16. What is the [H$^+$] in the following solutions?
  a. 0.010 M HCl
  b. 0.31 M HNO$_3$
  c. 0.090 M HClO$_4$

17. What is the [OH$^-$] in the following solutions?
  a. 0.20 M NaOH
  b. 0.13 M KOH
  c. 0.063 M RbOH

15.3 pH

18. Calculate the missing values for $[H^+]$, $[OH^-]$, pH, and pOH in the following table. Assume that the aqueous solutions are at 25°C.

	$[H^+]$	$[OH^-]$	pH	pOH
a.	10^{-7}	10^{-7}	7	7
b.	10^{-3}	10^{-11}		
c.			4.0	10
d.		3.2×10^{-8}		
e.			5.4	8.6
f.	6.8×10^{-11}			4.3
g.			9.7	
h.			10.6	3.4
i.		5.0×10^{-2}		
j.			1.2	12.8

19. How many moles of KOH must be dissolved in 3.00×10^2 ml of solution to give a pH of 10.0?

*20. The ion-product constant of water at 60°C is 1.0×10^{-13}. What is the pH of a neutral solution at this temperature?

15.4 Weak Acids and Bases

21. Calculate the ionization constants for the following solutions:

Solution	Molarity	pH
a. Formic acid	0.100	2.4
b. Benzoic acid	0.100	3.1
c. Hydrocyanic acid	0.100	5.1
d. Aqueous ammonia	0.100	11.1

22. Lactic acid is a weak monoprotic acid found in sour milk. The ionization constant of this acid is 1.4×10^{-4}. What is the concentration of H^+ in a 2.0 M solution of this acid?

23. The $[H^+]$ in a 0.020 M solution of the weak base BOH is 3.0×10^{-9} M. What is the ionization constant of BOH?

24. Calculate the percent ionization in 1.0 M, 0.10 M, 0.010 M, and 0.0010 M aqueous ammonia. What generalization can be made with regard to concentration and percent ionization?

25. The pH of a 1.0 M solution of the hypothetical acid HZ is 3.8. Calculate the ionization constant of HZ.

26. What concentration of nitrous acid is 3.1 percent ionized?

27. What concentration of aqueous NH_3 has $[OH^-] = 1.9 \times 10^{-3}$?

28. What is the $[H^+]$ in the following solutions?
 a. 1.0 M HCN   b. 0.23 M $HC_2H_3O_2$ c. 0.018 M HCOOH
 d. 0.41 M NH_3 e. 2.1 M $HC_2H_3O_2$

15.5 Common-Ion Effect in Acid-Base Reactions

29. Calculate the weight of potassium acetate that must be added to 8.00×10^2 ml of a $0.20\ M\ HC_2H_3O_2$ solution to change its pH to 5.5.

30. Calculate the $[OH^-]$ in solutions containing the following concentrations of solutes:
   a. $0.51\ M\ NH_3$ and $0.23\ M\ NH_4Cl$
   b. $0.042\ M\ HC_2H_3O_2$ and $0.19\ M\ NaC_2H_3O_2$
   c. $0.082\ M\ HCOOH$ and $0.12\ M\ HCOONa$
   d. $1.2\ M\ HCN$ and $0.93\ M\ NaCN$
   e. $1.4\ M\ NH_3$ and $1.8\ M\ NH_4Cl$

31. How is the pH of $0.10\ M$ acetic acid affected by the addition of the following?
   a. HCl   b. H_2O c. $NaNO_2$ d. NaOH

15.6 Buffers

*32. A buffer solution is prepared by dissolving 2.00 mol of ammonium chloride in 1.00 liter of $2.0\ M$ aqueous ammonia.
   a. What is the pH of the buffer?
   b. What will be the change in pH if 0.050 mol of sodium hydroxide is added to 8.00×10^2 ml of the buffer solution?
   c. What will be the change in pH if 0.50 mol of hydrochloric acid is added to 8.00×10^2 ml of the original buffer solution?

33. Calculate the ratio of salt to acid or base that will give the indicated pH for the following buffers:

Buffer	pH
a. NaCN + HCN	9.7
b. $NaNO_2$ + HNO_2	3.5
c. NH_4NO_3 + aqueous ammonia	9.2

15.7 Polyprotic Acids

34. What is the sulfide ion concentration in $0.30\ M\ HCl$ saturated with H_2S? In $0.30\ M\ HC_2H_3O_2$ saturated with H_2S? (A saturated solution of H_2S at room conditions is $0.10\ M$.)

35. What is the concentration of each ionic and molecular species in the following solutions?
   a. $0.20\ M\ H_2CO_3$ b. $0.13\ M\ H_2SO_3$ c. $0.080\ M\ H_2S$

15.8 Hydrolysis

36. Calculate the hydrolysis constants for the following:
   a. $NaC_2H_3O_2$ b. NaCN   c. NH_4Cl

37. A hypothetical acid ionizes according to the following equation:

$$HA \rightleftharpoons H^+ + A^-$$

a. If a $0.20\ M$ solution is 2.0 percent ionized, what is the ionization constant of the acid?

b. What is the $[H^+]$ in a 2.00 M HA solution?
c. What is the percent ionization in part b?
d. What is $[OH^-]$ in part b?
e. What is the pH of a 1.00 M NaA solution?
f. What is the percent hydrolysis in part e?

38. What concentration of $NaNO_2$ has a pH of 7.6?

*39. The hypothetical salt NaZ is 4.6 percent hydrolyzed in a 0.050 M solution. What is the ionization constant of HZ?

40. Calculate the pH of the following solutions:
    a. 0.13 M $NaC_2H_3O_2$ b. 1.0 M NaCN
    c. 0.32 M NH_4Cl d. 0.090 M NaCN

15.1–15.8

41. What is the $[H^+]$ (or $[OH^-]$ if the solution is basic) in each of the following solutions?
    a. 0.020 M $HC_2H_3O_2$ b. 0.070 M $NaNO_3$ c. 0.10 M HCl
    d. 0.10 M H_2CO_3 e. 0.010 M H_2SO_4
    f. 0.20 M $HC_2H_3O_2$ and 0.40 M $NaC_2H_3O_2$ g. 1.0 M HNO_2
    h. 0.70 M NaCN   i. 0.60 M NH_3 j. 1.0×10^{-4} M $HC_2H_3O_2$
    k. 0.80 M NH_3 and 1.0 M NH_4Cl
    l. 0.300 liter of 3.0 M KOH diluted to 1.70 liters
    m. 60.0 g of NaCN dissolved in 1.20×10^2 liters of solution
    n. 0.50 M NH_4Cl o. 0.20 M HNO_3 p. 0.80 M $Ca(C_2H_3O_2)_2$
    q. 5.02 percent NH_3 (specific gravity = 0.980)   r. 2.0 M Na_2CO_3
    s. 10.0 percent $HC_2H_3O_2$ (specific gravity = 1.010)
    t. A solution that results from mixing 5.00×10^2 ml of 0.50 M HCl with 7.00×10^2 ml of 0.60 M KOH
    u. 0.060 M H_2S v. 0.17 M NaOH   w. 0.92 M H_3PO_4
    x. A mixture of 3.00×10^2 ml of 0.50 M HOAc and 2.00×10^2 ml of 0.40 M KOH
    y. The solution that results from mixing 4.00×10^2 ml of 0.40 M NH_3 with 1.00×10^2 ml of 0.60 M HCl
    z. 1.0 M HCN
    aa. 4.00×10^2 ml of 0.30 M $HC_2H_3O_2$ diluted to 1.3 liters
    bb. 1.0 M $NaC_2H_3O_2$ cc. 0.20 M H_2SO_3
    dd. 5.0 liters of 3.0 M NaCN diluted to 8.0 liters
    ee. 0.0050 M $Ca(OH)_2$ (a strong base)

42. What is the concentration of each of the ions and molecules of the following?
    a. 0.10 M HCl   b. water   c. 0.30 M HNO_2
    d. 0.020 M H_2CO_3 e. 0.15 M $NaC_2H_3O_2$

15.9 Indicators

43. What colors would the following indicators have in each of the solutions in Problem 18?
    a. methyl violet   b. congo red   c. bromothymol blue
    d. litmus   e. thymol blue   f. phenolphthalein

*44. An indicator has an ionization constant of 1.0×10^{-8}. Its acid form is blue, whereas its base form is yellow. An obvious blue color demands a ratio of acid form to base form of 20 to 1. The yellow color is evident when the ratio of acid form to base form is 3 to 1. Calculate the pH range for the color change of this indicator.

15.10 Titrations

*45. Match the acid-base titration curves in Figure 15.7 with the following titrations. Assume that all solutions are 0.10 M.
   a. HOAc titrated with NH_3
   b. HCl titrated with NaOH
   c. NH_3 titrated with HOAc
   d. NH_3 titrated with HCl
   e. NaOH titrated with HOAc
   f. HCl titrated with NH_3
   g. NaOH titrated with HCl
   h. HOAc titrated with NaOH

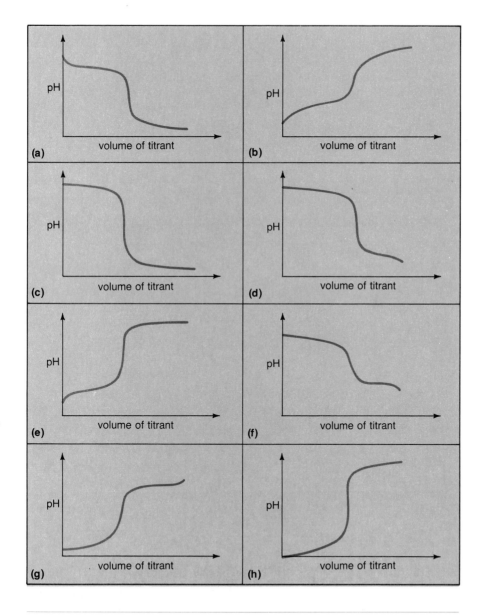

Figure 15.7 Figure to accompany Question 45.

Discussion Starters

46. Explain or account for the following:
   a. The pH of strong monoprotic acids with identical molarities is always the same; however, the pH of weak monoprotic acids with the same molarities varies from acid to acid.
   b. Dilute sulfuric acid is often referred to as a strong acid; yet the hydrogen ion concentration is not equal to twice the molarity of the acid.
   c. The concentration of the divalent ion in an aqueous solution of any weak diprotic acid is equal to K_{a_2} for the acid.
   d. There is no denominator in the expression for K_w.
   e. The hydrogen ion concentration is equal to K_a in a solution containing equal numbers of moles of a weak acid and the sodium salt of the weak acid.
   f. A solution made by adding 1.0 mol of hydrogen chloride to 1.0 liter of 1.0 M aqueous ammonia has the same pH as 1.0 M ammonium chloride.

Precipitation and Solubility Products

16

We turn now from reactions of acids and bases to a much more general study of reactions in aqueous solution. Suppose, for example, that we want to produce CuS from the reaction of Na_2S and $CuSO_4$:

$$Na_2S + CuSO_4 \longrightarrow Na_2SO_4 + CuS$$

Mixing the two solid reactants, as we might two gases, will not do; the ions of both Na_2S and $CuSO_4$ are held firmly in a crystal lattice and thus have little chance to react as long as the compounds remain in a solid state. Here is where water plays its crucial role. Both of these compounds are soluble in water since the solvation energy is great enough to overcome the interionic forces in the crystals. The result is a homogeneous mixture of solvated Na^+, S^{2-}, Cu^{2+}, and SO_4^{2-} ions, in which the copper(II) and sulfide ions have a chance to interact. Thus aqueous solution is a necessary medium for this and most other reactions of ionic compounds. Indeed, most of the metabolic reactions essential to life demand an aqueous medium.

We must now ask what happens when the copper ions encounter the sulfide ions in solution. Unlike Na_2S and $CuSO_4$, copper(II) sulfide is not appreciably soluble in water: The interionic forces binding it into a crystal lattice are too strong. Thus when Cu^{2+} and S^{2-} ions meet in solution, they form an insoluble compound, and CuS *precipitates* from the reaction mixture. If we have taken care to mix stoichiometric amounts of Na_2S and $CuSO_4$, we can now filter out all the black CuS, leaving a solution (*filtrate*) containing only Na^+ and SO_4^{2-} ions. If we then remove the water by evaporation, we have crystals of Na_2SO_4.

This simple example was intended mainly to emphasize the essential role that water plays in most reactions of ionic compounds. We can now look at other such reactions, asking which ions usually form insoluble precipitates and which remain in solution and asking especially just what we mean by "soluble" and "insoluble."

16.1 Equations for Precipitation Reactions

To see the general form for the equations of a precipitation reaction, let us consider the reaction between AgF and NaI. If we mix solutions of these two compounds, we find that we can recover the insoluble precipitate AgI and then recover NaF by evaporating the remaining water. Thus we write the overall equation

$$AgF(aq) + NaI(aq) \longrightarrow AgI(s) + NaF(aq)$$

The ionic equation for the same reaction is

$$(Ag^+ + F^-) + (Na^+ + I^-) \longrightarrow AgI + (Na^+ + F^-)$$

As we did in Chapter 14, we write the formulas for soluble strong electrolytes in the open form and the formulas for insoluble, as well as un-ionized, compounds in the closed form. Finally, we can write the net ionic equation for the reaction:

$$Ag^+(aq) + I^-(aq) \longrightarrow AgI(s)$$

The Na^+ and F^- ions do not appear in this last equation since they undergo no chemical change. These ions are called *spectator ions;* they remain in solution and do not form a new chemical substance. The compound NaF can be isolated only by evaporating the solvent.

Such equations are easy to write once we have experimental evidence as to which compounds precipitate and which ions remain in solution. To be able to write an equation in advance of the experiment, however, we need to be able to predict which pairs of ions form insoluble products and which do not. The five following generalizations will serve as useful guidelines.

1. The Na^+, K^+, and NH_4^+ ions will exist in solution with almost any common negative ion; thus practically all sodium, potassium, and ammonium compounds are soluble.

2. The NO_3^-, ClO_4^-, ClO_3^-, and $C_2H_3O_2^-$ ions will exist in solution with most positive ions; thus almost all nitrates, perchlorates, chlorates, and acetates are soluble.

3. The Cl^-, Br^-, and I^- ions will exist in solution with all common positive ions except Pb^{2+}, Ag^+, and Hg_2^{2+} (the nature of the Hg_2^{2+} ion is explained on p. 157). Therefore, all chlorides, bromides, and iodides are soluble except those of lead(II), silver, and mercury(I).

4. The SO_4^{2-} ion will exist in solution with most cations except Ba^{2+}, Pb^{2+}, Ca^{2+}, and Hg_2^{2+}. In other words, all sulfates are soluble except $BaSO_4$, $PbSO_4$, $CaSO_4$, and Hg_2SO_4.

5. The O^{2-}, S^{2-}, OH^-, CO_3^{2-}, PO_4^{3-}, and SO_3^{2-} ions almost never exist in solution with common cations except Na^+, K^+, and NH_4^+. Thus, most oxides, sulfides, hydroxides, carbonates, phosphates, and sulfites are insoluble, except those of ammonium and the Group IA metals. (We shall not find the oxide ion in solution even with Na^+ or other ions of Group IA metals since it reacts with water to form the hydroxide ion. Also, we do not find many NH_4^+ and OH^-

ions in the same solution since they combine to form aqueous NH_3. We discussed both of these reactions in Chapters 14 and 15.)

With these few chemical facts we can now predict the outcome of many precipitation reactions. For example, consider what happens when we mix the solutions of $Ca(NO_3)_2$ and K_3PO_4. The reaction mixture contains K^+, Ca^{2+}, NO_3^-, and PO_4^{3-} ions (as well as H_3O^+ and OH^- ions from the ionization of H_2O). From the guidelines above we find that Ca^{2+} ions and PO_4^{3-} ions will form $Ca_3(PO_4)_2$, whereas K^+ ions and NO_3^- ions remain in solution. We can therefore write the overall equation for the reaction

$$2\ K_3PO_4(aq) + 3\ Ca(NO_3)_2(aq) \longrightarrow Ca_3(PO_4)_2(s) + 6\ KNO_3(aq)$$

The ionic equation is

$$2(3\ K^+ + PO_4^{3-}) + 3(Ca^{2+} + 2\ NO_3^-) \longrightarrow Ca_3(PO_4)_2 + 6(K^+ + NO_3^-)$$

Note that this equation has two types of coefficients. Those outside the parentheses refer to everything enclosed in the parentheses and are the usual coefficients necessary to balance the chemical equation. Those inside the parentheses refer only to the ions they immediately precede. These coefficients are equal to the subscripts for each ion in the formula for the compound. Thus $2\ K_3PO_4$ becomes $2(3\ K^+ + PO_4^{3-})$.

From the ionic equation we can now write the net ionic equation by canceling the K^+ and NO_3^- ions from both sides:

$$2\ PO_4^{3-}(aq) + 3\ Ca^{2+}(aq) \longrightarrow Ca_3(PO_4)_2(s)$$

Question

1. Using the solubility rules listed in this section, write overall, ionic, and net ionic equations for any reaction that would occur in a solution of the following compounds. If you would expect no reaction, say so.
   a. NaI and $Pb(NO_3)_2$
   b. $KClO_3$ and $BaCl_2$
   c. $Fe_2(SO_4)_3$ and NaOH
   d. $Ca(NO_3)_2$ and $(NH_4)_2SO_4$
   e. $AgNO_3$ and K_2CO_3
   f. NaCl and $(NH_4)_3PO_4$
   g. ammonium chloride and mercury(I) nitrate
   h. iron(III) bromide and potassium hydroxide
   i. barium chloride and sodium phosphate
   j. silver perchlorate and sodium sulfide

16.2 Separation of Ions by Precipitation Reactions

By applying what we know about the solubilities of ionic compounds (either by using the generalizations in Section 16.1 or by consulting more extensive compilations of solubility data), we can often neatly separate the ions of a complex mixture. Let us say that we have a solution of nitrate salts of barium, silver, zinc, and ammonium. We might first remove only the Ag^+ cations by adding NaCl, precipitating AgCl, and then filtering the solution. We could then precipitate in turn $BaSO_4$ and $ZnCO_3$, again by adding the appropriate sodium salts. Only NH_4^+ would then remain of the original four cations.

This means of separating and identifying ions is one of the tools of *qualitative analysis*. The usual systematic procedure used in qualitative analyses can be used to identify about 25 cations, separated into five major groups according to their solubilities. The first of these groups consists of the cations that form insoluble chlorides. Group II contains cations that form soluble chlorides but are precipitated by hydrogen sulfide in a dilute acid solution. Group III includes cations that are not precipitated with Groups I and II cations but precipitate either as hydroxides or sulfides in ammonium hydroxide and ammonium sulfide. Group IV consists of cations that are not precipitated in any of the first three groups but form insoluble carbonates. Finally, Group V consists of the cations that do not precipitate with any of the first four groups. Figure 16.1 shows this separation scheme along with the cations that make up each of the five groups. Two important aspects of this scheme are not explained by the solubility rules alone. First, not all the insoluble sulfides are precipitated in Group II. We must explain this in terms of relative solubility: The ions of Group II form less soluble sulfides than do the ions of Group III. By controlling the hydrogen ion concentration with HCl, we can keep the S^{2-} concentration high enough to precipitate the Group II cations but not high enough to precipitate the more soluble sulfides of the Group III ions. Second, Mg^{2+} does not precipitate with $BaCO_3$ and $CaCO_3$, even though we classified $MgCO_3$ as insoluble. This too is explained in terms of relative solubilities. The CO_3^{2-} ion concentration is controlled

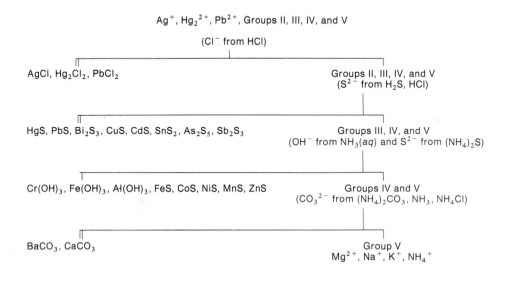

Figure 16.1 Separation of the qualitative analysis groups (double vertical lines indicate insoluble compounds and single vertical lines indicate soluble compounds).

(using NH_3 and NH_4Cl to control the pH) so that it is large enough to precipitate the Ca^{2+} and Ba^{2+} ions but not large enough to precipitate Mg^{2+}.

Questions

2. What characteristic distinguishes the ions of Group I from the other ions in the qualitative analysis scheme?

3. What is the basis for separating the ions of Group II and the ions of Group III?

Sidelight
Water Softening

We can find one important application of precipitation reactions in the processes used to remove unwanted ions from natural waters. Water often contains inorganic ions in solution, including Na^+, K^+, Ca^{2+}, Mg^{2+}, Fe^{3+}, NH_4^+, H^+, OH^-, NO_3^-, Cl^-, SO_4^{2-}, HCO_3^-, and CO_3^{2-}. Of these, the Ca^{2+}, Mg^{2+}, and Fe^{3+} ions especially are nuisances since they form precipitates with soap. Water containing these ions is called *hard water*, and substances designed to remove them are known as water *softeners*.

Soaps are sodium or potassium salts of high-molecular-weight organic acids such as stearic acid [$CH_3(CH_2)_{16}COOH$]. When these salts are added to hard water, insoluble precipitates form. For example,

$$Ca^{2+}(aq) + 2\,CH_3(CH_2)_{16}COO^-(aq) \longrightarrow [Ca(CH_3(CH_2)_{16}COO)_2](s)$$

These sticky, insoluble precipitates of calcium, magnesium, and iron(III) salts are responsible for "tattletale gray" and "bathtub ring." Furthermore, soap does not become an effective cleansing agent until all the calcium, magnesium, and iron ions are precipitated. If enough soap is added to hard water, it will eventually remove these offending ions, but this is an inefficient and expensive way to soften water. It is more practical to use inexpensive precipitating agents such as Na_2CO_3 (soda). For example,

$$Ca^{2+}(aq) + CO_3^{2-}(aq) \longrightarrow CaCO_3(s)$$

Another method of softening water makes use of *ion exchangers* such as zeolites. These are complex sodium aluminosilicate compounds, either natural or synthetic, consisting of huge three-dimensional porous anion lattices and sodium ions (Chapter 9). As hard water passes through columns packed with zeolite granules, the unwanted ions in the water replace the sodium ions in the silicate lattice; for example,

$$Na_2(\text{zeolite})(s) + Ca^{2+}(aq) \rightleftharpoons Ca\,(\text{zeolite})(s) + 2\,Na^+(aq)$$

In addition, since this reaction is reversible, the zeolite water softener can be regenerated by flushing the column with a concentrated solution of NaCl and thus driving the reaction to the left. A number of other reusable synthetic resins have been developed that serve as ion exchangers much like zeolites.

Questions

i. What distinguishes hard water from soft water?
ii. Write two net ionic equations illustrating the softening of hard water.
iii. What is soap?
iv. Why are calcium ions undesirable in water used for cleaning?
v. What causes "bathtub ring"?

16.3 Stoichiometry of Precipitation Reactions

When working out stoichiometry problems for practical precipitation reactions, we shall often find it useful to express the quantity of a reactant or product as the volume of a solution with a specified concentration rather than in grams or moles. Otherwise, the following examples are governed by the same principles we learned in Chapter 3.

Example 16.1 How many grams of Ag_2CO_3 can be precipitated from 10.0 ml of 0.100 M $AgNO_3$ by using an excess of Na_2CO_3?

Solution

a. Write the overall equation for the reaction:

$$2\ AgNO_3(aq) + Na_2CO_3(aq) \longrightarrow Ag_2CO_3(s) + 2\ NaNO_3(aq)$$

b. Determine the number of moles of $AgNO_3$ that react:

$$(10.0\ ml) \left(\frac{1\ liter}{1000\ ml}\right) \left(\frac{0.100\ mol\ AgNO_3}{1.00\ liter}\right)$$

c. The chemical equation specifies that 1 mol of Ag_2CO_3 is produced from 2 mol of $AgNO_3$; hence the number of moles of Ag_2CO_3 is

$$(10.0\ ml) \left(\frac{1\ liter}{1000\ ml}\right) \left(\frac{0.100\ mol\ AgNO_3}{1.00\ liter}\right) \left(\frac{1\ mol\ Ag_2CO_3}{2\ mol\ AgNO_3}\right)$$

d. Therefore the number of grams of Ag_2CO_3 produced is

$$(10.0\ \cancel{ml}) \left(\frac{1\ \cancel{liter}}{1000\ \cancel{ml}}\right) \left(\frac{0.100\ \cancel{mol\ AgNO_3}}{1.00\ \cancel{liter}}\right)$$
$$\times \left(\frac{1\ \cancel{mol\ Ag_2CO_3}}{2\ \cancel{mol\ AgNO_3}}\right) \left(\frac{276\ g\ Ag_2CO_3}{1.00\ \cancel{mol\ Ag_2CO_3}}\right)$$

$$= 0.138\ g\ Ag_2CO_3$$

Example 16.2 What weight of PbI_2 will precipitate if 253 ml of 0.943 M $Pb(NO_3)_2$ solution is mixed with 491 ml of 0.998 M NaI solution?

Solution

a. Write the overall equation for the reaction:

$$Pb(NO_3)_2 + 2\ NaI \longrightarrow PbI_2 + 2\ NaNO_3$$

b. This example is like Example 3.8 in that we must determine which of the two reactants is the limiting one. To do this we need to know the number of moles of each reactant in the reaction mixture. This in turn is done by first finding the number of moles per milliliter and then multiplying by the number of milliliters specified in the problem:

$$\left(\frac{0.943 \text{ mol } Pb(NO_3)_2}{1.00 \text{ liter } Pb(NO_3)_2 \text{ solution}}\right)\left(\frac{1 \text{ liter}}{1000 \text{ ml}}\right)(253 \text{ ml})$$
$$= 0.238 \text{ mol } Pb(NO_3)_2$$

$$\left(\frac{0.998 \text{ mol NaI}}{1.00 \text{ liter NaI solution}}\right)\left(\frac{1 \text{ liter}}{1000 \text{ ml}}\right)(491 \text{ ml}) = 0.490 \text{ mol NaI}$$

c. Calculate the ratio of moles of NaI to moles of $Pb(NO_3)_2$ in the mixture:

$$\left(\frac{0.490 \text{ mol NaI}}{0.238 \text{ mol } Pb(NO_3)_2}\right) = \left(\frac{2.05 \text{ mol NaI}}{1.00 \text{ mol } Pb(NO_3)_2}\right)$$

d. Since this ratio is larger than the 2:1 ratio indicated in the balanced equation, NaI is in excess and the amount of PbI_2 precipitated will depend on the amount of $Pb(NO_3)_2$ present.

e. Calculate the number of moles of PbI_2 precipitated:

$$(0.238 \text{ mol } Pb(NO_3)_2)\left(\frac{1 \text{ mol } PbI_2}{1 \text{ mol } Pb(NO_3)_2}\right)$$

f. Thus the number of grams of PbI_2 precipitated is

$$(0.238 \text{ mol } Pb(NO_3)_2)\left(\frac{1 \text{ mol } PbI_2}{1 \text{ mol } Pb(NO_3)_2}\right)\left(\frac{461 \text{ g } PbI_2}{1.00 \text{ mol } PbI_2}\right)$$
$$= 1.10 \times 10^2 \text{ g } PbI_2$$

16.4 Solubility Equilibria and Solubility Products

So far in this chapter we have been content with labeling most ionic compounds as either soluble or insoluble. For example, we have said that all common sodium salts are water-soluble, whereas most carbonates are insoluble and thus precipitate from aqueous solutions. In Chapter 10, however, we learned that even highly soluble salts, NaCl for example, form saturated solutions in which the dissolved ions are in equilibrium with undissolved solid. Likewise, we intimated that strictly speaking even "insoluble" salts must be regarded as slightly soluble. This is not unlike the case of favorable and unfavorable reactions, each of which proceeds to an extent indicated by an equilibrium constant. We can thus expect to extend the principles of equilibrium to solubility, though we shall find one important modification necessary.

For the general equilibrium between undissolved and dissolved solute, namely

$$\text{solute (solid)} \rightleftharpoons \text{solute (solution)}$$

we can write an expression for an equilibrium constant:

$$K = \frac{[\text{solute}]_{\text{solution}}}{[\text{solute}]_{\text{solid}}}$$

Though we have derived this expression in the usual way, it represents the first example we have encountered of a **heterogeneous equilibrium.** The components of this equilibrium are present in more than one phase (one a solid, one in aqueous solution), whereas the equilibria for which we have calculated equilibrium constants in previous chapters involved substances in only one phase. This raises the question of how we measure the concentration of a solid. If we think of concentration as related to the density of the solid, we conclude that however we measure it, it can be thought of as a constant for a given solid. Only very large changes in temperature or pressure have an effect on the density of a solid. Therefore we can multiply both sides of our expression for the equilibrium constant by $[\text{solute}]_{\text{solid}}$:

$$K[\text{solute}]_{\text{solid}} = [\text{solute}]_{\text{solution}}$$

or

$$[\text{solute}]_{\text{solution}} = K'$$

where K' represents the new constant.

To make this theoretical exercise more concrete, consider the equilibrium between the "insoluble" crystalline solid AgI and a solution of its ions:

$$\text{AgI}(s) \rightleftharpoons \text{Ag}^+(aq) + \text{I}^-(aq)$$

The expression for the equilibrium constant for this reaction is

$$\frac{[\text{Ag}^+][\text{I}^-]}{[\text{AgI}]} = K$$

or, using the argument we have already presented,

$$[\text{Ag}^+][\text{I}^-] = K[\text{AgI}] = K_{sp}$$

The product of the two constants K and [AgI] is referred to as the **solubility product** and is designated by K_{sp}. We can verify this expression experimentally. Regardless of the amount of solid silver iodide in the equilibrium mixture, the concentration of the Ag^+ and I^- ions dissolved in the solution does not vary. As with other equilibria we have studied, this is a dynamic equilibrium: Dissolution and precipitation continue to take place at equal rates.

To write the expressions for the solubility products of ionic compounds containing unequal numbers of positive and negative ions, we follow the same guidelines we used with other equilibrium constants. The exponents in the expression for K_{sp} correspond to the coefficients in the balanced chemical equation. For the equilibrium

$$PbI_2(s) \rightleftharpoons Pb^{2+}(aq) + 2\,I^-(aq)$$

the expression for the equilibrium constant is

$$\frac{[Pb^{2+}][I^-]^2}{[PbI_2]} = K$$

Since $[PbI_2]$ is a constant, the expression for K_{sp} is

$$[Pb^{2+}][I^-]^2 = K[PbI_2] = K_{sp}$$

Similarly, for the equilibrium

$$Ca_3(PO_4)_2 \rightleftharpoons 3\,Ca^{2+} + 2\,PO_4^{3-}$$

we have

$$\frac{[Ca^{2+}]^3[PO_4^{3-}]^2}{[Ca_3(PO_4)_2]} = K$$

and

$$[Ca^{2+}]^3[PO_4^{3-}]^2 = K[Ca_3(PO_4)_2] = K_{sp}$$

The most straightforward calculation of the K_{sp} of a compound is based on a measurement of the solubility of the compound. A large amount of water is added to an excess of the compound and the container is shaken until the solution is saturated. Any undissolved solute can then be filtered out and a liter of the saturated solution evaporated to dryness. Finally, the solid residue remaining after evaporation is weighed. The following examples illustrate how we can calculate the solubility product from this kind of data.

Example 16.3 An aqueous suspension of barium sulfate, ingested by the patient, is used as an opaque medium in X-ray photography of the gastrointestinal tract.

The solubility of this salt is 2.56×10^{-3} g/liter at 25°C. Calculate the solubility product of $BaSO_4$.

Solution

a. Write the expression for the solubility product:

$$[Ba^{2+}][SO_4^{2-}] = K_{sp}$$

b. Determine the concentration of the dissolved salt in moles per liter:

$$\left(\frac{2.56 \times 10^{-3} \text{ g BaSO}_4}{1.00 \text{ liter}}\right)\left(\frac{1.00 \text{ mol}}{233 \text{ g BaSO}_4}\right)$$

c. Determine $[Ba^{2+}]$ and $[SO_4^{2-}]$, remembering that the square brackets stand for concentration in moles per liter. Since one mole of Ba^{2+} and one mole of SO_4^{2-} enter solution for every mole of $BaSO_4$ that dissolves, the concentration of each ion is the same as the concentration of dissolved $BaSO_4$; that is,

$$[SO_4^{2-}] = [Ba^{2+}] = \left(\frac{2.56 \times 10^{-3} \text{ g}}{1.00 \text{ liter}}\right)\left(\frac{1.00 \text{ mol}}{233 \text{ g}}\right) = 1.10 \times 10^{-5} \text{ M}$$

d. Calculate the solubility product by substitution:

$$K_{sp} = (1.10 \times 10^{-5})^2 = 1.2 \times 10^{-10}$$

Solubility products, like most equilibrium constants, are not normally expressed with units.

Example 16.4 Mercury(I) chloride (Hg_2Cl_2), often called calomel, has been used in laxatives. If the solubility of the compound is 3.4×10^{-4} g/liter at 25°C, what is the solubility product of Hg_2Cl_2 at this temperature? The mercury-mercury bond discussed in Chapter 6 remains intact in solution; hence

$$Hg_2Cl_2(s) \rightleftharpoons Hg_2^{2+}(aq) + 2 \text{ Cl}^-(aq)$$

Solution

a. Write the expression for the solubility product:

$$[Hg_2^{2+}][Cl^-]^2 = K_{sp}$$

b. Calculate the solubility of Hg_2Cl_2 in moles per liter:

$$\left(\frac{3.4 \times 10^{-4} \text{ g Hg}_2\text{Cl}_2}{1.00 \text{ liter}}\right)\left(\frac{1.00 \text{ mol Hg}_2\text{Cl}_2}{472 \text{ g Hg}_2\text{Cl}_2}\right) = 7.2 \times 10^{-7} \text{ mol Hg}_2\text{Cl}_2/\text{liter}$$

c. Determine $[Hg_2^{2+}]$ and $[Cl^-]$. Since one mole of Hg_2^{2+} and two moles of Cl^- go into solution for every mole of Hg_2Cl_2 that dissolves,

$$[Hg_2^{2+}] = \left(7.2 \times 10^{-7}\, \frac{\text{mol } \cancel{Hg_2Cl_2}}{1.0 \text{ liter}}\right)\left(\frac{1 \text{ mol } Hg_2^{2+}}{1 \text{ mol } \cancel{Hg_2Cl_2}}\right)$$

$$= 7.2 \times 10^{-7} \text{ mol } Hg_2^{2+}/\text{liter}$$

and

$$[Cl^-] = 2[Hg_2^{2+}] = \left(\frac{7.2 \times 10^{-7} \cancel{\text{ mol } Hg_2^{2+}}}{1.0 \text{ liter}}\right)\left(\frac{2 \text{ mol } Cl^-}{1 \cancel{\text{ mol } Hg_2^{2+}}}\right)$$

$$= 1.4 \times 10^{-6} \text{ mol } Cl^-/\text{liter}$$

d. Calculate K_{sp}

$$K_{sp} = [Hg_2^{2+}][Cl^-]^2 = (7.2 \times 10^{-7})(1.4 \times 10^{-6})^2 = \mathbf{1.4 \times 10^{-18}}$$

We should now expect that solubility products will serve as a precise way of describing whether a compound is "soluble" or "insoluble." Ionic compounds that we have called insoluble will have small solubility products (Table 16.1 lists several), whereas we would predict soluble compounds to have large solubility products. In fact, however, the principle of solubility products is seldom applied to solutions with concentrations greater than 0.02 M. As the concentration of ions in a solution increases, the mutual attraction of positive and negative ions becomes significant, thus reducing the effective ionic concentrations and giving rise to erroneous solubility products. Furthermore, we must be careful, even with less soluble compounds, when comparing their solubilities on the basis of their solubility products alone. The numerical value of a solubility product depends not only on the solubility of a compound but also on the number of ions in each formula unit. Thus the relative values of solubility products reflect the relative solubilities of two different compounds *only* if they contain the same number of ions. The following problem serves as an illustration.

Example 16.5 Compare the solubilities of $AgCl$, Ag_2CrO_4, and AgI. What is the concentration of each ion in a saturated solution of each salt? Use the solubility products in Table 16.1.

Solution For $AgCl$:

a. Write the expression of the solubility product of $AgCl$:

$$[Ag^+][Cl^-] = 1.8 \times 10^{-10}$$

Table 16.1 Solubility Products (at 25°C)

Compound	K_{sp}
$Al(OH)_3$	3.7×10^{-15}
$BaCO_3$	8.1×10^{-9}
$BaCrO_4$	2.4×10^{-10}
BaF_2	1.7×10^{-6}
$BaSO_4$	1.3×10^{-10}
$CaCO_3$	8.7×10^{-9}
CaF_2	1.7×10^{-10}
CdS	3.6×10^{-29}
CuS	8.5×10^{-45}
$Fe(OH)_2$	1.6×10^{-14}
FeS	3.7×10^{-19}
PbI_2	1.4×10^{-8}
$PbSO_4$	1.1×10^{-8}
$MgCO_3$	2.6×10^{-5}
$Mg(OH)_2$	1.2×10^{-11}
$Mn(OH)_2$	4.0×10^{-14}
MnS	1.4×10^{-15}
HgS	4×10^{-53}
NiS	1.4×10^{-24}
$AgBr$	7.7×10^{-13}
Ag_2CO_3	6.6×10^{-12}
$AgCl$	1.8×10^{-10}
Ag_2CrO_4	1.1×10^{-12}
AgI	8.3×10^{-17}
Ag_2S	1.6×10^{-49}
$Zn(OH)_2$	1.8×10^{-14}

b. Since for every mole of AgCl dissolved, one mole of Ag^+ and one mole of Cl^- are in solution,

$$[Ag^+] = [Cl^-] = \text{molar solubility of AgCl}$$

c. Calculate the concentrations of Ag^+ and Cl^- ions in a saturated solution. Let

$$x = [Ag^+] = [Cl^-]$$

and then

$$x^2 = 1.8 \times 10^{-10}$$
$$x = \sqrt{1.8 \times 10^{-10}} = 1.3 \times 10^{-5}$$
$$[Ag^+] = [Cl^-] = 1.3 \times 10^{-5} \, M$$

d. Thus the molar solubility of AgCl is $1.3 \times 10^{-5} \, M$.

For Ag_2CrO_4:

a. From the balanced equation

$$Ag_2CrO_4(s) \rightleftharpoons 2\,Ag^+(aq) + CrO_4^{2-}(aq)$$

Write the expression for the solubility product of Ag_2CrO_4:

$$[Ag^+]^2[CrO_4^{2-}] = 1.1 \times 10^{-12}$$

b. Determine the relative concentrations of Ag^+ and CrO_4^{2-} in a saturated solution of Ag_2CrO_4. For every mole of Ag_2CrO_4 dissolved, one mole of CrO_4^{2-} and two moles of Ag^+ are in solution; hence

$$[Ag^+] = 2[CrO_4^{2-}]$$

and

$$[CrO_4^{2-}] = \text{molar solubility of } Ag_2CrO_4$$

c. Calculate the concentrations of Ag^+ and CrO_4^{2-} in a saturated solution. Let

$$x = [CrO_4^{2-}]$$

and then

$$[Ag^+] = 2x$$
$$(x)(2x)^2 = 1.1 \times 10^{-12}$$
$$4x^3 = 1.1 \times 10^{-12} = 1100 \times 10^{-15}$$
$$x^3 = 275 \times 10^{-15}$$
$$x = 6.5 \times 10^{-5}$$
$$[CrO_4^{2-}] = 6.5 \times 10^{-5} \, M$$
$$[Ag^+] = 1.3 \times 10^{-4} \, M$$

d. The molarity of a saturated solution of Ag_2CrO_4 is therefore $6.5 \times 10^{-5} \, M$. For AgI, the solution is similar to that for AgCl:

$$[Ag^+][I^-] = 8.3 \times 10^{-17}$$
$$[Ag^+] = [I^-] = x$$
$$x^2 = 8.3 \times 10^{-17} = 83 \times 10^{-18}$$
$$x = 9.1 \times 10^{-9}$$
$$[Ag^+] = [I^-] = 9.1 \times 10^{-9} \, M$$

A saturated solution of AgI is $9.1 \times 10^{-9} \, M$.

We would therefore have correctly concluded that AgCl is more soluble than AgI merely by comparing their solubility products; however, Ag_2CrO_4 is more soluble than either, even though its solubility product is smaller than that of AgCl.

Questions

4. Write the expressions for K_{sp} for each of the following:
   a. AgCN   b. Bi_2S_3 c. $Mg_3(PO_4)_2$
   d. $Fe_4[Fe(CN)_6]_3$, where $Fe_4[Fe(CN)_6]_3(s) \rightleftharpoons 4\ Fe^{3+}(aq) + 3\ Fe(CN)_6^{4-}(aq)$
   e. $MgNH_4PO_4$, where $MgNH_4PO_4(s) \rightleftharpoons Mg^{2+}(aq) + NH_4^+(aq) + PO_4^{3-}(aq)$

5. Explain why the solubility product is independent of the quantity of solid in equilibrium with the saturated solution.

16.5 Solubility Products and the Common-Ion Effect

In our discussion of solubility products, we have so far considered only solutions of a single slightly soluble salt in water. Thus the ratio of the ions in solution is equal to the ratio of the ions in the pure solid. We now want to look at the case where the solvent already contains a common ion, say from a highly soluble ionic salt. We have already seen (in Chapter 15) that, in the presence of a common ion, chemical equilibria shift to reduce the concentration of the common species. Thus if we measure the solubility of AgCl in a solution of NaCl, one would expect to find the equilibrium

$$AgCl(s) \rightleftharpoons Ag^+(aq) + Cl^-(aq)$$

shifted to the left because of the Cl^- ions from NaCl.

Example 16.6 What is the solubility of AgCl in a 0.0010 M solution of NaCl?

Solution

a. We must first determine the relative contributions of AgCl and NaCl to the concentration of the common ion, Cl^-. We found in Example 16.5 that $[Cl^-]$ in a saturated solution of AgCl is $1.3 \times 10^{-5}\ M$. In the presence of NaCl the solubility of AgCl will be even less; thus we expect the concentration of Cl^- from AgCl in this problem to be very small compared to that from NaCl.

b. From the conclusion above,

$$[Cl^-] = 0.0010\ M$$

c. Calculate $[Ag^+]$ in a solution where $[Cl^-] = 0.0010\ M$:

$$[Ag^+][Cl^-] = 1.8 \times 10^{-10}$$

$$[Ag^+] = \frac{1.8 \times 10^{-10}}{[Cl^-]}$$

$$[Ag^+] = \frac{1.8 \times 10^{-10}}{1.0 \times 10^{-3}} = 1.8 \times 10^{-7}\ M$$

d. Since all the Ag^+ ion in solution comes from the AgCl that dissolves, the solubility of AgCl in 0.0010 M NaCl is **1.8×10^{-7} mol/liter.** Note that this is about one-hundredth the solubility in pure water and that it justifies our assumption in part a.

Example 16.7 What is the solubility of Ag_2CrO_4 in a 0.0010 M solution of Na_2CrO_4?

Solution

a. In Example 16.5, we found that $[CrO_4^{2-}]$ in a saturated aqueous solution of Ag_2CrO_4 is 6.5×10^{-5} M. Because of the common-ion effect, we expect the solubility of Ag_2CrO_4 to be even less in a solution of Na_2CrO_4. We can therefore ignore the concentration of chromate ions from Ag_2CrO_4 (which will be less than 6.5×10^{-5} M) compared to that from Na_2CrO_4 (which is 0.0010 M):

$$[CrO_4^{2-}] = 0.0010 \ M$$

b. Calculate the $[Ag^+]$ in a 0.0010 M solution of Na_2CrO_4 saturated with Ag_2CrO_4:

$$[Ag^+]^2[CrO_4^{2-}] = 1.1 \times 10^{-12}$$

$$[Ag^+] = \sqrt{\frac{1.1 \times 10^{-12}}{[CrO_4^{2-}]}}$$

$$= \sqrt{\frac{1.1 \times 10^{-12}}{1.0 \times 10^{-3}}}$$

$$= 3.3 \times 10^{-5} \ M$$

c. All the Ag^+ in the solution comes from Ag_2CrO_4, but two moles of Ag^+ go into solution for every mole of Ag_2CrO_4 dissolved; hence the concentration of Ag_2CrO_4 is one-half $[Ag^+]$ or **1.6×10^{-5} M.**

16.6 Predicting the Formation of Precipitates

The solubility product is often used to predict whether or not a precipitate will form. First the *ion product* must be determined; it is the product of the concentrations of each ion raised to the power of its coefficient in the equation. The ion product is then compared with the solubility product. If the ion product exceeds the solubility product, precipitation will take place. If the ion product is equal to or less than the solubility product, precipitation will not occur. The following problems illustrate this practical use of solubility products.

Example 16.8 Will precipitation occur if we mix 500 ml each of 0.00020 M NaCl and 0.00020 M $AgNO_3$?

Solution

a. The first step in all such problems is to determine the concentrations of any ions that form insoluble compounds, assuming that no precipitation occurs. In this case we can ignore the Na^+ and NO_3^- ions, since they form a highly soluble compound. Since both solutions are diluted to twice their original volumes when combined, the concentrations of Ag^+ and Cl^- are reduced by one-half; hence

$$[Cl^-] = 1.0 \times 10^{-4} \, M$$
$$[Ag^+] = 1.0 \times 10^{-4} \, M$$

b. Determine the ion product, again assuming no precipitation:

$$[Ag^+][Cl^-] = (1.0 \times 10^{-4})^2$$
$$= 1.0 \times 10^{-8}$$

c. Compare the ion product with the solubility product:

$$1.0 \times 10^{-8} > 1.8 \times 10^{-10}$$

Since the calculated ion product exceeds the solubility product, **AgCl will precipitate until $[Ag^+][Cl^-] = 1.8 \times 10^{-10}$**.

Example 16.9 How many grams of crystalline NaOH can be added to 10.0 liters of 0.010 M $MgCl_2$ before precipitation begins to take place?

Solution

a. Magnesium hydroxide will begin to precipitate as soon as the product of $[Mg^{2+}]$ and $[OH^-]^2$ exceeds the solubility product (1.2×10^{-11}). Thus we can add NaOH until

$$[Mg^{2+}][OH^-]^2 = 1.2 \times 10^{-11}$$

$$[OH^-]^2 = \frac{1.2 \times 10^{-11}}{1.0 \times 10^{-2}}$$

$$[OH^-] = 3.5 \times 10^{-5} \, M$$

b. Calculate the weight of NaOH that will produce 10.0 liters of a $3.5 \times 10^{-5} \, M$ solution:

$$\left(\frac{3.5 \times 10^{-5} \, \text{mol NaOH}}{1.0 \, \text{liter}}\right)\left(\frac{40.0 \, \text{g NaOH}}{1.00 \, \text{mol NaOH}}\right)(10.0 \, \text{liters}) = 0.014 \, \text{g NaOH}$$

Example 16.10 A solution is 0.10 M in Ba^{2+} and 0.10 M in Ca^{2+}. If Na_2SO_4 is added gradually, will $BaSO_4$ or $CaSO_4$ precipitate first? When the second sulfate begins to precipitate, what concentration of the first cation remains in solution? Is this an effective method of separating Ca^{2+} and Ba^{2+}?

Solution

a. Determine $[SO_4^{2-}]$ when $CaSO_4$ begins to precipitate:

$$[Ca^{2+}][SO_4^{2-}] = 1.2 \times 10^{-6}$$

$$[SO_4^{2-}] = \frac{1.2 \times 10^{-6}}{1.0 \times 10^{-1}} = 1.2 \times 10^{-5} \, M$$

b. Determine $[SO_4^{2-}]$ when $BaSO_4$ starts to precipitate:

$$[Ba^{2+}][SO_4^{2-}] = 1.3 \times 10^{-10}$$

$$[SO_4^{2-}] = \frac{1.3 \times 10^{-10}}{1.0 \times 10^{-1}} = 1.3 \times 10^{-9} \, M$$

Therefore $BaSO_4$ precipitates first.

c. When $CaSO_4$ begins to precipitate, $[SO_4^{2-}] = 1.2 \times 10^{-5} \, M$. Therefore

$$[Ba^{2+}] = \frac{1.3 \times 10^{-10}}{1.2 \times 10^{-5}} = 1.1 \times 10^{-5} \, M$$

d. To two significant figures, 100 percent of the barium is precipitated before $CaSO_4$ begins to precipitate; therefore this procedure effectively separates the Ba^{2+} from the Ca^{2+}.

Example 16.11 When solving Example 16.1, we assumed that Ag_2CO_3 was *completely* insoluble and that it all precipitated. Was this assumption justified?

Solution

a. Calculate the concentration of Ag_2CO_3 in a saturated solution:

$$[Ag^+]^2[CO_3^{2-}] = 6.6 \times 10^{-12}$$

Let

$$x = [CO_3^{2-}]$$

then

$$2x = [Ag^+]$$
$$4x^3 = 6.6 \times 10^{-12}$$
$$x = 1.2 \times 10^{-4}$$

Thus a saturated solution of Ag_2CO_3 is $1.2 \times 10^{-4} \, M$.

b. Calculate the number of grams of Ag_2CO_3 dissolved in 10.0 ml of saturated solution:

$$\left(\frac{1.2 \times 10^{-4} \text{ mol}}{1.0 \text{ liter}}\right)\left(\frac{1 \text{ liter}}{1000 \text{ ml}}\right)(10.0 \text{ ml})\left(\frac{216 \text{ g}}{1.0 \text{ mol}}\right) = 0.00026 \text{ g}$$

c. Since this answer does not affect the third significant figure of the answer to Example 16.1 (0.138 g), our assumption was justified. Furthermore, since Example 16.1 specified an excess of Na_2CO_3, the amount of Ag_2CO_3 dissolved would be less than 0.00026 g.

Questions and Problems

16.2 Separation of Ions by Precipitation Reactions

6. What distinguishes the Group V ions from the other ions of the qualitative analysis scheme?

7. During the separation of the ions of Group II from the ions of Group III of the qualitative analysis scheme, what is the role of hydrochloric acid? Write the net ionic equation for the reaction involved.

16.3 Stoichiometry of Precipitation Reactions

8. Calculate the number of grams of NiS that can be prepared from 3.00×10^2 ml of a 2.00 M solution of $NiCl_2$ and an excess of Na_2S.

9. A mixture of 10.3 g of KBr and 28.9 g of $AgNO_3$ was added to 5.00×10^2 ml of water. Identify and calculate the weight of any precipitate formed.

10. How many milliliters of 4.03 M NaOH are needed to prepare 9.06 g of $Ba(OH)_2$? Assume an excess of $BaCl_2$.

11. How many moles of PbI_2 can be precipitated from 567 ml of a 3.22 M $Pb(NO_3)_2$ solution and an excess of KI?

12. Calculate the weight of Ag_3PO_4 formed by mixing 1.30×10^2 ml of 4.91 M $AgNO_3$ and 2.56×10^2 ml of 0.892 M Na_3PO_4.

13. When 312 ml of $(NH_4)_2CO_3$ solution was treated with an excess of calcium chloride, 4.96 g of $CaCO_3$ precipitated. What was the molarity of the $(NH_4)_2CO_3$ solution?

16.4 Solubility Equilibria and Solubility Products

14. Calculate the solubility products for the following compounds of silver, given the molarities of the saturated solutions:

a. AgBr $\quad 8.8 \times 10^{-7} M$ b. $Ag_2C_2O_4 \quad 1.1 \times 10^{-5} M$
c. $Ag_3PO_4 \quad 1.2 \times 10^{-6} M$ d. AgCN $\quad 1.7 \times 10^{-7} M$

15. Calculate the solubility products of the following compounds from their solubilities:
  a. $Mn(OH)_2$ 0.00020 g/100 cc
  b. NiS 0.00036 g/100 cc
  c. TlI 0.00060 g/100 cc
  d. SrF_2 0.011 g/100 cc

16. A saturated solution of the slightly soluble hypothetical salt RZ_2 has $[R^{2+}] = 0.0021\ M$. What is the solubility product for this salt?

For Problems 17 to 33, refer to Table 16.1 for solubility products. In solving these problems, assume that there is no reaction between the ions and water.

17. Calculate the molarity of saturated solutions of each of the following compounds:
  a. $PbSO_4$ b. BaF_2 c. Ag_2CO_3 d. $Al(OH)_3$

18. Calculate the solubilities of the following carbonates, and express them in grams per 100 ml:
  a. $BaCO_3$ b. $CaCO_3$ c. Ag_2CO_3

19. Calculate the number of sulfide ions in 1 liter of a saturated solution of Ag_2S.

20. Which has the greater molar solubility, $CaCO_3$ or CaF_2? Justify your answer.

21. Which is more soluble, $Al(OH)_3$ or $Zn(OH)_2$? Justify your answer.

22. The two compounds RA_2 and ZA both have solubility products of 1.0×10^{-18}. Which compound is more soluble? Justify your answer.

16.5 Solubility Products and the Common-Ion Effect

23. What is the molar solubility of $BaCrO_4$ in $0.000012\ M\ Na_2CrO_4$?

24. What is the solubility of $BaCO_3$ in
  a. $0.010\ M\ Ba(NO_3)_2$?   b. $0.0010\ M\ Na_2CO_3$?

25. What concentrations of Ag^+ ion are necessary to initiate precipitation of a silver salt from the following solutions?
  a. $1.0 \times 10^{-5}\ M\ NaBr$ b. $2.3 \times 10^{-4}\ M\ K_2CrO_4$ c. $3.6 \times 10^{-3}\ M\ NaI$

16.6 Predicting the Formation of Precipitates

26. Will $Mg(OH)_2$ precipitate when 1.3 liters of $1.0 \times 10^{-4}\ M$ NaOH is mixed with 9.0×10^2 ml of $1.3 \times 10^{-3}\ M\ MgCl_2$? Justify your answer with calculations.

27. What concentration of Ca^{2+} ion will precipitate CaF_2 from 2.3 liters of a solution that contains 0.53 g of NaF?

28. At what concentration of NaI will precipitation begin from a saturated solution of AgCl?

29. How many moles of lead iodide can be dissolved in 105 liters of $0.012\ M$ NaI?

*30. A city's water contains 3.0×10^{-3} g Ca^{2+}/liter. Suppose the city wants to reduce the concentration of Ca^{2+} to 1.0×10^{-6} g/liter. What weight of $Na_2CO_3 \cdot 10\ H_2O$ must be added to 100,000 gal of water?

31. If 3.0×10^2 ml of $0.040\ M\ Ba(NO_3)_2$ and 8.0×10^2 ml of $0.030\ M$ NaF are mixed,
  a. what would be the concentrations of Ba^{2+} and F^- ions if no precipitation took place?

b. what are the concentrations of the same ions after precipitation?
c. what is the weight of BaF_2 precipitated?

*32. A solution is 0.0050 M in CO_3^{2-} and 0.0030 M in F^-. If Ca^{2+} solution is added gradually, which calcium salt will precipitate first? What concentration of the first anion will remain in solution when the second begins to precipitate?

*33. Which of the following solutions has the highest pH?
   a. A saturated solution of $Mg(OH)_2$.
   b. A saturated solution of $Mn(OH)_2$.
   c. A solution that is 1.00×10^{-3} M in NH_3 and 1.00×10^{-2} M in NH_4Cl.
   d. A 0.010 M solution of sodium acetate.

Discussion Starters

34. Criticize the following statements:
   a. Silver iodide is an insoluble compound.
   b. Silver nitrate reacts with sodium iodide to form silver iodide.

17 Oxidation and Reduction

The element hydrogen figured prominently in our discussion of acid-base reactions in Chapter 14. But we saw also that the early, limited definitions of acids and bases could be extended, so that we can now consider many reactions that involve no hydrogen at all as acid-base reactions. In much the same way the terms *oxidation* and *reduction* were originally used in a narrow sense—referring only to reactions where a substance either combined with oxygen or lost oxygen—but have now been generalized to encompass a much wider range of reactions.

17.1 Reactions Involving Oxygen

The simplest oxidation reactions are those to which the word was first applied—reactions where a substance combines with oxygen. Weathering and corrosion processes are often reactions of this type. For example, the reddish rust that forms on iron is iron(III) oxide:

$$4\ Fe(s) + 3\ O_2(g) \longrightarrow 2\ Fe_2O_3(s)$$

Other simple oxidation reactions are much more vigorous. We call these strongly exothermic processes *combustion* reactions. The combustion of coal and coke (carbon), natural gas (principally methane), propane (C_3H_8), and butane (C_4H_{10}) are all examples that play an important role in energy production:

$$C(s) + O_2(g) \longrightarrow CO_2(g) + \text{heat}$$
$$CH_4(g) + 2\ O_2(g) \longrightarrow CO_2(g) + 2\ H_2O + \text{heat}$$
$$C_3H_8(g) + 5\ O_2(g) \longrightarrow 3\ CO_2(g) + 4\ H_2O + \text{heat}$$
$$2\ C_4H_{10}(g) + 13\ O_2(g) \longrightarrow 8\ CO_2(g) + 10\ H_2O + \text{heat}$$

In each of these equations, we have assumed that there is an ample supply of oxygen, assuring that the combustion is complete. Thus we find that the characteristic product of the complete combustion of carbon is CO_2 and those of hydrocarbons (compounds composed only of carbon and hydrogen) are CO_2

and H_2O. However, if there is not enough oxygen to assure complete combustion, the products may include carbon monoxide. For example, we must write two equations to represent the combustion of gasoline in the cylinders of an automobile engine:

$$2\ C_8H_{18}(l) + 25\ O_2(g) \longrightarrow 16\ CO_2(g) + 18\ H_2O$$
(octane)

and

$$2\ C_8H_{18}(l) + 17\ O_2(g) \longrightarrow 16\ CO(g) + 18\ H_2O$$

Likewise, a smoker inhales the combustion products of his cigarette before complete combustion can take place; hence the smoke contains considerable carbon monoxide.

The simplest example of a reduction reaction is little more than the reverse of a simple oxidation reaction. Practical examples include some of the important steps in metal production. Natural minerals suitable for refining, which we call *ores*, are often metal oxides. They can frequently be reduced with carbon (coke) or carbon monoxide:

$$Fe_2O_3(s) + 3\ C(s) \longrightarrow 3\ CO(g) + 2\ Fe(s)$$
$$SnO_2(s) + 2\ C(s) \longrightarrow 2\ CO(g) + Sn(s)$$
$$PbO(s) + CO(s) \longrightarrow Pb(s) + CO_2(g)$$

In each of these reactions, of course, the carbon or carbon monoxide is being oxidized at the same time that the metal oxide is being reduced. In the same way, chromium metal is produced by the reduction of chromium(III) oxide in a process that also oxidizes aluminum:

$$Cr_2O_3(s) + 2\ Al(s) \longrightarrow 2\ Cr(s) + Al_2O_3(s)$$

Questions

1. What evidence of corrosion have you seen recently?
2. What is meant by complete combustion?
3. Why are some minerals called ores and others not?
4. What three reagents for removing oxygen from oxides are mentioned in this section?

17.2 Oxidation and Reduction

We shall now turn back to the first example we cited as an oxidation reaction and use it to expand our definitions of oxidation and reduction. When iron combines with oxygen to form iron(III) oxide,

$$4\ Fe(s) + 3\ O_2(g) \longrightarrow 2\ Fe_2O_3(s)$$

changes occur in the electron configurations of each element. We shall discuss this reaction as if we could isolate individual atoms of iron and individual formula units of iron(III) oxide. In fact, both exist in crystalline arrays where discrete units cannot be identified, as we described in Chapter 9. Each neutral iron atom loses three electrons, producing a triply charged cation. And each neutral oxygen atom gains two electrons to produce a doubly charged oxide ion. We can thus think of the oxidation of Fe not as its combination with oxygen but more generally as its loss of electrons. Therefore, **oxidation** *is any process by which a substance loses electrons, and* **reduction** *is any process by which a substance gains electrons.* When iron rusts, it loses electrons and thus is oxidized; the oxygen gains electrons and thus is reduced. *Every oxidation is accompanied by a reduction,* because an oxidation-reduction reaction involves a partial or complete *transfer* of electrons.

When magnesium metal is burned in air, the magnesium reacts with both oxygen and nitrogen to produce magnesium oxide and magnesium nitride:

$$2\ Mg(s) + O_2(g) \longrightarrow 2\ MgO(s)$$
$$3\ Mg(s) + N_2(g) \longrightarrow Mg_3N_2(s)$$

According to the more restrictive definition of oxidation, the magnesium in the first equation but not in the second is oxidized. But, more broadly, we can see that magnesium undergoes the same change (from an oxidation state of 0 to one of +2) in both. The only difference is the substance that causes the oxidation, which we call the **oxidizing agent.** In the first case, it is oxygen; in the second, nitrogen. In all cases the oxidizing agent is reduced during the reaction. Conversely, the substance that provides the electrons (Mg in these examples) and is therefore oxidized by the reaction is called the **reducing agent.**

Many chemical reactions involve sharing rather than transferring electrons. The combustion of sulfur, for example, produces covalent sulfur dioxide, in which neither element has gained or lost electrons:

$$S_8(s) + 8\ O_2(g) \longrightarrow 8\ SO_2(g)$$

However, since oxygen is more electronegative than sulfur, the oxygen atoms in each SO_2 molecule more strongly attract the shared electron pairs than does the sulfur atom. Thus we say that the sulfur atom has been oxidized (has "lost" electrons) and that the oxygen atoms have been reduced (have "gained" electrons).

Question

5. What is the oxidizing agent in each of these reactions?
   a. $Zn + 2\ HCl \longrightarrow ZnCl_2 + H_2$ b. $2\ NaBr + Cl_2 \longrightarrow Br_2 + 2\ NaCl$

17.3 Oxidation Numbers

We introduced the concept of *oxidation numbers* as a bookkeeping device when discussing chemical nomenclature in Chapter 6. We can now expand the rules given there and perhaps make more sense of the term itself. The **oxidation number** of an element indicates the extent to which the element has been oxidized and thus can be thought of as a measure of how many electrons the element has gained or lost compared to its neutral state. For compounds comprising only ions, the oxidation number of each ion is the same as the ionic charge. In KCl, the oxidation number of potassium is $+1$ and that of chlorine is -1. Thus we say that potassium is in a $+1$ *oxidation state* and that chlorine is in a -1 *oxidation state*. Similarly, in magnesium nitride, Mg_3N_2, the oxidation number of magnesium is $+2$, and the oxidation number of nitrogen is -3.

For compounds with covalent bonds, we know that none of the atoms have gained or lost electrons in the strict sense, but we assign oxidation numbers as if all the bonds were ionic. The shared electrons are given to the more electronegative atom. Thus in ammonia

$$H:\overset{..}{\underset{..}{N}}:H$$
$$H$$

the shared electrons are assigned to the nitrogen atom, giving it an oxidation number of -3 and each hydrogen an oxidation number of $+1$. Fortunately, however, it is not necessary to memorize electronegativities to assign oxidation numbers to elements in covalent compounds. Instead we can rely on the following set of seven rules, which include those already listed in Chapter 6.

Rule 1: An element in the free state has an oxidation number of 0, regardless of the type of unit in which we find it. Thus the oxidation number of each element in the following natural substances is 0: H_2, Ar, Fe, Cu, O_2, O_3, S_2, S_8, P_4, C_n (graphite or diamond).

Rule 2: The oxidation number of a simple (monatomic) ion is the same as its ionic charge. Thus Group IA and IIA elements, whenever found in a compound or in solution, have oxidation numbers of $+1$ and $+2$, respectively, and aluminum has an oxidation number of $+3$.

Rule 3: The algebraic sum of the oxidation numbers of the atoms in the formula of a compound (taking note of the subscripts) must equal zero. Likewise, the sum of the oxidation numbers in the formula of a polyatomic ion must equal the net charge on the ion.

Rule 4: When combined with other elements, hydrogen always has an oxidation number of $+1$ unless Rule 2 or 3 requires otherwise. Thus hydrogen has a $+1$ oxidation number in compounds such as HCl, H_2O, and NH_3, but in NaH and CaH_2 (the metallic hydrides) Rules 2 and 3 require it to have a -1 oxidation number.

Rule 5: Oxygen in combined form always has an oxidation number of -2, unless Rule 2, 3, or 4 requires otherwise. Thus oxygen has an oxidation number of -2 in such compounds as Fe_2O_3, SO_2, and $NaOH$; an oxidation number of -1 in BaO_2, Na_2O_2, and H_2O_2 (the peroxides); and $-\frac{1}{2}$ in KO_2 (a superoxide).

Rule 6: The oxidation number of the halogens in combined form is always -1, unless one of the rules above requires otherwise. Thus chlorine has an oxidation number of -1 in $FeCl_3$, $SnCl_2$, Hg_2Cl_2, and HCl, but it has an oxidation number of $+1$ in $HClO$, $+3$ in $HClO_2$, $+5$ in $HClO_3$, and $+7$ in $HClO_4$.

Rule 7: Unless one of the first six rules requires otherwise, Group VIA elements in combined form have oxidation numbers of -2 and Group VA elements have -3.

If we apply these rules in their proper sequence, they provide an unambiguous scheme for assigning oxidation numbers to the elements of most compounds. (One exception is oxygen difluoride, OF_2, where the more electronegative fluorine has a -1 oxidation number and oxygen has a $+2$ oxidation number.) We should remember, however, that oxidation numbers are merely convenient labels, whose uses we shall soon discover, and not experimental quantities. Thus we should not be surprised to find, for example, that lead in Pb_3O_4 (red lead oxide) has an oxidation number of $+2\frac{2}{3}$. This number does not imply that each lead atom loses $2\frac{2}{3}$ electrons. (This oxide, incidentally, can be better represented as $2\ PbO \cdot PbO_2$, a formula that yields less awkward oxidation numbers.)

Example 17.1 Determine the oxidation number of Mn in $KMnO_4$.

Solution

a. Rule 1 does not apply. Rule 2 requires that the oxidation number of potassium (a Group IA metal) be $+1$.

b. Rule 3 requires that the sum of the three oxidation numbers be zero, but since we're still missing two numbers, we can't apply the rule.

c. Rule 4 is irrelevant but Rule 5 requires that the oxidation number of oxygen be -2.

d. We can now return to Rule 3 and calculate the oxidation number for Mn. Let $x =$ the oxidation number of Mn. Then

$$(+1) + (x) + 4(-2) = 0$$
$$x = +7$$

e. The oxidation number of Mn in $KMnO_4$ is $+7$.

Oxidation numbers usually range from -3 to $+7$, although $+8$ (for example, for osmium in OsO_4) and -4 (for carbon in Al_4C_3) are possible. Since $+7$ is the highest oxidation number found for manganese, the manganese in $KMnO_4$ can only be reduced (gain electrons) in subsequent oxidation-reduction reactions. Stated another way, the MnO_4^- ion can serve only as an oxidizing agent.

Example 17.2 Determine the oxidation numbers of the elements in Na_2O, Na_2O_2, and NaO_2.

Solution

a. According to Rule 2 sodium (Group IA) must have an oxidation number of $+1$ in each compound.

b. Rule 3 requires that the sum of the oxidation numbers be zero in each case. Therefore the oxidation number for oxygen is

-2 in Na_2O

-1 in Na_2O_2

$-\frac{1}{2}$ in NaO_2

This series of compounds provides an example of an *oxide*, a *peroxide*, and a *superoxide*. Binary compounds where oxygen has a -2 oxidation state are oxides; binary compounds where oxygen has a -1 oxidation state are peroxides; and binary compounds where oxygen has a $-\frac{1}{2}$ oxidation state are superoxides.

We can now turn from the routine calculation of oxidation numbers to the question of where they fit into our general discussion of oxidation-reduction reactions. We can begin with a basic observation: *A substance that is oxidized during a reaction will show an increase in oxidation number, whereas a substance that is reduced will show a decrease in oxidation number.*

Example 17.3 Assuming that the reaction below takes place in aqueous solution, determine which *element* is oxidized and which is reduced, and determine which *reagent* is the oxidizing agent and which is the reducing agent.

$$Zn(s) + 2\ HCl(aq) \longrightarrow ZnCl_2(aq) + H_2(g)$$

Solution

a. As a first step, assign oxidation numbers to each element:

$$\underset{Zn}{(0)} + \underset{2\ HCl}{(+1)(-1)} \longrightarrow \underset{ZnCl_2}{(+2)(-1)} + \underset{H_2}{(0)}$$

b. The oxidation number of zinc changes from 0 to +2; thus the zinc is oxidized by the reaction. The oxidation number of hydrogen, on the other hand, changes from +1 to 0; thus hydrogen is reduced during the reaction. The chlorine remains unchanged.

c. HCl, the reagent that causes the oxidation of the zinc and that contains the element reduced, is the oxidizing agent. Zinc, which reduces the hydrogen and is itself oxidized, is the reducing agent.

The net ionic equation for the reaction in Example 17.3 shows the oxidation-reduction process more clearly:

$$Zn(s) + 2\,H^+(aq) \longrightarrow Zn^{2+}(aq) + H_2(g)$$

In general, it is easier to discover which *element* has been oxidized and which has been reduced from the net ionic equation than from the overall equation, but from the net ionic equation above we know only that the oxidizing agent was some substance that contributed hydrogen ions. It might have been, say, HBr or HI. To identify HCl we need the overall equation.

Example 17.4 Identify the oxidizing and reducing agents and derive the net ionic equation for this reaction:

$$4\,HNO_3(aq) + Cu(s) \longrightarrow Cu(NO_3)_2(aq) + 2\,NO_2(g) + 2\,H_2O(l)$$

Solution

a. Assign oxidation numbers:

$$\underset{4\,HNO_3}{(+1)(+5)(-2)} + \underset{Cu}{(0)} \longrightarrow \underset{Cu(NO_3)_2}{(+2)(+5)(-2)} + \underset{2\,NO_2}{(+4)(-2)} + \underset{2\,H_2O}{(+1)(-2)}$$

b. The oxidation number of copper changes from 0 to +2; thus the copper is oxidized and serves as the reducing agent.

c. The oxidation number of nitrogen decreases from +5 to +4; thus nitrogen is reduced. HNO_3 is the oxidizing agent.

d. The net equation for this reaction is

$$4\,H^+(aq) + 2\,NO_3^-(aq) + Cu(s) \longrightarrow Cu^{2+}(aq) + 2\,NO_2(g) + 2\,H_2O$$

Several common strong oxidizing agents are listed in Table 17.1 together with the product of the reduction they undergo. Table 17.2 provides a similar list for reducing agents. We shall discuss a number of these substances, including HNO_3 with its several reduction products, in the following sections.

Table 17.1 Common Oxidizing Agents and Their Reduction Products

Oxidizing Agent	Reduction Product
F_2	F^-
Cl_2	Cl^-
O_2	O^{2-}
HNO_3	$NO_2, HNO_2, NO, N_2O, N_2, NH_3, NH_4^+$
conc. H_2SO_4	SO_2, S_8, H_2S
$K_2Cr_2O_7$	Cr^{3+}
K_2CrO_4	Cr^{3+}
$KMnO_4$ (acid solution)	Mn^{2+}
$KMnO_4$ (basic solution)	MnO_2
H_2O_2	H_2O
Fe^{3+}	Fe^{2+}

Table 17.2 Common Reducing Agents and Their Oxidation Products

Reducing Agent	Oxidation Product
Na	Na^+
K	K^+
Mg	Mg^{2+}
Ca	Ca^{2+}
Zn	Zn^{2+}
Fe	Fe^{2+}, Fe^{3+}
H_2S	S_8, SO_2
H_2	H^+
C	CO, CO_2
CO	CO_2
H_2SO_3	H_2SO_4
SO_2	SO_3
I^-	I_2
Br^-	Br_2
Cl^-	Cl_2
S^{2-}	S_8, SO_2
$SnCl_2$	$SnCl_4$

Questions

6. Calculate the oxidation number of the underlined element in each of the following compounds:

a. $\underline{N}H_4NO_3$ b. $Ca\underline{C}O_3$ c. $K\underline{H}$
d. $K_2\underline{O}_2$ e. $K\underline{I}O_3$ f. $\underline{Cr}_2(SO_4)_3$

7. The calculated oxidation number for carbon in acetic acid ($HC_2H_3O_2$) is zero. The electron formula for acetic acid is given below. By assigning each pair of shared electrons to the more electronegative element, determine the oxidation number of each individual atom in the molecule. Where a pair of electrons is shared equally, assign one electron to each atom.

$$\text{H : C : C} \begin{matrix} \text{H} \\ \text{H} \end{matrix} \begin{matrix} \ddot{\text{O}}: \\ \ddot{\text{O}}: \\ \text{H} \end{matrix}$$

8. Determine which *element* is oxidized, which element is reduced, which *reagent* acts as an oxidizing agent, and which reagent acts as a reducing agent in each equation below:
  a. $Br_2(aq) + 2\ KI(aq) \longrightarrow I_2(aq) + 2\ KBr(aq)$
  b. $16\ H_2S(g) + 8\ SO_2(g) \longrightarrow 3\ S_8(s) + 16\ H_2O(g)$

17.4 Half-Reactions

One shortcut to analyzing *redox* reactions (as oxidation-reduction reactions are often called) is to consider separate **half-reactions;** that is, to separate the equation for the reaction into two more elementary equations, one describing the reduction and the other, the oxidation. Bear in mind, however, that neither of these equations can represent a complete chemical reaction, because neither an oxidation nor a reduction can occur in the absence of the other. The reason for this is clear when we realize that any equation for a half-reaction shows electrons as reactants or products, whereas in any complete redox reaction electrons are merely transferred from one substance to another. Thus any complete reaction comprises, at least, two half-reactions. For example, we can think of the oxidation of copper by nitric acid (Example 17.4) as the sum of two half-reactions:

Oxidation half-reaction: $Cu \longrightarrow Cu^{2+} + 2e^-$

Reduction half-reaction: $NO_3^- + 2\ H^+ + e^- \longrightarrow NO_2 + H_2O$

Since the electrons produced in the first half-reaction must be consumed by the second, the reduction half-reaction must be multiplied by two before the equations are added:

$$Cu \longrightarrow Cu^{2+} + 2e^-$$
$$\underline{2NO_3^- + 4H^+ + 2e^- \longrightarrow 2NO_2 + 2H_2O}$$
$$Cu(s) + 2NO_3^-(aq) + 4H^+(aq) + 2e^- \longrightarrow 2NO_2(g) + 2H_2O(l) + 2e^- + Cu^{2+}(aq)$$

The sum is the net ionic equation for the reaction.

Just as we have listed acids according to their strengths and found the list useful in predicting the direction of acid-base reactions, we can now list a number of substances according to their strengths as oxidizing and reducing agents (Table 17.3). The elements that appear on the table behave much as we would

Table 17.3 Relative Strengths of Oxidizing and Reducing Agents
(A) indicates $[H^+] = 1\,M$; (B) indicates $[OH^-] = 1\,M$.

Oxidizing Agent		Reducing Agent
$F_2 + 2e^-$		$\rightleftarrows 2\,F^-$
$H_2O_2 + 2\,H^+ + 2e^-$	(A)	$\rightleftarrows 2\,H_2O$
$Au^+ + e^-$		$\rightleftarrows Au$
$MnO_4^- + 8\,H^+ + 5e^-$	(A)	$\rightleftarrows Mn^{2+} + 4\,H_2O$
$Au^{3+} + 3e^-$		$\rightleftarrows Au$
$ClO_4^- + 8\,H^+ + 8e^-$	(A)	$\rightleftarrows Cl^- + 4\,H_2O$
$Cl_2 + 2e^-$		$\rightleftarrows 2\,Cl^-$
$Cr_2O_7^{2-} + 14\,H^+ + 6e^-$	(A)	$\rightleftarrows 2\,Cr^{3+} + 7\,H_2O$
$O_2 + 2\,H_2O + 4e^-$		$\rightleftarrows 4\,OH^-$
$Br_2 + 2e^-$		$\rightleftarrows 2\,Br^-$
$NO_3^- + 4\,H^+ + 3e^-$	(A)	$\rightleftarrows NO + 2\,H_2O$
$Hg^{2+} + 2e^-$		$\rightleftarrows Hg$
$2\,NO_3^- + 4\,H^+ + 2e^-$	(A)	$\rightleftarrows N_2O_4 + 2\,H_2O$
$Ag^+ + e^-$		$\rightleftarrows Ag$
$Fe^{3+} + e^-$		$\rightleftarrows Fe^{2+}$
$O_2 + 4\,H^+ + 4e^-$	(A)	$\rightleftarrows 2\,H_2O$
$MnO_4^- + 2\,H_2O + 3e^-$	(B)	$\rightleftarrows MnO_2 + 4\,OH^-$
$I_2 + 2e^-$		$\rightleftarrows 2\,I^-$
$8\,H_2SO_3 + 32\,H^+ + 32e^-$	(A)	$\rightleftarrows S_8 + 24\,H_2O$
$Cu^{2+} + 2e^-$		$\rightleftarrows Cu$
$Co(NH_3)_6^{3+} + e^-$	(B)	$\rightleftarrows Co(NH_3)_6^{2+}$
$2\,H^+ + 2e^-$	(A)	$\rightleftarrows H_2$
$Fe^{3+} + 3e^-$		$\rightleftarrows Fe$
$CrO_4^{2-} + 4\,H_2O + 3e^-$	(B)	$\rightleftarrows Cr(OH)_3 + 5\,OH^-$
$Pb^{2+} + 2e^-$		$\rightleftarrows Pb$
$Sn^{2+} + 2e^-$		$\rightleftarrows Sn$
$Ni^{2+} + 2e^-$		$\rightleftarrows Ni$
$Co^{2+} + 2e^-$		$\rightleftarrows Co$
$Cd^{2+} + 2e^-$		$\rightleftarrows Cd$
$Fe^{2+} + 2e^-$		$\rightleftarrows Fe$
$Cr^{2+} + 2e^-$		$\rightleftarrows Cr$
$Fe(OH)_3 + e^-$	(B)	$\rightleftarrows Fe(OH)_2 + OH^-$
$Cr^{3+} + 3e^-$		$\rightleftarrows Cr$
$Zn^{2+} + 2e^-$		$\rightleftarrows Zn$
$SO_4^{2-} + H_2O + 2e^-$	(B)	$\rightleftarrows SO_3^{2-} + 2\,OH^-$
$Mn^{2+} + 2e^-$		$\rightleftarrows Mn$
$Be^{2+} + 2e^-$		$\rightleftarrows Be$
$Al^{3+} + 3e^-$		$\rightleftarrows Al$
$Sc^{3+} + 3e^-$		$\rightleftarrows Sc$
$Mg^{2+} + 2e^-$		$\rightleftarrows Mg$
$Na^+ + e^-$		$\rightleftarrows Na$
$Ca^{2+} + 2e^-$		$\rightleftarrows Ca$
$Sr^{2+} + 2e^-$		$\rightleftarrows Sr$
$Ba^{2+} + 2e^-$		$\rightleftarrows Ba$
$Cs^+ + e^-$		$\rightleftarrows Cs$
$K^+ + e^-$		$\rightleftarrows K$
$Li^+ + e^-$		$\rightleftarrows Li$

← decreasing strength as an oxidizing agent

increasing strength as a reducing agent →

predict from their electronegativities. Metals tend to lose electrons and are thus reducing agents; nonmetals tend to gain electrons and are thus oxidizing agents. Compounds, on the other hand, are often more difficult to predict. Their behavior can usually be judged only after careful experiments. In Table 17.3, substances appear at the left in order of decreasing strength as oxidizing agents, together with the half-reactions by which they are reduced. At the right of the table are the reduction products, which fall into a list of reducing agents arranged in order of increasing strength. This is as we would predict since if the reduction reactions near the top of the table proceed vigorously, the reverse reactions (by which the reducing agents are oxidized) must be very unfavorable. By the same reasoning, the reactions near the bottom of the table are more likely to proceed from right to left, reflecting weak oxidizing agents and strong reducing agents. We should emphasize, however, that oxidizing and reducing strengths are relative and, except for the reactions at the very top and the very bottom of the table, all the reactions can go either way, depending on the other half-reaction in the complete equation. For example, bromine is a fairly strong oxidizing agent and thus we expect to find it reduced to Br^- in most redox reactions; however, in a reaction with a stronger oxidizing agent (such as Cl_2), Br^- is oxidized to Br_2.

The common metals from the right-hand column of Table 17.3, if listed with the strongest reducing agent at the top, form the **electrochemical series** (Table 17.4). Since this list compares metals according to how easily they lose electrons in solution, we might compare it to a similar list derived from ionization energies. The differences are due to the role played by hydration energy in the

TABLE 17.4 The Electrochemical Series (Activity Series) of the Metals

Lithium
Potassium
Sodium
Magnesium
Scandium
Aluminum
Manganese
Zinc
Chromium
Iron
Cobalt
Nickel
Tin
Lead
(Hydrogen)
Copper
Arsenic
Bismuth
Antimony
Mercury
Silver
Platinum
Gold

electrochemical series. Ionization energies are measured for elements in the gas phase. Potassium is therefore the most easily ionized metal (of those listed in Table 17.4), but lithium with a high hydration energy is a more *active* metal in solution.

Questions

9. List the three strongest oxidizing agents and the three strongest reducing agents from Table 17.3.

10. How is it possible for a substance to function well as an oxidizing agent in one reaction and a reducing agent in another?

17.5 Predicting the Direction of Oxidation-Reduction Reactions

Before we attempt to use Table 17.3 to predict the directions of oxidation-reduction reactions, careful note must be taken of the conditions for which the table was derived. Each reaction is assumed to take place between the substances with each one in its standard state at 25°C (see Section 11.4) and all reactants at 1 M concentration. (To be precise we should use activities here rather than molarities, as discussed briefly in Chapter 13. But using molarities introduces little error.) Furthermore, the reactions marked (A) occur in acidic solutions with $[H^+] = 1\ M$; those marked (B) take place in basic solutions with $[OH^-] = 1\ M$.

With this understanding, we can now say that in general a substance to the left of the arrows will oxidize any substance to the right that lies below it. Thus chlorine will react with zinc, oxidizing it to Zn^{2+}. The zinc loses electrons and the chlorine gains electrons. To show this, we combine the two appropriate half-reactions, reversing the one for zinc. Since the two half-reactions involve the same number of electrons, the reactions can simply be added:

$$Cl_2 + 2e^- \rightleftharpoons 2\ Cl^-$$
$$Zn \rightleftharpoons Zn^{2+} + 2e^-$$
$$\overline{Cl_2(aq) + Zn(s) + \cancel{2e^-} \rightleftharpoons 2\ Cl^-(aq) + Zn^{2+}(aq) + \cancel{2e^-}}$$

The result is the net ionic equation for the formation of zinc chloride from zinc and chlorine (in aqueous solution).

Also from the table, we expect copper(II) ion to react with zinc metal:

$$Zn \rightleftharpoons Zn^{2+} + 2e^-$$
$$Cu^{2+} + 2e^- \rightleftharpoons Cu$$
$$\overline{Zn(s) + Cu^{2+}(aq) + \cancel{2e^-} \rightleftharpoons Zn^{2+}(aq) + Cu(s) + \cancel{2e^-}}$$

This equation shows what happens when a piece of metallic zinc is immersed in a solution of a soluble copper(II) compound: The zinc dissolves and the copper is deposited as a coating on the zinc. A typical overall equation for this reaction might be

$$Zn(s) + CuSO_4(aq) \longrightarrow ZnSO_4(aq) + Cu(s)$$

In this reaction we say that zinc has *replaced* the copper, meaning that it has replaced it in the combined state.

In a similar reaction aluminum replaces zinc:

$$\frac{\begin{array}{r} 2(Al \rightleftarrows Al^{3+} + 3e^-) \\ 3(Zn^{2+} + 2e^- \rightleftarrows Zn) \end{array}}{2\,Al(s) + 3\,Zn^{2+}(aq) + \cancel{6e^-} \rightleftarrows 2\,Al^{3+}(aq) + 3\,Zn(s) + \cancel{6e^-}}$$

Here, however, we notice a difference. Before the two half-reaction equations can be added, they must be adjusted so that the number of electrons gained equals the number lost. Therefore, we multiply the aluminum equation by two and the zinc equation by three. An overall equation for this reaction might be

$$2\,Al(s) + 3\,ZnSO_4(aq) \rightleftarrows Al_2(SO_4)_3 + 3\,Zn(s)$$

This reaction would take place if we placed a strip of aluminum metal in a solution of zinc sulfate: The aluminum would dissolve and the zinc would be deposited. The aluminum metal is the reducing agent and the zinc ion is the oxidizing agent. These last two reactions illustrate the relative nature of reducing or oxidizing ability. In a reaction with copper, zinc metal is oxidized, whereas, in a reaction with aluminum, Zn^{2+} is reduced.

It should be apparent from the relative positions in the table of copper, zinc, and aluminum that we could expect no spontaneous reaction if we placed a strip of copper metal in a zinc sulfate solution or a strip of zinc metal in an aluminum sulfate solution. (It might also be obvious that the use of the term *spontaneous* will eventually involve us in the principles of thermodynamics as we look further into redox reactions.)

In Example 17.5, we shall look at another series of replacement reactions, this time involving nonmetals.

Example 17.5 Determine all the reactions that will take place spontaneously among the following pairs: Cl_2 and Cl^-, Br_2 and Br^-, and I_2 and I^-. Write typical overall equations for each reaction.

Solution

a. We expect a spontaneous reaction between a substance to the left of the arrows and any substance below it and to the right of the arrows. Thus Cl_2 will react with Br^- and I^-, and Br_2 will react with I^-.

b. When Cl_2 reacts with Br^-, Cl_2 is the oxidizing agent and Br^- is the reducing agent:

$$Cl_2 + 2e^- \rightleftharpoons 2\ Cl^-$$
$$2\ Br^- \rightleftharpoons Br_2 + 2e^-$$
$$\overline{Cl_2(aq) + 2\ Br^-(aq) + \cancel{2e^-} \rightleftharpoons Br_2(aq) + 2\ Cl^-(aq) + \cancel{2e^-}}$$

The chlorine has replaced the bromide ion. We saw a typical equation for this reaction earlier:

$$Cl_2(aq) + 2\ KBr(aq) \rightleftharpoons Br_2(aq) + 2\ KCl(aq)$$

c. The reaction between Cl_2 and I^- is exactly parallel to that above. An overall equation for the reaction might be

$$Cl_2(aq) + 2\ KI(aq) \rightleftharpoons 2\ KCl(aq) + I_2(aq)$$

d. Br_2 reacts with I^- in a similar way. An overall equation for the reaction might be

$$Br_2(aq) + CaI_2(aq) \rightleftharpoons CaBr_2(aq) + I_2(aq)$$

e. No reaction will take place between I_2 and Br^-, I_2 and Cl^-, or Br_2 and Cl^-. These reactions, incidentally, form the basis for an identification test that distinguishes among the three colorless ions, Cl^-, Br^-, and I^-. Chlorine water (water saturated with chlorine gas), along with a small amount of an organic solvent such as carbon tetrachloride is added to a solution of the unknown halide ion. Any free halogen produced will dissolve in the organic solvent and its characteristic color will be concentrated. In carbon tetrachloride, Cl_2 is almost colorless, Br_2 is orange, and I_2 is purple.

Question

11. Which of the following pairs of reagents will react according to the data in Table 17.3?
    a. Cl_2 with a solution of NaI.
    b. Cobalt metal with the solution of a strong acid.
    c. Aluminum metal with a copper nitrate solution.
    d. Nickel metal with a calcium acetate solution.

17.6 The Preparation of Hydrogen

In Chapters 19 and 20 we shall discuss some of the properties and reactions of the nonmetals. But it is more appropriate to consider the preparation of one of these nonmetals, namely hydrogen, here. The reason is that all the reactions by which hydrogen is produced, both in the laboratory and commercially, are oxidation-reduction reactions. Furthermore, we can use these reactions to illustrate several general principles that have been introduced in this chapter and in earlier ones.

Since hydrogen most often occurs in the +1 state, the problem of preparing hydrogen gas would appear simply to be one of selecting the proper reagent to reduce it to the zero oxidation state. All metals below hydrogen in Table 17.3 will reduce the H^+ ion, so there would seem to be a wide choice. However, not all of these metals react with all hydrogen-containing compounds in solution to produce hydrogen gas, and others react too slowly or too vigorously to be useful. The reaction between hydrochloric acid and potassium, for example, generates so much heat that the gas generated is likely to ignite or explode. Less active metals, such as aluminum, zinc, iron, and manganese, can be used more successfully with hydrochloric acid in laboratory preparations of hydrogen.

The story is very much the same with *dilute* sulfuric acid (about 6 M): Aluminum, zinc, iron, and manganese will produce hydrogen gas at a convenient rate. There is, however, an important difference between the reactions with these two acids. Whereas the reaction rate *increases* as the concentration of hydrochloric acid increases, it *decreases* markedly in concentrated H_2SO_4. The reason is that sulfuric acid is a strong acid only in dilute solutions; 6 M H_2SO_4 is almost completely ionized, but 18 M H_2SO_4 is only about 3 percent ionized. And of course it is the hydrogen ions, not the un-ionized acid, that the metal reduces.

We might also compare the rates of reaction of metals with a strong acid like 6 M HCl and a weak acid like 6 M acetic acid ($HC_2H_3O_2$). In the acetic acid the hydrogen ion concentration is less than 0.12 M; thus it reacts very slowly with metals such as zinc or iron. Magnesium (a metal somewhat lower in Table 17.4) still produces hydrogen at a reasonable rate. It is important to emphasize, however, that even with zinc a liter of 6 M acetic acid will produce as much hydrogen gas as a liter of 6 M hydrochloric acid. It will just take longer. As the zinc reduces the H^+ ions in the acetic acid, a stress is applied to the system at equilibrium,

$$HC_2H_3O_2(aq) \rightleftharpoons H^+(aq) + C_2H_3O_2^-(aq)$$

causing it to shift to the right. Thus although the hydrogen ion concentration in 6 M acetic acid is never so great as it is in 6 M HCl, the amount of H^+ eventually reduced is the same.

A fourth common laboratory acid is nitric acid. Like hydrochloric acid it is a strong acid, but as a source of reducible hydrogen ions the resemblance ends there. If we add metals to nitric acid, the product is likely to be an orange-brown gas rather than colorless hydrogen gas. Since the nitrate ion is above the hydrogen ion in Table 17.3, the nitrate ion is more readily reduced. Any metal which reacts with nitric acid will therefore produce not hydrogen gas but one (or more) of the reduction products of NO_3^-: NO_2, HNO_2, NO, N_2O, N_2, NH_3, or NH_4^+. As we would expect, the extent of reduction depends on the metal and on the concentration of the nitric acid, but nitrogen dioxide and nitrogen oxide are the most common products. Curiously, however, some very active metals, such as chromium and aluminum, which we would expect to reduce nitrate ions vigorously, do not react strongly with nitric acid. Unlike most metals, which form metallic ions when they react with nitric acid, these metals form metallic oxides, which insulate the metals from further contact with the acid. Incidentally, by forming oxides, aluminum and chromium show that they have considerable nonmetallic character: Nonmetals such as carbon, sulfur, and phosphorus also react with nitric acid to form their oxides. Still other elements, such as tin, behave as metals when they react with dilute nitric acid and nonmetals when they react with concentrated nitric acid:

In dilute acid:

$$3\,Sn(s) + 8\,H^+(aq) + 2\,NO_3^-(aq) \longrightarrow 3\,Sn^{2+}(aq) + 2\,NO(g) + 4\,H_2O(l)$$

In concentrated acid:

$$Sn(s) + 4\,H^+(aq) + 4\,NO_3^-(aq) \longrightarrow SnO_2(s) + 4\,NO_2(g) + 2\,H_2O(l)$$

These reactions are a good illustration of the generalization that when an element has more than one oxidation state, it behaves more like a metal in the lower state.

Sometimes temperature is an important factor in reactions between metals and acids. Zinc, for example, will reduce the hydrogen ion in *cold* concentrated sulfuric acid; but if zinc is added to *hot* concentrated sulfuric acid, it reduces H_2SO_4 molecules to SO_2:

$$2\,H_2SO_4(l) + Zn(s) \longrightarrow ZnSO_4(aq) + SO_2(g) + 2\,H_2O(l)$$

Nitric acid and hot concentrated sulfuric acid, among others, are called *oxidizing acids*. These acids contain species more readily reduced than H^+; hence they react with metals to produce products other than hydrogen.

A final common source of hydrogen ions is water. Although the concentration of hydrogen ions in water is very low, even cold water will react explosively with the metals of Group IA. Barium and calcium, both in Group IIA, react at a more moderate rate and are convenient laboratory sources of hydrogen gas:

$$Ca(s) + 2\,H_2O(l) \longrightarrow Ca(OH)_2(s) + H_2(g)$$

Less active metals will not liberate hydrogen from cold water but will react with hot water or steam:

$$3\,Fe(s) + 4\,H_2O(g) \longrightarrow Fe_3O_4(s) + 4\,H_2(g)$$

$$Zn(s) + H_2O(g) \longrightarrow ZnO(s) + H_2(g)$$

Questions

12. Why does zinc react less vigorously with concentrated sulfuric acid than with moderately dilute sulfuric acid?

13. Each of these pairs of reagents is *unsuitable* for a laboratory preparation of hydrogen. Explain why.
    a. $Cr + HNO_3$ b. $Cu + HCl$ c. $Zn + $ conc. H_2SO_4 d. $Na + HCl$
    e. $Fe + HNO_3$ f. $Zn + H_2O$ g. $K + H_2O$

17.7 Writing Equations for Redox Reactions

We have already discussed the most basic way of writing equations for redox reactions: *the method of half-reactions*, for which we needed a list like Table 17.3. We simply added the appropriate half-reaction equations to produce the net ionic equation, ensuring that no electrons appeared as reactants or products. We can now derive the overall equation from this net ionic equation if we observe one important precaution: No changes can be made that will upset the equality of the electron exchange. This generally means that we leave unchanged the coefficients of substances oxidized and reduced, as well as the coefficients of their products. Occasionally we must multiply them by some integer to avoid fractional coefficients elsewhere in the equation. When we do, we must multiply *all* coefficients of substances involved in the electron exchange by the same number. As an example, let us derive the overall equation for the reaction between copper metal and nitric acid from the net ionic equation:

$$Cu(s) + 2\,NO_3^-(aq) + 4\,H^+(aq) \longrightarrow Cu^{2+}(aq) + 2\,NO_2(g) + 2\,H_2O(l)$$

An overall equation for this reaction must identify the soluble, ionic compound that contains the Cu^{2+} ions, as well as account for the two extra hydrogen ions on the left side of the equation. In this example there is little difficulty identifying the negative ion we need. Whereas some nitrate ions are reduced by the reaction, others remain unchanged and do not appear in the net ionic equation. These spectator ions (see Chapter 16) are the species that were canceled out of a complete ionic equation to derive a net ionic equation. We can thus add two nitrate ions to *each* side of the equation to produce the overall equation:

$$Cu(s) + 4\,HNO_3(aq) \longrightarrow Cu(NO_3)_2(aq) + 2\,NO_2(g) + 2\,H_2O(l)$$

Two other means at our disposal for writing equations for redox reactions do not require the use of Table 17.3. These are the *electron-transfer method*, which can be applied to all redox reactions but is sometimes misleading, and *the ion-electron method*, which applies only to reactions in aqueous solutions and which furnishes more explicit chemical information. For reactions in aqueous solution, each method can be used as a check on the other.

Ion-Electron Method This method requires only that we know whether the reaction takes place in an acidic or basic solution. The following steps will then produce the overall equation:

Step a: Write the formula for each reactant and each product as it would appear in an ionic equation.

Step b: Determine which substances are oxidized and which are reduced. If it is necessary to use oxidation numbers, be sure not to confuse them with ionic charges. Begin writing the equation for the oxidation half-reaction by writing the formulas of substances containing the elements oxidized along with the formulas of the oxidation products. Do the same for the reduction half-reaction. Write formulas only as they appear in the statement of Step a.

Step c: Complete the equations for the two half-reactions as follows:

1. In each equation, balance the elements oxidized or reduced with appropriate coefficients.

2. Balance all other elements except hydrogen and oxygen. Remember, no formula should be written that does not appear in the ionic statement.

3. Balance hydrogen and oxygen by using hydrogen ions and water molecules if the reaction takes place in acid solution, and by using hydroxide ions and water molecules if in basic solution.

4. Balance the charge on the two sides of an equation by adding electrons as required.

Step d: Multiply each half-reaction equation by the lowest factor that will make the number of electrons lost equal to the number gained.

Step e: Add the two half-reaction equations to obtain the net ionic equation. Derive the overall equation as before.

Example 17.6 Obtain the overall equation for the reaction in dilute HCl between $KMnO_4$ and $FeCl_2$ to produce $MnCl_2$, $FeCl_3$, KCl, and H_2O.

Solution

a. Write the formulas for the reactants and products as they would appear in an ionic equation. In this problem all salts are soluble and HCl is a strong acid; therefore

$$(K^+ + MnO_4^-) + (Fe^{2+} + 2\ Cl^-) + (H^+ + Cl^-) \longrightarrow$$
$$(Mn^{2+} + 2\ Cl^-) + (Fe^{3+} + 3\ Cl^-) + (K^+ + Cl^-) + H_2O$$

b. Write the equation for the oxidation half-reaction. It needs no coefficients and can be balanced by adding one electron:

$$Fe^{2+} \longrightarrow Fe^{3+} + e^-$$

c. Write the equation for the reduction half-reaction. The species reduced is the permanganate ion, which is reduced to the manganese(II) ion:

$$MnO_4^- \longrightarrow Mn^{2+}$$

No coefficients are required to balance the manganese. Since the reaction takes place in acid solution, water molecules must be added to the right side to balance the oxygen:

$$MnO_4^- \longrightarrow Mn^{2+} + 4\ H_2O$$

The hydrogen can be balanced by adding hydrogen ions to the left side:

$$8\,H^+ + MnO_4^- \longrightarrow Mn^{2+} + 4\,H_2O$$

Finally, add five electrons to balance the charge:

$$8\,H^+ + MnO_4^- + 5e^- \longrightarrow Mn^{2+} + 4\,H_2O$$

d. To match the five electrons in the reduction half-reaction, the oxidation half-reaction must be multiplied by five. The two equations can then be added:

$$5\,Fe^{2+} \longrightarrow 5\,Fe^{3+} + 5e^-$$
$$\underline{MnO_4^- + 8\,H^+ + 5e^- \longrightarrow Mn^{2+} + 4\,H_2O}$$
$$MnO_4^-(aq) + 8\,H^+(aq) + 5\,Fe^{2+}(aq) + 5e^- \longrightarrow 5\,Fe^{3+}(aq) + Mn^{2+}(aq) + 4\,H_2O + 5e^-$$

e. The overall equation requires one additional K^+ ion and 18 Cl^- ions on each side:

$$\mathbf{KMnO_4(aq) + 8\,HCl(aq) + 5\,FeCl_2(aq) \longrightarrow 5\,FeCl_3(aq)}$$
$$\mathbf{+ \,MnCl_2(aq) + KCl(aq) + 4\,H_2O}$$

Example 17.7 Derive the overall equation for the reaction in aqueous NaOH between $MnCl_2$ and Br_2 to produce MnO_2, NaCl, NaBr, and H_2O.

Solution

a. $MnCl_2$, NaCl, NaBr, and NaOH are soluble ionic compounds. MnO_2 is insoluble, and bromine and water are essentially molecular.

$$(Mn^{2+} + 2\,Cl^-) + Br_2 + (Na^+ + OH^-) \longrightarrow MnO_2$$
$$+ (Na^+ + Cl^-) + (Na^+ + Br^-) + H_2O$$

b. The reduction half-reaction equation is

$$Br_2 + 2e^- \longrightarrow 2\,Br^-$$

c. In the oxidation half-reaction, manganese(II) ion is oxidized to manganese(IV) oxide:

$$Mn^{2+} \longrightarrow MnO_2$$

Because the reaction takes place in basic solution, we can use only H_2O and OH^- to balance the elements on each side of the equation:

$$Mn^{2+} + 4\,OH^- \longrightarrow MnO_2 + 2\,H_2O$$

Finally, two electrons balance the charge:

$$Mn^{2+} + 4\,OH^- \longrightarrow MnO_2 + 2\,H_2O + 2e^-$$

d. Add the two half-reaction equations:

$$Br_2 + 2e^- \longrightarrow 2\,Br^-$$
$$4\,OH^- + Mn^{2+} \longrightarrow MnO_2 + 2\,H_2O + 2e^-$$
$$\overline{4\,OH^- + Mn^{2+} + Br_2 \longrightarrow MnO_2 + 2\,H_2O + 2\,Br^-}$$

e. Derive the overall equation by adding two Cl^- ions and four Na^+ ions to each side:

$$4\,NaOH(aq) + MnCl_2(aq) + Br_2(aq) \longrightarrow MnO_2(s)$$
$$+ 2\,NaBr(aq) + 2\,NaCl(aq) + 2\,H_2O(l)$$

Electron-Transfer Method This method is quicker than the ion-electron method since it employs a sort of shorthand notation for representing the changes that take place in the reaction. Furthermore, its applicability is not restricted to aqueous solutions. However, the electron-transfer method often distorts the chemical facts since it treats all reactants as if they were discrete atoms, molecules, or ions. The steps in this method are

Step a: Write an unbalanced expression for the overall reaction and indicate all oxidation numbers.

Step b: Use a bracket to connect each element oxidized to its product and each element reduced to its product.

Step c: Indicate with the bracket the number of electrons gained or lost *per atom*.

Step d: Check the subscripts on the formulas of substances containing the elements oxidized and reduced and if necessary change the notation above the brackets to indicate the gross number of electrons gained or lost per molecule or per formula unit. Use temporary coefficients to balance the number of atoms of the element oxidized with the number of atoms in its product. Do the same for the element reduced.

Step e: Multiply the coefficients of the substances connected by brackets by numbers that will make the gain and loss of electrons equal.

Step f: Complete the equation by determining the remaining coefficients. In this last step, do not alter the coefficients of the substances oxidized and reduced or the coefficients of their products so as to upset the electron exchange.

Example 17.8 Derive an overall equation for a reaction in which CdS, I_2, and HCl react to form $CdCl_2$, HI, and S_8.

a. Write an unbalanced expression for the overall reaction and indicate all oxidation numbers.

$$\underset{\text{CdS}}{(+2)(-2)} + \underset{\text{I}_2}{(0)} + \underset{\text{HCl}}{(+1)(-1)} \longrightarrow \underset{\text{CdCl}_2}{(+2)(-1)} + \underset{\text{HI}}{(+1)(-1)} + \underset{\text{S}_8}{(0)}$$

b. Use a bracket to connect each element oxidized to its product and each element reduced to its product. Indicate with each bracket the number of electrons gained or lost *per atom*.

$$\overset{(+1e^- \text{ per atom})}{\overbrace{\underset{\text{CdS}}{(+2)(-2)} + \underset{\text{I}_2}{(0)} + \underset{\text{HCl}}{(+1)(-1)} \longrightarrow \underset{\text{CdCl}_2}{(+2)(-1)} + \underset{\text{HI}}{(+1)(-1)} + \underset{\text{S}_8}{(0)}}}$$
$$\underset{(-2e^- \text{ per atom})}{\underbrace{}}$$

c. Use temporary coefficients to balance the number of atoms of the element oxidized with the number of atoms of its product. Do the same for the element reduced.

$$\overset{(+1e^- \text{ per atom})}{\overbrace{(8)\ \text{CdS} + \text{I}_2 + \text{HCl} \longrightarrow \text{CdCl}_2 + (2)\ \text{HI} + \text{S}_8}}$$
$$\underset{(-2e^- \text{ per atom})}{\underbrace{}}$$

d. Indicate with each bracket the number of electrons gained or lost per reacting (or produced) unit; then multiply the coefficients of the substances connected by brackets by numbers that will make the gain and loss of electrons equal.

$$\overset{\substack{+2e^- \text{ per I}_2 \text{ molecule} \\ (+1e^- \text{ per atom})}}{\overbrace{(8)\ \text{CdS} + 8\ \text{I}_2 + \text{HCl} \longrightarrow \text{CdCl}_2 + (16)\ \text{HI} + \text{S}_8}}$$
$$\underset{\substack{(-2e^- \text{ per atom}) \\ -16e^- \text{ per S}_8 \text{ molecule}}}{\underbrace{}}$$

e. Complete the equation by determining the remaining coefficients.

8 CdS + 8 I$_2$ + 16 HCl $\longrightarrow$ 8 CdCl$_2$ + 16 HI + S$_8$

17.8 Equivalent Weights in Redox Reactions

The **equivalent weight** *of a substance in an oxidation-reduction reaction is that weight of the substance which gains or loses* 6.02×10^{23} *electrons.* Thus one equivalent of an oxidizing agent reacts exactly with one equivalent of a reducing agent. With this definition we can now approach the stoichiometry of redox reactions just as we have the stoichiometry of other chemical reactions. To calculate the weight of $KMnO_4$ necessary to oxidize 2.30×10^2 g of $Fe(NO_3)_2$ in solution, we need only to determine the number of equivalents of $Fe(NO_3)_2$ that react, for that will also be the number of equivalents of $KMnO_4$ that react. Of course we must know the products of the reaction, since they tell us how many electrons have been transferred.

Example 17.9 Calculate the weight of $KMnO_4$ (formula weight = 158) required to oxidize 2.30×10^2 g of $Fe(NO_3)_2$ (formula weight = 180) to $Fe(NO_3)_3$. The other product of the reaction is $Mn(NO_3)_2$.

Solution

a. By looking at the oxidation numbers of Mn and Fe, we first calculate the equivalent weight of each reacting substance. Each formula unit of $KMnO_4$ gains $5e^-$ as the Mn is reduced:

$$\underset{KMnO_4}{\overset{+7}{\vphantom{|}}} \xrightarrow{+5e^-} \underset{Mn(NO_3)_2}{\overset{+2}{\vphantom{|}}}$$

Thus there are 5 equivalents/mol of $KMnO_4$, and the weight of one equivalent is

$$\left(\frac{158 \text{ g}}{1.00 \text{ mol}}\right)\left(\frac{1 \text{ mol}}{5 \text{ equivalents}}\right) = 31.6 \text{ g/equivalent}$$

Each ion of Fe^{2+} loses one electron as it is oxidized:

$$\underset{Fe^{2+}}{\vphantom{|}} \xrightarrow{-1e^-} \underset{Fe^{3+}}{\vphantom{|}}$$

The equivalent weight of $Fe(NO_3)_2$ is thus the same as the gram formula weight (180 g).

b. 2.30×10^2 g of $Fe(NO_3)_2$ is

$$(230 \text{ g } Fe(NO_3)_2)\left(\frac{1.00 \text{ equivalent}}{180 \text{ g}}\right) = 1.28 \text{ equivalents } Fe(NO_3)_2$$

c. Each equivalent of $Fe(NO_3)_2$ reacts with one equivalent of $KMnO_4$, and each equivalent of $KMnO_4$ weighs 31.6 g; therefore, the weight of $KMnO_4$

required is

$$(1.28 \text{ equivalents KMnO}_4) \left(\frac{31.6 \text{ g}}{1.00 \text{ equivalent KMnO}_4} \right) = 40.4 \text{ g}$$

Example 17.10 In an acid solution $KMnO_4$ oxidizes H_2S to free sulfur, and the manganese is reduced to the $+2$ state. What weight of free sulfur will be produced if 75.0 g of $KMnO_4$ and 24.0 liters (STP) of H_2S are allowed to react?

Solution

a. Determine the equivalent weight (or volume) of each of the two reactants. As in Example 17.9 each formula unit of $KMnO_4$ gains five electrons; thus its equivalent weight is 31.6 g. Each molecule of H_2S loses two electrons as the sulfur is oxidized; thus there are two equivalents of H_2S per mole:

$$\left(\frac{22.4 \text{ liters H}_2\text{S}}{1.00 \text{ mol H}_2\text{S}} \right) \left(\frac{1 \text{ mol H}_2\text{S}}{2 \text{ equivalents H}_2\text{S}} \right) = 11.2 \text{ liters H}_2\text{S/equivalent}$$

b. 75.0 g of $KMnO_4$ is

$$(75.0 \text{ g KMnO}_4) \left(\frac{1.00 \text{ equivalent KMnO}_4}{31.6 \text{ g KMnO}_4} \right) = 2.37 \text{ equivalents KMnO}_4$$

24.0 liters (STP) of H_2S is

$$(24.0 \text{ liters H}_2\text{S}) \left(\frac{1.00 \text{ equivalent H}_2\text{S}}{11.2 \text{ liters H}_2\text{S}} \right) = 2.14 \text{ equivalents H}_2\text{S}$$

c. Since an equivalent of one substance reacts with exactly an equivalent of another, the entire sample of H_2S will react with 2.14 equivalents of the $KMnO_4$, producing 2.14 equivalents of free sulfur.

d. Calculate the equivalent weight of free sulfur. Recalling that the molecular formula of sulfur is S_8, we find that 16 electrons must be transferred to produce each molecule:

$$\overset{8(-2e^-)}{\underset{8 \text{ H}_2\text{S}}{\overline{-2}} \underset{S_8}{\overline{0}}}$$

There are 16 equivalents of sulfur in each mole of S_8.

e. Thus the weight of sulfur produced is

$$(2.14 \text{ equivalents S}_8) \left(\frac{1 \text{ mol S}_8}{16 \text{ equivalents S}_8} \right) \left(\frac{256 \text{ g S}_8}{1.00 \text{ mol S}_8} \right) = 34.2 \text{ g S}_8$$

Questions and Problems

17.1 Reactions Involving Oxygen

14. Write balanced chemical (overall) equations for the complete combustion of each of the following:

$$C_2H_6 \quad C_6H_6 \quad C_4H_8$$

15. Write a chemical equation for the combustion of cigarette paper to produce carbon monoxide. Assume that the paper is composed mainly of cellulose $(C_6H_{10}O_5)_x$.

16. Write overall equations for the roasting of copper(I) sulfide ore and the subsequent reduction of copper(I) oxide with carbon.

17. Draw electron formulas for the two principal products formed when magnesium metal burns in air.

17.3 Oxidation Numbers

18. Determine the oxidation number of each element in the following compounds:
a. HClO   b. $Pb(NO_3)_2$ c. NaH   d. KH_2PO_4 e. NH_3
f. P_4O_{10} g. BaO_2 h. $H_2S_2O_8$ i. OF_2

19. Identify each of the following as an oxide, a peroxide, or a superoxide:
a. BaO   b. K_2O c. H_2O_2 d. Na_2O_2 e. CaO_2 f. KO_2

20. In each of the reactions below, determine which *element* is oxidized and which is reduced, and determine which *reagent* is the oxidizing agent and which is the reducing agent:
a. $PbS + 4 H_2O_2 \rightarrow PbSO_4 + 4 H_2O$
b. $KClO + H_2O_2 \rightarrow KCl + H_2O + O_2$
c. $Cr_2O_3 + 2 Na_2CO_3 + 3 KNO_3 \rightarrow 2 Na_2CrO_4 + 2 CO_2 + 3 KNO_2$
d. $2 HNO_3 + 6 HI \rightarrow 2 NO + 3 I_2 + 4 H_2O$

21. Among H^+, Mg, Cl_2, Cl^-, and Mg^{2+}, which can function as oxidizing agents? List them in order of increasing strength as oxidizing agents.

17.4 Half-Reactions

22. Use Table 17.3 to write equations for the reactions that take place between the species listed below. If no reaction occurs, say so.
a. Cl_2 and Cu   b. Br^- and I_2 c. H^+ and Zn
d. I_2 and Ni^{2+} e. Al and Br_2

17.5 Predicting the Direction of Oxidation-Reduction Reactions

23. Use Table 17.3 to predict whether a reaction will take place between the reagents listed below. If a reaction takes place, write a net ionic equation for it.
a. Chlorine water and a sodium iodide solution.
b. Dilute HCl and a sodium chloride solution.
c. A potassium nitrate solution and a sodium iodide solution, made 1 M in H^+.
d. Tin metal and a manganese(II) nitrate solution.

e. Oxygen and a potassium fluoride solution.
f. Aluminum metal and cobalt(II) nitrate.

17.6 The Preparation of Hydrogen

24. Which of the following would be suitable for the laboratory preparation of hydrogen? For those that would be suitable, write overall, ionic, and net ionic equations. Why are the other reactions unsuitable?
  a. Zn + dil. HCl
  b. Na + H_2O
  c. Cu + dil. HCl
  d. K + dil. H_2SO_4
  e. Fe + dil. $HC_2H_3O_2$
  f. Fe + conc. H_2SO_4
  g. Ca + H_2O
  h. Mg + dil. HCl
  i. Mg + dil. $HC_2H_3O_2$
  j. Fe + dil. HNO_3
  k. Pb + H_2O
  l. Ag + dil. HNO_3
  m. Au + dil. HCl
  n. Cu + dil. H_2SO_4

17.7 Writing Equations for Redox Reactions

25. Balance the following by the ion-electron method. Assume all reactions take place in aqueous solution.
  a. $Na_2SO_3 + NaMnO_4 + NaOH \rightarrow Na_2SO_4 + Na_2MnO_4 + H_2O$
  b. $KMnO_4 + HCl + FeCl_2 \rightarrow MnCl_2 + KCl + FeCl_3 + H_2O$
  c. $CuS + HNO_3 \rightarrow Cu(NO_3)_2 + S_8 + H_2O + NO$
  d. $Zn + HNO_3 \rightarrow N_2 + Zn(NO_3)_2 + H_2O$
  e. $Cl_2 + NaCN + NaOH \rightarrow NaCNO + NaCl + H_2O$
  f. $K_2Cr_2O_7 + KBr + H_2SO_4 \rightarrow K_2SO_4 + Cr_2(SO_4)_3 + Br_2 + H_2O$
  g. $As_2S_5 + HNO_3 \rightarrow H_3AsO_4 + H_2SO_4 + H_2O + NO_2$
  h. $FeS + H_2SO_4 \rightarrow Fe_2(SO_4)_3 + SO_2 + H_2O$
  i. $Sb + HNO_3 \rightarrow Sb_2O_5 + NO_2 + H_2O$
  j. $K_2CrO_4 + H_2S + H_2SO_4 \rightarrow K_2SO_4 + Cr_2(SO_4)_3 + S_8 + H_2O$
  k. $FeCl_2 + K_2Cr_2O_7 + HCl \rightarrow FeCl_3 + KCl + CrCl_3 + H_2O$
  l. $Bi_2S_3 + HNO_3 \rightarrow Bi(NO_3)_3 + NO + S_8 + H_2O$
  m. $NaMnO_4 + Na_3AsO_3 + HNO_3 \rightarrow H_3AsO_4 + NaNO_3 + Mn(NO_3)_2 + H_2O$
  n. $NiS + HCl + HNO_3 \rightarrow NiCl_2 + S_8 + NOCl + H_2O$

26. Balance the following by the electron-transfer method:
  a. $F_2 + H_2O \rightarrow HF + O_3$
  b. $P_4 + HClO_4 + H_2O \rightarrow H_3PO_4 + HCl$
  c. $FeCl_3 + H_2S \rightarrow FeCl_2 + HCl + S_8$
  d. $Bi(OH)_3 + Na_2SnO_2 \rightarrow Bi + Na_2SnO_3 + H_2O$
  e. $SnCl_2 + H_2SO_3 + HCl \rightarrow SnCl_4 + H_2S + H_2O$
  f. $KIO_3 + KI + HC_2H_3O_2 \rightarrow KC_2H_3O_2 + H_2O + I_2$
  g. $KMnO_4 + AsH_3 + H_2SO_4 \rightarrow H_3AsO_4 + MnSO_4 + H_2O + K_2SO_4$
  h. $CuO + NH_3 \rightarrow N_2 + H_2O + Cu$
  i. $H_2S + HNO_3 \rightarrow S_8 + NO + H_2O$
  j. $Sn + HNO_3 \rightarrow SnO_2 + NO_2 + H_2O$
  k. $PH_3 + NO_2 \rightarrow H_3PO_4 + N_2$
  l. $NH_3 + O_2 \rightarrow NO_2 + H_2O$
  m. $Cr_2O_3 + Na_2CO_3 + KNO_3 \rightarrow Na_2CrO_4 + CO_2 + KNO_2$
  n. $SF_6 + H_2S \rightarrow HF + S_8$
  *o. $HPO_3 + C \rightarrow H_2 + CO + P_4$
  *p. $NaAsO_2 + I_2 + NaHCO_3 \rightarrow Na_2HAsO_4 + NaI + CO_2 + H_2O$
  *q. $K_3Fe(CN)_6 + Cr_2O_3 + KOH \rightarrow K_4Fe(CN)_6 + K_2CrO_4 + H_2O$

27. Balance the following by either method. If you use the ion-electron method, assume aqueous solutions.
  a. $MnSO_4 + Br_2 + NaOH \rightarrow MnO_2 + NaBr + Na_2SO_4 + H_2O$
  b. $NaI + H_2SO_4 \rightarrow H_2S + I_2 + Na_2SO_4 + H_2O$
  c. $V + KClO_3 \rightarrow V_2O_5 + KCl$
  d. $NH_4Cl + Cl_2 \rightarrow NCl_3 + HCl$
  e. $NH_3 + Br_2 + NaOH \rightarrow N_2 + NaBr + H_2O$
  f. $AuCl_3 + Sb_2O_3 + H_2O \rightarrow Au + Sb_2O_5 + HCl$
  g. $KI + NaClO + H_2O \rightarrow KOH + NaCl + I_2$
  h. $HNO_3 + HCl \rightarrow Cl_2 + NO + H_2O$
  i. $H_3SbO_3 + I_2 + H_2O \rightarrow H_3SbO_4 + HI$
  j. $SnCl_2 + HgCl_2 \rightarrow Hg_2Cl_2 + SnCl_4$
  k. $HNO_3 + HI \rightarrow NO + I_2 + H_2O$
  l. $MnCl_2 + NaOH + Br_2 \rightarrow MnO_2 + NaCl + NaBr + H_2O$
  m. $Fe_2(SO_4)_3 + Zn \rightarrow FeSO_4 + ZnSO_4$
  n. $S_8 + HNO_3 \rightarrow H_2SO_4 + NO$
  o. $H_2SO_3 + I_2 + H_2O \rightarrow H_2SO_4 + HI$
  p. $KClO + HCl \rightarrow Cl_2 + KCl + H_2O$
  q. $C + H_2SO_4 \rightarrow CO_2 + SO_2 + H_2O$
  r. $C + HNO_3 \rightarrow CO_2 + NO + H_2O$
  s. $H_2SO_3 + HNO_3 \rightarrow H_2SO_4 + H_2O + NO$
  t. $AsH_3 + AgNO_3 + H_2O \rightarrow H_3AsO_3 + Ag + HNO_3$
  u. $Cu + HNO_3 \rightarrow Cu(NO_3)_2 + NO_2 + H_2O$
  v. $Cu + HNO_3 \rightarrow Cu(NO_3)_2 + NO + H_2O$
  w. $CrI_3 + KOH + Cl_2 \rightarrow K_2CrO_4 + KIO_4 + KCl + H_2O$
  * x. $K_2Cr_2O_7 + FeSO_4 + H_2SO_4 \rightarrow Cr_2(SO_4)_3 + K_2SO_4 + Fe_2(SO_4)_3 + H_2O$
  * y. $Cu_2S + HNO_3 \rightarrow Cu(NO_3)_2 + NO + S_8 + H_2O$
  z. $Sb_2S_5 + HNO_3 \rightarrow Sb_2O_5 + H_2SO_4 + NO + H_2O$
  * aa. $As_2S_3 + HNO_3 + H_2O \rightarrow H_3AsO_4 + H_2SO_4 + NO$
  * bb. $K_4Fe(CN)_6 + KMnO_4 + H_2SO_4 \rightarrow Fe_2(SO_4)_3 + K_2SO_4 + CO_2 + HNO_3 + MnSO_4 + H_2O$

28. Use Tables 17.1 and 17.2 to predict the probable products of reactions between the reagents below. Write overall equations for each reaction.
  a. $F_2(g) + Zn(s)$
  b. $H_2SO_4(l) + H_2S(g)$
  c. $KMnO_4(aq) + H_2SO_3(aq)$ (yields Mn^{2+})
  d. $HNO_3(aq) + Zn(s)$ (to produce NO)
  e. $FeCl_3(aq) + SnCl_2(aq)$ (assume that Cl^- is not oxidized)
  f. $C(s) + H_2SO_4(l)$
  g. $HI(aq) + H_2SO_4(l)$
  h. $FeCl_3(aq) + H_2S(aq)$
  i. $KMnO_4(aq) + SO_2(g) + H_2O$

* **29.** Write net ionic equations for the following reactions. All products are not necessarily indicated. Assume aqueous solutions.
  a. $P_4 + OH^- \rightarrow H_2PO_2^- + PH_3$
  b. $I_2 + NO_3^- \rightarrow NO_2 + IO_3^-$
  c. $Mn^{2+} + BiO_3^- \rightarrow MnO_4^- + Bi^{3+}$
  d. $Co(OH)_2 + O_2^{2-} \rightarrow Co(OH)_3 + OH^-$
  e. $H_2O_2 + MnO_4^- \rightarrow O_2 + Mn^{2+}$ (acidic solution)
  f. $I_2 + H_2O + Cl_2 \rightarrow IO_3^- + H^+ + Cl^-$
  g. $Zn + CNS^- \rightarrow Zn^{2+} + H_2S + HCN$

17.8 Equivalent Weights in Redox Reactions

30. Calculate the number of equivalent weights in
   a. 23.4 g of $KMnO_4$ (if reduced to Mn^{2+}).
   b. 6.04 g formula wt of $Co(OH)_3$ (if reduced to $Co(OH)_2$).
   c. 3.51 g of H_2SO_4 (if reduced to S_8).
   d. 24.0 ml of 3.00 M HNO_3 (if reduced to NO_2).
   e. 1.25×10^2 ml of 0.100 M $SnCl_2$ solution (if oxidized to $SnCl_4$).

31. Calculate the equivalent weight of HNO_3 for reactions that produce the following:
   a. NO_2 b. NO c. N_2O d. N_2 e. NH_3

32. Calculate the equivalent weight of A if 20.0 g of A reacts with 0.385 equivalent of B.

33. What is the equivalent weight of a metal if 0.216 g of it combines with 0.192 g of oxygen?

34. 8.33 liters of H_2 gas at 27°C and 1.10 atm was displaced from aqueous HCl by 12.1 g of an unknown metal. What is the equivalent weight of the metal?

35. How many equivalents of hydrogen are needed to reduce 46.0 g of CuO to free copper?

Use equivalents rather than equations to work problems 36–38.

36. How many grams of magnesium will react with 3.00 g of chlorine to form magnesium chloride?

37. How many grams of chlorine can be prepared from excess HCl by treating it with 1.00×10^2 g of $K_2Cr_2O_7$?

38. What volume of hydrogen (STP) can be prepared from 10.0 g of zinc and excess HCl?

Discussion Starters

39. Explain or account for the following:
   a. Many oxidation reactions do not involve oxygen at all.
   b. The metal lithium appears at the bottom of the listing of oxidizing agents but at the top of the activity series.
   c. Zinc reacts more vigorously with dilute sulfuric acid than with concentrated sulfuric acid.
   d. Magnesium reacts more vigorously with hydrochloric acid than with acetic acid.
   e. The reaction of zinc with nitric acid would not be a successful laboratory preparation of hydrogen.
   f. Radium reacts more vigorously with water to produce hydrogen than calcium does.
   g. It is difficult if not impossible to extinguish a fire of magnesium with water.

18 Electrochemistry

The special feature of redox reactions—that they involve the transfer of electrons—links them inevitably to electricity. As in all chemical reactions the currency of exchange is still free energy but, unlike, say, a spontaneous precipitation reaction, a spontaneous redox reaction (a **galvanic reaction**) can be harnessed in a galvanic cell to produce a flow of electrons—electricity. If on the other hand Table 17.3 tells us that a redox reaction is not spontaneous (thus is an **electrolytic reaction**), we can set up an *electrolytic cell* and drive the reaction with a battery.

18.1 Galvanic Cells

We have already discussed the reaction that takes place when we immerse a strip of zinc in a copper(II) sulfate solution (p. 524). The zinc metal dissolves by forming Zn^{2+} ions, and the copper plates out as free copper. The reaction is spontaneous because copper(II) ions attract electrons more strongly than zinc ions do, that is, they are stronger oxidizing agents. The equations for the two half-reactions are

$$Zn(s) \rightleftharpoons Zn^{2+}(aq) + 2e^-$$

and

$$Cu^{2+}(aq) + 2e^- \rightleftharpoons Cu(s)$$

To construct a *galvanic cell* from these two half-reactions, we must separate them so that the transfer of electrons takes place *through a wire*, thus producing an electric current. One of the earliest commercial galvanic cells was the Daniell cell, which used this overall reaction to produce electricity for early telegraph stations.

The Daniell Cell Figure 18.1 shows a Daniell cell. The lower liquid layer is a solution of copper(II) sulfate, which remains saturated because of the excess

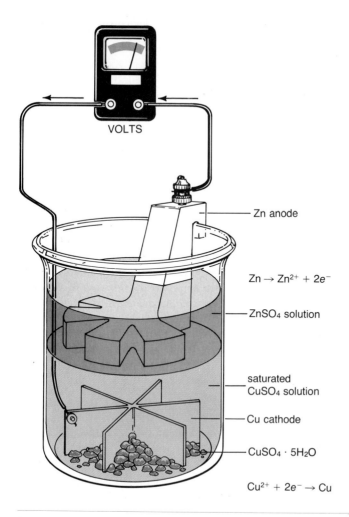

Figure 18.1 The Daniell (gravity) cell.

$CuSO_4 \cdot 5H_2O$ crystals. A copper electrode is immersed in this lower solution. The less dense upper layer is a dilute solution of $ZnSO_4$ containing a heavy zinc electrode. As the zinc dissolves, electrons travel through the external circuit to the copper electrode, where Cu^{2+} ions are reduced to free copper. As this transfer of charge occurs, of course, there must be some compensating change so that both solutions remain electrically neutral. What we find is that negative ions (mainly SO_4^{2-}) migrate from the $CuSO_4$ solution across the boundary between the solutions to the $ZnSO_4$ solution and that some positive ions (principally Zn^{2+}) migrate the other way.

We must now take care of some problems of nomenclature. Earlier we called the negatively charged source of electrons in a cathode-ray tube the *cathode* and the positively charged electrode the *anode*. In galvanic cells we must be careful how we use these two terms. *The electrode where oxidation takes place is the* **anode.** As metal atoms are oxidized to metal ions and enter the solution, their electrons are left behind on the anode; thus the anode in all galvanic cells is designated as *negative*. Negative ions migrate toward the anode—not because of its charge, of course, but to preserve the charge neutrality of the solution, which would otherwise be unbalanced as the oxidation produces (in this case) positive zinc ions in solution at this electrode. *The electrode at which reduction takes place is called the* **cathode** and is designated as *positive*. Positive ions migrate toward this electrode—again, not because of its charge but to preserve the charge neutrality in the solution, which would be unbalanced because (in this case) Cu^{2+} ions leave the solution as they are reduced to Cu atoms.

Before going on to look at other specific examples, it might be helpful to reinforce the concepts and definitions we just summarized by looking at a general model of the galvanic cell. Figure 18.2 is a schematic representation of just such a cell. It comprises two *half-cells*, each of which consists of an electrode and an electrolyte solution. The two electrodes are connected by a wire, and the two half-cells are separated by a porous barrier, which permits the migration of ions between the two electrolytes. For metal A to function as the anode, as we have

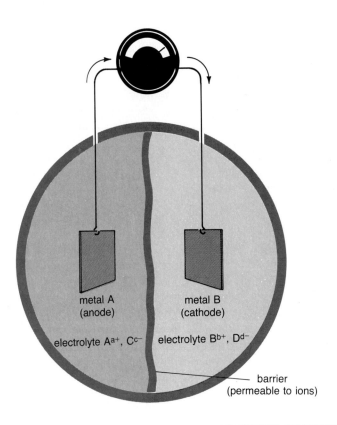

Figure 18.2 Schematic diagram of a galvanic cell.

designated it in Figure 18.2, it must lie nearer the bottom right-hand side of Table 17.3 than does metal B; that is, it must be a stronger reducing agent. With this assumption, we can now write the two half-reactions for this cell.

Oxidation half-reaction: $A(s) \rightleftharpoons A^{a+}(aq) + ae^-$

Reduction half-reaction: $B^{b+}(aq) + be^- \rightleftharpoons B(s)$

The equation for the *cell reaction* is of course just the sum of these two half-reactions. To ensure that no electrons appear in the final equation we can use a and b as coefficients for the two half-reaction equations:

$$\frac{\begin{array}{c} b[A(s) \rightleftharpoons A^{a+}(aq) + ae^-] \\ a[B^{b+}(aq) + be^- \rightleftharpoons B(s)] \end{array}}{aB^{b+}(aq) + bA(s) + \cancel{abe^-} \rightleftharpoons bA^{a+}(aq) + aB(s) + \cancel{abe^-}}$$

(Spectator ions C^{c-} and D^{d-}, shown in Figure 18.2, do not appear in the net ionic equation.) Finally, we can describe this cell concisely by resorting to a shorthand notation where we list first the metal of the anode, then its corresponding ion, and then the ion and metal of the cathode:

$A|A^{a+}|B^{b+}|B$

Each straight line indicates a phase boundary.

Many galvanic cells fit this model exactly. However, few can be set up like the Daniell cell where gravity and different solution densities are sufficient to separate the half-cells. A more common setup uses a *salt bridge*, as shown schematically in Figure 18.3. The salt bridge consists of a U-tube filled with concentrated salt solution and usually a gel. The ions of the salt solution are chosen so that they will not react in either half-cell. As the galvanic reaction proceeds, the positive ions of the salt bridge move out of the tube and into the cathode half-cell, and the negative ions move into the anode half-cell. Neutrality is thus preserved in both half-cells. A salt bridge in a galvanic cell is denoted by a double bar:

$A|A^{a+}||B^{b+}|B$

Example 18.1 For the galvanic cell, $Cu|Cu^{2+}||Ag^+|Ag$, determine (a) which metal is the anode, (b) the equations for the half-reactions and for the cell reaction, (c) the direction of electron flow in the wire, and (d) the ions that migrate toward each electrode.

Solution

a. Table 17.3 shows that copper metal is a stronger reducing agent than silver metal; thus copper will act as the anode:

Oxidation half-reaction: $Cu(s) \rightleftharpoons Cu^{2+}(aq) + 2e^-$

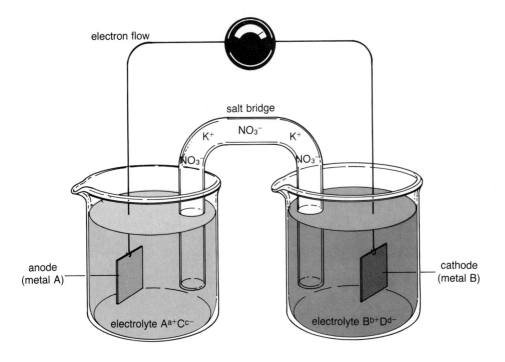

Figure 18.3 Schematic representation of a galvanic cell utilizing a salt bridge.

b. Current will flow in the external circuit from the copper anode to the silver cathode. Silver ions will then be reduced to free silver:

Reduction half-reaction: $Ag^+(aq) + e^- \rightleftharpoons Ag(s)$

c. The equation for the cell reaction is the sum of the two half-reaction equations. The reduction half-reaction equation needs a coefficient of 2 to ensure that no electrons appear in the result:

$$2[Ag^+(aq) + e^- \rightleftharpoons Ag(s)]$$
$$Cu(s) \rightleftharpoons Cu^{2+}(aq) + 2e^-$$
$$\overline{2\,Ag^+(aq) + Cu(s) + \cancel{2e^-} \rightleftharpoons Cu^{2+}(aq) + 2\,Ag(s) + \cancel{2e^-}}$$

d. All negative ions migrate toward the copper anode. This includes the negative ions of both electrolytes and of the salt bridge and OH^- ions from the water. Conversely, all positive ions migrate toward the silver cathode. This includes the Ag^+ and Cu^{2+} ions, the cation of the salt bridge, and H^+ ions from the water.

e. This cell is diagrammed in Figure 18.4.

So far all the cells we have described have had reactive electrodes—electrodes that actually take part in the oxidation and reduction half-reactions.

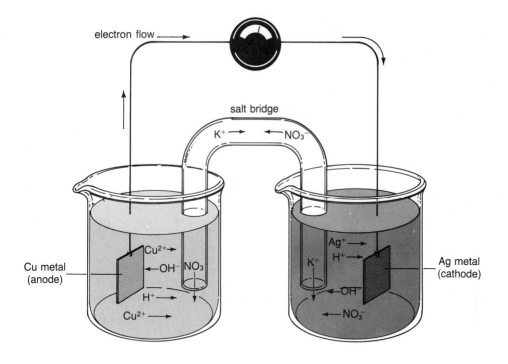

Figure 18.4 Schematic representation of the Cu|Cu^{2+}||Ag$^+$|Ag cell:
Electrolytes: Cu(NO$_3$)$_2$(aq), AgNO$_3$(aq)
Salt bridge: KNO$_3$(aq)

However, it is equally possible to construct a cell, similar, say, to the one in Example 18.1, with an inert electrode. With platinum as the passive cathode, the shorthand notation now becomes

$$\text{Cu}|\text{Cu}^{2+}||\text{Ag}^+|\text{Ag}|\text{Pt}$$

Note that metallic Ag still appears. It is still the cathode product, which forms at the Pt electrode. We have also limited our discussion to half-cells in which metals are oxidized or reduced. However, with a little more effort we might just as well have constructed half-cells where, say, hydrogen is oxidized or chlorine reduced. The **standard hydrogen electrode,** for example, consists of hydrogen gas at 1 atm bubbling over the surface of a platinum electrode immersed in an acid solution with unit activity (approximately 1 M in H$^+$; see p. 414). The half-reaction for this electrode, when it functions as an anode, is an oxidation half-reaction:

$$\text{H}_2(g) \rightleftharpoons 2\,\text{H}^+(aq) + 2e^-$$

When this electrode serves as the cathode, the reverse reaction takes place.

Example 18.2 For the galvanic cell $Pt|H_2|H^+\|Ag^+|Ag$, determine (a) which half-cell contains the anode and (b) the equation for the cell reaction.

Solution

a. Table 17.3 shows that Ag^+ is a stronger oxidizing agent than H^+. Thus Ag^+ will be reduced at the cathode, and H_2 will be oxidized at the anode.

b. The equation for the reduction half-reaction can be written just as it appears in Table 17.3, but the equation for the oxidation half-reaction must be reversed:

$$\text{Reduction half-reaction:} \quad Ag^+(aq) + e^- \rightleftharpoons Ag(s)$$
$$\text{Oxidation half-reaction:} \quad H_2(g) \rightleftharpoons 2\,H^+(aq) + 2e^-$$

c. Multiply the reduction half-reaction by two to ensure that the gain and loss of electrons cancel in the cell reaction; then add the two half-reactions:

$$2\,Ag^+(aq) + H_2(g) \rightleftharpoons 2\,H^+(aq) + 2\,Ag(s)$$

d. This cell is diagrammed in Figure 18.5.

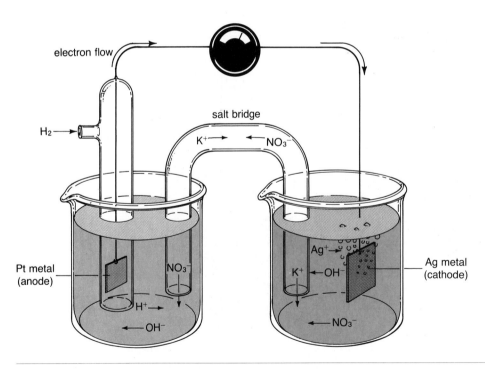

Figure 18.5 Schematic representation of the $Pt|H_2|H^+\|Ag^+|Ag$ cell:
Electrolytes: $AgNO_3(aq)$, $HNO_3(aq)$
Salt bridge: $KNO_3(aq)$

Example 18.3 Determine which half-cell will be the anode, and write the equation for the cell reaction for a cell comprising the standard hydrogen electrode and a $Zn|Zn^{2+}$ half-cell.

Solution

a. In Table 17.3 the $H^+|H_2$ half-reaction appears above the $Zn^{2+}|Zn$ half-reaction. This tells us that H^+ will be the oxidizing agent which is reduced at the cathode and that Zn will be the reducing agent which is oxidized at the anode:

$$\text{Reduction half-reaction:} \quad 2\,H^+(aq) + 2e^- \rightleftharpoons H_2(g)$$
$$\text{Oxidation half-reaction:} \quad Zn(s) \rightleftharpoons Zn^{2+}(aq) + 2e^-$$

b. The cell reaction equation is

$$Zn(s) + 2\,H^+(aq) \rightleftharpoons Zn^{2+}(aq) + H_2(g)$$

c. This cell (Figure 18.6) would be described as

$$Zn|Zn^{2+}\|H^+|H_2|Pt$$

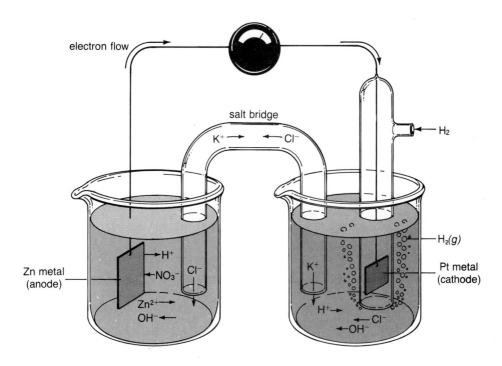

Figure 18.6 Schematic representation of the $Zn|Zn^{2+}\|H^+|H_2|Pt$ cell:
Electrolytes: $Zn(NO_3)_2(aq)$, $HCl(aq)$
Salt bridge: $KCl(aq)$

Questions

1. Distinguish between an electrolytic cell and a galvanic cell.
2. How are the $CuSO_4$ and $ZnSO_4$ solutions in the Daniell cell separated?
3. What is the purpose of a salt bridge in a galvanic cell?

18.2 Commercial Cells

Except for the Daniell cell, the galvanic cells we have discussed so far have little use outside the laboratory. It would be cumbersome indeed to design a transistor radio around one of these bulky cells. Fortunately, other galvanic cells can be packaged more conveniently. Perhaps the most common commercial cells are *primary cells*, which eventually become exhausted and cannot be reused. *Secondary* or *rechargeable cells*, on the other hand, can be recharged, at least several times, after they become exhausted. We shall describe the recharging process later in this chapter. Finally, some commercial cells are true *fuel cells*, where the anode and cathode materials are constantly replenished as the cells operate. We shall look at examples of each of these cells below. Incidentally, the word *battery*, when applied to a single cell, is a misnomer. A battery is a *series* of cells, where the anode of one is connected to the cathode of the next, and so on. The electron flow in a battery is thus continuous through all the connected cells.

The Leclanché Dry Cell The dry cell, which is not dry but damp and which we usually call a battery, was devised more than 100 years ago. Nevertheless, it is still the most common power source for low voltage appliances, such as flashlights and mechanical toys. The container for this cell is a zinc can that also acts as the anode (Figure 18.7). The can is lined with porous paper, which isolates the zinc from the other reactants but permits diffusion of ions. In the center of the can, an inert carbon rod serves as the cathode. Between the cathode and the paper lining is an aqueous paste of ammonium chloride, manganese (IV) oxide, and zinc chloride. Finally, the top of the cell is sealed to prevent evaporation of water, and a steel or cardboard case encloses the entire cell to reduce corrosion. Curiously, even after 100 years, the chemistry of the dry cell is still not completely understood, because the reactions that occur depend on the demands imposed on the cell. Essentially, however, the dry cell is a zinc-hydrogen cell, in which the source of hydrogen ions is the acidic solution of ammonium chloride. At the anode, zinc atoms give up electrons:

$$Zn \longrightarrow Zn^{2+} + 2e^-$$

The zinc ions then migrate toward the cathode, where ammonium ions are being converted to ammonia and hydrogen:

$$2\,NH_4^+ + 2e^- \longrightarrow 2\,NH_3 + 2\,H$$

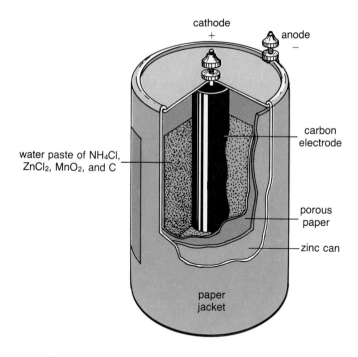

Figure 18.7 The "dry" cell.

The zinc then combines with the ammonia to form ammoniated zinc ions $[Zn(NH_3)_4]^{2+}$. This reaction not only removes the ammonia gas from around the cathode, but it also ensures that the Zn^{2+} concentration remains constant, which in turn ensures a fairly constant cell voltage of about 1.5 volts. The hydrogen atoms formed at the cathode are oxidized by the manganese (IV) oxide:

$$2\,MnO_2 + 2\,H \longrightarrow Mn_2O_3 + H_2O$$

Alkaline Cells Alkaline cells have been developed and marketed in the past 20 years or so and have to some extent replaced the Leclanché-type cell. They are similar to the dry cell except that the electrolyte includes a high concentration of KOH or NaOH. These alkaline cells can deliver higher currents and have longer lives than dry cells and thus are often used in electric shavers, tape recorders, and radios. Alkaline cells function well at temperatures as low as $-20°C$.

Mercury Cell The mercury cell has several distinct advantages over most galvanic cells. It is small, it maintains a constant voltage over a long life, and it can supply high currents for brief periods. Mercury cells are most frequently used in hearing aids, heart pacemakers, and other kinds of electronic equipment where a constant voltage is essential. As shown in Figure 18.8, these cells contain an electrolyte

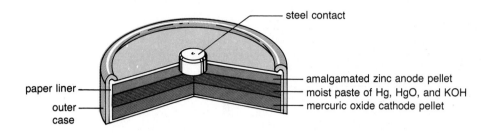

Figure 18.8 The mercury cell.

of KOH and ZnO, sandwiched between a zinc-mercury anode and a mercury(II) oxide cathode. In simplest form we can write the electrode reactions for the mercury cell as follows:

Anode: $\quad Zn + 2\,OH^- \longrightarrow ZnO + H_2O + 2e^-$
Cathode: $\quad HgO + H_2O + 2e^- \longrightarrow Hg + 2\,OH^-$
Cell reaction: $\quad HgO + Zn \longrightarrow ZnO + Hg$

Silver Oxide Cell The silver oxide cell is similar to the mercury cell, except that silver oxide forms the cathode instead of mercury(II) oxide.

Lead-Acid Storage Battery Theoretically we should be able to regenerate the active materials in a galvanic cell by forcing a current through it from the cathode to the anode. In practice, however, this works with only a few commercial cells, the most conspicuous example being the lead storage cell. One electrode in this cell is metallic lead; the other is lead(IV) oxide pressed onto a lead grid (Figure 18.9). The electrolyte is a solution of sulfuric acid. When the cell is producing a current, the reactions are

Anode: $\quad Pb \longrightarrow Pb^{2+} + 2e^-$
$\quad\quad\quad\quad\ Pb^{2+} + SO_4^{2-} \longrightarrow PbSO_4$
Cathode: $\quad 2e^- + PbO_2 + 4\,H^+ \longrightarrow Pb^{2+} + 2\,H_2O$
$\quad\quad\quad\quad\ Pb^{2+} + SO_4^{2-} \longrightarrow PbSO_4$
Cell reaction: $\quad Pb + PbO_2 + 2\,H_2SO_4 \longrightarrow 2\,PbSO_4 + 2\,H_2O$

As the cell discharges, insoluble lead(II) sulfate accumulates at both the anode and the cathode, and sulfuric acid is consumed. As most automobile owners know, the density of the sulfuric acid solution can thus be used as a measure of the charge on the battery. The battery is adequately charged if the specific gravity is between 1.25 and 1.30 and is in need of recharging when the specific

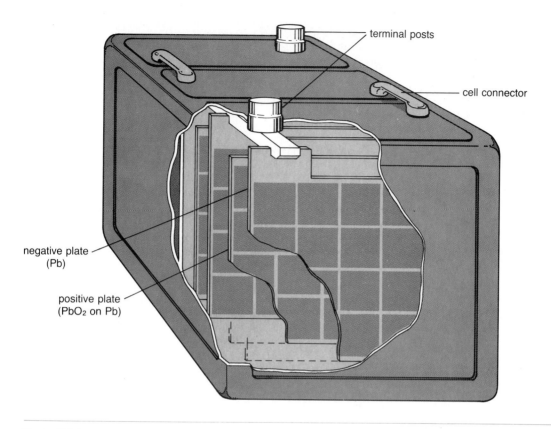

Figure 18.9 The lead storage battery.

gravity falls below 1.20. Eventually the lead-acid cell deteriorates as lead and lead(IV) oxide are lost from the electrodes, and it can no longer be recharged. Most automobile batteries consist of six lead-acid cells in series, each of which supplies about 2 volts.

Nickel-Cadmium Cells Batteries of nickel-cadmium cells are replacing the lead storage battery in many applications where light weight is a virtue. Nickel-cadmium cells also have long shelf lives (they keep their charge even when not used) and can be recharged repeatedly without deteriorating. They are used extensively in space vehicles, as well as in more mundane household appliances and electronic calculators. In the nickel-cadmium cell, the cathode is nickel oxide and the anode is cadmium. The electrode reactions and the cell reactions are

$$\text{Anode:} \quad Cd + 2\,OH^- \longrightarrow Cd(OH)_2 + 2e^-$$
$$\underline{\text{Cathode:} \quad NiO_2 + 2\,H_2O + 2e^- \longrightarrow Ni(OH)_2 + 2\,OH^-}$$
$$\text{Cell reaction:} \quad Cd + NiO_2 + 2\,H_2O \longrightarrow Cd(OH)_2 + Ni(OH)_2$$

Table 18.1 Primary and Secondary (Rechargeable) Cells

Primary Cells

Type	Electrodes, Electrolyte	Voltage (V)	Uses
Standard (Leclanché) dry cell	Zinc anode; carbon (inert) cathode; electrolyte: paste of NH_4Cl, MnO_2, $ZnCl_2$, and graphite	1.5	Flashlights, toys, novelties, radios
Alkaline cell	Zinc anode; steel or carbon (inert) cathode; electrolyte: paste of MnO_2 and KOH or NaOH	1.5	Bicycle lights, toys, tape recorders, radios, walkie-talkies, cameras
Mercury cell	Zn-Hg amalgam anode; mercury(II) oxide (HgO) cathode; electrolyte: KOH	1.3	Hearing aids, heart pacemakers, radios, transistorized equipment, "electric eye" cameras, nonrechargeable calculators
Silver oxide cell	Zinc anode; Ag_2O cathode; electrolyte: KOH or NaOH	1.5	Instruments, hearing aids, electric watches, photoelectric exposure devices

Secondary (Rechargeable) Cells

Type	Electrodes, Electrolyte	Voltage (V)	Uses
Lead-acid storage battery	Lead anode; PbO_2 cathode; electrolyte: $H_2SO_4(aq)$	2.0	Automobile electrical systems
Nickel-cadmium cell	Anode material: Cd; cathode material: NiO(OH); electrolyte: KOH(*aq*) and LiOH(*aq*)	1.3	Cordless electric appliances (toothbrushes, carving knives, grass trimmers), amplifiers, portable television sets, movie cameras, electronic calculators

Table 18.1 summarizes the properties and uses of the primary and secondary commercial cells we've discussed above.

Fuel Cells To convert a conventional fuel to electricity, it is usually necessary to change chemical energy to heat energy, heat energy to mechanical energy, and finally mechanical energy to electrical energy. A galvanic cell designed around the oxidation of a conventional chemical fuel would of course provide a highly efficient shortcut. This is the basis of the fuel cell. A second characteristic of the fuel cell, one that distinguishes it from other galvanic cells, is that the reactants are constantly replenished and the products withdrawn, so that it provides constant voltage and power. The first fuel cells were H_2—O_2 cells developed for the Apollo space program. Most still use H_2 and O_2 as reactants. One type, the molten-carbonate fuel cell (Figure 18.10), employs an electrolyte that is a mixture of alkali metal carbonates and ceramic particles. At room temperature, this electrolyte looks like a solid tile, but at the cell-operating temperature of 700°C it is a molten ionic conductor. The electrodes are porous nickel. We can write the reactions for this fuel cell as

Anode:	$2(H_2 + CO_3^{2-} \longrightarrow H_2O + CO_2 + 2e^-)$
Cathode:	$O_2 + 2\,CO_2 + 4e^- \longrightarrow 2\,CO_3^{2-}$
Cell reaction:	$2\,H_2 + O_2 \longrightarrow 2\,H_2O$

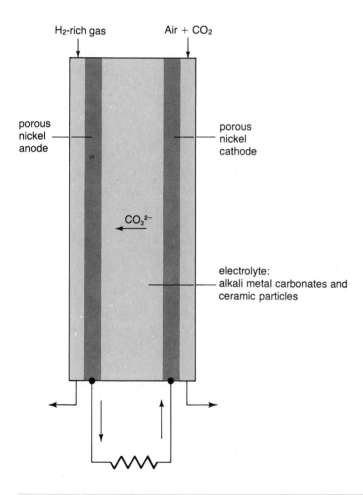

Figure 18.10 Schematic representation of a molten carbonate fuel cell.

Questions

4. Distinguish between a cell and a battery.

5. What is the cathode of a dry cell made of?

6. What are some advantages of the mercury cell?

7. What is the special advantage of the nickel-cadmium cell?

8. How do fuel cells differ from other commercial cells?

18.3 Electrode Potentials

In our discussion of commercial cells and batteries, we alluded to the characteristic voltages of several galvanic cells. You might also have correctly inferred that the voltage of a battery of cells in series is just the sum of the voltages of its constituent cells. To put a firmer base beneath this notion of cell voltages, we shall now look first at how charge is measured in oxidation-reduction reactions and then at how we measure (and predict) the driving force of these reactions.

The charge on a single electron is 4.803×10^{-10} esu (electrostatic units), where 2.998×10^9 esu $= 1$ coulomb. Thus the charge on a mole of electrons is

$$(6.022 \times 10^{23} \text{ electrons}) \left(\frac{4.803 \times 10^{-10} \text{ esu}}{1 \text{ electron}} \right) \left(\frac{1 \text{ coul}}{2.998 \times 10^9 \text{ esu}} \right)$$

$$= 9.65 \times 10^4 \text{ coul}$$

This quantity of electric charge has been given the name *faraday*, in honor of the scientist who first described how the amount of electrical current produced in a redox reaction is related to how much of the reactants is oxidized or reduced. We can review what he discovered by considering a cell in which zinc is oxidized at the anode and chlorine reduced at the cathode. The anode half-reaction, $Zn \longrightarrow Zn^{2+} + 2e^-$, liberates two electrons; thus the oxidation of 1 gram atomic weight of zinc requires the transfer of $2 \times 6.02 \times 10^{23}$ electrons, or 2 faradays of charge. At the cathode, 2 faradays will reduce 1 mol of chlorine to 2 mol of chloride ions, $Cl_2(g) + 2e^- \longrightarrow 2\,Cl^-(aq)$.

Example 18.4 How many faradays of electricity are produced if 365 g of zinc metal reacts in the cell, $Zn\,|\,Zn^{2+}\,||\,Cl_2\,|\,Cl^-\,|\,Pt$?

Solution

a. According to Table 17.3, chlorine gas is the strongest oxidizing agent of the four reagents in the cell (Pt is inert); thus we know it will be reduced at the cathode, while zinc metal is oxidized at the anode.

Reduction half-reaction: $Cl_2(g) + 2e^- \longrightarrow 2\ Cl^-(aq)$

Oxidation half-reaction: $Zn(s) \longrightarrow Zn^{2+}(aq) + 2e^-$

Cell reaction: $Zn(s) + Cl_2(g) \longrightarrow Zn^{2+}(aq) + 2\ Cl^-(aq)$

b. The number of moles of zinc that will be oxidized is

$$(365\ g\ Zn)\left(\frac{1.00\ mol\ Zn}{65.4\ g\ Zn}\right)$$

c. Every mole of zinc oxidized requires the transfer of two moles of electrons, or 2 faradays of charge. Thus the amount of electricity produced is

$$(365\ g\ Zn)\left(\frac{1.00\ mol\ Zn}{65.4\ g\ Zn}\right)\left(\frac{2\ mol\ e^-}{1\ mol\ Zn}\right)\left(\frac{1\ faraday}{1\ mol\ e^-}\right) = \textbf{11.2 faradays}$$

Example 18.5 What volume (STP) of chlorine will have been reduced in the cell described in Example 18.4?

Solution

a. Two moles of electrons are needed to reduce each mole of chlorine gas:

$$Cl_2(g) + 2e^- \longrightarrow 2\ Cl^-(aq)$$

b. Therefore, 11.2 faradays of electricity will reduce 5.60 moles of chlorine:

$$(11.2\ faradays)\left(\frac{1\ mol\ e^-}{1\ faraday}\right)\left(\frac{1\ mol\ Cl_2}{2\ mol\ e^-}\right) = 5.60\ mol\ Cl_2$$

c. At STP, 5.60 mol of chlorine has a volume of

$$(5.60\ mol\ Cl_2)\left(\frac{22.4\ liters}{1.00\ mol}\right) = \textbf{125 liters of } Cl_2 \textbf{ (STP)}$$

Another galvanic cell could be made with the $Zn|Zn^{2+}$ half-cell by coupling it with, say,

$$Br_2(l) + 2e^- \longrightarrow 2\ Br^-(aq)$$

The only obvious difference between this cell and the first is that bromine would now be reduced in the place of chlorine. But chlorine is a stronger oxidizing agent than bromine; that is, it has a stronger attraction for electrons so we might expect a less obvious difference as well. In each case the halogen's attraction for electrons, compared to the lesser attraction of zinc ions for electrons, is the *driving force* for the complete redox reaction. Thus the zinc-chlorine reaction proceeds more vigorously and releases more energy than the zinc-bromine reaction. We say that the *electromotive force* (E) is different in the two cases, although the same amount of charge is transferred in each. We measure E in volts and can relate it to charge and energy with the following equation:

$$\text{energy (joules)} = \text{electrical charge (coulombs)} \times E \text{ (volts)}$$

We'll have more to say about this relationship later.

We can now modify Table 17.3 by assigning each reduction half-reaction a **standard reduction potential, $E°$**, which is a measure of how readily the reaction proceeds with each reactant in its standard state at 25°C and 1 atm. Table 18.2 shows the result. Of course, the values of $E°$ are meaningful only when compared to one another, since we can only measure the voltage of a complete galvanic cell—never the voltage of a single half-cell. To establish a reference value, $E°$ for the standard hydrogen electrode has been established by convention as 0.00 V. The standard reduction potential for any other half-cell can thus be established by pairing it with a hydrogen electrode and measuring the voltage of the complete cell. If hydrogen gas is oxidized in the complete reaction [$H_2(g) \longrightarrow 2\ H^+(aq) + 2e^-$], we give the other cell a *positive* $E°$; if hydrogen ions are reduced [$2\ H^+(aq) + 2e^- \longrightarrow H_2(g)$], we give the other cell a *negative* $E°$. *Standard reduction potentials are always written for reduction half-reactions, regardless of which way the half-reaction goes in the complete cell.* For example, the cell potential for the $Zn\,|\,Zn^{2+}\,||\,H^+\,|\,H_2\,|\,Pt$ cell, in which Zn is oxidized, is 0.7628 V; the standard reduction potential for the Zn, Zn^{2+} half-cell ($Zn^{2+} + 2e^- \longrightarrow Zn$) is thus listed as -0.7628 V. On the other hand, the cell potential for the $Pt\,|\,H_2\,|\,H^+\,||\,Ag^+\,|\,Ag$ cell, where Ag^+ is reduced, is 0.7996 V; thus $E°$ for the Ag, Ag^+ half-cell [$Ag^+(aq) + e^- \longrightarrow Ag(s)$] is 0.7996 V. *The larger (more positive) the standard reduction potential for any reduction half-reaction, the greater its tendency to proceed as written.*

If Table 18.2 allowed us only to calculate the cell voltages of cells that included a hydrogen electrode, it would not find much practical use. In fact, however, we can calculate the cell potential of any cell from the standard reduction potentials of the two half-cells. The cell voltage, $E°$ cell, is the *potential difference* between the cathode and the anode:

$$E°_{cell} = E°_{cathode} - E°_{anode}$$

standard

If $E°_{cell}$ is positive, current flows from the anode to the cathode.

Example 18.6 Use $E°$ values in Table 18.2 to determine the potential for a zinc-copper cell in which all reactants are in their standard states at 25°C and 1 atm.

Table 18.2 Standard Reduction Potentials
(A) indicates $[H^+] = 1\,M$; (B) indicates $[OH^-] = 1\,M$.

Oxidizing Agent		Reducing Agent	$E°$ (V)
$F_2 + 2e^-$		$\rightleftarrows 2\,F^-$	+2.87
$H_2O_2 + 2H^+ + 2e^-$	(A)	$\rightleftarrows 2\,H_2O$	+1.776
$Au^+ + e^-$		$\rightleftarrows Au$	+1.68
$MnO_4^- + 8H^+ + 5e^-$	(A)	$\rightleftarrows Mn^{2+} + 4\,H_2O$	+1.491
$Au^{3+} + 3e^-$		$\rightleftarrows Au$	+1.42
$ClO_4^- + 8H^+ + 8e^-$	(A)	$\rightleftarrows Cl^- + 4\,H_2O$	+1.37
$Cl_2 + 2e^-$		$\rightleftarrows 2\,Cl^-$	+1.36
$Cr_2O_7^{2-} + 14H^+ + 6e^-$	(A)	$\rightleftarrows 2\,Cr^{3+} + 7\,H_2O$	+1.33
$O_2 + 2H_2O + 4e^-$		$\rightleftarrows 4\,OH^-$	+1.23
$Br_2 + 2e^-$		$\rightleftarrows 2\,Br^-$	+1.087
$NO_3^- + 4H^+ + 3e^-$	(A)	$\rightleftarrows NO + 2\,H_2O$	+0.96
$Hg^{2+} + 2e^-$		$\rightleftarrows Hg$	+0.851
$2\,NO_3^- + 4H^+ + 2e^-$	(A)	$\rightleftarrows N_2O_4 + 2\,H_2O$	+0.81
$Ag^+ + e^-$		$\rightleftarrows Ag$	+0.7996
$Fe^{3+} + e^-$		$\rightleftarrows Fe^{2+}$	+0.770
$O_2 + 4H^+ + 4e^-$	(A)	$\rightleftarrows 2\,H_2O$	+0.682
$MnO_4^- + 2H_2O + 3e^-$	(B)	$\rightleftarrows MnO_2 + 4\,OH^-$	+0.58
$I_2 + 2e^-$		$\rightleftarrows 2\,I^-$	+0.535
$8\,H_2SO_3 + 32H^+ + 32e^-$	(A)	$\rightleftarrows S_8 + 24\,H_2O$	+0.45
$Cu^{2+} + 2e^-$		$\rightleftarrows Cu$	+0.3402
$Co(NH_3)_6^{3+} + e^-$	(B)	$\rightleftarrows Co(NH_3)_6^{2+}$	+0.10
$2H^+ + 2e^-$	(A)	$\rightleftarrows H_2$	0.00
$Fe^{3+} + 3e^-$		$\rightleftarrows Fe$	−0.036
$CrO_4^{2-} + 4H_2O + 3e^-$	(B)	$\rightleftarrows Cr(OH)_3 + 5\,OH^-$	−0.12
$Pb^{2+} + 2e^-$		$\rightleftarrows Pb$	−0.126
$Sn^{2+} + 2e^-$		$\rightleftarrows Sn$	−0.1364
$Ni^{2+} + 2e^-$		$\rightleftarrows Ni$	−0.23
$Co^{2+} + 2e^-$		$\rightleftarrows Co$	−0.28
$Cd^{2+} + 2e^-$		$\rightleftarrows Cd$	−0.4026
$Fe^{2+} + 2e^-$		$\rightleftarrows Fe$	−0.409
$Cr^{2+} + 2e^-$		$\rightleftarrows Cr$	−0.557
$Fe(OH)_3 + e^-$	(B)	$\rightleftarrows Fe(OH)_2 + OH^-$	−0.56
$Cr^{3+} + 3e^-$		$\rightleftarrows Cr$	−0.74
$Zn^{2+} + 2e^-$		$\rightleftarrows Zn$	−0.7628
$SO_4^{2-} + H_2O + 2e^-$	(B)	$\rightleftarrows SO_3^{2-} + 2\,OH^-$	−0.92
$Mn^{2+} + 2e^-$		$\rightleftarrows Mn$	−1.029
$Be^{2+} + 2e^-$		$\rightleftarrows Be$	−1.70
$Al^{3+} + 3e^-$		$\rightleftarrows Al$	−1.706
$Sc^{3+} + 3e^-$		$\rightleftarrows Sc$	−2.08
$Mg^{2+} + 2e^-$		$\rightleftarrows Mg$	−2.375
$Na^+ + e^-$		$\rightleftarrows Na$	−2.67
$Ca^{2+} + 2e^-$		$\rightleftarrows Ca$	−2.76
$Sr^{2+} + 2e^-$		$\rightleftarrows Sr$	−2.89
$Ba^{2+} + 2e^-$		$\rightleftarrows Ba$	−2.90
$Cs^+ + e^-$		$\rightleftarrows Cs$	−2.923
$K^+ + e^-$		$\rightleftarrows K$	−2.924
$Li^+ + e^-$		$\rightleftarrows Li$	−3.02

↓ decreasing strength as an oxidizing agent

↑ increasing strength as a reducing agent

Solution

a. Table 18.2 shows that copper ions have a greater tendency to be reduced than zinc ions. Thus the copper half-cell will be the cathode and the zinc half-cell, the anode.

b. The half-reaction equations and the cell equation are

$$\begin{array}{ll}\text{Cathode:} & Cu^{2+}(aq) + 2e^- \longrightarrow Cu(s) \\ \text{Anode:} & Zn(s) \longrightarrow Zn^{2+}(aq) + 2e^- \\ \hline \text{Cell reaction:} & Cu^{2+}(aq) + Zn(s) \longrightarrow Zn^{2+}(aq) + Cu(s)\end{array}$$

c. The cell potential is

$$\begin{aligned} E°_{cell} &= E°_{cathode} - E°_{anode} \\ &= +0.3402 - (-0.7628) \\ &= +1.1030 \text{ V} \end{aligned}$$

If we had mistakenly assumed that copper would be the anode and zinc the cathode, a negative $E°_{cell}$ would have signaled our mistake. A cell cannot produce a negative voltage if we consistently stick to the conventions we have adopted.

Example 18.7 Determine the cell potential for the $Zn|Zn^{2+}||Ag^+|Ag$ cell.

Solution

a. Silver will form the cathode and zinc, the anode.

b. The half-reaction equations and the cell reaction equation are

$$\begin{array}{ll}\text{Cathode:} & 2[Ag^+ + e^- \longrightarrow Ag(s)] \\ \text{Anode:} & Zn(s) \longrightarrow Zn^{2+}(aq) + 2e^- \\ \hline \text{Cell reaction:} & Zn(s) + 2\,Ag^+(aq) \longrightarrow Zn^{2+}(aq) + 2\,Ag(s)\end{array}$$

c. The cell potential is

$$\begin{aligned} E°_{cell} &= E°_{cathode} - E°_{anode} \\ &= +0.7996 - (-0.7628) \\ &= +1.5624 \text{ V} \end{aligned}$$

Notice that although we multiplied the reduction half-reaction by two, we did not tamper with $E°_{cathode}$.

Question

9. How many faradays of electricity must be transferred to deposit 1.00 mol of Ag metal from $AgNO_3$ solution? To reduce 1.00 mol of bromine to bromide ions?

18.4 The Nernst Equation

In Chapter 17 we pointed out that the hierarchy of oxidizing and reducing agents established by Table 17.3 depended on the reactants in the half-cells having unit activity. Naturally the same warning must be appended to Table 18.2. There is, however, a way to make adjustments to the standard electrode potentials for reactants whose activities are not equal to one. To do so we shall begin with some approximations that allow us to estimate accurately the activities of substances:

activity of an ion in solution $\cong$ molarity

activity of a gas $\cong$ partial pressure (atm)

activity of a pure solid or liquid and of water in dilute solutions $\cong 1$

With these approximations we can now make use of the **Nernst equation**, the derivation of which we shall defer to the end of this section:

$$E = E° - \frac{RT}{nF} \ln Q$$

In this equation, E is the adjusted electrode potential of the half-cell, R is the gas constant, T is the absolute temperature, n is the number of electrons that appear in the half-reaction; F is the value of the faraday in coulombs, and Q is the reaction quotient. (This is the same reaction quotient we first encountered in Chapter 13: It has the same form as an equilibrium constant, but it is evaluated with nonequilibrium concentrations.) We can rewrite the Nernst equation in a more useful form by substituting $R = 8.32 \text{ J}/(\text{K})(\text{mol})$, $T = 298 \text{ K}$, and $F = 9.65 \times 10^4 \text{ coul/mol } e^-$ and by converting to base 10 logarithms:

$$E = E° - \frac{0.05916}{n} \log Q$$

$$M^{2+} + 2e^- \longrightarrow M \qquad Q = \frac{[M]}{[M^{2+}]}$$

Example 18.8 Calculate the half-cell potential for $Zn^{2+}|Zn$ if the concentration of Zn^{2+} is 0.10 M instead of the standard 1.0 M.

Solution

a. The half-reaction (from Table 18.2) is

$$Zn^{2+}(aq) + 2e^- \longrightarrow Zn(s)$$

b. The value of the reaction quotient is

$$Q = \frac{[Zn(s)]}{[Zn^{2+}(aq)]} = \frac{1}{0.10} = 10$$

Electrochemistry

c. Two electrons appear in the half reaction, so $n = 2$.

d. Thus the revised half-cell potential is

$$E = -0.7628 - \left(\frac{0.05916}{2}\right) \log 10$$
$$= -0.7628 - 0.02958$$
$$= \mathbf{-0.7924 \text{ V}}$$

The effect of a tenfold reduction in the Zn^{2+} ion concentration is thus fairly small; yet it is the effect we would have predicted on the basis of Le Chatelier's principle: Zinc metal is more easily oxidized to Zn^{2+}. This phenomenon can be exploited in a *concentration cell*, where the two half-cells contain the same reactants but at different concentrations (Figure 18.11). The electrode surrounded by the dilute solution will be the anode, and the concentration of the dilute solution will increase as the electrode is oxidized. The ions in the more concentrated solution, on the other hand, will be reduced, plating out on the cathode. Finally the concentrations of the two solutions will become equal, and current will cease to flow. As one might infer from the result of Example 18.8, the voltages of concentration cells are usually very low.

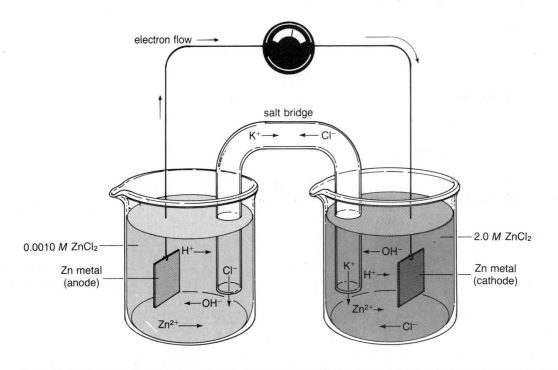

Figure 18.11 The concentration cell: $Zn|Zn(0.0010\ M)||Zn^{2+}(2.0\ M)|Zn$:
Electrolyte: $ZnCl_2(aq)$
Salt bridge: $KCl(aq)$

Although we introduced the Nernst equation in the context of half-cell potentials, it is equally applicable to the task of calculating the potentials of galvanic cells:

$$E = E°_{cell} - \frac{0.05916}{n} \log Q$$

Here we use the cell reaction, not the reduction half-reaction, to calculate Q. A close look at this equation is revealing. It shows that just as the concentration cell "runs down" when $Q = 1$, any galvanic cell will cease to operate ($E = 0$) when Q reaches a certain characteristic value. At that point, where

$$E°_{cell} = \frac{0.05916}{n} \log Q$$

Q is equal to the familiar equilibrium constant K.

Example 18.9 Calculate the equilibrium constant for the reaction

$$Zn(s) + Cu^{2+}(aq) \longrightarrow Cu(s) + Zn^{2+}(aq)$$

at 25°C, and 1 atm.

Solution

a. The half-reactions are

$$Zn(s) \longrightarrow Zn^{2+}(aq) + 2e^-$$
$$Cu^{2+}(aq) + 2e^- \longrightarrow Cu(s)$$

b. The standard cell potential is 1.10 V (Example 18.6), and n (the number of moles of electrons transferred in the reaction as written) is 2.

c. At equilibrium (no net reaction occurring), $E = 0$:

$$0 = +1.10 - \frac{0.05916}{2} \log K$$

$$\log K = (1.10)\left(\frac{2}{0.05916}\right) = 37.2$$

$$K = 10^{37}$$

The size of this constant is typical of redox-reaction equilibria; thus we usually treat redox reactions as if they went to completion. Nonetheless, we shouldn't lose sight of the fact that they are true dynamic equilibria.

Earlier in this chapter we expressed a general relationship among energy, charge, and electromotive force. We can now be more specific. The cell voltage of a galvanic cell, E, is directly related to the free energy of the cell reaction, ΔG:

$$\Delta G = -nFE$$

where n is the number of faradays transferred, $F = 9.65 \times 10^4$ coul, and ΔG is expressed in joules. As we expect, a positive cell voltage produces a negative change in free energy, which we recognize as the requirement for spontaneity. If the cell voltage is measured for a cell with standard electrodes (all activities equal to unity), this relationship becomes

$$\Delta G° = -nFE°$$

where $\Delta G°$ is the standard free energy change for the reaction.

Now we can see the origin of the Nernst equation. On p. 415, we wrote an equation relating the free energy change for a gas phase reaction to the standard free energy change and the reaction quotient:

$$\Delta G = \Delta G° + RT \ln Q$$

If we now recognize that this equation is applicable to all chemical reactions, provided we express Q in terms of activities, the Nernst equation is a direct result. Substituting for ΔG and $\Delta G°$,

$$-nFE = -nFE° + RT \ln Q$$

$$E = E° - \frac{RT}{nF} \ln Q$$

Question

10. In a concentration cell with one active metal electrode and one inert platinum electrode, would the inert electrode be immersed in the dilute or in the concentrated solution?

18.5 Electrolysis

An electrolytic cell differs from a galvanic cell in two important respects. First, both electrodes (which may be active or inert) are immersed in the same electrolyte solution rather than in separate half-cells; second, an external source of current drives the cell reaction in a direction it would not spontaneously go. A few examples should make clear the importance of these two differences, as well as point out the special problems of predicting the products in electrolytic reactions.

Electrolysis of Molten Sodium Chloride An electrolyte need not be an aqueous solution; any substance containing ions that are free to move under the influence of an electrical potential will do. Molten salts are good examples. For example, if we place two platinum electrodes, connected to a battery, in molten NaCl, the result is an electrolytic cell (Figure 18.12). The battery moves electrons toward the cathode, which becomes negatively charged. Sodium ions thus migrate toward the cathode, pick up electrons, and are reduced to free sodium metal:

$$Na^+ + e^- \longrightarrow Na$$

Meanwhile, since the battery has removed electrons from the anode, it bears a positive charge and chloride ions drift toward it, where they give up electrons

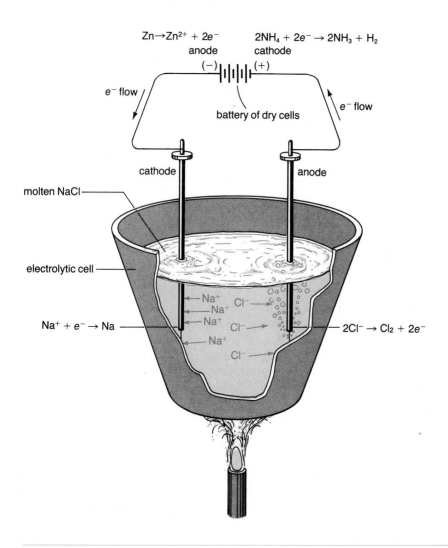

Figure 18.12 Electrolysis of molten NaCl.

and are oxidized to chlorine atoms. Pairs of chlorine atoms then unite to form molecules of chlorine gas:

$$2(\text{Cl}^- \longrightarrow \text{Cl} + e^-)$$
$$\underline{\text{Cl} + \text{Cl} \longrightarrow \text{Cl}_2}$$
$$2\,\text{Cl}^- \longrightarrow \text{Cl}_2 + 2e^-$$

As in a galvanic cell, the cathode is the site of reduction and the anode is the site of oxidation. However, the sign convention must be reversed. Since the external circuit imposes a positive charge on the anode and a negative charge on the cathode, *in an electrolytic cell the anode is designated positive and the cathode, negative.*

Electrolysis of Sodium Chloride Solution The electrolysis of molten sodium chloride is the commercial source of sodium metal, with chlorine a by-product. We might expect the electrolysis of a sodium chloride solution to be more complicated than the electrolysis of the molten salt. After all, the aqueous solution contains, in addition to sodium and chloride ions, hydrogen ions, hydroxide ions, and water molecules. Under the influence of an imposed potential, both cations (Na^+ and H^+) migrate toward the cathode, and both anions (Cl^- and OH^-) migrate toward the anode. (The water molecules tend only to orient themselves in the electric field, as all molecules with a permanent dipole do.) Thus there are two possible reactions at each electrode:

At the cathode: $Na^+(aq) + e^- \longrightarrow Na(s)$
 $2\,H^+(aq) + 2e^- \longrightarrow H_2(g)$

At the anode: $2\,Cl^-(aq) \longrightarrow Cl_2(g) + 2e^-$
 $4\,OH^-(aq) \longrightarrow O_2(g) + 2\,H_2O(l) + 4e^-$

We must now ask which are the most likely. At the cathode the answer is fairly simple: As Table 18.2 shows, H^+ is a much stronger oxidizing agent than Na^+; thus despite its lower concentration H^+ will be reduced at the cathode, producing hydrogen gas. As this happens, water molecules will ionize to replenish, at least partly, the supply of hydrogen ions. The net reaction at the cathode is therefore

$$2\,H^+(aq) + 2e^- \longrightarrow H_2(g)$$
$$\underline{2\,H_2O(l) \longrightarrow 2\,H^+(aq) + 2\,OH^-(aq)}$$
$$2\,H_2O(l) + 2e^- \longrightarrow H_2(g) + 2\,OH^-(aq)$$

At the anode the situation is not so straightforward. Hydroxide ions are somewhat stronger reducing agents than Cl^-, but the disparity is not so great as between H^+ and Na^+. Therefore, because the Cl^- concentration is much greater than that of OH^- in most NaCl electrolysis cells, we find Cl_2 rather than O_2 evolved at the anode. The net cell reaction thus evolves hydrogen at the cathode and chlorine at the anode:

Anode: $2\ Cl^-(aq) \longrightarrow Cl_2(g) + 2e^-$

Cathode: $2\ H_2O(l) + 2e^- \longrightarrow H_2(g) + 2\ OH^-(aq)$

Cell reaction: $2\ H_2O(l) + 2\ Cl^-(aq) \longrightarrow H_2(g) + 2\ OH^-(aq) + Cl_2(g)$

The complete cell equation,

$$2\ NaCl(aq) + 2\ H_2O \longrightarrow Cl_2(g) + H_2(g) + 2\ NaOH(aq)$$

shows that the cell changes a sodium chloride solution to a sodium hydroxide solution. This cell is used for the commercial preparation of sodium hydroxide, hydrogen, and chlorine.

Electrolysis of Hydrochloric Acid Hydrochloric acid contains water molecules, hydrogen ions, chloride ions, and a much smaller concentration of hydroxide ions. Since hydrogen ions are the only cations in the solution, the cathode reaction is

$$2\ H^+(aq) + 2e^- \longrightarrow H_2(g)$$

From what we found in the NaCl cell, we expect the plentiful chloride ions rather than the hydroxide ions to be oxidized at the anode:

$$2\ Cl^-(aq) \longrightarrow Cl_2(g) + 2e^-$$

The cell equation is

$$2\ H^+(aq) + 2\ Cl^-(aq) \longrightarrow H_2(g) + Cl_2(g)$$

Since both hydrogen and chloride ions are removed from s operates, the hydrochloric acid solution becomes more d⋮

Electrolysis of Sulfuric Acid Another common ¹ complicated case. Dilute sulfuric acid cont⋯ hydrogen sulfate ions, sulfate ions, and ⋯ Again, there is only one possible cat⋯ ions. But there are now three ani⋯ the anode:

$$SO_4^{2-}(aq) \longrightarrow SO_4(a_4$$
$$2\ HSO_4^-(aq) \longrightarrow H_2S_2O_8(a_⋯$$
$$4\ OH^-(aq) \longrightarrow O_2(g) + 2\ H_2$$

The hydroxide ion is oxidized *much* more ⋯ sulfate ions so, despite the very low concentra⋯

at the anode. Since the supply of OH⁻ ions is then replenished by the ionization of water, the net anode reaction is the sum of two equations:

$$4\ OH^-(aq) \longrightarrow O_2(g) + 2\ H_2O(l) + 4e^-$$
$$\underline{4\ H_2O(l) \longrightarrow 4\ H^+(aq) + 4\ OH^-(aq)}$$
$$2\ H_2O(l) \longrightarrow O_2(g) + 4\ H^+(aq) + 4e^-$$

The cell reaction is therefore

$$2\ H_2O(l) \longrightarrow O_2(g) + 4\ H^+(aq) + 4e^-$$
$$\underline{4\ H^+(aq) + 4e^- \longrightarrow 2\ H_2(g)}$$
$$2\ H_2O(l) \longrightarrow 2\ H_2(g) + O_2(g)$$

Thus this electrolytic process produces hydrogen and oxygen by decomposing water. It is sometimes used commercially to prepare oxygen of high purity, but since it is a costly process, other methods are used when less pure oxygen will suffice.

The electrolysis of *concentrated* sulfuric acid at low temperatures is a different story entirely. The cathode reaction must of course remain the same, but the anode reaction is

$$2\ H_2SO_4(l) \longrightarrow H_2S_2O_8(l) + 2\ H^+(aq) + 2e^-$$

The resulting hydrogen peroxydisulfate molecule can be hydrolyzed to form hydrogen peroxide in a commercial preparation for the latter substance.

$$H_2S_2O_8(l) + H_2O(l) \longrightarrow H_2O_2(aq) + H^+(aq) + HSO_4^-(aq)$$

From these examples it should now be clear that it is not always easy to predict the products of electrolysis reactions. Table 18.2 remains a useful guide, but it provides few hard and fast rules. The main complication is concentration. As we know from the Nernst equation, reaction potentials depend on the concentrations of the reactants, and it is especially difficult to know the concentrations of competing reactants near the electrodes of an operating electrolytic cell. A second problem is temperature, which can affect different reactions differently. Finally, reaction *rates* may favor one product, even when Table 18.2 suggests another. Gases especially are slow to evolve unless considerable *overvoltage* is applied to the cell. For example, to produce hydrogen at a practical rate at the cathode of an electrolytic cell, we must apply a much higher voltage than we would calculate naively from Table 18.2. But when we do this, other reduction half-reactions become possible, even though they appear lower in the table. In fact metals in Table 18.2 down to and including zinc can be recovered by electrolysis from aqueous solutions.

In the case of reactions at the anode, it is easier to remember the order in which common anions are oxidized than to try to rationalize the several variables. In solutions of moderate concentration, common anions are usually in the order I^-, Br^-, Cl^-, OH^-, NO_3^-, SO_4^{2-}.

Electrolysis of Nickel Sulfate Solution Aqueous nickel sulfate consists of water molecules, nickel ions, sulfate ions, hydrogen ions, and hydroxide ions. Since the hydroxide ion is oxidized more readily than the sulfate ion, O_2 is evolved at the anode, with the net anode reaction as previously derived:

$$2\,H_2O(l) \longrightarrow O_2(g) + 4\,H^+(aq) + 4e^-$$

Hydrogen ion is more easily reduced than is nickel ion, but remembering the fact that excessive overvoltage is necessary to produce hydrogen at a practical rate we realize that the predominant cathode product will be nickel metal:

$$Ni^{2+}(aq) + 2e^- \longrightarrow Ni(s)$$

Electrolysis of Copper(II) Sulfate Solution Since the Cu^{2+} ion is above the H^+ ion in Table 18.2, copper(II) ions have a greater affinity for electrons than do hydrogen ions. Hence, copper is reduced at the cathode. As in several earlier examples, hydroxide ions are oxidized at the anode. Thus the solution in the cell changes from copper(II) sulfate to sulfuric acid. When the copper(II) ion is exhausted, the cell continues to operate, liberating hydrogen at the cathode. Note in this example that matters would be different if the anode were copper instead of inert graphite or platinum. Then it would be copper metal oxidized at the anode to Cu^{2+}. The result would be to reduce the size of the anode and to plate copper onto the cathode. The concentrations of species in solution would remain unchanged.

Questions

11. Define electrolyte and overvoltage.

12. Identify the predominant products at the anode and the cathode for the electrolysis of the following (with inert electrodes):

    a. molten potassium bromide   b. aqueous potassium bromide
    c. aqueous potassium sulfate

18.6 Applications of Electrolysis

Refining of Metals Many metals can be refined electrolytically using a process similar to that we described at the end of Section 18.5. Copper, in fact, is a good example. Copper ore is roasted and reduced to produce a crude copper (about 99.4 percent pure) known as *blister copper*. This crude product is then used as the anode in an electrolytic cell. The cathode is a thin sheet of pure copper, and the electrolyte contains copper(II) sulfate, sulfuric acid, and sodium chloride. The copper is dissolved from the anode, along with more active metals like zinc and iron. Less active metals (such as gold and silver) are not oxidized at the anode and thus drop to the bottom of the solution forming an inert *anode mud*. Meanwhile, Cu^{2+}, in preference to the impurities oxidized at the anode, is plated out at the cathode. This refining process is especially important, since very small impurities have an enormous effect on the usefulness of copper in its major application: electrical wiring and appliances. For example, a few hundredths of 1 percent of arsenic can reduce the conductivity of copper by 15 percent.

Aluminum is another metal produced electrolytically. The chief ore of aluminum is bauxite, a hydrated aluminum oxide ($Al_2O_3 \cdot n\,H_2O$). But to use this as a source of pure aluminum is not an easy matter. Because Al^{3+} is a very poor oxidizing agent, bauxite cannot be reduced by carbon or other common chemical reducing agents nor can aluminum be purified by conventional electrolytic methods. It reacts with water; hence aqueous solutions cannot be used. Also, aluminum compounds either have prohibitively high melting points or are poor conductors; thus electrolysis of molten aluminum compounds is generally not feasible. Because of these problems aluminum long remained an unexploited resource, in spite of its being the most abundant metal in the earth's crust. In 1886, however, Charles Hall (1863–1914) a very young American scientist, developed the modern electrolytic process for the production of aluminum.

In the Hall process, purified aluminum oxide is dissolved in molten cryolite (Na_3AlF_6), where it ionizes into aluminum ions and oxide ions. When this mixture is electrolyzed, molten aluminum is produced at the cathode and oxygen at the anode. (In a second reaction at the anode, the carbon electrode is attacked by oxygen and gradually used up.) Thus

Cathode: $Al^{3+} + 3e^- \longrightarrow Al$

Anode: $2\,O^{2-} \longrightarrow O_2 + 4e^-$

The aluminum produced by this process is more than 99 percent pure.

Electroplating Electroplating is merely an electrolytic process where the object to be plated is the cathode and the metal used for coating it is the anode. Originally used only to improve the appearance of objects by coating them with silver, electroplating is now used for more practical purposes. It can be used to protect surfaces against corrosion or to provide harder, smoother surfaces for moving parts. Worn parts can also be rebuilt to their original dimensions. Also it provides a thinner, more uniform coating than dipping, painting, or spraying processes.

Charging the Lead Storage Cell In Section 18.2 we described the lead storage cell in detail. We also intimated that it could be recharged. It should now be clear how this is done: It is in effect converted to an electrolytic cell (the anode becomes the cathode, and vice versa) and electrons are forced through it in the opposite direction. The cell reaction is reversed and lead, lead(IV) oxide, and sulfuric acid are produced.

Corrosion Although corrosive processes have been studied extensively, there are still many things we do not understand about them. But we do believe that corrosion is essentially an electrochemical phenomenon. After we have studied the metals in greater detail, we shall return to this topic.

Questions

13. Why can't aluminum be produced by a reduction of its oxide by carbon?

14. Why can't aluminum be produced by electrolysis of an aqueous solution of an aluminum compound or of a molten aluminum compound?

18.7 Quantitative Aspects of Electrolysis

We first summarized the laws of Michael Faraday in Chapter 4. After our discussion of electrolysis, we are now in a position to make more sense of them. The laws are

1. The weight of a substance produced by electrolysis is directly proportional to the amount of electricity passed through the substance.

2. Weights of different substances produced by the same amount of current are proportional to the equivalent weights of those substances.

The following two examples serve as illustrations.

Example 18.10 What weight of copper can be produced by a 3.00-A (ampere) current flowing through a copper(II) sulfate solution for 5.00 hours?

Solution

a. Calculate the quantity of electricity (in faradays) used.

$$1 \text{ faraday} = 9.65 \times 10^4 \text{ coulomb}$$

$$1 \text{ coulomb} = (1 \text{ ampere})(1 \text{ second})$$

$$(3.00 \text{ A})(5.00 \text{ hr})\left(\frac{3600 \text{ s}}{1 \text{ hr}}\right)\left(\frac{1 \text{ faraday}}{9.65 \times 10^4 \text{ A·s}}\right) = 0.560 \text{ faraday}$$

b. The half-reaction, $Cu^{2+} + 2e^- \rightarrow Cu$, shows that two electrons are required to reduce one copper(II) ion; thus 2 faradays of electricity are required to produce 1 gram atomic weight of copper. Therefore 0.560 faraday would produce

$$(0.560 \text{ faraday})\left(\frac{1 \text{ g at. wt}}{2 \text{ faradays}}\right)\left(\frac{63.5 \text{ g Cu}}{1 \text{ g at. wt}}\right) = 17.8 \text{ g of Cu}$$

Example 18.11 How long must a 3.00-A current flow through a solution of dilute H_2SO_4 to liberate 4.00 liters of oxygen measured at 27°C and 2.60 atm?

Solution

a. Calculate the number of moles of oxygen liberated:

$$(4.00 \text{ liters}) \left(\frac{2.60 \text{ atm}}{1.00 \text{ atm}}\right) \left(\frac{273 \text{ K}}{300 \text{ K}}\right) \left(\frac{1.00 \text{ mol}}{22.4 \text{ liters}}\right) = 0.422 \text{ mol}$$

b. Oxygen has been changed by the electrolysis from an oxidation state of -2 to 0; thus each atom of oxygen has lost two electrons, and each molecule of O_2 has lost four. Therefore the number of equivalents of oxygen liberated is four times the number of moles liberated:

$$(0.422 \text{ mol } O_2) \left(\frac{4 \text{ equivalents } O_2}{1 \text{ mol } O_2}\right) = 1.69 \text{ equivalents } O_2$$

c. Each equivalent requires 1 faraday or 9.65×10^4 coul (or 9.65×10^4 A s). The number of ampere-seconds required is

$$(1.69 \text{ equivalents}) \left(\frac{9.65 \times 10^4 \text{ As}}{1.00 \text{ equivalent}}\right) = 1.63 \times 10^5 \text{ A s}$$

d. And, finally, the time required is

$$\frac{1.63 \times 10^5 \text{ As}}{3.00 \text{ A}} = 5.45 \times 10^4 \text{ s}$$

$$(5.43 \times 10^4 \text{ s}) \left(\frac{1.00 \text{ hr}}{3.60 \times 10^3 \text{ s}}\right) = \textbf{15.1 hr}$$

Questions and Problems

18.1 Galvanic Cells

15. For each of the following cells (with standard electrodes), write the equations for the reactions at each electrode and for the cell reaction; identify the anode, cathode, positive electrode, and negative electrode; and indicate the direction of electron flow of positive ion movement and of negative ion movement and the change in mass of the electrodes, if any.

a. $Mg|Mg^{2+}||Co^{2+}|Co$ b. $Pt|I^-|I_2||Cl_2|Cl^-|Pt$

c. $Al|Al^{3+}||Au^+|Au$ d. $Pt|Cl^-|Cl_2||Mg^{2+}|Mg$

e. $Cr|Cr^{3+}||Sn^{2+}|Sn$ f. $Mn|Mn^{2+}||Ag^+|Ag$

18.2 Commercial Cells

16. Why does a measurement of the specific gravity of the liquid in a lead storage battery indicate whether the battery needs to be charged?

17. Would adding concentrated sulfuric acid correct the problem indicated by the low specific gravity of a lead storage battery?

18. What is the purpose of the manganese(IV) oxide in the dry cell?

18.3 Electrode Potentials

19. Calculate the voltage delivered by each cell in Problem 15.

18.4 Free Energy Changes in Galvanic Cells

20. Write the equation for the reactions at each electrode; identify the anode, cathode, positive electrode and negative electrode; and indicate the direction of electron flow, the direction of positive ion movement and of negative ion movement, and the change in the mass of the electrodes, if any, for the cell:

$$Ag\,|\,Ag^+(dil.)\,\|\,Ag^+(conc.)\,|\,Ag$$

21. Calculate the potential for the concentration cell in Problem 20, if the concentrations of silver nitrate in the two half-cells are 0.0010 M in one and 2.0 M in the other.

22. Calculate the potential for the cells in Problem 15 assuming the electrolytes have concentrations of 0.010 M.

23. Use Table 18.2 to calculate the value of K for the following reactions:
   a. $Br_2 + 2\,I^- \longrightarrow 2\,Br^- + I_2$ b. $Mn + 2\,H^+ \longrightarrow H_2 + Mn^{2+}$

18.5 Electrolysis

24. Write the equations for the reactions that take place at both electrodes during the electrolysis of each of the following. Assume inert electrodes unless otherwise indicated.
   a. dilute solution of sulfuric acid
   b. solution of sodium sulfate
   c. solution of copper(II) sulfate
   d. solution of copper(II) sulfate (copper electrodes)
   e. solution of sodium hydroxide
   f. solution of lead(II) nitrate
   g. solution of potassium chloride
   h. solution of magnesium iodide

18.7 Quantitative Aspects of Electrolysis

25. How many coulombs are
   a. equivalent to 0.300 faraday?
   b. equivalent to 1.20×10^{23} electrons?
   c. produced by a 10.0-A current in 3.00 min?
   d. required to produce 1.00 liter of hydrogen at STP?
   e. required to produce 1.00 liter of oxygen from H_2O at STP?
   f. required to produce 1.00 g of silver from Ag^+ solution?

g. required to produce 1.00 g of zinc from Zn^{2+} solution?
h. required to produce 1.00 g of chromium from Cr^{3+} solution?

26. How many grams, gram atomic weights, and atoms of each of the following can be produced by 5.00 faradays of electricity?
   a. Cr from Cr^{3+} b. Cr from Cr^{2+} c. Ag from Ag^{+}

27. A current is passed through a solution of copper(II) sulfate long enough to deposit 14.0 g of copper. What volume (STP) of oxygen is produced at the same time?

28. Two electrolytic cells are connected so that the same current goes through each. If 1.30 g of chromium is produced from a Cr^{2+} solution in the first cell, what weight of chromium will be produced from a Cr^{3+} solution in the second?

29. How many faradays are needed to produce a kilogram of sodium metal by electrolysis of molten sodium chloride?

*30. For 30.0 min a 5.00-A current passes through all the following solutions: sodium sulfate, dilute sulfuric acid, copper(II) sulfate, and sodium chloride. How many grams of copper, oxygen, and hydrogen are produced?

31. If 5.00×10^4 coul is needed to produce 15.0 g of a tripositive metal, what is the atomic weight of the metal?

Discussion Starters

32. Explain or account for the following:
   a. In the galvanic cell, $Zn|Zn^{2+}||Cl^-|Cl_2|Pt$, we say the zinc electrode is negative; yet negative ions migrate toward this electrode.
   b. Only oxidation-reduction reactions are suitable for the direct production of electrical energy.
   c. In a magnesium-silver cell, silver is positive; but in a silver-gold cell, silver is negative.
   d. The electrolysis of $1\ M\ Sn(NO_3)_2$ would produce free tin at the cathode, even though (according to Table 18.2) H^+ is a better oxidizing agent than Sn^{2+}.
   e. The same products are obtained at the electrodes in the electrolysis of aqueous sodium sulfate as in the electrolysis of dilute sulfuric acid.
   f. The same products are obtained at the electrodes in the electrolysis of aqueous sodium chloride as in the electrolysis of hydrochloric acid.
   g. A used dry cell does not smell highly of ammonia, even though this gas is produced at the carbon electrode.
   h. In the electrolysis of dilute sulfuric acid, the volumes of gases produced at the electrodes are not the same.
   i. Hydrogen rather than sodium is formed at the cathode during the electrolysis of aqueous sodium chloride, in spite of the fact that sodium ions are present in the solution in much greater quantities than hydrogen ions.

The Chemistry of the Nonmetals of Groups VIA and VIIA

The next three chapters of this book are devoted largely to what many people would regard as the foremost mission of chemistry—a study of the properties of the elements and their compounds. As the layout of this text suggests, most modern chemists regard the theoretical framework more important than such descriptive details, but it would be hard to convince many laymen that thermodynamics and kinetics are more basic than the colors and densities of the alkali metals. And in fact it is important to discuss the properties and reactions of elements and compounds, especially when they point to family relationships or to practical applications.

We can begin by once more dividing the elements into their two most general classifications: metals and nonmetals. We have already described many of the differences between these two. Distinctions have been drawn in terms of electron configurations, ionization energies, electronegativities, positions in the periodic table, chemical bonding, acid and base properties, oxidizing and reducing characteristics, and physical properties. A summary of these differences appears in Table 19.1.

We should also remember that despite these differences the line between metals and nonmetals is a fuzzy one. For example, carbon behaves chemically like a typical nonmetal, but in the form of graphite it is shiny and it conducts electricity. Gold and copper, on the other hand, are typical metals, but neither have the characteristic silver-white color associated with metals. Likewise the metallic elements of Group IA lack the tensile strength typical of most familiar metals. And of course the properties of the metalloids are even more ambiguous. Nonetheless, Table 19.1 provides some useful guidelines—as we shall see in the coming chapters.

We shall now tackle the elements more or less systematically, beginning with the nonmetals of Groups VIA and VIIA. In general, we won't repeat things we have already covered, so you may wish to refer especially to Chapters 6, 7, and 9 as we discuss the chemistry of the nonmetals.

19.1 The Halogens

Although the halogens get their name ("formers of sea salt") from the chlorides, bromides, and iodides in seawater, two of them must be sought elsewhere.

Table 19.1 Summary of the Properties of Metals and Nonmetals

Property	Metals	Nonmetals
Electron configurations (Chapter 5)	Three or fewer electrons in the outermost energy level (bismuth, tin, and lead are exceptions).	Four or more electrons in the outermost energy level (hydrogen and boron are exceptions).
Ionization energies (Chapter 5)	Low	High
Electronegativities (Chapter 6)	Low	High
Positions in the periodic table (Chapter 5)	All areas of the periodic table except the upper right-hand corner.	Upper right-hand corner (excluding the noble gases).
Chemical bonding (Chapter 6)	Form ionic bonds by transferring electrons to atoms of nonmetals.	Form ionic bonds by accepting electrons from atoms of metals; also form covalent bonds with each other.
Acid-base properties (Chapter 14)	Metallic oxides are basic.	Nonmetallic oxides are acidic.
Oxidizing-reducing characteristics (Chapter 17)	Often act as reducing agents (donate electrons).	Often act as oxidizing agents (accept electrons).
Physical properties (Chapter 21)	Lustrous, malleable, ductile, hard, good electrical and thermal conductors, generally high melting and boiling points; most are solids at room conditions.	If solid, they are dull and brittle; insulators, poor conductors, generally low melting and boiling points; one is a liquid, five are gases under ordinary conditions.

Fluorine is most abundant as insoluble fluorides in minerals such as fluorspar (CaF_2) and cryolite (Na_3AlF_6); and astatine, which has been prepared by nuclear reactions (Chapter 25), is rarely found in nature at all. There is good reason for beginning our study of the nonmetals with the halogens: They constitute the only group in the periodic table made up entirely of nonmetals (though some chemists regard astatine as a metalloid). In fact, within any period, the halogen is the most typically nonmetallic element: It has the highest electronegativity, is the strongest oxidizing agent, and forms the strongest acid. Thus fluorine, the most nonmetallic halogen, is also the most nonmetallic of all elements.

Each of the electron configurations for the halogens concludes with s^2p^5; hence their most characteristic chemical property is their tendency to gain or

	IA																	0
	H	IIA											IIIA	IVA	VA	VIA	VIIA	He
	Li	Be											B	C	N	O	F	Ne
	Na	Mg	IIIB	IVB	VB	VIB	VIIB		—VIIIB—		IB	IIB	Al	Si	P	S	Cl	Ar
	K	Ca	Sc	Ti	V	Cr	Mn	Fe	Co	Ni	Cu	Zn	Ga	Ge	As	Se	Br	Kr
	Rb	Sr	Y	Zr	Nb	Mo	Tc	Ru	Rh	Pd	Ag	Cd	In	Sn	Sb	Te	I	Xe
	Cs	Ba	La	Hf	Ta	W	Re	Os	Ir	Pt	Au	Hg	Tl	Pb	Bi	Po	At	Rn
	Fr	Ra	Ac															

Ce	Pr	Nd	Pm	Sm	Eu	Gd	Tb	Dy	Ho	Er	Tm	Yb	Lu
Th	Pa	U	Np	Pu	Am	Cm	Bk	Cf	Es	Fm	Md	No	Lr

share one electron per atom. They all exist as diatomic molecules, and no allotropic forms are known. As Table 19.2 shows, the melting points, boiling points, and color intensities of the halogens increase with increasing atomic weight. Conversely, electronegativities and oxidizing strengths decrease as one goes down the group.

Table 19.2 Properties of the Halogens

	Fluorine	Chlorine	Bromine	Iodine	Astatine
Atomic number	9	17	35	53	85
Atomic weight	18.998	35.453	79.909	126.904	(210)
Physical state	Gas	Gas	Liquid	Solid	Solid
Melting point (°C)	−218	−101	−7	+113	—
Boiling point (°C)	−188	−34	+59	+184	—
Density (g/cm^3)	1.108 (liq. at b.p.)	1.57 (liq. at b.p.)	3.12 (at 20°C)	4.93 (at 20°C)	—
Color	Pale yellow-green	Yellow-green	Reddish brown	Black(s) violet(g)	—
Crystal structure	Cubic	Tetragonal	Orthorhombic	Orthorhombic	—
Heat of vaporization (kcal/mol)	1.51	4.88	7.34	11.14	—
Electronegativity	4.0	3.0	2.8	2.5	—

Preparation Except for traces of iodine, the halogens do not exist as free elements in nature. But as we saw in Chapter 18, they can be prepared by electrolysis. Indeed, electrolysis is the *only* way to prepare fluorine: Since fluorine is the strongest of all oxidizing agents, there is no other way to oxidize the fluoride ion. The other halide ions are more easily oxidized. The following oxidizing agents are strong enough to oxidize the chloride, bromide, and iodide ions: NO_3^-, $Cr_2O_7^{2-}$, CrO_4^{2-}, MnO_4^-, $S_2O_8^{2-}$, MnO_2, and PbO_2. Weaker oxidizing agents, such as sulfuric acid and free chlorine itself, will oxidize bromide and iodide but not chloride ions. The ferric ion and free bromine will oxidize only iodide ions. Some typical net ionic equations appear below.

$$2\, MnO_4^-(aq) + 10\, Br^-(aq) + 16\, H^+(aq) \longrightarrow$$
$$2\, Mn^{2+}(aq) + 5\, Br_2(l) + 8\, H_2O(l)$$

$$2\, I^-(aq) + Cl_2(g) \longrightarrow I_2(s) + 2\, Cl^-(aq)$$

$$2\, Fe^{3+} + 2\, I^-(aq) \longrightarrow I_2(s) + 2\, Fe^{2+}(aq)$$

$$MnO_2(s) + 2\, Cl^-(aq) + 4\, H^+(aq) \longrightarrow Mn^{2+}(aq) + Cl_2(g) + 2\, H_2O(l)$$

We would not have predicted some of these reactions from the data in Table 18.2, but under the appropriate conditions they occur readily.

Physical and Chemical Properties At room temperature, chlorine and fluorine are gases, bromine is a liquid, and iodine is a solid. (Bromine is one of the two elements that are liquid at room temperature.) The vapor pressure of iodine is considerable even at room temperature; thus it sublimes readily at temperatures below its melting point. Chlorine, bromine, and iodine dissolve in organic solvents such as alcohol, carbon tetrachloride, chloroform, ether, and carbon disulfide. Since fluorine reacts with water and with all organic solvents, there are no solutions of fluorine in common solvents. Incidentally, iodine is unique among the halogens in that it changes colors (from violet to brown) in many polar solvents.

Fluorine is the most reactive substance known. It reacts directly with every element except oxygen, nitrogen, helium, neon, and argon, and even stable nitrogen fluorides and oxygen fluorides exist. For example, the reaction between fluorine and water produces oxygen difluoride:

$$H_2O(l) + 2\, F_2(g) \longrightarrow 2\, HF(g) + OF_2(g)$$

This equation illustrates the ability of fluorine to remove elements for which it has great affinity (hydrogen in this case) even from very stable compounds. All fluorine reactions are very rapid. Nonmetals with low or average electronegativities, such as hydrogen, phosphorus, and sulfur, are likely to react explosively when placed in an atmosphere of fluorine, and most hot metals burn vigorously in fluorine. Even asbestos and glass can be made to burn in fluorine. (However, at moderate temperatures some metals, such as iron, nickel, and cobalt, form adherent metal-fluoride coatings that protect the metals from further reaction.) All of these examples of reactivity depend on fluorine's high electronegativity and the weakness of the fluorine-to-fluorine bond. This last observation may be surprising, but we can attribute it to the repulsion among unshared electron pairs in the small fluorine atoms of F_2. This repulsion is a less significant factor in the other halogen molecules, since the atoms are larger and the bond distances greater.

Since the reactivities of halogens decrease with increasing atomic weight, each halogen is less reactive than the one just above it. For example, chlorine is clearly less reactive than fluorine. Chlorine still reacts with all the metals, but high temperatures and moisture are necessary for many of these reactions and the list of elements chlorine does not react with must be expanded to include carbon and the rest of the noble gases. Bromine and iodine are even less reactive: The reactions are slower and less exothermic. Neither of these elements reacts with the less active metals, such as gold and platinum.

When metals such as mercury, tin, and iron, which exhibit more than one oxidation number in binary compounds, combine with a halogen, they usually assume the higher oxidation number:

$$2\ Fe(s) + 3\ Cl_2(g) \longrightarrow 2\ FeCl_3(s)$$

$$Hg(l) + Cl_2(g) \longrightarrow HgCl_2(s)$$

$$Sn(s) + 2\ Cl_2(g) \longrightarrow SnCl_4(l)$$

On the other hand, nonmetals with the same property often react with the halogens to form more than one compound. In a limited supply of chlorine, phosphorus reacts to form PCl_3, but, in a plentiful supply of chlorine, the product is PCl_5.

All the halogens react with water. Fluorine reacts violently, even with cold water, in several simultaneous reactions. The other halogens react much less vigorously to form hypohalous (HOCl, HOBr, HOI) and hydrohalic (HCl, HBr, HI) acids. For example,

$$Cl_2(g) + H_2O(l) \rightleftharpoons HOCl(aq) + HCl(aq)$$

(Note that we usually write the formulas of hypohalous acids with the oxygen before the halogen to emphasize the fact that hydrogen is bonded to oxygen.) This is an example of autooxidation-reduction or *disproportionation*. One reactant serves as both oxidizing and reducing agent and is thereby both oxidized and reduced. In this case Cl_2 is both oxidized to HOCl (oxidation state $+1$) and reduced to HCl (oxidation state -1).

The hypohalous acids are weak, unstable acids and strong oxidizing agents. The ionization and decomposition reactions of these acids are illustrated by the following equations:

$$HOX \rightleftharpoons H^+ + XO^- \quad \text{(ionization)}$$

$$\left. \begin{array}{ll} \text{Overall:} & 2\ HOX \longrightarrow 2\ HX + O_2 \\ \text{Ionic:} & 2\ HOX \longrightarrow 2(H^+ + X^-) + O_2 \end{array} \right\} \quad \text{(decomposition)}$$

Since both the reaction of the halogen with water and the ionization of the resulting hypohalous acid are incomplete, as indicated by the double arrows, an aqueous solution of chlorine, bromine, or iodine contains the following ions and molecules: $H_2O, X_2, H^+, X^-, HOX, OH^-$, and OX^-. Since light promotes the decomposition of the hypohalous acids, solutions of the halogens are usually stored in brown bottles.

All the halogens react with an enormous number of organic compounds. We discussed one of the simplest examples—the reaction between methane and chlorine—in Chapter 12 (p. 393). Depending on the ratio of reactants, the reaction can produce four different compounds. (In each case light or heat is necessary to make the reaction favorable.)

$$CH_4(g) + Cl_2(g) \longrightarrow HCl(g) + CH_3Cl(g) \quad \text{(methyl chloride or chloromethane)}$$

$$CH_4(g) + 2\,Cl_2(g) \longrightarrow 2\,HCl(g) + CH_2Cl_2(l) \quad \text{(methylene chloride or dichloromethane)}$$

$$CH_4(g) + 3\,Cl_2(g) \longrightarrow 3\,HCl(g) + CHCl_3(l) \quad \text{(chloroform or trichloromethane)}$$

$$CH_4(g) + 4\,Cl_2(g) \longrightarrow 4\,HCl(g) + CCl_4(l) \quad \text{(carbon tetrachloride or tetrachloromethane)}$$

We shall discuss other reactions between organic compounds and halogens in Chapter 22. Another halogen reaction we have already mentioned (Chapter 17) is the displacement of a less active halogen; thus chlorine displaces bromine and iodine but not fluorine from metal halides.

From what we've said so far, it should be clear that fluorine holds a special place among the halogens—for example, it is by far the most reactive. Indeed, the differences between fluorine and chlorine, the element below it, are greater than between any other pair of adjacent halogens. We'll often find that the second period element is unique within a group, and we've already alluded to one of the reasons: Atoms of the second period elements are usually much smaller than those below them. Another important difference between fluorine and the other halogens is the presence of d orbitals in chlorine, bromine, and iodine. Thus all but fluorine are able to expand their outermost energy levels to hold more than eight electrons. As we shall see later, this means that there is a ClF_3 molecule but no FCl_3. This characteristic is even more pronounced in bromine and iodine, and it explains the observation that iodine is more soluble in aqueous KI than in water alone:

$$I_2(s) + I^-(aq) \longrightarrow I_3^-(aq)$$

Hydrogen Halides The hydrogen halides can be prepared by (1) the direct reaction of the elements, (2) the reaction of concentrated sulfuric or phosphoric acid and metal halides, (3) the hydrolysis of phosphorus halides, or (4) the reaction of halogens and hydrocarbons.

The vigor of the direct reaction between hydrogen and the halogens varies from an explosive reaction with fluorine to a very slow and incomplete reaction with iodine. A mixture of hydrogen and chlorine will not react at all in the dark but will explode when exposed to a bright light. The reaction between hydrogen and bromine is incomplete. Since none of these reactions is both thermodynamically favorable and easy to control, we seldom see them in the laboratory.

The second method listed above is much more useful, especially for the preparation of hydrogen chloride and hydrogen fluoride. Calcium fluoride and sodium chloride usually take the role of the metal halide, since both are abundant and inexpensive:

$$CaF_2(s) + H_2SO_4(l) \longrightarrow CaSO_4(s) + 2\ HF(g)$$

$$NaCl(s) + H_2SO_4(l) \longrightarrow NaHSO_4(s) + HCl(g)$$

$$NaHSO_4(s) + NaCl(s) \longrightarrow Na_2SO_4(s) + HCl(g)$$

Either concentrated sulfuric acid or phosphoric acid is suitable for these preparations, since both are nonvolatile and since HCl and HF are practically insoluble in both. However, since sulfuric acid is cheaper and more readily available, it is generally chosen. The first stage in the reaction of sodium chloride with concentrated sulfuric acid $[NaCl(s) + H_2SO_4(l) \longrightarrow NaHSO_4(s) + HCl(g)]$ takes place at room temperature, but the second stage

$$[NaHSO_4(s) + NaCl(s) \longrightarrow Na_2SO_4(s) + HCl(g)]$$

requires elevated temperatures. If we try to extend the use of H_2SO_4 to the preparation of hydrogen bromide or hydrogen iodide from metal halides, we find the product contaminated by the free halogen. Both HBr and HI are strong enough reducing agents to reduce H_2SO_4 to H_2SO_3 or H_2S:

$$H_2SO_4(l) + 2\ HBr(g) \longrightarrow Br_2(l) + H_2SO_3(aq) + H_2O(l)$$

$$H_2SO_4(l) + 8\ HI(g) \longrightarrow 4\ I_2(s) + H_2S(g) + 4\ H_2O(l)$$

However, since phosphoric acid is a nonoxidizing acid, it does not give rise to such undesirable side reactions:

$$NaBr(s) + H_3PO_4(l) \longrightarrow NaH_2PO_4(s) + HBr(g)$$

The hydrolysis of phosphorus halides is a third practical source of hydrogen bromide. The usual laboratory procedure is to place red phosphorus in a limited amount of water and then add bromine slowly in small amounts. The bromine combines readily with the phosphorus to form phosphorus tribromide, which in turn reacts with water to form hydrogen bromide:

$$2\ P(s) + 3\ Br_2(l) \longrightarrow 2\ PBr_3(l)$$

$$PBr_3(l) + 3\ H_2O(l) \longrightarrow H_3PO_3(aq) + 3\ HBr(g)$$

All hydrogen halides are pungent, colorless, water-soluble gases. (Figure 19.1 illustrates a dramatic demonstration of the solubility of HCl.) Their boiling points are 19.5°C (HF), −85°C (HCl), −67°C (HBr), and −36°C (HI). We can explain the unusually high boiling point of hydrogen fluoride by recalling that it readily forms hydrogen bonds (Chapter 9)—an insignificant phenomenon in the other hydrogen halides.

Since the hydrogen halides do not conduct electricity appreciably (no more than water), we usually regard them as covalent compounds. Anhydrous hydrogen chloride is a very stable and unreactive substance. It does not react with any of the metals at low temperatures, and at higher temperatures it reacts with only the most active metals to liberate hydrogen:

$$2 \, K(s) + 2 \, HCl(g) \longrightarrow 2 \, KCl(s) + H_2(g)$$

It does not react with carbonates or oxidizing agents. As mentioned before, however, it does react with ammonia:

$$NH_3(g) + HCl(g) \longrightarrow NH_4Cl(s)$$

Even this reaction appears to require a small amount of water as a catalyst, since if both gases are absolutely dry, the reaction proceeds very slowly.

Hydrogen fluoride differs from the other hydrogen halides in that it reacts with silicon dioxide and silicates. Thus glass and porcelain, which are silicates, are attacked by hydrogen fluoride.

$$CaSiO_3(s) + 6 \, HF(g) \longrightarrow SiF_4(g) + CaF_2(s) + 3 \, H_2O(l)$$

Hydrofluoric acid is likewise unique among the hydrohalic acids in being a weak electrolyte. It is therefore a weak acid:

$$HF(g) + H_2O(l) \rightleftharpoons H_3O^+(aq) + F^-(aq)$$

The fluoride ions can then form hydrogen bonds with un-ionized hydrogen fluoride molecules:

$$HF(g) + F^-(aq) \rightleftharpoons FHF^-(aq)$$

Hydrofluoric acid thus contains water molecules, hydrogen fluoride molecules, hydrogen ions, bifluoride ions (FHF^-), and small numbers of fluoride ions and hydroxide ions. Salts containing the bifluoride ion can be prepared by adding a stoichiometric amount of base to hydrofluoric acid.

Interhalogen Compounds The halogens react with each other to form a group of compounds that we call the *interhalogen compounds*. Each pair of halogens forms several compounds whose properties are similar to those of the halogens themselves. Some of these compounds, such as ClF_3, ClF, and BrF_3, react explosively with water and with organic compounds.

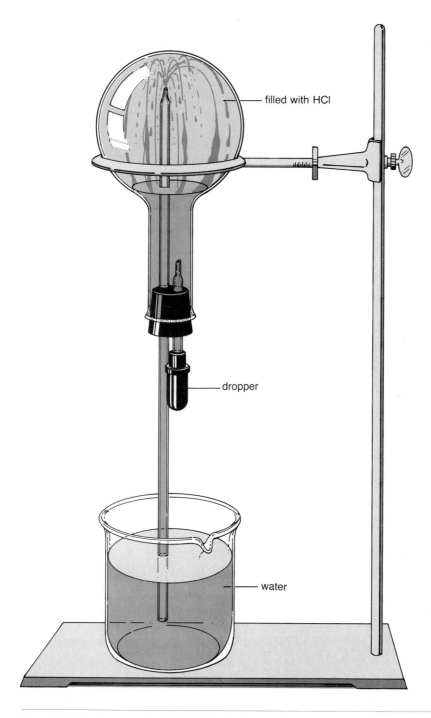

Figure 19.1 The hydrogen chloride fountain. The fountain begins when a few drops of water are forced from the dropper into the flask filled with hydrogen chloride gas. The hydrogen chloride dissolves in the water and pressure in the flask is reduced. Consequently, the water rises rapidly up the tube, forming a fountain.

Table 19.3 The Interhalogen Compounds

Formula	Shape	Formula	Shape	Formula	Shape
ClF	Linear	BrCl	Linear	IBr	Linear
ClF_3	T-shaped	BrF	Linear	ICl	Linear
ClF_5	Square-pyramidal	BrF_3	T-shaped	ICl_3	T-shaped
		BrF_5	Square-pyramidal	IF_3	T-shaped
				IF_5	Square-pyramidal
				IF_7	Pentagonal-bipyramidal

The formulas of the interhalogen compounds appear in Table 19.3. This list makes several points. First, each compound consists of one larger atom and an odd number of smaller atoms. Second, in compounds of more than two atoms the central atom contains more than eight electrons in its outer energy level. Finally, the greater the disparity in atomic radii, the more small atoms the larger central atom can accommodate (see Figure 5.19). The molecular shapes listed in Table 19.3 follow logically from the formulas, if we use the methods described in Chapters 6 and 7. (It may be useful to review the material on p. 204.)

Applications of Halogen Chemistry Sodium chloride, one of the earth's most abundant mineral resources, is the source of most commercially produced chlorine. In Chapter 18, we described several electrolytic processes by which chlorine could be produced from NaCl. On an industrial scale, however, most of these simple cells are impractical. For one thing, the products of the half-cells often react with each other; for example, Cl_2 reacts with sodium hydroxide to form sodium hypochlorite (NaOCl) or sodium chlorate ($NaClO_3$) and can react explosively with hydrogen to form HCl. There have been a number of ingenious commercial solutions to this problem, and the principles we discussed for the simple cells are still valid. Once produced, chlorine is one of the few elements, apart from the common metals, that are shipped in huge quantities. It is used to sterilize drinking water, as a bleach, and in the production of other chemicals.

Since chlorine is inexpensive and potent, it is the most common water disinfectant in the country. Typically, 1 part of chlorine to 1 million parts of water will make the water safe for drinking. Chlorine's effectiveness as a disinfectant (and as a bleach) depends on oxidation. But according to most chemists it is hypochlorous acid, a product of the reaction between chlorine and water ($Cl_2 + H_2O \longrightarrow HCl + HOCl$), that is the active oxidizing agent. In fact, anhydrous chlorine is not an effective bleach. Aqueous chlorine can be used to bleach paper pulp and cotton, but it badly oxidizes fabrics like silk and wool.

Chlorine is finding increased use in the chemical industry, including the production of other halogens.

The concentration of bromide ions in seawater is only 70 parts per million, but it is feasible to extract even this small amount on a commercial scale. The seawater is acidified with sulfuric acid and then the bromide ions are oxidized with chlorine. The bromine is driven out by air blown through the solution and collected as a liquid. The major use of bromine is in the production of 1,2-dibromoethane ($C_2H_4Br_2$) for the manufacture of antiknock gasolines (see Chapter 22).

Fluorine, the most reactive halogen, finds many applications in relatively inert fluorides. Their inactivity, of course, is explained by an eighth electron in the outer energy level—the same electron that elemental fluorine reacts so vigorously to obtain. A good example is the household polymer, Teflon (Chapter 22). We can think of Teflon (a trade name for this fluorocarbon polymer) as a high-molecular-weight hydrocarbon in which all the hydrogen atoms have been replaced with fluorine atoms:

$$F-\left[\begin{array}{cc} F & F \\ | & | \\ C-C \\ | & | \\ F & F \end{array}\right]_n-F$$

and as most cooks know, it is practically unreactive at normal cooking temperatures. Other fluorocarbons (fluorine-containing hydrocarbons) are used as refrigerants (Freon-12, CCl_2F_2), insecticides (CCl_3F), and inert propellants in cans of hair spray and deodorants).

Another important, if less widespread, fluoride is uranium(VI) fluoride. Nuclear weapons, as well as many nuclear reactors, demand fuel that is *enriched* in fissionable isotopes. Since isotopes have identical chemical properties, enrichment methods depend on physical rather than chemical processes. The most successful of these methods is based on Graham's law of gaseous diffusion (Chapter 8). One of the few compounds of uranium that is a gas under workable conditions is uranium(VI) fluoride. The lighter $^{235}_{92}UF_6$ molecule has a slightly greater velocity than the $^{238}_{92}UF_6$ molecule and hence diffuses faster through a porous barrier. If the process is repeated, it produces increasingly pure samples of gas that contains fissionable uranium-235.

A final use of fluoride salts, and perhaps the best known, is as an additive to drinking water and toothpastes to reduce dental caries.

Iodine in the form of iodide salts is an essential nutrient. Normally, we receive enough iodides in water and in certain foods such as green beans, spinach, and milk, but to ensure an adequate intake manufacturers of table salt add a small amount of potassium iodide (0.023 percent) to their product. Iodine deficiency in the diet leads to goiter, a disease of the thyroid. Tincture of iodine, an alcohol solution of the element, is used as an antiseptic, though less frequently today than in the past.

A final common use of the halogens—as silver halides—is in photographic emulsions, which we shall discuss more in Chapter 21.

A comparison of fluorine with the other halogens is given in Table 19.4.

Table 19.4 A Comparison of Fluorine with the Other Halogens

Atomic radius (nm)	F (0.072) Cl (0.0994) Br (0.1142) I (0.1334)	Note that the most dramatic difference between adjacent halogens is between F and Cl.
Electron configuration	The outer energy level electron configuration for all halogen atoms is s^2p^5.	
Electronegativity	F (4.0) Cl (3.0) Br (2.8) I (2.5) At (2.2)	Again, the most dramatic difference is between F and Cl. Each halogen is the most electronegative element in its period; fluorine is the most electronegative of all elements.
Ionic valence and oxidation number	The most characteristic property of all the halogens is their tendency to gain or share one electron. All halogens thus exhibit the ionic charge and oxidation number of -1. In addition, all halogens below fluorine show positive oxidation numbers. Fluorine is the only element that *never* shows a positive oxidation number, since it is the most electronegative element.	
Expansion of the outermost energy level	Since there are no d orbitals in the outermost energy level of fluorine, it can never accommodate more than eight electrons in this level. The other halogens, however, can accommodate more than eight electrons in their outer energy levels by forming hybrid orbitals involving d orbitals. In many such cases, interhalogen compounds form in which fluorine is the peripheral element (see Table 19.3).	
Bond strength	All halogens occur naturally as diatomic molecules. The fluorine-to-fluorine bond is by far the weakest of the homonuclear halogen bonds, since the small fluorine atoms allow greater repulsion among the unshared electron pairs.	
Ease of oxidation of the halide ion	The heavier the halogen, the more easily can its halide ion be oxidized. Fluorine is the most powerful oxidizing agent known; hence the fluoride ion cannot be oxidized by any chemical oxidizing agent. It can be oxidized only by electrolysis.	
Hydrogen bonding	Among halogen compounds, only hydrogen fluoride exhibits hydrogen bonding. Only the fluorine atom combines the two requirements for hydrogen bonding: small size and high electronegativity.	
Strength of hydrohalic acids	Hydrochloric, hydrobromic, and hydroiodic acids are strong acids and hence are completely ionized in water at all concentrations. Hydrofluoric acid is a weak acid, which differs from most weak acids in that it forms a hydrogen bond with its anion: $$HF + F^- \rightleftharpoons FHF^-$$	

Example 19.1 List seven different methods for preparing zinc chloride. In designing procedures for preparing this salt, keep in mind the problem of how to isolate the zinc chloride from other products of the reaction. It is best if the other products are insoluble, volatile, or both.

Solution

a. Recall that acids and bases react to form salts and water:

$$Zn(OH)_2(s) + 2\ HCl(aq) \longrightarrow ZnCl_2(aq) + 2\ H_2O(l)$$

b. Metallic oxides react with acids to form salts and water:

$$ZnO(s) + 2\ HCl(aq) \longrightarrow ZnCl_2(aq) + H_2O(l)$$

c. Carbonates react with acids to form weak, unstable carbonic acid, which dissociates to CO_2 and H_2O:

$$ZnCO_3(s) + 2\ HCl(aq) \longrightarrow ZnCl_2(aq) + H_2O(l) + CO_2(g)$$

d. Similarly, sulfites react with acids to form the weak, unstable acid H_2SO_3, which dissociates to SO_2 and H_2O:

$$ZnSO_3(s) + 2\ HCl(aq) \longrightarrow ZnCl_2(aq) + H_2O(l) + SO_2(g)$$

e. Chlorine will oxidize either the bromide ion or the iodide ion, forming the chloride ion and free bromine or iodine. Both bromine and iodine are easily volatilized.

$$ZnBr_2(aq) + Cl_2(g) \longrightarrow ZnCl_2(aq) + Br_2(l)$$
$$ZnI_2(aq) + Cl_2(g) \longrightarrow ZnCl_2(aq) + I_2(s)$$

f. Zinc will displace hydrogen from hydrochloric acid:

$$Zn(s) + 2\ HCl(aq) \longrightarrow ZnCl_2(aq) + H_2(g)$$

g. In each of the preparations above, zinc chloride can be isolated by evaporating the water or other volatile products. If the reaction produces an insoluble product, it can be removed by filtration before the water is evaporated. For example,

$$ZnSO_4(aq) + BaCl_2(aq) \longrightarrow BaSO_4(s) + ZnCl_2(aq)$$

Questions

1. Contrast the physical properties of metals and nonmetals.
2. Indicate the electron configuration of each of the halogens.

3. Which halogen
    a. never shows a positive oxidation state?
    b. is a liquid at room temperatures?
    c. cannot be produced by a chemical oxidizing agent?
    d. has the darkest color?
    e. is the most electronegative?
    f. has the highest atomic weight?
    g. is the strongest oxidizing agent?
    h. is the most reactive?
    i. is the least reactive?
    j. has the smallest ionization energy?

4. Which hydrogen halide
    a. has the highest boiling point?
    b. is the best reducing agent?
    c. reacts with sand?
    d. is the poorest reducing agent?

5. Which hydrohalic acid
    a. is the best reducing agent?
    b. is the weakest acid?
    c. is the poorest electrical conductor?
    d. has the highest hydroxide ion concentration?
    e. has the lowest hydrogen ion concentration?
    f. shows hydrogen bonding?

19.2 Hydrogen

Hydrogen occupies a unique position in the periodic table. Since the hydrogen atom has only one electron in its outermost energy level and forms compounds analogous to the compounds of the Group IA metals, we often see it placed at the top of Group IA. On the other hand, since it lacks only one electron of having a noble gas electron configuration and since it exists as a diatomic molecule and forms several compounds analogous to the halides, it sometimes appears at the top of Group VIIA. However, neither spot is entirely appropriate. In contrast to the elements of Group IA, hydrogen behaves chemically and physically as a nonmetal and it has a very high ionization energy. But neither is the parallel with the halogens convincing; its electronegativity is much lower and it is much less reactive. In the end, we can best consider it as belonging in a group by itself, placing it at the top and center of the periodic table with lines drawn to both Groups IA and VIIA.

Nor does hydrogen's uniqueness end with its position in the periodic table. It is by far the most abundant element in the universe: In our galaxy it is 10 times more common than the next most abundant element, helium. In fact, some have gone so far as to characterize the universe as consisting entirely of hydrogen, with the other elements constituting only impurities. Unlike the hydrogen in the stars and in interstellar space, most terrestrial hydrogen is found in compounds, the

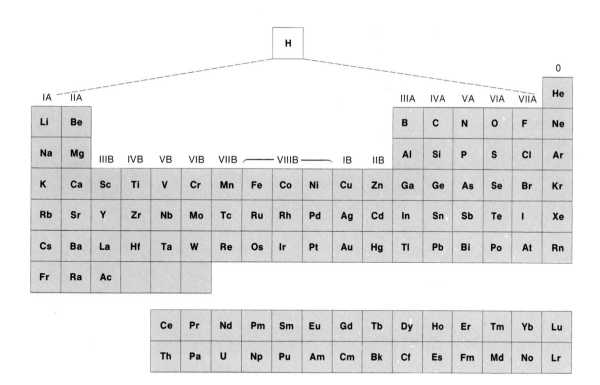

most abundant of which is water. Indeed, there are more compounds of hydrogen than of any other element, with the possible exception of carbon.

Isotopes There are three isotopes of hydrogen, each with its own name: protium, with mass number 1, deuterium (sometimes given the symbol D) with mass number 2, and tritium (sometimes given the symbol T), with mass number 3. Protium is by far the most common; it is 5000 times as abundant as deuterium and 10^{17} times as abundant as tritium. Although the three isotopes exhibit the same chemical properties, the large differences in mass manifest themselves in important ways. For example, at room temperature protium combines with chlorine nearly three times as fast as deuterium does.

Physical and Chemical Properties We have already looked at several of the properties of hydrogen in different contexts. We discussed hydrogen bonding in Chapter 8, the hydration of hydrogen ions in Chapter 14, the generation of hydrogen in electrochemical reactions in Chapter 17, and the reactions of hydrogen with nitrogen and iodine in Chapter 13. Now we shall take a more systematic, though brief, look at some other important properties.

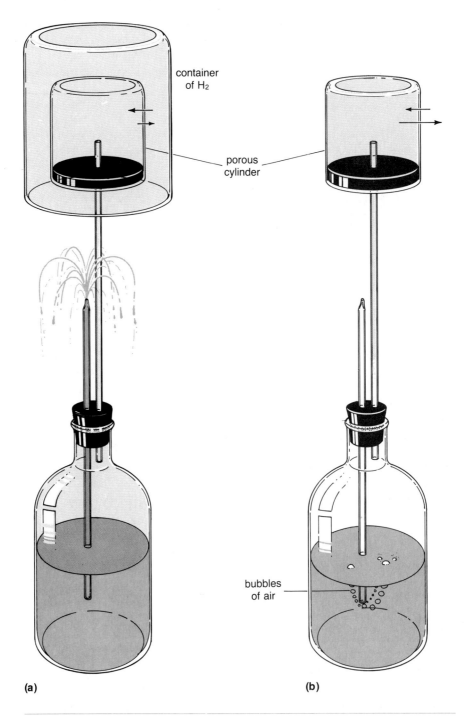

Figure 19.2 The "hydrogen fountain" showing the rapid diffusion of hydrogen through the porous walls of a cylinder. In **(a)** the hydrogen diffuses into the porous cylinder faster than air diffuses out. The resulting increase in pressure in the cylinder, tube, and bottle forces water out through the jet tube. In **(b)** the container of hydrogen has been removed. The porous cylinder now contains both hydrogen and air. Since the hydrogen in the cylinder diffuses out faster than air diffuses in, the pressure in the cylinder, tube, and bottle decreases causing air to enter the bottle through the jet tube.

Hydrogen has no color, taste, or odor. At room temperature it is a diatomic gas, and it is the lightest of the elements. Because of its lightness, hydrogen diffuses rapidly, a fact that can be demonstrated neatly as shown in Figure 19.2. The velocity of a hydrogen molecule is about 1 mps at 0°C. Table 19.5 summarizes several other important physical properties of hydrogen.

Table 19.5 Properties of Hydrogen

Atomic number	1
Atomic weight	1.00797
Melting point	−259.14°C
Boiling point	−252.8°C
Critical temperature	−239.9°C
Critical pressure	12.8 atm
Density	0.08988 g/liter at 0°C and 1 atm
Specific gravity of liquid	0.070 at −252°C
Solubility	1.9 ml/100 g of H_2O at 20°C and 1 atm
Electronegativity	2.1

Chemically, hydrogen is a relatively inactive molecule at room temperature. The reason is the considerable energy needed to break the H—H bond. Nonetheless, under proper conditions, hydrogen will react with all the nonmetals and many of the metals. It combines with oxygen to form water, but the reaction must be started with an input of energy. If a mixture of these two gases is ignited, the reaction proceeds with explosive violence (see Chapter 12); if oxygen is added slowly, pure hydrogen will burn quietly with a hot and nearly colorless flame. In both cases the reaction proceeds more readily in the presence of finely divided platinum acting as a catalyst.

The metals of Groups IA and IIA (except beryllium and magnesium) react with hydrogen at high temperatures. For example,

$$2\ Na(s) + H_2(g) \longrightarrow 2\ NaH(s)$$

The products of these reactions are very different from the more familiar compounds formed between nonmetals and hydrogen. The hydrides of the metals of Groups IA and IIA are white crystalline solids, with each hydrogen atom having acquired an extra electron to form a hydride ion ($:H^-$). By contrast, compounds of hydrogen and nonmetals are liquid or gaseous molecular compounds.

There is some evidence to indicate that hydrogen also reacts with the transition metals. The products, however, are quite unlike the hydrides of the Group IA and IIA metals; instead they resemble metals more than compounds. Indeed, the nature of these substances has led some chemists to argue that a physical rather than a chemical change is involved and that the results are not true chemical compounds.

Applications of Hydrogen Chemistry In Chapter 17 we discussed several ways to produce hydrogen by reducing hydrogen ions in solution. Although several of these methods are practical for the laboratory, all are too expensive when carried out on an industrial scale. A more economical source of hydrogen is water. Some methods of industrial production of hydrogen follow.

At high temperatures and in the presence of a suitable catalyst, superheated steam reacts with hydrocarbons to produce carbon dioxide and hydrogen:

$$C_nH_{2n+2} + 2n\ H_2O(g) \longrightarrow n\ CO_2(g) + (3n+1)\ H_2(g)$$

Under similar conditions steam will react with carbon alone:

$$C(s) + H_2O(g) \xrightarrow{1000°C} CO(g) + H_2(g)$$

The carbon monoxide-hydrogen mixture produced in this second reaction is often called *water gas*. Since both these gases are excellent reducing agents and fuels, the mixture can be used as it is in many applications. If, however, the carbon monoxide must be removed, more steam can be added:

$$CO(g) + H_2O(g) \xrightarrow{500°C} CO_2(g) + H_2(g)$$

The carbon dioxide is then separated from the hydrogen by dissolving it in an alkaline solution.

The reaction of steam with iron is one of the oldest industrial sources of hydrogen:

$$4\ H_2O(g) + 3\ Fe(s) \xrightarrow{800°C} Fe_3O_4(s) + 4\ H_2(g)$$

Another important source of commercial hydrogen is the catalyzed thermal decomposition of hydrocarbons to produce gasoline, where both hydrogen and carbon black (which is used in the manufacture of automobile tires) are produced as by-products. Other important sources, especially of very pure hydrogen, are electrolytic processes, where again hydrogen is often merely a by-product.

Two other processes with more specialized applications are the catalyzed thermal decomposition of ammonia and the reaction of calcium hydride with water. The equation for the second of these can be written

$$CaH_2(s) + 2\ H_2O(l) \longrightarrow Ca(OH)_2(s) + 2\ H_2(g)$$

Neither of these processes can compete on a large scale with the production of hydrogen from steam, but each has its unique advantage. Where relatively small amounts of hydrogen are required, it is cheaper to ship liquid ammonia and then produce hydrogen at the site where it is needed than to ship tanks of hydrogen gas. Likewise, calcium hydride can be easily transported and can be kept indefinitely in airtight cans. One application that exploits these advantages is inflating meteorological balloons in remote areas.

About half of the hydrogen produced commercially is used in the manufacture of ammonia by the Haber process. It is also used in the industrial synthesis of hydrogen chloride, methanol, formaldehyde, and aniline. As a reducing agent, hydrogen is sometimes used to produce free metals, such as tungsten and molybdenum, from metal oxides. For example,

$$WO_3(s) + 3\,H_2(g) \longrightarrow W(s) + 3\,H_2O(g)$$

In Chapter 23 we shall look at yet another commercially important reaction involving hydrogen —*hydrogenation*, by which vegetable oils can be converted to shortening and coal to gasoline. In light of dwindling oil supplies, this last process is becoming increasingly important.

Hydrogen is also used to produce high temperature flames. The oxyhydrogen torch, which burns hydrogen in pure oxygen, produces temperatures as high as 5000°C. The atomic hydrogen torch burns even hotter by first dissociating hydrogen molecules in an electric field and then converting the energy of both recombination and oxidation to temperatures up to 6000°C.

Since the disastrous burning of the dirigible Hindenburg in 1937, hydrogen has not been used in large lighter-than-air craft; however, it is still used in unmanned meteorological balloons. And it was used to inflate barrage balloons in England during World War II.

A comparison of hydrogen with the halogens is given in Table 19.6.

Table 19.6 A Comparison of Hydrogen with the Halogens

Molecular structure	Both hydrogen and the halogens occur naturally as diatomic molecules.
Physical properties	Hydrogen fits well in Group VIIA. For example, it has a lower boiling point than fluorine and has no color. Thus with hydrogen at the head of the group, there remains a smooth progression from colorless gas to a black solid (iodine).
Electronegativity	Hydrogen is less electronegative than even the heaviest of the halogens; thus it does not fit well above fluorine.
Ionic charge and oxidation state	Hydrogen, like the halogens, reacts with the metals of Group IA to form ionic compounds. For example, we have both $[Na^+ + :H^-]$ and $[Na^+ + :\ddot{\underset{..}{F}}:^-]$. However, hydrogen is more likely than the halogens to form covalent compounds.

Questions

6. Write an equation for the reaction of hydrogen with each of the following:
   a. Na   b. Cl_2 c. O_2 d. WO_3 e. N_2

7. Write the electron formulas for the following:
   a. hydrogen ion   b. hydrogen molecule   c. hydride ion   d. hydrogen atom

8. What are the differences among the atomic structures of protium, deuterium, and tritium?

19.3 Oxygen

According to the principles we have learned to apply to the periodic table, we expect the elements of Group VIA to be more metallic than the halogens. However, they still manifest very little metallic behavior. The first three elements of the group, oxygen, sulfur, and selenium, are nonmetallic, and tellurium and polonium are usually classified as metalloids. Because of its unique importance we'll turn first to oxygen.

Although hydrogen is by far the most abundant element in the universe, oxygen is the most abundant in that part of the universe accessible to man. The earth's crust, oceans, and atmosphere are almost 50 percent oxygen by weight. In its elemental form oxygen usually exists as a diatomic molecule (Chapter 7), but a second allotrope, ozone (O_3), can be found in the higher reaches of the atmosphere. Under normal conditions ozone is very unstable and converts rapidly to O_2; however, it can be prepared in the laboratory by using electrical discharges or ultraviolet radiation. We can best describe O_3 as a hybrid of two resonance forms:

$$\left[\ddot{:\underset{..}{O}}: \quad \underset{..}{\overset{..}{O}} \quad \underset{..}{O}: \quad \longleftrightarrow \quad :\underset{..}{O} \quad \underset{..}{\overset{..}{O}} \quad :\underset{..}{\ddot{O}}: \right]$$

The bonding is sp^2-hybridized; thus the molecular shape is bent.

IA																	0
H	IIA											IIIA	IVA	VA	VIA	VIIA	He
Li	Be											B	C	N	O	F	Ne
Na	Mg	IIIB	IVB	VB	VIB	VIIB	____VIIIB____			IB	IIB	Al	Si	P	S	Cl	Ar
K	Ca	Sc	Ti	V	Cr	Mn	Fe	Co	Ni	Cu	Zn	Ga	Ge	As	Se	Br	Kr
Rb	Sr	Y	Zr	Nb	Mo	Tc	Ru	Rh	Pd	Ag	Cd	In	Sn	Sb	Te	I	Xe
Cs	Ba	La	Hf	Ta	W	Re	Os	Ir	Pt	Au	Hg	Tl	Pb	Bi	Po	At	Rn
Fr	Ra	Ac															

Ce	Pr	Nd	Pm	Sm	Eu	Gd	Tb	Dy	Ho	Er	Tm	Yb	Lu
Th	Pa	U	Np	Pu	Am	Cm	Bk	Cf	Es	Fm	Md	No	Lr

Preparation As we saw in Chapter 18, oxygen is produced during several common electrolytic reactions, but the easiest laboratory preparation is by thermal decomposition of potassium chlorate ($KClO_3$):

$$2\ KClO_3(s) \longrightarrow 2\ KCl(s) + 3\ O_2(g)$$

Potassium chlorate melts at 368.4°C. At about 400°C it appears to boil, but the frothing is in fact caused by the evolution of oxygen gas. At 600°C the reaction will continue until decomposition is complete. The reaction also proceeds at a convenient rate at temperatures as low as 250°C if some MnO_2 or Fe_2O_3 is present. These oxides do not themselves decompose at these lower temperatures; instead they serve as catalysts, increasing the rate of oxygen evolution. Although it is a popular source of oxygen, molten potassium chlorate must always be treated as a hazardous compound. It is a powerful oxidizing agent and in contact with organic materials such as rubber stoppers and wooden splints can cause violent explosions.

Other compounds that yield oxygen when heated with a Bunsen flame are metallic nitrates and oxides and carbonates of heavy metals (mercury and the elements below it in the activity series). For example,

$$2\ KNO_3(s) \longrightarrow 2\ KNO_2(s) + O_2(g)$$
$$2\ HgO(s) \longrightarrow 2\ Hg(l) + O_2(g)$$
$$2\ Pb(NO_3)_2(s) \longrightarrow 2\ PbO(s) + 4\ NO_2(g) + O_2(g)$$
$$2\ Ag_2CO_3(s) \longrightarrow 4\ Ag(s) + 2\ CO_2(g) + O_2(g)$$

Notice in the third and fourth equations that oxygen is not the only gas produced; this somewhat limits the use of $Pb(NO_3)_2$ and Ag_2CO_3 as practical sources of oxygen.

Physical and Chemical Properties In the gaseous state, oxygen is colorless, odorless, and tasteless, but liquid and solid oxygen are pale blue. Some of the quantitative properties of oxygen are listed in Table 19.7. There are three isotopes of oxygen having mass numbers of 16, 17, and 18. The lightest of these is by far the most abundant.

Despite its chemical similarity to diatomic oxygen, ozone has come to play a unique role in the earth's atmosphere. Between 12 and 50 km above the earth's surface, the atmosphere maintains (through a series of complex reactions) a fairly constant concentration of ozone. This ozone effectively shields the earth from intense ultraviolet radiation by absorbing the harmful light, at the same time dissociating to form an oxygen atom and an oxygen molecule:

$$O_3 \longrightarrow O_2 + O$$

Table 19.7 Properties of Oxygen

Atomic number	8
Atomic weight	15.9994
Melting point	−218.4°C
Boiling point	−183.0°C
Critical temperature	−118°C
Critical pressure	50 atm
Density	1.429 g/liter at 0°C
Specific gravity of liquid	1.14 at −182.96°C
Solubility	3.1 cc/100 g of water at 20°C and 1 atm
Electronegativity	3.5

Were it not for this protective ozone layer, the incidence of skin cancer and genetic mutations would be much higher, and terrestrial climate would be dramatically different. Current concern over compounds that could deplete the ozone layer are thus easy to understand. Two of the suspected culprits are nitrogen oxides, which are produced by jet aircraft, and fluorocarbons such as CF_2Cl_2 and $CFCl_3$, which have been used as propellants in aerosol cans. Interestingly, it is the very inertness of the fluorocarbons that is the source of the threat they pose. Whereas many potentially harmful atmospheric pollutants are washed out by rain or removed by other atmospheric processes at low altitudes, the insoluble and unreactive fluorocarbons diffuse freely upward to the ozone layer. There they are exposed to intense uv radiation, which liberates chlorine atoms. These in turn catalyze a chain raction that decomposes ozone:

$$Cl + O_3 \longrightarrow ClO + O_2$$

$$ClO + O \longrightarrow Cl + O_2 \quad (\text{O from uv decomposition of } O_3, \text{ p. 593})$$

Since the chlorine atoms are catalysts, a single chlorine atom can cause the dissociation of several ozone molecules.

Paradoxically, man's technological advancement has produced not only compounds that threaten the ozone layer at high altitudes but also airborne pollutants that *increase* the ozone concentration nearer the ground. There the ozone kills vegetation and promotes the deterioration of textiles and rubber. The attack on rubber probably takes place at the double bonds (Chapter 22).

The more common diatomic allotrope of oxygen combines with all metals except the noble metals, and even oxides of the noble metals can be synthesized by indirect methods. These metal-oxygen reactions can produce not only simple oxides, where oxygen has a −2 oxidation number, but also peroxides (containing O_2^{2-}) and superoxides (containing O_2^-). The oxidation number of each oxygen atom in the peroxide ion is −1, and the average oxidation number of each oxygen atom in the superoxide ion is $-\frac{1}{2}$. The electron formulas for each appear below.

$:\ddot{\underset{..}{O}}:^{2-}$ $\quad$ $:\ddot{\underset{..}{O}}:\ddot{\underset{..}{O}}:^{2-}$ $\quad$ $:\ddot{\underset{..}{O}}:\dot{\underset{..}{O}}:^{-}$

oxide $\qquad$ peroxide $\quad$ superoxide
ion $\qquad\quad$ ion $\qquad\quad$ ion

The metals of Group IA react spontaneously and rapidly with oxygen. The product depends on the metal and the reaction conditions:

$$4\,Li(s) + O_2(g) \longrightarrow 2\,Li_2O(s) \quad \text{(an oxide)}$$
$$2\,Na(s) + O_2(g) \longrightarrow Na_2O_2(s) \quad \text{(a peroxide)}$$
$$K(s) + O_2(g) \longrightarrow KO_2(s) \quad \text{(a superoxide)}$$

Self-contained breathing units often make use of potassium superoxide as a source of oxygen. Moisture from the breath reacts with the KO_2 to form oxygen and potassium hydroxide:

$$4\,KO_2(s) + 2\,H_2O(l) \longrightarrow 4\,KOH(s) + 3\,O_2(g)$$

The potassium hydroxide, in turn, reacts with exhaled CO_2, removing it from the system:

$$CO_2(g) + KOH(s) \longrightarrow KHCO_3(s)$$

Hence during rescue work in poisonous atmosphere, workers can be completely independent of the surrounding air.

The reactions between oxygen and less active metals usually proceed slowly at room temperature, are catalyzed by moisture, and produce oxides. At higher temperatures and in atmospheres of pure oxygen, the reactions are often much faster. In the case of iron, the product too depends on the conditions. At room temperature and in a moist environment, iron slowly rusts to form iron(III) oxide:

$$4\,Fe(s) + 3\,O_2(g) \longrightarrow 2\,Fe_2O_3(s)$$

At higher temperatures and in pure oxygen, the product is a more unusual iron oxide, one in which both oxidation numbers are exhibited. This is iron(II, III) oxide, more commonly called *magnetic oxide*.

$$3\,Fe(s) + 2\,O_2(g) \longrightarrow Fe_3O_4(s)$$

Oxygen does not combine with the noble gases nor with the halogens. It does combine with all other nonmetals, however, and even oxides of the halogens can be synthesized indirectly. The reaction with sulfur is typical:

$$S_8(s) + 8\,O_2(g) \longrightarrow 8\,SO_2(g)$$

In other cases, as in the reaction with iron above, the amount of available oxygen determines the product. We have made this observation earlier when we looked at the combustion products of carbon, and the case is similar with phosphorus:

$$P_4(s) + 3\,O_2(g) \longrightarrow P_4O_6(s) \quad \text{(limited supply of } O_2\text{)}$$
$$P_4(s) + 5\,O_2(g) \longrightarrow P_4O_{10}(s) \quad \text{(plentiful supply of } O_2\text{)}$$

Oxygen reacts with many compounds, often producing simple oxides of the constituent elements. Thus if hydrogen sulfide burns in a plentiful supply of oxygen, the products are water and sulfur dioxide:

$$2\,H_2S(g) + 3\,O_2(g) \longrightarrow 2\,H_2O + 2\,SO_2(g)$$

Similarly, zinc sulfide reacts with oxygen to yield zinc oxide and sulfur dioxide:

$$2\,ZnS(s) + 3\,O_2(g) \longrightarrow 2\,ZnO(s) + 2\,SO_2(g)$$

However, when silver sulfide is heated with oxygen, the products are free silver and sulfur dioxide:

$$Ag_2S(s) + O_2(g) \longrightarrow 2\,Ag(s) + SO_2(g)$$

Silver is one of the noble metals that do not react directly with oxygen; hence this reaction produces no silver oxide.

As a footnote to the discussion of the chemical properties of oxygen, we should mention the unstable gas oxygen difluoride (OF_2), the single compound in which oxygen has an oxidation number of $+2$.

Example 19.2 Show by a series of equations how sulfurous acid might be prepared from methane, potassium chlorate, and copper sulfide.

Solution

a. Probably the best approach to a synthesis problem of this sort is to start with the product and work backward. The only method of preparing sulfurous acid we have described is the reaction between sulfur dioxide and water:

$$SO_2(g) + H_2O(l) \longrightarrow H_2SO_3(aq)$$

b. Since neither sulfur dioxide nor water is among the starting materials, however, we must prepare them too. We can do this by burning copper sulfide and methane separately:

$$2\,CuS(s) + 3\,O_2(g) \longrightarrow 2\,CuO(s) + 2\,SO_2(g)$$

$$CH_4(g) + 2\,O_2(g) \longrightarrow CO_2(g) + 2\,H_2O(l)$$

c. Now we need only produce the oxygen used in step b. This we do by heating potassium chlorate:

$$2\,KClO_3(s) \longrightarrow 2\,KCl(s) + 3\,O_2(g)$$

Industrial Aspects of the Chemistry of Oxygen The two principal industrial methods of oxygen production are the electrolysis of water and the fractional distillation of liquid air. The electrolytic process is economically feasible only when hydrogen is also being marketed. Industrial electrolytic cells typically contain

10–25 percent solutions of sodium hydroxide. By far the most common industrial source of oxygen is liquid air. Since air is a mixture, not a compound, it can be separated into its components by a physical process. On an industrial scale this is done by compressing and cooling air, which has been freed of water and carbon dioxide, until it condenses. The liquid air is then separated into its two major components by the differences in their boiling points ($-183°C$ for oxygen and $-196°C$ for nitrogen). This distillation process yields nitrogen and oxygen that are better than 99.5 percent pure, as well as several noble gases.

A major use of oxygen is in the manufacture of steel and in other metallurgical processes; several efficient production methods depend on air enriched with oxygen. A second modern application is the use of liquid oxygen, *lox*, as an oxidizer in rockets. Oxygen is also used in oxyacetylene and oxyhydrogen torches, the manufacture of chemicals such as nitric acid and methanol, and self-contained breathing apparatus.

Questions

9. Write equations illustrating the following:
    a. The burning of a hydrocarbon.
    b. The formation of a peroxide.
    c. The production of an acid anhydride.
    d. A laboratory preparation of oxygen from a chlorate.
    e. A laboratory preparation of oxygen from a nitrate.

10. Which of the three anions containing only oxygen is paramagnetic?

11. Write the electron formulas for the following:
    a. A molecular oxide.   b. An ionic oxide.   c. A molecular peroxide.
    d. An ionic peroxide.   e. A hydroxide.   f. A superoxide.

12. What are the two allotropic forms of oxygen?

19.4 Sulfur

Since sulfur is directly below oxygen in the periodic table, we should not be surprised to find many similarities between the two elements. For example, we shall find that both exhibit allotropism. And like other elements in this group, both oxygen and sulfur have an electron configuration that ends with s^2p^4, and both tend to gain or share two electrons per atom. Consequently, a common oxidation number for all Group VIA elements is -2, and there are often sulfur-containing analogs to common oxygen-containing compounds and anions. For example,

H_2O H_2S
Na_2O Na_2S
Al_2O_3 Al_2S_3
AsO_4^{3-} AsS_4^{3-}

IA																	0
H	IIA											IIIA	IVA	VA	VIA	VIIA	He
Li	Be											B	C	N	O	F	Ne
Na	Mg	IIIB	IVB	VB	VIB	VIIB	\multicolumn{3}{c}{VIIIB}	IB	IIB	Al	Si	P	S	Cl	Ar
K	Ca	Sc	Ti	V	Cr	Mn	Fe	Co	Ni	Cu	Zn	Ga	Ge	As	Se	Br	Kr
Rb	Sr	Y	Zr	Nb	Mo	Tc	Ru	Rh	Pd	Ag	Cd	In	Sn	Sb	Te	I	Xe
Cs	Ba	La	Hf	Ta	W	Re	Os	Ir	Pt	Au	Hg	Tl	Pb	Bi	Po	At	Rn
Fr	Ra	Ac															

Ce	Pr	Nd	Pm	Sm	Eu	Gd	Tb	Dy	Ho	Er	Tm	Yb	Lu
Th	Pa	U	Np	Pu	Am	Cm	Bk	Cf	Es	Fm	Md	No	Lr

Among the conspicuous differences between sulfur and oxygen is the readiness with which sulfur forms compounds in which it has a positive oxidation number. Another difference is sulfur's greater tendency to form homonuclear bonds (bonds between two atoms of the same element). And unlike oxygen, the sulfur atom has d orbitals in its outermost energy level; hence it can expand the number of valence electrons to more than eight. Finally, sulfur forms double bonds much less readily than oxygen. (The smaller oxygen atom can more easily approach other atoms closely enough to allow p-orbital overlap.) In the context of these chemical differences between oxygen and sulfur, it is worthwhile to mention that the differences between sulfur and selenium, the element below sulfur, are much less pronounced. This is merely an extension of the observation we made when studying the halogens: The greatest contrast between adjacent members of a periodic group is likely to be between the first two.

Although sulfur is not nearly so abundant as oxygen, it is common in nature not only in compounds but also in elemental form. Nearly all the sulfur used in this country comes from the Gulf Coast states, where it is found in swampy areas that make conventional mining impossible. A simple but ingenious method for obtaining this sulfur was developed in 1891 by Herman Frasch (1851–1914). The Frasch process, which is illustrated in Figure 19.3, requires that three concentric pipes be sunk into the sulfur deposits. Superheated water (165°C) is pumped down the outer pipe, and compressed air is forced down the inner pipe. The water melts the sulfur (m.p. = 112.8°C), and a frothy mixture of sulfur, air, and water rises to the surface in the middle (outlet) pipe. The frothy mixture is necessary since the sulfur, which is about twice as dense as water, would not otherwise rise to the surface. As it rises, the sulfur remains molten because it is surrounded by

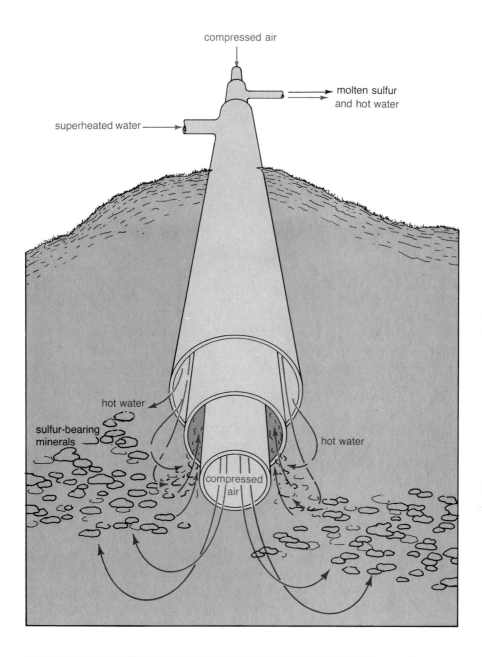

Figure 19.3 The Frasch process.

the superheated water in the outer pipe. When it reaches the surface, the sulfur quickly solidifies. The product is better than 99.5 percent pure.

About three-quarters of the sulfur produced is used to manufacture sulfuric acid. Much of the rest is used in the manufacture of carbon disulfide, plastics, gunpowder, and vulcanized rubber.

Allotropes of Sulfur At room temperature sulfur is a tasteless, odorless, yellow solid, insoluble in water but soluble in carbon disulfide. This stable form of sulfur, called *rhombic sulfur*, comprises cyclic S_8 molecules arranged in an orthorhombohedral crystal lattice (see Figure 9.19). More than 20 different allotropes of sulfur have been identified, two of which are easily observed at room temperature. Several of these allotropes, and the processes by which they can be produced, appear in Figure 19.4. As the figure illustrates, between 95.6°C and 119°C the

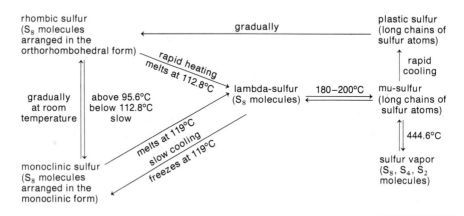

Figure 19.4 Allotropic forms of sulfur.

S_8 molecules are arranged in a monoclinic lattice, hence the allotrope *monoclinic sulfur*. Likewise the molecular structure of the liquid depends on the temperature: Between about 180°C and 200°C the cyclic S_8 molecules break into *diradicals* (atoms or groups of atoms containing two unpaired electrons), which in turn link into long chains, as shown in Figure 19.5.

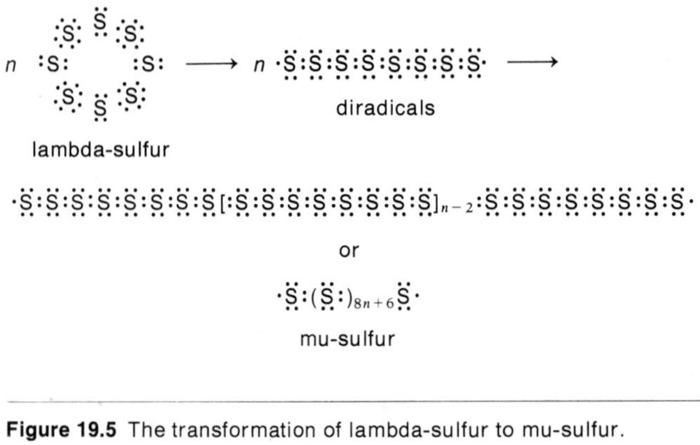

Figure 19.5 The transformation of lambda-sulfur to mu-sulfur.

General Chemical Properties Like oxygen, sulfur is relatively inactive at room temperature, but it combines with most elements at elevated temperatures. Among the exceptions are gold, platinum, iridium, tellurium, nitrogen, iodine, and the noble gases. In general, sulfur reacts more vigorously with metals than with nonmetals. When a mixture of powdered zinc and sulfur is heated, for example, the reaction is exothermic enough to heat the mass to incandescence.

When sulfur reacts with active metals (those with low electronegativities), it acts as an oxidizing agent, gaining two electrons per atom, and it acquires an electrovalence and an oxidation number of -2. When it reacts with less active metals or with inactive nonmetals, it still usually acquires a -2 oxidation number, but the bonds are more likely to be covalent. Finally, if it reacts with active nonmetals, it behaves as an electron-pair donor, acquiring a positive oxidation number and forming covalent bonds.

Sulfides (Oxidation State -2) Binary compounds in which sulfur is the more electronegative element are called *sulfides*. In Chapter 15 we mentioned one of the most important—hydrogen sulfide. This common reagent is often prepared from the reaction between hydrochloric acid and iron(II) sulfide:

$$FeS(s) + 2\ HCl(aq) \longrightarrow FeCl_2(aq) + H_2S(g)$$

Hydrogen sulfide is a colorless, foul-smelling, poisonous gas. As stated in Chapter 14, H_2S dissolves in water to form a weak, diprotic acid called hydrosulfuric acid. In this form it is a common precipitating agent for metallic sulfides. Hydrogen sulfide is also a strong enough reducing agent to reduce nitric acid, concentrated sulfuric acid, iodine, the permanganate ion, the dichromate ion, the iron(III) ion, and sulfur dioxide.

When aqueous solutions of soluble hydroxides or sulfides are heated with sulfur, the latter dissolves to form solutions of polysulfides. These compounds consist of metallic or ammonium ions and negative ions in which two or more sulfur atoms form chains by sharing electrons. These chains vary in length from two to five sulfur atoms. The tendency of sulfur atoms to form covalent links (among themselves) is responsible for the formation of the polysulfide ions:

$$:\!\ddot{\underset{..}{S}}\!:\!\ddot{\underset{..}{S}}\!:^{2-} \text{ to } :\!\ddot{\underset{..}{S}}\!:\!\ddot{\underset{..}{S}}\!:\!\ddot{\underset{..}{S}}\!:\!\ddot{\underset{..}{S}}\!:\!\ddot{\underset{..}{S}}\!:^{2-}$$

The $+4$ Oxidation State Sulfur forms several compounds and ions in which it has an oxidation number of $+4$. The common examples are sulfur dioxide, sulfurous acid, and the sulfite and hydrogen sulfite ions. Sulfurous acid is a weak, unstable acid, which decomposes to produce sulfur dioxide and water. Thus when the salts of sulfurous acid (sulfites and hydrogen sulfites) react with strong acids, the final products are SO_2 and H_2O:

$$SO_3^{2-}(aq) + 2\,H^+(aq) \longrightarrow H_2SO_3(aq)$$
$$H_2SO_3(aq) \longrightarrow H_2O(l) + SO_2(g)$$

These reactions can be used to identify sulfites and hydrogen sulfites, since SO_2 has a distinct pungent odor. This procedure is also used to prepare SO_2 in the laboratory.

Sulfur dioxide is one of the noxious pollutants produced by the combustion of sulfur-containing fuels. Unless present in very large concentrations, it has little effect on healthy adults, but it is exceedingly harmful to vegetation and it can pose a health hazard to the elderly and to those who suffer from respiratory diseases.

Since the $+4$ oxidation number is an intermediate oxidation state for sulfur, the sulfites and related substances can act as either oxidizing agents or reducing agents. For example, if we mix hydrogen sulfide and sulfur dioxide (or sulfurous acid), the result is free sulfur:

$$8\,SO_2(g) + 16\,H_2S(g) \longrightarrow 16\,H_2O(l) + 3\,S_8(s)$$

Sulfur dioxide thus acts as an oxidizing agent. On the other hand, both sulfur dioxide and sulfurous acid can serve as reducing agents (thus being oxidized to sulfate ions) in reactions with strong oxidizing agents such as nitric acid, permanganate ions, dichromate ions, and iron(III) ions.

The +6 Oxidation State The highest common oxidation number of sulfur is $+6$, which we find in sulfur trioxide, sulfuric acid, the sulfates, and the hydrogen sulfates. Sulfur trioxide, whose polymeric structure we described in Chapter 9, is a colorless liquid at room temperature. It combines with water in a highly exothermic reaction to form sulfuric acid. Concentrated (98 percent) sulfuric acid, in turn, is a colorless, oily liquid with a specific gravity of 1.84. We have already characterized it as a weak nonvolatile acid, which becomes a strong acid when diluted. (The second hydrogen atom ionizes only weakly, however.) In addition, hot concentrated sulfuric acid is an effective oxidizing agent. It oxidizes metals, carbon, sulfur, bromide ions, and iodide ions, being reduced at the same time to sulfurous acid, sulfur dioxide, or (in the presence of a strong reducing agent) hydrogen sulfide.

Concentrated sulfuric acid has a great affinity for water; thus it serves as an effective drying agent for gases and liquids that do not react with it chemically. It will even react with compounds like sugar ($C_{12}H_{22}O_{11}$), extracting from them the component elements of water. And in other cases it will extract a hydrogen atom from one molecule and OH from another, thus initiating a reaction that would not otherwise occur:

$$H_2SO_4 + HX + YOH \longrightarrow XY + H_2SO_4 \cdot H_2O$$

This powerful affinity for water is the reason that when concentrated H_2SO_4 is diluted in the laboratory, the acid must always be added to the water. If water is added to the acid, the solution boils violently.

Thiosulfates The prefix *thio-* is used to indicate that sulfur has been substituted for oxygen in an oxyacid of sulfur or in one of its salts. The most common example is sodium thiosulfate ($Na_2S_2O_3$), which can be prepared by boiling a solution of sodium sulfite with free sulfur:

$$8\ Na_2SO_3(aq) + S_8(s) \longrightarrow 8\ Na_2S_2O_3(aq)$$

This solution of sodium thiosulfate will then react with a strong acid to produce thiosulfuric acid. This weak, unstable acid subsequently decomposes, producing sulfur, sulfur dioxide, and water:

$$Na_2S_2O_3(aq) + 2\ HCl(aq) \longrightarrow H_2S_2O_3(aq) + 2\ NaCl(aq)$$
$$8\ H_2S_2O_3(aq) \longrightarrow 8\ H_2O(l) + S_8(s) + 8\ SO_2(g)$$

The thiosulfate ion can also be oxidized by I_2 to produce the tetrathionate ion:

$$2\ [S_2O_3]^{2-}(aq) + I_2(s) \longrightarrow [O_3S\text{-}S\text{-}S\text{-}SO_3]^{2-}(aq) + 2\ I^-(aq)$$

Example 19.3 How can each of the steps below be carried out?

$$H_2SO_4 \longrightarrow H_2S \longrightarrow S_8 \longrightarrow Na_2S_2O_3$$

Solution

a. The first step requires that we change the oxidation number of sulfur from $+6$ to -2. We can do this by heating concentrated sulfuric acid with a strong reducing agent such as zinc metal:

$$4\ Zn(s) + 5\ H_2SO_4(l) \longrightarrow 4\ ZnSO_4(s) + H_2S(g) + 4\ H_2O(l)$$

This reaction must not be confused with the reaction of zinc with cold, dilute sulfuric acid, where hydrogen ion is reduced to hydrogen gas.

b. To oxidize the sulfur in H_2S to free sulfur, we might choose any of several oxidizing agents. For example,

$$8\ SO_2(g) + 16\ H_2S(g) \longrightarrow 16\ H_2O(l) + 3\ S_8(s)$$

c. Finally, to produce sodium thiosulfate, we might boil aqueous sodium sulfite with the free sulfur:

$$8\ Na_2SO_3(aq) + S_8(s) \longrightarrow 8\ Na_2S_2O_3(aq)$$

Example 19.4 What sulfur-containing compounds can you make from the following starting materials: S_8, O_2, NaCl, H_2SO_4, and Zn? How would you prepare them?

Solution

a. $Zn(s) + H_2SO_4(l) \longrightarrow ZnSO_4(aq) + H_2(g)$

b. $S_8(s) + 8\ O_2(g) \longrightarrow 8\ SO_2(g)$

c. In Chapter 14 we found that concentrated sulfuric acid reacts with salts of volatile acids to form the volatile acid and either sodium sulfate (if the reaction mixture is heated) or sodium hydrogen sulfate (if the reaction mixture is not heated). Therefore,

$$2\ NaCl(s) + H_2SO_4(l) \xrightarrow{heat} Na_2SO_4(aq) + 2\ HCl(g)$$

$$NaCl(s) + H_2SO_4(l) \longrightarrow NaHSO_4(aq) + HCl(g)$$

d. If we burn the hydrogen from step a to form water, we can then synthesize sulfurous acid:

$$SO_2(g) + H_2O(l) \longrightarrow H_2SO_3(aq)$$

e. $Zn(s) + H_2SO_3(aq) \longrightarrow H_2(g) + ZnSO_3(aq)$

f. Hydrogen sulfide can be made by bubbling hydrogen through boiling sulfur:

$$n\ H_2(g) + S_n(l) \longrightarrow n\ H_2S(g)$$

(Since liquid sulfur exists in a number of different molecular forms, we represent it here as S_n.)

g. $ZnSO_4(aq) + H_2S(aq) \longrightarrow ZnS(s) + H_2SO_4(aq)$

Questions

13. What is the oxidation number of sulfur in each of the following compounds?
    a. H_2S b. H_2SO_3 c. H_2SO_4 d. $H_2S_2O_3$

14. Name each of the compounds in Question 13.

15. Classify the following substances as oxidizing agents, reducing agents, or both:
    a. S_8 b. H_2SO_3 c. H_2SO_4 d. SO_2 e. Na_2S f. Na_2SO_3

Table 19.8 A Comparison of Oxygen and Sulfur

Electron configuration	All elements in Group VIA, including oxygen and sulfur, have electron configurations ending with s^2p^4.
Atomic radius (nm)	O (0.074) As we saw among the halogens, the most S (0.104) dramatic difference is between the first two Se (0.117) members of the group. Te (0.137)
Allotropism	Both elements exhibit allotropism, sulfur more dramatically than oxygen. Indeed, sulfur has more allotropes than any other element.
Expansion of the outermost energy level	Since oxygen is in the second period, it has no d orbitals in its outermost energy level; thus it can never have more than eight valence electrons. Sulfur, on the other hand, does have d orbitals; hence it can accommodate more than eight electrons in its outermost energy level.
Ionic valence and oxidation numbers	Both oxygen and sulfur tend to gain two electrons per atom to form -2 ions. Both will also share two electrons to form two covalent bonds per atom. Oxygen usually exhibits an oxidation number of -2, less often -1 (peroxides), $-\frac{1}{2}$ (superoxides), and $+2$ (OF_2). Sulfur has oxidation states of -2, $+2$, $+4$, and $+6$.
Double bonds	Oxygen commonly forms double bonds with another oxygen atom as well as with other elements. The larger sulfur atom forms double bonds far less easily.
Homonuclear bonds	Oxygen forms homonuclear bonds only in O_2, O_2^{2-}, O_2^-, and O_3. Sulfur, on the other hand, forms homonuclear bonds in polysulfides, thiosulfates, tetrathionates, and proteins; it forms cyclic S_8 molecules in rhombic, monoclinic, and lambda-sulfur; and in mu-sulfur it forms long chains containing hundreds of sulfur-sulfur bonds.
Hydrogen bonding	The oxygen atom meets both requirements for forming hydrogen bonds: high electronegativity and small size. Sulfur does not.

16. What property of sulfuric acid allows it to
    a. dissolve metals below hydrogen in the activity series?
    b. dry gases and liquids that do not react with it?
    c. produce nitric acid from sodium nitrate?
    d. produce carbon dioxide from calcium carbonate?

17. Classify solutions of the following compounds as acidic, basic, or neutral. Give the reasons for your choices.
    a. $NaHSO_4$ b. Na_2SO_3 c. $NaHSO_3$ d. Na_2S

18. How do the members of each of the following pairs differ?
    a. lambda-sulfur and mu-sulfur
    b. monoclinic and rhombic sulfur
    c. sulfate and thiosulfate ions

Questions and Problems

19.1 The Halogens

19. List six reagents that can be used to oxidize chloride ions to free chlorine.

20. If an oxidizing agent will oxidize the bromide ion but not the chloride ion, which other halide ion will it oxidize? Which will it not oxidize?

21. Starting with a mixture of sodium halides, how would you liberate in sequence iodine, bromine, and chlorine?

22. Write equations illustrating the following changes in the oxidation number of chlorine:
 a. -1 to 0 b. 0 to -1 c. 0 to $+1$

23. Compare the properties of anhydrous hydrogen chloride to those of aqueous hydrogen chloride.

24. Write equations for reactions between chlorine and the following:
 a. water   b. iron   c. CH_4 d. $Br^-(aq)$ e. $I^-(aq)$ f. H_2

25. How would you carry out each of the following steps?
 a. $CaI_2 \longrightarrow CaBr_2 \longrightarrow CaCl_2 \longrightarrow Cl_2$ b. $HI \longrightarrow KI \longrightarrow KI_3$

26. Write balanced net ionic equations for the reactions between
 a. hydrobromic acid and sodium carbonate.
 b. hydrochloric acid and sodium carbonate.
 c. hydroiodic acid and sodium hydroxide.
 d. hydrofluoric acid and sodium hydroxide.
 e. sodium chloride and concentrated sulfuric acid.
 f. sodium bromide and concentrated sulfuric acid.
 g. hydrochloric acid and potassium permanganate.
 h. hydrofluoric acid and silicon dioxide.
 i. hydrogen iodide and water.
 j. sodium fluoride and water.
 k. bromine and water.
 l. chlorine and phosphorus.
 m. lead(IV) oxide and hydrochloric acid.
 n. phosphorus, bromine, and water.

*__27.__ List the ions and molecules present in aqueous solutions of the following:
 a. HCl   b. HF   c. Cl_2

Discussion Starters

28. Explain or account for the following:
 a. Hydrogen fluoride has the highest boiling point of all the hydrogen halides, even though it has the smallest molecular weight.
 b. Fluorine cannot be prepared by electrolysis of an aqueous solution containing fluoride ions.

c. Within a periodic group the first two elements differ more dramatically than any other adjacent pair.
d. Fluorine is extremely active chemically.
e. The reaction between concentrated sulfuric acid and sodium iodide is not a suitable laboratory preparation of hydrogen iodide.
f. Fluorine cannot be liberated from fluorides by any chemical reagent.
g. Hydrofluoric acid attacks glass, but hydrochloric acid does not; however, we call hydrochloric acid a strong acid and hydrofluoric acid a weak acid.
h. The formula for hydrofluoric acid is sometimes written H_2F_2.
i. Chlorine reacts with water; yet we say that chlorine forms a solution in water.
j. Solutions of sodium iodide are neutral, but solutions of sodium fluoride are basic.
k. The solubility of iodine in water increases in the presence of potassium iodide.
l. Direct union of the elements is an unsatisfactory method of preparaing hydrogen fluoride.
m. Direct union of the elements is an unsatisfactory method of preparing hydrogen iodide.
n. Chlorine water gradually becomes more acidic during storage.
o. The reaction between concentrated phosphoric acid and sodium bromide is a suitable method of preparing HBr in the laboratory, but the reaction between concentrated sulfuric acid and sodium bromide is not suitable.
p. Hydrogen chloride is very soluble in water, whereas nitrogen and oxygen are only slightly soluble in water.
q. Astatine was the last halogen to be isolated.
r. There are no solutions of fluorine in ordinary solvents.
s. Elemental chlorine is never found in nature.
t. Fluorine would generally not be used to liberate bromine from aqueous solutions of sodium bromide.
u. Hydrofluoric acid cannot be stored in glass bottles.
v. Fluorine is the only element in Group VIIA that never expands the outermost energy level beyond eight electrons.

19.2 Hydrogen

29. By chemical means, how would you distinguish hydrogen from each of the following gases?
    a. water vapor   b. oxygen   c. nitrogen   d. air

30. What is the percentage (by weight) of hydrogen in each of the following compounds?
    a. H_2O b. H_2O_2 c. C_2H_2 d. CaH_2 e. NaOH   f. NH_3

31. Write the electron formulas of the compounds listed in Question 31.

32. 40.0 liters of hydrogen and 19.0 liters of oxygen (measured at STP) were mixed in a strong container and then ignited.
    a. What substances were present in the container after the explosion?
    b. What mass of each was present?

Discussion Starters

33. Explain or account for the following statements:
    a. The hydrogen molecule is less reactive than the fluorine molecule.
    b. A hydrogen ion is a proton.

c. Among the elements of the earth's crust, hydrogen ranks fourth in percent by number of atoms but eleventh in percent by weight.

d. In some periodic tables, hydrogen appears in Group VIIA, above fluorine.

19.3 Oxygen

34. Write the electron configuration of each element in Group VIA of the periodic table.

35. a. Which has the higher melting point, selenium or tellurium?
b. Which is more metallic, sulfur or selenium?
c. Which is more electronegative, sulfur or tellurium?
d. Which is larger, the sulfur atom or the oxygen atom?
e. Which is more nonmetallic, sulfur or oxygen?

36. In what ways does oxygen differ from the other elements in Group VIA of the periodic table?

37. For each of the following descriptions, write the formula of an appropriate compound:
a. A ternary compound that decomposes when heated to liberate all its oxygen.
b. A metallic oxide that decomposes when heated to give the free metal and oxygen.
c. A ternary compound that liberates one-third of its oxygen when heated.
d. A compound that catalyzes the decomposition of $KClO_3$.

38. By recalling the chemistry we have discussed in this chapter, show how you could distinguish between the following pairs:
a. air and oxygen
b. oxygen and nitrogen
c. air and nitrogen
d. KNO_3 and KCl
e. KNO_2 and $KClO_3$
f. $KClO_3$ and KCl
g. CuO and Ag_2O
h. iron and silver
i. carbon dioxide and carbon monoxide

39. Write equations for the burning of the following. Assume an unlimited supply of oxygen in each case.
a. CuS
b. Al
c. HCl
d. Fe (rapid burning)
e. Fe (slow burning)
f. H_2S
g. C_6H_6
h. C_5H_{12}
i. $C_6H_{10}O_5$
j. $C_{12}H_{22}O_{11}$

40. Show by a series of equations how carbonic acid, sulfurous acid, and phosphoric acid might be prepared from KNO_3, ZnS, P_4, and C_2H_6.

*__41.__ When a mixture of $KClO_3$ and MnO_2 is heated, oxygen comes from the $KClO_3$ rather than from the MnO_2. How could you prove this?

19.4 Sulfur

42. Write electron formulas for the following:
a. Na_2S
b. $(NH_4)_2S_5$
c. Na_2SO_3
d. H_2SO_4
e. $S_2O_3^{2-}$
f. $S_4O_6^{2-}$

43. How could one distinguish between the following pairs of substances?
a. Na_2S and ZnS
b. Na_2SO_4 and $NaNO_3$
c. $BaSO_4$ and $BaCO_3$

44. Name the following compounds:
a. Na_2S
b. $NaHSO_4$
c. $Na_2S_2O_3$

45. Write the equations for the reactions between the following:
   a. hot concentrated H_2SO_4 and zinc   b. dilute H_2SO_4 and zinc

46. Write two equations illustrating the use of sulfuric acid in the preparation of other strong acids from their salts. What property is important here?

47. Show by equations how the following conversions might be made:
   a. $Na_2S_2O_3$ to S_8 b. H_2S to S_8 c. H_2S to SO_2
   d. H_2SO_4 to SO_2 e. S_8 to ZnS f. S_8 to H_2SO_3
   g. SO_3 to Na_2SO_4 h. FeS to H_2S i. S_8 to H_2S
   j. S_8 to $Na_2S_2O_3$

48. Write equations illustrating the following changes in the oxidation number of sulfur:
   a. 0 to -2 b. -2 to $+4$ c. -2 to 0
   d. 0 to $+4$ e. $+6$ to $+4$ f. $+6$ to -2

49. Complete and balance the following:
   a. $H_2S + KMnO_4 + H_2SO_4 \longrightarrow S_8 + MnSO_4$ ($KMnO_4$ is the oxidizing agent)
   b. $C + H_2SO_4 \longrightarrow CO_2 + SO_2$
   c. $S_8 + H_2SO_4 \longrightarrow SO_2$
   d. $H_2S + H_2SO_3 \longrightarrow S_8$
   e. $Ag + H_2SO_4 \longrightarrow Ag_2SO_4 + SO_2$
   f. $H_2S + K_2Cr_2O_7 + H_2SO_4 \longrightarrow S_8 + Cr_2(SO_4)_3 + K_2SO_4$
   g. $HBr + H_2SO_4 \longrightarrow Br_2 + SO_2$
   h. $H_2S + HNO_3 \longrightarrow S_8 + NO_2$

50. How can each of the following steps be carried out?
   a. $S_8 \longrightarrow SO_2 \longrightarrow H_2SO_3 \longrightarrow Na_2SO_3$
   b. $S_8 \longrightarrow H_2S \longrightarrow SO_2 \longrightarrow SO_3 \longrightarrow H_2SO_4 \longrightarrow Na_2SO_4$

51. List all the molecules and ions mentioned in this chapter that contain homonuclear, covalent bonds.

52. What volume of 85 percent sulfuric acid (specific gravity 1.83) is needed to oxidize 50.0 g of carbon? Assume sulfuric acid is reduced to SO_2.

53. What volume of air is needed to burn 1.00×10^2 liters of H_2S at STP? Assume that the products are water and SO_2.

54. How many liters of hydrogen sulfide, measured at 25°C and 7.40×10^2 mm of Hg, are required to precipitate 50.0 g of ZnS from solution?

55. What volume of H_2S, measured at 27°C and 2.00 atm will be produced when 1.00×10^2 g of FeS is treated with an excess of hydrochloric acid?

56. Write equations showing three different methods of preparing sodium sulfide.

57. What is the volume of 1.00×10^2 g of SO_2, measured at 13°C and 7.20×10^2 mm of Hg?

58. Write equations illustrating three ways of preparing sulfur dioxide.

59. How many liters of oxygen are required to produce 3.00×10^2 liters of SO_2 by burning sulfur? Consider the gases at the same conditions of temperature and pressure.

Discussion Starters

60. Explain or account for the following:
  a. Sulfur, but not oxygen, may have more than eight electrons in its outermost energy level.
  b. The Frasch process would not be successful if the temperature of the water used was 200°C.
  c. Sulfur exhibits less tendency to form double bonds than oxygen.
  d. Sulfur and sulfur dioxide can act as either oxidizing or reducing agents, but hydrogen sulfide can act only as a reducing agent and sulfuric acid can act only as an oxidizing agent.
  e. Sulfur has two melting points.
  f. When liquid sulfur is heated, the viscosity increases until the temperature reaches 200°C. Upon further heating, the viscosity decreases.
  g. Solutions of sodium sulfite are basic.
  h. Even though copper is a metal below hydrogen in the activity series, it dissolves in hot concentrated sulfuric acid.
  i. Sulfuric acid cannot be prepared by treating sodium sulfate with hydrochloric acid.
  j. Solutions of sodium sulfide are strongly basic.
  k. If hydrogen sulfide were accidentally released on the second floor of a three-story building, the gas would likely be detected on the first floor before it was detected on the third floor.
  l. Zinc reduces concentrated sulfuric acid to H_2S, whereas copper reduces it only to SO_2.
  m. Water has a relatively high boiling point compared to hydrogen sulfide.
  n. Hydrochloric acid, rather than nitric acid, is used in the laboratory preparation of H_2S from FeS.

The Chemistry of the Nonmetals of Groups IVA and VA and the Noble Gases

20.1 Nitrogen

The elements of Group VA, each characterized by an s^2p^3 ending to its electron configuration, show all the trends we have come to expect in a periodic group: From top to bottom, the electronegativities and ionization energies decrease, whereas boiling points, melting points, atomic radii, and metallic character generally increase. Table 20.1 lists the properties of the Group VA elements, showing at the same time that the trend in melting points is the least reliable. The trend from nonmetal to metal, however, is especially pronounced in this group: Nitrogen is a typical nonmetal; phosphorus shows some metallic characteristics, which become even more obvious in arsenic and antimony; and bismuth is a metal, showing only vestiges of nonmetallic traits. Table 20.1 also confirms the observations we made in Groups VIA and VIIA that the most pronounced differences between adjacent elements in a group are between the first two (nitrogen and phosphorus, in this case).

Nitrogen, a colorless, odorless, tasteless gas, is the only gas in Group VA. It is most common as a very stable diatomic molecule, having no allotropes. In this form, it constitutes 78 percent of the earth's atmosphere by volume and 75 percent by weight. It has only two natural isotopes, having mass numbers of 14 and 15. The lighter isotope makes up 99.64 percent of all nitrogen. Combined nitrogen is much less abundant than the elemental form. The only major source is extensive deposits of saltpeter ($NaNO_3$) in the deserts of Chile. Protein is about 16 percent nitrogen; hence all living matter contains some of this element. When organic material decays, the nitrogen from the proteins can be freed as elemental nitrogen or can remain fixed in ammonia, nitrites, or nitrates.

Nitrogen is usually prepared commercially by distilling liquid air (see Chapter 19). In the laboratory, small quantities of nitrogen can be prepared by heating nitrogen-containing compounds such as ammonium nitrite. Since this unstable compound is not usually found among laboratory reagents, the laboratory procedure uses instead an aqueous solution of ammonium and nitrite ions:

$$NH_4Cl(aq) + NaNO_2(aq) \longrightarrow N_2(g) + NaCl(aq) + 2\ H_2O(l)$$

Table 20.1 Properties of Group VA Elements

Property	Nitrogen	Phosphorus	Arsenic	Antimony	Bismuth
Atomic number	7	15	33	51	83
Atomic weight	14.0067	30.97376	74.9216	121.75	208.9804
Melting point (°C)	−210	44.1	814 (at 36 atm)	630	271
Boiling point (°C)	−196	280	613 (sublimes)	1325	1560
Atomic radius (nm)	0.070	0.110	0.121	0.141	0.146
First ionization energy (kcal/g at. wt)	336	242	231	199	185
Electronegativity	3.0	2.1	2.0	1.9	1.9
Classification	Nonmetal	Nonmetal	Metalloid	Metalloid	Metal

The product of this preparation is relatively pure except for water vapor, which can be easily removed by passing the gas mixture over calcium chloride. Notice also that nitrogen is both oxidized and reduced in this reaction: The ammonium ion is a reducing agent, and the nitrite ion is an oxidizing agent. Ammonium dichromate is a second compound that decomposes when heated to produce N_2:

$$(NH_4)_2Cr_2O_7(s) \longrightarrow N_2(g) + 4\ H_2O(g) + Cr_2O_3(s)$$

Chemical Properties Considering its position in the periodic table and its high electronegativity, we might expect nitrogen to be quite an active element: It is adjacent to two very active elements, oxygen and phosphorus, and its electronegativity (3.0) is exceeded only by oxygen and fluorine and is equal to chlorine. However, the most striking chemical property of nitrogen is its inactivity.

To explain this observation we can turn back to Chapter 6, where we first mentioned the highly stable triple bond in N_2. In contrast, remember that we attributed the great chemical activity of fluorine to the weak fluorine-to-fluorine bond in F_2. Similarly, in discussing oxygen and hydrogen we noted that they were not particularly active at room temperatures but were quite reactive when small amounts of energy were supplied to break or distort the oxygen-oxygen or hydrogen-hydrogen bond. In nitrogen the bond is so strong that the molecule is unreactive under most laboratory conditions; indeed, N_2 is the most stable diatomic molecule known (see Table 11.4). As we learned in Chapter 6, the bond between the two atoms is a triple covalent bond, involving three orbitals from each atom. One of these bonds is a sigma bond formed by the endwise overlap of hybridized *sp* orbitals (see Figure 20.1), and the other two bonds are pi bonds formed by the lengthwise overlap of the two $2p$ orbitals in each atom.

Just as we found that oxygen was the only member of Group VIA that readily formed multiple bonds, we'll find that nitrogen is unique among Group VA elements. Only the elements of the second period are small enough to form pi bonds easily; these pi bonds require that the bonding atoms be close enough together to allow *p*-orbital overlap.

Despite the stability of N_2, nitrogen will react with hydrogen, oxygen, and the metals of Group IA and IIA, although only at high temperatures or in the presence of catalysts. The binary compounds produced by these reactions can be used to produce more complex nitrogen compounds. In fact, there are probably more compounds containing nitrogen than any element except hydrogen, oxygen, and carbon. Even in simple compounds, nitrogen shows all possible oxidation numbers between -3 and $+5$, the most common being -3, $+3$, and $+5$. Table 20.2 lists some common compounds containing nitrogen in each of its oxidation states. Note that all common compounds containing nitrogen in a positive oxidation state have N—O bonds.

Hydrogen Compounds of Nitrogen Ammonia has come up several times in previous chapters. We have discussed its structure, its property as a weak base, its role as a source of fixed nitrogen, and its manufacture. But we have yet to present its physical and chemical properties systematically. Ammonia is a colorless, toxic, choking gas at room temperature. It melts at $-77.7°C$ and boils at $-33.4°C$. Its heat of vaporization at the boiling point is 322 cal/g, higher than for any other common liquid except water. All of these values are considerably higher than we would expect for a compound having a molecular weight of only 17. But as we have done with water, we can explain these surprising observations in terms of hydrogen bonding. Since the hydrogen bonds are neither as strong nor as extensive as they are in water, however, the surprises are not so great. The critical temperature and pressure of ammonia are $132.4°C$ and

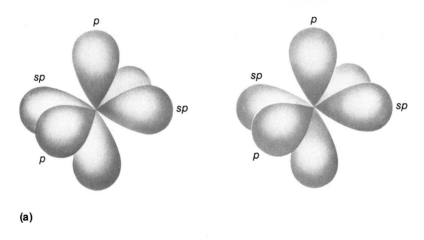

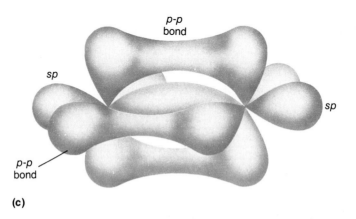

Figure 20.1 Formation of the N₂ molecule. (**a**) Orbitals involved in bonding two nitrogen atoms. (**b**) Formation of the sigma bond. (**c**) Formation of the two pi bonds.

Table 20.2 Oxidation Numbers of Nitrogen

Oxidation Number	Example	
−3	NH_3	Ammonia
	NH_4^+	Ammonium ion
	NH_2^-	Amide ion
	N^{3-}	Nitride ion
−2	N_2H_4	Hydrazine
−1	NH_2OH	Hydroxylamine
0	N_2	Nitrogen
+1	N_2O	Nitrous oxide (dinitrogen oxide)
+2	NO	Nitric oxide (nitrogen oxide)
+3	N_2O_3	Dinitrogen trioxide
	HNO_2	Nitrous acid
	NO_2^-	Nitrite ion
+4	NO_2	Nitrogen dioxide
	N_2O_4	Dinitrogen tetraoxide
+5	N_2O_5	Dinitrogen pentaoxide
	HNO_3	Nitric acid
	NO_3^-	Nitrate ion

111.5 atm, respectively. Ammonia is highly soluble in water (see Figure 20.2); nearly 700 liters of ammonia will dissolve in 1 liter of water at room temperature and one atmosphere. At 0°C the solubility increases to about 1300 liters of ammonia to 1 liter of water.

The most interesting chemical properties of NH_3 might be listed concisely as follows:

1. At temperatures above 700°C, ammonia decomposes rapidly into nitrogen and hydrogen. At 1000°C this decomposition is nearly complete.

2. Ammonia burns in oxygen to produce water and nitrogen or water and nitric oxide:

$$4\ NH_3(g) + 3\ O_2(g) \longrightarrow 2\ N_2(g) + 6\ H_2O(g)$$

or

$$4\ NH_3(g) + 5\ O_2(g) \longrightarrow 4\ NO(g) + 6\ H_2O(g)$$

The second reaction proceeds at a reasonable rate only at high temperatures and in the presence of a platinum catalyst.

3. Ammonia undergoes rapid oxidation in chlorine to give off heat and light:

$$2\ NH_3(g) + 3\ Cl_2(g) \longrightarrow 6\ HCl(g) + N_2(g)$$

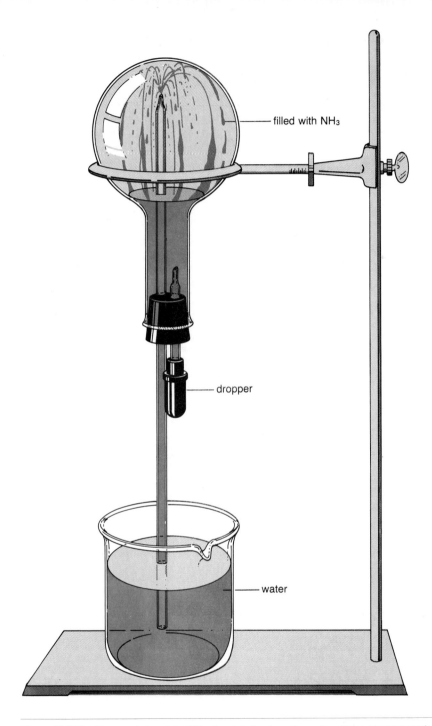

Figure 20.2 The "ammonia fountain." The fountain begins when a few drops of water are forced from the dropper into the flask filled with ammonia gas. The ammonia dissolves in the water and the pressure in the flask is reduced. Consequently, the water rises rapidly up the tube, forming a fountain.

4. Hot ammonia will reduce the oxides of some metals:

$$3\ CuO(s) + 2\ NH_3(g) \longrightarrow 3\ Cu(s) + N_2(g) + 3\ H_2O(g)$$

5. Ammonia acts as an oxidizing agent when passed over heated metals. With magnesium the product is magnesium nitride:

$$3\ Mg(s) + 2\ NH_3(g) \longrightarrow Mg_3N_2(s) + 3\ H_2(g)$$

6. Metals of Group IA react with liquid NH_3 to form amides:

$$2\ NH_3(l) + 2\ Na(s) \longrightarrow 2\ NaNH_2(\text{solvated}) + H_2(g)$$

Like the ammonia molecule, the ammonium ion (NH_4^+) is sp^3-hybridized. And since all four orbitals contain a pair of shared electrons, both the electron-pair geometry and the molecular shape are tetrahedral. As we've seen before, all ammonium salts are water-soluble, and all hydrolyze to form ammonia:

$$NH_4^+(aq) + H_2O(l) \rightleftharpoons H_3O^+ + NH_3(aq)$$

Ammonium salts are also unstable at high temperatures. When dry ammonium salts of nonoxidizing acids are heated, ammonia and the corresponding acids are formed. For example,

$$NH_4Cl(s) \longrightarrow NH_3(g) + HCl(g)$$
$$NH_4HCO_3(s) \longrightarrow NH_3(g) + H_2O(g) + CO_2(g)$$

Ammonium salts of oxidizing acids, on the other hand, undergo autooxidation when heated:

$$NH_4NO_3(s) \longrightarrow N_2O(g) + 2\ H_2O(g)$$
$$(NH_4)_2Cr_2O_7(s) \longrightarrow N_2(g) + 4\ H_2O(g) + Cr_2O_3(s)$$

In these autooxidation reactions, the ammonium ion is the reducing agent, and the negative ion is the oxidizing agent. Thus the ammonium salt is both oxidized and reduced in the same reaction. Some of these reactions can proceed with explosive violence.

Less common than ammonia and its salts is hydrazine (N_2H_4), in which nitrogen has the oxidation number -2. We can represent its electron formula as

$$\begin{array}{cc} H & H \\ H\!:\!\overset{..}{N}\!:\!\overset{..}{N}\!:\!H & \end{array}$$

Hydrazine and its derivatives have been used extensively as rocket fuels.

Nitrogen Oxides The most common compounds containing nitrogen in its positive oxidation states are the oxides—of which there are a number. One of the most common is nitrous oxide (more correctly but less frequently called dinitrogen oxide), N_2O, in which the oxidation number of nitrogen is $+1$. This colorless, unreactive gas can be prepared by the decomposition of ammonium nitrate. At about 500°C, N_2O decomposes into nitrogen and oxygen; hence we say that it supports combustion. As shown in Figure 20.3, the N_2O molecule is linear and is isoelectronic with carbon dioxide. Its best known application has been as a general anesthetic, whence came the nickname "laughing gas" because of the mild hysteria it produces when inhaled. It was also once used extensively as the propellant in cans of "instant whipped cream," but it has now been largely replaced by fluorocarbons. It is soluble in cream and in other fats and oils.

Nitrogen exhibits the $+2$ oxidation state in the colorless, insoluble, poisonous gas called nitric oxide, NO. (The less common, systematic name for this compound is nitrogen oxide.) We can best represent this oxide as a resonance hybrid of two molecular structures. This representation (Figure 20.3) also reveals that the molecule contains an odd number of electrons, one of which cannot be properly assigned to either the nitrogen or the oxygen. Instead it occupies an orbital that encompasses both atoms. This odd electron means that NO is paramagnetic (see Chapter 6). Nitric oxide can be prepared by oxidizing nitrogen in

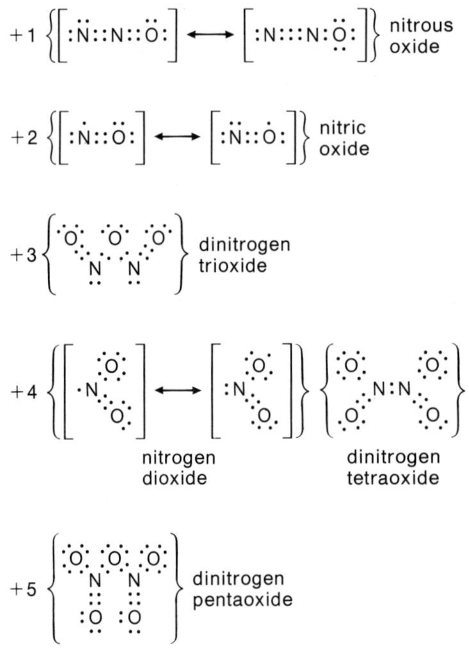

Figure 20.3 The electron formulas of the nitrogen oxides, showing the resonance forms of nitrous oxide, nitric oxide, and nitrogen dioxide.

an electrical discharge, by oxidizing ammonia in the presence of a platinum catalyst, or by reducing dilute nitric acid with copper or mercury. We can represent the reduction of HNO_3 as follows:

$$3\ Cu(s) + 8\ HNO_3(aq) \longrightarrow 3\ Cu(NO_3)_2(aq) + 4\ H_2O(l) + 2\ NO(g)$$

Since nitric oxide combines readily with oxygen to form nitrogen dioxide, it is difficult to prepare pure nitric oxide; however, the two gases can be separated by passing the mixture through a sodium hydroxide solution. The nitrogen dioxide reacts with the base to form NO_3^- and NO_2^- in solution. Nitric oxide can also be oxidized, for example by Cl_2, to form the nitrosyl ion (NO^+), which is isoelectronic with N_2.

Nitrogen appears in the $+4$ oxidation state in nitrogen dioxide, which like nitric oxide is a paramagnetic molecule. It is a brown gas at room temperature, but when cooled its brown color as well as its magnetic properties disappear. The explanation is that the NO_2 molecules *dimerize* to form N_2O_4 molecules. The equilibrium between the *monomer* and the *dimer* can be written

$$2\ NO_2(g) \rightleftharpoons N_2O_4(g)\quad + \text{energy}$$

$$\text{brown}\qquad\quad\ \text{colorless}$$
$$\text{paramagnetic}\quad\text{diamagnetic}$$

Since the reaction is exothermic as written, high temperatures favor the reverse reaction (at 140°C dissociation is almost complete) and low temperatures favor the forward reaction (at 0°C dimerization is almost complete). At room temperature nitrogen dioxide is a mixture of NO_2 and N_2O_4.

The NO_2 and N_2O_4 mixture can be prepared by reacting concentrated nitric acid and copper:

$$Cu(s) + 4\ HNO_3(aq) \longrightarrow Cu(NO_3)_2(aq) + 2\ NO_2(g) + 2\ H_2O(l)$$

It is also produced when the nitrates of heavy metals decompose at high temperatures:

$$2\ Pb(NO_3)_2(s) \longrightarrow 2\ PbO(s) + 4\ NO_2(g) + O_2(g)$$

As we mentioned above, nitric oxide reacts readily with O_2 to produce nitrogen dioxide. Perhaps the most significant chemical property of nitrogen dioxide is its reaction with cold water to form nitric acid and nitric oxide:

$$3\ NO_2(g) + H_2O(l) \rightleftharpoons 2\ HNO_3(aq) + NO(g)$$

This reaction is one of the reasons for the concern over nitrogen oxides in the atmosphere. Where their concentrations are high, the air takes the distinctive brown color of NO_2.

Nitrogen trioxide (dinitrogen trioxide), N_2O_3, and nitrogen pentaoxide (dinitrogen pentaoxide), N_2O_5, are the anhydrides of nitrous and nitric acids, respectively. These oxides, containing nitrogen in its $+3$ and $+5$ oxidation states, are unstable and of little practical importance.

Nitrous Acid and Nitrites The nitrites of the Group IA metals are most conveniently prepared by heating the corresponding nitrates:

$$2\ KNO_3(s) \longrightarrow 2\ KNO_2(s) + O_2(g)$$

(We've already mentioned this reaction as a source of oxygen.) Solutions of these nitrites are basic, since the metal hydroxides are strong bases, whereas HNO_2 (nitrous acid) is a weak acid. This weak, unstable acid can be prepared by reacting nitrites with sulfuric acid:

$$2\ KNO_2(s) + H_2SO_4(aq) \longrightarrow 2\ HNO_2(aq) + K_2SO_4(aq)$$

This reaction mixture must be kept cool to minimize the decomposition of the product.

Since the oxidation number of nitrogen in nitrous acid and the nitrites is the middle one of the three most common ($-3, +3, +5$), these compounds can serve as either oxidizing or reducing agents. For example, the nitrite ion and nitrous acid will oxidize the iodide ion but reduce the permanganate ion:

$$2\ HNO_2(aq) + 2\ HI(aq) \longrightarrow 2\ H_2O(l) + I_2(s) + 2\ NO(g)$$

$$5\ HNO_2(aq) + 2\ KMnO_4(aq) + 3\ H_2SO_4(aq) \longrightarrow 2\ MnSO_4(aq) + K_2SO_4(aq) + 5\ HNO_3(aq) + 3\ H_2O(l)$$

Nitric Acid and Nitrates Far more common and more stable than nitrous acid is the common laboratory reagent nitric acid, HNO_3. This strong, volatile acid can be prepared by reacting sodium nitrate with concentrated sulfuric acid:

$$NaNO_3(s) + H_2SO_4(l) \longrightarrow NaHSO_4(s) + HNO_3(g)$$

This reaction is convenient in the laboratory, but on a commercial scale it is more economical to use the Ostwald process, which begins with the oxidation of ammonia:

$$4\ NH_3(g) + 5\ O_2(g) \longrightarrow 4\ NO(g) + 6\ H_2O(g) + \text{heat}$$

This first step requires heat and a platinum catalyst, but once begun the reaction itself generates enough heat to maintain the temperature at about $400°C$. The nitric oxide produced in the first reaction reacts with excess oxygen to form nitrogen dioxide:

$$2\ NO(g) + O_2(g) \longrightarrow 2\ NO_2(g)$$

This nitrogen dioxide then passes through a water spray, producing nitric acid:

$$3\ NO_2(g) + H_2O(l) \longrightarrow 2\ HNO_3(l) + NO(g)$$

Finally, the nitric oxide produced by this last reaction becomes the source of still more nitrogen dioxide.

Pure nitric acid is a colorless liquid that boils at 83°C and freezes at −42°C. At room temperature it decomposes slightly,

$$4\ HNO_3(l) \rightleftharpoons 2\ H_2O(l) + 4\ NO_2(g) + O_2(g)$$

and this decomposition becomes increasingly evident at higher temperatures. Concentrated nitric acid, as used in the laboratory, is a 70 percent aqueous solution. It boils at 121°C and has a specific gravity of 1.4. The most important applications of nitric acid are in the preparation of both inorganic and organic nitrates and nitro compounds. Some of these compounds are used as explosives and fertilizers; others are intermediates for the manufacture of dyes, plastics, and lacquers; 100 percent nitric acid is also sometimes used as an oxidizing agent for rocket propellants. It is not, however, often used solely for its property as a strong acid, since both sulfuric acid and hydrochloric acid are cheaper.

Nitric acid behaves as a typical strong acid in a wide range of reactions, but since it is an oxidizing acid (Chapter 17), it usually produces something other than hydrogen in reactions with metals. Most nonmetals do not react with nitric acid, although carbon, sulfur, and phosphorus, (among others) do. In these reactions, concentrated nitric acid is usually reduced to nitrogen dioxide and dilute nitric acid, to nitric oxide. The nonmetallic element is, at the same time, oxidized to an oxide or oxyacid. For example,

$$3\ C(s) + 4\ HNO_3(l) \longrightarrow 3\ CO_2(g) + 4\ NO(g) + 2\ H_2O(l)$$

Nitric acid also oxidizes some nonmetallic ions, among them Cl^-, Br^-, I^-, and S^{2-}. The reaction conditions often determine whether the oxidation stops at the free element or goes on to the oxide or acid. For example, the sulfide ion can be oxidized to free sulfur, sulfur dioxide, or sulfuric acid.

As we've already seen, all salts of nitric acid (the nitrates) are soluble, and since HNO_3 is a strong acid, the nitrate ion never hydrolyzes in solution. Nitrate ions are also colorless; thus the color of any nitrate solution can be attributed to the cation. All nitrates decompose when heated. As noted previously, ammonium nitrate decomposes to produce nitrous oxide and water.

The nitrates of the Groups IA and IIA metals decompose to produce stable nitrites and oxygen. Nitrates of other metals, however, yield nitrogen dioxide, oxygen, and either a metallic oxide or the free metal. Iron and lead nitrates thus produce metal oxides when heated:

$$4\ Fe(NO_3)_3(s) \longrightarrow 2\ Fe_2O_3(s) + 12\ NO_2(g) + 3\ O_2(g)$$

$$2\ Pb(NO_3)_2(s) \longrightarrow 2\ PbO(s) + 4\ NO_2(g) + O_2(g)$$

And since the oxide of silver is unstable, $AgNO_3$ yields the free metal:

$$2\ AgNO_3(s) \longrightarrow 2\ Ag(s) + 2\ NO_2(g) + O_2(g)$$

Example 20.1 Write equations for reactions illustrating the following changes in the oxidation state of nitrogen:
a. 0 to $+2$
b. 0 to -3
c. $+2$ to $+4$
d. -3 to 0
e. $+5$ to $+4$
f. $+5$ to $+2$

Solution

a. $N_2(g) + O_2(g) \longrightarrow 2\ NO(g)$
This reaction requires an electric discharge, and since the NO reacts with O_2 to form NO_2, the product is certain to be impure. However, the mixture can be separated by bubbling it through water.

b. $3\ Mg + N_2 \longrightarrow Mg_3N_2$
Even when magnesium is burned in air, some magnesium nitride is formed. You can demonstrate this by adding water to the oxidation products of magnesium: The odor of ammonia will be distinct. The same change in oxidation state occurs in the Haber process, $N_2 + 3\ H_2 \rightleftharpoons 2\ NH_3$, but this is not a convenient laboratory procedure.

c. $2\ NO(g) + O_2(g) \longrightarrow 2\ NO_2(g)$
This reaction takes place spontaneously when NO and O_2 are mixed.

d. $2\ NH_3(g) + 3\ Cl_2(g) \longrightarrow 6\ HCl(g) + N_2(g)$

e. When nitrates of heavy metals are heated, nitrogen is reduced to the $+4$ oxidation state:

$$2\ Pb(NO_3)_2(s) \longrightarrow 2\ PbO(s) + 4\ NO_2(g) + O_2(g)$$

The same change occurs when concentrated nitric acid reacts with an inactive metal such as Cu:

$$Cu(s) + 4\ HNO_3(aq) \longrightarrow Cu(NO_3)_2(aq) + 2\ NO_2(g) + 2H_2O(l)$$

f. $3\ Cu(s) + 8\ HNO_3(aq) \longrightarrow 3\ Cu(NO_3)_2(aq) + 2\ NO(g) + 4\ H_2O(l)$
In this reaction, which should be compared with the one immediately above it, the nitric acid must be dilute.

Questions

1. Compare ammonia and water with regard to the following:
   a. molecular shape
   b. heat of vaporization
   c. hydrogen bonding
   d. action on Group IA metals

2. Write equations showing the effect of heat on the following:
   a. $(NH_4)_2CO_3$
   b. NH_4Cl
   c. NH_4NO_2
   d. NH_4NO_3
   e. $(NH_4)_2Cr_2O_7$
   f. $NaNO_3$
   g. $AgNO_3$
   h. $Cu(NO_3)_2$

3. What effect will increased pressure have on the following equilibrium? What about increased temperature?

   $$2\ NO_2 \rightleftharpoons N_2O_4 + \text{heat}$$

4. What is a commercial use of hydrazine and its derivatives?

5. Classify solutions of the following as acidic, basic, or neutral:
   a. $NaNO_3$
   b. $NaNO_2$
   c. NH_4NO_3
   d. NH_4Cl
   e. $(NH_4)_2SO_4$

20.2 Phosphorus

Much can be learned about phosphorus by comparing it to the elements adjacent to it in the periodic table. We'll find it especially useful to note the parallels and contrasts with sulfur and nitrogen. Both phosphorus and nitrogen are nonmetals, and both have an outer energy level electron configuration of s^2p^3. In addition, they share common oxidation states and have similar formulas for many compounds. However, there are important differences, as we've come to expect between second and third period elements. As usual we can account for most on the basis of the heavier element's larger size and much smaller electronegativity. For example, the homonuclear bonds in free phosphorus are weaker than in N_2; hence phosphorus is far more reactive. Thus phosphorus reacts more readily with highly electronegative elements (such as oxygen and chlorine) and forms more stable binary compounds with them. However, covalent binary compounds of phosphorus and nonmetals with low electronegativities (for example, hydrogen) are less stable than their nitrogen analogs.

Both elemental sulfur and elemental phosphorus form extended polyatomic molecules, both have several allotropes, and (unlike the elements above them) both are solids. They also often violate the octet rule. There is some similarity in the formulas of the compounds of these elements. For example, consider Na_2SO_4 and Na_3PO_4 or $NaHSO_3$ and Na_2HPO_3.

Unlike nitrogen, phosphorus never occurs in the free state, but phosphorus compounds are abundant and widely distributed. The principal natural form is calcium phosphate, which is a common mineral as well as a major constituent of bones and teeth. Phosphorus compounds are also found in muscles and the nervous system. They appear in the diet in beans, eggs, and milk.

Physical Properties Table 20.3 lists the physical properties of the two most common allotropes of phosphorus. The tetrahedral molecular structure of white phosphorus is shared by liquid phosphorus and by phosphorus vapor at temperatures below 800°C (see Chapter 9). At temperatures above 800°C, the P_4

Table 20.3 The Common Allotropic Forms of Phosphorus

Property	White	Red
Appearance	White waxy solid, turns yellow in air	Red-violet powder
Solubility in water	Insoluble	Insoluble
Solubility in organic solvents	Dissolves in CS_2, ether, benzene, and turpentine	Insoluble
Melting point	44°C	Neither melts nor boils at 1 atm; sublimes at 416°C
Boiling point	280°C	
Specific gravity	1.83	2–2.3
Molecular formula	P_4	Large molecules (P_n)
Kindling temperature	30°C–40°C	About 240°C
Reactivity	Very reactive	Much less reactive than white phosphorus

molecules dissociate into P_2 molecules. Red phosphorus is believed to consist of large interlinked chains and networks of phosphorus atoms. This polymeric form accounts for the insolubility, low vapor pressure, and reduced reactivity of red phosphorus.

White phosphorus can be isolated by condensing, under water, the phosphorus vapors obtained by heating phosphates with carbon and silicon dioxide. We can write the equation for this process as

$$Ca_3(PO_4)_2(s) + 3\ SiO_2(s) + 5\ C(s) \longrightarrow 3\ CaSiO_3(s) + 5\ CO(g) + P_2(g)$$

$$2\ P_2(g) \longrightarrow P_4(g) \longrightarrow P_4(s)$$

There is no transition temperature between white and red phosphorus such as there is between rhombic and monoclinic sulfur. However, if white phosphorus is heated to 400°C in the absence of air and in the presence of an iodine catalyst, solid red phosphorus will condense from the vapor.

Chemical Properties Although red phosphorus is much less reactive than white phosphorus, the reaction products for the two are identical. And as Table 20.4 shows, these products can contain phosphorus in a wide range of oxidation states. As with nitrogen, the most important are -3, $+3$, and $+5$. Among the characteristic reactions of free phosphorus is the reaction with active metals to form ionic phosphides. For example,

$$6\ Mg\ (s) + P_4(s) \longrightarrow 2\ Mg_3P_2(s)$$

The phosphide ion, however, does not exist in aqueous solution since it hydrolyzes completely to form phosphine (PH_3).

Phosphorus also reacts with oxygen, but the conditions necessary to promote the reaction depend greatly on the allotrope. White phosphorus oxidizes slowly in moist air at room temperature, emitting the familiar glow associated with the name of the element. At slightly above room temperature, white phosphorus

Table 20.4 Oxidation Numbers of Phosphorus

Oxidation Number	Example	
-3	PH_3	Phosphine
-2	P_2H_4	Diphosphine
0	P_4	Phosphorus
$+1$	H_3PO_2	Hypophosphorous acid
$+3$	H_3PO_3	Phosphorous acid
	P_4O_6	Tetraphosphorus hexaoxide
$+5$	H_3PO_4	Phosphoric acid
	P_4O_{10}	Tetraphosphorus decaoxide

ignites, making it necessary to store it under water. Red phosphorus, on the other hand, does not ignite unless heated to 240°C; hence it is much easier and safer to work with. Phosphorus also reacts with sulfur and with the halogens. It is oxidized by oxidizing acids, and it reacts with strong solutions of sodium hydroxide. But it will not react with carbon, nitrogen, or hydrogen.

Phosphine and Phosphonium Compounds In a reaction analogous to the preparation of ammonia from metallic nitrides, the colorless, odorous gas phosphine (PH_3) can be prepared by the hydrolysis of phosphides:

$$Mg_3P_2(s) + 6\ H_2O(l) \longrightarrow 3\ Mg(OH)_2(s) + 2\ PH_3(g)$$

A more common laboratory preparation of phosphine, however, involves warming small pieces of white phosphorus in a concentrated solution of sodium hydroxide:

$$3\ NaOH(aq) + 3\ H_2O(l) + P_4(s) \longrightarrow 3\ NaH_2PO_2(aq) + PH_3(g)$$

This is another example of an autooxidation reaction since P_4 is both oxidized and reduced. It also requires special precautions since phosphine ignites spontaneously on contact with air. The products of the combustion are tetraphosphorus decaoxide and water. [Some chemists attribute this spontaneous flammability to the presence of small amounts of diphosphine (P_2H_4), an unstable compound analogous to hydrazine (N_2H_4).]

The phosphine molecule, like the ammonia molecule, is trigonal pyramidal. However, it is only slightly polar since the electronegativities of phosphorus and hydrogen are nearly the same. Because of this smaller polarity, phosphine is less basic than ammonia, is only slightly soluble in water, and does not form hydrogen bonds in the liquid state. In a final analogy to ammonia, phosphine reacts with hydrogen chloride or hydrogen iodide to form phosphonium salts:

$$PH_3(g) + HI(g) \longrightarrow PH_4I(s)$$

Unlike the ammonium ion, however, the phosphonium ion does not exist in aqueous solution since it hydrolyzes completely on contact with water to form insoluble PH_3.

$$PH_4I(s) + H_2O(l) \rightleftharpoons PH_3(g) + H_3O^+ + I^-(aq)$$

Phosphorus Halides Phosphorus combines directly with all the halogens to form trihalides (PX_3). As we might predict from the generalizations in Chapter 7 (and by analogy to PH_3), the molecules of these compounds are trigonal pyramidal. Phosphorus trifluoride is a colorless gas, the trichloride and tribromide are colorless liquids, and the triiodide is a solid. All but the trifluoride are completely and irreversibly hydrolyzed on contact with water:

$$PX_3 + 3\ H_2O(l) \longrightarrow H_3PO_3(aq) + 3\ HX(aq)$$

Because of this hydrolysis, all phosphorus halides produce fumes of the hydrogen halide in moist air.

Unlike nitrogen, with which we've drawn numerous parallels, phosphorus reacts with excess chlorine or bromine to produce pentahalides. The pentachloride (PCl_5) in the liquid or vapor state is a trigonal bipyramidal molecule. As a solid it appears to consist of tetrahedral PCl_4^+ ions and octahedral PCl_6^- ions. The participation of d orbitals in the PCl_5 molecule and the PCl_6^- ion, of course, is the feature that bars nitrogen from forming a similar class of compounds. In the gas phase, PCl_5 dissociates reversibly:

$$PCl_5(g) \rightleftharpoons PCl_3(g) + Cl_2(g)$$

In water it hydrolyzes to form phosphoric acid and hydrogen chloride:

$$PCl_5(s) + 4\ H_2O(l) \longrightarrow H_3PO_4(aq) + 5\ HCl(aq)$$

Phosphorus Oxides The combustion of phosphorus in air results in a mixture of P_4O_6 and P_4O_{10}. We often represent these compounds by their empirical formulas, P_2O_5 and P_2O_3, and refer to them as phosphorus pentaoxide and phosphorus trioxide, respectively. More accurate names are tetraphosphorus hexaoxide and tetraphosphorus decaoxide.

We can derive the structure of the P_4O_6 molecule schematically from that of the P_4 molecule (Figure 20.4) by placing an oxygen atom between each pair of phosphorus atoms in the P_4 tetrahedron. The result appears in Figure 20.5. Four more oxygen atoms, one at each corner of the tetrahedron, produces P_4O_{10} (Figure 20.6). Both oxides are white solids and dissolve in water to form acids:

$$P_4O_6(s) + 6\ H_2O(l) \longrightarrow 4\ H_3PO_3(aq)$$
$$P_4O_{10}(s) + 6\ H_2O(l) \longrightarrow 4\ H_3PO_4(aq)$$

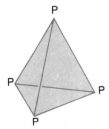

Figure 20.4 The P_4 molecule.

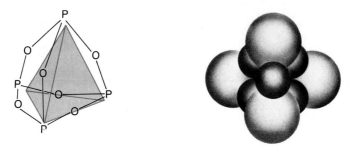

Figure 20.5 The P$_4$O$_6$ molecule.

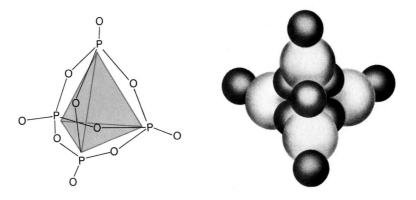

Figure 20.6 The P$_4$O$_{10}$ molecule.

The reaction of tetraphosphorus decaoxide with water is especially vigorous; hence P$_4$O$_{10}$ is often used as a dehydrating agent. Its affinity for water is so great that it will even dehydrate sulfuric acid, which is an excellent dehydrating agent itself:

$$2\ H_2SO_4(l) + P_4O_{10}(s) \longrightarrow 4\ HPO_3(l) + 2\ SO_3(l)$$

Acids of Phosphorus Phosphorus—like sulfur, the element adjacent to it—forms a number of oxygen acids. Table 20.5 is only a partial list. Except for hypophosphorous acid, the *-ous* acids contain phosphorus in a +3 oxidation state, and all the *-ic* acids contain phosphorus in a +5 state. All these acids are weak acids, and only the hydrogen atoms bonded to oxygen are ionizable. Except for hypophosphorous acid, which has no acid anhydride, the compounds in Table 20.5 are hydration products of P$_4$O$_6$ or P$_4$O$_{10}$. Tetraphosphorus hexaoxide is the acid anhydride of the *-ous* acids, and tetraphosphorus decaoxide is the anhydride of the *-ic* acids. The prefixes *meta-*, *pyro-*, and *ortho-* indicate increasing

Table 20.5 Oxyacids of Phosphorus

Oxidation Number of Phosphorus	Formula	Name	Electron Formula	Number of Ionizable Hydrogen Atoms per Molecule
+1	H_3PO_2	Hypophosphorous acid		1
+3	H_3PO_3	Orthophosphorous or phosphorous acid		2
+3	$H_4P_2O_5$	Pyrophosphorous acid		2
+3	$(HPO_2)_n$	Metaphosphorous acid		n
+5	H_3PO_4	Orthophosphoric or phosphoric acid		3
+5	$H_4P_2O_7$	Pyrophosphoric acid		4
+5	$(HPO_3)_n$	Metaphosphoric acid		n

degrees of hydration. Thus

$$P_4O_6(s) + 6\ H_2O(l) \longrightarrow 4\ H_3PO_3(aq) \quad \text{orthophosphorous acid}$$

$$P_4O_6(s) + 4\ H_2O(l) \longrightarrow 2\ H_4P_2O_5(aq) \quad \text{pyrophosphorous acid}$$

$$n\ P_4O_6(s) + 2n\ H_2O(l) \longrightarrow 4\ (HPO_2)_n(aq) \quad \text{metaphosphorous acid}$$

$$P_4O_{10}(s) + 6\ H_2O(l) \longrightarrow 4\ H_3PO_4(aq) \quad \text{orthophosphoric acid}$$

$$P_4O_{10}(s) + 4\ H_2O(l) \longrightarrow 2\ H_4P_2O_7(aq) \quad \text{pyrophosphoric acid}$$

$$n\ P_4O_{10}(s) + 2n\ H_2O(l) \longrightarrow 4\ (HPO_3)_n(aq) \quad \text{metaphosphoric acid}$$

As we might expect, when orthophosphoric acid is heated gently, it loses water to form pyrophosphoric acid, which in turn loses more water at 200°C to produce metaphosphoric acid:

$$2\ H_3PO_4 \longrightarrow H_4P_2O_7 + H_2O$$

$$n\ H_4P_2O_7 \longrightarrow 2\ (HPO_3)_n + n\ H_2O$$

By far the most important acid of phosphorus is orthophosphoric acid or, more simply, *phosphoric acid*. It is a white, water-soluble solid. Commercial concentrated phosphoric acid is an 85 percent aqueous solution. Since it is a triprotic acid, H_3PO_4 ionizes in three stages, the degree of ionization decreasing from the first stage to the third. Likewise, there are three families of salts of phosphoric acid. For example, if we add sodium hydroxide to the acid, the reaction that takes place depends on the quantities of the reactants:

$$H_3PO_4 + NaOH \longrightarrow NaH_2PO_4 + H_2O$$

$$H_3PO_4 + 2\,NaOH \longrightarrow Na_2HPO_4 + 2\,H_2O$$

$$H_3PO_4 + 3\,NaOH \longrightarrow Na_3PO_4 + 3\,H_2O$$

To complicate matters more, each of these salts has three names.

NaH_2PO_4 sodium dihydrogen phosphate
                monosodium phosphate
                primary sodium phosphate

Na_2HPO_4 sodium monohydrogen phosphate
                disodium phosphate
                secondary sodium phosphate

Na_3PO_4 sodium phosphate
                trisodium phosphate
                tertiary sodium phosphate

Since phosphoric acid is a weak acid at all three stages of ionization, each of these sodium salts hydrolyzes in solution to produce hydroxide ions:

$$H_2PO_4^-(aq) + H_2O(l) \rightleftharpoons H_3PO_4(aq) + OH^-(aq)$$

$$HPO_4^{2-}(aq) + H_2O(l) \rightleftharpoons H_2PO_4^-(aq) + OH^-(aq)$$

$$PO_4^{3-}(aq) + H_2O(l) \rightleftharpoons HPO_4^{2-}(aq) + OH^-(aq)$$

At the same time, any $H_2PO_4^-$ and HPO_4^{2-} ions in solution ionize to produce hydrogen ions:

$$H_2PO_4^-(aq) \rightleftharpoons H^+(aq) + HPO_4^{2-}(aq)$$

$$HPO_4^{2-}(aq) \rightleftharpoons H^+(aq) + PO_4^{3-}(aq)$$

As we've seen before (Chapter 15), this complex situation makes it difficult to predict beforehand whether a salt solution will be acidic or basic. The answer depends on several ionization constants. Solutions of sodium dihydrogen phosphate are slightly acidic, which means that the ionization of the $H_2PO_4^-$ ion predominates over its hydrolysis. On the other hand, monohydrogen phosphate solutions are slightly basic, and phosphate solutions are strongly basic. In each of these last two, dominant hydrolysis reactions are responsible for the basicity.

None of the acids of phosphorus, nor their salts, function as oxidizing agents; however, phosphorous acid and hypophosphorous acid, as well as their salts, can serve as reducing agents.

Phosphorus, like nitrogen, is an essential nutrient for plants and animals. Fertilizers thus often contain soluble compounds rich in phosphorus. One of the most common is monocalcium orthophosphate [$Ca(H_2PO_4)_2$], which is a derivative of calcium phosphate, a cheap but insoluble source of phosphorus. If $Ca_3(PO_4)_2$ is heated with phosphoric acid, the result is $Ca(H_2PO_4)_2$, which when used as a fertilizer goes by the uninformative name of triple superphosphate. A fertilizer rich in both nitrogen and phosphorus is $(NH_4)_2HPO_4$, which can be made by dissolving ammonia in phosphoric acid.

An unfortunate side effect of these phosphate fertilizers, as well as phosphate water softeners, has been the pollution of many lakes and streams. Soluble phosphorus- and nitrogen-containing compounds promote the growth of aquatic plants; the plants are innocuous enough while they are alive but when they die, decomposition processes seriously deplete the supply of oxygen dissolved in the water, thus endangering fish and other marine life. The properties of nitrogen and phosphorus are compared in Table 20.6.

Example 20.2 How could you distinguish among the following sodium salts?

NaI Na_2SO_4 Na_2SO_3 Na_2CO_3
Na_3PO_4 $NaNO_3$ $Na_2S_2O_3$ $NaCl$

Solution

a. All of these salts are white, crystalline, water-soluble salts; hence neither color nor solubility is a helpful clue.

b. We can, however, test the aqueous solutions with litmus paper to divide the salts into two categories:

Neutral: NaI Na_2SO_4 $NaNO_3$ $NaCl$

Basic: Na_2SO_3 Na_2CO_3 Na_3PO_4 $Na_2S_2O_3$

c. Of the salts forming neutral solutions, only $NaNO_3$ decomposes when heated:

$$2\ NaNO_3(s) \longrightarrow 2\ NaNO_2(s) + O_2(g)$$

d. We can then distinguish NaI from Na_2SO_4 and $NaCl$ by treating it with concentrated sulfuric acid, producing free iodine:

$$5\ H_2SO_4(l) + 8\ NaI(s) \longrightarrow 4\ I_2(s) + H_2S(g) + 4\ Na_2SO_4(aq) + 4\ H_2O(l)$$

Table 20.6 A Comparison of Nitrogen and Phosphorus

Atomic radius (nm)	N (0.074) P (0.110)	As in other groups of the periodic table, there is a large difference in size between the atoms of the first two elements of Group VA.
Electronegativity	N (3.0) P (2.1)	Again, the first two elements in a group differ greatly in electronegativity.
Electron configuration		All elements in Group VA have an electron configuration ending in s^2p^3.
Pi bonding		Their small size allows nitrogen atoms to approach one another closely enough to form strong pi bonds. The same is not true of the larger phosphorus atoms.
Ionic valence and oxidation numbers		Both nitrogen and phosphorus react with active metals to form -3 ions, but both the nitride and the phosphide ion hydrolyze in solution. The common oxidation numbers of both elements are -3, $+3$, and $+5$. Binary compounds of these two elements often have analogous formulas, for example, PH_3 and NH_3, N_2H_4 and P_2H_4, Mg_3N_2 and Mg_3P_2. However, ternary compounds of nitrogen and phosphorus frequently do not have analogous formulas because only nitrogen forms strong pi bonds with oxygen.
Allotropism		Phosphorus exists in several allotropic forms; nitrogen does not exhibit allotropism.
Homonuclear bonds		Red phosphorus is characterized by extensive homonuclear bonding, but in compounds neither element readily forms homonuclear bonds.
Hydrogen bonding		The small size and high electronegativity of the nitrogen atom make it a likely candidate for hydrogen bonding. Thus ammonia exhibits extensive hydrogen bonding. The phosphorus atom is too large to participate in hydrogen bonding.
Acid-base properties of compounds		Since phosphorus is much less electronegative than nitrogen, the phosphine molecule is less polar than the ammonia molecule and it is a much weaker base. For the same reason, nitric acid is a strong acid, whereas all the phosphorus oxyacids are weak.
Redox reactions		Nitric acid is a strong oxidizing agent. Nitrous acid and its salts can be either reducing or oxidizing agents. Acids of phosphorus never function as oxidizing agents, but phosphorous acid and hypophosphorous acid, along with their salts, can serve as reducing agents.

e. Sodium sulfate and sodium chloride can then be distinguished from each other by treating aqueous solutions of each with a $BaCl_2$ solution. Barium sulfate will precipitate from the Na_2SO_4 solution.

f. Finally, we can distinguish the salts in the basic group from one another by adding hydrochloric acid to a sample of each:

$$Na_2SO_3(s) + 2\ HCl(aq) \longrightarrow 2\ NaCl(aq) + H_2O(l) + SO_2(g)$$
$$Na_2CO_3(s) + 2\ HCl(aq) \longrightarrow 2\ NaCl(aq) + H_2O(l) + CO_2(g)$$
$$Na_3PO_4(s) + 3\ HCl(aq) \longrightarrow 3\ NaCl(aq) + H_3PO_4(aq)$$
$$8\ Na_2S_2O_3(s) + 16\ HCl(aq) \longrightarrow 16\ NaCl(aq) + S_8(s) + 8\ SO_2(g) + 8\ H_2O(l)$$

The Na_3PO_4 forms a stable weak acid (H_3PO_4) and evolves no gas; sodium carbonate produces unstable carbonic acid, which decomposes to produce an odorless gas (CO_2). Both Na_2SO_3 and $Na_2S_2O_3$ form unstable sulfurous acid, which then decomposes to evolve foul-smelling SO_2. The $Na_2S_2O_3$ also produces free sulfur.

Questions

6. Contrast the chemical properties of
   a. nitrogen and phosphorus.   b. sulfur and phosphorus.
   c. white phosphorus and red phosphorus.

7. Write equations for the derivations of the phosphoric acids from their anhydride.

8. Compare the structures of P_4, P_4O_6, and P_4O_{10}.

9. Write equations for the successive stages of ionization of orthophosphoric acid.

10. Write formulas for all the sodium salts of orthophosphoric acid.

20.3 Carbon

In each periodic group we've looked at so far, there has been a noticeable decrease in nonmetallic properties as we've gone from top to bottom. Group IVA is no different; indeed, the trend is more pronounced here than elsewhere. Both carbon and silicon are nonmetals and will be considered in this chapter, whereas the two elements at the bottom of the group, tin and lead, are both metals.

Though it is certainly among the earth's most important elements, carbon is not among the earth's most abundant. (See Table 21.1.) It is, nevertheless, common and widespread in both free and combined forms. The most abundant source of free carbon is coal, which contains compounds of carbon in addition to the free element. Diamond and graphite, the two allotropes of carbon (Chapter 9), are much less abundant forms of free carbon. In the combined form, carbon

IA																	0
H	IIA											IIIA	IVA	VA	VIA	VIIA	He
Li	Be											B	C	N	O	F	Ne
Na	Mg	IIIB	IVB	VB	VIB	VIIB	⎯	VIIIB	⎯	IB	IIB	Al	Si	P	S	Cl	Ar
K	Ca	Sc	Ti	V	Cr	Mn	Fe	Co	Ni	Cu	Zn	Ga	Ge	As	Se	Br	Kr
Rb	Sr	Y	Zr	Nb	Mo	Tc	Ru	Rh	Pd	Ag	Cd	In	Sn	Sb	Te	I	Xe
Cs	Ba	La	Hf	Ta	W	Re	Os	Ir	Pt	Au	Hg	Tl	Pb	Bi	Po	At	Rn
Fr	Ra	Ac															

Ce	Pr	Nd	Pm	Sm	Eu	Gd	Tb	Dy	Ho	Er	Tm	Yb	Lu
Th	Pa	U	Np	Pu	Am	Cm	Bk	Cf	Es	Fm	Md	No	Lr

can be found in all living matter as well as in substances like coal and petroleum that have their origins in living matter. Carbon is also found in the metal carbonates, the most common being calcium carbonate, which we know as limestone.

The carbonates are examples of *inorganic* compounds of carbon, whereas most carbon compounds derived from living matter are regarded as *organic* compounds. The distinction between these two classes is not always a sharp one, but we can make some useful generalizations. Organic compounds usually burn, contain more than one carbon atom per molecule, contain hydrogen, and are nonelectrolytes. Inorganic compounds generally do not burn, usually contain only one carbon atom per molecule or formula unit, may or may not contain hydrogen, and are often electrolytes. In this section we'll concern ourselves only with inorganic compounds; we'll discuss organic chemistry in Chapter 22.

Chemical Properties The outer energy level electron configuration of carbon, as well as the other elements in Group IVA, is s^2p^2. In inorganic compounds, carbon usually has an oxidation number of $+4$, but it can also exhibit oxidation numbers of -2 or $+2$.

Elemental carbon is not highly reactive (witness the stability of diamond), but it does participate in several characteristic reactions at high temperatures. For example, as we saw in Chapter 19, hot carbon reacts with steam to produce carbon monoxide and hydrogen:

$$C(s) + H_2O(g) \longrightarrow CO(g) + H_2(g)$$

In similar reactions, hot carbon reduces other nonmetallic oxides such as carbon dioxide and silicon dioxide:

$$CO_2(g) + C(s) \longrightarrow 2\ CO(g)$$
$$SiO_2(s) + C(s) \longrightarrow Si(s) + CO_2(g)$$

Hot carbon will also reduce the oxides of metals above aluminum in Table 17.3. The products are the metal and either carbon dioxide or carbon monoxide:

$$2\ Fe_2O_3(s) + 3\ C(s) \longrightarrow 4\ Fe(s) + 3\ CO_2(g)$$
$$CuO(s) + C(s) \longrightarrow Cu(s) + CO(g)$$

But attempts to reduce the oxides of more active metals produce metal carbides:

$$2\ Al_2O_3(s) + 9\ C(s) \longrightarrow 6\ CO(s) + Al_4C_3(s)$$

Carbon Monoxide Carbon monoxide can be produced by (1) burning carbon in a limited supply of oxygen or at high temperatures, (2) burning carbon-containing compounds in a limited supply of oxygen, (3) reducing carbon dioxide, or (4) reducing metal oxides with carbon. The second of these sources—specifically the incomplete combustion of gasoline in automobile engines—is responsible for most of the carbon monoxide in urban areas. In downtown parking areas and in traffic jams, the concentration of carbon monoxide can, in fact, reach dangerously high levels. Inhaled carbon monoxide reacts with the iron atom in hemoglobin (the oxygen-carrying substance in the blood) to form a stable complex called carboxyhemoglobin. This complexed hemoglobin molecule thus becomes unavailable for transporting oxygen through the body. Some other gases, such as hydrogen cyanide and hydrogen sulfide, form similar complexes, and like carbon monoxide they can be deadly. Unlike carbon monoxide, however, they have strong odors and can be easily detected.

Carbon monoxide is a colorless, odorless, tasteless gas with a density of 1.250 g/liter at STP. It is slightly soluble in water (3.5 ml/100 g at 0°C and 1.00 atm). The carbon monoxide molecule is isoelectronic with the nitrogen molecule:

$$:N:::N:\quad \text{and} \quad :C:::O:$$

and thus we might expect similar chemical properties. However, our expectations are borne out only at ordinary temperatures, where both molecules are relatively inert. At higher temperatures carbon monoxide is much more reactive. The reason is the difference in electronegativities between carbon and oxygen, a difference that polarizes and thus weakens the triple bond in carbon monoxide.

Carbon monoxide burns readily to form carbon dioxide; thus it is an excellent reducing agent for metallic oxides:

$$CuO(s) + CO(g) \longrightarrow Cu(s) + CO_2(g)$$
$$FeO(s) + CO(g) \longrightarrow Fe(s) + CO_2(g)$$

It also reacts with chlorine to form carbonyl chloride, a compound more commonly called phosgene:

$$Cl_2(g) + CO(g) \rightleftharpoons COCl_2(l)$$

Phosgene is a highly toxic liquid used in the manufacture of pesticides and herbicides. In reactions similar to the formation of carboxyhemoglobin, carbon monoxide also reacts with some free metals by sharing the electrons on carbon to form metal carbonyls such as $Fe(CO)_5$, $Ni(CO)_4$, and $Cr(CO)_6$. We shall discuss these compounds in more detail in Chapter 24.

Carbon Dioxide Carbon dioxide usually constitutes about 0.03 percent of the earth's atmosphere, although this percentage varies a good deal. In a crowded room it can be as high as 1 percent. Carbon dioxide is a colorless, odorless gas about one and one-half times as dense as air. It is soluble in water (90.1 ml/100 g at 20°C and 1.00 atm). Carbon dioxide does not exist as a liquid at 1 atm, but it freezes at −78.5°C. At 5.2 atm it melts at −56.6°C.

A number of simple reactions produce carbon dioxide: (1) burning carbon or carbon-containing materials in a plentiful supply of oxygen, (2) reducing oxides with carbon or carbon monoxide, (3) heating metal carbonates (other than those of Group IA metals), (4) heating bicarbonates, and (5) treating carbonates or bicarbonates with acids. This last reaction is often used in the laboratory as a source of carbon dioxide. Both carbonates and bicarbonates hydrolyze in the acid solution, forming carbonic acid:

$$CO_3^{2-}(aq) + H^+(aq) \rightleftharpoons HCO_3^-(aq)$$
$$HCO_3^-(aq) + H^+(aq) \rightleftharpoons H_2CO_3(aq)$$

The unstable carbonic acid then decomposes to evolve carbon dioxide:

$$H_2CO_3(aq) \longrightarrow H_2O(l) + CO_2(g)$$

Although carbon dioxide is more than 70 percent oxygen, it is not a very good oxidizing agent. It will, however, oxidize hot carbon and metals above aluminum in the activity series.

Carbonic Acid, Carbonates, and Bicarbonates We have already had several encounters with the unstable acid, carbonic acid. An equilibrium exists between it and water and carbon dioxide:

$$CO_2(g) + H_2O(l) \rightleftharpoons H_2CO_3(aq)$$

Like all weak, diprotic acids, the first hydrogen atom of H_2CO_3 ionizes more strongly than the second, but even the first ionizes only slightly. Furthermore, the existence of the H_2CO_3 molecule itself is very tenuous; recent work suggests that the equilibrium above lies well to the left. Carbonic acid exists only in solution.

Since it contains two ionizable protons, carbonic acid forms two series of salts—the bicarbonates and the carbonates. All bicarbonates decompose when heated to form carbonates, water, and carbon dioxide:

$$2\ NaHCO_3(s) \longrightarrow Na_2CO_3(s) + H_2O(g) + CO_2(g)$$

Bicarbonates both ionize and hydrolyze:

Ionization: $HCO_3^-(aq) \rightleftharpoons H^+(aq) + CO_3^{2-}(aq)$

Hydrolysis: $HCO_3^-(aq) + H_2O(l) \rightleftharpoons H_2CO_3(aq) + OH^-(aq)$

But since the hydrolysis predominates, solutions of bicarbonates of strong bases, such as $NaHCO_3$, are basic.

Carbonates, except those of the Group IA metals, evolve carbon dioxide when heated:

$$CuCO_3(s) \longrightarrow CuO(s) + CO_2(g)$$

Or if the metal oxide, too, is unstable:

$$2\ Ag_2CO_3(s) \longrightarrow 2\ CO_2(g) + 4\ Ag(s) + O_2(g)$$

Only ammonium carbonate and the carbonates of the Group IA metals are soluble in water. Since the carbonate ion hydrolyzes, the soluble metal carbonates form basic solutions. All carbonates are soluble in acid since the high H^+ ion concentration promotes the formation of bicarbonate ions.

Questions

11. What physical property do carbon dioxide and iodine have in common?

12. What is the effect of heat on each of the following?
   a. $CaCO_3$ b. K_2CO_3 c. $KHCO_3$
   d. $HgCO_3$ e. $CuCO_3$ f. Ag_2CO_3

13. What molecules and ions are present in an aqueous solution of carbon dioxide? Account for each.

20.4 Silicon

Like carbon, silicon often forms sp^3-hybridized orbitals, and a number of simple silicon compounds have the characteristic tetrahedral shape. It is no surprise, then, that there are many silicon analogs of common carbon compounds. However, considering the proximity of the two elements, the differences between carbon and silicon are striking. We'll look at examples of two important differences.

[Periodic table highlighting Si]

First, the silicon atom is larger than the carbon atom and it has *d* orbitals available for bonding; thus as we've seen in other periodic groups, this third period element can accommodate more than eight electrons in its valence shell. As a result the two colorless, volatile, nonpolar liquids, silicon tetrachloride and carbon tetrachloride, exhibit important differences. For example, unlike carbon tetrachloride, silicon tetrachloride hydrolyzes readily. A *d* orbital of the silicon atom first accepts an unshared pair of electrons from the water molecule, forming a trigonal bipyramidal intermediate with five bonds. This intermediate then loses a molecule of hydrogen chloride:

$$\text{:}\overset{..}{\underset{..}{Cl}}\text{:}\overset{\overset{\text{:}\overset{..}{\underset{..}{Cl}}\text{:}}{|}}{\underset{\underset{\text{:}\overset{..}{\underset{..}{Cl}}\text{:}}{|}}{Si}}\text{:}\overset{..}{\underset{..}{Cl}}\text{:} \; + \; \xrightarrow{H_2O} \; \text{:}\overset{..}{\underset{..}{Cl}}\text{:}\overset{\overset{\text{:}\overset{..}{\underset{..}{Cl}}\text{:}\;\text{:}\overset{..}{\underset{..}{Cl}}\text{:}}{}}{\underset{\underset{\text{:}\overset{..}{\underset{..}{Cl}}\text{:}\;H}{}}{Si}}\text{:}\overset{..}{\underset{..}{O}}\text{:}H \; \xrightarrow{-HCl} \; \text{:}\overset{..}{\underset{..}{Cl}}\text{:}\overset{\overset{\text{:}\overset{..}{\underset{..}{Cl}}\text{:}}{|}}{\underset{\underset{\text{:}\overset{..}{\underset{..}{Cl}}\text{:}}{|}}{Si}}\text{:}\overset{..}{\underset{..}{O}}\text{:}H$$

This process is repeated until all the chlorine atoms have been replaced. The resulting $Si(OH)_4$ then loses two water molecules, and SiO_2 precipitates. We can thus write the overall equation as

$$SiCl_4(l) + 2\,H_2O(l) \longrightarrow SiO_2(s) + 4\,HCl(g)$$

A second major difference between carbon and silicon is the strength of homonuclear bonds: a weak silicon-silicon bond in contrast to a stable carbon-carbon bond. This important difference is emphasized by comparing the hydrides of these two elements. The hydrides of carbon (hydrocarbons) are extremely numerous, and there seems to be no limit to the length of the carbon chains that can be synthesized. In contrast, only a few hydrides of silicon (silanes) have been isolated, the longest of which contains only six silicon atoms (Si_6H_{14}). Furthermore, the hydrocarbons frequently contain stable double or triple bonds, whereas silicon invariably forms only single bonds. Finally, the silanes are much more reactive than the hydrocarbons: They burn spontaneously at room temperature and are readily hydrolyzed. The chemistry of silicon, unlike the chemistry of carbon, is therefore primarily concerned with the silicon-oxygen bond. This bond, unlike the silicon-silicon or silicon-hydrogen bonds, is very strong (see Chapter 9).

20.5 The Noble Gases

In 1785 Sir Henry Cavendish (1731–1810), one of chemistry's great experimentalists, made an observation that could have led to the discovery of the Group 0 elements. He had subjected samples of air to electrical sparks, causing the nitrogen and oxygen to react:

$$N_2(g) + 2\,O_2(g) \longrightarrow 2\,NO_2(g)$$

He removed the resulting NO_2 from the reaction container by dissolving it in a sodium hydroxide solution:

$$2\,NO_2(g) + 2\,OH^-(aq) \longrightarrow NO_3^-(aq) + NO_2^-(aq) + H_2O(l)$$

Cavendish then added more oxygen to react with the remaining nitrogen, again passed a spark through the mixture, and again dissolved the product in aqueous sodium hydroxide. Finally he removed the excess oxygen from the reaction vessel by passing the remaining gas over hot copper:

$$2\ Cu(s)\ +\ O_2(g) \longrightarrow 2\ CuO(s)$$

Cavendish recorded that a small bubble of gas remained whose volume was about 1 percent of the original volume of air, but he did not speculate about the nature of the residual gas. Presumably he, as well as others who studied his data, attributed it to experimental error. In retrospect we know that Cavendish was a very careful, skillful worker and that the small amount of gas that refused to react should not have been overlooked.

Almost a hundred years later, in 1868, when helium was discovered in the solar spectrum, no thought was given to a new group of elements since there was no evidence that helium was to be found on earth. Even Mendeleev left no room for the noble gases in his periodic table of 1869. But in 1894 a pair of experiments confirmed Cavendish's earlier observations and established the existence of a new family of elements.

In that year another Englishman, Lord Rayleigh (1842–1919), observed that the gas that remained after he removed the water, carbon dioxide, and oxygen from air was slightly more dense than the pure nitrogen isolated from ammonia. Suspecting that this slight difference was significant, one of Lord Rayleigh's co-workers, William Ramsey (1852–1916), performed a second experiment similar to Cavendish's. He allowed the gas isolated from air to react with hot magnesium:

$$3\ Mg\ +\ N_2 \longrightarrow Mg_3N_2$$

Like Cavendish, Ramsey found that a small amount of gas remained. With a spectroscope, a tool unavailable to Cavendish, Ramsey showed that this residue was a mixture of undiscovered gases. These elements have since been referred to as the inert gases, the rare gases, or the noble gases. The first name has fallen into disfavor with many, since chemists have now prepared a number of compounds containing these elements. We shall use the third name, which might be thought of as a less emphatic way of describing their unreactive properties. Some of the physical properties of the noble gases appear in Table 20.7.

Until the 1960s only a few unstable compounds of the Group 0 elements had been reported in the scientific literature. And since these compounds existed only under extreme conditions, little attention was paid to them. In 1933 Linus Pauling had predicted that fluorides of xenon and krypton should be stable, but nobody had successfully prepared them. But in 1962 Neil Bartlett exploited the oxidizing power of platinum hexafluoride (PtF_6) to synthesize the first stable compound of the noble gases. After preparing O_2PtF_6, which consists of O_2^+ and PtF_6^- ions, Bartlett surmised that if PtF_6 was a powerful enough oxidizing agent to remove electrons from O_2, it might also react with xenon.

Table 20.7 Properties of the Noble Gases

Property	Helium	Neon	Argon	Krypton	Xenon	Radon
Atomic number	2	10	18	36	54	86
Atomic weight	4.0	20.18	39.95	83.8	131.3	222
Melting point (°C)	−272.1 (25 atm)	−248.6	−189.4	−156.6	−111.5	−71
Boiling point (°C)	−269.0	−246.4	−185.9	−152.9	−107.1	−65
Critical temperature (°C)	−267.9	−228.7	−122.3	−63.8	−16.6	105
Critical pressure (atm)	2.26	26.9	48.3	54.3	57.6	62.4
Atmospheric concentration (ppm)	5.2	18	9340	1.1	0.09	6×10^{-14}

A few months later he prepared the yellow crystalline $XePtF_6$ in a reaction between xenon and PtF_6 at room temperature. This compound consists of Xe^+ and PtF_6^- ions and is called xenon hexafluoroplatinate(V).

Within a few months chemists at Argonne National Laboratory had prepared the first stable binary compound of a noble gas by heating a mixture of xenon and fluorine:

$$Xe(g) + 2 F_2(g) \longrightarrow XeF_4(s)$$

Since then, several other compounds of xenon have been prepared, and compounds of krypton and radon have been reported.

The discovery of these noble gas compounds came as a surprise to most scientists. Chemists had become so comfortable with the octet rule that they had accepted a few unsuccessful attempts to prepare compounds of the Group 0 elements as a confirmation of their intuition. That there are few safe assumptions is a lesson that scientists must learn and relearn. Once the noble gas compounds were prepared, however, their structure was quickly explained in terms of expanded outer energy levels (see Chapter 7). Table 20.8 lists the properties of some noble gas compounds.

Table 20.8 Some Properties of Noble Gas Compounds

Compound	Properties
$XePtF_6$	Yellow ionic crystals.
XeF_2	Colorless crystals that melt at 140°C; linear molecules; stable in dry air; reacts with water.
XeF_4	Colorless crystals that melt at 114°C; square planar molecules.
XeF_6	Colorless crystals that melt at 48°C; distorted octahedral molecules; stable in dry air; reacts with water.

Questions and Problems

20.1 Nitrogen

14. Write the electron formula for each of the following, indicating all resonance forms:
 a. NH_3
 b. N_2H_4
 c. ammonium sulfate
 d. nitrite ion
 e. nitrate ion
 f. nitrous oxide
 g. nitric oxide
 h. nitrogen dioxide

15. Describe the shapes of the following molecules and ions:
 a. SO_2
 b. H_2S
 c. SO_3^{2-}
 d. NO_2^-
 e. NO_3^-
 f. $S_2O_3^{2-}$

16. List the ions and molecules present in aqueous ammonia.

17. Make a list of molecules and ions that contain a nitrogen-nitrogen bond.

18. Balance the following:
 a. $HNO_3 + H_2S \longrightarrow S_8 + NO_2 + H_2O$
 b. $HNO_3 + HCl \longrightarrow Cl_2 + NO_2 + H_2O$
 c. $HNO_3 + CuS \longrightarrow S_8 + Cu(NO_3)_2 + NO + H_2O$
 d. $HNO_3 + Zn \longrightarrow Zn(NO_3)_2 + NH_4NO_3 + H_2O$
 e. $HNO_3 + Mg \longrightarrow Mg(NO_3)_2 + N_2O + H_2O$
 f. $HNO_3 + C \longrightarrow CO_2 + NO_2 + H_2O$
 g. $HNO_3 + Fe \longrightarrow N_2 + Fe(NO_3)_3 + H_2O$

19. Show how the following conversions might be made:
 a. $NO \longrightarrow NO_2 \longrightarrow HNO_3 \longrightarrow NaNO_3 \longrightarrow NaNO_2 \longrightarrow HNO_2$
 b. $NH_3 \longrightarrow NH_4NO_3 \longrightarrow N_2O \longrightarrow N_2 \longrightarrow Mg_3N_2 \longrightarrow NH_3$

20. Write equations for the reactions in which the oxidation number of nitrogen changes from
 a. $+3$ to 0
 b. $+5$ to $+1$
 c. $+4$ to $+5$
 d. -3 to $+1$
 e. -3 to $+2$
 f. $+1$ to 0

21. Write the equations for the industrial preparation of nitric acid from nitrogen.

22. How would you distinguish between members of the following pairs of compounds?
 a. $NaNO_3$ and $NaNO_2$
 b. NH_4Cl and $NaCl$
 c. NH_4NO_2 and $NaNO_2$
 d. NH_4NO_3 and NH_4NO_2
 e. NH_4NO_2 and NH_4Cl
 f. $NaNO_3$ and $Pb(NO_3)_2$

23. How would you distinguish between the four colorless gases, NH_3, N_2O, NO, and N_2?

Discussion Starters

24. Explain or account for the following:
 a. Sodium nitrate solutions are neutral, but sodium nitrite solutions are basic.
 b. There are no precipitation tests for the nitrate ion.
 c. Nitric oxide is paramagnetic.
 d. Nitrogen is a relatively inactive element in spite of its high electronegativity.

e. Nitrous acid is both an oxidizing agent and a reducing agent, but nitric acid is only an oxidizing agent.

f. Nitric acid is not used for the laboratory preparation of hydrogen.

g. Although ammonia has a lower molecular weight than hydrogen sulfide, it boils at a higher temperature.

h. A binary compound of nitrogen and iodine is properly referred to as a nitride, but a binary compound of nitrogen and fluorine is referred to as a fluoride.

i. A burning magnesium ribbon continues to burn when thrust into a container of nitrogen.

*j. When ammonium chloride is heated at the closed end of a horizontal tube with a piece of moist red litmus paper at the open end, the litmus first turns blue and then red.

k. Hydrochloric acid cannot be substituted for sulfuric acid in the laboratory preparation of nitric acid.

l. NH_4NO_2 is not commonly found in the laboratory.

m. Copper dissolves in nitric acid but not in hydrochloric acid.

n. Nitric oxide, a colorless gas, turns brown when mixed with air. The brown color disappears when the gas is cooled.

*o. When lead is treated with concentrated nitric acid, a brown gas is evolved. If an attempt is made to collect this gas by water displacement, the gas that is collected is colorless. The colorless gas turns brown if exposed to air.

p. Ammonia is a better Lewis base than nitrogen trifluoride.

20.2 Phosphorus

25. Which of the following formulas represent sodium salts of acids of phosphorus?
    a. Na_2HPO_2 b. $Na_4P_2O_7$ c. Na_2HPO_4
    d. Na_3PO_2 e. $Na_4P_2O_5$ f. $Na_2H_2P_2O_7$
    g. NaH_2PO_2 h. Na_2HPO_3 i. NaH_2PO_3

26. Name the following:
    a. Na_3PO_4 b. Na_2HPO_4 c. NaH_2PO_4 d. H_3PO_4 e. $H_4P_2O_7$
    f. $(HPO_3)_3$ g. H_3PO_2 h. H_3PO_3 i. $H_4P_2O_5$ j. $(HPO_2)_n$
    k. PH_4Cl l. P_4O_6 m. P_4O_{10}

27. Write the electron formulas for each of the following:
    a. PH_3 b. PH_4^+ c. P_2H_4 d. PCl_5 e. PCl_3
    f. $P_2O_7^{4-}$ g. H_3PO_2 h. H_3PO_3 i. H_3PO_4 j. Mg_3P_2

*28. List at least four ions of phosphorus that exist as solids but not in solution.

29. Write equations illustrating the reaction of water with each of the following:
    a. Mg_3P_2 b. PCl_3 c. PCl_5 d. P_4O_6
    e. P_4O_{10} f. $H_2PO_4^-$ g. HPO_4^{2-} h. PO_4^{3-}

30. Write equations for reactions illustrating the following changes in the oxidation number of phosphorus:
    a. 0 to -3 b. 0 to $+1$ c. 0 to $+3$ d. 0 to $+5$ e. -3 to $+5$

31. Write the equations for the reactions that take place between the following pairs of substances:
    a. P_4 and HNO_3 b. Mg and P_4 c. P_4 and excess air

32. How could the following changes be effected?

$$P_4 \longrightarrow P^{3-} \longrightarrow PH_3 \longrightarrow PH_4I$$

$$P_4 \longrightarrow P_4O_{10} \longrightarrow H_3PO_4 \longrightarrow NaH_2PO_4$$

$$P_4 \longrightarrow PCl_3 \longrightarrow H_3PO_3 \longrightarrow Na_2HPO_3$$

33. How would you separate a mixture of white and red phosphorus?

34. How can the following be prepared from white phosphorus?
    a. red phosphorus   b. PH_3 c. H_3PO_4 d. H_3PO_3 e. H_3PO_2

Discussion Starters

35. Explain or account for the following:
    a. H_3PO_4, H_3PO_3, and H_3PO_2 are triprotic, diprotic, and monoprotic acids, respectively.
    b. Solutions of NaH_2PO_4 are acidic, but solutions of Na_2HPO_4 are alkaline.
    c. P_4O_{10} is sometimes referred to as phosphorus pentaoxide.
    *d. There is a CN^- ion but no CP^- ion.
    e. There is no PO_3^- ion analogous to the NO_3^- ion.
    f. Phosphorus chlorides fume when exposed to moist air.
    g. Phosphorus pentachloride is a saltlike substance.
    h. White phosphorus is more soluble in organic solvents, has a higher vapor pressure, and is chemically more active than red phosphorus.

20.3 Carbon

36. Write electron formulas for the following ions or molecules:
    a. carbon monoxide   b. carbon tetrachloride
    c. carbonate ion   d. bicarbonate ion

37. Write electron formulas for three pairs of isoelectronic ions or molecules. At least one member of each pair should contain carbon.

38. What is the spatial arrangement of each of the following?
    a. CO_2 b. CO_3^{2-} c. CO d. CCl_4

39. Write the net ionic equations for the reaction of sulfuric acid with each of the following:
    a. Na_2CO_3 b. $BaCO_3$ c. $NaHCO_3$

40. Write equations for the complete combustion of C_2H_6, CO, and CH_4.

41. Write the equations for the reactions of carbon with the following substances. Assume any temperature necessary to get the reaction to proceed.
    a. carbon dioxide   b. calcium oxide   c. aluminum oxide
    d. steam   e. copper(II) oxide

42. Write equations for the reactions of water with the following:
    a. carbonate ion   b. bicarbonate ion   c. carbon dioxide

43. Write equations for reactions in which the oxidation number of carbon changes from
   a. 0 to +4   b. 0 to −4   c. 0 to +2
   d. +2 to +4   e. 0 to −1   f. +4 to +2

44. Indicate how you would make the following conversions:
   a. $CO_2 \longrightarrow H_2CO_3 \longrightarrow NaHCO_3 \longrightarrow Na_2CO_3 \longrightarrow CaCO_3$
   b. $CaCO_3 \longrightarrow CO_2 \longrightarrow CO \longrightarrow COCl_2$

*45. How can you prove that diamond consists entirely of carbon?

46. Write five equations illustrating *different* ways of forming carbon dioxide.

Discussion Starters

47. Explain or account for the following:
   a. The bicarbonate ion both ionizes and hydrolyzes.
   b. Na_2CO_3 solutions are basic.
   c. An attempt to obtain sodium hydrogen carbonate crystals by boiling off the water from an aqueous solution would not be successful.
   *d. Carbon dioxide could not be used effectively to extinguish a magnesium fire.
   e. Carbon dioxide is a nonpolar molecule even though the difference in electronegativity between carbon and oxygen is 1.0.
   f. Automobile exhaust contains carbon monoxide.
   g. The molecules of CO and N_2 are isoelectronic; however, carbon monoxide boils at a slightly higher temperature than nitrogen.

20.4 Silicon

Discussion Starters

48. Explain or account for the following:
   a. Silicon compounds analogous to C_2H_4 and C_2H_2 do not exist.
   b. Silicon tetrachloride hydrolyzes readily, but carbon tetrachloride does not.

20.5 The Noble Gases

49. Write the electron formulas and indicate the type of hybrid bonding, electron-pair geometry, and molecular shape for the three xenon fluorides. Refer to Chapter 7 if necessary.

50. What are the names for the three xenon fluorides?

*51. So far no compounds of helium or neon have been prepared. Why do you think compounds of these elements would be less stable than the compounds of xenon?

21 The Chemistry of Metals

Our systematic discussion of the nonmetals in Chapters 19 and 20 has reflected the regular trends in nonmetallic properties, trends that culminate in fluorine—unarguably the most nonmetallic of the elements. Aside from the nonmetallic corner of the periodic table, however, things are not so simple. To illustrate, we might try to select the most metallic element. The properties that distinguish metals from nonmetals are luster, thermal and electrical conductivity, high melting points and densities, malleability and ductility, low ionization energies, and large (negative) electrode potentials. Is the most metallic element then cesium, which has the lowest ionization energy, or lithium, which has the most negative electrode potential? Or is it silver, the most lustrous and most conductive element; osmium, the densest; gold, the most malleable; or tungsten, with the highest melting point? To confuse matters more, not only are these properties found in different elements, but also, as Figure 21.1 shows, these elements are scattered widely about the periodic table.

Part of the reason that the metals form such a complex class is that there are so many of them. Three-fourths of the elements are metals; only about 30 are noble gases, nonmetals, or metalloids. But a more important reason is that the fundamental atomic properties—atomic radii, nuclear charge, electron configuration—that account for chemical and physical properties contribute in complicated and sometimes contradictory ways to what we've called metallic properties. Thus the trend in ionization energy, which depends on little else than atomic radius, may be very different from the trend in electrical conductivity, which depends on the structure of the metallic crystal (which depends in turn on more than one atomic property). One of the missions of this chapter will be to sort out these trends as we study the different groups of metals: the alkali metals and alkaline earths; the representative metals of Groups IIIA, IVA, and VA; the transition metals; and the rare earths.

21.1 Abundance, Availability, and Usefulness

Table 21.1 lists the most abundant elements in the earth's crust and thus demonstrates that there is little correlation between familiarity and abundance. Except for aluminum (actually a metalloid) and iron, most of the metals near the top

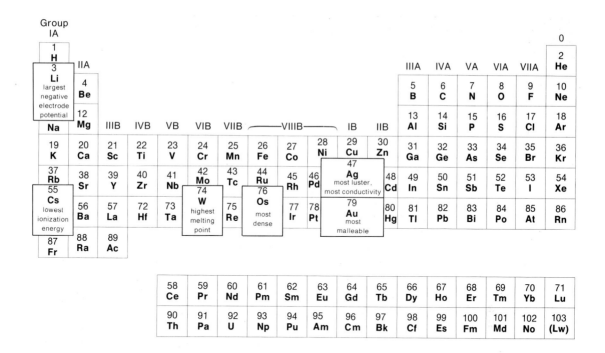

Figure 21.1 Periodic table locations of elements with extreme values of various metallic properties.

Table 21.1 Relative Abundances of the Elements in the Earth's Crust Expressed as Percent by Weight

Element	%	Element	%
O	46.6	Cr	0.020
Si	27.7	V	0.015
Al	8.13	Zn	0.013
Fe	5.00	Ni	0.008
Ca	3.63	Cu	0.007
Na	2.83	W	0.007
K	2.59	Li	0.007
Mg	2.09	N	0.005
Ti	0.44	Ce	0.005
H	0.14	Sn	0.004
P	0.12	Y	0.003
Mn	0.10	Nd	0.002
S	0.052	Nb	0.002
C	0.032	Co	0.002
Cl	0.031	La	0.002
Rb	0.031	Pb	0.002
F	0.030	Ga	0.002
Sr	0.030	Mo	0.002
Ba	0.025	Th	0.002
Zr	0.022		

The Chemistry of Metals

of the list are familiar only to chemists. And the elements common in everyday life, zinc, nickel, chromium, tin, appear well down the list; gold and silver don't appear at all. Instead of necessarily being the abundant ones, the familiar metals are instead the ones that are both useful and easily mined and refined. Aluminum, for example, despite its useful properties and abundance, was practically unknown a century ago, since no economical method was known for refining it. Like titanium, another valuable lightweight structural material, aluminum is most commonly found combined with silicon and oxygen in the silicates (Chapter 9). These stable compounds, resistant to erosion and decay, are the most abundant compounds in the earth's crust. But their very stability makes it impractical, even today, to extract aluminum and titanium from them. Instead, we depend on far less abundant compounds as ores for these important elements: *bauxite*, $Al_2O_3 \cdot nH_2O$, and *cryolite*, Na_3AlF_6, for aluminum, and *rutile*, TiO_2, and *ilmenite*, $FeTiO_3$, for titanium. In contrast, lead is easy to refine; thus despite there being no more than 1 gram in each 60 kilograms of the earth's crust, it has been a familiar metal for centuries. It was used by the ancient Romans in their water systems and it is from the Latin name for lead, *plumbum*, that we get our word *plumbing*. Silver and gold, with calculated abundances of 1×10^{-5} percent and 5×10^{-7} percent, respectively, were known for centuries before the Romans. Both of these rare elements are most commonly found in uncombined form.

Questions

1. Why are "abundance" and "availability" of a metal not the same thing?

2. Add the percent abundances of the first three elements in Table 21.1. In what forms are these elements principally found in the earth's crust?

21.2 The Metals of Groups IA and IIA

Since the right-hand side of the periodic table contains the most nonmetallic elements, we might expect Groups IA and IIA to contain the most metallic. If we consider only chemical properties, we shall not be disappointed. Physically, however, the alkali and alkaline-earth metals are not nearly so metallic as we might expect.

Chemical Properties As we first discussed in Chapter 5, within any period or group the ionization energy decreases with increasing atomic radius. Thus within any period the alkali metal (Group IA) has the lowest ionization energy, and within any group the ionization energy decreases from top to bottom. Cesium then has the lowest ionization energy of all. (We omit from consideration the element francium, since it does not exist in nature and since all its isotopes undergo rapid radioactive decay. The work that has been done on francium, however, suggests that its properties are those we would predict from its position in the periodic table.)

IA																	0
H	IIA											IIIA	IVA	VA	VIA	VIIA	He
Li	Be											B	C	N	O	F	Ne
Na	Mg	IIIB	IVB	VB	VIB	VIIB	⎡——VIIIB—— ⎤			IB	IIB	Al	Si	P	S	Cl	Ar
K	Ca	Sc	Ti	V	Cr	Mn	Fe	Co	Ni	Cu	Zn	Ga	Ge	As	Se	Br	Kr
Rb	Sr	Y	Zr	Nb	Mo	Tc	Ru	Rh	Pd	Ag	Cd	In	Sn	Sb	Te	I	Xe
Cs	Ba	La	Hf	Ta	W	Re	Os	Ir	Pt	Au	Hg	Tl	Pb	Bi	Po	At	Rn
Fr	Ra	Ac															

Ce	Pr	Nd	Pm	Sm	Eu	Gd	Tb	Dy	Ho	Er	Tm	Yb	Lu
Th	Pa	U	Np	Pu	Am	Cm	Bk	Cf	Es	Fm	Md	No	Lr

A second chemical measure of metallic character is the reduction potential. By this criterion (Table 18.2), fluorine is again the most nonmetallic element, and lithium, with the largest (negative) reduction potential, is the most metallic. We might pause here to note that fluorine has both the highest ionization energy and the most positive reduction potential, whereas cesium and lithium must share the honors at the other end of the scale. We have discussed the reason before. The ionization energy is the energy needed to remove a valence electron from a neutral gaseous atom. Thus the largest of the alkali metals has the lowest ionization energy. Reduction potentials, on the other hand, are related to the energies needed to add an electron to the atom or ion *in aqueous solution*. Here we cannot overlook the role of water. Reduction potentials depend somewhat on hydration energies, since ion-water bonds must be broken before electrons can be restored to the ions. The interaction between the small lithium ion and water is strong; that between cesium and water is weak.

Low ionization energies and large reduction potentials may seem impractical measures of metallic character, but they produce two important properties as direct consequences: the tendency to form ionic compounds and high reactivity. All compounds of the alkali and alkaline earth metals (remember that beryllium is a metalloid, not a metal) are essentially ionic, conducting electric current both in aqueous solution and as molten salts. And all the metals of Groups IA and IIA are highly reactive. The IA metals react with the oxygen of the air or the moisture in the air so that they must be stored under oil or kerosene. Group IIA metals oxidize much more slowly in air. In fact, magnesium is very stable in air, apparently because of a thin, tightly adhering oxide layer that forms and then prevents further reaction. All metals of both groups react with water to produce hydrogen:

$$2\ Na(s) + 2\ H_2O(l) \longrightarrow 2\ NaOH(aq) + H_2(g)$$

$$Ca(s) + 2\ H_2O(l) \longrightarrow Ca(OH)_2(aq) + H_2(g)$$

Reactions of the Group IA metals range from vigorous to explosive; Group IIA metals react at a more controllable rate. This difference between the two groups is not always obvious from thermodynamic data. For example, a comparison of free energies suggests that the reaction of calcium with water is more favorable than the reaction of sodium with water. But rate is another factor, and the sodium reaction is many times faster. One reason for this rate difference is the disparity between the melting points of IA and IIA metals. The heat of the reaction between sodium and water melts the sodium, causing it to spread out in a thin layer on the water. Thus the sodium presents more reactive surface to the water than does a lump of solid calcium.

Physical Properties The strength of the bonds in a metallic crystal depend largely on the number of delocalized electrons contributed by each atom. Physical properties that depend on strong metallic bonds are therefore not so obvious among the Group IA and IIA metals as among others. The bond in Group IA metals, especially, is relatively weak, and the metals are quite soft. A lump of sodium metal can easily be cut into pieces with an ordinary pocketknife. Group IA metals are also the lightest (least dense) of the metals—lithium is the lightest of all—and have relatively low melting points—cesium has the lowest melting point of all the metals except mercury. Group IIA metals, because they have twice as many valence electrons per atom, are harder and denser and have higher melting points than the IA metals. For both groups, densities increase with increasing atomic number, except that potassium and calcium are less dense than the metals immediately above them. All metals of both groups exhibit the characteristic luster of metals.

Hydrates The chemical formulas for many solid salts include a specific number of water molecules. Thus gypsum, the natural mineral form of calcium sulfate, has the formula $CaSO_4 \cdot 2\ H_2O$, and plaster of Paris has the formula $CaSO_4 \cdot \frac{1}{2}\ H_2O$. Despite the fact that the water molecules are set off from the rest of the formula, there is always a consistent stoichiometric ratio among all the atoms of these compounds. Gypsum, whatever its source, has a ratio of one atom of sulfur, six atoms of oxygen, and four atoms of hydrogen for each atom of calcium. However, the bonds between the water molecules and the ions of the salt are weaker than true ionic bonds. (We'll say more about these molecule-ion bonds in Chapter 24.) Thus most hydrates can be *dehydrated* by increasing the temperature.

Since hydrates consist of water molecules as well as ions, most of them have measurable vapor pressures at ordinary temperatures. And of course the vapor pressures increase with temperature. Certain hydrates, such as $Na_2CO_3 \cdot 10\ H_2O$, spontaneously give up water when exposed to dry air:

$$Na_2CO_3 \cdot 10\ H_2O(s) \longrightarrow Na_2CO_3(s) + 10\ H_2O(g)$$

This is an example of an **efflorescent** hydrate. Other hydrates, as well as many anhydrous salts, absorb water from the atmosphere to form hydrates or higher hydrates. Such substances are said to be **hygroscopic.** If a solid continues to absorb water until it dissolves, it is **deliquescent** as well as hygroscopic. An example is anhydrous calcium chloride, which absorbs water from the atmosphere to form $CaCl_2 \cdot 6\ H_2O$ and then continues to absorb water until the hydrate dissolves. Anhydrous copper(II) sulfate merely absorbs water from the air until it forms $CuSO_4 \cdot 5\ H_2O$. Thus it is hygroscopic but not deliquescent.

The tendency to form stable hydrates depends mainly on a large hydration energy. Since hydration energies typically increase as the ionic charge increases and as the ionic radius decreases, they are small for most Group IA and IIA cations. Most compounds of these cations thus have little tendency to form crystals containing water molecules. Notable exceptions are lithium and magnesium, whose hydration energies are relatively large.

Binary Oxygen Compounds When burned under ordinary conditions, the Group IA and IIA metals yield products as shown in Table 21.2. Each of the binary oxygen compounds of Group IA metals hydrolyzes completely and irreversibly:

$$M_2O(s) + H_2O \longrightarrow 2\ M^+(aq) + 2\ OH^-(aq)$$

$$M_2O_2(s) + 2\ H_2O \longrightarrow 2\ M^+(aq) + 2\ OH^-(aq) + H_2O_2(aq)$$

$$2\ MO_2(s) + 2\ H_2O \longrightarrow O_2(g) + 2\ M^+(aq) + 2\ OH^-(aq) + H_2O_2(aq)$$

(Temperature must be carefully controlled for the latter two reactions to prevent the decomposition of H_2O_2 to H_2O and O_2.) Oxides of Group IIA metals react with water to form hydroxides. The hydroxides of all Group IA and IIA metals except $Mg(OH)_2$ are strong bases.

Table 21.2 Products Obtained When the IA and IIA Metals are Burned Under Ordinary Conditions

	IA	IIA	
oxide	Li_2O		
peroxide	Na_2O_2	MgO	
	KO_2	CaO	oxides
superoxides	RbO_2	SrO	
	CsO_2	BaO	

Colors of Compounds The alkali metal and alkaline earth cations are colorless in solution but can readily be made to emit visible energy. According to the quantum theory, atoms must emit energy in discrete amounts when electrons drop from one energy level to a lower one (Chapter 5). If a compound containing a Group IA or IIA cation is heated (excited) with a Bunsen burner, it emits light with a characteristic color when the excited ions relax to the ground state. These colors, listed in Table 21.3, form the basis of identification by flame tests (Figure 21.2).

Table 21.3 Flame Colors

Ion	Color
Li^+	Carmine (vivid purplish red)
Na^+	Yellow
K^+, Rb^+, Cs^+	Lavender
Ca^{2+}	Yellow-red
Sr^{2+}	Scarlet (vivid orangish red)
Ba^{2+}	Yellow-green

Natural Occurrence Compounds of the Group IA metals are so soluble that they are seldom found in large deposits. Wherever rainwater can reach them, they are carried into streams, rivers, lakes, and finally the oceans. Solid deposits of the Group IIA metal compounds are more common; yet they too can be found in seawater in relatively high concentrations. Because of their solubility, compounds of both groups have been deposited where large bodies of water have evaporated. Such is the origin of the great salt flats in Utah and of large underground salt deposits.

Uses These metals are so reactive that most find only limited use as free metals. For example, lithium and magnesium are used in the metal industry as *scavengers*, reacting with and removing traces of oxygen and nitrogen from molten metals. Sodium is common in the chemistry laboratory, where it is a useful reagent, especially for organic reactions. Cesium is used in photoelectric cells because the cesium atom can be ionized with visible light (see "The Photoelectric Effect and the Garage Door," in Chapter 5). Radium, the rarest of the alkaline earths, is used as a source of radiation in radiotherapy. Only one of the metals in these groups finds wide use. Magnesium is a lightweight structural material in airplanes; it is the fine wire in many photoflashes and in flares; it is an important laboratory reagent in organic chemistry; it is used for anodic protection against corrosion (see p. 670), and it can be added to aluminum to form several different alloys (p. 657), a familiar one being *duralumin*.

The compounds of these reactive elements are a different story from the rarely used elements. They find wide applicability, and there is room here to mention only a few of their most important uses. Sodium chloride is common table salt. It is used as a seasoning and a preservative, and in industry it is the

Figure 21.2 Performing a flame test.

starting point for almost all sodium compounds as well as for hydrochloric acid and chlorine. (We discussed the electrolysis of sodium chloride solutions as well as molten sodium chloride in Chapter 18.) Sodium hydroxide finds wide use in industry: It is used in the production of textiles and paper, in soap and detergent manufacturing, in the purification of fats, and in the refining of petroleum. Sodium carbonate is also important in industry, about half of it being used in the production of glass. Sodium bicarbonate is used in baking powders and antacids; sodium nitrate in fertilizers. Potassium superoxide is used in inhalators as described in Chapter 19. Potassium nitrate is the oxidizing agent for many explosives; it is the chief constituent of black gunpowder. Among the compounds of the alkaline earths, calcium chloride is among the most common: It is the salt that is sprinkled on icy streets. Calcium hydroxide is an ingredient of builder's mortar. Plaster of Paris is calcium sulfate in the hemihydrate form $(CaSO_4 \cdot \frac{1}{2} H_2O)$. When water is added, it "sets" and becomes permanent in the gypsum form

Table 21.4 A Comparison of Group IA and Group IIA Metals

Property	IA	IIA	Notes
Atomic radius (nm)	Li (0.152) Na (0.185) K (0.231) Rb (0.246) Cs (0.263)	Mg (0.160) Ca (0.197) Sr (0.215) Ba (0.217)	Atomic radii increase with atomic number in each group, and each Group IIA atom has a smaller radius than the Group IA atom of the same period.
Ionization energy (kcal/g at. wt)	Li (0.060) Na (0.095) K (0.133) Rb (0.148) Cs (0.169)	Mg (0.065) Ca (0.099) Sr (0.113) Ba (0.135) Ra (0.140)	Ionization energies decrease from top to bottom in a group but increase from left to right in a period.
Ionic radius (nm)	Li^+ (0.060) Na^+ (0.095) K^+ (0.133) Rb^+ (0.148) Cs^+ (0.169)	Mg^{2+} (0.065) Ca^{2+} (0.099) Sr^{2+} (0.113) Ba^{2+} (0.135) Ra^{2+} (0.140)	Ionic radii follow the same trends as atomic radii.
Hydration energy	\multicolumn{3}{l}{Hydration energies are low for all ions in these two groups, except for the very small Li^+ and Mg^{2+} ions.}	
Reduction potential $E°$ (volts)	Li (−3.045) Na (−2.711) K (−2.924) Rb (−2.925) Cs (−2.923)	Mg (−2.375) Ca (−2.76) Sr (−2.89) Ba (−2.90) Ra (−2.92)	Generally, large (negative) reduction potentials reflect low ionization energies. Lithium upsets the trend because of its uncharacteristically large hydration energy.
Density (g/cm³)	Li (0.534) Na (0.97) K (0.86) Rb (1.53) Ca (1.88)	Mg (1.74) Ca (1.54) Sr (2.6) Ba (3.51)	Densities generally increase with increasing atomic number, except for the step from the third to the fourth period. Here the change in atomic weight is not relatively so great as the change in atomic radius.
Melting point (°C)	Li (179) Na (97.9) K (63.7) Rb (38.5) Cs (28.5)	Mg (651) Ca (851) Sr (770) Ba (710)	Melting points reflect the strength of the metal bond. Group IIA metals have much higher melting points than the Group IA metals. Cesium has the lowest melting point of any metal except mercury.
Compounds	\multicolumn{3}{l}{Most compounds of IA metals are very soluble in water. Compounds of IIA metals are somewhat less soluble, because greater ionic charges and smaller ionic radii give them higher lattice energies. Group IA hydroxides are all soluble, strong bases. Group IIA hydroxides, except for $Ba(OH)_2$, are insoluble, but all except $Mg(OH)_2$ are strong bases.}	
Abundance	\multicolumn{3}{l}{Ca, Na, K, and Mg are among the 10 most abundant elements in the earth's crust. Rb, Sr, and Ba are among the next 10, and Li is twenty-seventh. Cs and Ra are much rarer, and Fr does not occur naturally since all of its isotopes undergo rapid radioactive decay.}	

($CaSO_4 \cdot 2\ H_2O$). Strontium and barium nitrates are often used to provide the colors in fireworks. And as described in Chapter 16, barium sulfate is used as an opaque intestinal coating for diagnostic X rays. The properties of Group IA and Group IIA metals are compared in Table 21.4.

Questions

3. Explain why cesium has the lowest ionization potential but not the largest (negative) reduction potential of the elements.

4. Which term is the opposite of efflorescent: hygroscopic or deliquescent?

5. Draw electron formulas for Li_2O, Na_2O_2, and KO_2.

21.3 Transition Elements

Chemically, if not physically, the metals of Groups IA and IIA are ideal metals. If on the other hand we apply physical criteria, it is among the transition metals that we find the most metallic elements. In general, the transition metals are harder, denser, and more malleable and have higher melting points than the alkali and alkaline earth metals, but they are less chemically reactive, and many compounds of transition metals have considerable covalent character. In addition, whereas the hydroxides of all IA and IIA metals except magnesium are strong bases, the hydroxides of the transition elements are either weakly basic or amphoteric.

The transition elements are the elements that occupy the B groups of the periodic table. Each is characterized by filled or partly filled d orbitals. Atoms and ions of the transition metals are generally smaller than those of the IA and IIA metals in the same period. Figure 9.34 shows this trend in ionic radii among metals that form strongly ionic compounds. The transition metals are also uniformly less chemically active than the IA and IIA metals; consequently, they appear well above the bottom of Table 18.2. In fact, some of the transition metals are so unreactive that they have been accorded the name **noble metals.** This category includes gold, silver, mercury, palladium, platinum, rhodium, ruthenium, iridium, and osmium. Except for mercury, none of the noble metals react with oxygen when heated in air, and most are usually found uncombined in nature. In contrast, the **base metals,** which include all metals except the noble ones, oxidize when heated in air.

Although silver will not react with molecular oxygen alone, it does form a dark tarnish when exposed to sulfur compounds and oxygen:

$$4\ Ag(s) + 2\ H_2S(g) + O_2(g) \longrightarrow 2\ Ag_2S(s) + 2\ H_2O(l)$$

The hydrogen sulfide necessary for this reaction is often present in the atmosphere as the result of the combustion of sulfur-containing fuels. A less obvious source of combined sulfur is eggs, which contain an abundance of sulfur-containing proteins. As any housekeeper can attest, nothing tarnishes silverware faster than eggs! Unlike silver, gold does not tarnish, nor does it dissolve in nitric acid as

IA																	0
H	IIA											IIIA	IVA	VA	VIA	VIIA	He
Li	Be											B	C	N	O	F	Ne
Na	Mg	IIIB	IVB	VB	VIB	VIIB	⎯	VIIIB	⎯	IB	IIB	Al	Si	P	S	Cl	Ar
K	Ca	Sc	Ti	V	Cr	Mn	Fe	Co	Ni	Cu	Zn	Ga	Ge	As	Se	Br	Kr
Rb	Sr	Y	Zr	Nb	Mo	Tc	Ru	Rh	Pd	Ag	Cd	In	Sn	Sb	Te	I	Xe
Cs	Ba	La	Hf	Ta	W	Re	Os	Ir	Pt	Au	Hg	Tl	Pb	Bi	Po	At	Rn
Fr	Ra	Ac															

Ce	Pr	Nd	Pm	Sm	Eu	Gd	Tb	Dy	Ho	Er	Tm	Yb	Lu
Th	Pa	U	Np	Pu	Am	Cm	Bk	Cf	Es	Fm	Md	No	Lr

silver does. On the other hand, gold will dissolve in *aqua regia* (three parts concentrated hydrochloric acid and one part concentrated nitric acid), whereas silver will not. In this case the silver is protected by a tightly adhering silver chloride coating.

Many transition elements form ions by losing only s electrons. However, the elements in Group IIIB and some of the other transition metals lose one d electron along with two s electrons to form simple ions. The $+3$ ions of the IIIB elements thus have a noble gas configuration, whereas those of Ti, V, Cr, Fe, and Co are d^n ions (see Chapter 6).

Most of the transition metals can be found in a variety of oxidation states, but most of these states do not reflect true ionic charges. Instead they result from bonds that have some covalent character. In some cases the transition metal shares its d electrons to form covalent bonds, in others empty d orbitals share electrons contributed by the anions. The larger the anion and the more loosely held its electrons, the more likely this latter type of bond. A good example of an essentially covalent compound is $HgCl_2$. Although the Hg^{2+} ion exists in $Hg(NO_3)_2$ and in other compounds and although its radius has been measured, it does not exist in $HgCl_2$. Evidence for this includes the observations that $HgCl_2$ is three times as soluble in alcohol as it is in water and that it is also soluble in covalent diethyl ether. Furthermore, the conductivity of a solution containing Hg^{2+} ions decreases as chloride ions are added, demonstrating that the formation of $HgCl_2$ molecules reduces the number of ions in solution. Finally, $HgCl_2$ does not produce hydrogen chloride when treated with concentrated H_2SO_4 nor chlorine when treated with HNO_3. If we look at the chlorides of the metals above mercury in Group IIB, we find an example of the general observation that covalent character increases from top to bottom in a group of transition metals. Zinc chloride is essentially ionic in aqueous solution, and $CdCl_2$ ionizes partially.

Color A compound appears colored if it absorbs energy in the visible range. The light reflected or transmitted then lacks certain wavelengths and thus takes on a characteristic color. Ions of the representative elements are colorless, since the electronic energy levels are so far apart that the electrons can't be excited by visible light. Additional energy—sometimes a Bunsen burner is sufficient—is needed to excite them. It is important to realize that although visible energy is often not enough to excite the electrons, light sometimes appears when excited electrons return to their ground state energy levels. The flame tests summarized in Table 21.2 are examples. The explanation is that the electrons return to the ground state in several smaller steps, emitting energy with each step. The gaps between the energy levels of *d* electrons are much smaller, thus many transition metal ions absorb visible light. For example, the energy of red light raises a $3d$ electron in the cobalt ion to a higher energy level. The red light is absorbed and the remaining transmitted light gives the ion its blue color.

Alloys The physical properties we associate with useful metals—strength, hardness, resistance to corrosion, tensile strength, ductility, malleability, conductivity, and high melting points—can often be enhanced by producing alloys. A notable exception is electrical conductivity, which diminishes markedly in copper if even a trace of impurity is present. An alloy is formed by combining two or more elements into a material that has metallic properties. Generally, alloys are prepared by melting together two or more metals—although alloys can also contain nonmetals—then allowing the homogeneous mixture to cool and solidify. We can divide the resulting materials into three classes:

1. *Simple mixtures* The crystals of two or more insoluble metals can form an intimate though heterogeneous mixture. Tin and lead in solder form this type of alloy.

2. *Solid solutions* Solid solutions can be of two kinds. In a *substitutional solid*, such as those formed of zinc with copper and of chromium with nickel, ions of one metal replace the ions of another in its crystal lattice. In an *interstitial solid*, smaller atoms, such as carbon, fit into the spaces between the ions of a metal lattice.

3. *Intermetallic compounds* Some metals combine in stoichiometric ratios to form such unlikely compounds as Ag_3Al and Cu_5Zn_8.

The properties of an alloy are rarely the simple sums or averages of the properties of the constituent elements. Steel, that most useful and most widely used of all metallic substances, known for its strength and hardness, is an alloy of iron and carbon. Pure iron is a soft metal, and carbon, in the form of graphite, is crumbly and weak; yet an interstitial solid of iron containing 0.5 percent carbon is hard enough to be used for railway tracks. Likewise the ancient alloys brass (copper and zinc) and bronze (copper and tin) are far sturdier than the elements they comprise.

Although the reasons for the differences between the properties of alloys and those of their component metals are almost as numerous as the alloys themselves, it is worthwhile to take a closer look at the example of brass, an alloy of copper, with a face-centered cubic lattice, and zinc, with a hexagonal lattice. When small amounts of zinc are added to copper, a substitutional solid is formed, with zinc ions replacing the similarly sized copper ions in the face-centered lattice. Since the intruding ions are not exactly the same size as the copper ions, however, the copper lattice becomes increasingly distorted as more zinc is added. This distorted crystal resists deformation; thus brass is harder than copper. When the proportion of zinc reaches 36 percent, the structure of the lattice changes from face-centered to body-centered, and the properties of the alloy change abruptly. As the proportion of zinc continues to increase, the hardness of the alloy increases even more dramatically. Then a third, a fourth, and finally a fifth phase of the alloy appear. Thus this alloy system is fairly complex, though others are even more so. Copper and nickel, on the other hand, with atoms of similar size and identical lattice structures, form a continuous range of solid solutions from 100 percent copper to 100 percent nickel. One nickel-copper alloy is Monel, an extremely corrosion-resistant material.

Iron and Steel Iron and steel, the *ferrous metals*, are produced in much larger quantities than all the nonferrous metals put together. Steel especially, which is produced at a rate of over 700 million tons annually, has become part of almost every activity in our lives. There are various steels, each an alloy of iron with other elements. A small percentage of carbon dramatically increases the hardness of steel, as does magnesium. Nickel and chromium increase steel's resistance to corrosion, and tungsten or molybdenum can be added to produce steel tools that hold their edges at high temperature. Common stainless steel contains carbon, chromium, and nickel.

Steel production begins, of course, with the mining of iron ore. The principal ores are *magnetite* (Fe_3O_4) and *hematite* (Fe_2O_3). Occasionally carbonate and sulfide ores are used, and recent advances in extraction methods have permitted the use of *taconite*, a low-grade oxide ore. The ore is then reduced in a blast furnace by subjecting the ore, coke, and a limestone flux to a blast of oxygen-enriched air. The result is molten *pig iron*, which contains many impurities, including up to 4 percent carbon.

The next step is the *basic oxygen furnace*. This furnace may be computer controlled, with the emission spectrum of the steel being continuously analyzed. A jet of very pure oxygen is directed into the molten pig iron, burning away impurities which combine with the limestone slag. This modern process is both rapid and efficient, taking less than one-half hour to complete and, as a result, is replacing older refining methods. The final step in the production of steel is to add the other elements that comprise the alloy being made.

Halides Many of the halides of transition metals are among the most important and most common compounds, and we must devote a fair share of attention to them. We can begin by tracing a few trends in covalency among the transition metal halides. For example, within a periodic group of transition metals, covalent character usually increases as the size of the metal ion increases. We have already noted that in Group IIB, $ZnCl_2$ is ionic, $CdCl_2$ is partially covalent, and $HgCl_2$ is covalent. Within a period, however, covalency increases with the charge of the cation, regardless of the trend in ionic radius. Thus KCl, with the largest cation in the fourth period, is essentially ionic: $CaCl_2$ and $ScCl_3$ with smaller cations are progressively less ionic; and $TiCl_4$ is essentially covalent. As the charge on the cation increases, the polarization of the anion increases; that is, the electrons on the anion become more distorted from their normal positions by the attractive charge of the cation. The greater the distortion, the more the covalent character. Indeed we can look at distortion as a sort of sharing. An additional factor in this series of fourth period chlorides is the involvement of d electrons in the bonding of $ScCl_3$ and $TiCl_4$.

The size of the halide ion is another factor in determining the amount of covalence in a metal halide. Larger halide ions are more easily polarized than smaller ones. Thus covalence increases in the series AgCl, AgBr, AgI—a fact manifest in the colors of these compounds. As the electrons on the anions are polarized, the electronic energy levels are shifted so that the gaps among them become smaller. Therefore, although the bromide and iodide ions are normally colorless, they are able to absorb visible light when they are polarized as in AgBr and AgI. Thus AgCl is white, but AgBr is very pale yellow and AgI is light yellow. Indeed, if exposed to light long enough, all three of these compounds will absorb visible and ultraviolet light, which can cause the transfer of an electron from the deformed halide ion to the small silver ion. The result is free silver and free halogen atoms. Thus all these compounds darken when exposed to light because of the accumulation of free silver.

Some transition metals form more than one halide. In these cases, the higher the oxidation state of the metal, the more covalent the compound. Thus chromium(II) chloride, iron(II) chloride, and copper(I) chloride are essentially ionic, whereas chromium(III) chloride, iron(III) chloride, and copper(II) chloride are covalent. In fact, because of their covalent character, it is difficult to represent the solid-state structures of copper(II) chloride and iron(III) chloride with simple formulas. Solid $CuCl_2$ actually consists of chainlike molecules in which each copper atom is surrounded by four chlorine atoms:

$$\left[Cu \begin{array}{c} ..\ddot{C}l.. \\ ..\ddot{C}l.. \end{array} Cu \begin{array}{c} ..\ddot{C}l.. \\ ..\ddot{C}l.. \end{array} Cu \begin{array}{c} ..\ddot{C}l.. \\ ..\ddot{C}l.. \end{array} \right]_n$$

Iron(III) chloride is commonly represented by its empirical formula, $FeCl_3$, but this simple molecule probably exists only in the vapor state at high temperatures. At lower temperatures the vapor consists mostly of Fe_2Cl_6 molecules, and the solid consists of huge sheetlike structures in which each iron atom is surrounded by six chlorine atoms and each chlorine atom is bonded to two iron atoms. The anhydrous gas is easily prepared by passing chlorine over heated iron. The product sublimes readily and is soluble in many common organic solvents.

Sidelight
Photography

The instability of silver halides when exposed to light is the key to the photographic process. The four steps in the process are

Exposure Ordinary photographic film consists of a mixture of very small silver halide crystals suspended in a gelatin that is spread on transparent cellulose acetate. When exposed to light the silver ions in this *emulsion* are "sensitized," that is, made susceptible to reduction.

Development The exposed film is immersed in a solution of a mild reducing agent. In the Polaroid (a trademark) cameras the reducing agent is automatically spread on the film. In either case the result is silver atoms:

$$Ag^+ + e^- \text{ (reducing agent)} \longrightarrow Ag$$

The number of silver ions that are reduced is directly related to the amount of light that struck the area when the film was exposed. The more light, the more silver halide ions that were sensitized, and the blacker that area of the film becomes.

Fixation After development the film is treated with a thiosulfate ion solution (called *hypo*) which removes the unreduced silver halide:

$$Ag^+ + 2\,S_2O_3^{2-} \longrightarrow [Ag(S_2O$$

(More is said about this kind of complex ion reaction in Chapter 24.) The result of fixation is the finished negative.

Printing As we've all seen and as the name suggests, the negative represents light objects as dark and dark objects as light. The reason is that the brightly lighted objects have produced the greatest amounts of free black silver. Thus the three steps above must be repeated, this time using the negative as the object of the photograph and using photographic paper instead of photographic film. The result is the print.

Among the most interesting of the transition metal compounds are the halides of mercury. The mercury atom has an electron configuration ending with $5d^{10}6s^2$. We therefore expect the mercury(I) halides to be paramagnetic since an unpaired s electron should remain in each Hg^+ ion. However, mercury(I) compounds show no paramagnetism. The widely accepted reason is that there is no Hg^+ ion but rather a Hg_2^{2+} ion, which is a mercury(II) ion and a mercury atom bonded together by a covalent linkage. Thus we write Hg_2Cl_2, Hg_2Br_2, and Hg_2I_2 but never HgCl, HgBr, or HgI.

In many ways the mercury(I) ion (Hg_2^{2+}) is similar to the silver ion. For example, most of the mercury(I) salts, like most of the silver salts, are insoluble. Among the common salts, only the nitrate and perchlorate compounds of these ions are soluble in water. The mercury(I) ion, like the silver ion, is colorless in solution but forms some colored solids (Hg_2Br_2 and Hg_2I_2) with colorless anions. Also, the solubilities of the mercury(I) halides, like the silver halides, decrease as the atomic number of the halide increases. Finally, mercury(I) chloride, bromide, and iodide darken when exposed to light. [But whereas the silver halides produce free silver and the free halogen, the mercury(I) halides yield mercury(II) halides and mercury.]

A solution containing the mercury(I) ion can be prepared by the reaction of excess mercury with cold dilute nitric acid. The halides can then be precipitated from this solution. However, attempts to precipitate other mercury(I) compounds such as the sulfides produce only free mercury and the mercury(II) compound:

$$Hg_2^{2+}(aq) + S^{2-}(aq) \longrightarrow Hg(l) + HgS(s)$$

One way to look at this reaction is to see it as the precipitation of mercury(I) sulfide and then its subsequent autooxidation. However, it seems more likely to be a direct reaction between sulfide ions and Hg(II) ions, which are in equilibrium with Hg(I) ions:

$$Hg_2^{2+} \rightleftharpoons Hg + Hg^{2+}$$

There are a number of other reactions of the same sort. For example;

$$Hg_2^{2+}(aq) + 2\,OH^-(aq)\ [\text{pH above 4}] \longrightarrow Hg(l) + HgO(s) + H_2O(l)$$

$$Hg_2^{2+}(aq) + CO_3^{2-}(aq) \longrightarrow Hg(l) + HgO(s) + CO_2(g)$$

$$Hg_2Cl_2(s) + 2\,NH_3(aq) \longrightarrow Hg(l) + HgNH_2Cl(s) + NH_4^+(aq) + Cl^-(aq)$$

Transition Metal Anions Metals generally form cations rather than anions and bases rather than acids, but some of the transition metals can be found in the anions of acids. Common examples are the chromate (CrO_4^{2-}) and dichromate ($Cr_2O_7^{2-}$) ions and the permanganate (MnO_4^-) and manganate (MnO_4^{2-}) ions. The metal in each of these ions is in one of its highest oxidation states, an observation that illustrates a generalization we made in Chapter 14, namely, that when an element exhibits more than one oxidation state, it is less metallic and its oxides are less basic in the higher oxidation states. Thus in its +2 oxidation state manganese appears in MnO and Mn(OH)$_2$. In its +7 oxidation state it is found in MnO_4^-, Mn_2O_7, and the acid $HMnO_4$. As we might expect, anions containing these metals in high oxidation states are strong oxidizing agents.

Among the anions we are using as examples, the two manganese anions contain the metal in different oxidation states, but the two chromium ions do not. Chromates and dichromates can be transformed reversibly from one to the other by adjusting the hydrogen ion concentration. The first hydrogen atom in chromic acid (H_2CrO_4) ionizes fairly strongly, but the second ionizes only weakly. Hence, when hydrogen ions are added to a chromate solution, bichromate ions ($HCrO_4^-$) are formed. The bichromate ions in turn condense into dichromate ions. We can represent both of these changes by the following equilibria:

$$2\,H^+ + 2\left[\begin{array}{c}:\ddot{O}:^{2-}\\:\ddot{O}:\ddot{C}r:\ddot{O}:\\:\ddot{O}:\end{array}\right] \rightleftharpoons 2\left[\begin{array}{c}:\ddot{O}:^-\\H:\ddot{O}:\ddot{C}r:\ddot{O}:\\:\ddot{O}:\end{array}\right] \rightleftharpoons \left[\begin{array}{c}:\ddot{O}:\quad:\ddot{O}:^{2-}\\:\ddot{O}:\ddot{C}r:\ddot{O}:\ddot{C}r:\ddot{O}:\\:\ddot{O}:\quad:\ddot{O}:\end{array}\right] + H_2O$$

Incidentally, a second illustration of decreasing metallic character with higher oxidation states is the amphoteric nature of some transition metal hydroxides. For instance, $Cr(OH)_2$ dissolves only in acid, but $Cr(OH)_3$ dissolves in either acid or base.

Questions

6. Write an equation for the reaction of NaCl with concentrated H_2SO_4. How does $HgCl_2(s)$ act with the same acid? What does this show?

7. What is an alloy? Are all alloys metallic?

8. What are the constituents of brass? Of bronze?

9. What similar properties of copper and nickel are responsible for their forming a continuous range of solid solutions from 100 percent copper to 100 percent nickel?

21.4 Rare Earths: Inner Transition Metals

The rare earths are sometimes called the inner transition elements. Those in Period 6 are called the *lanthanides* because they follow lanthanum, and those in Period 7 are known as the *actinides*. The members of each series are so similar to one another in all their properties that it is difficult to separate them. The lanthanides occur chiefly as a mixture of complex phosphates in *monazite sand*. Most

IA																	0
H	IIA											IIIA	IVA	VA	VIA	VIIA	He
Li	Be											B	C	N	O	F	Ne
Na	Mg	IIIB	IVB	VB	VIB	VIIB	———	VIIIB	———	IB	IIB	Al	Si	P	S	Cl	Ar
K	Ca	Sc	Ti	V	Cr	Mn	Fe	Co	Ni	Cu	Zn	Ga	Ge	As	Se	Br	Kr
Rb	Sr	Y	Zr	Nb	Mo	Tc	Ru	Rh	Pd	Ag	Cd	In	Sn	Sb	Te	I	Xe
Cs	Ba	La	Hf	Ta	W	Re	Os	Ir	Pt	Au	Hg	Tl	Pb	Bi	Po	At	Rn
Fr	Ra	Ac															

Ce	Pr	Nd	Pm	Sm	Eu	Gd	Tb	Dy	Ho	Er	Tm	Yb	Lu
Th	Pa	U	Np	Pu	Am	Cm	Bk	Cf	Es	Fm	Md	No	Lr

of the actinides, on the other hand, do not occur in nature at all—they have only been synthesized in nuclear reactions.

The differentiating electron for all the rare earths is in an inner f sublevel. As a result the atomic radii of all these elements are very similar. The most stable oxidation state for all the lanthanides is $+3$, but the $+2$ and $+4$ states also exist for some—an observation we explain in terms of the stability of a filled or half-filled f sublevel. The actinides exhibit oxidation states from $+2$ to $+7$ and in general have a slightly more complicated chemistry.

21.5 The Representative Metals of Groups IIIA, IVA, and VA

We are now left with six metals to discuss: gallium, indium, and thallium in Group IIIA; tin and lead in Group IVA; and bismuth in Group VA.

Gallium is a soft silvery white metal with a very low melting point (30°C) and a very high boiling point (>2000°C). This wide range (indeed the widest for any known substance) has made it useful as a thermometer liquid. Indium is relatively scarce and very expensive, but it plays an important role as a corrosion-resistant coating for bearings. Thallium, like the other two members of Group IIIA, is soft, white, and relatively reactive.

Tin and lead are, of course, familiar metals. An unfamiliar form of tin, however, is one of its allotropes—gray tin. Gray tin is a nonmetallic powder with no metallic strength at all. It is stable below -12.3°C, and the slow conversion of metallic tin to its crumbly allotrope at low temperatures is known as "tin disease." The chief use of tin is in a structural material called tinplate, a thin sheet of steel coated with an even thinner sheet of tin for protection. Because of their sonorous tone, tin alloys are used in church bells, organ pipes, and cymbals.

Lead has only one form—the familiar soft, dense, and malleable metal. Its low melting point allows it to be melted and recast for such diverse uses as bullets and printers' type. It's also used in storage batteries (Chapter 18) and as a shielding against radioactivity. Finally, a use that is a source of current concern is in the manufacture of gasoline additives (Chapter 22). Over a quarter of a million tons of lead are used annually for such additives, and most of it becomes part of automobile exhaust. Although the extent of the present danger is still being debated, we do know that in large doses lead, like many of the heavy metals, is toxic, both as a free element and in combined form (Chapter 23).

Solders, which are low-melting alloys, are frequently alloys of tin and lead, with a little antimony added for strength. Tin and tin-lead alloys are also used to coat electrical or mechanical components that are later to be soldered.

Bismuth is another low-melting metal. It is used chiefly in *fusible alloys*, such as *Woods metal*. Woods metal melts at 70°C and is made by combining four parts (by weight) bismuth, two parts lead, one part tin, and one part cadmium. Fusible alloys are an important part of heat-sensitive fire extinguishers, which are activated when the temperature rises enough to melt the alloy. Bismuth has another useful and very unusual property: Like water, it expands when it freezes. Combined with metals that are to be cast, it ensures that the metal takes a good impression.

Other systematic properties of these six metals give us an opportunity to review and confirm some of the generalizations we've already made.

1. *Metallic character within a periodic group increases with atomic number.* Tin has several allotropic forms, one of which is a gray, nonmetallic powder. Lead has no such nonmetallic form.

2. *Heavy metal atoms with two s electrons in the outer energy level often form inert-pair ions.* The electron configuration endings for Ga, In, and Tl are

Ga: $3d^{10}, 4s^2, 4p^1$

In: $4d^{10}, 5s^2, 5p^1$

Tl: $5d^{10}, 6s^2, 6p^1$

All three of these elements form $+3$ ions predominantly, but the $+1$ oxidation state becomes more stable as atomic number increases.

The electron configuration endings for tin and lead are $4d^{10}, 5s^2, 5p^2$ and $5d^{10}, 6s^2, 6p^2$, respectively. Both metals have $+2$ and $+4$ oxidation states but only a $+2$ ionic state. The two s electrons are sometimes shared but never lost.

Although bismuth has five valence electrons ($5d^{10}, 6s^2, 6p^3$), we find the $+5$ oxidation state only in the highly reactive bismuthate ion (BiO_3^-). In all other compounds, the oxidation state of bismuth is $+3$.

3. *If an element exhibits more than one oxidation state, it is less metallic in the higher*

oxidation states, and the oxides and hydroxides of the higher states are more acidic. $SnCl_2$ and $PbCl_2$ are ionic, whereas $SnCl_4$ and $PbCl_4$ are essentially covalent. Lead(II) hydroxide exists but is amphoteric; lead(IV) hydroxide does not exist.

In its +3 state, bismuth exists exclusively as a cation. In its +5 state, however, we find it only in an anion. [The bismuthate ion (BiO_3^-) is a strong oxidizing agent.]

4. *Color in ions usually means partially filled and closely spaced electronic energy levels.* Outside an inner core of filled energy levels, these representative elements and their ions have electrons only in s and p orbitals, where the spacing is larger than in d and f orbitals. Thus all the ions of these six metals are colorless.

5. *Metallic character decreases from left to right within a period.* Bismuth has a positive reduction potential and thus will not reduce the hydrogen ion. The two Group IVA metals, tin and lead, have negative potentials, but they are so low that the metals are not practical for the preparation of hydrogen. The remaining three metals have larger (negative) reduction potentials.

Table 21.5 A Comparison of the Representative Metals and Their Compounds with the Transition Metals and Their Compounds

Atomic radius	In any period, the transition metals have smaller atomic radii than the metals of Group IA and IIA, but they do not differ markedly from the atomic radii of the elements of Groups IIIA, IVA and VA.
General physical properties	Physical metallic properties such as density, high melting point, malleability, and conductivity are more pronounced among the transition elements than among the representative metals (see Figure 21.1).
Color of ions and compounds	All representative metal ions are colorless, whereas many transition metal ions are colored. When colorless representative metal cations combine with colorless anions to form colored compounds, we can usually infer that the ions have been polarized and that the compound is partially covalent.
Solubilities of compounds	Most compounds of the IA and IIA metals are water-soluble. The representative elements of Groups IIIA, IVA, and VA have fewer soluble compounds, and most compounds of the transition metals are insoluble.
Ionic nature of compounds	Almost all compounds of the IA and IIA metals are essentially ionic. Compounds of the metals in Groups IIIA, IVA, and VA show increasing covalent character. Aside from some compounds of the fourth period transition metals, transition metal compounds show considerable covalency.
Uses	Transition metals are often used as free elements, either pure or in alloys. With some exceptions, notably tin and lead, the representative metals are more frequently useful as parts of compounds.

We should add that the reaction of lead with sulfuric acid is unlike the usual behavior of metals with this acid. In contrast to most metals, lead liberates hydrogen faster from concentrated sulfuric acid than from dilute sulfuric acid. In the dilute acid, an insoluble lead(II) sulfate coating forms on the metal, inhibiting the reaction. In concentrated sulfuric acid, soluble $Pb(HSO_4)_2$ forms instead, and a fresh lead surface is always available to the acid.

A comparison of the representative metals and their compounds with the transition metals and their compounds appears in Table 21.5.

Questions

10. In what sense can we apply the generalization, "Metallic character increases with atomic number within a periodic group," to Groups IA and IIA?

11. Can you provide some examples from among the transition metals of the generalization, "If an element exhibits more than one oxidation state, it is more nonmetallic in the higher oxidation states, and the oxides and hydroxides of the higher states are more acidic"?

21.6 The Metalloids

We have now given some attention to all the elements of the periodic table except that family of elements whose job seems to be to keep the metals and the nonmetals apart—the metalloids. They are beryllium in Group IIA, boron and aluminum in Group IIIA, silicon and germanium in Group IVA, arsenic and antimony in Group VA, and tellurium and polonium in Group VIA. Since we discussed boron briefly in Chapters 6, 7, and 9 and looked at silicon in Chapters 9 and 20, we'll devote ourselves to the remaining seven here.

Beryllium Beryllium is very unlike the other elements of Group IIA, and aside from its oxidation state it more closely resembles aluminum than magnesium. This similarity is analogous to the resemblance of lithium to magnesium and boron to silicon. These resemblances are accounted for by the fact that elements tend to become more nonmetallic as one goes across the periodic table and more metallic as one goes down the periodic table. Elemental beryllium is very light and is quite strong for its weight; thus it would seem to have possibilities as a structural material. However, it is very difficult to extract from its ores.

The beryllium atom contains only four s electrons tightly bound to the nucleus. As a consequence it is relatively nonreactive, has a small atomic radius, and has a high ionization energy. The high ionization energy means that all beryllium compounds are at least partially covalent. We discussed some of these compounds, the beryllium halides, in Chapter 7. However, these linear molecules probably exist only in the vapor state at high temperatures. At lower temperatures, they dimerize and polymerize. As we would expect for a metalloid, beryllium hydroxide is amphoteric. On the other hand, when alloyed with copper, beryllium behaves like a metal, giving the alloy a strength and hardness comparable to steel. This alloy has little tendency to produce sparks; tools made of beryllium-copper are thus often used in explosive atmospheres. Its use in industry is limited by the toxicity of its compounds.

Aluminum We discussed some of the properties and uses of elemental aluminum among the metals, since in the free state there is no doubt that aluminum is a metal—a lightweight structural material, highly resistant to corrosion, and a good electrical and thermal conductor. But among its compounds we begin to recognize aluminum as a metalloid. Aluminum hydroxide is amphoteric. It dissolves in acids to give salts in which aluminum is the cation, but it also dissolves in strong bases to form salts in which aluminum is part of the anion. We'll say more about these reactions in Chapter 24. Also, the aluminum halides are not the ordinary ionic salts we would expect of a true metal. Aluminum fluoride is best represented as an extended sheet of aluminum and fluorine atoms. And the other halides are dimers in most states, better represented by formulas such as Al_2Cl_6 than the empirical formulas.

Aluminum reacts with nonoxidizing acids to produce hydrogen but with nitric acid (p. 527) to form a tightly adhering oxide coating which effectively protects it from further reaction.

Aluminum oxide containing traces of other metal ions occurs in nature as gem stones. The contaminant determines the gem. Ruby is aluminum oxide with Cr^{3+} ions, and blue sapphire is aluminum oxide with titanium ions. White sapphire is pure aluminum oxide. All of these gem stones are now produced synthetically in large quantities.

Germanium Germanium, element 32, was one of the missing elements for which Mendeleev allowed room in his periodic table. He predicted that it would have properties similar to silicon, and he designated it "eka-silicon." Germanium remains rare even today, but it does indeed resemble silicon closely. It crystallizes in a diamondlike structure.

Arsenic and Antimony Among the nonmetallic characteristics of these two elements is allotropism. Both exist in more than one form, though the forms that are stable at room temperature are bright and metallic. Neither element finds much use in its elemental form, although both are added to lead as hardeners.

The stability of binary hydrogen compounds is a chemical property that can be used to illustrate the range of metallic character among the elements of Group VA, including arsenic and antimony. The nonmetals, nitrogen and phosphorus, form stable ammonia and phosphine. Arsine (AsH_3), an extremely poisonous compound, readily undergoes thermal decomposition to free arsenic. Stibine (SbH_3) is even less stable, and the binary hydrogen compound of bismuth (BiH_3) has been synthesized only in trace amounts. Another property that shows the same trend is the basicity of oxide hydrolysis products. H_3PO_3 is a moderately weak acid with $K_{a_1} = 1 \times 10^{-2}$; H_3AsO_3 is a very weak acid with $K_{a_1} = 6 \times 10^{-10}$; and H_3SbO_3 is amphoteric, so that we might equally well write it as $Sb(OH)_3$; $Bi(OH)_3$ is a base.

Arsenic compounds such as As_2O_3, As_2S_3, $AsCl_3$, As_2O_5, and As_2S_5 are all at least partially covalent. As_2O_3 is a dimer in the vapor phase and hence better written As_4O_6. Its structure is much like the analogous oxide of phosphorus (see Figure 20.5). The structure of As_2O_5 (empirical formula) is as yet unknown. Arsenic almost never forms a cation in solution; instead it is hydrolyzed completely to the AsO_3^{3-} ion. Antimony, on the other hand, does form +3 ions, but in the +5 state it is always found in an anion. Both of these elements have +3 oxidation states—a result of their outer s electrons being an "inert pair" (see Chapter 6).

Some of the compounds of arsenic are used in insecticides and pesticides. Allowed to accumulate in the body, arsenic is also poisonous to humans. In fact, from an analysis of hair from Napoleon's disinterred remains, we now know that it was arsenic, not syphilis, that was primarily responsible for his death.

Tellurium and Polonium Although it is a metalloid, tellurium is in many ways similar to selenium, which in turn resembles sulfur. The chief use of tellurium is as a hardener in steel and in copper alloys. Telluric acid, H_6TeO_6 or $Te(OH)_6$, is amphoteric.

Polonium is radioactive and is the most metallic member of Group VIA. It was the first radioactive element identified by Marie Curie and was named for her native Poland.

Questions

12. What is a metalloid?

13. What is amphoteric behavior?

21.7 Corrosion

We apply the general term *corrosion* to any disintegration of a metal caused by its reaction with substances in the environment. But since iron is such an important resource and yet is so susceptible to corrosion, most practical and theoretical studies of corrosion have centered around this metal. The corrosion of iron is the process we commonly call *rusting*, the formation of a crumbly, non-adhering oxide that readily peels off, constantly exposing a fresh surface to the attack of the atmosphere. Iron's susceptibility to this corrosive process causes losses that run into billions of dollars annually in this country.

The formation of FeO (or Fe_2O_3) on the surface of iron may seem an easy process to explain, but in fact there are still many unanswered questions about it. To explain it adequately we must account for the following observations.

1. Iron will not rust in dry air nor will it rust in water free of air; that is, both water and air are necessary for corrosion.

2. The presence of electrolytes greatly enhances corrosion. This is evident when we look at automobiles that have been driven on salted streets.

3. Strained metal corrodes faster than unstrained metal; thus metal corrodes faster when bent or distorted.

4. Heated metals corrode faster than unheated metals.

5. A metal in contact with a less active metal corrodes faster then when alone.

We now recognize corrosion as essentially an electrochemical phenomenon. In contact with an electrolyte, miniature galvanic cells are established on the surface of the iron. Strained areas, especially, are likely to become anodes:

$$Fe \longrightarrow Fe^{2+} + 2e^-$$

The electrons move through the metal to other areas which act as cathodes. Here oxygen is reduced in the presence of hydrogen ions that have been contributed by moisture in the air.

$$O_2 + 4H^+ + 4e^- \longrightarrow 2H_2O$$

The oxygen oxidizes the iron(II) ion at the anode to the iron(III) ion:

$$4\ Fe^{2+} + O_2 + 8\ OH^- \longrightarrow 2\ Fe_2O_3 + 4\ H_2O$$

The corrosion of iron can be prevented by alloying it with other elements to form corrosion-resistant alloys or by coating it with materials less susceptible to corrosion. Coatings include paint, lacquer, grease, other metals, ceramic enamels, and magnetic oxide (Fe_3O_4), which is formed by treating hot iron with steam.

Zinc is a more active metal than iron, as a comparison of their reduction potentials (Table 18.2) shows:

$$Fe \longrightarrow Fe^{2+} + 2e^- \qquad +0.409\ V$$
$$Zn \longrightarrow Zn^{2+} + 2e^- \qquad +0.7628\ V$$

It is surprising, therefore, that zinc can be used as a protective coating for iron. The explanation is the unusual corrosion product of zinc. When exposed to the atmosphere, zinc forms a thin layer of basic zinc carbonate [$Zn(OH)_2 \cdot ZnCO_3$], which adheres tightly to the surface and protects the metal from further corrosion. Furthermore, the greater activity of zinc provides an added measure of protection for iron. If a break should occur in the zinc coating, exposing iron to the atmosphere, zinc becomes a sacrificial metal, that is, the zinc rather than an area of the iron becomes the anode and the iron serves as the cathode. On the other hand, if a break occurs in a protective coating provided by a less active metal, such as tin, the iron is the anode in the galvanic cell and it is corroded even faster than if tin were not present. In fairness, tinplate—iron coated with tin—does have some advantages over galvanized iron—iron coated with zinc. For example, its chemical inactivity makes it more suitable for storing foods.

To illustrate the electrochemical nature of corrosive processes further, another example is in order. Magnesium is often used as a sacrificial metal in steel pipelines and on ship hulls. Pieces of magnesium, wired to the pipe at suitable intervals, act as anodes in the corrosive reaction, thus protecting the steel from oxidation.

A number of active metals can be made "passive" to oxidation by allowing a thin layer of oxidation product to form on them. Aluminum, chromium, iron, cobalt, and nickel are all active metals that react with most strong acids to produce hydrogen. But to all appearances these metals do not react with concentrated nitric acid. Instead, an insoluble, tightly adhering oxide coating forms, protecting the metal from further reaction. The same kind of coating protects some of these metals, notably aluminum, nickel, and chromium, from normal corrosion. This passive coating is often produced before the metal is used by exposing it to a strong oxidizing agent, such as a dichromate solution.

Questions

14. What is the chemical composition of the rust on iron?

15. Write equations illustrating the production of H^+ ions in water from the CO_2 of the atmosphere.

Questions and Problems

21.1 Abundance, Availability, and Usefulness

16. What percent of the elements are metals?

17. What is the difference between a mineral and an ore?

21.2 The Metals of Groups IA and IIA

18. Explain the following facts in terms of atomic or ionic properties.
    a. None of the metals of Groups IA or IIA is found free in nature.
    b. Compounds of the IA and IIA metals are found in high concentration in seawater.
    c. Some of the IA and IIA metal ions produce colored flames.
    d. Lithium has the largest (negative) reduction potential, but cesium has the lowest ionization energy.

19. What is responsible for the similar properties of lithium and magnesium (and their compounds) and of aluminum and beryllium?

20. Draw Lewis structures for the following:
    a. Na_2O b. KO_2 c. BaO_2 d. MgO e. H_2O_2

21. In what ways does the metal crystal resemble (a) an ionic crystal, (b) a graphite crystal, and (c) a network crystal?

21.3 Transition Elements

22. Contrast the properties of the transition metals and their compounds with the properties of the Group IA and IIA metals and their compounds.

23. Arrange these oxides in order of increasing acidity:

    MnO_2 Mn_2O_7 MnO Mn_3O_4 MnO_3

24. Using only nitric and hydrochloric acids, how would you distinguish among members of each of the following groups?
    a. gold, silver, and copper
    b. silver, aluminum, and magnesium
    c. chromium, gold, and copper

25. What property has led to the use listed below for each metal?
    a. silver   coinage
    b. nickel   nickel-plating
    c. aluminum   aircraft construction
    d. copper   electrical wiring
    e. lithium   scavenger in the steel industry
    f. bismuth   casting alloys
    g. lead   printers' type
    h. mercury   thermometer liquid
    i. bismuth   fusible alloys

26. List all the factors that should be considered when explaining the difference in covalence between two transition metal halides.

27. What explanation can you offer for the observation that mercury(II) sulfide, not mercury(I) sulfide, is precipitated when mercury(I) nitrate is treated with H_2S?

28. If Hg_2O could be prepared, would it likely be more or less acidic than HgO?

29. What weight of iron could be prepared from 1.00 ton of ore that is 45 percent Fe_2O_3?

30. If a current of 2.00 A is used in the electrorefining of copper, what weight of copper will be deposited at the cathode in 1.00 day?

*31. Explain the following facts about color:
    a. Group IA and IIA metal ions are colorless.
    b. Transition metal ions frequently are highly colored.
    c. A cobalt solution that absorbs red light appears blue.
    d. AgBr and AgI are colored compounds, even though they are formed from colorless ions.

32. Explain what is meant by polarization of an anion.

33. Compare and contrast the following terms:
    a. interstitial solution   b. solid solution
    c. substitutional solution   d. intermetallic compound

21.4 Rare Earths

34. Account for the similarity of properties among the members of each of the two groups of elements known as lanthanides and actinides.

21.5 The Representative Metals of Groups IIIA, IVA, and VA

35. Explain why Pb, Sn, and Bi, though representative metals, each exhibit more than one oxidation state.

36. Zinc and other active metals react with dilute sulfuric acid faster than with the concentrated acid. Lead, however, reacts faster with concentrated sulfuric acid than with dilute. Explain.

37. Which metal
    a. is the best light reflector?
    b. has the greatest density?
    c. is the most malleable?
    d. is the best electrical conductor?
    e. has the lowest melting point?
    f. has the highest melting point?
    g. has the lowest ionization energy?
    h. has the largest (negative) oxidation potential?
    i. has the lowest density?
    j. is the most abundant in the earth's crust?

38. Which metals
    a. are known as alkali metals?
    b. are called the noble metals?
    c. are the base metals?
    d. are known as the alkaline earth metals?
    e. will not combine with oxygen when heated in air?

39. Explain why clay, a naturally occurring compound of aluminum, is not an ore of aluminum.

21.6 The Metalloids

40. In what respects is aluminum like a metal? In what ways is it like a nonmetal?

21.7 Corrosion

41. Explain how zinc, a more active metal than iron, is used to protect iron from corrosion.

42. Explain why iron corrodes faster when in contact with tin than when alone.

Discussion Starters

43. Explain or account for the following:
    a. There are more hydrates of lithium and magnesium than of other Group IA and IIA metals.
    b. Although silver does not combine directly with oxygen from the air, it does tarnish.
    c. Some alloys are pure substances; some are not.
    d. Tin appears to disintegrate when exposed to very cold temperatures for a long time.
    e. Metals are electrical conductors.
    f. Graphite shows a number of metallic characteristics.
    g. Silver dissolves readily in nitric acid but is attacked very slowly by aqua regia.
    h. Silver is a better heat conductor than diamond.
    i. Aluminum, the most abundant metal, was not developed commercially until recently.
    j. Iron is the most widely used metal.
    k. Iron was one of the first metals to be used by man.
    l. Metals are not necessarily priced in relation to their abundance.
    m. Aluminum, a very active element, is resistant to corrosion.
    n. Calcium, which is as readily available as iron, is not important commercially.
    o. It is essential that the copper used in electrical conductors be pure.
    p. For many uses a pure metal is not the most desirable.

22 Organic Chemistry

"I must tell you that I can make urea without the use of kidneys, either man or dog." So wrote Friedrich Wöhler, a German chemist, in a letter to the great Swedish chemist J. J. Berzelius. The year was 1828, and this triumphant statement heralded the beginning of modern organic chemistry. Wöhler had successfully synthesized urea (H_2NCONH_2), a waste product eliminated from the body in urine, from the inorganic compound ammonium cyanate (NH_4CNO). Before this achievement it was generally believed that compounds produced by plants or animals (organic compounds) could not be synthesized without a special "vital force" that only a living organism could supply. Ordinary laboratory methods were thought to be useless.

Because the terms *organic* and *inorganic* represent convenient divisions of chemistry, they survive even today; however, they no longer refer strictly to whether a compound originated in an organism or not. Instead, organic chemistry is now often defined as the chemistry of carbon. Better yet, we shall define it as the chemistry of **hydrocarbons** (binary compounds of hydrogen and carbon) and their derivatives since some carbon compounds, such as carbonates, bicarbonates, carbides, carbon dioxide, and carbon monoxide, are traditionally regarded as inorganic.

Over 2 million organic compounds have been synthesized, identified, and described in the scientific literature, far more than the number of compounds classified as inorganic. And in recent years more than 30,000 new organic compounds have been added to that list annually. Because of the importance of many of these compounds, more than one-third of all chemists specialize in organic chemistry.

In beginning our study of organic chemistry, we might first ask what unique property of carbon enables it to combine with other elements to form so many different compounds. Primarily it is the ability of carbon-carbon bonds to form and persist under normal conditions. It is true that homonuclear bonds are not the exclusive property of carbon (in Chapters 19 and 20 we mentioned homonuclear bonds of nitrogen, oxygen, sulfur, phosphorus, and silicon), but the carbon-carbon bond is by far the most versatile and most stable. Silicon is the element closest to carbon in its chemical properties, and next to carbon it forms the greatest array of compounds containing homonuclear bonds. These compounds are not commonly encountered because the hydrolysis of silanes occurs readily and completely. Conversely, the hydrocarbons are practically immune to hydrolysis. The larger silicon atom uses available d orbitals to accept a pair of electrons from a water molecule in the first step of hydrolysis. Thus, in our normal wet environment, carbon-carbon bonds persist but silicon-silicon bonds do not.

Although organic chemistry is a vast subject, its study can be approached logically. We have already suggested a way to begin: The organic compounds can first be separated into two large categories—hydrocarbons on the one hand and their derivatives on the other. Next, the hydrocarbons can be divided into two categories according to whether the carbon atoms are arranged in chains or in cyclic structures (Table 22.1). Chain hydrocarbons are further classified by the type of linkage between carbon atoms. Those bonded together exclusively by single bonds are known as *saturated hydrocarbons* (also as the methane series, paraffins, or *alkanes*). Those containing at least one multiple bond are referred to as *unsaturated hydrocarbons* and are further divided into *alkenes* and *alkynes*. Alkenes, sometimes called olefins or the ethylene series, are characterized by one or more double bonds between carbon atoms. Alkynes, known as the acetylene series, have one or more triple bonds.

We can make similar distinctions among the cyclic compounds. First, they are classified as either *aromatic* or *nonaromatic*. (All nonaromatic hydrocarbons, both chain and cyclic, are referred to collectively as *aliphatic hydrocarbons*.) The term *aromatic* originally referred to the agreeable odor often associated with these compounds, but it now has a more precise meaning that we shall discuss

Table 22.1 Classification of Hydrocarbons

Category	Synonyms	Bonding Between Carbon Atoms	Example
		Chain	
Alkanes	Saturated hydrocarbons, methane series, paraffins	Single bonds only	C_2H_6
Alkenes	Ethylene series, olefins	Contains at least one double bond	C_2H_4
Alkynes	Acetylene series	Contains at least one triple bond	C_2H_2
		Cyclic	
Cycloalkanes		Single bonds only	C_6H_{12}
Cycloalkenes		Contains at least one double bond	C_6H_{10}
Cycloalkynes		Contains at least one triple bond	C_6H_8
Aromatic (there are several aromatic series—only the benzene series is represented here)		Pi bonding that extends over a sigma-bonded six-carbon atom ring	C_6H_6

in detail later. The nonaromatic cyclic compounds are classified as *cycloalkanes*, *cycloalkenes*, or *cycloalkynes*, again according to the linkages that join the carbon atoms.

22.1 Structure and Nomenclature of Alkanes

The alkanes have the general formula C_nH_{2n+2}. The simplest compound that fits this general formula is methane (CH_4), where $n = 1$. Next in line is ethane (C_2H_6), and then propane (C_3H_8)—see Table 22.2. The list can be extended by continuing to add CH_2 groups, as demanded by the general formula. Since the formulas of all noncyclic alkanes differ only by one or more CH_2 groups, they are said to constitute a *homologous series*, and each member of the series is called a *homolog*.

Table 22.2 Formulas for the First Three Members of the Alkane Series

Hydrocarbon	Empirical Formula	Molecular Formula	Electron Formula	Structural Formula	Condensed Structural Formula	Carbon Skeleton Formula
Methane	CH_4	CH_4	H:C:H with H above and below	H—C—H with H above and below	CH_4	C
Ethane	CH_3	C_2H_6	H:C:C:H with H's	H—C—C—H with H's	CH_3—CH_3	C—C
Propane	C_3H_8	C_3H_8	H:C:C:C:H with H's	H—C—C—C—H with H's	CH_3—CH_2—CH_3	C—C—C

Structural Formulas We said previously that empirical formulas express only the ratio of the numbers of atoms in a compound and not the absolute number of atoms of each element in each molecule. Molecular formulas are needed to provide this missing piece of information. As Table 22.2 shows, the empirical formula and the molecular formula are the same for methane but not for ethane. The empirical formula for ethane is CH_3, but the molecular formula is C_2H_6.

For most ionic compounds, empirical formulas are sufficient, but most molecular inorganic compounds require molecular formulas. With most organic compounds we must go even a step further, specifying electron formulas or structural formulas to avoid ambiguity. In structural formulas, a bond is indicated by a line rather than a pair of dots, and unshared pairs of electrons are not always shown. Table 22.2 shows each of these formulas for the three simplest alkanes, along with two types of formulas we haven't encountered yet. The *condensed structural formulas* save space by showing only the C—C bonds, and the carbon skeleton formulas omit the hydrogen atoms as well.

Models Three-dimensional models are helpful in representing the spatial arrangement of atoms in molecules. Figures 22.1 through 22.5 illustrate two methods of representing the spatial arrangement of atoms in molecules of lower-molecular-weight hydrocarbons. The models on the left illustrate the so-called

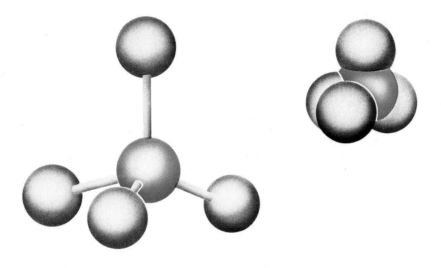

Figure 22.1 Methane models.

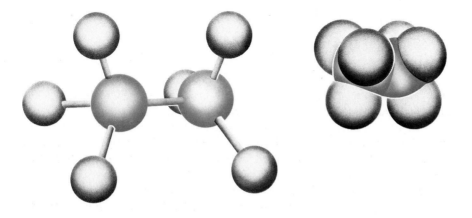

Figure 22.2 Ethane models.

Organic Chemistry 677

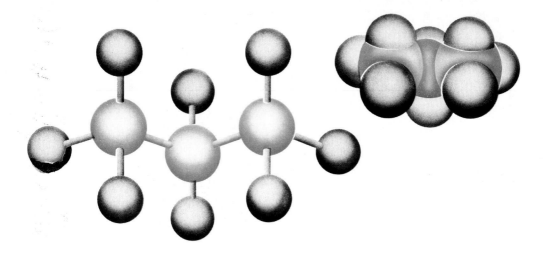

Figure 22.3 Propane models.

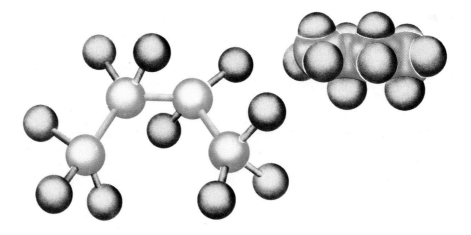

Figure 22.4 Butane models.

"ball-and-stick" method, where each stick represents one covalent bond, and black balls and light balls represent carbon and hydrogen atoms, respectively. In the models on the right, each atom is represented by a sphere from which one segment for each bond has been removed. Both kinds of models have their advantages and disadvantages. The ball-and-stick models correctly and clearly indicate bond angles and relative positions of atoms in the molecule. Of course, they are not constructed to scale. On the other hand, the space-filling models on the right are much closer to scale and give a much clearer impression of the surface of the molecule, indicating that molecules are not likely to become entangled—an impression that might be conveyed by the ball-and-stick models.

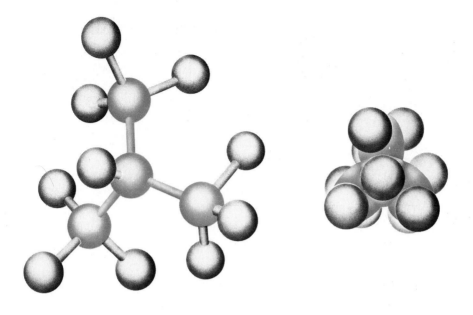

Figure 22.5 Structural isomer of model in Figure 22.4 (methylpropane models).

We shall now look in more detail at several of the simple alkanes. Since we covered methane in Chapter 7, we'll begin with ethane.

Ethane As in all the noncyclic alkanes, the orbitals of the carbon atoms in ethane are sp^3-hybridized. To visualize the structure of the ethane molecule, therefore, we can think of it as a derivative of the tetrahedral methane molecule in which one of the hydrogen atoms has been replaced with a second methyl group ($\cdot CH_3$). The result is two tetrahedra joined at one corner, with hydrogen atoms at each of the six remaining corners and a carbon atom at the center of each tetrahedron (Figure 22.6). This geometrical construction somewhat misrepresents bond lengths, but it is a useful way to picture the molecule (see also Figure 22.2).

Propane The third member of the alkane series, propane, adds one CH_2 group to the ethane molecule. We can again derive this structure from one we recognize—in this case, ethane—by replacing a hydrogen atom with a tetrahedral methyl group. Since we customarily write such structural formulas with as many carbon atoms as possible in a straight line, the result for propane usually appears as

$$\begin{array}{c} H H H \\ | | | \\ H-C-C-C-H \\ | | | \\ H H H \end{array}$$

Organic Chemistry

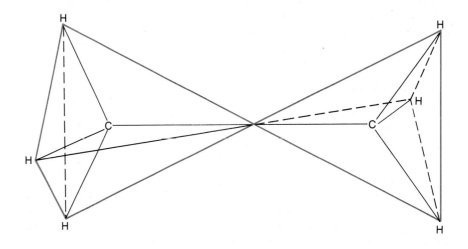

Figure 22.6 Spatial arrangement of the two carbon atoms and six hydrogen atoms in the ethane molecule.

rather than

$$\begin{array}{c} \text{H} \quad \text{H} \\ | \quad | \\ \text{H}-\text{C}-\text{C}-\text{H} \\ | \\ \text{H} \\ \\ \text{H}-\text{C}-\text{H} \\ | \\ \text{H} \end{array}$$

However, the two formulas are equally correct. And both equally misrepresent the bond angles, which are *not* 90° but 109°28′ (see Figure 22.3). We should also note that because of the free rotation around every carbon-carbon single bond all hydrogen atoms bonded to a given carbon atom are structurally equivalent. In ethane the two carbon atoms are equivalent; therefore, all six hydrogen atoms are equivalent. In propane the six hydrogen atoms on the terminal carbon atoms are equivalent to one another, as are the two hydrogen atoms on the middle carbon atom. Thus

$$\begin{array}{ccc} \text{H H H} & \text{H H Cl} & \text{H H H} \\ | \; | \; | & | \; | \; | & | \; | \; | \\ \text{H}-\text{C}-\text{C}-\text{C}-\text{Cl} \quad \text{H}-\text{C}-\text{C}-\text{C}-\text{H} \quad \text{and} \quad \text{Cl}-\text{C}-\text{C}-\text{C}-\text{H} \\ | \; | \; | & | \; | \; | & | \; | \; | \\ \text{H H H} & \text{H H H} & \text{H H H} \end{array}$$

represent the same chlorinated derivative of propane, and

$$\begin{array}{c} \text{H} \quad \text{H} \quad \text{H} \\ | \quad\; | \quad\; | \\ \text{H—C—C—C—H} \\ | \quad\; | \quad\; | \\ \text{H} \quad \text{Cl} \quad \text{H} \end{array}$$

represents a second, chemically distinct derivative.

The Butanes Since there is only one structure for each of the three simplest alkanes, we can safely identify these compounds by using their molecular formulas. However, beyond propane, structural formulas become necessary. For example, there are two structures that correspond to butane (C_4H_{10}):

$$\begin{array}{c} \text{H} \;\; \text{H} \;\; \text{H} \;\; \text{H} \\ | \;\;\; | \;\;\; | \;\;\; | \\ \text{H—C—C—C—C—H} \\ | \;\;\; | \;\;\; | \;\;\; | \\ \text{H} \;\; \text{H} \;\; \text{H} \;\; \text{H} \end{array} \quad \text{and} \quad \begin{array}{c} \text{H} \;\; \text{H} \;\; \text{H} \\ | \;\;\; | \;\;\; | \\ \text{H—C—C—C—H} \\ | \;\;\; | \;\;\; | \\ \text{H} \quad | \quad \text{H} \\ \;\;\;\;\; \text{H—C—H} \\ \;\;\;\;\;\;\;\; | \\ \;\;\;\;\;\;\;\; \text{H} \end{array}$$

These structural formulas represent two different compounds, each with its own physical and chemical properties (see Table 22.3). When two or more such compounds have the same molecular formula but different structural formulas, we call them *structural isomers*, and we call their relationship *structural isomerism*. We should emphasize that there are *only two* structural isomers of C_4H_{10}. One

Table 22.3 Properties of the Two Butanes

	H—C—C—C—C—H (with H's)	H—C—C—C—H with H—C—H branch
Molecular formula	C_4H_{10}	C_4H_{10}
Molecular weight	58.1	58.1
Melting point (°C)	−138	−145
Boiling point (°C)	−1	−10

might think that

$$\begin{array}{c} \text{H} \quad \text{H} \quad \text{H} \\ | \quad \ | \quad \ | \\ \text{H}-\text{C}-\text{C}-\text{C}-\text{H} \\ | \quad \ | \quad \ | \\ \text{H} \quad \text{H} \quad \ \ \\ \quad \quad \quad \text{H}-\text{C}-\text{H} \\ \quad \quad \quad \quad | \\ \quad \quad \quad \quad \text{H} \end{array}$$

is another structure and another structural isomer. However, if one remembers that the four bonds of the carbon atom are identical and tetrahedrally directed (Chapter 7), it is obvious that this structure is the same as the one represented by a structural formula with the four carbon atoms arranged in a straight line (see Figures 22.4 and 22.5).

The Pentanes When we come to pentane (C_5H_{12}), we can write structural formulas for three chemically distinct compounds:

$$\begin{array}{cc}
\begin{array}{c}
\text{H H H H H} \\
| \ | \ | \ | \ | \\
\text{H}-\text{C}-\text{C}-\text{C}-\text{C}-\text{C}-\text{H} \\
| \ | \ | \ | \ | \\
\text{H H H H H}
\end{array}
&
\begin{array}{c}
\text{H H H H} \\
| \ | \ | \ | \\
\text{H}-\text{C}-\text{C}-\text{C}-\text{C}-\text{H} \\
| \ | \ \ \ | \\
\text{H H} \quad \text{H} \\
\quad \quad \text{H}-\text{C}-\text{H} \\
\quad \quad \quad \text{H}
\end{array}
\end{array}$$

$$\begin{array}{c}
\text{H} \\
| \\
\text{H}-\text{C}-\text{H} \\
\text{H} \quad | \quad \text{H} \\
| \quad \quad | \\
\text{H}-\text{C}-\text{C}-\text{C}-\text{H} \\
| \quad \quad | \\
\text{H} \quad | \quad \text{H} \\
\text{H}-\text{C}-\text{H} \\
| \\
\text{H}
\end{array}$$

The situation becomes even more complicated with larger alkanes: There are 4347 structural isomers with the molecular formula of $C_{15}H_{32}$.

Nomenclature Assigning names to the vast array of alkanes obviously requires a carefully designed system. Such a system was established by an international group of chemists in 1892. The original system has subsequently been revised and enlarged several times, and today it permits unambiguous and unique names to be assigned to all organic compounds. It is referred to as the IUPAC (International Union of Pure and Applied Chemistry) system or the Geneva system. We shall not try to cover all the details of this complex system in this chapter but as we discuss each group of compounds, we'll present enough rules to make it possible to name the simpler compounds unambiguously.

The following rules apply to the naming of alkanes:

1. All alkanes end in -*ane*.

2. Unbranched alkanes—through decane—have the names listed in Table 22.4. The first four names do not indicate the number of carbon atoms in the molecule, but beyond that point the name of the compound designates the number of carbon atoms in each molecule.

3. If the carbon atoms are not arranged in an unbranched chain, the longest chain is selected as the parent hydrocarbon. The compound is then named as a derivative of the parent hydrocarbon. For example, the following compound is named as a derivative of pentane since the longest chain contains five carbon atoms:

```
    H   H   H   H   H
    |   |   |   |   |
H—C⁵—C⁴—C³—C²—C¹—H
    |   |   |   |   |
    H   H   H   |   H
                |
            H—C—H
                |
                H
```

Note that this alkane is a hexane since it contains six carbon atoms, but that it will be named as a derivative of pentane to specify which hexane it is.

Table 22.4 Unbranched Alkanes through Decane

Name	Molecular Formula
Methane	CH_4
Ethane	C_2H_6
Propane	C_3H_8
Butane	C_4H_{10}
Pentane	C_5H_{12}
Hexane	C_6H_{14}
Heptane	C_7H_{16}
Octane	C_8H_{18}
Nonane	C_9H_{20}
Decane	$C_{10}H_{22}$

Organic Chemistry

4. The carbon atoms in the parent hydrocarbon are numbered so as to give the lowest possible number to the carbon atoms to which other substituents are attached. The numbering in the example above is from right to left, giving the number 2 to the substituted carbon atom.

5. The names of the substituent groups are attached directly to the name of the parent hydrocarbon, with a number to indicate the carbon atom in the parent hydrocarbon to which they are attached. The numbers precede the names of the substituted groups and are separated from them by a hyphen.

6. The names of the substituent groups containing only carbon and hydrogen (*alkyl* groups) are derived by dropping *-ane* from the name of the appropriate hydrocarbon and adding *-yl*. Thus the complete name of the compound above is 2-methylpentane. In some cases the number of the substituted carbon atom can be dropped without introducing any ambiguity. For example, there is only one methylbutane:

7. The prefixes *di-*, *tri-*, *tetra-*, etc., are used when the same substituent appears more than once. If two substituents are on the same carbon, the number must be repeated. Consecutive numbers are separated by commas. For example,

2,2,3,3-tetramethylpentane

Questions

1. Write the carbon skeleton, electron, molecular, structural, condensed structural, and empirical formulas of methylpropane.

2. Write the structural formulas for all the structural isomers of heptane and assign an IUPAC name to each.

3. Write the structural formula and name for a homolog of methylpropane. Write the structural formula and name for a structural isomer of methylpropane.

22.2 Optical Isomerism

We've seen how compounds with the same molecular formulas can have very different structural formulas because of the different ways that the same atoms can be put together. Now we shall see that there is a more subtle form of isomerism where molecules have exactly the same bonds but differ in the way their atoms are arranged in space. We call such molecules **stereoisomers,** and we can often depict their differences only with three-dimensional models or by carefully illustrating their three-dimensional properties in two dimensions. One form of stereoisomerism is **geometric isomerism.** As we shall see in Section 22.4, some geometric isomers can be distinguished from one another in a two-dimensional drawing. The second form of stereoisomerism is **optical isomerism,** and optical isomers (*enantiomers*) can never be distinguished without examining the three-dimensional arrangement of their atoms.

Put simply, an optical isomer is a molecule with a nonsuperimposable mirror image; that is, if a molecule and its mirror image are not identical in every respect, they constitute a pair of enantiomers. An example of this idea from the everyday world is the human hand. A right hand is not superimposable on its mirror image, which in fact is identical to a left hand (Figure 22.7). Therefore a right and a left hand form a pair of macroscopic enantiomers.

No such optical isomers are found in the alkane series until we reach the heptanes. Among the heptanes, however, are two pairs of optical isomers; there are two forms of 3-methylhexane,

```
  H  H  H  H  H  H
  |  |  |  |* |  |
H—C—C—C—C—C—C—H
  |  |  |  |  |  |
  H  H  H  |  H  H
           |
         H—C—H
           |
           H
```

and two forms of 2,3-dimethylpentane,

```
  H  H  H            H  H
  |  |  |*           |  |
H—C—C—C————————————C—C—H
  |  |  |            |  |
  H  H  |            |  H
        |            |
      H—C—H        H—C—H
        |            |
        H            H
```

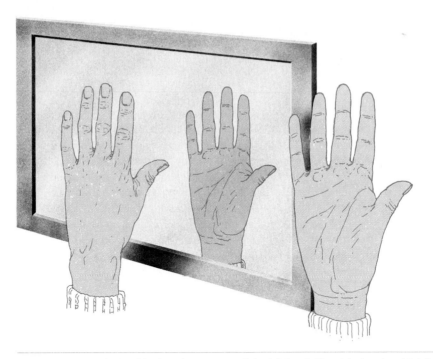

Figure 22.7 The right hand is the mirror image of the left hand. The right hand and left hand are nonsuperimposable and hence asymmetric.

Each of these compounds contains the trademark of most optical isomers: an asymmetric carbon atom. In each structural formula, an asterisk marks a carbon atom that is bonded to four different groups of atoms. We can show that this asymmetric carbon atom ensures optical isomerism by placing it at the center of a tetrahedron and then putting the four groups at the corners. For 3-methylhexane we can draw two nonsuperimposable forms, one the mirror image of the other (Figure 22.8). Figure 22.8 also shows that a molecule having no asymmetric atom can be superimposed on its mirror image.

Optical isomers are identical in most respects. They have the same color, the same solubilities, and the same melting point. However, there are three important ways by which they can be distinguished. First, they differ in their reactions with other optical isomers. Second, there may be important biological differences. For example, they usually differ in taste, smell, toxicity, and effectiveness as medicinal agents. In some cases, bacteria may feed on one enantiomer and ignore the other. We'll say more about the importance of optical isomerism in biology in Chapter 23.

The third important difference between optical isomers is their opposite rotation of *plane-polarized light*, the property by which they are usually identified and the one for which they were named. To explain this phenomenon, we must first recall some of the properties of light itself.

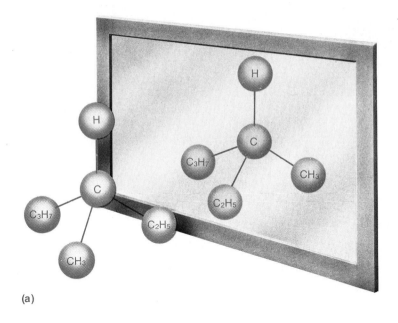

(a)

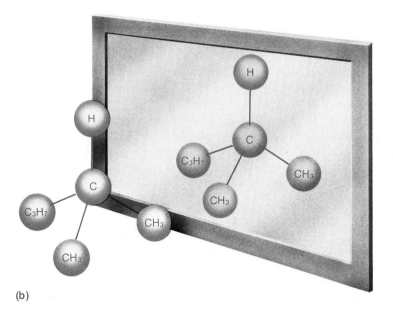

(b)

Figure 22.8 **(a)** The model of 3-methylhexane and its mirror image are nonsuperimposable. Thus the model and the mirror image represent optical isomers. **(b)** The model of 2-methylpentane and its mirror image are superimposable. Hence the model and the mirror image represent the same compound.

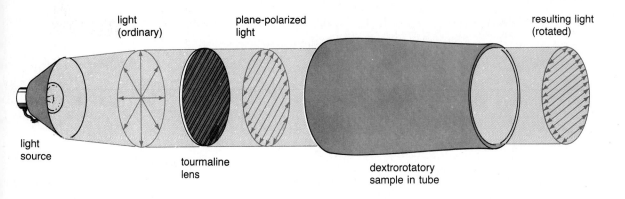

Figure 22.9 Schematic representation of production and rotation of plane-polarized light.

Light can be thought of as a wave phenomenon (Chapter 5) in which each quantum is characterized by its wavelength. In addition, each quantum can be described by the plane in which it vibrates as it moves through space. Ordinarily light consists of waves vibrating in all planes; hence it is said to be *unpolarized*. However, when light passes through certain materials, such as tourmaline (a complex silicate) or calcite (a crystalline form of calcium carbonate), the light that emerges is plane-polarized—all the waves vibrate in the same plane (Figure 22.9).

If this plane-polarized light then passes through a solution containing only one of a pair of enantiomers, the plane of polarization will be rotated. Thus optical isomers are said to be *optically active*. One isomer rotates the plane of polarization to the left and the other rotates the plane of polarization to the right, the angle of rotation being the same in each case. We call the former isomer *levorotatory* (from the Latin *laevus*, meaning *left*) and the latter *dextrorotatory* (from the Latin *dexter*, meaning *right*). A solution containing equal concentrations of the two has no net effect on plane-polarized light and is called a *racemic mixture*.

Questions

4. Distinguish between the members of the following pairs of terms:
   a. structural isomer and stereoisomer
   b. plane-polarized light and ordinary light
   c. enantiomer and structural isomer

5. Write the IUPAC names for the optically active octanes.

22.3 Structure and Nomenclature of Alkenes

The homologous series of hydrocarbons containing one double bond has the general formula C_nH_{2n}. We shall look at several of these compounds as examples of alkenes. The names of all the compounds in this series are derived from the name of the analogous alkane by simply changing the *a* to *e*.

Ethene The simplest alkene is ethene (often called ethylene). Its structure was discussed in Chapter 7.

Propene The name of the alkene containing three carbon atoms is propene. Its condensed structural formula is $CH_3-CH=CH_2$.

Butenes A new type of structural isomerism appears in C_4H_8. The structural isomerism we encountered in the alkanes is called *chain isomerism* and has to do with the arrangement of the carbon atoms. In alkenes we again find chain isomerism, but we also discover *position isomerism*, which arises because the double bond can be at different places along the same carbon chain. There is only one unbranched butane but there are two structurally isomeric unbranched butenes: $CH_3-CH_2-CH=CH_2$ and $CH_3-CH=CH-CH_3$.

Since these two compounds must have different names, we must make an obvious addition to the IUPAC rules of nomenclature. When alkenes are being named, the parent hydrocarbon is the longest chain that contains the double bond, and it is numbered by starting at the end nearer the double bond. The position of the double bond is then indicated by the smaller of the two numbers adjacent to it. Thus the two isomers above are 1-butene and 2-butene, respectively.

The third butene is a chain isomer of the first two:

$$\begin{array}{c} CH_3 \\ \diagdown \\ C=CH_2 \\ \diagup \\ CH_3 \end{array}$$

The name of this compound is methylpropene. Since there is only one possible position for the double bond and only one for the methyl substituent, no numbers are needed to locate either.

22.4 Geometric Isomerism

A second type of stereoisomerism is geometric isomerism (see Section 22.2) and, as with position isomerism, it is with the butenes that we first encounter it. Whenever *each* of two doubly bonded carbon atoms carries two different substituents, different geometric arrangements of the substituents are possible. This is the case with 2-butene, where each of the doubly bonded carbon atoms is attached to both a hydrogen atom and a methyl group. Thus we have *cis*-2-butene if the two methyl groups are adjacent and *trans*-2-butene if they are not:

$$\underset{cis\text{-2-butene}}{\overset{H}{\underset{CH_3}{\diagdown}}C=C\overset{H}{\underset{CH_3}{\diagup}}} \qquad \underset{trans\text{-2-butene}}{\overset{H}{\underset{CH_3}{\diagdown}}C=C\overset{CH_3}{\underset{H}{\diagup}}}$$

Each of these geometric isomers is a distinct chemical compound, and one cannot be converted to the other without breaking the rigid double bond. The other two structural isomers of butene do not exhibit geometric isomerism since they do not meet the requirements established above:

$$\begin{array}{cc} \underset{\text{1-butene}}{\overset{H}{\underset{H}{>}}C=C\overset{CH_2-CH_3}{\underset{H}{<}}} & \underset{\text{methylpropene}}{\overset{H}{\underset{H}{>}}C=C\overset{CH_3}{\underset{CH_3}{<}}} \end{array}$$

Questions

6. Distinguish between members of the following pairs of terms:
   a. optical isomer and geometric isomer
   b. chain isomer and position isomer
   c. *cis*-3-hexene and *trans*-3-hexene

7. Indicate which of the following exhibit geometric isomerism:
   a. $CH_3-CH_2-CH=CH_2$
   b. $CH_3-CH=C(CH_3)(CH_2-CH_3)$
   c. $CH_3-CH_2-CH=C(CH_3)_2$
   d. $CH_3-CH=CH-CH_3$

8. Which pentenes show geometric isomerism?

Sidelight
Insect Pheromones

It has been estimated that 12 percent of the world's crop production is lost to insects. In a world where many people suffer from malnutrition, this is a startling figure, and the need to reduce it is obvious. Yet some effective means of insect control are unsuitable for ecological reasons. DDT is a well-known example (see p. 711).

One especially exciting current approach to insect control is the use of insect *sex pheromones*. These are compounds emitted by insects to attract mates. They are highly specific, and naturally enough insects respond to them strongly. Therefore, if the sex pheromone of an insect can be identified and synthesized, it can be placed in traps and used to remove insects from the reproductive cycle.

As a surprisingly simple example, the sex pheromone of the female housefly has been identified as the *cis* isomer of an unbranched alkene containing 23 carbon atoms:

$$CH_3(CH_2)_{11}CH_2\overset{H}{\underset{}{>}}C=C\overset{H}{\underset{CH_2(CH_2)_6CH_3}{<}}$$

It is often called muscalure, but its IUPAC name is *cis*-9-tricosene.

22.5 Structure and Nomenclature of Alkynes

The series of alkynes that contain only one triple bond is described by the formula C_nH_{2n-2}. The compounds are named in a way exactly analogous to the alkenes in Section 22.3, except that here we replace the *a* in the corresponding alkane with a y. Ethyne, more often called acetylene, is the first member of the series and was discussed in Chapter 7. Next is propyne:

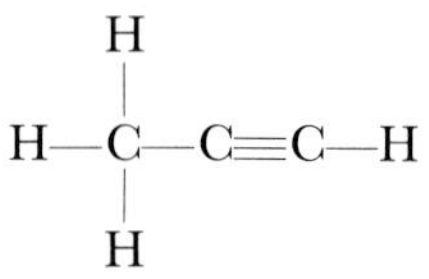

$$\text{H}-\underset{\underset{\text{H}}{|}}{\overset{\overset{\text{H}}{|}}{\text{C}}}-\text{C}\equiv\text{C}-\text{H}$$

With butyne, the third alkyne, we again encounter isomerism, though it is never as extensive as among the alkenes—the triple bond offers no opportunity for geometric isomerism. The two structural isomers of butyne are position isomers:

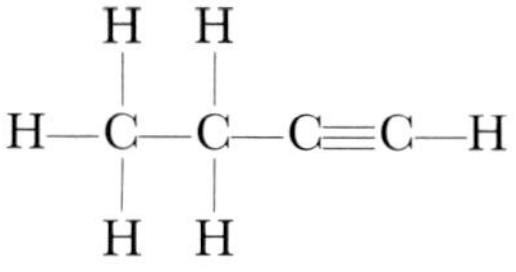

1-butyne 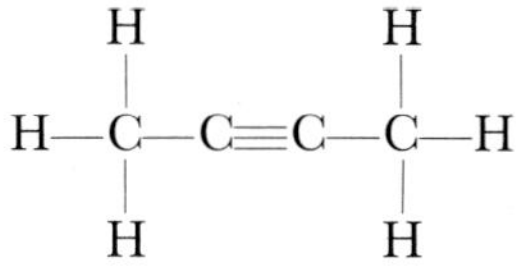   2-butyne

Questions

9. Why are there no chain isomers of C_4H_6?

10. Write structural formulas for the seven hexynes.

22.6 Structure and Nomenclature of the Cycloalkanes

The cycloalkanes can be thought of as chain alkanes in which two nonadjacent carbon atoms have been linked together to form a ring, at the same time losing two hydrogen atoms. Their general formula is, therefore, C_nH_{2n}. The four simplest unsubstituted cycloalkanes are

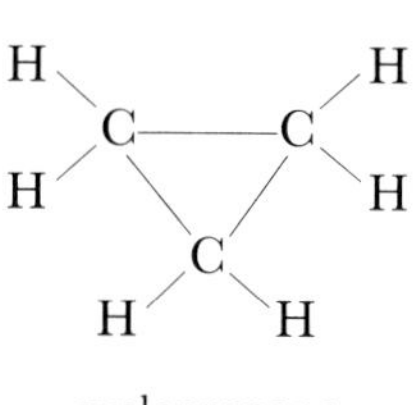

cyclopropane 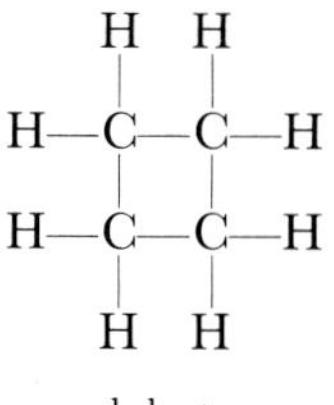   cyclobutane

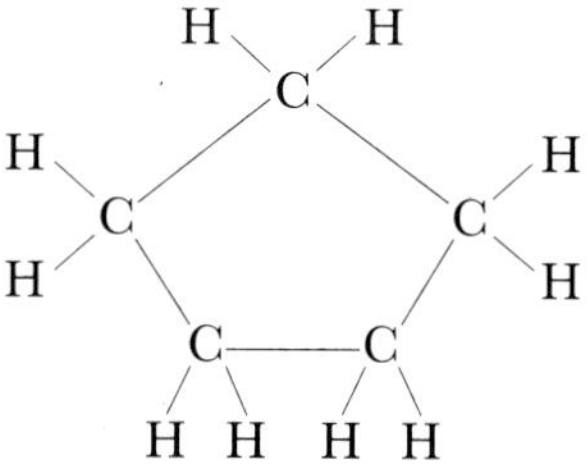

cyclopentane 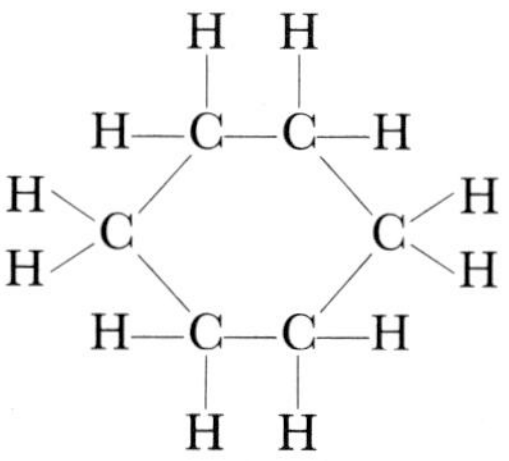   cyclohexane

Organic Chemistry

The geometric structures of both cyclopentane and cyclohexane permit the bond angles around carbon to be close to the usual 109°28′. However, the bond angles in the cyclopropane and cyclobutane rings (60° and 90°, respectively) deviate markedly from 109°28′. As a result, the three- and four-membered rings are much less stable than the larger ones. It should not surprise us to find that five- and six-membered rings are by far the most common in organic chemistry.

Question

11. Write structural formulas for the structural isomers of C_5H_{10}. Include cyclic and chain hydrocarbons.

22.7 Structure and Nomenclature of Cycloalkenes

Introducing double bonds into the structure of a cycloalkane produces a cycloalkene. For example,

cyclopentene   cyclohexene

All positions of the double bond are equivalent in these simple examples; hence no position isomers exist, and no numbers are needed to name them. When we add a second double bond, however, we must be more careful. For example,

1,3-cyclohexadiene   1,4-cyclohexadiene

These compounds are again named by a simple extension of the IUPAC rules. The carbon atoms of the ring are numbered to give the double bonds the lowest numbers possible; then the compounds are given the suffix -*diene*. Introducing a third double bond into cyclohexane produces

which we might expect to call 1,3,5-cyclohexatriene. Instead we produce benzene, the simplest of the aromatic compounds and one of the most important of all organic compounds. As we shall see in Section 22.8, it is not merely a cycloalkene with three double bonds.

22.8 Structure and Nomenclature of Aromatic Hydrocarbons

The reason benzene holds such a special place in organic chemistry is suggested by its structure. The picture we have drawn for the mythical cyclohexatriene molecule implies that it must contain both single and double bonds and that pi electrons are localized between pairs of carbon atoms. Experiments show that in benzene neither of these is the case. *All the bonds between the carbon atoms in the benzene molecule are equivalent*—neither single nor double but something in between.

We can account for this finding in terms of resonance (see Chapter 7), writing the two contributing forms as

The result is an unexpectedly stable structure lacking many of the properties usually associated with unsaturated hydrocarbons. For example, unlike alkenes, benzene will not readily react with hydrogen or bromine nor will it decolorize an aqueous solution of potassium permanganate (see Section 22.12).

These resonance structures provide a useful picture of the bonding in benzene, but a more precise physical explanation is possible by comparing it to the bonding in graphite (see Chapter 9). As in graphite, the carbon atoms in each benzene molecule lie in the same plane, suggesting sp^2 bonding. The sp^2 bonds form the skeleton of the benzene ring, providing sigma bonds between adjacent carbon atoms and contributing to the carbon-hydrogen bonds (Figure 22.10). The remaining $2p$ orbitals, which are perpendicular to the plane of the ring, overlap one another, forming a pi cloud. The electrons in this pi cloud are not localized between two carbon atoms but move freely about the ring (see Figure 22.11).

Since it is difficult to represent the benzene molecule adequately with a normal structural formula, many chemists prefer to show it simply as a hexagon containing a circle:

It is understood that one carbon and one hydrogen reside at each corner and that the circle represents the six delocalized pi electrons.

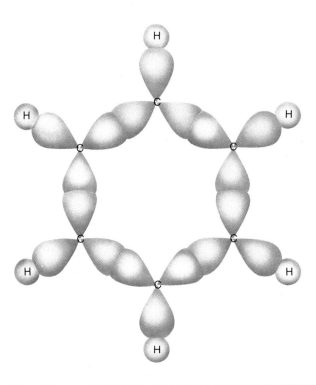

Figure 22.10 The sigma-bond framework of benzene.

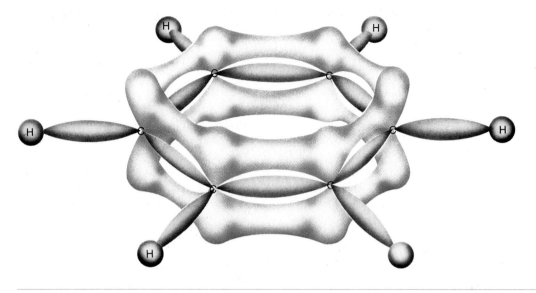

Figure 22.11 A representation of the benzene molecule illustrating both sigma and pi bonds.

In addition to benzene, there are several other aromatic hydrocarbons, many of which we can represent as fused benzene rings. For example,

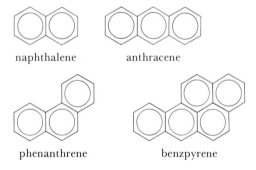

naphthalene   anthracene

phenanthrene   benzpyrene

Naphthalene is commonly sold as moth repellant, and it gives its distinctive aroma to "moth balls." Benzpyrene has been identified as a *carcinogenic* (cancer-producing) compound and has been detected in cigarette smoke.

Many important hydrocarbons can be thought of as combinations of aromatic compounds and alkanes. The simplest of these is a homolog of benzene, methylbenzene (commonly called toluene):

We can think of toluene either as a derivative of benzene, in which a hydrogen atom on the ring has been replaced by a methyl group, or as a derivative of

Organic Chemistry   695

methane, in which a hydrogen atom has been replaced by a *phenyl* group. The phenyl group is simply a benzene ring minus a hydrogen atom.

There is only one methylbenzene since all positions on the benzene ring are equivalent, but when we add a second methyl group to benzene, three structural isomers are possible.

ortho-xylene
1,2-dimethylbenzene

meta-xylene
1,3-dimethylbenzene

para-xylene
1,4-dimethylbenzene

Like toluene, xylene is a widely accepted unsystematic name, and the prefixes *ortho-*, *meta-*, and *para-* indicate that the two substituents are adjacent, separated by one carbon, and opposite, respectively. To arrive at the systematic names, we must number the carbons of the benzene rings so that the substituents have the lowest numbers possible. Thus the systematic name of *meta*-xylene is 1,3-dimethylbenzene and not 1,5-dimethylbenzene.

Questions

12. Write structural formulas for the following:
    a. 3,3-dimethylhexane   b. *cis*-2-pentene   c. 2-methyl-4-ethyloctane
    d. 1-hexyne   e. toluene   f. *meta*-xylene   g. anthracene

13. Write structural and molecular formulas for
    a. 1,2,4-trimethylbenzene   b. ethylbenzene
    c. 2-phenylbutane   d. cyclopentadiene

22.9 Physical Properties of Hydrocarbons

Since most alkenes, alkynes, and aromatic hydrocarbons closely resemble the alkanes, we can discuss the properties of the hydrocarbons in terms of the alkanes alone. They are colorless substances with their boiling points increasing regularly with molecular weight (Figure 22.12). Since the electronegativities of hydrogen and carbon are very similar, the bonds in most hydrocarbons are essentially nonpolar. This together with the symmetrical arrangement of the bonds ensures that most hydrocarbons are nonpolar molecules and that molecular interactions are limited almost completely to London dispersion forces. Since these forces are weaker among the more compact branched alkanes (because of their smaller surface areas), they have slightly lower boiling points than their unbranched structural isomers.

Since they are nonpolar molecules, alkanes are insoluble in water and other polar solvents. For the same reason, they dissolve nonpolar compounds but not ionic or polar substances.

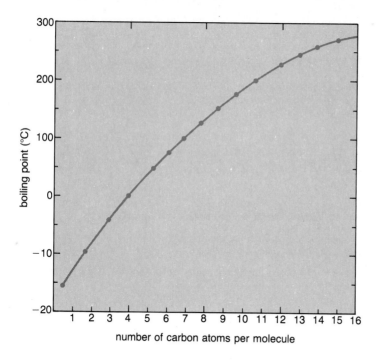

Figure 22.12 Boiling points of unbranched alkanes.

22.10 Combustion of Hydrocarbons

Among the most important reactions in chemistry are those of hydrocarbons with oxygen. Because these reactions are highly exothermic, hydrocarbons are widely used as fuels, as Table 22.5 amply demonstrates.

The combustion products vary with the relative amounts of the hydrocarbon and oxygen. The following equations, for example, represent three possible reactions between methane and oxygen:

$$CH_4(g) + O_2(g) \longrightarrow C(s) + 2\,H_2O(g)$$
$$2\,CH_4(g) + 3\,O_2(g) \longrightarrow 2\,CO(g) + 4\,H_2O(g)$$
$$CH_4(g) + 2\,O_2(g) \longrightarrow CO_2(g) + 2\,H_2O(g)$$

The ratio of oxygen to methane is 1 to 1 in the first reaction, 1.5 to 1 in the second, and 2 to 1 in the last. Of course the extent to which the carbon is oxidized varies with this ratio.

We can write similar equations for other hydrocarbons. A practical illustration is the complete combustion of an octane:

$$2\,C_8H_{18}(l) + 25\,O_2(g) \longrightarrow 16\,CO_2(g) + 18\,H_2O(g)$$

Organic Chemistry

Table 22.5 Hydrocarbon Fuels

Name of Fuel	Composition	Use
Natural gas	Mixture of hydrocarbons with 1 to 4 carbon atoms per molecule (usually over 80 percent CH_4)	Home and industrial use
Propane	C_3H_8	Heating farm homes, drying crops, camp stoves, bottled gas fuel
Butane	C_4H_{10}	Home and industrial fuel
Gasoline	Complex mixture of hydrocarbons, mostly octanes, heptanes, and hexanes	Internal combustion engines
Kerosene	Mixture of hydrocarbons containing 10 to 16 carbon atoms per molecule	Jet fuel, lighting in remote rural areas
Paraffins	Mixture of hydrocarbons containing 16 to 28 carbon atoms per molecule	Candles (also greases)
Acetylene	C_2H_2	High-temperature work, welding and cutting metals, illuminant

Since octanes are among the main ingredients of gasoline, this is one of the many reactions that take place in internal combustion engines. Unfortunately, most automobiles perform best if the carburetor is adjusted so that combustion is not complete. As a result, appreciable quantities of carbon monoxide are expelled through the exhaust system.

Question

14. Write an equation for the complete combustion of heptane.

22.11 Substitution Reactions

Substitution reactions are characteristic of alkanes. In these reactions hydrogen is replaced by another element in the presence of light. The following series of equations illustrates the four substitution products possible when chlorine reacts with methane:

$$CH_4 + Cl_2 \longrightarrow HCl + CH_3Cl \quad \text{chloromethane (methyl chloride)}$$

$$CH_4 + 2\,Cl_2 \longrightarrow 2\,HCl + CH_2Cl_2 \quad \text{dichloromethane (methylene chloride)}$$

$$CH_4 + 3\,Cl_2 \longrightarrow 3\,HCl + CHCl_3 \qquad \text{trichloromethane (chloroform)}$$

$$CH_4 + 4\,Cl_2 \longrightarrow 4\,HCl + CCl_4 \qquad \text{tetrachloromethane (carbon tetrachloride)}$$

The relative amounts of chlorine and methane determine which product will predominate. Analogous equations can be written for the reactions of bromine with methane, but iodine will not replace hydrogen in hydrocarbons. With fluorine on the other hand, hydrocarbons ignite spontaneously in a reaction too vigorous to be controlled.

We discussed the mechanism for the substitution of chlorine in methane in Chapter 12.

Substitution reactions in other alkanes proceed by similar mechanisms. But beyond ethane, the situation is slightly complicated by the possibility of isomers, even when only one chlorine atom is substituted. And the more complex the alkane molecule, the greater the number of possible isomers.

We mentioned earlier in this chapter that the alkanes are sometimes referred to as paraffins. This word comes from the Latin words meaning *little affinity*. The term is appropriate since, aside from participating in combustion and substitution reactions, alkanes are relatively inactive substances. They do not react with strong acids, bases, or common aqueous oxidizing agents. The same holds true for most cycloalkanes, though cyclopropane and cyclobutane are special cases because of their strained and unstable ring structures. Among the reactions that only they undergo is the following:

$$\underset{CH_2-CH_2}{\overset{CH_2}{\triangle}} + H_2 \xrightarrow[\text{catalyst}]{\text{heat}} CH_3-CH_2-CH_3$$

Question

15. Why are alkanes called paraffins?

22.12 Reactions of Unsaturated Hydrocarbons

Unsaturated hydrocarbons are considerably more reactive than the alkanes—a reflection of the greater electron density in molecules containing multiple bonds. Characteristically, therefore, the unsaturated hydrocarbon behaves as a Lewis base or *nucleophile* (from the Greek, *nucleus loving*), seeking a nucleus with which to share its electrons. The other reactant then is a Lewis acid or *electrophile* (*electron loving*), seeking electrons to share.

Among the most common such reactions are addition reactions across a double bond, where one pair of the electrons of the double bond is used to form new covalent bonds. Among the reagents that will "add" to alkenes are hydrogen, chlorine, bromine, hydrogen chloride, hydrogen bromide, and sulfuric acid. Thus ethene reacts with hydrogen to form ethane:

$$\text{H}_2\text{C}=\text{CH}_2 + \text{H}_2 \xrightarrow{\text{Ni}} \text{H}_3\text{C}-\text{CH}_3$$
ethene → ethane

(It is customary to indicate a catalyst by writing it over the arrow as we've done here.) Other typical addition reactions can be written as follows:

$$\text{H}_2\text{C}=\text{CH}_2 + \text{Cl}_2 \longrightarrow \text{ClH}_2\text{C}-\text{CH}_2\text{Cl}$$
ethene → 1,2-dichloroethane

$$\text{H}_2\text{C}=\text{CH}_2 + \text{HCl} \longrightarrow \text{H}_3\text{C}-\text{CH}_2\text{Cl}$$
ethene → chloroethane (ethyl chloride)

$$\text{H}_2\text{C}=\text{CH}_2 + \text{H}_2\text{SO}_4 \longrightarrow \text{H}_3\text{C}-\text{CH}_2\text{OSO}_3\text{H}$$
ethene → ethylsulfuric acid

Alkynes undergo similar addition reactions, but as we might expect, twice as much of the electrophile is needed to *saturate* a triple bond:

$$\text{H}-\text{C}\equiv\text{C}-\text{H} + 2\,\text{H}_2 \xrightarrow{\text{Ni}} \text{H}_3\text{C}-\text{CH}_3$$
ethyne → ethane

$$\text{H}-\text{C}\equiv\text{C}-\text{H} + 2\,\text{Cl}_2 \longrightarrow \text{Cl}_2\text{HC}-\text{CHCl}_2$$
ethyne → 1,1,2,2-tetrachloroethane

Alkenes and alkynes also react with common aqueous oxidizing agents. For example, solutions of potassium permanganate are decolorized by unsaturated hydrocarbons. Thus a common test for unsaturation is to shake a small quantity of the hydrocarbon with aqueous permanganate. If the hydrocarbon is unsaturated, the purple color will disappear. As we've already mentioned, potassium permanganate does not attack alkanes. Another test for unsaturation is to treat the hydrocarbon with a solution of bromine in carbon tetrachloride. Unsaturated hydrocarbons combine with the bromine quickly, and the solution is soon decolorized. Although alkanes also react with bromine, the substitution reaction is much slower than the addition reaction, so the orange color of the bromine solution persists much longer.

Questions

16. By what mechanism do olefins typically react?

17. How would you prepare monochloroethane from
 a. ethane? b. ethene?

22.13 Polymers

The word *polymer* comes from the Greek word meaning *many parts*. In chemistry, therefore, it is applied to huge molecules, comprising many smaller parts, that are formed by the **polymerization** of simpler molecules called *monomers* (*single part*). Molecular weights of polymers range from several thousand to as high as 15 million.

Polymers can be classified in a number of different ways, but the two most obvious classes are natural and synthetic. Some natural polymers are natural rubber, cellulose, starch, and proteins, each of which we'll discuss later in this chapter or in the next chapter. Among the synthetic polymers that everyone has encountered are plastics and synthetic fibers such as Orlon. We can also classify polymers as inorganic or organic. We're concerned only about the more common and much more useful organic polymers here, but an example of an inorganic polymer is *mu*-sulfur, where the monomer is the S_8 molecule.

Polymerization can take place by simple self-addition of the monomers, or it can occur by the elimination of a small molecule (often water) during the reaction of monomers. We shall discuss only the *addition polymers* here, reserving discussion of *condensation polymers* for later in this chapter.

Polyethylene is one of the simplest examples of an addition polymer. More of it is produced in the United States than any other synthetic polymer. One of the several methods for polymerizing ethylene (ethene) uses hydrogen peroxide to initiate the reaction:

$$n\ CH_2\!\!=\!\!CH_2 \xrightarrow{\text{hydrogen peroxide}} -(CH_2-CH_2)_n-$$

A second method for polymerizing ethylene is the *Ziegler process*, named after the German chemist Karl Ziegler (1898–1973), who won the Nobel prize in 1963 for developing it. A typical catalyst for this process is formed from aluminum alkyls (which are compounds such as triethylaluminum, where aluminum atoms are bonded to alkyl groups) and titanium(IV) chloride ($TiCl_4$). The mechanism for this reaction is believed to involve the formation of bonds between ethene and titanium.

The properties of polyethylene can be modified somewhat by controlling the length of the chain, but in general we can describe this plastic as a light, translucent solid that is highly resistant to chemical attack. It appears in such products as insulating material, carpeting, rope, flexible piping, and containers.

Numerous other addition polymers are made by using derivatives of ethene. Some of these are listed in Table 22.6, along with their formulas, common names, properties and uses and the structural formulas of the monomers from which they are made.

A third way to classify polymers is by their properties. One class would then be the *plastics*—synthetic polymers that can be molded into stable forms by heat and pressure. Plastics that can be remolded when heated are *thermoplastic*; those that cannot are *thermosetting*. The chemical reason for the difference between these two types of plastics is that heating changes the molecular structure of the latter but not the former.

A second class of polymers distinguished by their properties is the *elastomers*—linear polymers that return to their original shape after being stretched to more than 150 percent of their original length. Until World War II, natural rubber was the only important elastomer. It consists of unsaturated hydrocarbons of high molecular weight and is found in the coagulated latex of certain tropical trees. The monomeric unit is commonly known as isoprene. The structural formulas for isoprene and natural rubber are

$$\underset{\text{isoprene}}{\begin{array}{c}\text{H} \\ | \\ \text{H}-\text{C}-\text{H} \\ \text{H} \quad\quad | \quad\quad \text{H} \quad \text{H} \\ | \quad\quad | \quad\quad | \quad\quad | \\ \text{H}-\text{C}=\text{C}-\text{C}=\text{C}-\text{H}\end{array}} \quad \text{and} \quad \left[\begin{array}{c}\text{H} \\ | \\ \text{H} \quad \text{H}-\text{C}-\text{H} \quad \text{H} \quad \text{H} \\ | \quad\quad | \quad\quad\quad | \quad\quad | \\ -\text{C}-\quad\text{C}=\text{C}-\text{C}- \\ | \quad\quad\quad\quad\quad\quad\quad\quad | \\ \text{H} \quad\quad\quad\quad\quad\quad\quad \text{H}\end{array}\right]_n$$

The molecular weights of rubber molecules may vary considerably, but they probably range as high as 400,000.

Natural rubber, as such, is not a very useful material because it is soft, sticky, and lacks resilience. The useful commercial product is made by heating natural rubber with sulfur in a process called *vulcanization*. The sulfur links carbon atoms in parallel adjacent chains by reacting at the double bonds. It thus acts as a bridge among the polymer chains. A soft rubber contains about 5 percent sulfur, whereas hard rubbers contain as much as 30 percent sulfur.

During World War II, when supplies of natural rubber were disrupted, several important synthetic elastomers were produced. Among them were addition polymers of 1,3-butadiene and 2-chloro-1,3-butadiene. Rubber sub-

Table 22.6 Addition Polymers of Ethene Derivatives

Name of Polymer	Common Name of Monomer	Structural Formula of Monomer	Formula of Polymer	Properties	Uses
Saran	Vinylidene chloride	$H_2C{=}CCl_2$	$[-CH_2-CCl_2-]_n$	Chemically resistant, low water absorption	Wrapping material, upholstery, curtains, and drapery
Teflon*	Tetrafluoroethylene	$F_2C{=}CF_2$	$[-CF_2-CF_2-]_n$	Inert	Radar parts, gaskets for high-temperature uses, cooking-pan coating
Vinylite	Vinyl chloride	$H_2C{=}CHCl$	$[-CH_2-CHCl-]_n$	Chemically resistant	Floor tiling
Orlon*	Vinyl cyanide (acrylonitrile)	$H_2C{=}CHCN$	$[-CH_2-CHCN-]_n$	High tensile strength, light	Textiles
Styrofoam*	Styrene	$C_6H_5CH{=}CH_2$	$[-CH(C_6H_5)-CH_2-]_n$	Light weight	insulation, "foam" drinking cups, boats, toys, packing material
Polyvinyl acetate	Vinyl acetate	$H_2C{=}CH-O-C(=O)-CH_3$	$[-CH_2-CH(O-C(=O)-CH_3)-]_n$	Expands on solidification	Adhesive, emulsion, paints

*Trademarks

stitutes are now available that have properties nearly identical to and in some cases superior to those of natural rubber.

Questions

18. Give an example of each of the following:
    a. inorganic polymer   b. organic polymer
    c. natural polymer   d. synthetic polymer

19. Distinguish between the following terms:
    a. addition polymer and condensation polymer   b. monomer and polymer

22.14 Aromatic Substitution

In Section 22.8 we mentioned that aromatic compounds often fail to behave chemically like other unsaturated hydrocarbons. The reason given was the extra stability afforded by delocalized electrons. If benzene were to undergo an addition reaction, the benefit of delocalization would be largely lost; hence benzene does not often react in this way. (In the presence of a catalyst, however, benzene can be hydrogenated to form cyclohexane.) Many of the characteristic reactions of benzene, therefore, are substitution reactions, making it more akin to alkanes than alkenes. For example,

$$C_6H_6 + Cl_2 \xrightarrow{FeCl_3} C_6H_5Cl \text{ (chlorobenzene)} + HCl$$

$$C_6H_6 + HNO_3 \xrightarrow{H_2SO_4} C_6H_5NO_2 \text{ (nitrobenzene)} + H_2O$$

$$C_6H_6 + H_2SO_4 \xrightarrow{SO_3} C_6H_5SO_3H \text{ (benzenesulfonic acid)} + H_2O$$

Question

20. What chemical reaction do all hydrocarbons have in common?

22.15 Petroleum

It is difficult to overestimate the importance of petroleum to modern civilization. The headlines of the 1970s have, perhaps for the first time, made us all aware of its crucial role as a source of energy, but its use as a source of raw materials for plastics, drugs, insecticides, and synthetic fibers is still not widely appreciated or understood.

Composition of Petroleum Petroleum is a highly complex mixture consisting of hundreds of compounds having molecular weights that range from 16 (methane) to over 6000. Furthermore, the composition of petroleum varies widely, depending on its source. However, hydrocarbons—predominantly unbranched chain alkanes—make up the major portion of any sample of petroleum. Other components include compounds containing oxygen, nitrogen, and sulfur.

Refining Petroleum The separation of pure components from a large sample of petroleum would be a gigantic task. Fortunately, complete separation is not necessary to obtain usable products. Instead, petroleum is separated into various fractions, each with a characteristic boiling point range. It is this process that we call *refining*. The names of the various fractions, their hydrocarbon content, and their boiling point ranges appear in Table 22.7.

Table 22.7 Petroleum Fractions

Name	Hydrocarbon Content	Boiling Point Range (°C)	Uses
Natural gas	Methane–butanes	Below room temperature	Fuel, production of gasoline
Petroleum ether	Pentanes, hexanes	20–60	Solvent, dry cleaning
Ligroin	Hexanes, heptanes	60–100	Solvent
Gasoline	Hexanes–dodecanes	50–200	Motor fuel
Kerosene	Dodecanes–hexadecanes	175–275	Diesel fuel, jet fuel, rocket fuel, illuminant, production of gasoline
Fuel oil, diesel fuel	Pentadecanes–octadecanes	250–400	Fuel, production of gasoline
Lubricating oils	Hydrocarbons above hexadecanes	350 and above	Lubricant
Asphalt	Variable and complex mixture of derivatives of hydrocarbons	Residue	Paving, roofing

All the fractions in Table 22.7 have important uses, but the most important of them by far is gasoline. The gasoline fraction obtained directly from petroleum distillation is called *straight-run* gasoline. It is neither sufficient in quantity nor of high enough quality to meet modern demands. Fortunately, processes are available to improve its quality as well as to produce more from the natural gas, kerosene, and fuel oil fractions.

Gasoline The efficiency of a gasoline engine is governed largely by its compression ratio—the ratio of the volume of the gas and air mixture before compression to the volume of the mixture after compression. The larger the compression ratio, the greater the power that can be obtained from the engine. However, if the compression ratio is raised too high, the engine starts to *knock*. This phenomenon, which many drivers of older high-compression cars have heard all too often, is a result of sudden violent explosions in the cylinder instead of smooth combustion.

The tendency of a gasoline to resist knocking is indicated by its *octane number*, which is determined by comparing its antiknock quality to the antiknock quality of standard fuels. Since branched hydrocarbons tend to knock less than unbranched hydrocarbons, 2,2,4-trimethylpentane (incorrectly referred to as *isooctane* in the petroleum industry) was originally selected as the standard by which others would be compared. It was arbitrarily given an octane number of 100. At the other end of the scale, heptane, which produces severe knocking, was assigned an octane number of zero. Gasolines can therefore be rated according to how they compare with various mixtures of 2,2,4-trimethylpentane and heptane. A fuel showing the same tendency to knock as a 90/10 percent mixture of 2,2,4-trimethylpentane and heptane is given an octane number of 90. Since the standard is arbitrary, some gasolines may have octane numbers greater than 100.

The gasolines of the early days of the automobile were very poor fuels compared to our modern gasolines. The most important reasons for this have been the various additives that petroleum research has developed—foremost among them tetraethyl lead. It was observed in 1922 that a few milliliters of this compound added to each gallon of gasoline greatly increased the octane number of gasoline. Since then, until very recently, it was a common constituent of gasoline. To prevent the formation of lead(II) oxide (which would foul the cylinders) during the oxidation of ethyl gasoline, it is also necessary to add 1,2-dibromoethane or 1,2-dichloroethane to the gasoline. The halogens in these additives form volatile lead(II) halides, which are expelled with the exhaust gases. Thus formation of the undesirable PbO is eliminated or greatly reduced. Recently, environmental concerns have dictated a drastic reduction in the use of lead compounds in gasoline. Many late-model cars now require the use of unleaded gasoline.

Questions

21. How is straight-run gasoline produced?

22. What causes engine knock?

23. What is the octane rating of a gasoline with the same antiknock properties as a mixture of 48 percent heptane and 52 percent 2,2,4-trimethylpentane?

22.16 Other Sources of Hydrocarbons

Although petroleum is by far the leading industrial source of hydrocarbons, coal is another source worth our attention. Most coal mined in this country is converted into *coke* by heating it in the absence of air. The volatile materials driven off in this process are complex mixtures, referred to as coal gas and coal tar (a liquid), depending on their physical states. Coal tar contains mostly aromatic hydrocarbons, including simple ones such as benzene, toluene, and the xylenes. These compounds can then be isolated by distillation.

By far the major use of coke is in the production of iron and steel (see Chapter 21); another is in the production of ethyne (acetylene). This process involves heating lime (CaO) and coke together at about 2000°C. The resulting calcium carbide (CaC_2) hydrolyzes to form acetylene:

$$3\ C(s) + CaO(s) \longrightarrow CaC_2(s) + CO(g)$$

$$CaC_2(s) + 2\ H_2O(l) \longrightarrow Ca(OH)_2(s) + C_2H_2(g)$$

Coal is not currently being used as a source of gasoline but as petroleum reserves dwindle, we may have to consider this alternative. Our petroleum reserves have been estimated as sufficient for only a few more decades at the present rate of consumption, whereas coal reserves in the United States have been calculated to be sufficient for hundreds of years. Several methods have already been developed to produce gasoline from coal, but none of them is economically competitive yet.

Question

24. Write equations illustrating the following:
  a. Synthesis of hydrocarbons from inorganic compounds.
  b. Synthesis of an alkane from an alkyne.

22.17 Air Pollution from Hydrocarbon Fuels

The development of the internal combustion engine has been one of the great technological advances of our time. But as so often happens with technological advances in our society, we failed for too long to be concerned with its by-products. As a result, some of our major population centers now suffer from one of these intolerable side effects—air pollution.

The major pollutants from the gasoline engine are carbon monoxide, nitrogen oxides, lead compounds, and unburned hydrocarbons. We've already mentioned all but the last as pollutants, but we now want to look at each more closely.

Man's activities account for only a small portion of these airborne compounds (except the lead compounds) on a global scale; however, man's activities do account for the dangerously high concentrations of them around large cities. In the case of carbon monoxide, the concentrations in urban areas may be 50 to 100 percent above the global average. Although inhaling carbon monoxide deprives the body of oxygen (see Chapter 21), the long-term effect on healthy people of sublethal doses of carbon monoxide has not been established. It appears that in most cases the effects of carbon monoxide are reversible, but some fragmentary evidence suggests adverse effects on the nervous system when even low concentrations of CO are inhaled. High concentrations of CO present the greatest hazard to people who suffer from respiratory or heart ailments. It has been shown, for example, that there is a greater danger of heart failure when the CO concentration is high.

The unburned hydrocarbons emitted by the automobile engine do not remain in the atmosphere long and are harmful only in high concentrations. The major problem with hydrocarbons is that in bright sunlight they react with nitrogen oxides to form more dangerous pollutants.

The reaction $N_2(g) + O_2(g) \longrightarrow NO(g)$ proceeds to an insignificant degree at normal temperatures. However, the high temperatures in the internal-combustion engine promote this reaction and result in considerable NO being emitted in automobile exhaust. The NO subsequently reacts with O_2 to form NO_2, which is largely responsible for the reddish brown color of smog. Nitrogen dioxide is a lung irritant, but again the major dangers are the products of reactions between NO_2 and hydrocarbons.

Nitrogen dioxide can dissociate photochemically (that is, by absorbing light) to yield nitric oxide and a highly reactive oxygen atom:

$$NO_2 \xrightarrow{\text{sunlight}} NO + O$$

The atomic oxygen ($:\!\ddot{O}\!\cdot$) can then react with a hydrocarbon to form an organic *radical* (a molecular fragment having one or more unpaired electrons):

$$CH_3\!-\!CH_2\!-\!CH_2\!-\!CH_2\!-\!CH_3 + :\!\dot{\ddot{O}}\!\cdot \longrightarrow$$

$$CH_3\!-\!CH_2\!-\!CH_2\!-\!CH_2\!-\!CH_2\!\cdot + \cdot\ddot{O}\!:\!H$$

It is these radicals that are believed to be primarily responsible for the irritation caused by smog and, worse, they are suspected of causing cancer. Atomic oxygen can also react with diatomic oxygen to form ozone (Chapter 19), which in turn can react with hydrocarbons to form radicals.

A large portion of the lead halides (from tetraethyl lead) in automobile exhaust is in aerosol form; hence these lead compounds remain airborne for long periods. Lead is a cumulative poison and lingers in the body, damaging the central nervous system and the brain. Also, Pb^{2+} ions can replace the Ca^{2+} ions in bone structures, and lead compounds can interfere with the action of certain enzymes (Chapter 23). As we said in Chapter 21, there is no consensus on the seriousness of low-level lead poisoning from exhaust fumes, but it seems prudent to continue making efforts to develop efficient automobile engines that do not require even small amounts of lead compounds in gasoline.

22.18 Hydrocarbon Derivatives

As numerous as the hydrocarbons are, they are far outnumbered by their derivatives, each of which can be thought of as a hydrocarbon with one or more hydrogen atoms replaced with atoms or groups of atoms of some other element. We call this newly substituted portion of the molecule a *functional group* and the unaltered *hydrocarbon residue* an *alkyl* or *aryl* group. Aryl derivatives are those in which there has been substitution on an aromatic ring; all others are alkyl derivatives.

The capital letter R is commonly used to represent hydrocarbon residues; thus the general formula for a class of derivatives consists of the letter R plus the functional group that identifies the class. Table 22.8 lists the most common of these functional groups, along with the names given to the resulting derivatives. We shall not try to discuss all of these functional groups in this text.

Although hydrocarbon residues can undergo the same reactions as the original hydrocarbons, the functional group is the most reactive site in most derivatives. As a result, the study of hydrocarbon derivatives is primarily the study of functional groups. And although the size of the hydrocarbon residue can affect the chemistry of the functional group, the effect is usually one of degree rather than kind. There are, of course, many organic compounds that contain more than one functional group, in which case the chemistry can be more complicated.

22.19 Halides

The simpler halides are often referred to by the name of the alkyl or aryl residue. For example, CH_3—Br is methyl bromide, and C_2H_5—Cl is ethyl chloride. For larger halides, however, we must use the proper IUPAC names. This system can be complicated (though unambiguous) for complex molecules, but usually we can simply replace the *-ine* halogen ending with an *-o* and then locate the substituent with a number:

```
    H   H   H   H   H
    |   |   |   |   |
H—C—C—C—C—C—H
    |   |   |   |   |
    H   H   Cl  H   H
```
3-chloropentane

The polarity of the carbon-halogen bond in these compounds results in boiling points somewhat higher than those of hydrocarbons having the same molecular weight. However, in spite of their polarity, most organic halides are insoluble in water.

Table 22.8 Classes of Derivatives

Functional Group	General Formula	Name
—X (X = F, Cl, Br, I)	R—X	Halides
—O—H (—OH)	R—O—H	R = alkyl group: alcohols R = aryl group: phenols
—CHO (H—C=O)	R—CH=O	Aldehydes
—CO— (>CO)	R—CO—R	Ketones*
—COOH	R—COOH	Carboxylic acids
—COO⁻M⁺	R—COO⁻M⁺	Carboxylate salts (M^+ = metal cation)
—COO—	R—COO—R'	Esters*
—O—	R—O—R	Ethers*
—NH (—NH₂)	R—NH—H	Amines
—NH⁺ (—NH₃⁺)	R—NH₂⁺—H X⁻	Derived ammonium salts
—CONH₂	R—CONH₂	Amides
—S—H (—SH)	R—S—H	Thiols
—C≡N (—CN)	R—C≡N	Nitriles
—NO₂	R—NO₂	Nitro compounds

*R′, R″, etc., represent alkyl or aryl groups that may or may not be the same as the group represented by R.

There are many commercially important hydrocarbon halides, some of which are listed in Table 22.9. Of particular interest is the last entry in the table, DDT (an acronym derived from the common name *dichlorodiphenyltrichloroethane*). This compound first came to the public's attention in the 1960s following the publication in 1962 of Rachel Carson's book, *Silent Spring*. At that time DDT was one of the most widely used agents for controlling insects. DDT appeared to be relatively harmless to man, and it resisted degradation by either environmental agents or organisms, which meant that it persisted long enough to control more than one generation of insects. Ironically, however, this last quality proved to have disastrous consequences. DDT's persistence, together with its solubility in fatty tissues, has resulted in harmful accumulations in animal systems. In a study of wildlife in Long Island (New York) marshes (published in 1967), for example, it was found that plankton had a DDT content of 4×10^{-5} percent; minnows that eat the plankton had a DDT concentration of 1×10^{-4} percent; and birds that eat the minnows had a DDT concentration of 7.5×10^{-3} percent. One effect of this increase of DDT along the food chain was that some birds suffered a calcium deficiency and produced only soft-shelled eggs that were

Table 22.9 Commercially Important Hydrocarbon Halides

Formula	Common Name	IUPAC Name	Use
$CHCl_3$	Chloroform	Trichloromethane	Solvent
CCl_2F_2	Freon-12	Dichlorodifluoromethane	Refrigerant, propellant in spray cans
CH_2BrCH_2Br	Ethylene bromide	1,2-Dibromoethane	Gasoline additive (see p. 706)
(H₂C=CHCl structure)	Vinyl chloride	Chloroethene	Manufacture of polymers (see p. 703)
(p-dichlorobenzene structure)	Paradichlorobenzene	p-Dichlorobenzene or 1,4-dichlorobenzene	Moth repellant
(DDT structure)	DDT	1,1,1-trichloro-2,2-bis(p-chlorophenyl)ethane	Insecticide

unlikely to remain intact long enough to produce young. The near demise of birds such as the bald eagle and the brown pelican can be explained in large part by the presence of DDT and other chlorinated insecticides in the food chain. Since 1972 the use of DDT in the United States has been severely restricted, and many birds are making significant comebacks.

On the positive side, it must be added that DDT was responsible for preventing the spread of typhus in Europe after World War II and that it is still used in many parts of the world to control malaria. Nevertheless, it seems clear that we must continue to seek compounds that control insects without profoundly upsetting the ecological balance. To this end several effective insecticides have been developed that disintegrate rapidly in water, but all of them remain more expensive than DDT.

Since the halogens are among the very few things that react with alkanes, the production of alkyl halides is often the first step in the multistep manufacture of common organic compounds from raw materials such as petroleum and coal. The typical reaction of alkyl halides is the substitution of another atom or group of atoms for the halogen. These substitution reactions, which should be carefully distinguished from the reactions we discussed in Section 22.11, are acid-base reactions in which the alkyl halide (a Lewis acid) reacts with a group containing an unshared pair of electrons (a Lewis base). Because of its unshared pair of electrons, the base is attracted to the partial positive charge on the carbon atom bonded to the halogen atom in an alkyl halide. We can represent this type of substitution by the following general equation, where B is the base:

$$R:\ddot{X}: + :B^- \longrightarrow R:B + :\ddot{X}:^-$$

Typical bases and the products they form are listed in Table 22.10. These reactions are examples of an important type of reaction called *nucleophilic substitution*. The base is a nucleophile since it seeks a center of positive charge. The displaced group, in this case a halide ion, is called a *leaving group*.

Most nucleophilic substitutions are reversible. The direction of the reaction often depends on the concentrations of the reactants, and it can often be driven forward by removing the product by distillation. Aryl halides do not generally undergo nucleophilic substitution.

Table 22.10 Nucleophilic Substitution of Alkyl Halides

Nucleophile	General Formula of Product	Type of Product
OH⁻	R	Alcohol
OR⁻	R—O—R	Ether
CN⁻	R—C≡N	Nitrile
SH⁻	R—S—H	Thiol

NH₃	R—N—H
	      |
	      H
		Amine

Question

25. Distinguish between an alkyl halide and an aryl halide.

22.20 Alcohols

Like the simple alkyl halides, the lower-molecular-weight alcohols are often given informal names according to the hydrocarbon residue. Thus the first two members of the series are methyl alcohol and ethyl alcohol. The IUPAC names, however, are obtained by substituting *-ol* for *-e* in the names of the corresponding alkanes. Therefore, the systematic names for the two simplest alcohols are methanol and ethanol. When a distinction must be drawn between position isomers, the position of the —OH group is located with a number. The longest chain containing the —OH group is the parent compound, and the numbering begins at the end nearer the —OH group:

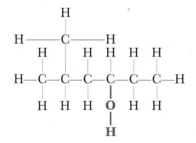

5-methyl-3-hexanol

Another way to look at alcohols is as derivatives of water, where an alkyl group is substituted for a hydrogen atom. And, in fact, alcohols do have a number of properties in common with water. Among these is hydrogen bonding, which we can represent as shown below. Here the short solid line represents a covalent bond and the long dotted line, a hydrogen bond.

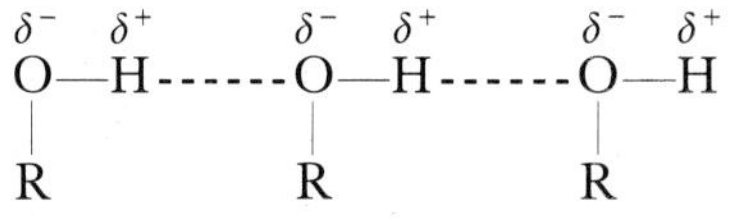

$$\overset{\delta^-}{O}-\overset{\delta^+}{H}------\overset{\delta^-}{O}-\overset{\delta^+}{H}------\overset{\delta^-}{O}-\overset{\delta^+}{H}$$
$$\mid \qquad\qquad \mid \qquad\qquad \mid$$
$$R \qquad\qquad R \qquad\qquad R$$

In addition, the lower-molecular-weight alcohols are miscible with water. (The higher-molecular-weight alcohols, however, are only slightly soluble or insoluble in water.) A final similarity is that, like water, alcohols react with active metals to liberate hydrogen:

$$2\ R-O-H + 2\ Na \longrightarrow 2\ (Na^+ + O-R^-) + H_2$$

The evaporation of the excess alcohol leaves a saltlike alkoxide $(Na^+ + O-R^-)$.

Table 22.11 lists four of the most common alcohols, including two with more than one —OH group. The simplest of the alcohols, methanol (also called wood alcohol because it was once commercially prepared by the destructive distillation of wood), is a poison. It attacks the optic nerve first, causing blindness, and can be fatal.

Ethanol (also called grain alcohol since it is the product of grain fermentation or sometimes just *alcohol*) is one of the most important of all synthetic chemicals, both in terms of its commercial value and of the quantity produced. It is the least toxic of all monohydric (one —OH group per molecule) alcohols and is the *only* one that can be consumed by human beings in moderate amounts without serious danger. The human body is equipped to metabolize ethyl alcohol at a restricted rate, but in high concentrations it too can have toxic and even lethal effects.

Alcoholic beverages are made by promoting the fermentation of sugar, a natural product of fruits and grains. Man's ingenuity and thirst for intoxicants have produced drinkable products from grapes, rye, barley, dandelions, cactus pulp, berries, corn, and honey. The percentage of alcohol in these products varies widely: Beer is roughly 4 percent alcohol; wines about 10–12 percent; and gin, brandies, and whiskies are 40–50 percent. Because the alcohol in these beverages is highly taxed, certain precautions are taken to prevent alcohol manufactured for industrial purposes from being consumed. Usually this involves adding a *denaturant*, a substance that renders ethanol unpalatable or poisonous. One of the most common denaturants is methanol, which is similar to ethanol in physical and chemical properties.

Compounds that contain more than one —OH group are referred to as polyhydric alcohols. Since two —OH groups on the same carbon is an unstable arrangement, these —OH groups are always found on different carbon atoms. Thus the simplest polyhydric alcohol is ethylene glycol:

```
     H   H
     |   |
 H—C—C—H
     |   |
    OH  OH
```

Table 22.11 Some Uses of Alcohols

Formula	Common Name	IUPAC Name	Use
CH_3OH	Methyl alcohol	Methanol	Solvent
CH_3CH_2OH	Ethyl alcohol	Ethanol	Solvent, synthetic intermediate, beverage, gasoline additive
CH_2OHCH_2OH	Ethylene glycol	1,2-Ethanediol	Antifreeze
$CH_2OHCHOHCH_2OH$	Glycerol (glycerine)	1,2,3-Propanetriol	Manufacture of drugs, food products

It can be prepared by treating 1,2-dichloroethane with aqueous KOH. The two —OH groups promote more extensive hydrogen bonding than in monohydric alcohols; thus its boiling point is high (197°C). It is miscible with water.

Glycerol, or glycerine, which occurs naturally in the form of its esters, has the structural formula

$$\begin{array}{c} \text{H} \quad \text{H} \quad \text{H} \\ | \quad | \quad | \\ \text{H}-\text{C}-\text{C}-\text{C}-\text{H} \\ | \quad | \quad | \\ \text{OH} \; \text{OH} \; \text{OH} \end{array}$$

It is a viscous high-boiling (290°C) liquid, miscible with water.

Alcohols are the first stage in the oxidation of hydrocarbons, even though hydrocarbons cannot normally be oxidized directly by inorganic oxidizing agents. The next step, however—the oxidation of alcohols—occurs more readily. The products of this oxidation vary, depending on the location of the —OH group. If the —OH group is bonded to a carbon atom that is bonded in turn to only one other carbon atom, the alcohol is a *primary alcohol* and is oxidized to an aldehyde:

$$\text{R}-\text{CH}_2-\text{OH} + \text{proper oxidizing agent} \longrightarrow \text{R}-\overset{\text{H}}{\underset{}{\text{C}}}=\text{O}$$

(Methanol is a special example of this case, where R = H.) Since aldehydes are readily oxidized to acids, this reaction must be carefully controlled. If the —OH group is bonded to a carbon atom that is bonded to two other carbon atoms, the alcohol is a *secondary alcohol* and is oxidized to a ketone:

$$\text{R}-\underset{\text{H}}{\overset{\text{R}'}{\text{C}}}-\text{OH} + \text{proper oxidizing agent} \longrightarrow \text{R}-\underset{\text{O}}{\overset{}{\text{C}}}-\text{R}'$$

Further oxidation of the ketone is not possible without breaking C—C bonds. Finally, if the —OH group is bonded to a carbon atom that is bonded to three other carbon atoms, the alcohol is a *tertiary alcohol* and cannot be oxidized without breaking C—C bonds:

$$\text{R}-\underset{\text{R}''}{\overset{\text{R}'}{\text{C}}}-\text{OH} + \text{oxidizing agent} \longrightarrow \text{no reaction}$$

If the —OH group is bonded to an aryl group rather than to an alkyl group, the compound is called a *phenol*. An important difference between alcohols and phenols is the greater acidity of phenols.

Questions

26. In what ways do alcohols resemble water?

27. Write the structural formulas of all the monohydric alcohols containing five carbon atoms. Which of these alcohols may be oxidized to an aldehyde? An acid? A ketone? Which cannot be oxidized without breaking a carbon-carbon bond?

22.21 Ethers

Ethers are compounds with the general formula of R—O—R′, where the R groups may or may not be the same. Ethers are named simply by identifying the two alkyl groups attached to the oxygen and adding the word *ether*. The following are examples:

```
    H   H          H H   H H          H   H H
    |   |          | |   | |          |   | |
H—C—O—C—H      H—C—C—O—C—C—H      H—C—O—C—C—H
    |   |          | |   | |          |   | |
    H   H          H H   H H          H   H H

  dimethyl ether     diethyl ether     methyl ethyl ether
```

Each of these ethers is a structural isomer of a simple alcohol; for example, dimethyl ether is a structural isomer of ethanol. This type of structural isomerism is called *functional-group isomerism*: Isomers in this case have different functional groups.

Diethyl ether (simply called *ether*) has been used as an anesthetic since 1846. Because of its high vapor pressure (b.p., 35°C) and its flammability, it must be administered under strictly controlled conditions. It is also a common laboratory solvent.

The preferred laboratory preparation of ethers is nucleophilic substitution of a halide ion in an alkyl halide with an anion of an alkoxide:

$$R\text{—}X + (Na^+ + {}^-O\text{—}R') \longrightarrow R\text{—}O\text{—}R' + (Na^+ + X^-)$$

Ethers are generally unreactive compounds: They are not affected by most oxidizing and reducing agents. They will burn, however. Their high volatility and flammability make them extremely hazardous materials.

Question

28. Write the structural formula of
   a. two functional-group isomers of methyl ethyl ether.
   b. the ethers that are structural isomers of 1-pentanol.

22.22 Carbonyl Compounds

Carbonyl compounds are characterized by an oxygen atom doubly bonded to a carbon atom ($-\overset{\overset{O}{\|}}{C}-$). This double bond, like the others we've encountered, consists of a sigma bond and a pi bond. Since the orbitals of the carbon atom are sp^2-hybridized, the bond angles are about 120°, and the atoms bonded to the carbonyl carbon are located at the corners of a planar triangle. If the carbonyl carbon is bonded to two other carbon atoms, the compound is a ketone; if it is bonded to one (or two) hydrogen atom(s), the compound is an aldehyde.

The IUPAC ending for the aldehydes is -*al*. Since the carbonyl group is always at the end of an aldehyde molecule, it is not necessary to use a number to specify its location. The systematic names for the first two members of this series are methanal and ethanal, but they are usually referred to by their common names, formaldehyde and acetaldehyde, respectively.

Most of the approximately 1 million tons of formaldehyde (CH_2O) manufactured in the United States each year is used in the manufacture of plastics such as Bakelite (trademark). Formaldehyde, which is a gas (b.p., $-21°C$), is probably best known in an aqueous solution, called *formalin*. Since formaldehyde is toxic to lower forms of life, it is used in this form as a disinfectant and as a preservative for biological specimens. Another use of formaldehyde is in the treatment of durable-press textiles. Acetaldehyde (CH_3CHO) is also a gas at room temperature (b.p., 21°C) and is used primarily as an intermediate in the manufacture of other organic compounds, such as acetic acid. Although these lower-molecular-weight aldehydes have sharp irritating odors, higher-molecular-weight aldehydes (especially aromatic aldehydes) are fragrant substances often used as flavoring agents and in perfumes. For example, benzaldehyde, vanillin, and cinnamaldehyde smell like almonds, vanilla, and cinnamon, respectively.

benzaldehyde   vanillin   cinnamaldehyde

Since the carbonyl carbon in the ketones is bonded to two other carbon atoms, the simplest ketone is the common laboratory solvent acetone:

The IUPAC ending for ketones is *-one*; thus the systematic name for this compound is propanone. Beyond butanone we must use numbers to locate the carbonyl carbon atom. For example,

$$\text{H}-\underset{\underset{\text{H}}{|}}{\overset{\overset{\text{H}}{|}}{\text{C}}}-\underset{\underset{\text{H}}{|}}{\overset{\overset{\text{H}}{|}}{\text{C}}}-\underset{\underset{\text{O}}{\|}}{\text{C}}-\underset{\underset{\text{H}}{|}}{\overset{\overset{\text{H}}{|}}{\text{C}}}-\underset{\underset{\text{H}}{|}}{\overset{\overset{\text{H}}{|}}{\text{C}}}-\text{H}$$

is 3-pentanone, and

$$\text{H}-\underset{\underset{\text{H}}{|}}{\overset{\overset{\text{H}}{|}}{\text{C}}}-\underset{\underset{\text{H}}{|}}{\overset{\overset{\text{H}}{|}}{\text{C}}}-\underset{\underset{\text{H}}{|}}{\overset{\overset{\text{H}}{|}}{\text{C}}}-\underset{\underset{\text{O}}{\|}}{\text{C}}-\underset{\underset{\text{H}}{|}}{\overset{\overset{\text{}}{}}{\text{C}}}-\text{H}$$

is 2-pentanone.

Acetone is used commercially in the paint and varnish industry as a solvent and in the home as a fingernail polish remover. It plays a role in biological processes as well. Ketosis, a condition that results when the body must break down fatty tissue for energy, is characterized by the presence of acetone and other ketones in the breath. The most common cause of ketosis is severe diabetes; another is starvation.

Ketones and aldehydes are among the irritating substances present in smog. They are formed by the oxidation of radicals (see p. 708) by ozone or by oxygen atoms.

Since oxygen is much more electronegative than carbon, the double bond in carbonyl groups (unlike the double bonds in alkenes) is strongly polar:

$$\overset{\delta+}{\underset{}{\text{C}}}=\overset{\delta-}{\text{O}}$$

The partial negative charge on the oxygen atom makes it subject to electrophilic attack, and the electron deficiency of the carbonyl carbon atom renders it subject to nucleophilic attack. Thus the typical reaction of the carbonyl group is an *addition* across the double bond, where the carbon atom acts as a Lewis acid, accepting an electron pair from an electron-pair donor, and the oxygen atom behaves as a Lewis base:

$$\overset{\delta+}{\text{C}}=\overset{\delta-}{\text{O}} \quad (\text{B:}^-) \ (\text{H}^+)$$

We can illustrate this mechanism with the reaction between a ketone and hydrogen cyanide:

$$\underset{R}{\overset{R'}{\diagup}}C\!\!=\!\!O^{\delta^+\;\delta^-} + HCN \longrightarrow R'\!-\!\underset{R}{\overset{CN}{\underset{|}{C}}}\!-\!OH$$

Alcohols add to carbonyls in a similar fashion, forming molecules in which an —OH group and an —OR group (from the alcohol) are bonded to the same carbon atom. These compounds are called *hemiacetals* if derived from aldehydes and *hemiketals* if derived from ketones:

$$R\!-\!\overset{O}{\overset{\|}{C}}\!-\!H + R'\,OH \rightleftharpoons R\!-\!\underset{H}{\overset{OH}{\underset{|}{\overset{|}{C}}}}\!-\!OR'$$

a hemiacetal

$$R\!-\!\overset{O}{\overset{\|}{C}}\!-\!R + R'\,OH \rightleftharpoons R\!-\!\underset{R}{\overset{OH}{\underset{|}{\overset{|}{C}}}}\!-\!OR'$$

a hemiketal

The polarity of the carbonyl functional group also explains the observation that aldehydes and ketones have higher boiling points than nonpolar compounds of the same molecular weights. But all the hydrogen atoms in aldehydes and ketones are bonded to carbon atoms and hence do not form hydrogen bonds; aldehydes and ketones have lower boiling points than corresponding alcohols. Finally, again because of the polarity of the carbonyl group, the smaller aldehydes and ketones are soluble in water.

Although aldehydes and ketones share many physical and chemical properties, ketones are generally less reactive. Thus aldehydes can be readily oxidized to form carboxylic acids, but ketones can be oxidized only with difficulty and only if C—C bonds are broken. As a practical example, aldehydes are oxidized by the silver ion in an aqueous ammonia solution, and the free silver thus produced appears as a characteristic mirrorlike deposit on the walls of the reaction container. Since ketones do not undergo this reaction (called *Tollen's test*), it is often used as a test for aldehydes.

Since primary and secondary alcohols can be oxidized to aldehydes and ketones, it is not surprising that the reverse reactions also occur:

$$H\!-\!\underset{H}{\overset{H}{\underset{|}{\overset{|}{C}}}}\!-\!\overset{H}{\underset{|}{C}}\!\!=\!\!O + \text{proper reducing agent} \longrightarrow H\!-\!\underset{H}{\overset{H}{\underset{|}{\overset{|}{C}}}}\!-\!\underset{H}{\overset{H}{\underset{|}{\overset{|}{C}}}}\!-\!OH$$

$$\begin{array}{c} \text{H} \quad \text{H} \\ | \quad\quad | \\ \text{H}-\text{C}-\text{C}-\text{C}-\text{H} \\ | \quad\; \| \quad\; | \\ \text{H} \quad \text{O} \quad \text{H} \end{array} + \text{proper reducing agent} \longrightarrow \begin{array}{c} \text{H} \quad \text{H} \quad\; \text{H} \\ | \quad\quad | \quad\quad | \\ \text{H}-\text{C}-\text{C}-\text{C}-\text{H} \\ | \quad\quad | \quad\quad | \\ \text{H} \quad \text{OH} \quad \text{H} \end{array}$$

Questions

29. Distinguish between an aldehyde and a ketone.

30. Assign names to the following compounds:

a. CH_3-CH_2-CHO

b. $CH_3-CH_2-\underset{\underset{O}{\|}}{C}-CH_3$

c. $CH_3-CH_2-\underset{\underset{O}{\|}}{C}-CH_2-CH_2-CH_3$

22.23 Carboxylic Acids

Carboxylic acids represent the third stage in the oxidation of hydrocarbons—a step beyond the carbonyls and two steps beyond the alcohols. They are characterized by the carboxyl group:

$$-C\overset{\displaystyle O}{\underset{\displaystyle OH}{\diagup}}$$

The IUPAC ending for acids is *-oic*, and since the carboxyl group is always at the end of the molecule, it is not necessary to identify its position with a number. The systematic names and formulas for the two simplest carboxylic acids are

$$H-C\overset{\displaystyle O}{\underset{\displaystyle OH}{\diagup}} \quad \text{and} \quad H-\underset{\underset{\text{H}}{|}}{\overset{\overset{\text{H}}{|}}{C}}-C\overset{\displaystyle O}{\underset{\displaystyle OH}{\diagup}}$$

methanoic acid   ethanoic acid

Since many carboxylic acids were isolated from natural sources long before systematic nomenclature was established, we still often refer to them by their common names, which usually indicate their sources. For example, methanoic acid is often called formic acid, from the Latin word for ant (*formica*). It was first isolated by distilling a species of red ant, and it is at least partly responsible for the irritation of insect bites. Ethanoic acid is common acetic acid, which takes its name from the Latin word for vinegar (*acetum*). Vinegar is simply a dilute solution of acetic acid. The heavier carboxylic acids have distinctly unpleasant odors. The odors of goats, locker rooms, and certain aromatic cheeses can be attributed to carboxylic acids. Butyric acid (butanoic acid) is found in rancid butter; caproic (hexanoic), caprylic (octanoic), and capric (decanoic) acids all take their names from the Latin word for goat (*caper*).

We have already mentioned one way of preparing carboxylic acids: the oxidation of primary alcohols or aldehydes. Another is the hydrolysis of nitriles (organic compounds containing the CN^- group), which in turn are prepared by treating alkyl halides with aqueous sodium cyanide:

$$R-Cl + NaCN \longrightarrow R-CN + NaCl$$

$$H^+ + R-CN + 2\,H_2O \xrightarrow{\text{aqueous acid}} R-C\begin{smallmatrix}\diagup O \\ \diagdown OH\end{smallmatrix} + NH_4^+$$

Note that in this reaction series the length of the carbon chain has been increased by one carbon atom.

Organic acids exhibit all the properties normally associated with the inorganic acids we've already discussed. They react with bases to form salts; they react with active metals to liberate hydrogen; and they react with carbonates to form carbon dioxide. Nearly all organic acids are classified as weak acids, and although they often contain a large number of hydrogen atoms, the acid properties of these compounds are due exclusively to the hydrogen atom of the carboxyl group. It is, of course, possible for a molecule to contain more than one carboxyl group. In fact, the acid taste of some fruit juices is due to such acids or their salts.

In addition to characteristic reactions in which the acidic hydrogen atom is ionized, carboxylic acids also undergo reactions in which the carbon-oxygen single bond is broken and the whole —OH group removed. For example, when heated with alcohols in the presence of an inorganic acid catalyst, carboxylic acids are converted to *esters*:

$$R-C\begin{smallmatrix}\diagup O \\ \diagdown O-H\end{smallmatrix} + R'-O-H \underset{}{\overset{H^+}{\rightleftharpoons}} R-C\begin{smallmatrix}\diagup O \\ \diagdown O-R'\end{smallmatrix} + H_2O$$

The oxygen in the eliminated water molecule comes from the acid. The reaction works best with primary alcohols and since it is reversible, the reaction conditions must be carefully chosen.

Although they are oxidation products of aldehydes or primary alcohols, carboxylic acids are generally difficult to reduce. One of the very few effective reducing agents is lithium aluminum hydride ($LiAlH_4$).

Formic acid, the first member of this homologous series, is the only acid that can be readily oxidized by the usual aqueous oxidizing agents:

$$5\,HCOOH + 2\,MnO_4^- + 6\,H^+ \longrightarrow 5\,CO_2 + 2\,Mn^{2+} + 8\,H_2O$$

This reaction is the fourth and final stage in the oxidation of methane. We can represent all four stages, along with the oxidation numbers of the carbon atom, as follows:

$$\underset{-4}{CH_4} \longrightarrow \underset{-2}{CH_3-O-H} \longrightarrow \underset{0}{H-\underset{\underset{H}{|}}{C}=O} \longrightarrow \underset{+2}{H-C\underset{OH}{\overset{O}{\nearrow}}} \longrightarrow \underset{+4}{CO_2}$$

Question

31. Assign IUPAC names to the following:
   a. $CH_3-CH_2-CH_2-COOH$
   b. $CH_3-\underset{\underset{}{\overset{\overset{CH_3}{|}}{}}}{CH}-CH_2-CH_2-COOH$
   c. formic acid

22.24 Esters

Esters are represented by the general formula $R-\overset{\overset{O}{\|}}{C}-O-R'$. They are named by identifying the acid from which they are derived, along with the hydrocarbon radical that replaces the acidic hydrogen. The *-ic* ending of the acid is replaced with *-ate*; thus;

methyl acetate or methyl ethanoate

ethyl formate or ethyl methanoate

Unlike many organic compounds, esters usually have pleasant tastes and odors. The scents of fruits and flowers are often the aromas of esters, and jams and jellies are often flavored with them, sometimes artificially. Table 22.12 lists some common esters, along with the fruit flavor and odor we associate with them. Esters are also used in industry as solvents, and a bifunctional ester is one of the monomeric units of Dacron (trademark) which we shall discuss in Section 22.25. Finally, among the most significant of all esters are *fats*, esters of the trihydric alcohol, glycerol (see Chapter 23).

Table 22.12 Fruit Odor or Flavor Associated with Esters

Structural Formula	Common Name	Fruit Odor or Flavor
$H-\overset{O}{\overset{\|\|}{C}}-O-C_2H_5$	Ethyl formate	Rum
$CH_3(CH_2)_2-\overset{O}{\overset{\|\|}{C}}-O-C_2H_5$	Ethyl butyrate	Pineapple
$CH_3(CH_2)_2-\overset{O}{\overset{\|\|}{C}}-O-CH_3$	Methyl butyrate	Apple
$CH_3-\overset{O}{\overset{\|\|}{C}}-O-(CH_2)_2-\overset{}{\underset{CH_3}{C}}H-CH_3$	Isoamyl acetate	Banana

Esters are hydrolyzed in either acidic or basic solutions. In acidic solutions the products are an alcohol and a carboxylic acid:

$$R-C(=O)(O-R') + H_2O \underset{}{\overset{H^+}{\rightleftharpoons}} R-C(=O)(OH) + R'-O-H$$

In basic solutions the reaction is called *saponification*, and the products are the same alcohol and a carboxylate anion:

$$R-C(=O)(O-R') + H_2O \xrightarrow{OH^-} R'-O-H + R-C(=O)(OH)$$

$$R-C(=O)(OH) + OH^- \longrightarrow R-C(=O)(O^-) + H_2O$$

Upon evaporation of the solvent, the carboxylate anion forms crystals of a carboxylate salt with the metallic cation from the base. Unlike carboxylic acids, most esters can be readily reduced. The products are alcohols:

$$R-C(=O)(O-R') + \text{proper reducing agent} \longrightarrow R-\underset{H}{\overset{H}{C}}-OH + R'-OH$$

Questions

32. Write structural formulas for the following compounds:
    a. methyl butanoate   b. isopropyl propanoate   c. ethyl ethanoate

33. Write structural formulas for the esters that are structural isomers of normal propyl acetate.

22.25 Synthetic Fibers

In Section 22.13 we defined condensation polymers as polymers formed by reactions between monomers in which small molecules are eliminated. Synthetic fibers such as Dacron (a trademark) and nylon are familiar examples of condensation polymers. These fibers, like all other condensation polymers, are formed from monomers that contain at least two functional groups. Dacron is a condensation polymer formed from the dihydric alcohol ethylene glycol and dimethyl terephthalate. Dimethyl terephthalate is a diester of methanol and terephthalic acid (1,4-benzenedicarboxylic acid):

$$CH_3-O-\underset{\underset{O}{\|}}{C}-C_6H_4-\underset{\underset{O}{\|}}{C}-O-CH_3$$

Dacron is formed by a condensation reaction that eliminates methanol:

$$H-O-\underset{\underset{H}{|}}{\overset{\overset{H}{|}}{C}}-\underset{\underset{H}{|}}{\overset{\overset{H}{|}}{C}}-O-H + CH_3-O-\underset{\underset{O}{\|}}{C}-C_6H_4-\underset{\underset{O}{\|}}{C}-O-CH_3 + H-O-\underset{\underset{H}{|}}{\overset{\overset{H}{|}}{C}}-\underset{\underset{H}{|}}{\overset{\overset{H}{|}}{C}}-O-H$$

The following repeating structure results:

$$\left[-O-\underset{\underset{H}{|}}{\overset{\overset{H}{|}}{C}}-\underset{\underset{H}{|}}{\overset{\overset{H}{|}}{C}}-O-\underset{\underset{O}{\|}}{C}-C_6H_4-\underset{\underset{O}{\|}}{C}-\right]_n$$

Nylon is formed from 1,6-diaminohexane,

$$H-N(H)-CH_2-CH_2-CH_2-CH_2-CH_2-CH_2-N(H)-H$$

and the dicarboxylic acid, adipic acid,

$$H-O-\underset{\underset{O}{\|}}{C}-CH_2-CH_2-CH_2-CH_2-\underset{\underset{O}{\|}}{C}-O-H$$

These compounds react to form a salt that, upon being heated, forms *peptide linkages*—bonds in which an amine group (–NH$_2$) and a carboxyl group link by eliminating a water molecule:

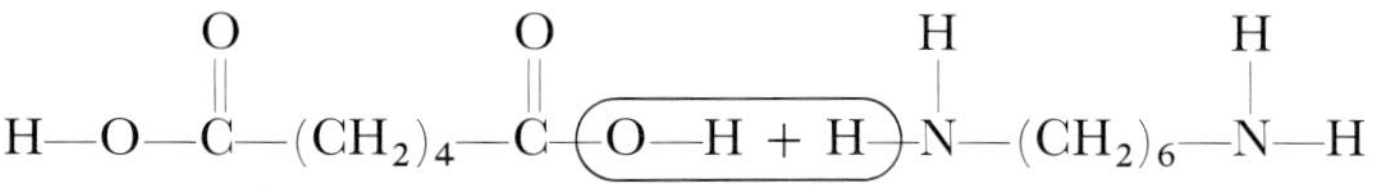

$$\text{H–O–}\overset{\text{O}}{\underset{\|}{\text{C}}}\text{–(CH}_2)_4\text{–}\overset{\text{O}}{\underset{\|}{\text{C}}}\text{–(O–H + H)–}\overset{\text{H}}{\underset{|}{\text{N}}}\text{–(CH}_2)_6\text{–}\overset{\text{H}}{\underset{|}{\text{N}}}\text{–H}$$

Nylon thus has the following structure:

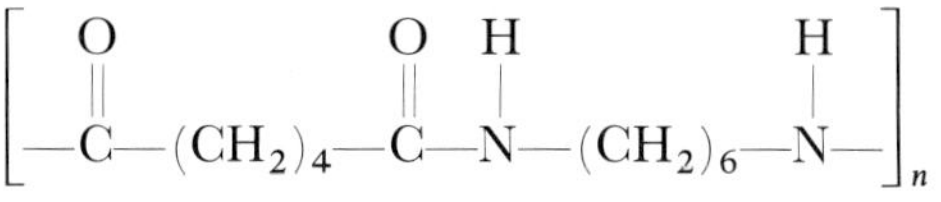

$$\left[\text{–}\overset{\text{O}}{\underset{\|}{\text{C}}}\text{–(CH}_2)_4\text{–}\overset{\text{O}}{\underset{\|}{\text{C}}}\text{–}\overset{\text{H}}{\underset{|}{\text{N}}}\text{–(CH}_2)_6\text{–}\overset{\text{H}}{\underset{|}{\text{N}}}\text{–}\right]_n$$

As we'll see in Chapter 23, these peptide linkages are also characteristic of proteins.

Questions and Problems

22.1 Structure and Nomenclature of Alkanes

34. Consider the following carbon skeleton of a saturated hydrocarbon:

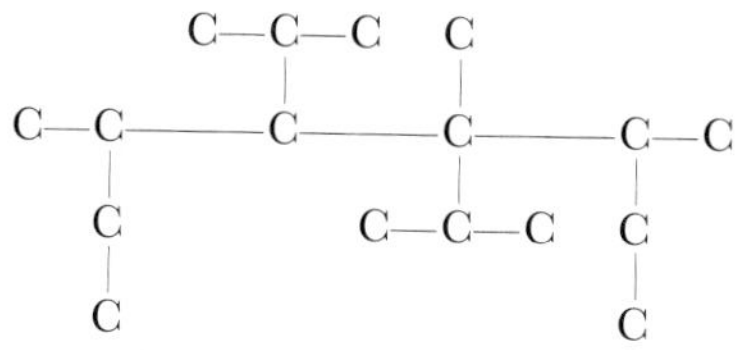

```
          C—C—C    C
          |        |
C—C———C———C———C—C
    |        |
    C    C—C—C  C
    |           |
    C           C
```

a. Write the general formula for the class of compounds to which this example belongs.
b. What is the molecular formula of this compound?
c. What is the empirical formula of this compound?
d. What type of hybrid bonding is manifested in this compound (sp, sp^2, etc.)?

35. What is the density of propane at 38.0°C and 0.500 atm?

22.4 Geometric Isomerism

36. Consider the following formula:

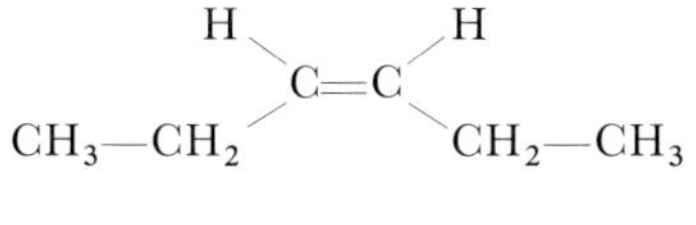

$$\text{CH}_3\text{–CH}_2\diagdown_\text{H}\overset{\diagup^\text{H}}{\text{C=C}}\diagdown_{\text{CH}_2\text{–CH}_3}$$

a. What is the name of the compound above?
b. What type of stereoisomerism is manifested in this compound?
c. Write the structural formula for a chain isomer of this compound.
d. Write the structural formula for a position isomer of this compound.
e. Write the structural formula for a stereoisomer of this compound.

37. Which members of each of the following sets of formulas represent the same compound?

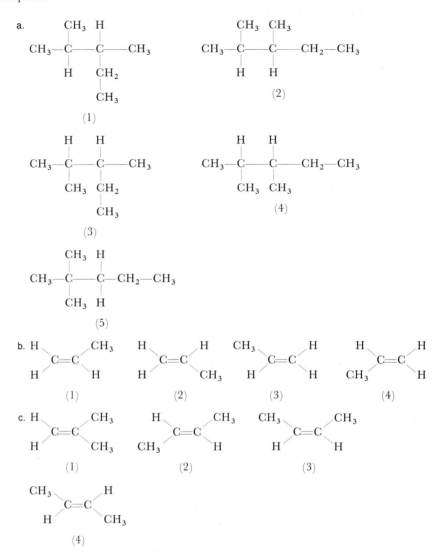

38. What is wrong with the following IUPAC names? Name each compound correctly.
  a. 1-methylpropane b. 2-ethylbutane c. 4,5-dimethylhexane
  d. 4-butene e. 3-pentene

39. What structural characteristics do you associate with
  a. optical isomers? b. geometric isomers?
  c. unrestricted rotation about a bond?

40. What is the molecular formula for the lowest-molecular-weight hydrocarbon for which
  a. structural isomers exist? b. optical isomers exist?
  c. geometric isomers exist?

41. Write the IUPAC names for the following:

a. H—C—C—C—C≡C—H (with H H H on top, H H H on bottom)

b. acetylene

c. H—C—C—C=C—H (with H H H H on top, H on bottom of C2, and H—C—H with H below attached to C2)

d. H—C—C—C—C—C—H with methyl branches (H—C—H groups) on the middle carbon

42. What is the C—C bond angle for each carbon atom in the following compound?

(benzene ring with positions a, b, c, d, e, f)—C^g—C^h=C^i—C^j—C^k≡C^l—H

43. The formula for a compound used in oral contraceptives is given below. Indicate the type of hybridization involved for each lettered carbon atom.

(steroid structure with labeled carbons a, b, c, d, e, f and a C≡C—H group)

22.7 Structure and Nomenclature of Cycloalkanes

44. Which is more stable, cyclopropane or cyclohexane? Why?

22.8 Structure and Nomenclature of Aromatic Hydrocarbons

45. Write structural formulas for the three isomers formed by substituting two chlorine atoms on the benzene ring. Which of these would you expect to be nonpolar?

46. What is the general formula for
    a. cycloalkenes?   b. homologs of benzene?

47. Phenanthrene is an isomer of anthracene and also contains three benzene rings. Write its structure.

48. Write the molecular formulas for the following:
    a. An alkane with 30 carbon atoms per molecule.
    b. An alkene with 9 carbon atoms per molecule.
    c. An aromatic hydrocarbon with 8 carbon atoms per molecule.
    d. An alkyne with 13 carbon atoms per molecule.
    e. Naphthalene.

49. Write the structural formula of
    a. a homolog of benzene   b. an isomer of benzene

22.10 Combustion of Hydrocarbons

50. Write an equation for the complete combustion of $C_{39}H_{80}$. What volume of oxygen at STP is required to burn 1.00 mol of this compound? How many liters of carbon dioxide at STP are produced?

*51. When a sample of gas containing ethane and hydrogen was burned, it produced 60.0 g of water and 20.0 g of carbon dioxide. What was the volume of the original gas mixture at STP?

22.11 Substitution Reactions

52. What is the typical reaction type of the paraffins?

53. How many dichloro- and trichloro-structural isomers are there of each of the following?
    a. ethane   b. propane

54. Which of the structural isomers above are optically active?

55. Write the structural formulas for alkanes that contain the following numbers of carbon atoms but that each form only one monochloro isomer.
    a. five   b. eight

*56. 46.4 ml of vapor of an unknown alkane weighed 0.106 g at $(1.00 \times 10^2)°C$ and 7.40×10^2 mm of Hg. Reaction with chlorine produced four monochloro isomers. Write the structural formula for the compound.

22.12 Reactions of Unsaturated Hydrocarbons

57. How would you distinguish between members of the following pairs of compounds by using simple chemical tests?
    a. ethane and ethene   b. ethane and ethyne

58. Distinguish between members of the following pairs of terms.
    a. saturated hydrocarbon and unsaturated hydrocarbon

b. position isomer and chain isomer
c. stereoisomerism and structural isomerism
d. alkene and alkyne
e. geometric isomerism and optical isomerism
f. alkene and alkane
g. organic compounds and inorganic compounds
h. alkanes and cycloalkanes

22.13 Polymers

59. A certain polymer has a molecular weight of 7.00×10^5. How many molecules are there in 1.00 g of this polymer?

22.14 Aromatic Substitution

60. What is the typical reaction type for benzene?

61. What simple chemical tests would you use to distinguish between cyclohexane and cyclohexene?

22.15 Petroleum

62. Which do you think would have better antiknock qualities, 2-methylheptane or 2,2,3,3-tetramethylpentane?

63. Write a name and structural formula for each of the following:
  a. an addition polymer
  b. a monomer of an elastomer
  c. a compound added to gasoline to improve its antiknock quality
  d. an inorganic polymer

64. Considering gasoline to be essentially a mixture of octanes, write an equation for its combustion. Assume an unlimited supply of oxygen. What volume of CO_2, measured at $(2.0 \times 10^1)°C$ and 7.40×10^2 mm of Hg, is produced when 1.00 liter of gasoline is burned? Assume a specific gravity of 0.700 for the gasoline.

22.16 Other Sources of Hydrocarbons

65. Indicate the series of reagents you would need to make the following conversions:
  a. $CH_4 \longrightarrow CH_3Cl$ b. $CaC_2 \longrightarrow H-C\equiv C-H \longrightarrow C_2H_6$

66. Write the structural formulas for the organic products resulting from the following reactions:
  a. ethene + hydrogen chloride
  b. benzene + nitric acid in the presence of sulfuric acid
  c. 1 mol ethyne + 2 mol chlorine
  d. ethyne + excess hydrogen in the presence of a catalyst
  e. 1 mol ethane + 1 mol chlorine in the presence of light
  f. 1 mol ethane + 2 mol chlorine in the presence of light
  g. calcium carbide + water
  h. 1 mol cyclohexane + 1 mol bromine in the presence of light

67. How would you prepare 1,1,2,2-tetrabromoethane, starting with CaC_2?

22.19 Halides

68. Consider the reaction: $R\text{---}Cl + NaOH \longrightarrow R\text{---}OH + NaCl$
  a. What is the nucleophile?
  b. What is the leaving group?
  c. If NaOH is replaced by NaCN, what type of product is formed?

69. What is the general characteristic of all nucleophiles?

70. Write formulas for the products formed when 2-bromopropane reacts with each of the following (see Table 22.10):
  a. NaSH   b. NaCN   c. $NaOC_2H_5$ d. NaOH   e. NH_3

22.20 Alcohols

*71. Write the structural formulas for all possible oxidation products of ethylene glycol.

72. Both ethyl alcohol and water exhibit hydrogen bonding. Why does water boil at a considerably higher temperature?

22.22 Carbonyl Compounds

73. What will be the main product in each of the following reactions?
  a. 1 mol Cl_2 + 1 mol CH_4
  b. carefully controlled oxidation of 2-propanol
  c. hydrogenation of ethene
  d. hydrolysis of ethyl chloride with aqueous KOH
  e. hydrolysis of CaC_2

22.23 Carboxylic Acids

74. What products are formed in the successive stages of oxidation in methane? (The first oxidation step must be carried out indirectly.) What is the oxidation number of carbon in each product?

75. Classify the following compounds as to whether they are derivatives of (i) methane, (ii) ethane, or (iii) some other hydrocarbon. What is the systematic name of each compound?
  a. acetic acid   b. chloroform   c. formic acid
  d. acetaldehyde   e. acetone   f. formaldehyde

76. Which of the following compounds show optical isomerism?
  a. oxalic acid   b. 2-pentanol   c. glycerine   d. 3-pentanol

22.24 Esters

77. What classes of compounds are formed by the following reactions?
  a. an alcohol with a carboxylic acid
  b. oxidation of a secondary alcohol
  c. partial oxidation of a primary alcohol
  d. partial hydrogenation of an alkyne
  e. complete hydrogenation of an alkyne
  f. halogenation of ethene
  g. reduction of esters
  h. hydrolysis of esters

*78. Show how methyl acetate might be prepared from methane.

79. Distinguish between members of the following pairs of terms:
  a. a primary alcohol and a secondary alcohol

b. a secondary alcohol and a tertiary alcohol
c. an ester and an ether

80. Give the structural formulas for the products indicated by the capital letters:

$$C_2H_5Br \xrightarrow{NaCN} A \quad A \xrightarrow[\text{hydrolysis}]{\text{acid}} B \quad B \xrightarrow[H^+]{\text{ethyl alcohol}} C \quad C \xrightarrow{\text{basic hydrolysis}} D \text{ and } E$$

81. What class of compounds is represented by each of the following general formulas?

a. C_nH_{2n+2} b. C_nH_{2n} c. R—O—H   d. $R-\overset{\overset{\displaystyle O}{\|}}{C}-OH$

e. R—O—R   f. $R-\overset{\overset{\displaystyle O}{\|}}{C}-H$ g. R—X   h. $R-\overset{\overset{\displaystyle O}{\|}}{C}-R$

i. C_nH_{2n-2} j. $R-\overset{\overset{\displaystyle O}{\|}}{C}-O-R$

82. Starting with ethyl acetate, show how the following compounds may be prepared:
a. ethyl alcohol   b. sodium acetate   c. ethyl chloride

22.25 Synthetic Fibers

83. Write a name or structural formula for each of the following:
a. a compound used as an antifreeze
b. a condensation polymer
c. a moth repellent
d. a flavoring extract
e. a denaturant
f. a bactericide

Discussion Starters

84. Explain or account for the following:
a. Si—Si bonds are easily broken during hydrolysis, but C—C bonds are not.
b. Alkenes and cycloalkanes have the same general formula.
c. No numbers are necessary in naming

$$CH_3-\underset{\underset{\displaystyle CH_3}{|}}{\overset{\overset{\displaystyle CH_3}{|}}{C}}-CH_3$$

d. There are position isomers but no chain isomers for C_4H_6.
e. There are no geometric isomers of alkynes.
f. The two compounds below have never been isolated and identified.

g. No numbers are necessary in the name methylpropene.
h. Cyclohexane is a more stable molecule than cyclopropane.
i. Branched alkanes have slightly lower boiling points than unbranched alkanes.
j. The octane number for 2,3,4-trimethylpentane is considerably higher than the octane number for octane.
k. Cyclohexatriene is not a good name for C_6H_6.
l. Reactions between alcohols and carboxylic acids cannot be regarded as equivalent to neutralization reactions.
m. Alcohols boil at higher temperatures than alkanes with similar molecular weights.
n. Polyhydric alcohols have higher boiling points than monohydric alcohols with approximately the same molecular weights.
o. Methyl alcohol is soluble in water, but ethane is not.
p. In the preparation of wine, care must be taken to exclude air.
q. 1-propanol is more water-soluble than 1-chloropropane.

The Structure and Properties of Biomolecules

A logical and essential extension of organic chemistry is biological chemistry—the study of the molecules of living organisms. Often, simple organic molecules of the kind we've already encountered play critical roles in life processes, but other biomolecules can be extraordinarily complex and can have molecular weights as high as 10^9. Fortunately, these *macromolecules* are usually composed of relatively simple molecules, which serve as building blocks in a larger structure. Our approach here will be to describe these building-block molecules and then to show how they are linked to form the macromolecules.

The macromolecules themselves can be conveniently divided into three classes: **lipids** (fats, oils, and waxes), **carbohydrates,** and **proteins.** Together these three kinds of molecules make up most of the solid, nonbony bulk of plants and animals. (In considering biomolecules we shouldn't lose sight of the importance of water in life processes: Plants and animals are 60 to 90 percent water.)

Most lipids are glyceryl esters of carboxylic acids that contain an even number (between 4 and 22) of carbon atoms. They include fats and oils (and waxes, which have simpler structures) and are soluble in carbon tetrachloride, chloroform, ether, and other water-immiscible solvents; but they are insoluble in water. Lipids, in the form of fat, are stored by the body when food intake exceeds energy requirements. They help insulate the body, act as the major source of reserve energy, and protect the vital organs from heavy blows. Most plants do not store fats or oils as energy sources, but some seeds and nuts are rich in them. Fats and oils make up about 40 percent of the American diet.

Carbohydrates (in the form of starch and sugars) are the main sources of energy for many animals, including man. *Cellulose* (another carbohydrate) is the chief structural material in plants.

The most important nonbony structural element of most animal bodies is protein. Fibrous or structural proteins are constituents of the cell membrane, fingernails, cartilage, muscle, and tendons. Proteins also comprise the *antibodies* that protect the body from the effects of foreign substances such as bacteria and viruses. As *enzymes*, proteins catalyze biochemical reactions; as *hormones*, they regulate metabolic processes (though not all hormones are proteins). Finally, hemoglobin is an example of a *transport protein;* it provides for the transfer of oxygen from the lungs to the rest of the body.

23.1 Lipids

Materials such as lard, tallow, butter, and vegetable oils are made up largely of naturally occurring esters of glycerol. These compounds are fats and have the general formula

$$\begin{array}{c} H \\ | \\ H-C-O-C-R \\ | \quad \parallel \\ \quad O \\ H-C-O-C-R' \\ | \quad \parallel \\ \quad O \\ H-C-O-C-R'' \\ | \\ H \end{array}$$

where R, R', and R" may or may not be the same hydrocarbon group. In most natural fats and oils, they are not; hence these natural lipids are called *mixed triglycerides*. Tristearin, on the other hand, which is a glyceryl ester of stearic acid (which has 18 carbon atoms), is an example of a *simple triglyceride*:

$$\begin{array}{c} H \quad\quad O \quad \left[\begin{array}{c}H\\|\\C\\|\\H\end{array}\right]_{16} H \\ | \quad\quad \parallel \quad\quad | \\ H-C-O-C-C-H \\ | \\ \\ \quad\quad O \quad \left[\begin{array}{c}H\\|\\C\\|\\H\end{array}\right]_{16} H \\ \quad\quad \parallel \quad\quad | \\ H-C-O-C-C-H \\ | \\ \\ \quad\quad O \quad \left[\begin{array}{c}H\\|\\C\\|\\H\end{array}\right]_{16} H \\ \quad\quad \parallel \quad\quad | \\ H-C-O-C-C-H \\ | \\ H \end{array}$$

In general, natural solid fats are animal products and are esters of saturated organic acids.

The oils are simply unsaturated fats. These liquid fats usually are plant products. Triolein illustrates the structure of a typical oil (all three carbon-carbon double bonds have *cis* geometry):

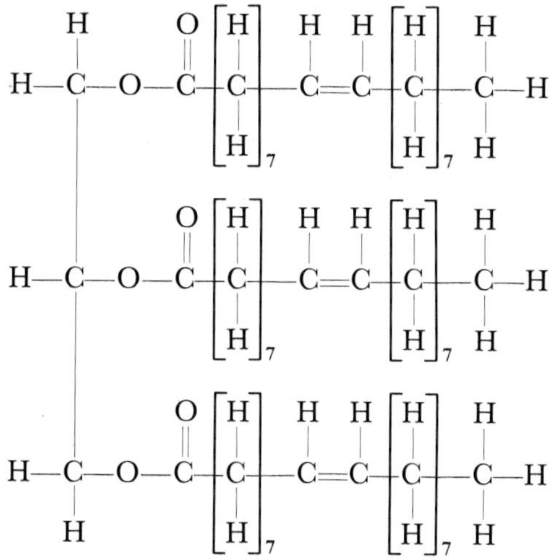

Oils may be converted into solid fats by hydrogenation. For example, when hydrogen is added across the double bond in triolein, the product is tristearin.

If fats or oils are hydrolyzed, the ester linkage is broken and *fatty acids* result. Table 23.1 lists the most common of these natural fatty acids. Where geometric isomerism is possible, it is usually only one of the isomers that occurs naturally. For example, consider

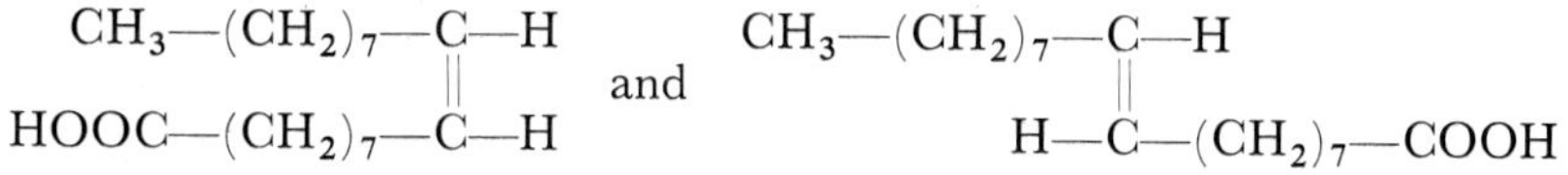

oleic acid (*cis* form) and elaidic acid (*trans* form)

Oleic acid is found in natural fats; elaidic acid is not.

Table 23.1 Examples of Common Fatty Acids

Name	Formula	Source
Stearic	$CH_3(CH_2)_{16}COOH$	Tallow, lard, butter
Palmitic	$CH_3(CH_2)_{14}COOH$	Butter, lard, cottonseed oil
Oleic	$CH_3(CH_2)_7CH\overset{(cis)}{=}CH(CH_2)_7COOH$	Olive oil, corn oil
Linoleic	$CH_3(CH_2)_4CH\overset{(cis)}{=}CHCH_2CH\overset{(cis)}{=}CH(CH_2)_7COOH$	Soybean oil, safflower oil
Linolenic	$CH_3CH_2CH\overset{(cis)}{=}CHCH_2CH\overset{(cis)}{=}CHCH_2CH\overset{(cis)}{=}CH(CH_2)_7(COOH)$	Linseed oil

As we mentioned above, most natural fats and oils are mixed triglycerides; hence they contain more than one fatty acid. However, most of these natural fatty acids, including the ones in Table 23.1, share the following characteristics:

1. They usually contain an even number of carbon atoms.

2. The carbon atoms are arranged in a continuous (unbranched) chain.

3. If unsaturated they usually contain 18 or more carbon atoms and have *cis* geometry.

4. If saturated they usually contain 10 to 18 carbon atoms.

Waxes are the third type of lipid. They are essentially simple esters, but both the alcohol and acid portions of waxes contain long carbon chains. Waxes are widely distributed among both plants and animals; in the case of plants, they often serve as protective coatings. Beeswax and the wax from our ears are examples of waxes produced by animals; and carnauba wax, which is used as a floor and automobile wax, is a natural product of certain Brazilian palm trees.

As foodstuffs waxes are inconsequential, whereas fats are one of the three principal components of our diets (proteins and carbohydrates are the other two). Fats and oils are used by the body to synthesize essential constituents of tissues and to provide energy. They produce about 9.5 kcal/g when oxidized to carbon dioxide and water, which is considerably more energy than the 4 kcal obtained from a gram of either carbohydrate or protein. Unfortunately for some of us, much more fat than necessary can be stored in the body for future use. This stored fat includes not only triglycerides from fatty foods but also fats that the body produces from unused carbohydrates and proteins.

Most of us have seen TV commercials extolling the virtues of unsaturated shortenings. The rationalization for these ads is some statistical evidence linking saturated fats with atherosclerosis, a condition characterized by chronic hardening and thickening of the arterial walls, often caused by fat deposits, and resulting in restricted circulation of the blood. This is the reason that some people prefer polyunsaturated margarine to butter.

Linoleic and linolenic acids are *essential* fatty acids; that is, they are required by the body but cannot be synthesized by it. Hence they must be part of our diet. Oleic acid, on the other hand, can be synthesized from stearic acid.

Questions

1. What are the principal nonbony constituents of living organisms?

2. What are the products of the hydrolysis of
   a. tristearin?   b. triolein?

3. Distinguish between
   a. fat and oil.   b. simple triglyceride and mixed triglyceride.   c. wax and fat.

Sidelight
Soaps and Other Detergents

Any compound that reduces the surface tension of water or reduces the interfacial tension between two liquids or between a liquid and a solid is referred to as a *surface-active agent* or a *surfactant*. One of the most familiar categories of surfactants comprise the *detergents*, a term we generally apply to surface-active agents that aid in loosening and removing dirt. The oldest and most common of these agents is soap: a sodium or potassium salt of a high-molecular-weight organic acid such as stearic acid, $CH_3-(CH_2)_{16}-COOH$. Soaps are produced commercially by heating fats and fatty oils with either NaOH or KOH. This process, which we first encountered in Chapter 22, is called *saponification*. In this case it produces soap and glycerol, a by-product of the soap industry. The equation for the saponification of tristearin is

tristearin + 3 NaOH → glycerol + 3 sodium stearate

Saponification with sodium hydroxide produces *hard soaps* (solids), whereas saponification with potassium hydroxide produces *soft soaps*. The potassium soaps are more soluble in water; hence they produce softer lathers.

The cleansing action of soap is illustrated in Figure 23.1. If we take the stearate anion $[CH_3-(CH_2)_{16}-COO^-]$ as our example, we find in Figure 23.1b that the carboxylate portion $(-COO^-)$, which is soluble in water, is at or slightly below the surface of the water, whereas the hydrocarbon portion $[CH_3-(CH_2)_{16}-]$, which is soluble in oil or grease but not in water, extends slightly above the surface of the water. When this partial solution is brought in contact with oily materials attached to a surface (skin, clothing, wood, etc.), the hydrocarbon end dissolves in the water-insoluble oil (Figure 23.1c). This changes the properties of the interface between the oil and the water and makes these two immiscible materials compatible. The oil thus moves into the body of the liquid as a *micelle*, an oil particle surrounded by fatty acid anions. Whenever dirt is associated with oil (skin oil, lubricating oils, foods, etc.), soap and water can be effective cleansers where water alone is useless.

However, soap does have a number of limitations as a detergent. Chief among them is its reaction with hard water (which contains Ca^{2+} and Mg^{2+} ions) to produce insoluble calcium and magnesium salts:

$$2\ CH_3-(CH_2)_{16}-COO^-(aq) + Ca^{2+}(aq) \longrightarrow Ca[CH_3-(CH_2)_{16}-COO]_2(s)$$

These sticky, insoluble compounds contribute to bathtub ring and the infamous "ring around the collar." Because of this undesirable trait, soap has, to a large extent, been replaced by synthetic detergents (*syndets*) that do not have this property.

Soaps and Other Detergents

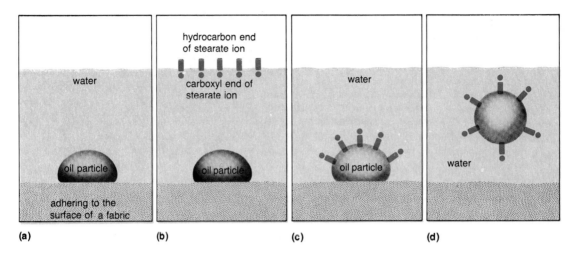

Figure 23.1 The cleansing action of soap. **(a)** Water alone has no effect on the oil particle adhering to the surface of a fabric. **(b)** The carboxylate portion (small circle) of the stearate ion at the surface of a soap and water mixture is below the surface, but the hydrocarbon portion (rectangle) extends above the surface. **(c)** When the soap and water mixture is brought into contact with the oil particle adhering to the surface, the hydrocarbon end is dissolved in the oil and the carboxylate portion remains dissolved in the water. **(d)** The oil particle is pulled from the surface and moved into the body of the liquid.

One of the most popular commercial synthetic detergents is manufactured according to the following steps:

Coconut oil is reduced to produce alcohols. This step yields the alcohols without producing carboxylic acids as intermediates. For example, trilaurin, one of the prominent fats in coconut oil, yields glycerol and lauryl alcohol [$CH_3-(CH_2)_{10}-CH_2-O-H$].

Lauryl alcohol is esterified with sulfuric acid to lauryl hydrogen sulfate:

$$CH_3-(CH_2)_{10}-CH_2 \quad + \; H_2SO_4 \longrightarrow$$
$$CH_3-(CH_2)_{10}-CH_2 \quad\quad\quad + \; H_2O$$

The desired product, sodium lauryl sulfate, is produced by neutralization:

$$CH_3(CH_2)_{10}-CH_2- \quad\quad + \; NaOH \longrightarrow$$
$$CH_3-(CH_2)_{10}-CH_2- \quad\quad + \; H_2O$$

The anion of this product, like that of all soaps, is a long, nonpolar hydrocarbon chain attached to a polar group. In addition, sodium lauryl sulfate is a biodegradable compound; that is, bacteria and other microorganisms present in sewage systems are able to break it down into small nonpolluting components. This virtue became widely recognized in the 1960s when foam from nonbiodegradable detergents began to appear in freshwater supplies. These detergents were alkyl benzene sulfonates with branched alkyl chains. For example,

$$CH_3-CH(CH_3)-CH_2-CH(CH_3)-CH_2-CH(CH_3)-CH_2-CH(C_6H_4SO_3^-Na^+)-CH_3$$

On the other hand, if the alkyl chain attached to the ring is unbranched, the detergent is largely biodegradable:

$$CH_3-CH_2-CH_2-CH_2-CH_2-CH_2-CH_2-CH_2-CH_2-CH_2-CH_2-CH_2-C_6H_4SO_3^-Na^+$$

Questions

i. What does the term *detergent* refer to?

ii. Distinguish between a soap and a synthetic detergent. Give an example of each.

iii. What are the similarities between a soap and a synthetic detergent?

iv. Why are small amounts of synthetic detergent effective in hard water, where large quantities of soap are needed?

v. What causes "bathtub ring"?

vi. What is the mechanism by which a detergent acts as a cleansing agent?

23.2 Carbohydrates

Carbohydrates include sugars, starch, and cellulose, as well as a number of related compounds. The term itself was first applied to these compounds because it was observed that they decomposed into carbon and water when heated. Indeed, the formulas of most of these compounds can be written $C_y(H_2O)_x$. However, both the name and this general formula are misleading. Unlike the hydrates described in Chapter 21, carbohydrates do not contain discrete molecules of water. Nor can water be added to carbon to form a carbohydrate, as, for example, we can form copper(II) sulfate pentahydrate by adding water to the anhydrous salt.

Monosaccharides We shall look first at the simple sugars, or *monosaccharides*. These monosaccharides are polyhydroxy carbonyls and can be further classified as either aldoses (hydroxyaldehydes) or ketoses (hydroxyketones). The most important of the monosaccharides, in fact the most important carbohydrate, is glucose ($C_6H_{12}O_6$), also called dextrose. It is found in fruits, honey, and sap and constitutes 0.08 to 0.1 percent of the blood of most mammals. It is the body's principal source of energy, and it is an inability to assimilate glucose that is characteristic of *diabetes mellitus*. Glucose is the only product of the complete physiological hydrolysis of maltose, glycogen, starch, or cellulose. It is one of the products, along with fructose, of the hydrolysis of sucrose (common cane sugar) in the body.

As a first step in adequately representing the structure of glucose, we can write the following:

$$HO-CH_2-\underset{OH}{CH}-\underset{OH}{CH}-\underset{OH}{CH}-\underset{OH}{CH}-\overset{O}{\underset{\|}{C}}-H$$

This is the general formula for an aldohexose—a carbohydrate containing six carbon atoms and an aldehyde group in each molecule. (Note that we customarily use the *-ose* suffix to indicate a sugar or other carbohydrate.) Glucose is indeed an aldohexose, but since there are four asymmetric carbon atoms in the formula above, there are 16 stereoisomers, of which only one is glucose, represented by this formula. To distinguish among them we must determine the spatial arrangement of the atoms around the asymmetric carbon atoms. This was first done experimentally by Emil Fischer (1852–1919), who received the Nobel prize in 1902 for his monumental work. The glucose molecule is often represented in two dimensions by the following:

$$\begin{array}{c} H\diagdown\!\!\!\diagup O \\ C \\ | \\ H-C-OH \\ | \\ HO-C-H \\ | \\ H-C-OH \\ | \\ H-C-OH \\ | \\ H-C-OH \\ | \\ H \end{array}$$

An even better representation would indicate the true bond angles of 109°28′:

This shows the proximity of the hydroxyl group on the fifth carbon atom to the carbonyl group.

In Chapter 22 we saw that an addition reaction between an aldehyde and an alcohol can produce a hemiacetal. In just this way, glucose forms a cyclic hemiacetal, introducing another center of asymmetry at the former carbonyl carbon. Thus there are two naturally occurring cyclic forms of glucose.

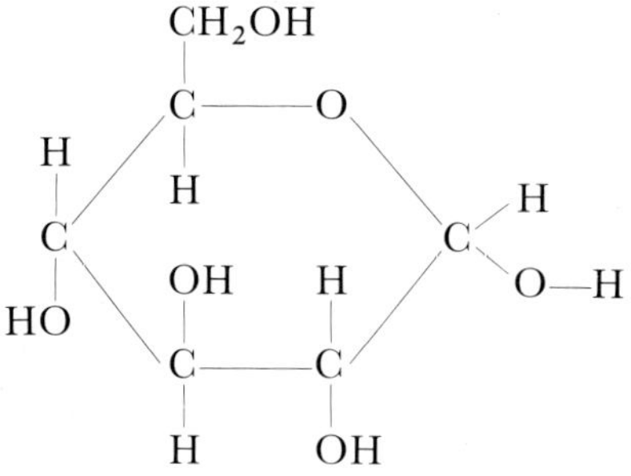

α-glucose 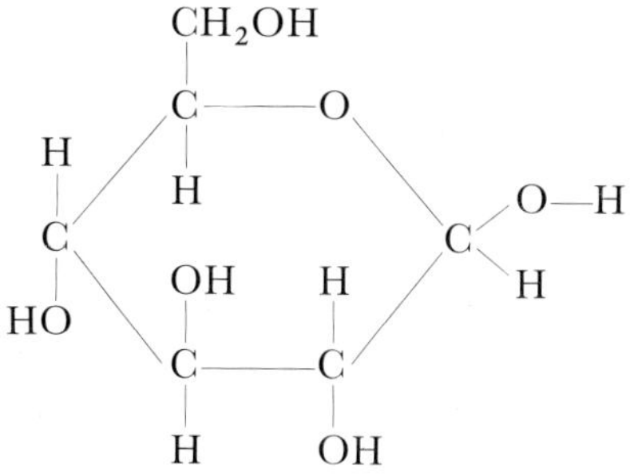   β-glucose

These two forms differ only in the orientation of a single —OH group.

When either α- or β-glucose is dissolved in water, it forms the other isomer via the open-chain (aldehyde) form until the mixture reaches equilibrium:

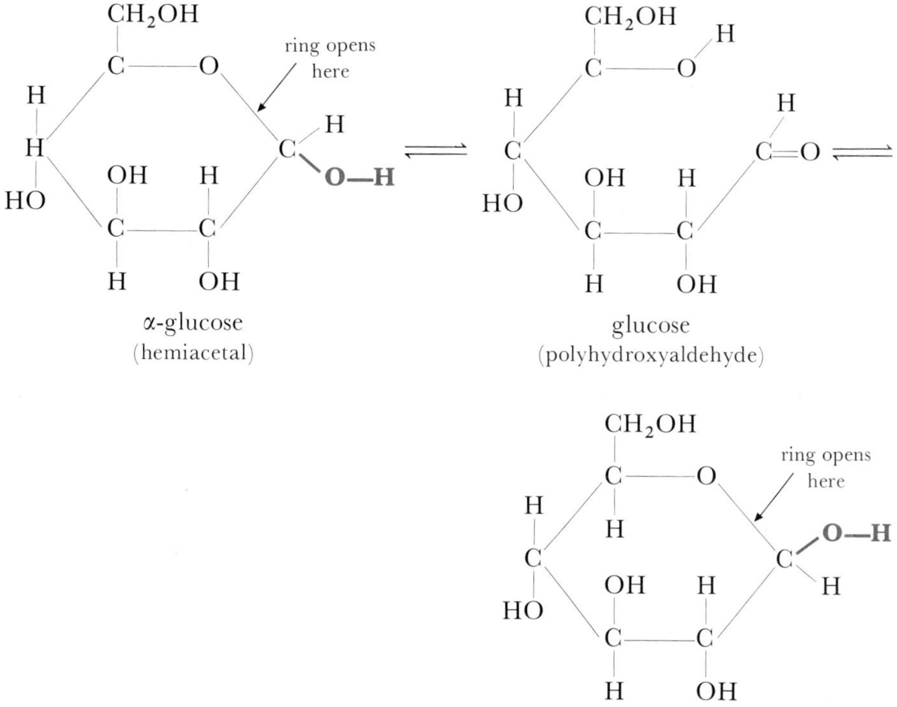

α-glucose
(hemiacetal)

glucose
(polyhydroxyaldehyde)

β-glucose
(hemiacetal)

The Structure and Properties of Biomolecules

At equilibrium only about 0.02 percent of the glucose is in the open-chain form; however, this is sufficient to ensure that aqueous solutions of glucose react as typical aldehydes—for example, reducing Ag(I) or Cu(II) ions.

Although the cyclic ring structures more adequately represent glucose than the open-chain formula, there are difficulties even with ring structures. The most important of these is that they represent the molecule as planar; whereas if we consider the bond angles, it is obvious that all the atoms in the ring are not in the same plane. Thus a fourth (and final) representation of glucose is a pair of zigzag ring structures:

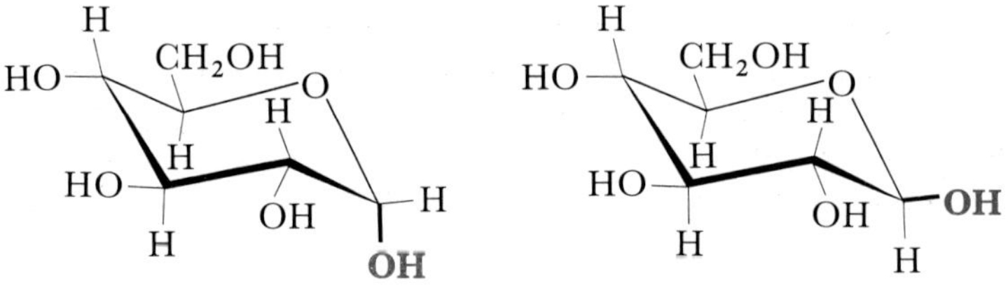

α-glucose

β-glucose

Another aldohexose, galactose, is formed by the hydrolysis of lactose (see p. 744). Galactose differs from glucose only in the orientation of the —OH group on the fourth carbon atom (considering the carbonyl carbon atom as number one). Like glucose, it exists in aqueous solution as an equilibrium mixture of three forms:

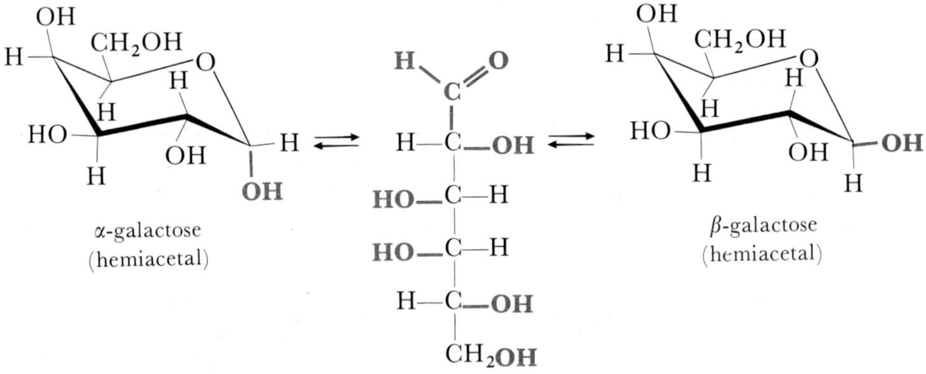

α-galactose
(hemiacetal)

galactose
(polyhydroxyaldehyde)

β-galactose
(hemiacetal)

Another common sugar, fructose (also called levulose and fruit sugar), is an example of a ketose. It is the sweetest of all sugars and is found in fruit and honey. Its three most common equilibrium forms can be represented as follows:

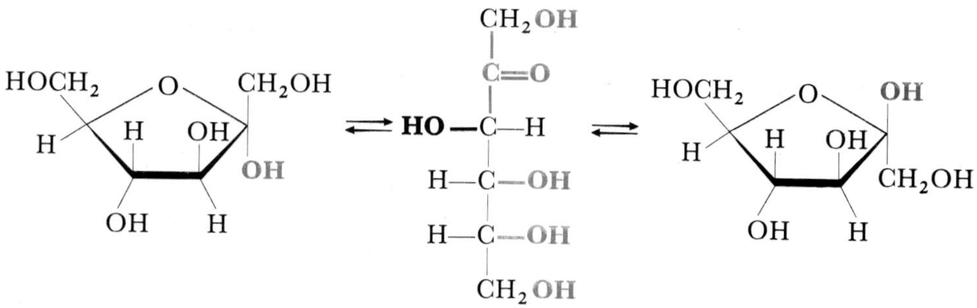

α-fructose   fructose β-fructose

Because of the position of the carbonyl carbon, fructose most readily forms five-membered rings, in contrast with the six-membered cyclic structures of glucose and galactose. Fructose can also exist as a six-membered ring, but it is less common. To indicate the size of the rings in sugars, the terms *pyranose* and *furanose* are often incorporated into the names. These terms are derived from pyran and furan:

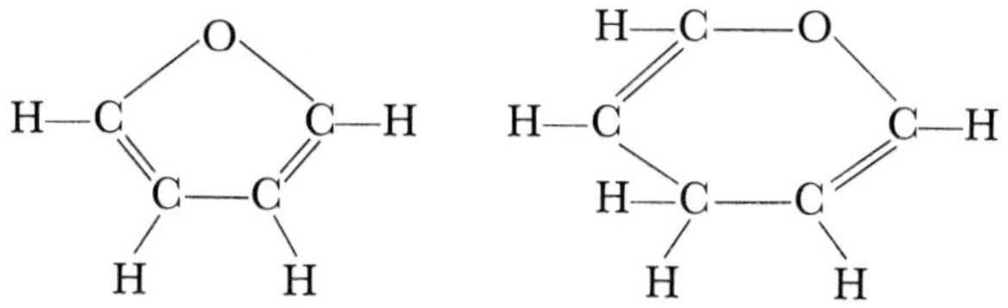

furan   pyran

(These are examples of *heterocyclic compounds*—compounds with ring structures containing more than one kind of atom.) Thus the names of the monosaccharides we have discussed become

α- or β-glucopyranose
α- or β-galactopyranose
α- or β-fructofuranose

Two final examples of sugars are pentoses of singular significance in biological processes—ribose and deoxyribose. Both are components of the vitally important nucleic acids, which control protein biosynthesis and the transfer of hereditary characteristics. We can represent the cyclic structures of these compounds as follows:

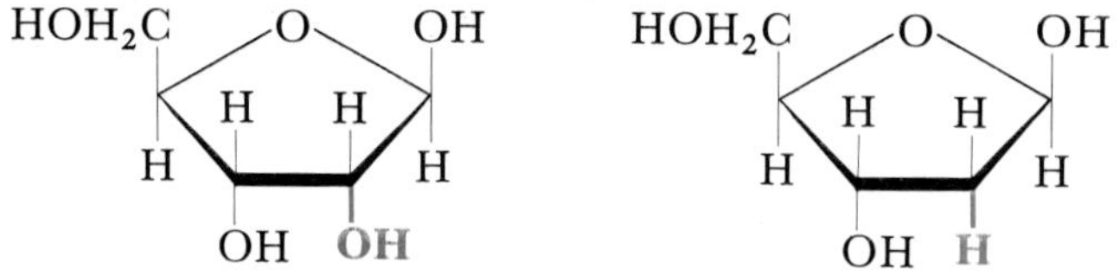

ribose

deoxyribose
(sometimes called β-2′-deoxyribose, where the 2′ indicates the carbon atom lacking an oxygen)

Disaccharides We can envision the *disaccharides* as linked pairs of monosaccharides, in which a molecule of water has been eliminated. The structures of three important disaccharides (sucrose, maltose, and lactose) are shown in Figure 23.2. Each of them has the molecular formula of $C_{12}H_{22}O_{11}$.

Sucrose is common table sugar, produced from sugarcane and sugar beets. Its hydrolysis yields glucose and fructose. Maltose (malt sugar) is obtained by the enzymatic hydrolysis of starch. It, in turn, hydrolyzes to form two molecules of glucose. Lactose comprises 5 to 8 percent of human milk and hydrolyzes to form glucose and galactose.

Properties of Sugars The disaccharides, like all other sugars, form hard crystals and are soluble in water. We can attribute both of these properties to the presence

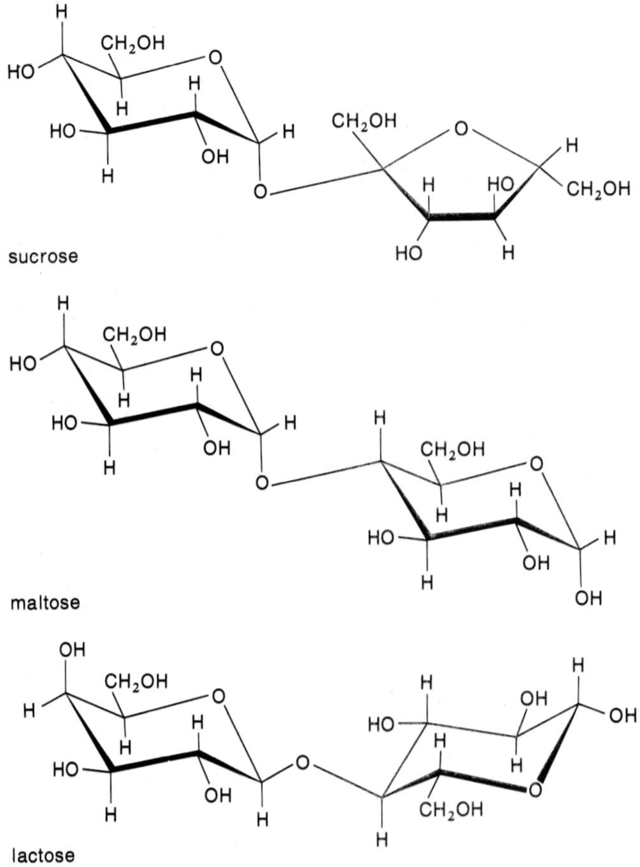

Figure 23.2 Structures of three important disaccharides.

of —OH groups that form hydrogen bonds with other sugar molecules in the crystal and with water molecules in solution.

Sugars are classified as reducing sugars or nonreducing sugars depending on whether they reduce metallic ions (usually copper or silver ions) in an alkaline solution. If they do, they are themselves oxidized in the process. Reducing sugars are characterized by hemiacetal linkages that can open in solution to form an aldehyde group. Since glucose, galactose, and fructose (changes to an aldohexose in alkaline solutions) possess such hemiacetal linkages, they are reducing sugars. Both maltose and lactose retain hemiacetal linkages and hence are also classified as reducing sugars. The carbon farthest to the right as drawn in Figure 23.2 becomes the aldehyde group when the ring opens. Sucrose, on the other hand, is formed by a linkage between carbon atom number 1 of glucose and carbon atom number 2 of fructose (the carbonyl carbon atom). Thus the sucrose molecule does not have a hemiacetal linkage that can open to form an aldehyde, and sucrose is a nonreducing sugar.

Polysaccharides Carbohydrates that are composed of large numbers of monosaccharide units are called *polysaccharides*. Starch, glycogen, and cellulose are important examples. In the case of starch and glycogen, complete hydrolysis produces α-glucose units, whereas hydrolysis of cellulose yields β-glucose units.

Starch serves as the reserve carbohydrate for most plants. It occurs chiefly in cereal grains and plant tubers and consists of two components, amylose and amylopectin. Amylose (10 to 25 percent starch) is a linear chain of approximately 200 to 1000 glucose units joined by α,1 → 4 linkages (see Figure 23.3). In solution the helically coiled chain of amylose gives an intense blue color with iodine. Amylopectin, the major component of plant starch (about 75 to 90 percent), consists of 500 to 2000 glucose units. Most of the linkages are α, 1 → 4 linkages, but amylopectin also contains branching chains that are interconnected by α, 1 → 6 linkages. Amylopectin yields a red-violet color with iodine.

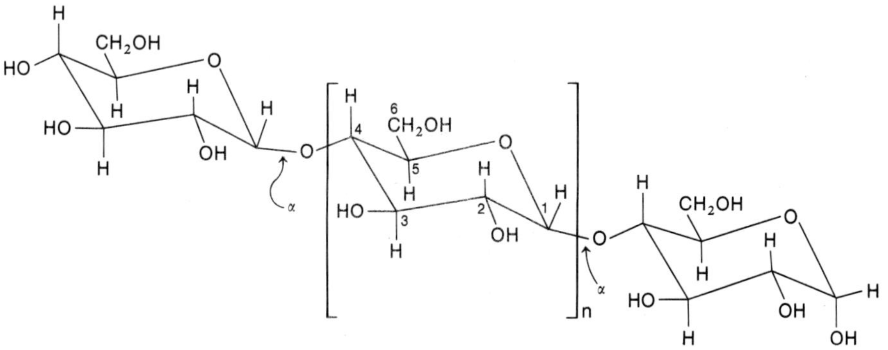

Figure 23.3 α-Linked polysaccharide (amylose component of starch).

Glycogen is similar to the amylopectin portion of starch except that it consists of shorter chains and appears to be more highly branched. It is the reserve carbohydrate for animals, serving as an energy source for the muscles. It is stored in small amounts in the muscles and liver, and it helps to maintain the correct level of glucose in the blood.

Cellulose is the principal structural material of plants; hence it is the chief component of plant products like wood and cotton. Cellulose is a linear chain of thousands of glucose units linked together by $\beta, 1 \rightarrow 4$ linkages (see Figure 23.4).

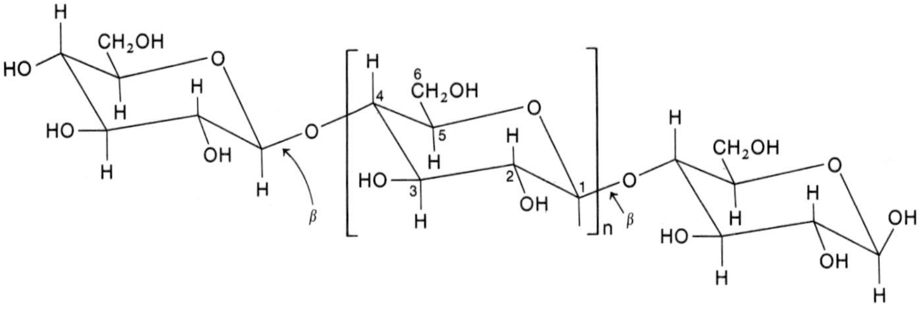

Figure 23.4 β-Linked polysaccharide (cellulose).

The subtle difference between $\alpha, 1 \rightarrow 4$ linkages and $\beta, 1 \rightarrow 4$ linkages may seem trifling, but it has a dramatic effect: Starch can be readily hydrolyzed during digestion, whereas cellulose can be hydrolyzed only with difficulty and in fact cannot be digested by humans. Cellulose can be digested by cattle and other herbivorous animals (including the lowly termite) only with the help of enzymes manufactured by symbiotic bacteria that inhabit the animals' gut. Both cellulose and starch are synthesized by plants as products of photosynthesis.

Questions

4. What is the historical origin of the term *carbohydrate*?

5. Give a specific example of each of the following:
   a. ketohexose   b. pentose   c. aldohexose
   d. disaccharide   e. monosaccharide   f. polysaccharide

6. Distinguish between α-glucose and β-glucose.

7. What is the name of a carbohydrate that does not fit the general formula of $C_y(H_2O)_x$ for carbohydrates?

8. Which carbohydrates occur as components of nucleic acids?

23.3 Amino Acids

Amino acids are so called because they contain both an amine ($-NH_2$) group and a carboxyl ($-COOH$) group. (They may, of course, contain other functional groups as well.) In naturally occurring amino acids, the amine group is always attached to the α-carbon—the one adjacent to the carboxyl group. The natural amino acids can thus be represented by the general formula

$$\begin{array}{c} H \\ | \\ R-C-C\begin{array}{c}\diagup O \\ \diagdown O-H\end{array} \\ | \\ H-N \\ | \\ H \end{array}$$

where R can be a hydrogen, a simple alkyl group, or a complex group.

Actually the properties of amino acids are somewhat inconsistent with this general formula. They differ from both the amines and the carboxylic acids in that they are both crystalline solids and generally soluble in water and other polar solvents but insoluble in nonpolar solvents. To explain these properties, which suggest strongly polar molecules, we might better represent the amino acids as

$$\begin{array}{c} H \\ | \\ R-C-C\begin{array}{c}\diagup O \\ \diagdown O^-\end{array} \\ | \\ H-N-H \\ | \\ H^+ \end{array}$$

The $-NH_2$ group is a stronger base than the $-COO^-$ group, and the result is a *dipolar ion* or *zwitterion* (an ion with both a positive and negative charge).

Twenty α-amino acids have been isolated from plant and animal proteins. Their formulas and common names—amino acids are nearly always referred to by common names—are listed in Table 23.2. (Notice that aspartic acid and glutamic acid can appear as the amides of asparigine and glutamine.) The eight acids marked with an asterisk are the *essential amino acids* for humans; these acids cannot be synthesized in sufficient quantities by the body and thus must be supplied in the diet. Except for glycine, in which R = H, all the amino acids in Table 23.2 contain an asymmetric carbon atom and hence are optically active. Most of the amino acids contain one amino group and one carboxyl group and are classified as neutral. Those that contain more than one of either group are classified as basic or acidic, depending on which group predominates.

Question

9. What two functional groups do all amino acids contain? What is the relative location of these two functional groups in naturally occurring amino acids?

Table 23.2 Amino Acids

Neutral Amino Acids

Name (abbreviation)	Comment	Formula
1. Glycine (Gly)	Glycine is the simplest amino acid. It is unique in that it is the only amino acid without an asymmetric carbon atom.	H₂N–CH₂–COOH
2. Alanine (Ala)		CH₃–CH(NH₂)–COOH
*3. Valine (Val)	Branched-chain amino acids are not usually synthesized by animals.	(CH₃)₂CH–CH(NH₂)–COOH
*4. Leucine (Leu)		(CH₃)₂CH–CH₂–CH(NH₂)–COOH
*5. Isoleucine (Ile)	Has two asymmetric carbon atoms.	CH₃–CH₂–CH(CH₃)–CH(NH₂)–COOH
6. Serine (Ser)	The OH group reacts in the usual way—ester formation, etc.	HO–CH₂–CH(NH₂)–COOH
*7. Threonine (Thr)	Has two asymmetric carbon atoms.	CH₃–CH(OH)–CH(NH₂)–COOH
*8. Phenylalanine (Phe)		C₆H₅–CH₂–CH(NH₂)–COOH

Table 23.2 (*continued*)

Neutral Amino Acids

Name (abbreviation)	Comment	Formula
9. Tyrosine (Tyr)		*(structure of tyrosine)*
*10. Tryptophan (Trp)	Heterocyclic	*(structure of tryptophan)*
11. Proline (Pro)	Proline is an *imino* acid. The nitrogen atom is part of a heterocyclic ring as well as joined to the α carbon atom. However, this does not prevent the nitrogen from taking part in protein formation.	*(structure of proline)*
12. Hydroxyproline (hyp)	Heterocyclic	*(structure of hydroxyproline)*

Basic Amino Acids

Name (abbreviation)	Comment	Formula
13. Histidine (his)	Heterocyclic	*(structure of histidine)*
*14. Lysine (lys)		*(structure of lysine)*
15. Arginine (arg)		*(structure of arginine)*

Table 23.2 (continued)

Acidic Amino Acids

Name (abbreviation)	Comment	Formula
16. Aspartic acid (asp)	Sometimes found in proteins as the amide asparigine (Asn)	[structural formula of aspartic acid and asparagine]
17. Glutamic acid (glu)	Sometimes formed in proteins as the amide glutamine (Gln). The monosodium salt of glutamic acid is sold commercially ("Accent") to accentuate the flavor of meat.	[structural formula of glutamic acid and glutamine]

Sulfur-Containing Amino Acids

Name (abbreviation)	Comment	Formula
*18. Methionine (met)		[structural formula of methionine]
19. Cysteine (cys)	The SH group is moderately reactive upon oxidation (dehydrogenation). A disulfide linkage is formed between two molecules to form cystine.	[structural formula of cysteine]
20. Cystine (cys-scy)	It is often part of two proteins or located in two widely separated parts of the same protein chain. Disulfide linkages play an important role in the structure of proteins. (See Figure 23.5.) Cysteine has two asymmetric carbon atoms per molecule.	[structural formula of cystine]

23.4 Proteins

Proteins are among the most complex substances known. Molecular weights range from 5000 into the millions, and hydrolysis of a single protein can produce as many as 20 different α-amino acids. However, this last observation can also be the key to understanding proteins, in spite of their complexities: Proteins are long chains of amino acids held together by peptide linkages. Much like the peptide linkages in nylon (see Chapter 22), these bonds are formed between the carboxyl group of one amino acid and the amine group of a second, with the elimination of a water molecule:

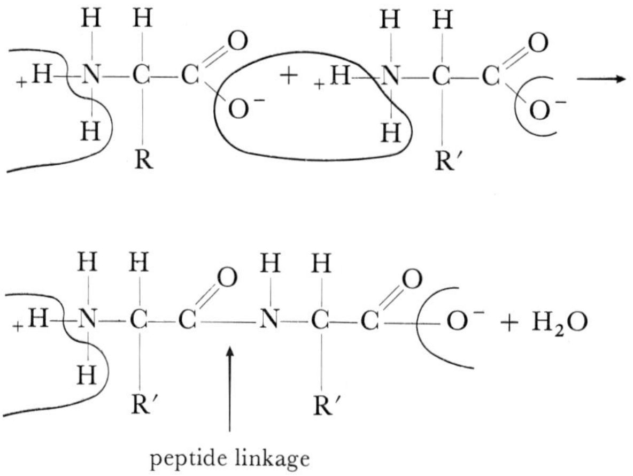

[Diagram showing two amino acids combining to form a peptide linkage with elimination of H_2O, with R and R' side chains, labeled "peptide linkage"]

In the living cell, the formation of each peptide linkage is more complicated than we've shown here, demanding many enzymes, which ensure the synthesis of highly specific proteins.

Since these peptide linkages produce unbranched chains of α-amino acids, we can represent the structure of any protein chain with a general formula, much as we have for less complex polymers:

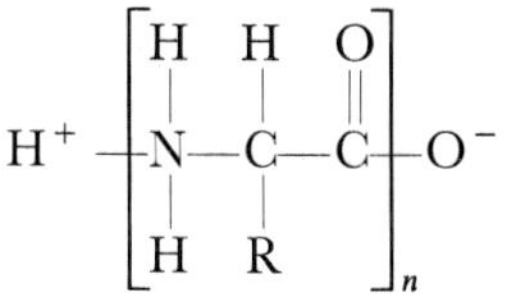

[General formula of protein chain with repeating unit $[-N(H)(H)-C(H)(R)-C(=O)-O^-]_n$ with H^+ at one end]

It is important to emphasize here that the group represented by R may vary from subunit to subunit in the protein molecule; indeed, this sequence of amino acids distinguishes one protein from another. The number (n) of amino acids in a single chain can vary greatly, reaching into the thousands. (The natural protein, insulin, comprises two chains, one with $n = 21$, and one with $n = 30$. (See Figure 23.5.)

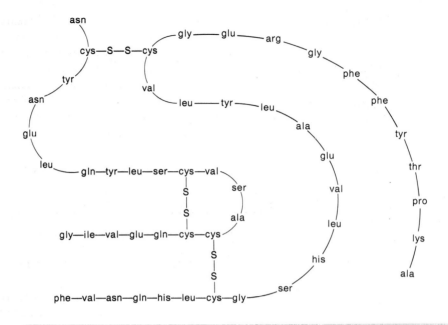

Figure 23.5 Primary structure of beef insulin.

Sometimes one amino acid can be substituted for another in a protein chain with little apparent effect. Other times the consequences are serious. For example, a hereditary disease known as sickle-cell anemia results when two valine molecules are substituted for two glutamic acid molecules among the hundreds of amino acids in hemoglobin (see Chapter 24).

Amino Acid Sequences: Primary Structure Until recently, determining the number and sequence of amino acids in a protein has been an extraordinarily difficult and painstaking task. Even today it is a demanding and time-consuming procedure. The first step is to hydrolyze a pure sample of the protein completely, breaking *all* the peptide linkages in the molecule. This makes it possible to determine the relative abundance of the constituent amino acids, and this information, together with the molecular weight of the protein, yields the absolute abundance of each amino acid.

The second step is to unravel the original sequence of these amino acids. One approach is to hydrolyze another pure sample of the protein partially, a procedure that yields fragments (called *peptides*) containing a small number of amino acids. Each peptide is then isolated and purified. The sequence of amino acids in each of the fragments can be determined by comparing them with known peptides. Having identified small segments of the whole chain, the biochemist can then work out the sequence for larger and larger segments. For example, if he identifies the following six fragments,

> gly-ser-his
> gly-ser
>   ser-his-leu
>   ser-his
>   his-leu-val
>   leu-val

he might tentatively conclude that somewhere in the protein is the sequence gly-ser-his-leu-val.

A more modern method of determining the sequence of amino acids involves tagging the NH_3^+ end of a large protein fragment with a special reagent. The tagged amino acid can then be preferentially split off and identified, and the procedure can be repeated again and again with the rest of the fragment. Specially designed instruments, called *sequencers*, now analyze protein fragments automatically using this method.

The amino acid cysteine plays a special role in the structure of proteins. A pair of cysteine molecules can join to form the disulfide bond of cystine:

$$2\ HSCH_2-\underset{\underset{H}{|}}{\overset{\overset{NH_2}{|}}{C}}-COOH \xrightarrow{-2\ H} HOOC-\underset{\underset{H}{|}}{\overset{\overset{NH_2}{|}}{C}}-\underset{\underset{H}{|}}{\overset{\overset{H}{|}}{C}}-S-S-\underset{\underset{H}{|}}{\overset{\overset{H}{|}}{C}}-\underset{\underset{H}{|}}{\overset{\overset{NH_2}{|}}{C}}-COOH$$

      cysteine   cystine

This permits linkage of two separate peptide chains in a complex protein or formation of a cyclic structure somewhere within a single chain. Three such bonds are found in bovine insulin (Figure 23.5). Together, the amino acid sequence of a protein and the presence of disulfide bridges determine the protein's *primary structure*.

Secondary and Tertiary Structure The primary structure of a protein yields only a partial solution to the riddle of its structure. Protein molecules are not simple linear chains; they fold and coil in highly specific ways. This spatial arrangement of the chain is the *secondary structure* of the protein. Whereas the primary structure involves covalent bonding, the secondary structure involves mostly hydrogen bonding.

One of the prevalent secondary structures of protein molecules is the spiral or helix. The helix may be right-handed (as are the threads on wood screws, for example) or left-handed (Figure 23.6). The helix is held together by hydrogen bonds that link the $-\overset{\overset{H}{|}}{N}-$ groups with double-bonded carbonyl oxygen atoms

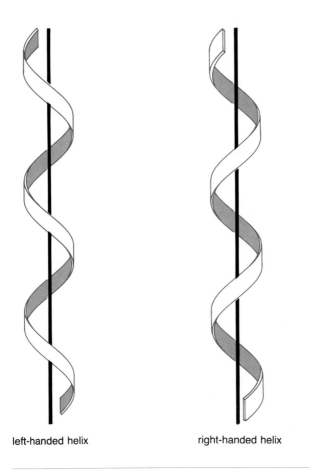

left-handed helix   right-handed helix

Figure 23.6 Left- and right-handed helices.

on a neighboring coil (Figure 23.7). The right-handed or α-*helix* is more common in proteins than the left-handed β-*helix*, since it ensures that the R groups are almost completely outside the helix, where they have the most room and the least effect on the stable structure of the helix.

Another type of secondary structure involves interchain hydrogen bonding (Figure 23.8), in contrast to the intrachain bonding of the helix. Unlike the α-helix, which is characteristic of proteins with large, bulky R groups, this *pleated-sheet* structure (Figure 23.9) is characteristic of proteins comprising amino acids with small R groups, such as glycine and alanine.

The *tertiary structure* of a protein is determined by the way the helix itself folds and coils. It depends on various kinds of intermolecular attractions—hydrogen bonding, electrovalent forces, London dispersion forces, and occasionally disulfide linkages. Often the helix folds into a structure that is nearly spherical;

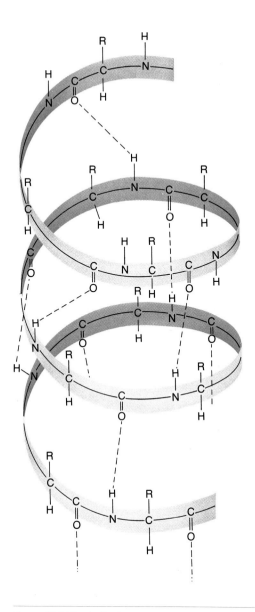

Figure 23.7 The helical structure of a protein. Hydrogen bonds are indicated by broken lines.

proteins with this tertiary structure are called *globular proteins* (Figure 23.10). Other proteins have fiberlike or sheetlike tertiary structures; these are called *fibrous proteins* (Figure 23.11). These more pliable fibrous proteins serve structural and supporting functions and are found in arteries, hair, and muscles.

Globular proteins, which are usually water-soluble, are especially susceptible to *denaturation*—the loss of secondary and tertiary structure. When exposed

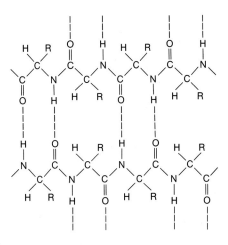

Figure 23.8 Interchain hydrogen bonding.

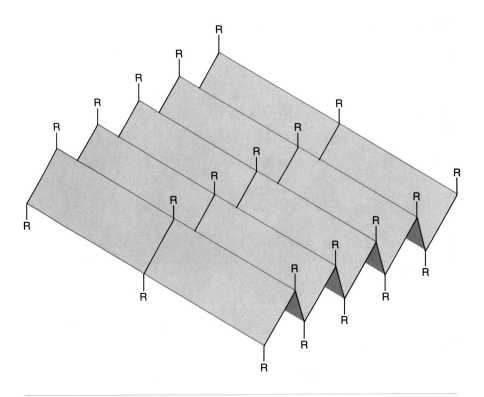

Figure 23.9 Pleated-sheet structure of a protein.

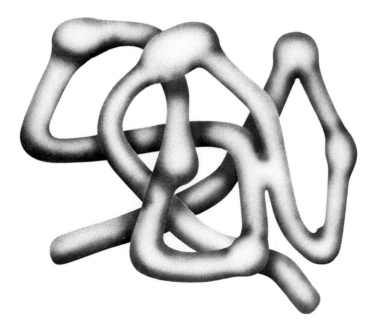

Figure 23.10 The tertiary structure of a globular protein.

to heat, extreme pH, or organic solvents, proteins often uncoil and unfold, even though the peptide linkages remain intact. After denaturation, globular proteins often become insoluble and lose their physiological properties. Boiling an egg is an example of an irreversible denaturing process.

We can now take the protein structure even one step further. Many complex proteins, such as hemoglobin, consist of more than one distinct subunit. These subunits may or may not be identical and can be bonded together by disulfide linkages, hydrogen bonds, or other intermolecular forces. They can usually be separated, often reversibly, under the right conditions. The nature of their interactions and the arrangement of these subunits determines the *quaternary structure* of the protein. The subunits alone are often inactive and cannot carry out the function of the protein.

Questions

10. What is a peptide linkage?

11. Write balanced equations for the two reactions between alanine and glycine.

12. What is meant by the primary, secondary, tertiary, and quaternary structures of a protein?

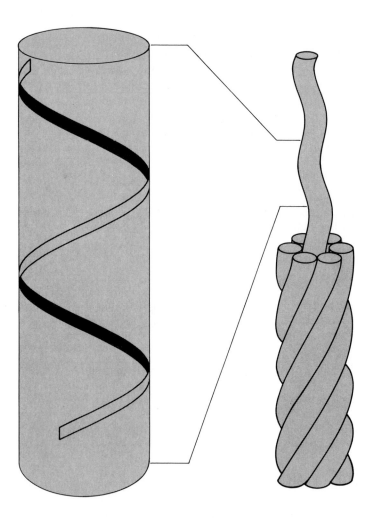

Figure 23.11 Structure of a fibrous protein consisting of seven strands arranged like a cable.

23.5 Medicinal Chemistry and Enzymes

We can define *medicinal* chemistry as the application of chemistry to biological systems with the specific aim of understanding how chemicals and medicines affect life processes. As a result of this kind of research, medicinal chemists have developed a multitude of antiseptics, antibiotics, and other medicines. We shall discuss only one of these here, as an illustration of the function of *enzymes*.

Although their function is simple in principle, enzymes are among the most remarkable of all substances. They are proteins that function as catalysts for chemical reactions in biological systems. However, the contrasts between enzymes and inorganic catalysts are striking. An inorganic catalyst may catalyze a number of reactions. For example, platinum catalyzes the hydrogenation of a double bond, the oxidation of SO_2 to SO_3, the combination of hydrogen and nitrogen to form ammonia, and other reactions as well. Enzymes, on the other hand, are very specific. We saw in Section 23.2 that despite only a subtle difference between starch and cellulose the human body can digest only starch. The reason is that amylase, an enzyme found in the saliva is able to hydrolyze the α-linkages between glucose units in starch but not the β-linkages in cellulose. Because of this kind of specificity, a great many enzymes are required to catalyze the enormous number of reactions that take place in the body. Thousands of different enzymes may be required in each cell of the body.

A second major difference between enzymes and inorganic catalysts is the rate at which they catalyze reactions. This rate is called the *turnover number*—the number of molecules of *substrate* (substance acted on by a catalyst) transformed per enzyme molecule per minute. Turnover numbers can range as high as 10^8; thus only a tiny amount of an enzyme is often needed to catalyze a reaction. Analogous numbers for inorganic catalysis rates are often a million times smaller.

Although enzymes are proteins, they are often not completely composed of amino acids. In fact, the protein portion of the enzyme (called the *apoenzyme*) by itself may be inactive. The apoenzyme is usually activated by a metal ion (*metal-ion activator*) or some small organic molecule (a *coenzyme*).

An active enzyme usually comprises an apoenzyme and a metal-ion activator or an apoenzyme and a coenzyme.

Coenzymes may be vitamins, or simple derivatives of vitamins, that cannot be synthesized by the body (thus the need to ensure an adequate daily vitamin intake). Thiamine, niacin, and riboflavin are among the vitamins that function as coenzymes. These vitamins, like all coenzymes, are relatively simple compounds (at least compared to the apoenzymes) and can be manufactured commercially.

The specificity of an enzyme is determined largely by its tertiary structure. It must be folded so that amino acids are positioned where they can interact loosely—perhaps by hydrogen bonding—with the substrate molecule. The location on the surface of an enzyme providing this singular arrangement of amino acids is referred to as the *active site*. The geometry of the active site ensures that the enzyme will interact only with specific substrate molecules. As a result of this specific interaction, an enzyme-substrate complex is formed, which enables the substrate to undergo a chemical change that it might not otherwise undergo. Once the product is formed, it is released from the enzyme, since it no longer meets the requirements of geometry and bonding at the active site. We can summarize the enzymatic reaction as follows:

enzyme + substrate $\rightleftharpoons$ enzyme-substrate complex $\longrightarrow$ enzyme + product

Figure 23.12 illustrates these principles in a schematic representation of a catalyzed hydrolysis reaction.

Although the chemistries of many chemotherapeutic agents are not completely understood, some are known to interfere with the synthesis of an essential

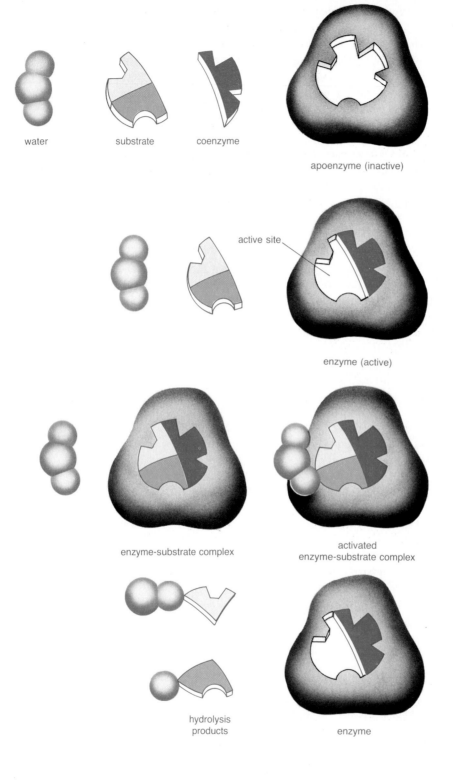

Figure 23.12 Hydrolysis reaction catalyzed by an enzyme.

molecule. This is the prevailing theory regarding the antibacterial action of the medicine we've set out to discuss, sulfanilamide (see Figure 23.13).

Folic acid is one of the B vitamins. In addition to being an essential element in the human diet where it functions as a coenzyme, it acts as a coenzyme in the metabolic processes of certain bacteria. These bacteria use *para*-aminobenzoic acid as an intermediate in folic acid synthesis. In the presence of the similar sulfanilamide molecule, however, the bacteria often select it instead of *para*-aminobenzoic acid to make the coenzyme. The molecule produced is inactive and bacterial reactions requiring folic acid cannot take place. Ultimately the growth and multiplication of the bacterial cells cease. Human cells, on the other hand, do not synthesize folic acid but must acquire it from the diet, so no human enzyme systems are inhibited by sulfanilamide. This antimetabolic action of sulfanilamide was first recognized in the 1930s. Within a few years, hundreds of derivatives of this compound, known collectively as *sulfa drugs*, had been synthesized and tested for their effectiveness as antimetabolites.

Since sulfa drugs interfere with enzyme systems in a desirable way, promoting the health and well-being of the human organism, they are called *medicines*. Substances that interfere with enzyme systems in an undesirable way are *poisons*. An example of this kind of interference might be to prevent the formation of an

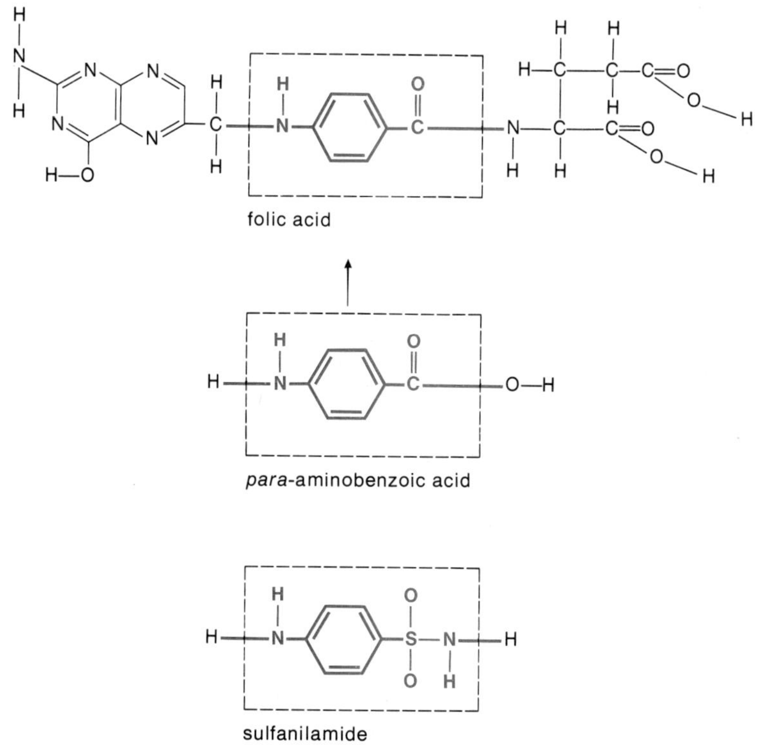

Figure 23.13 Formulas of folic acid, *para*-aminobenzoic acid and sulfanilamide.

essential enzyme from the apoenzyme and the activator. The cyanide ion (CN^-) is this kind of poison. It forms stable bonds with metal-ion activators (see Chapter 24) so that few metal ions remain to convert inactive apoenzymes into catalytically active enzymes. A second kind of poison behaves like a substrate but forms such stable complexes with the enzyme that the enzyme is permanently deactivated. Examples include certain heavy metal ions, such as Pb^{2+} or Hg^{2+}.

Since enzymes can be a million times as effective as inorganic catalysts, it is not surprising that man has attempted to apply them to industrial or commercial procedures. The difficulty is that industrial processes often require conditions that denature the enzymes, such as high temperatures and extreme pH's. Also, enzymes are much more expensive than simpler inorganic catalysts.

One recent application of enzymes has been in detergents. The idea was that added enzymes would function the same way they do in the digestive system, catalyzing the hydrolysis of large molecules into smaller, more readily removed ones. Thus stains from tea, fruit juices, eggs, and gravy would be more effectively removed. Unfortunately, these persistent enzymes began to appear in natural waters, where they threatened to disrupt the ecosystem. As a result, their use in detergents has been largely discontinued.

A more exotic use of enzymes is in the manufacture of candies with liquid centers. At first the center is a firm mixture containing sucrose and an enzyme that catalyzes its hydrolysis. After the center is covered with chocolate, the enzyme hydrolyzes the sucrose, producing products more soluble in the small amount of water initially present.

Questions

13. What accounts for the specificity of enzymes?

14. Why might an enzyme be effective in removing a gravy stain from cloth but have no effect on an ink stain?

15. What role do metal ions play in an enzyme system?

16. In what way might metal ions inhibit an enzyme system?

17. Why is the CN^- ion a poison?

Questions and Problems

23.1 Lipids

18. What are the essential fatty acids?

19. Write the formula and name of a geometric isomer of oleic acid.

23.2 Carbohydrates

20. Glucose is a very water-soluble compound. How do you explain this?

21. Identify a compound that is a
    a. structural isomer of glucose.   b. polymer of glucose.

22. How does the structure of starch differ from that of cellulose?

23. Bread tastes sweet after prolonged chewing. Why?

24. Distinguish between reducing sugars and nonreducing sugars.

23.3 Amino Acids

25. Identify an amino acid that fits each of the following descriptions:
    a. Simplest of the naturally occurring α-amino acids.
    b. A neutral amino acid.
    c. Essential for human beings.
    d. An amino acid whose monosodium salt is used to accentuate the flavor of meat.
    e. An amino acid that contains a sulfur-to-sulfur linkage.

26. Which of the following compounds show optical isomerism? Geometric isomerism?
    a. oleic acid   b. glycine   c. serine

23.4 Proteins

27. What chemical linkage is the basis for the primary structure of proteins?

28. Suppose the following fragments were identified in a mixture resulting from the partial hydrolysis of a protein. Determine a sequence of 10 amino acids that you might expect to be present in this protein.

    gly-his-leu
    phe-val-asp
    cys-gly-ser-his
    his-leu-cys-gly
    asp-gly-his
    gly-ser-his
    phe-val

29. What is formed by the complete hydrolysis of
    a. sucrose?   b. starch?   c. protein?   d. fat?   e. cellulose?

30. Identify the monomers of the following naturally occurring polymers:
    a. proteins   b. cellulose   c. starch

31. Show by equations how the polymers listed in Question 30 are related to their constituent monomers.

32. Distinguish between the members of each of the following pairs:
    a. essential and nonessential amino acids
    b. ribose and deoxyribose
    c. peptides and protein

24 Coordination Compounds

There is a large and colorful class of compounds that we have not considered. Structural units for these compounds appear at first to be a combination of two or more simpler units—atoms, ions, or molecules. Certain salts, for instance, are found to contain more than one type of metal ion, and sometimes more than one type of anion, which are always present in the same ratio by weight. At a loss for a satisfactory structural explanation for these salts, early chemists referred to them as *double salts* and wrote formulas such as $SnCl_4 \cdot 2\, NaCl$ and $Fe(CN)_2 \cdot 4\, KCN$. They used the dot to indicate a type of chemical bond they did not yet understand. The dot was employed, also, in the formulas for *ammines* such as $Zn(NO_3)_2 \cdot 4\, NH_3$ and hydrates such as $CuSO_4 \cdot 5\, H_2O$. Evidence that more than one type of bonding was involved in these compounds came from the observation that under proper conditions it was possible to drive off the ammonia (from ammines) or water (from hydrates), leaving a residue consisting of the simple salt. Nevertheless, these ammines and hydrates were recognized as true chemical compounds, because their compositions were the same wherever and whenever they were found.

There are two other groups of compounds that do not fit neatly into any of the classes we've encountered so far. *Metal carbonyls*, such as $Fe(CO)_5$ and $Ni(CO)_4$, appear to be combinations of neutral CO molecules and a metallic atom. In marked contrast to the properties of the components, however, metal carbonyls are volatile and are frequently low-melting solids or even liquids. Finally, *addition compounds* appear to be formed from the union of two stable molecular units. An example is the compound $BF_3 \cdot O(CH_3)_2$, where both a molecule of boron trifluoride and a molecule of dimethyl ether can be identified in its formula.

Our task in this chapter will be to shed some light on the special kind of bond that is common to all these compounds.

24.1 Coordination

The familiar ionic compound, ammonium chloride, has something in common with the compounds described above; namely, it can be visualized as a product of two stable, neutral substances. Covalent HCl reacts with covalent NH_3 to form the essentially ionic NH_4Cl:

$$\text{(H)}:\!\ddot{\underset{..}{Cl}}\!: \; + \; H:\!\underset{\underset{H}{|}}{\overset{\overset{H}{|}}{\ddot{N}}}\!:H \; \longrightarrow \; \left[H:\!\underset{\underset{H}{|}}{\overset{\overset{H}{|}}{\ddot{N}}}\!:H^+ \; + \; :\!\ddot{\underset{..}{Cl}}\!:^- \right]$$

For our present purposes, however, the most important feature of NH_4Cl is that the covalent bond that forms between the ammonia molecule and the proton from HCl uses the *lone pair* of electrons on the ammonia molecule and the empty $1s$ orbital on the proton. This illustrates the principle that will ultimately explain the double salts, ammines, hydrates, metal carbonyls, and addition compounds. Put simply, substances with an unshared pair of electrons often gain stability by sharing that pair, and substances with empty outer orbitals are often more stable if they can use those orbitals to form covalent bonds. (Recall from Chapter 6 the formation of NH_3BF_3, which we can now identify as an addition compound.) This kind of covalent bond-forming process, in which both of the bonding electrons are furnished by one member, is known as *coordination*, and the bond is sometimes called a coordinate covalent bond.

We now see that the ammines can be considered analogous to the ammonium compounds. In place of a proton, an ammine contains a metal ion (which, since it contains more available empty outer orbitals than a proton, can coordinate simultaneously with other ammonia molecules as well). As an example, we can look at the electron configuration of the zinc(II) ion. (See Chapter 6 for a discussion of the relative energies of the $3d$ and $4s$ sublevels.)

$$Zn^{2+} \quad \underset{3s}{\boxed{\uparrow\downarrow}} \quad \underset{3p}{\boxed{\uparrow\downarrow|\uparrow\downarrow|\uparrow\downarrow}} \quad \underset{3d}{\boxed{\uparrow\downarrow|\uparrow\downarrow|\uparrow\downarrow|\uparrow\downarrow|\uparrow\downarrow}} \quad \underset{4s}{\boxed{}} \quad \underset{4p}{\boxed{||}}$$

In the presence of electron-pair donors, such as NH_3, the four empty orbitals in the zinc ion may hybridize to form four sp^3 orbitals and then accept four pairs of electrons. The result is a $[Zn(NH_3)_4]^{2+}$ ion with tetrahedral geometry. Thus we write the formulas for ammines without the ambiguous dot; for example, $[Zn(NH_3)_4](NO_3)_2$. This formula fits the properties of the compound much better than the old formula of $Zn(NO_3)_2 \cdot 4\,NH_3$ because in aqueous solution the salt clearly dissociates into two kinds of ions, the $[Zn(NH_3)_4]^{2+}$ ion and the NO_3^- ion.

Historically the recognition of coordination compounds as a class (although not by that name) was the achievement of Alfred Werner (1866–1919), a Swiss chemist. At the age of 26, he proposed a theory to explain the formation of the double salts and similar compounds—a theory that, although subject to continual modification, still provides the basis of today's theories. Substances, said Werner, exhibit both a primary and a secondary valence. In $[Zn(NH_3)_4](NO_3)_2$, zinc is exhibiting a primary (ionic) valence of +2, which accounts for the two nitrate ions, and a secondary (covalent) valence of 4, which accounts for the four ammonia molecules.

Water is another example from a wide range of ions and molecules that can function as electron-pair donors. Hydrates form when a molecule of water shares one of the lone pairs of electrons on the oxygen atom. Unfortunately chemists continue to write hydrate formulas in the old style. For example, the copper sulfate hydrate is written $CuSO_4 \cdot 5\ H_2O$, which obscures an interesting pattern of bonds. In copper sulfate pentahydrate, four water molecules are coordinated to the Cu^{2+} ion, and the fifth is bonded by hydrogen bonds both to the water molecules attached to the Cu^{2+} ion and to the sulfate ion. A more descriptive formula might be $[Cu(H_2O)_4]SO_4 \cdot H_2O$.

An ion that readily coordinates to metals is the cyanide ion ($:C:::N:^-$). Thus the formula once written as $Fe(CN)_2 \cdot 4\ KCN$ becomes $K_4[Fe(CN)_6]$, showing the -4 ion produced when six cyanide ions (using the unshared electrons on the carbon atoms) are coordinated to an $Fe(II)$ ion. This example also illustrates another generalization: Group IA and IIA metal ions seldom form coordinate covalent bonds. In a similar example, six chloride ions coordinate to tin (in the $+4$ oxidation state) and we rewrite $SnCl_4 \cdot 2\ NaCl$ as $Na_2[SnCl_6]$.

The other compounds introduced in the opening paragraphs of this chapter can also be explained in terms of coordination chemistry. The carbon monoxide molecule ($:C:::O:$) is isoelectronic with the cyanide ion; thus the neutral metal carbonyls are analogous to anions such as $[Fe(CN)_6]^{-4}$. And the same principles apply to the addition compounds. Boron trifluoride contains three sigma bonds and no unshared electron pairs on the central boron atom; thus there are only six valence electrons in the outer energy level of the boron atom:

```
      F
      |
      B
     / \
    F   F
```

The oxygen atom in the dimethyl ether molecule has two unshared pairs of electrons:

```
    H    ..   H
    |   :O:   |
  H-C    \   C-H
    |     \  |
    H        H
```

Again we have the requirements for a coordination compound. The boron atom provides an empty orbital and the oxygen atom provides the electron pair for a coordinate covalent bond:

```
    F      H
     \     |
   F-B:Ö-C-H
     /   | |
    F    | H
       H-C-H
         |
         H
```

In one important way, this last example contrasts with the simple coordination of an ammonia molecule with a proton. The ammonium ion is a perfect tetrahedron. But the extreme size difference between the fluoride ion and the ether molecule distorts the geometry of $BF_3 \cdot O(CH_3)_2$ so that the ether molecule has more room. Effects such as this, due to the size or shape of a structural species rather than to its electron configuration, electronegativity, or some other property, are called *steric effects*.

Questions

1. What values would be assigned to Werner's primary valence and secondary valence for the metals within the brackets in these substances?
  a. $K_4[Fe(CN)_6]$ b. $[Ag(NH_3)_2]Cl$ c. $(NH_4)_2[Zn(SCN)_4]$

2. Rewrite the formulas for $MnF_4 \cdot 2\,KF$ and $CoCl_3 \cdot 6\,NH_3$ in the modern coordination form. Remember that IA and IIA metal ions seldom form coordinate covalent bonds.

3. Ammonia and boron trifluoride react to form an addition compound. Draw its electron formula. Which substance functions as a Lewis acid and which as a Lewis base?

24.2 Terminology

Before proceeding further, we'll define some of the terms we shall need as we discuss coordination compounds.

Lewis bases (atoms, ions, or molecules) that act as electron-pair donors in these compounds are called **ligands,** and the atom in the ligand that provides the electrons is the **donor atom.** The substance that has the necessary empty orbital(s) and that acts as an electron-pair acceptor is called the **center of coordination.** According to the Lewis acid-base theory (Chapter 14), therefore, ligands are Lewis bases and centers of coordination are Lewis acids. The center of coordination is usually a metal ion, although it is sometimes a neutral metal atom. The unit composed of the center of coordination plus all ligands is called the **coordination complex** or sometimes just the *complex*. If this unit has a net charge, as in $[Fe(CN)_6]^{4-}$ or $[Zn(NH_3)_4]^{2+}$, it may be referred to as a **complex ion.** The number of coordinate covalent bonds formed with the center of coordination is the **coordination number.**

A ligand such as NH_3 that forms only one bond with a center of coordination is called an *unidentate* ligand. Ligands such as $NH_2-CH_2-CH_2-NH_2$, with more than one donor atom, are called *polydentate* (*bidentate, tridentate*, etc.) ligands. Polydentate ligands are also called **chelating agents,** a term derived from the Greek word *chelos*, for claw. Thus the number of ligands can be smaller than the coordination number.

To illustrate these terms, let's consider the formula of the salt, potassium trichloroammineplatinate(II): $K[PtCl_3(NH_3)]$. The negative ion of this salt is a coordination complex, with a -1 charge. The center of coordination is the Pt^{2+} ion. There are four ligands in the complex, namely, three chloride anions and one ammonia molecule. Each of these is a unidentate ligand, so that there are four bonds formed with the center of coordination: The coordination number of the complex is 4. Each ligand contributes the necessary electron pair for a bond. In the ammonia molecule the nitrogen atom is the donor atom. Figure 24.1 illustrates this example.

In the salt $[CoCl(NO_2)(NH_2CH_2CH_2NH_2)_2]NO_2$, the coordination complex forms a positive ion having a $+1$ charge (Figure 24.2). The center of coordination is the Co^{3+} ion. There are four ligands, one chloride anion, one nitrite anion (in which the nitrogen atom is the donor atom), and two bidentate ethylenediamine molecules:

$$\overset{H}{\underset{H}{N}}-\overset{\overset{H}{|}}{\underset{\underset{H}{|}}{C}}-\overset{\overset{H}{|}}{\underset{\underset{H}{|}}{C}}-\overset{H}{\underset{H}{N}}$$

This molecule is one of the simpler chelating agents. Since each of the two ethylenediamine molecules forms two bonds with the Co^{3+} ion, the coordination number of the complex is 6.

Question

4. A coordination compound has the formula $(NH_4)_2[Zn(SCN)_4]$. Answer the following questions about this compound:
  a. Is this compound essentially ionic or covalent?
  b. What is the charge on the complex?

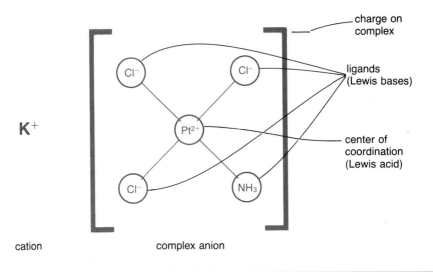

Figure 24.1 Some terminology applied to the compound:

$K[PtCl_3(NH_3)]$

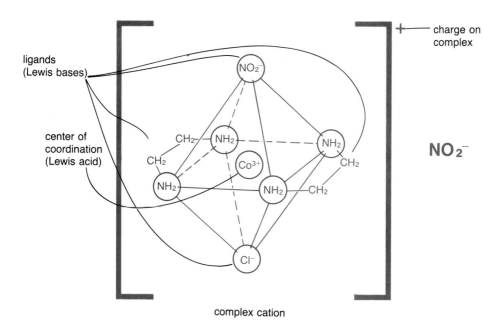

Figure 24.2 Some terminology applied to the compound:

[CoCl(NO$_2$)(NH$_2$CH$_2$CH$_2$NH$_2$)$_2$]NO$_2$

    c. What is the charge on each ligand?
    d. What is the coordination number of the complex?
    e. What is the oxidation state of the center of coordination?
    f. Identify a Lewis acid in the complex.
    g. Identify a Lewis base in the complex.

24.3 Nomenclature

At first glance, some of the complex formulas for coordination compounds would seem to defy simple rules of nomenclature. Indeed, the nomenclature rules applicable to the binary or ternary inorganic compounds are of little help here. However, unambiguous and informative names can be assigned after the following questions are answered:

1. Does the formula being named represent a covalent compound, an ionic compound, or an ion?
2. If the substance is ionic, what are the charges on the ions?
3. What are the ligands and what is the center of coordination?
4. What is the oxidation number of the center of coordination?

Because all of these questions must be answered, naming a coordination compound is not simply an empty exercise; it demands an understanding of structure.

Rules for Writing Formulas

1. If the compound is ionic, the formula of the cation is written first; that of the anion, second.

2. The formula for the entire coordination complex is enclosed in brackets.

3. The formula for the coordination complex begins with the symbol for the center of coordination, followed by the formulas of the ligands in this order:

a. Anionic ligands: O^{2-}; OH^-; one-element anions in order of increasing periodic group number or, within a group, of decreasing atomic number; polyatomic inorganic anions similarly ordered by the group number and atomic number of the donor atom; organic ligands in alphabetical order.

b. Neutral and cationic ligands: H_2O; NH_3; inorganic ligands ordered as above by the group number and atomic number of the donor atom; organic ligands in alphabetical order.

Rules for Naming Coordination Compounds and Ions

1. If the compound is ionic, the cation is named first. The remaining rules apply to any complex, whether it is an ion or molecule.

2. Ligands are named first; the center of coordination is named last. Names of ligands are read in order from left to right, using the prefixes *di-*, *tri-*, *tetra-*, etc., where appropriate. If the name of the ligand is a complicated, polysyllabic one, the prefixes *bis-*, *tris-*, and *tetrakis-*, are used instead, and the name of the ligand is enclosed in parentheses.

3. If the complex has no charge or if it is a cation, the center of coordination is given its usual name without an added ending. If the complex is an anion, the ending *-ate* is added to the name of the center. In the following cases this ending is added to the Latin rather than the English name of the element: Cu (cuprate), Au (aurate), Fe (ferrate), Pb (plumbate), Ag (argentate), and Sn (stannate). (See Table 24.1.)

4. The oxidation number of the center of coordination is indicated in Roman numerals (or an Arabic zero) in parentheses following its symbol.

5. Anionic ligands are given names ending in *o*. Some common ones are

F^-	fluoro	CN^-	cyano
Cl^-	chloro	SCN^-	thiocyanato
Br^-	bromo	NO_2^-	nitro
I^-	iodo	SO_3^{2-}	sulfito
O^{2-}	oxo	O_2^{2-}	peroxo
OH^-	hydroxo	$C_2O_4^{2-}$	oxalato
S^{2-}	thio	$S_2O_3^{2-}$	thiosulfato

Table 24.1 Names for Centers of Coordination of Anionic Complexes

Center of Coordination	Name
Al^{3+}	aluminate(III)
Ag^+	argentate(I)
Au^+	aurate(I)
Cr^{3+}	chromate(III)
Co^{3+}	cobaltate(III)
Co^{2+}	cobaltate(II)
Cu^{2+}	cuprate(II)
Fe^{3+}	ferrate(III)
Fe^{2+}	ferrate(II)
Pb^{2+}	plumbate(II)
Hg^{2+}	mercurate(II)
Ni^{2+}	nickelate(II)
Pt^{2+}	platinate(II)
Sn^{2+}	stannate(II)
Zn^{2+}	zincate(II)

6. Names of neutral or cationic ligands are used without change, except in the following specific cases:

H_2O aquo
NH_3 ammine
CO carbonyl

7. Some organic ligands are

CH_3 methyl (symbol: Me) ⎫
C_2H_5 ethyl (symbol: Et) ⎬ Oxidation numbers for these ligands in coordination complexes must be taken as -1.
C_6H_5 phenyl (symbol: Ph) ⎭

C_5H_5N pyridyl (symbol: py)   Oxidation number $= 0$.

8. Some common chelating ligands and their names are

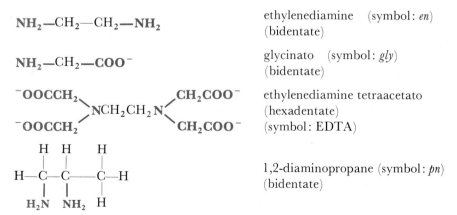

$NH_2-CH_2-CH_2-NH_2$ ethylenediamine (symbol: en) (bidentate)

$NH_2-CH_2-COO^-$ glycinato (symbol: gly) (bidentate)

ethylenediamine tetraacetato (hexadentate) (symbol: EDTA)

1,2-diaminopropane (symbol: pn) (bidentate)

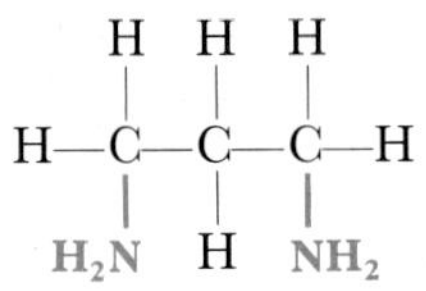

$$\begin{array}{c} \text{H} \quad \text{H} \quad \text{H} \\ | \quad \ | \quad \ | \\ \text{H}-\text{C}-\text{C}-\text{C}-\text{H} \\ | \quad \ | \quad \ | \\ \text{H}_2\text{N} \quad \text{H} \quad \text{NH}_2 \end{array}$$

1,3-diaminopropane (symbol: *tn*)
(trimethylenediamine)
(bidentate)

Example 24.1 Name the compound $[PtCl_2(NH_3)_2]$.

Solution

a. The entire formula appears inside the brackets with no indicated charge; thus this is a covalent compound whose molecule is a coordination complex.

b. The ligands are two Cl^- ions and two NH_3 molecules.

c. The center of coordination is a platinum ion. To determine its oxidation number, note that the charge on the whole complex is 0, the charge on each chloride ion is -1, and the ammonia molecules have no charge. Thus the center of coordination is the Pt(II) ion.

d. We name the ligands from left to right, followed by the center of coordination:

dichlorodiammineplatinum(II)

Example 24.2 Name the compound $(NH_4)_2[Sn(OH)_6]$.

Solution

a. The presence of the ammonium ions in the formula shows that this is an ionic compound. The coordination complex is an anion with a -2 charge.

b. The ligands within the complex are six OH^- ions.

c. The center of coordination is tin(IV). This we determine from the charge on the complex (-2) and the presence of six hydroxide ions (-1) within the complex.

d. We name the cation first and then the anion. Because the complex is an anion, the ending *-ate* is added to the (Latin) name of the center of coordination:

ammonium hexahydroxostannate(IV)

Example 24.3 Name the compound $[Co(H_2O)_2(NH_3)_4]Br_3$.

Solution

a. The compound is ionic. The cation is a coordination complex; the anions are three Br^- ions. The charge on the cation is $+3$.

b. The ligands are two H$_2$O molecules (neutral) and four ammonia molecules (neutral). Thus the charge on the center of coordination is the same as the charge on the complex, namely, $+3$.

c. Because the coordination complex is a cation, we give the name of the center of coordination no special ending.

d. The cation is named first, followed by the anion:

diaquotetraamminecobalt(III) bromide

Questions

5. Name the following substances:
   a. $[HgI_4]^{2-}$ b. $Na[Al(OH)_4]$ c. $[Co(CN)_5(Et)]^{3-}$

6. Write formulas for the following substances:
   a. nitropentaamminecobalt(III) ion
   b. tris(ethylenediamine)cobalt(III) nitrate
   c. potassium hexacyanoferrate(II)

24.4 The Stability of Coordination Complexes

Since we don't often encounter coordination complexes outside aqueous solutions, we shall discuss their stability in terms of their resistance to dissociation in water. Since this dissociation is an equilibrium reaction, we can take the equilibrium constant (usually called an *instability constant*) as a rough measure of stability. For similar complexes, the smaller the instability constant, the more stable the complex. (See Table 24.2.)

Table 24.2 Instability Constants at 25°C

$$K_{ins} = \frac{[\text{dissociation products}]}{[\text{coordination complex}]}$$

Complex	K_{ins}
$[AlF_6]^{3-}$	1.5×10^{-20}
$[Al(OH)_4]^-$	1.3×10^{-34}
$[Co(NH_3)_6]^{2+}$	1.3×10^{-5}
$[Co(NH_3)_6]^{3+}$	2.2×10^{-34}
$[Co(en)_3]^{2+}$	1.0×10^{-12}
$[Cu(NH_3)_4]^{2+}$	4.7×10^{-15}
$[Fe(CN)_6]^{4-}$	1×10^{-35}
$[Fe(CN)_6]^{3-}$	1×10^{-42}
$[Ag(NH_3)_2]^+$	6.0×10^{-8}
$[Ag(CN)_2]^-$	1×10^{-21}
$[Zn(NH_3)_4]^{2+}$	3.4×10^{-10}
$[Zn(OH)_4]^{2-}$	3.6×10^{-16}

Since equilibrium constants are measures of relative free energy, instability constants tell us only about thermodynamic stability. As we've seen before, kinetic factors may be even more important. Some complexes dissociate rapidly (often exchanging ligands for water molecules), thus attaining equilibrium as soon as they are mixed in water. These are called *labile* complexes. Others dissociate, or exchange ligands, very slowly and are called *inert*. It is therefore possible for a complex to be thermodynamically unstable and yet inert.

In principle, any metal ion can function as a center of coordination. In fact, not all are equally suited. The force that stabilizes a coordinate covalent bond is the electrostatic attraction between the positive nucleus of the center of coordination and the electron pair donated by the ligand; any factor tending to increase this attraction should make a metal ion more suitable as a center of coordination. Two of these factors are high ionic charge and small radius. If we divide the charge by the radius, the result (called the *charge density*) is a good measure of an ion's tendency to form complexes (Table 24.3).

Another important (in fact, essential) factor is the availability of empty orbitals at sufficiently low energy levels. Transition metal ions are thus more suitable than the representative metal ions. Some of the representative metal ions do form complexes but unless the ligands contribute unusual stability to the structure, these complexes are relatively unstable and usually unimportant.

Coordination compounds are known with coordination numbers from 1 to 9. By far the most common coordination numbers, however, are 2, 4, and 6. We can often predict these numbers from the number of low-energy orbitals available on the center of coordination, but we must also consider steric effects. With a small center of coordination and large ligands, it may simply be impossible to arrange the ligands around the center of coordination, even if the orbitals are available.

Table 24.3 The Charge Densities of Common Metal Ions

Ion	Radius (nm)	Charge Density (charge/radius)
K^+	0.133	7.5
Ag^+	0.126	7.9
Sn^{2+}	0.112	18
Hg^{2+}	0.110	18
Ca^{2+}	0.099	20
Cu^{2+}	0.096	21
Na^+	0.095	11
Mn^{2+}	0.080	25
Fe^{2+}	0.076	33
Co^{2+}	0.074	27
Zn^{2+}	0.074	27
Ni^{2+}	0.072	28
Cu^{2+}	0.069	29
Cr^{3+}	0.069	43
Fe^{3+}	0.064	47
Co^{3+}	0.063	48
Al^{3+}	0.050	60

The properties of the ligand can also influence the stability of the coordination complex. Volatile ligands such as NH_3 and H_2O are affected by temperature changes; thus, for example, almost all hydrates can be easily dehydrated by heating. Large negative charges and small radii, on the other hand, are stabilizing factors, much as the charge density of the center of coordination is. Probably the most strongly stabilizing factor of all, however, is the presence of polydentate ligands in the complex. Even representative metal ions form fairly stable complexes with the proper chelating agents. EDTA, the anion of ethylene-diaminetetraacetic acid, is thus an efficient water-softening agent because it forms a complex ion with the calcium(II) ion, one of the "hard" constituents of hard water. This complex prevents the precipitation of calcium salts such as calcium stearate (Chapter 16). The ethylenediaminetetraacetate ion has six donor atoms (Figure 24.3), and as a hexadentate ligand it forms six coordinate covalent bonds with the calcium ion. The calcium ion forms a similar soluble complex with the anion of metaphosphoric acid, $(HPO_3)_n$. Sodium metaphosphate is also used as a water softener.

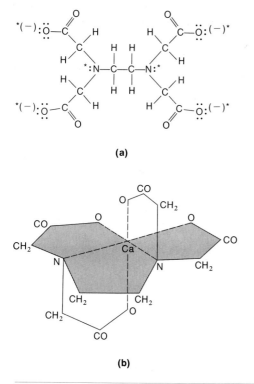

Figure 24.3 (a) The hexadentate ethylenediaminetetraacetate ion. Asterisks indicate donor atoms. **(b)** The chelate complex formed with the calcium ion and the anion of ethylenediaminetetraacetic acid. This coordination sphere contains five 5-membered rings. The three shaded rings are in the horizontal plane. One ring is above the horizontal and another is below the horizontal. Coordinate covalent bonds are indicated with dashed lines.

Questions

7. Which ion of each pair below would probably form the more stable complex (with the same ligand)? Consult Table 24.3 for comparative charge densities.
  a. Fe^{2+}, Fe^{3+} b. Ca^{2+}, Zn^{2+} c. Al^{3+}, Na^{+} d. Co^{3+}, Co^{2+}

8. What is a chelating agent?

24.5 Geometry

The geometry in coordination complexes can in most cases be deduced from the coordination number, with due consideration given to steric factors. If steric interactions or unbonded electron pairs do not distort the geometry, we can expect complexes with a coordination number of 2 to be linear, those with a coordination number of 3 (very rare) to be trigonal, and those with a coordination number of 6 to be octahedral. Complexes with coordination numbers of 4 can have either tetrahedral or square-planar geometry, and those with a coordination number of 5 can be either trigonal bipyramidal or square pyramidal. The most

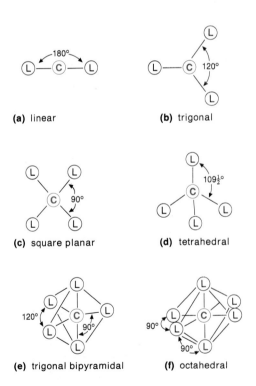

Figure 24.4 Geometries of some complexes: C = center of coordination, L = ligand.

common of these geometries is illustrated in Figure 24.4. Very few complexes with higher coordination numbers have been identified.

We can best understand these geometries, as well as the bonding arrangements that give rise to them, by again turning to the concept of hybrid orbitals (see Chapter 7). Thus the tetrahedral complexes employ four equivalent sp^3 orbitals on the center of coordination. This arrangement is especially common among the representative metals and the IB and IIB metals, where there are no d orbitals readily available for hybridization. (Remember that the IB and IIB metals form d^{10} ions—see Chapter 6.) Tetrahedral complexes also appear in other periodic groups, of which $[CoCl_4]^{2-}$ and $[FeCl_4]^-$ are examples. The Group VIII metals often form dsp^2 hybrid orbitals. These orbitals give rise to square-planar complexes such as $[Ni(CN)_4]^{2-}$. Ni(II) ion also forms complexes with coordination numbers of 5; their geometry is either trigonal bipyramidal or square pyramidal. Finally, many atoms or ions form octahedral complexes with either d^2sp^3 or sp^3d^2 hybridization. The important distinction between these two is the subject of Section 24.6.

The common hybridizations found for coordination numbers 4, 5, and 6 are summarized in Table 24.4.

Table 24.4 Hybridization and Shape of Coordination Complexes

Coordination Number	Hybridization	Shape	Example
4	sp^3	Tetrahedral	$[Zn(NH_3)_4]^{2+}$
	dsp^2	Square planar	$[Ni(CN)_4]^{2-}$
5	dsp^3	Trigonal bipyramidal	$[Fe(CO)_5]$
	dsp^3	Square pyramidal	$[Ni(CN)_5]^{3-}$
6	d^2sp^3	Octahedral	$[Fe(CN)_6]^{4-}$
	sp^3d^2	Octahedral	$[CoF_6]^{3-}$

24.6 Magnetic Properties and Orbitals

We first observed in Chapter 6 that atoms, ions, or molecules containing unpaired electrons responded differently to a magnetic field than those without. They are said to have a *magnetic moment*. And since the size of the magnetic moment is a direct measure of the number of unpaired electrons in an atom, ion, or molecule, it can be used to answer some questions about hybridization in coordination complexes. Consider the two complexes $[CoF_6]^{3-}$ and $[Fe(CN)_6]^{4-}$. The central ion in each is a d^6 metal ion, each has a coordination number of 6, and each has octahedral geometry. These observations suggest that each metal ion has formed bonds from one s, three p, and two d orbitals. However, the magnetic moments of the two complex ions indicate that the hexafluorocobaltate(III) ion has four unpaired electrons, whereas the hexacyanoferrate(II) ion has none. To explain this, let's look first at the electron configurations of the simple metal ions:

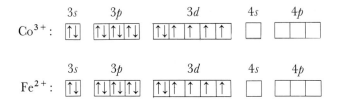

From these diagrams and the measurements of magnetic moment, we see that Co^{3+} has formed a complex without disturbing the electron configuration of the lone metal ion:

Co^{3+} (complex): [3d, 4s, 4p, 4d diagram with sp^3d^2 hybridization]

Since all the orbitals involved in the hybrid sp^3d^2 orbitals are from the same energy level, we call the resulting complex an *outer complex*.

The iron complex is a different story. Here all the electrons are paired, suggesting a second kind of hybridization:

Fe^{2+} (complex): [3d, 4s, 4p diagram with d^2sp^3 hybridization]

These hybrid d^2sp^3 orbitals involve both the third and fourth energy levels; thus the complex iron ion is an *inner complex*. Where all other factors are equal, an inner complex is more stable than an outer complex.

Example 24.4 Determine the likely geometry of $[CoCl_4]^{2-}$, given that it contains three unpaired electrons.

Solution

a. The center of coordination is the Co^{2+} ion. The electron configuration for the simple Co^{2+} ion is

Co^{2+}: [3s, 3p, 3d orbital diagram]

b. Square-planar geometry would require that the 3d electrons be rearranged so that one vacant orbital is available to form the dsp^2 hybrid orbitals. (The hybrid orbitals would arise from one 3d orbital, one 4s orbital, and two 4p orbitals.) Only one unpaired electron would remain.

c. Tetrahedral geometry demands sp^3 orbitals and thus allows the three unpaired 3d electrons to remain. The hybrid orbitals arise from one 4s orbital and three 4p orbitals. Experiments have confirmed that the geometry of $[CoCl_4]^{2-}$ is **tetrahedral**.

Question

9. Identify the orbitals used to bond the ligands in the following complexes. Describe the geometry of each.

	Number of unpaired electrons
a. $[Mn(H_2O)_6]^{2+}$	5
b. $[Os(CO)_5]$	0

Note that the center of coordination in b is a metal atom, not a metal ion.

24.7 Splitting of d Orbitals

We have not yet provided an adequate explanation of hybridization in coordination compounds. Why, for example, do the electrons in the 3d orbitals of the Fe(II) ion pair up while forming $[Fe(CN)_6]^{4-}$, whereas the electrons in the 3d orbitals of the Co(III) ion remain apparently undisturbed during the formation of $[CoF_6]^{3-}$? For an explanation we must look at the spatial arrangement of the five d orbitals. Though we have consistently represented them with a set of identical connected boxes, they by no means have identical shapes, which is apparent in Figure 24.5. It is then this set of five distinct orbitals that we must consider during the formation of a complex.

In the absence of any interactions, the five d orbitals have identical energies, despite their different shapes and orientations (Figure 24.6a). Furthermore, in the presence of a spherically symmetric electrical field, the energy of each orbital would be raised by the same amount (Figure 24.6b). However, during the formation of a complex, a metal ion is exposed to a nonuniform electric field (due to the unpaired electrons on the approaching ligands); the effect the field has upon the metal-ion orbitals depends on the coordination number the complex will ultimately have. For example, during the formation of an octahedral complex, the nonuniform field raises the d_{z^2} and $d_{x^2-y^2}$ orbitals to a higher energy than would a uniform field, and it raises the d_{xy}, d_{yz}, and d_{xz} orbitals less (Figure 24.6c). During the formation of a tetrahedral complex, the effect of the nonuniform ligand field is just the opposite (Figure 24.6d).

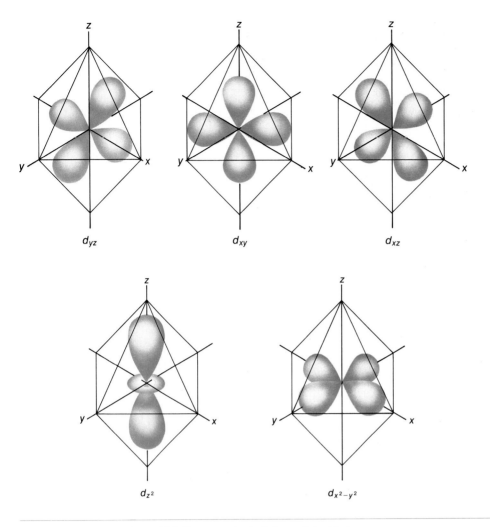

Figure 24.5 Shapes and orientations of d orbitals.

If this *field effect* causes the d orbitals to split enough, the electrons will pair up in the lower-energy orbitals rather than remain unpaired. This is exactly what we saw in the hexacyanoferrate(II) ion. In $[CoF_6]^{3-}$, however, the fluoride ligands did not split the d-orbital energy levels enough to cause the electrons to pair up. From this we can conclude that the cyanide ion produces a stronger field than the fluoride ion. The ligand is therefore the key, and we can list the common ligands in order of their decreasing ability to split d-orbital energy levels:

$$CO, \quad CN^- > NO_2^- > en > NH_3 > CNS^- > H_2O > OH^- >$$
$$F^- > Cl^- > Br^- > I^-$$

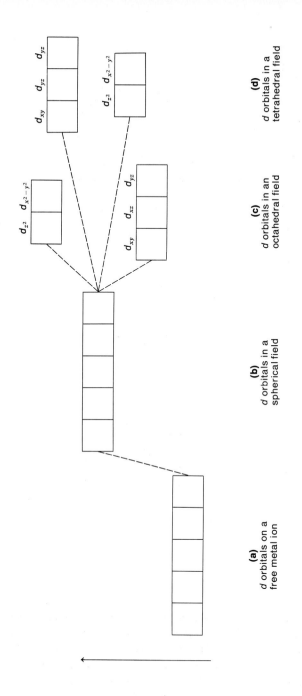

Figure 24.6 Splitting of d-orbitals.

Coordination Compounds 781

24.8 Isomerism

So far we have drawn examples of isomerism only from among organic compounds (Chapter 22). We'll now turn to isomerism among coordination compounds, where as a matter of fact there are more *types* of isomerism than among organic molecules, although not more examples. In some cases, variations in the arrangement of ligands about the center of coordination are responsible for compounds with very different properties. For instance, if some ligands in a complex are more electronegative than others, one arrangement of ligands can produce a nonpolar molecule, whereas another can produce a polar molecule. An example is the square-planar complex, $[PtCl_2(NH_3)_2]$. If the chloride ions are in adjacent positions, the complex is unsymmetric and polar:

$$\begin{bmatrix} NH_3 & & NH_3 \\ & \diagdown Pt \diagup & \\ Cl & & Cl \end{bmatrix}$$

If, on the other hand, the chloride ions are opposite from each other, the complex is symmetric and nonpolar:

$$\begin{bmatrix} Cl & & NH_3 \\ & \diagdown Pt \diagup & \\ NH_3 & & Cl \end{bmatrix}$$

We'll come back to this example of stereoisomerism after first considering some types of isomerism peculiar to coordination complexes.

Ionization Isomerism When an ion is bound by a covalent bond *within* the complex in one compound and by an ionic bond *to* the complex in a second, the result is a pair of ionization isomers. For example,

$[PtCl_2(NH_3)_4]Br_2$ $[PtBr_2(NH_3)_4]Cl_2$

dichlorotetraammineplatinum(IV) bromide   dibromotetraammineplatinum(IV) chloride

Coordination Isomerism Where both anion and cation are coordination complexes, a variation in the distribution of ligands between the two complexes gives rise to coordination isomerism. Two pairs of examples are

$[Co(NH_3)_6][Cr(CN)_6]$ $[Cr(NH_3)_6][Co(CN)_6]$

hexaamminecobalt(III) hexacyanochromate(III)   hexaamminechromium(III) hexacyanocobaltate(III)

$[Cu(NH_3)_4][PtCl_4]$ $[Pt(NH_3)_4][CuCl_4]$

tetraamminecopper(II) tetrachloroplatinate(II)   tetraammineplatinum(II) tetrachlorocuprate(II)

Attachment or Linkage Isomerism Some ligands, though consistently unidentate, have more than one atom that can function as a donor atom. The nitrite ion, for example, has lone pairs on an oxygen atom as well as on the nitrogen atom. When the nitrogen atom is the donor, the ligand is called the *nitro* ligand and is written NO_2. When the oxygen atom is the donor, it is called the *nitrito* ligand, and the formula is written (ONO). The following two compounds, then, are examples of attachment or linkage isomers:

$$[CoNO_2(NH_3)_5]^{2+} \qquad [Co(ONO)(NH_3)_5]^{2+}$$
nitropentaamminecobalt(III) ion nitritopentaamminecobalt(III) ion

Attachment isomers involving the thiocyanate (SCN^-) ligand are also known, with coordination taking place either through a lone pair of electrons on the sulfur atom or through the pair on the nitrogen atom.

Ligand Isomerism If the ligands themselves are large enough and complex enough to form structural isomers, the result can be ligand isomerism. Thus there is the ligand 1,2-diaminopropane (abbreviated *pn*) and the ligand 1,3-diaminopropane (called trimethylenediamine and written *tn*). Thus we have the following two isomeric complexes:

$$[CoCl_2(pn)_2]Cl$$
dichlorobis(1,2-diaminopropane)cobalt(III) chloride

$$[CoCl_2(tn)_2]Cl$$
dichlorobis(1,3-diaminopropane)cobalt(III) chloride

Geometric Stereoisomerism The most important isomers among coordination compounds are stereoisomers. As we saw in Chapter 22, stereoisomers have the same molecular formula but differ in the spatial arrangement of their components. They are of two types: geometric and optical.

Geometric isomerism occurs in complexes that have a square as part of their geometry. These include dsp^2 hybrids (with square-planar geometry) and sp^3d^2 or d^2sp^3 hybrids (with octahedral geometry). In a square, the two ligands diagonally opposite each other (*trans*) are farther apart than two ligands at adjacent corners (*cis*). Thus we can now name the two isomers of $[PtCl_2(NH_3)_2]$:

cis-dichlorodiammineplatinum(II)   *trans*-dichlorodiammineplatinum(II)

We began this section by pointing out the difference in polarity between these two isomers. It has been discovered only recently by cancer researchers that *cis*-dichlorodiammineplatinum(II) has significant value in fighting tumor growth, whereas the *trans* isomer has little or no effect.

Because the octahedron can be seen as consisting of three squares lying in mutually perpendicular planes (Figure 24.7), it opens even more possibilities for geometric isomerism than the square-planar complex. Two examples are given in Figures 24.8 and 24.9.

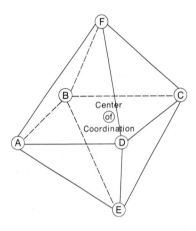

Figure 24.7 An octahedron consists of three mutually perpendicular squares with a common center. (Squares are defined by *ABCD*, *BEDF*, and *AECF*.)

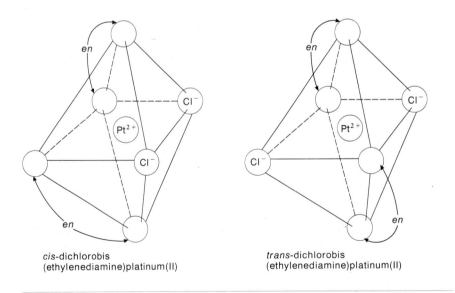

cis-dichlorobis(ethylenediamine)platinum(II)

trans-dichlorobis(ethylenediamine)platinum(II)

Figure 24.8 A pair of geometric isomers.

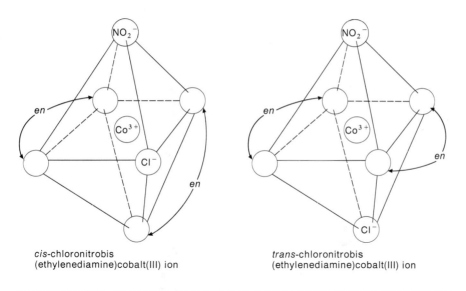

Figure 24.9 A pair of geometric isomers.

Optical Stereoisomerism Among the common geometries of coordination compounds, the tetrahedral and octahedral geometries allow the possibility of optical isomerism. Since octahedral complexes outnumber all others, examples of optically active octahedral complexes abound. It is interesting, however, that all involve chelates. (Two examples are illustrated in Figures 24.10 and 24.11.)

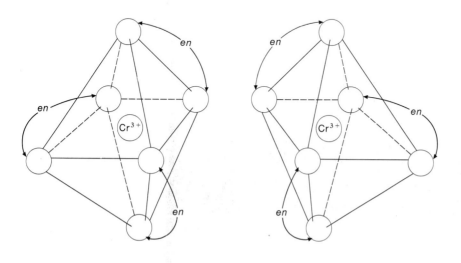

Figure 24.10 Optical isomers of the tris(ethylenediamine)chromium(III) ion.

Coordination Compounds

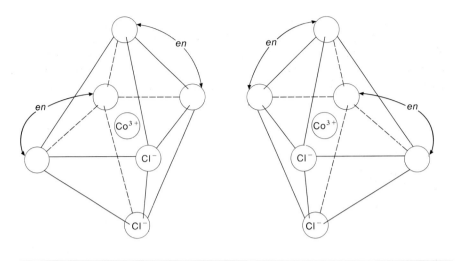

Figure 24.11 Optical isomers of the dichlorobis(ethylenediamine)cobalt(III) ion.

Questions

10. Draw and name the two square-planar isomers of the ion $[PtBr_2Cl_2]^{2-}$.

11. How many complex ions with the formula $[Ni(H_2O)_2(en)_2]^{2+}$ are possible?

24.9 Color and Some Interesting Complexes

The vivid and varied colors of coordination compounds require some explanation. As we saw in Chapter 5 (especially Figure 5.3), the eye is sensitive to only a small segment of the electromagnetic spectrum. Within this segment we see different wavelengths (or energies) as different colors. Thus a substance that selectively absorbs light of certain wavelengths within the visible spectrum will take on the color characteristic of the wavelengths that remain.

The critical question then is, why do some substances absorb visible light whereas others do not? And as we first saw in discussing the transition metals, the answer lies in the electronic energy levels of the atom. When light is absorbed, its energy is used to excite an electron—to raise the electron from one energy level to another. For many substances, such as representative metal ions, energy levels are so widely spaced that visible light lacks the energy to excite an electron from one to another. Consequently all visible light is reflected. Sodium chloride, potassium nitrate, and calcium carbonate are pure white since this reflected light is the combination of all visible wavelengths. On the other hand, transition metal ions, especially those in coordination complexes, often have closely spaced energy levels and hence often absorb visible light. The kind of d-orbital splitting we discussed in Section 24.7, for example, can produce orbitals (that is, energy levels) that differ only slightly.

Sometimes closely spaced energy levels can be provided by the ligand. An important example is chlorophyll, which must absorb light to be used by green plants, which in turn synthesize carbohydrates from CO_2 and water. Chlorophyll is a square-planar chelate of magnesium (Figure 24.12). The chelating agent is a tetradentate molecule of a class called *porphyrins*. The four donor atoms are

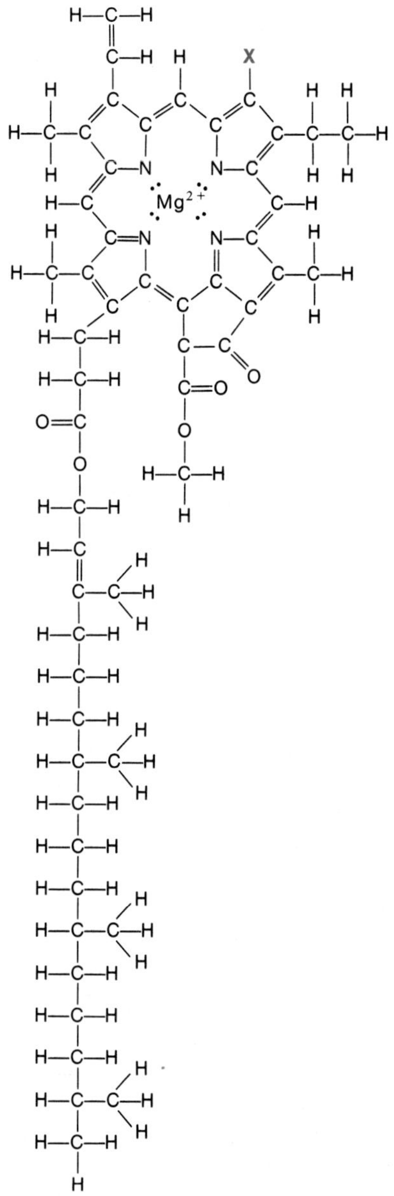

Figure 24.12 Chlorophyll. In chlorophyll *a*, X = $-CH_3$, in chlorophyll *b*, X = $-CHO$.

nitrogen atoms. The nature of the excited state of the chlorophyll molecule, which is the immediate result of light absorption, is a matter still under investigation. In any case, this excited molecule subsequently transfers its energy to another molecule or system of molecules, thus initiating the complex process of photosynthesis.

Another important porphyrin chelate, with iron as the center of coordination, appears in hemoglobin. The hemoglobin molecule itself (Figure 24.13) is a protein with a molecular weight in excess of 6×10^4. Each molecule contains four iron(II)-porphyrin complexes called *hemes* (Figure 24.14). Hemoglobin transports oxygen from the lungs to the body tissues by binding O_2 to each of these heme units. This oxygenated complex is called *oxyhemoglobin* and gives arterial blood its bright red color. When it releases the complexed oxygen, the hemoglobin reverts to a darker, purplish color. The fact that hemoglobin can also combine with CO, CN^-, and other ligands accounts for the toxicity of these substances.

An interesting application of coordination chemistry has been the use of chelating agents in cases of poisoning. For example, EDTA (see p. 775) is administered in cases of lead poisoning because it forms more stable complexes

Figure 24.13 The hemoglobin molecule. Four identical peptide chains, each surrounding a coordinated iron(II) ion, are tetrahedrally oriented in the molecule.

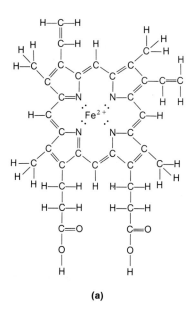

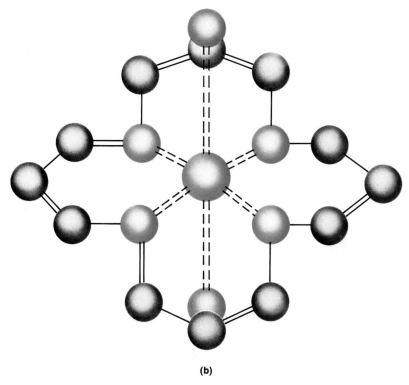

Figure 24.14 The heme group. **(a)** The iron ion forms a square-planar complex with a tetradentate protoporphyrin molecule. **(b)** The fifth octahedral position on the iron(II) ion is occupied by a nitrogen atom from an amino acid. The sixth is unoccupied in deoxyhemoglobin and occupied by an oxygen atom in oxyhemoglobin.

with the lead than do the proteins of the body. The lead is thus removed from the body as EDTA complexes. Other chelating agents have been used with success in other cases of heavy metal poisoning.

What might be thought of as the reverse process also has medical application. Metal chelates are sometimes an effective way to introduce needed metal ions into the body. A weak chelate of iron, for instance, is used in the treatment of anemia.

Commercial plant foods also make use of chelating agents. Thus chelated iron can be used to introduce iron into the soil, while avoiding the deleterious side effects that often accompany the use of soluble iron salts.

Sidelight
Reactions Involving Coordination Compounds

Some of the most interesting aqueous chemistry was omitted from earlier chapters because a discussion of it would have demanded an understanding of coordination compounds. We can now turn to some of these reactions.

For example, the solubility of the amphoteric hydroxides [Cr(OH)$_3$, Zn(OH)$_2$, Al(OH)$_3$, Sn(OH)$_2$, Sn(OH)$_4$, Pb(OH)$_2$, Be(OH)$_2$, Sb(OH)$_3$, and Te(OH)$_6$] involves the formation of coordination complexes. Each of these water-insoluble hydroxides dissolves in strong acid:

$$Al(OH)_3(s) + 3\,H^+(aq) \rightleftharpoons Al^{3+}(aq) + 3\,H_2O$$

(Following our usual custom, we do not show the Al^{3+} ion as hydrated, although it can be thought of as an aquo complex.) In addition, each of these hydroxides dissolves in strong bases because of the formation of hydroxo complexes:

$$Al(OH)_3(s) + OH^-(aq) \rightleftharpoons [Al(OH)_4]^-(aq)$$

$$Sn(OH)_2(s) + 4\,OH^-(aq) \rightleftharpoons [Sn(OH)_6]^{4-}(aq)$$

These ionic hydroxo complexes are all colorless.

The ammines are a colorful series of complex ions. Many water-insoluble hydroxides can be dissolved in aqueous ammonia—not because of the hydroxide ions present but because of the molecular NH$_3$. The products are ammines:

$$Cu(OH)_2(s) + 4\,NH_3(aq) \rightleftharpoons [Cu(NH_3)_4]^{2+}(aq) + 2\,OH^-(aq)$$

$$Co(OH)_3(s) + 6\,NH_3(aq) \rightleftharpoons [Co(NH_3)_6]^{3+}(aq) + 3\,OH^-(aq)$$

Some ammines and their colors are

[Cr(NH$_3$)$_6$]$^{3+}$	yellow	[Ni(NH$_3$)$_6$]$^{2+}$	violet-blue
[Co(NH$_3$)$_6$]$^{2+}$	pink to yellow-brown	[Pt(NH$_3$)$_4$]$^{2+}$	colorless
[Co(NH$_3$)$_6$]$^{3+}$	brown to red	[Ag(NH$_3$)$_2$]$^+$	colorless
[Cu(NH$_3$)$_4$]$^{2+}$	deep blue	[Zn(NH$_3$)$_4$]$^{2+}$	colorless

Another colorful example of coordination-complex chemistry centers around copper(II) chloride, a dark brown covalent compound. When it is dissolved in water, it forms several complexes with chloride ions and/or water molecules. The concentrated solution is yellow-green; the dilute solution, blue. If hydrochloric acid is added, the solution turns yellow. All of these changes in color are results of shifts in the various equilibria involving these complexes. Given below is the equilibrium between the completely chlorinated copper(II) ion and the completely hydrated copper(II) ion.

$$[CuCl_4]^{2-}(aq) + 4H_2O \rightleftharpoons [Cu(H_2O)_4]^{2+}(aq) + 4Cl^-(aq)$$
yellow   blue

A mixture of the two complex ions is yellow-green.

Almost all metal ions form complexes with the halide ions. The chloro complexes are seldom stable enough to shift an equilibrium (such as we saw above for the hydroxo and ammine complexes), but they sometimes contribute a little extra push. For instance, gold is a noble metal with a very large (negative) reduction potential. It is thus unaffected by concentrated hydrochloric acid. Nor does it react appreciably with concentrated nitric acid. But the combination of the two acids, *aqua regia*, will dissolve gold. The combination of the oxidizing power of HNO_3 and the complexing action of the Cl^- ion is responsible:

$$Au(s) + 3NO_3^-(aq) + 4Cl^-(aq) + 6H^+(aq) \longrightarrow 3NO_2(g) + [AuCl_4]^-(aq) + 3H_2O$$

A final example of the part coordination complexes play in solution chemistry is provided by iron(III) chloride, a covalent compound that forms large sheetlike structures. These giant two-dimensional structures dissolve in water to form $[Fe(H_2O)_6]^{3+}$ and Cl^- ions. If the water is evaporated from this solution, crystals of the ionic compound $[Fe(H_2O)_6]Cl_3$ remain. The original anhydrous iron(III) chloride cannot be recovered by heating the hydrated crystals since the heat causes other reactions, ultimately producing iron(III) oxide.

The hexaquoiron(III) ion is a very pale violet; however, its color is often obscured by yellow and brown hydroxo complexes formed by its hydrolysis. For example,

$$[Fe(H_2O)_6]^{3+}(aq) + H_2O \rightleftharpoons [Fe(OH)(H_2O)_5]^{2+}(aq) + H_3O^+(aq)$$
violet   brown

or, more simply,

$$Fe^{3+}(aq) + H_2O \rightleftharpoons [Fe(OH)]^{2+}(aq) + H^+(aq)$$

The extent of hydrolysis increases when the solution is boiled, causing a deepening of the color. Adding HNO_3 or H_2SO_4, on the other hand, reverses the reaction and causes lightening of the color. Since chloride complexes add to the intensity of the color, solutions of $FeCl_3$ are more intensely colored than solutions of $Fe(NO_3)_3$ or $Fe_2(SO_4)_3$. And unlike nitric and sulfuric acids, hydrochloric acid is not effective in reducing the color of solutions containing the iron(III) ion.

With these examples in hand, it is now possible to organize a fairly complete outline of the processes by which a solid, normally insoluble in water, can be dissolved by the addition of another reagent. Any "insoluble" substance in the presence of a solvent is in equilibrium with its dissolved phase, even though the concentrations of the dissolved species may be very low. Using AB as a general formula for an ionic compound, we can write this equilibrium as

$$AB(s) \rightleftharpoons A^+(aq) + B^-(aq)$$

To shift this equilibrium to the right, one of the ions must be "removed" so that the rate of the reverse reaction, $A^+(aq) + B^-(aq) \longrightarrow AB(s)$, is slowed. This can occur if a reagent is added that

1. reacts with one of the products to form a gas that escapes.
2. reacts with one of the products to form water.
3. reacts with one of the products to form a weak electrolyte (a weak acid, a weak base, or a complex ion).

4. reacts with one of the products to form a precipitate. This kind of reaction is not often employed in practical situations since it merely dissolves one substance by precipitating another, but we must include it for the sake of completeness.

Thus we can make several generalizations. All hydroxides dissolve in strong acid (since water is formed). Most insoluble salts of weak acids dissolve in strong acid (since a weak acid is formed). Amphoteric hydroxides dissolve in strong base (since a complex ion is formed). Some water-insoluble hydroxides dissolve in aqueous ammonia (since an ammine complex is formed).

Example 24.5 Describe how a mixture of the following solids might be separated, using only water and aqueous solutions of NaOH, NH_3, and HCl: $Fe(OH)_3$, $Al(OH)_3$, $Cu(OH)_2$, and $Zn(OH)_2$.

Solution

a. All the substances are insoluble in water. Treat them with NaOH and filter. The filtrate now contains hydroxo complexes of Al and Zn; the residue contains $Cu(OH)_2$ and $Fe(OH)_3$.

b. Neutralize the filtrate with HCl, and then treat it with aqueous NH_3. Filter. The zinc(II) ion forms $[Zn(NH_3)_4]^{2+}$ and remains in the filtrate. The residue is $Al(OH)_3$.

c. Treat the residue from step a with aqueous NH_3. Filter. The filtrate contains a copper(II) ammine complex, $[Cu(NH_3)_4]^{2+}$. The residue is $Fe(OH)_3$.

d. The zinc and copper ammine complexes can be neutralized with HCl and then the hydroxides carefully reprecipitated. Figure 24.15 outlines the complete procedures.

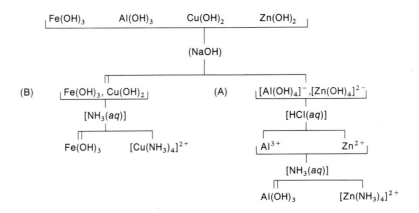

Figure 24.15 Separation scheme (Example 24.5). Double vertical lines indicate a precipitate (solid) as product. Single vertical lines indicate a soluble product that remains in solution.

Other analytical separation schemes make use of other solubility differences and other complexes. One such scheme of qualitative analysis based partly on the differences in the solubilities of the metal sulfides was first outlined in Chapter 16. The Group IA metal sulfides are soluble in water. ZnS, FeS, and MnS are insoluble in water but will dissolve in HCl because of the formation of the weak acid, HS^-. PbS, CuS, CdS, As_2S_5, Sb_2S_3, and SnS and SnS_2 are insoluble in dilute HCl but will dissolve in HNO_3 since the nitric acid oxidizes the sulfide ion to free sulfur. HgS, NiS, and CoS are so insoluble that only aqua regia will dissolve them. The aqua regia places three stresses on the equilibrium between the metal sulfide and its ions in solution: It promotes formation of the weak acid, HS^-; it promotes oxidation of the sulfide ion; and it promotes formation of a charged chloro complex with the metal ion. To represent these simultaneous reactions, we can write the following two equations:

$$8\,HgS(s) + 32\,H^+(aq) + 16\,NO_3^-(aq) + 32\,Cl^-(aq) \longrightarrow$$
$$8[HgCl_4]^{2-}(aq) + S_8(s) + 16\,NO_2(g) + 16\,H_2O$$

$$HgS(s) + H^+(aq) \longrightarrow Hg^{2+}(aq) + HS^-(aq)$$

Some sulfides can also be dissolved in solutions containing excess sulfide ion because of the formation of thio complexes. For example,

$$As_2S_3 + 3\,S^{2-} \longrightarrow 2[AsS_3]^{3-}$$
$$Sb_2S_3 + 3\,S^{2-} \longrightarrow 2[SbS_3]^{3-}$$
$$SnS_2 + S^{2-} \longrightarrow [SnS_3]^{2-}$$

Questions

i. In the manner of Example 24.5, devise a scheme for separating a mixture of K^+, Bi^{3+}, Cd^{2+}, and Zn^{2+} in aqueous solution, using only aqueous solutions of NaOH, NH_3, and HCl.

ii. Cadmium sulfide can be dissolved in a concentrated solution of NaCl. Can you suggest a reason?

iii. From among the reasons listed below, select the most important one for each process at the left.
  a. MnS dissolves in HCl.
  b. Ag_3PO_4 dissolves in HCl.
  c. $Zn(OH)_2$ dissolves in NaOH.
  d. $Zn(OH)_2$ dissolves in HCl.
  e. $Zn(OH)_2$ dissolves in $NH_3(aq)$.
  f. PbS dissolves in HNO_3 but not in HCl.
  g. NiS dissolves in aqua regia but not in HCl or HNO_3 alone.
  h. SnS_2 dissolves in Na_2S solution.

  1. the formation of water
  2. the formation of a weak acid
  3. the formation of a stable hydroxo complex.
  4. the formation of a stable ammine.
  5. the evolution of a gas or the formation of a free element
  6. the formation of a more insoluble compound
  7. the formation of a stable complex other than an ammine or a hydroxo complex

iv. Determine a method for separating K_2S, PbS, FeS, and HgS, using only water, HCl, and HNO_3.

Questions and Problems

24.1 Coordination

12. Determine the primary and secondary valences of the metal ion in each of these complexes:
    a. $[CoNO_2(NH_3)_5](NO_3)_2$ b. $Na_3[Ag(S_2O_3)_2]$ c. $[PtCl_2(NH_3)_2]Br_2$

13. Give an example of a coordination complex in which a bond is formed between
    a. an atom and a molecule.   b. an ion and a molecule.
    c. two molecules.   d. two ions.

14. Rewrite the formulas for the double salts and ammines listed below in the modern coordination form:
    a. $Fe(CN)_3 \cdot 3\,NaCN$ b. $SbCl_4 \cdot 2\,NH_4Cl$ c. $Cr(NO_3)_3 \cdot 6\,NH_3$

24.2 Terminology

15. For each of the complexes below, (i) identify it as an ionic compound, ion, or a molecular compound; (ii) determine the charges (if any); (iii) determine the coordination number; and (iv) determine the charge on the center of coordination.
    a. $K_4[Fe(CN)_6]$ b. $[Co(NO_2)_3(NH_3)_3]$
    c. $[Co(en)_3]_2(SO_4)_3$ d. $[Ag(NH_3)_2]Cl$
    e. $[CrF(H_2O)_5][SiF_6]$ f. $Na_3[Ag(S_2O_3)_2]$
    g. $[PtCl_2(NH_3)_4]Cl_2$ h. $(NH_4)_2[Zn(SCN)_4]$
    i. $[Co(ONO)(NH_3)_5]^{2+}$ j. $[Ni(CO)_4]$

16. Identify the donor atom of each ligand in Question 15.

24.3 Nomenclature

17. Name the substances in Question 15.

18. Write the formulas for the following substances:
    a. potassium dicyanoargentate(I)
    b. pentaaquoamminenickel(II) ion
    c. tris(ethylenediamine)chromium(III) ion
    d. *trans*-dichlorodiammineplatinum(II)
    e. hexaamminecobalt(III) hexacyanochromate(III)
    f. dichlorobis(ethylenediamine)cobalt(III) chloride
    g. pentacarbonyliron(0)
    h. *cis*-dichlorotetraammineplatinum(II)
    i. tetraammineplatinum(IV) tetrachlorocuprate(II)
    j. diamminesilver(I) tetraiodoplatinate(II)

19. a. List six ligands that are anions.
    b. List three ligands that are neutral molecules.
    c. List three ligands that contain more than one donor atom.

24.4 Stability of Coordination Complexes

20. List and discuss the factors that should be considered when predicting the relative stabilities of two complexes.

21. In each case below, predict which metal ion you would expect to form the more stable complex with the given ligand:
   a. With ammonia: Ca^{2+} or Cu^{2+} b. With chloride ion: Hg_2^{2+} or Hg^{2+}

22. The instability constant is one index of the stability of a complex ion. Compare the constants for these pairs of ions (Table 24.1) and explain their relative values.
   a. $[Co(NH_3)_6]^{2+}$ and $[Co(NH_3)_6]^{3+}$
   b. $[Fe(CN)_6]^{4-}$ and $[Fe(CN)_6]^{3-}$
   c. $[Al(OH)_4]^-$ and $[Zn(OH)_4]^{2-}$

24.6 Magnetic Properties and Orbitals

23. Determine the number of unpaired electrons in a molecule of $[PtICl(NH_3)_2]$ (square-planar geometry).

24. Making use of the coordination number, together with the number of unpaired electrons, determine for each of the following complexes (i) the hybridization, (ii) the geometry, and (iii) if an octahedron, whether an inner or outer complex.

Complex	Number of Unpaired Electrons
$[Fe(CN)_6]^{3-}$	1
$[CoCl_4]^{2-}$	3
$[Co(CN)_6]^{3-}$	0
$[CoF_6]^{3-}$	4
$[Ag(CN)_2]^-$	0
$[FeF_6]^{3-}$	5
$[Ni(CN)_4]^{2-}$	0

24.7 Isomerism

25. Draw structures that illustrate all geometric isomers of the planar iodobromochloroammineplatinum(II) ion.

26. Which of the two molecules in Figure 24.9 has an optical isomer? Draw the two optical isomers.

27. Draw the *cis-* and *trans-* forms of dichlorotetraamminechromium(III) ion.

28. How many different types of isomerism can the compound $[CoClNO_2(en)_2]NO_2$ exhibit? Illustrate with formulas or drawings.

29. The oxalate ion is bidentate. Draw the $[Co(C_2O_4)_3]^{3-}$ ion and its mirror image.

25 Nuclear Chemistry

The most important physical and chemical properties of substances depend on the electronic configurations of the atoms, ions, and molecules that make them up. Consequently, many chemists rarely give much thought to the properties of atomic nuclei. But no study of the structure of matter would be judged complete that did not include a brief description of nuclear phenomena and a discussion of the theories of the atomic nucleus.

Earlier we presented a simple picture of subatomic structure, in which atoms were made up of only three fundamental particles: the electron, the proton, and the neutron. In the last half-century, however, scientists in the field of high-energy physics, or particle physics, have identified several dozen different particles, all of which can be extracted from atomic nuclei and all of which can be thought of as **elementary particles.** The comprehensive theory that explains these many particles allows them to change from one to another through various interactions. Some elementary particles have very brief lives, measured in nanoseconds, whereas others are relatively stable. The neutron, for example, is stable within the nucleus, but outside the nucleus it has an average life of only about 1000 s, after which time it will likely change into one or more other elementary particles. It is the relative stability of the electron, proton, and neutron that makes them appropriate building blocks in a simple picture of the atom.

25.1 Theories of Nuclear Structure

Even today the questions facing physicists who seek to explain the structure of the atomic nucleus are much like those that confronted chemists before the advent of quantum theory. They are questions that simply can't be answered in terms of the classical theories that are adequate to the everyday world. For example, we cannot discuss the incredible densities of nuclear material in the same terms we used for elements and compounds, nor can we explain the stability of the highly charged nucleus with the same electromagnetic theory that justifies the formation of ions and molecules.

Example 25.1 Calculate the density of the material in the nucleus of the gold atom. The atomic weight of gold is 197 and the radius of its nucleus is approximately 8×10^{-13} cm.

Solution

a. The weight of an atom of gold to one significant figure is

$$\left(\frac{197 \text{ g}}{\text{g at. wt}}\right)\left(\frac{1.00 \text{ g at. wt}}{6.02 \times 10^{23} \text{ atoms}}\right) = 3 \times 10^{-22} \text{ g/atom}$$

b. The volume of the nucleus, if we take it to be a sphere, is

$$\tfrac{4}{3}\pi r^3 = \tfrac{4}{3}(3.14)(8 \times 10^{-13} \text{ cm})^3 = 2 \times 10^{-36} \text{ cm}^3$$

c. Thus the density of the nuclear material is

$$\frac{3 \times 10^{-22} \text{ g}}{2 \times 10^{-36} \text{ cm}^3} = 2 \times 10^{14} \text{ g/cm}^3$$

d. Or, put into more impressive units,

$$\left(\frac{2 \times 10^{14} \text{ g}}{1 \text{ cm}^3}\right)\left(\frac{1.00 \text{ lb}}{454 \text{ g}}\right)\left(\frac{1 \text{ ton}}{2000 \text{ lb}}\right) = 2 \times 10^8 \text{ tons/cm}^3$$

One cubic centimeter of nuclear material would weigh approximately 200 million tons! Clearly the world of the atomic nucleus is not governed by the same physical laws we have depended on so far. (As an aside, astronomers think they have found similar inconceivable densities on a much larger scale: *Neutron stars* are extinct stars that have collapsed as a result of their own gravity to densities as great as those of the atomic nucleus.)

To explain such densities, as well as the stability of the nucleus and the nature of particle interactions, physicists have advanced a theory of *nuclear forces*, totally different from electrostatic and gravitational forces. These forces predominate only at distances less than 10^{-12} cm, that is, within the dimensions of the atomic nucleus. Just as electrostatic forces dominate gravity under the right conditions, so two positively charged protons attract one another in the nucleus, in apparent contradiction of electromagnetic theory.

25.2 Natural Radioactivity

Natural radioactivity (see Chapter 4) provided the first important evidence about the structure of the nucleus. This phenomenon has, of course, been taking place since the beginning of the universe, but only at the beginning of this century did scientists begin to understand it. At the suggestion of Antoine Becquerel, Marie Curie (1867–1934) and her husband Pierre (1859–1906) began the analysis of a ton of pitchblende (uranium ore) to discover the source of a radiation emanating from it that was much greater than could be accounted for

Table 25.1 The Uranium-238 Radioactive Decay Series

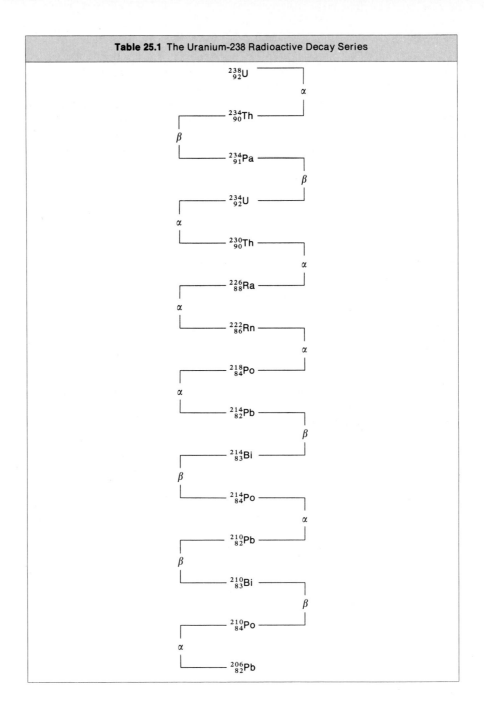

by the uranium it contained. They separated the components of the ore, by one chemical method after another, into smaller and smaller fractions. Eventually, they identified two highly radioactive portions, one containing predominantly bismuth salts and the other, predominantly barium salts. From the bismuth fraction the Curies isolated a new element, polonium, and from the barium fraction—with much more difficulty—a pure salt of a second new element, radium. From the ton of ore, they obtained 0.2 g of the radium salt—1 part in 4.5 million! For their tremendous labors and their discoveries, the Curies were honored together with the Nobel prize in 1903, and the widowed Marie Curie was honored again in 1911.

These and all other naturally occurring radioactive elements emit three kinds of radiation: alpha, beta, and gamma rays. Alpha radiation consists of particles identical to the helium nucleus, each consisting of two protons and two neutrons. Beta radiation consists of electrons, and gamma rays are high-frequency electromagnetic waves. Since each of these forms of radiation responds differently to electric and magnetic fields, they can be identified using the same principles that are the basis of the mass spectrograph (Figure 4.9).

We now know that beyond bismuth, element 83, all isotopes of all elements are radioactive and that each naturally occurring isotope is part of one of three natural decay series. Each of these series is named for its parent element (the first member of the series).

One such decay series, the uranium-238 series, appears in Table 25.1. This series is also designated as the $4n + 2$ series, since the mass number of each of its members is divisible by four with a remainder of 2. The other naturally occurring radioactive decay series are the uranium-235 series (the $4n + 3$ series) and the thorium-232 series (the $4n$ series). In all three series, decay takes place by successive emissions of alpha or beta particles; both are never emitted simultaneously from the same atom. Each series terminates with a stable isotope of lead.

In addition to the members of the three natural radioactive decay series, several lighter elements have naturally occurring radioactive isotopes: hydrogen, carbon, potassium, rubidium, indium, neodymium, samarium, lutetium, and rhenium. Most of these lighter elements decay by beta emission.

Question

1. What are the periodic table relationships of bismuth to polonium and of barium to radium? Suggest a reason that the separation of the first two might be easier than the second.

25.3 Detection of Radiation

Although none of the radiation emitted by radioactive materials is visible, several ways of detecting these rays have been developed. One, the effect on an exposed photographic plate, was accidentally discovered by Becquerel. (This effect is now the basis of the film badges carried by people who work with radioactive materials.) Since the three types of rays differ in their penetrating power, we can infer that Becquerel saw the effects of gamma or beta rays. A sheet of paper can stop alpha rays, whereas thin sheets of metal are necessary to stop beta rays, and thick layers of lead or concrete are required to stop gamma rays.

Today most radiation is detected by exploiting its tendency to ionize gases, a property shared by all three types of rays. One primitive, but dramatic, illustration of this property is the electroscope, which consists of two leaves of gold foil attached to a metal rod within a flask (Figure 25.1). The flask isolates the gold foil and protects it from air currents. When the electroscope is charged, the pieces of gold foil repel each other and fly apart. If the charged electroscope is then brought into the vicinity of radioactive materials, the rays ionize the air, causing it to become a conductor; the charge on the electroscope is therefore conducted away, and the gold leaves collapse.

The Geiger-Müller counter (Figure 25.2) is based on the same property of radiation as the electroscope, but it is much more sophisticated and is the most common detection device used today. The essential part of the Geiger counter is the ionization chamber, which consists of positive and negative electrodes

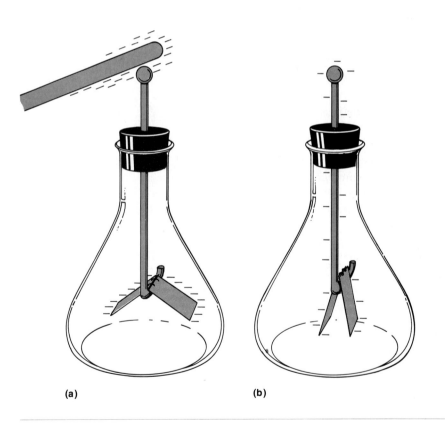

Figure 25.1 An electroscope (a) charged and (b) discharging.

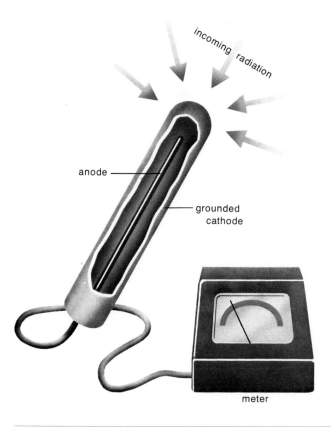

Figure 25.2 A Geiger counter.

sealed in a tube filled with an easily ionizable gas. When rays emitted by radioactive sources enter the ionization chamber, the gas is ionized. The positive ions and electrons formed by this ionization cause a small current to flow, which in turn is amplified and used to operate a recording device. Thus the Geiger counter measures as well as detects. Furthermore, by carefully choosing the ionizable gas and the material of the ionization chamber and by selecting the proper voltage across the electrodes, one can construct a counter that detects and measures a specific type of radiation.

The Wilson cloud chamber is a third device that depends on the ionization of a gas to detect radiation. It consists of a chamber containing a supersaturated solution of water vapor in air. The conditions are such that a fog of condensed vapor will form, but only in the presence of small dust particles or gaseous ions, which are necessary as the seeds of condensation. If radioactive materials are present, therefore, their rays will produce the gaseous ions that initiate condensation. Thin trails of water vapor, called *fog tracks* (Figure 25.3), mark the paths of particles. Photographed or seen under a microscope, tracks of the different types of particles display individual characteristics recognizable to a trained observer. The Wilson cloud chamber thus detects the presence of a single particle.

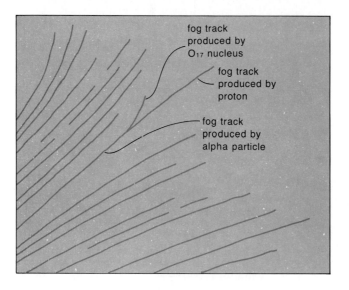

Figure 25.3 Fog tracks showing the formation of oxygen-17 by the bombardment of nitrogen-14 with alpha particles.

Certain substances known as *phosphors*, such as zinc sulfide, sodium iodide, and anthracene, fluoresce when radiation strikes them. This property is the basis of the *scintillation counter*. The flashes of light, or scintillations, that are produced when radiation strikes the phosphor are observed by a photoelectric cell, which in turn converts radiant energy to electrical energy (see Chapter 5). Each electrical pulse can then be counted separately.

Questions

2. What property common to all three types of radioactivity is the basis of the electroscope, the Geiger counter, and the Wilson cloud chamber?

3. In Figure 25.1a, the electroscope is charged by bringing a negatively charged object to the rod. What would have happened—in terms of electrons and the response of the gold foil—if a positively charged object had been brought to the rod?

25.4 Nuclear Equations

Equations for nuclear reactions convey a type of information different from that found in ordinary chemical equations. Only nuclear changes are indicated: changes in the number of nucleons or changes in the charge on the nucleus. A superscript to the left of the symbol for the element indicates the number of nucleons (the mass number) and a subscript to the left gives the number of protons (the nuclear charge or atomic number). Using this convention we can write symbols for the three fundamental particles we have discussed so far, plus the alpha particle:

$$\text{proton} \quad {}^{1}_{1}p \quad \text{or} \quad {}^{1}_{1}\text{H}$$
(hydrogen nucleus)

$$\text{neutron} \quad {}^{1}_{0}n$$

$$\text{electron} \quad {}^{0}_{-1}e \quad \text{or} \quad {}^{0}_{-1}\beta$$
(beta particle)

$$\text{alpha particle} \quad {}^{4}_{2}\text{He} \quad \text{or} \quad {}^{4}_{2}\alpha$$
(helium nucleus)

A complete nuclear equation must show the same total number of nucleons on each side and the same total nuclear charge. For example, the nuclear equation for the reaction in which a uranium isotope with a mass number of 234 emits an alpha particle is

$$^{234}_{92}\text{U} \longrightarrow {}^{4}_{2}\alpha + {}^{230}_{90}\text{Th}$$

The reduction of two in nuclear charge has produced the nucleus of a different element (thorium, atomic number 90), and the reduction of four in the number of nucleons has produced an isotope with a mass number of 230.

Emission of a beta particle has the effect of converting a neutron to a proton: The remaining nucleus gains a single positive charge, that is, one proton, but it retains the same total number of nucleons. For example, the nuclear equation for the emission of a beta particle from the nucleus of thorium-234 is

$$^{234}_{90}\text{Th} \longrightarrow {}^{0}_{-1}\beta + {}^{234}_{91}\text{Pa}$$

The increase in nuclear charge produces the nucleus of a different element, protoactinium, atomic number 91.

Example 25.2 The uranium isotope of mass number 238 decays by alpha emission. Write an equation for this reaction.

Solution

a. The symbols for the uranium isotope and the alpha particle are

$$^{238}_{92}\text{U} \qquad {}^{4}_{2}\alpha$$

b. The emission of an alpha particle reduces the number of nucleons in the nucleus by four and the number of protons by two. Thus the new nucleus must have a mass number of 234 and an atomic number of 90:

$$^{238}_{92}\text{U} \longrightarrow {}^{4}_{2}\alpha + {}^{234}_{90}\text{X}$$

c. Because an element is defined by its atomic number, we can identify the product as the nucleus of thorium-234:

$$^{238}_{92}U \longrightarrow {}^{4}_{2}\alpha + {}^{234}_{90}X$$

Example 25.3 Aluminum-27 decays by beta emission. Write the nuclear equation for this reaction and identify the product.

Solution

a. The symbols for aluminum-27 and the beta particle are

$$^{27}_{13}Al \qquad {}^{0}_{-1}\beta$$

b. The emission of a beta particle does not change the number of nucleons in a nucleus, but it increases nuclear charge by one:

$$^{27}_{13}Al \longrightarrow {}^{0}_{-1}\beta + {}^{27}_{14}X$$

c. We can check this equation by noting that the algebraic sums of the superscripts are the same on both sides of the arrow and that the same holds for the subscripts.

d. We can identify the element X by its atomic number as silicon. The complete equation is

$$^{27}_{13}Al \longrightarrow {}^{0}_{-1}\beta + {}^{27}_{14}Si$$

Questions

4. How is the mass number of a nucleus affected by alpha emission? How is the atomic number affected?

5. How are the mass number and the atomic number of a nucleus affected by beta emission? By gamma emission?

25.5 Nuclear Stability

Can the relative stability of a nucleus be determined in any way other than by looking at how rapidly it undergoes radioactive decay? It turns out that it is indeed possible to compare nuclear stabilities by applying Einstein's concept that the mass of a body is a measure of its energy content. His equation, probably as well known as any in the physical sciences, is

$$E = mc^2$$

where E is energy in joules, m is mass in kilograms, and c is the velocity of light in meters per second. This well-known equation might even be said to be *too* well known because it has led to the belief that, with the coming of the atomic age, unlimited sources of energy are available. Tremendous sources of energy are indeed becoming available as nuclear research continues, but it is only through certain reactions, which are very difficult to control and whose consequences are still unknown, that energy can be made available in the quantities implied by Einstein's equation. *If* all the mass of a railway ticket could be converted to energy, it would be sufficient to drive the train several times around the world. But this is a fantasy, and it is a fantasy that has done real harm, as unrealistic dreams often do. This chapter will conclude with a more realistic discussion of atomic energy production, the present state of the industry, and future possibilities.

Einstein's equation, which expresses the equivalence of mass and energy, can be used to calculate the energy holding a nucleus together. The masses of the proton, neutron, and electron and of the various atomic nuclei have been precisely determined. The mass of a nucleus is not easily measured, but it can be calculated with considerable accuracy by subtracting the total mass of the extranuclear electrons from the mass of the atom. But the total mass of the fundamental particles that make up a nucleus is always greater than the actual mass of the assembled nucleus. This difference in mass (called the *mass defect*), multiplied by c^2, is the *binding energy*. We can look upon it as the energy that would have to be supplied to separate the nucleus into its constituent parts. For nuclei with the same mass, the greater the binding energy, the more stable the nucleus.

Example 25.4 Calculate the binding energy of the nucleus of a carbon-12 atom (sometimes simply called C-12), first in joules and then in the unit usually employed for binding energies, million electron volts, MeV (1.000 MeV = 1.602×10^{-13} J). The following values will be needed:

Avogadro's number: 6.02252×10^{23}

Velocity of light: 2.99793×10^8 m/s

Atomic mass of $^{12}_{6}C$:   12.0000

Mass of the proton: 1.67252×10^{-27} kg

Mass of the neutron: 1.67482×10^{-27} kg

Mass of the electron: 9.1066×10^{-31} kg

Solution

a. The carbon nucleus contains six neutrons and six protons with a total mass of

$$6(1.67252 \times 10^{-27}) = 1.00351 \times 10^{-26} \text{ kg}$$
$$6(1.67482 \times 10^{-27}) = 1.00489 \times 10^{-26} \text{ kg}$$
$$\overline{\phantom{6(1.67482 \times 10^{-27}) =\ } 2.00840 \times 10^{-26} \text{ kg}}$$

b. The mass of the carbon atom is

$$\left(\frac{1.20000 \times 10^{-2} \text{ kg}}{1.00000 \text{ g at. wt}}\right)\left(\frac{1.00000 \text{ g at. wt}}{6.02252 \times 10^{23} \text{ atoms}}\right) = 1.99252 \times 10^{-26} \text{ kg/atom}$$

c. The mass of the carbon nucleus is the mass of the atom minus the mass of six electrons:

$$1.99252 \times 10^{-26} - 6(9.1066 \times 10^{-31}) = 1.99197 \times 10^{-26} \text{ kg}$$

d. The mass defect is

$$\begin{aligned} & 2.00840 \times 10^{-26} \text{ kg} \\ & -1.99197 \times 10^{-26} \text{ kg} \\ \hline & 0.01643 \times 10^{-26} \text{ kg} \end{aligned}$$

e. Using the relationship $E = mc^2$, this mass is equal to

$$(0.01643 \times 10^{-26} \text{ kg})\left(2.99793 \times \frac{10^8 \text{ m}}{\text{s}}\right)^2 = 1.477 \times 10^{-11} \text{ J}$$

or

$$(1.477 \times 10^{-11} \text{ J})\left(\frac{1.000 \text{ MeV}}{1.602 \times 10^{-13} \text{ J}}\right) = 92.20 \text{ MeV}$$

So that we might get some idea of how much energy this is, the next example compares it with the energy of combustion of one atom of carbon.

Example 25.5 Compare the binding energy of carbon with the energy released by the combustion of one atom of carbon. (1.000 cal = 2.609×10^{13} MeV.)

Solution

a. The binding energy of one carbon nucleus was found in Example 25.4 to be 92.20 MeV.

b. The combustion of a mole of carbon produces 94.0 kcal of heat (see Table 11.3). For one atom, this is

$$\left(\frac{94.0 \text{ kcal}}{1.00 \text{ mol}}\right)\left(\frac{1.00 \text{ mol}}{6.02 \times 10^{23} \text{ atoms}}\right) = 1.56 \times 10^{-22} \text{ kcal/atom}$$

c. Or, in MeV,

$$(1.56 \times 10^{-22} \text{ kcal}) \left(\frac{10^3 \text{ cal}}{1 \text{ kcal}} \right) \left(\frac{2.609 \times 10^{13} \text{ MeV}}{1.000 \text{ cal}} \right)$$
$$= 4.07 \times 10^{-6} \text{ MeV}$$

d. Thus the binding energy is more than 10^7 (10 million) times the heat of combustion for carbon:

$$\frac{92.20 \text{ MeV}}{4.07 \times 10^{-6} \text{ MeV}} = 2.27 \times 10^7$$

We are now close to our original goal—to assess relative nuclear stabilities. To do this we need only calculate other binding energies and then compare them with one another.

Example 25.6 Calculate the binding energy of the boron isotope with the mass number of 12 ($^{12}_{5}B$). Its atomic mass is 12.0143.

Solution

a. The mass of five protons and seven neutrons is

$$\begin{array}{rl} 5(1.67252 \times 10^{-27}) = & 8.36260 \times 10^{-27} \text{ kg} \\ 7(1.67482 \times 10^{-27}) = & 11.72374 \times 10^{-27} \text{ kg} \\ \hline & 20.08634 \times 10^{-27} \text{ kg} \\ = & 2.008634 \times 10^{-26} \text{ kg} \end{array}$$

b. The mass of one atom of this isotope is

$$\left(\frac{1.20143 \times 10^{-2} \text{ kg}}{1.00000 \text{ g at. wt}} \right) \left(\frac{1.00000 \text{ g at. wt}}{6.02252 \times 10^{23} \text{ atoms}} \right) = 1.99490 \times 10^{-26} \text{ kg/atom}$$

c. The mass of the nucleus is

$$(1.9949 \times 10^{-26} \text{ kg}) - 5(9.1066 \times 10^{-31} \text{ kg}) = 1.9944 \times 10^{-26} \text{ kg}$$

d. The mass defect is

$$\begin{array}{rl} & 2.008634 \times 10^{-26} \text{ kg} \\ - & 1.9944 \times 10^{-26} \text{ kg} \\ \hline & 0.0142 \times 10^{-26} \text{ kg} \end{array}$$

e. This mass is equal to

$$(0.0142 \times 10^{-26} \text{ kg})(3.00 \times 10^8 \text{ m/s})^2 = 1.28 \times 10^{-11} \text{ J}$$

$$(1.28 \times 10^{-11} \text{ J})\left(\frac{1.000 \text{ MeV}}{1.602 \times 10^{-13} \text{ J}}\right) = 79.9 \text{ MeV}$$

The carbon-12 nucleus, which consists of six neutrons and six protons, is about 12 MeV more stable than the boron isotope with the same mass number, which contains five protons and seven neutrons. Thus the boron nucleus can reach a more stable condition by emitting a beta particle:

$$^{12}_{5}\text{B} \longrightarrow {}^{0}_{-1}\beta + {}^{12}_{6}\text{C}$$

The comparison between the total binding energies of $^{12}_{6}\text{C}$ and $^{12}_{5}\text{B}$ is an effective way of comparing nuclear stability since they have the same mass number. In general, however, we must compare the binding energy *per nucleon*, since that indicates the strength with which each nucleon is held in the nucleus.

Figure 25.4 shows the relationship between the binding energy per nucleon and mass number for the stable isotopes. Since a large binding energy means

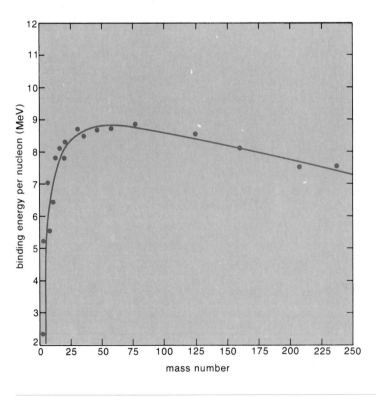

Figure 25.4 Binding energy per nucleon.

a stable nucleus, the most stable nuclei are found around mass number 60 (iron, cobalt, nickel, and copper). As we shall see later, this curve is the key to explaining the processes of *fission* and *fusion*. If a heavy, relatively unstable nucleus could be broken into two more stable fragments, the binding energy (hence the mass defect) per nucleon would increase, and large amounts of energy would be released. Likewise, if two very small nuclei could be combined, the increase in stability would be even more impressive and the energy release, even greater.

Question

6. Define binding energy and mass defect, and explain their relationship to each other.

25.6 Neutron-to-Proton Ratio

Calculating the mass defect and binding energy of a nucleus allows us to judge its relative stability, but only after we have information about the nuclear mass. These calculations tell us nothing about whether a hypothetical nucleus of, say, 82 protons and 124 neutrons is likely to be stable. As a step in the direction of understanding the more basic requisites for nuclear stability, we can turn to the information in Figure 25.5, where every known stable nucleus appears on a plot of number of neutrons versus number of protons. The general requirements for stability are immediately evident. For elements of low atomic number an equal number of protons and neutrons favors stability, whereas for heavier elements a higher ratio of neutrons to protons is required, reaching a little more than 1.5 for the heaviest elements. Nuclei with too high a neutron-to-proton ratio often attain stability through beta emission, which has the effect of converting a neutron to a proton. An example is the decay of protoactinium-234:

$$^{234}_{91}\text{Pa} \longrightarrow\ ^{0}_{-1}\beta + ^{234}_{92}\text{U}$$

$$\frac{143n}{91p} = 1.571 \qquad \frac{142n}{92p} = 1.543$$

Nuclei with too low a neutron-to-proton ratio, on the other hand, may achieve a more favorable balance by alpha emission. This reduces both the number of neutrons and the number of protons by two; thus for any isotope with more neutrons than protons the neutron-to-proton ratio is increased. The alpha decay of polonium-212 is an example:

$$^{212}_{84}\text{Po} \longrightarrow\ ^{208}_{82}\text{Pb} + ^{4}_{2}\alpha$$

$$\frac{128n}{84p} = 1.524 \qquad \frac{126n}{82p} = 1.536$$

Figure 25.6 is an expanded view of a small section of Figure 25.5. In it we find a further clue to nuclear stability. Between element 23 and element 33 there are

>16 stable isotopes with an even number of protons and an even number of neutrons

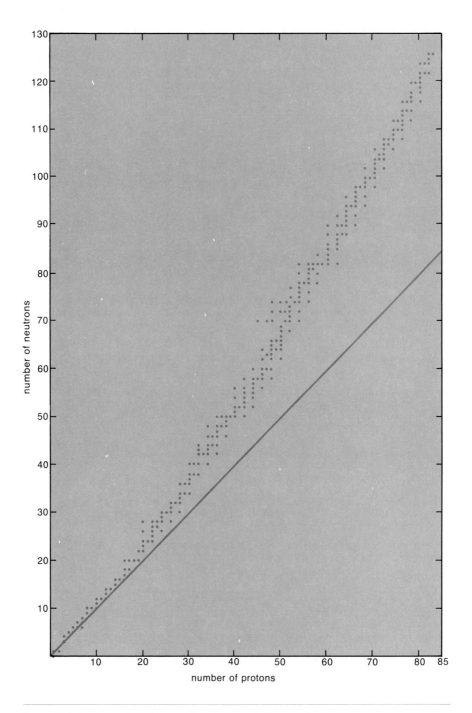

Figure 25.5 Numbers of neutrons and protons for stable nuclei.

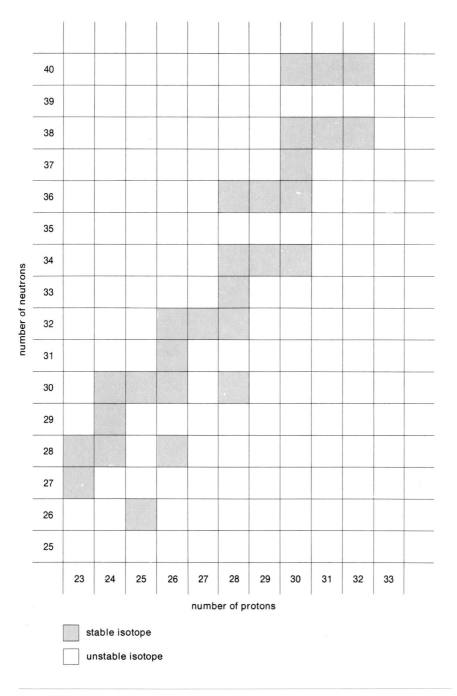

Figure 25.6 Stable isotopes with 23–33 protons.

7 stable isotopes with an odd number of protons and an even number of neutrons

4 stable isotopes with an even number of protons and an odd number of neutrons

1 stable isotope with an odd number of both protons and neutrons

If we extend this tabulation over the whole range of Figure 25.5, we find

164 isotopes with an even number of protons and an even number of neutrons

52 isotopes with an odd number of protons and an even number of neutrons

56 isotopes with an even number of protons and an odd number of neutrons

10 isotopes with an odd number of both protons and neutrons

If we go even further and take account of relative binding energies and abundances, we find that the most abundant and most stable nuclei have certain *magic numbers* of protons or neutrons: 2, 8, 20, 28, 50, 82, or 126. These numbers indicate something about the nuclear structure just as the numbers 2, 10, 18, 36, ... indicate something about the electronic structure of the atom. Nuclei like $^{208}_{82}\text{Pb}$ with two magic numbers are thus especially stable. Theories of nuclear stability that account for these magic numbers have been proposed, but we shall not discuss them here.

Questions

7. Calculate the change in neutron-to-proton ratio for these reactions:
   a. $^{214}_{82}\text{Pb} \longrightarrow {}^{214}_{83}\text{Bi} + {}^{\ \ 0}_{-1}\beta$
   b. $^{210}_{84}\text{Po} \longrightarrow {}^{206}_{82}\text{Pb} + {}^{4}_{2}\alpha$

8. What effect does alpha emission have on the neutron-to-proton ratio? What effect does beta emission have?

25.7 Half-Life

All radioactive decay processes are first-order reactions (Chapter 12); thus for a given nuclear reaction the length of time it takes for a certain fraction of material to decay is constant. Accordingly, each radioactive isotope is commonly characterized by its *half-life*, the time required for half of the starting material to decay. Half-lives of the isotopes in the uranium-238 radioactive decay series are given in Table 25.2. In this series half-lives vary from about 10^{-4} s to 4.5×10^{9} yr.

Table 25.2 Half-Lives of the Uranium-238 Radioactive Decay Series

Nuclide	Half-life
$^{238}_{92}U$	4.5×10^9 yr
$^{234}_{90}Th$	24.5 days
$^{234}_{91}Pa$	1.14 min
$^{234}_{92}U$	2.7×10^5 yr
$^{230}_{90}Th$	8.3×10^4 yr
$^{226}_{88}Ra$	1590 yr
$^{222}_{86}Rn$	3.82 days
$^{218}_{84}Po$	3.05 min
$^{214}_{82}Pb$	26.8 min
$^{214}_{83}Bi$	19.7 min
$^{214}_{84}Po$	1.4×10^{-4} s
$^{210}_{82}Pb$	22 yr
$^{210}_{83}Bi$	5.0 days
$^{210}_{84}Po$	140 days
$^{206}_{82}Pb$	Stable

Example 25.7 Bismuth-210 has a half-life of 5.0 days. How much of a 1.0-g sample remains after 20 (two significant figures) days?

Solution

a. Twenty days is equal to 4.0 half-lives:

$$(20 \text{ days}) \left(\frac{1.0 \text{ half-life}}{5.0 \text{ days}} \right) = 4.0 \text{ half-lives}$$

b. During each of these successive half-lives, the amount of Bi-210 is reduced to one-half that present when the half-life commenced. This means that we must multiply the original amount by $\frac{1}{2}$ for each half-life that elapses:

$$(1.0 \text{ g}) (\tfrac{1}{2})(\tfrac{1}{2})(\tfrac{1}{2})(\tfrac{1}{2}) = \mathbf{0.063 \text{ g}}$$

This leads to a general expression for A, the amount remaining after n half-lives, in terms of A_0, the original amount:

$$A = A_0 (1/2)^n$$

Example 25.8 Polonium-218 has a half-life of 3.05 min. How many atoms of a 5.00-g sample of Po-218 remain at the end of an hour?

Solution

a. In an hour there are

$$(60.0 \text{ min}) \left(\frac{1 \text{ half-life}}{3.05 \text{ min}} \right) = 19.7 \text{ half-lives}$$

b. Thus the sample of polonium will be reduced by one-half almost 20 times in an hour:

$$A = (5.00 \text{ g}) (\tfrac{1}{2})^{19.7}$$
$$\log A = \log 5.00 + 19.7 \log (\tfrac{1}{2})$$
$$= 0.699 - 19.7(0.301)$$
$$= -5.23$$

c. The amount of Po-218 remaining is, therefore, the antilog of -5.23 or 6.0×10^{-6} g.

d. One atom of Po-218 weighs

$$\left(\frac{218 \text{ g}}{1.00 \text{ g at. wt}} \right) \left(\frac{1.00 \text{ g at. wt}}{6.02 \times 10^{23} \text{ atoms}} \right) = 3.62 \times 10^{-22} \text{ g}$$

e. Therefore 6.0×10^{-6} g is

$$(6.0 \times 10^{-6} \text{ g}) \left(\frac{1 \text{ atom}}{3.62 \times 10^{-22} \text{ g}} \right) = 1.6 \times 10^{16} \text{ atoms}$$

Question

9. Why is it never possible to have a completely pure sample of a radioactive element?

25.8 Induced Nuclear Transformations

Only a limited amount of information and understanding can be gained from observation alone. We can hardly expect to produce a theory of nuclear structure simply by comparing masses and stabilities or by observing spontaneous reactions and their effects. Yet doing experiments on nuclei poses a number of practical problems. For example, we can't examine the nuclei directly with visible light because the wavelengths of visible light (10^{-7} m) are 10^7 times larger than an atomic nucleus. (We perceive objects by means of the light they reflect. Since visible light waves "miss" atomic nuclei altogether, there is no reflection.) Even the wavelengths of X rays (10^{-11} m) are a thousand times greater than the diameter of the largest nucleus. But beams of alpha particles or electrons have appropriate wavelengths and have been used since early in the century to probe the interior of the atomic nucleus.

In 1919 Ernest Rutherford achieved the first artificial nuclear *transmutation* (changing one element into another) by exposing nitrogen in a Wilson cloud chamber to the energetic alpha particles emitted by radium. He noted that most of the fog tracks produced by the alpha particles were straight lines. Occasionally, however, a branched track would appear (see Figure 25.3). Rutherford concluded that he was observing the direct collision of an alpha particle and a nitrogen nucleus, which combined to form a new nucleus with a very short life and high energy. This new nucleus then decayed to produce the branched fog track, one branch due to a proton and the other caused by a nucleus of oxygen-17. We can write this nuclear transmutation as follows:

$$^{14}_{7}N + ^{4}_{2}He \longrightarrow ^{1}_{1}H + ^{17}_{8}O$$

Observations of other transmutations of light elements soon followed.

In 1930 two German scientists, H. Becker (b. 1896) and Walter Bothe (1891–1957), discovered when they bombarded certain light nuclei with alpha particles, a new and highly penetrating type of radiation was emitted. Since this radiation was not deflected by electric or magnetic fields, they took it to be composed of high-energy X rays or gamma rays. In 1932 Irene Joliot-Curie (1897–1956), Marie Curie's daughter and her husband, Frederick Joliot (1900–1958), found that if a piece of paraffin were placed in front of the metal being bombarded, high-speed protons were ejected from the paraffin. It was James Chadwick, in England, who provided the explanation and who subsequently received the Nobel Prize for his "discovery" of the neutron. He used beryllium as the target nucleus and deduced that when an alpha particle struck or was absorbed by a beryllium nucleus, an unstable nucleus of 13 nucleons was produced. This nucleus quickly decayed by emitting a neutron:

$$^{9}_{4}Be + ^{4}_{2}\alpha \longrightarrow (^{13}_{6}C) \longrightarrow ^{12}_{6}C + ^{1}_{0}n$$

The neutron then penetrated the paraffin, ultimately colliding with a proton and ejecting it.

These discoveries provided two more tools for examining atomic nuclei—beams of protons and beams of neutrons. The neutron went undetected for so long because it is not charged. Since it has no ionizing effect as it passes through gases, it could not be detected by the methods in use early in the century. But this very lack of charge makes it especially valuable as a bombarding projectile. Unlike electrons, protons, or alpha particles, it experiences no electrostatic repulsion either from the extranuclear electrons or from the nuclear protons; thus it is able to penetrate the nucleus more easily.

Question

10. Why was the neutron the last of the fundamental particles (proton, neutron, electron) to be identified?

25.9 Accelerators

To produce nuclear reactions with nuclei of larger atoms, physicists needed more energetic bombarding particles than they found in natural radiation. Thus began the development of *accelerators*. The earliest such machines, called *voltage multipliers*, were developed in England by John Cockroft (1897–1967) and Ernest Walton (b. 1903). These machines accelerated protons to energies of several hundred thousand volts, and when fired at a lithium target they produced two alpha particles per lithium nucleus:

$$^{7}_{3}\text{Li} + ^{1}_{1}p \longrightarrow (^{8}_{4}\text{Li}) \longrightarrow 2^{4}_{2}\alpha$$

There has followed a long line of increasingly powerful *linear accelerators*. In each, particles are sent through a series of hollow cylindrical metal electrodes in an evacuated chamber. Figure 25.7 presents a simplified diagram of how the charges on these electrodes can be manipulated to accelerate a charged particle. When, for example, a positive particle is at A, the cylinder r is negatively charged so that the particle is drawn forward. When the particle reaches B, cylinder r is given a positive charge, and cylinder s is given a negative charge. Thus the positive particle is repelled from r and drawn toward s. When the particle reaches C, the charges are again reversed and the process continues. As the figure shows, the distances between the points of charge reversal grow as the particle accelerates.

Other accelerators (*cyclotrons*, *synchrotrons*, and *betatrons*) confine particles to circular paths by imposing powerful magnetic fields on them. But these machines still use oscillating electric charges to accelerate the particles.

Today's most powerful accelerators can accelerate not only electrons and protons but also alpha particles, deuterons ($^{2}_{1}\text{H}$), tritons ($^{3}_{1}\text{H}$), and even heavier ions. Currently the world's most powerful heavy-ion accelerator is the Bevalac at the University of California's Lawrence Berkeley Laboratory. It is a combination of the Super-HILAC (Heavy-Ion Linear Accelerator) and the Bevatron, a powerful cyclotron. Early experiments with the Bevalac produced a carbon-12 beam of 2.1 GeV (1 GeV = 10^{9} electron volts). By using such highspeed charged particles and neutrons, physicists have produced transmutations in practically

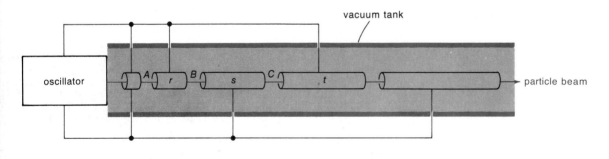

Figure 25.7 Schematic diagram of the principle of the linear accelerator.

all naturally occurring elements. In addition, they have synthesized elements that probably do not exist in nature: technetium, astatine, francium, promethium, and the **transuranic elements** (those with atomic numbers above 92).

Reactions that have been used to synthesize the missing elements technetium (43) and promethium (61) are

$$^{93}_{41}\text{Nb} + ^{4}_{2}\alpha \longrightarrow ^{96}_{43}\text{Tc} + ^{1}_{0}n$$

$$^{144}_{60}\text{Nd} + ^{2}_{1}p \longrightarrow ^{144}_{61}\text{Pm} + 2\,^{1}_{0}n$$

The first transuranic element was prepared in 1940. When uranium-238 was bombarded with neutrons, the product was neptunium:

$$^{238}_{92}\text{U} + ^{1}_{0}n \longrightarrow ^{239}_{93}\text{Np} + ^{0}_{-1}\beta$$

Elements through 105 have now been created, with claims from both Russian and American scientists for element 106 and from Russian scientists for 107. Equations for the production of some of the isotopes of some transuranic elements are given below:

$$^{238}_{92}\text{U} + ^{2}_{1}p \longrightarrow ^{240}_{94}\text{Pu} + ^{0}_{-1}\beta$$

$$^{239}_{94}\text{Pu} + ^{4}_{2}\alpha \longrightarrow ^{242}_{96}\text{Cm} + ^{1}_{0}n$$

$$^{241}_{95}\text{Am} + ^{4}_{2}\alpha \longrightarrow ^{243}_{97}\text{Bk} + 2\,^{1}_{0}n$$

$$^{253}_{99}\text{Es} + ^{4}_{2}\alpha \longrightarrow ^{256}_{101}\text{Md} + ^{1}_{0}n$$

$$^{250}_{98}\text{Cf} + ^{11}_{5}\text{B} \longrightarrow ^{257}_{103}\text{Lw} + 4\,^{1}_{0}n$$

Tons of plutonium (element 94) are now produced by the atomic energy industry, and appreciable amounts (kilograms) of curium are available, but most of the transuranic elements have been produced only in trace amounts—sometimes only a few atoms. Much interest is now focused on element 114, although it is several spaces beyond the end of the present periodic table. Theoretical calculations suggest that it may be much more stable than most transuranic elements. A reaction by which element 114 might be produced in given below, using the symbol X for element 114 and XX for an unstable intermediate.

$$^{238}_{92}\text{U} + ^{238}_{92}\text{U} \longrightarrow (^{476}_{184}\text{XX}) \longrightarrow ^{298}_{114}\text{X} + ^{170}_{70}\text{Yb} + 8\,^{1}_{0}n$$

Scientists at a laboratory outside of Frankfurt, Germany, are currently working to produce such reactions with the UNILAC, a linear accelerator expected to be capable of accelerating heavy nuclei such as uranium to 10 MeV/nucleon and lighter nuclei such as argon to 14 MeV/nucleon.

Question

11. Element 114 is expected to fall in Periodic Group IVA. Predict some of its physical and chemical properties.

25.10 Artificial Radioactivity

Induced nuclear transformations can produce both stable and unstable isotopes. Of the many nuclei that have been created by man, most (over 1200) are radioactive. This artifical radioactivity was first observed in 1934 by Irene Joliot-Curie and Frederick Joliot. They prepared unstable phosphorus-30 by bombarding aluminum-27 with alpha particles:

$$^{27}_{13}\text{Al} + ^{4}_{2}\alpha \longrightarrow ^{30}_{15}\text{P} + ^{1}_{0}n$$

Since then, radioactive isotopes of most of the elements have been produced.

Example 25.9 In an experiment, a sodium-23 nucleus absorbed a deuteron. The resulting nucleus was unstable and emitted a proton. Write nuclear equations for both steps.

Solution

a. The formulas for the particles involved are

   Sodium-23: $^{23}_{11}\text{Na}$

   Deuteron: $^{2}_{1}\text{H}$

   Proton: $^{1}_{1}p$

b. The equation for the absorption of a deuteron is

$$^{23}_{11}\text{Na} + ^{2}_{1}\text{H} \longrightarrow ^{25}_{12}\text{XX}$$

c. Or, since the element with atomic number of 12 is magnesium;

$$^{23}_{11}\text{Na} + ^{2}_{1}\text{H} \longrightarrow ^{25}_{12}\text{Mg}$$

d. The equation for the second step is

$$^{25}_{12}\text{Mg} \longrightarrow ^{1}_{1}p + ^{24}_{11}\text{XX}$$

e. Or, supplying the missing element;

$$^{25}_{12}\text{Mg} \longrightarrow ^{1}_{1}p + ^{24}_{11}\text{Na}$$

f. Therefore, the net effect of the whole process is an increase in the mass number of the sodium atom:

$$^{23}_{11}\text{Na} + ^{2}_{1}\text{H} \longrightarrow (^{24}_{12}\text{Mg}) \longrightarrow ^{1}_{1}p + ^{24}_{11}\text{Na}$$

Some processes of decay occur among artificial isotopes that are not usually found among natural radioactive isotopes. These processes are *positron emission* and *K-electron capture*. The first describes the emission of a particle with the same mass as an electron but a positive charge ($_{+1}^{0}\beta$). We can illustrate this kind of decay as follows:

$$_{15}^{30}P \longrightarrow {_{14}^{30}Si} + {_{+1}^{0}\beta}$$

The second process involves the capture of an electron from the $n = 1$ energy level by the nucleus. Electron shifts follow, so that the vacancy in the first energy level is filled by an electron from a higher level. (The name of this decay process comes from the designation of the $n = 1, 2, 3, \ldots$ energy levels as the $K, L, M, \ldots$ levels.) For example,

$$_{19}^{40}K + {_{-1}^{0}e} \longrightarrow {_{18}^{40}Ar}$$

Both of these processes have the effect of converting a proton into a neutron, thus decreasing the atomic number by one.

Most artificially radioactive isotopes decay by beta emission or positron emission. Alpha emission is relatively rare, except among heavier elements. There is also evidence that some decay processes produce a *neutrino*, a neutral particle much lighter than the electron. Neutrino emission is believed to occur in radioactive processes where there is a conversion of a proton to a neutron or vice versa, that is, in beta emission, positron emission, and K-electron capture. The neutron itself, unstable outside the nucleus, is believed to decay as follows:

$$_{0}^{1}n \longrightarrow {_{1}^{1}p} + {_{-1}^{0}\beta} + \text{neutrino}$$

Questions

12. Which types of radioactive emission alter the atomic number of the nucleus?
13. Which types alter the mass number?
14. Which types alter both? Which alter neither?

25.11 Nuclear Fission

The nuclear transformations we have discussed so far have been changes in which only a small part of the nucleus has been ejected and the major portion remained intact. By contrast, fission reactions split the nucleus into fragments of similar mass. A reaction of this type was first observed in 1939 by the German scientists Otto Hahn (1879—1968) and Fritz Strassman (b. 1902). They identified krypton and barium among the products formed when they bombarded uranium with neutrons. Natural uranium is a mixture of three isotopes: 99.27 percent U-238,

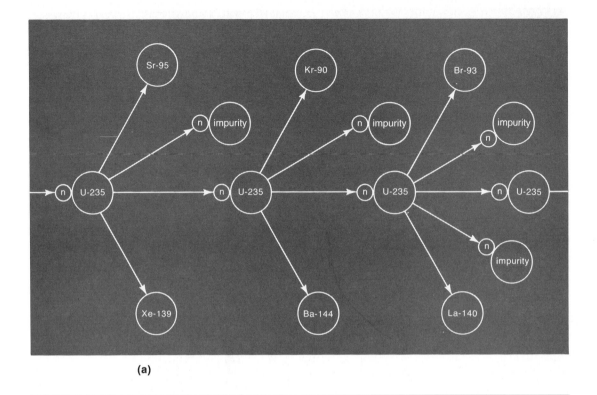

(a)

Figure 25.8 (a) Self-sustaining chain reaction. Only one neutron per fission is absorbed by a fissionable U-235 nucleus (power plant-type reaction). **(b)** Detonation reaction. All ejected neutrons are absorbed by U-235 nuclei with the result that the reaction is not only self-sustaining but accelerating.

0.72 percent U-235, and a few parts per million U-234. It was soon learned that it was the U-235 isotope that had undergone fission.

When fragmented by neutron bombardment, U-235 can form a number of different products, among which over 40 different elements have been identified. Most of these elements fall into two groups: One of the fragments will be a nucleus with a mass number between 85 and 105 and the other will be a nucleus with a mass number between 130 and 150. The following are typical fission reactions:

$$^{235}_{92}U + ^{1}_{0}n \longrightarrow ^{90}_{36}Kr + ^{144}_{56}Ba + 2\,^{1}_{0}n$$

$$^{235}_{92}U + ^{1}_{0}n \longrightarrow ^{95}_{38}Sr + ^{139}_{54}Xe + 2\,^{1}_{0}n$$

$$^{235}_{92}U + ^{1}_{0}n \longrightarrow ^{93}_{35}Br + ^{140}_{57}La + 3\,^{1}_{0}n$$

The fragments of nuclear fission are unstable and undergo radioactive decay.

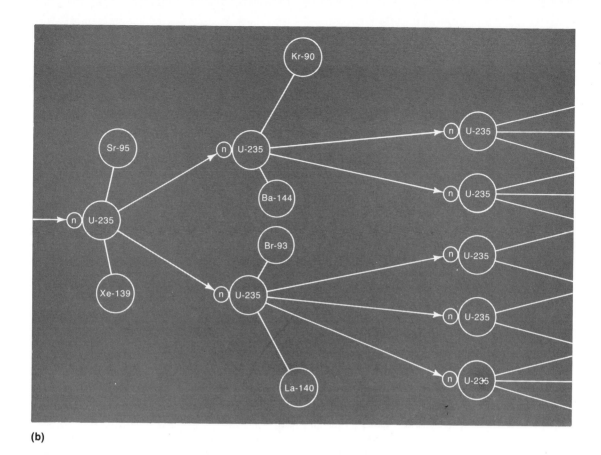

(b)

Figure 25.8 *(Continued)*

   As the typical equations above illustrate, two or more neutrons are released for each atom of U-235 that undergoes fission. These neutrons can then split other U-235 nuclei, thus propagating the reaction. Under the right conditions, therefore, the reaction is self-perpetuating (Figure 25.8). In exactly the same way as the branched-chain reactions of Chapter 12, it can proceed explosively. On the other hand, this chain reaction can be broken if neutrons are absorbed by impurities in the U-235 sample or if too many neutrons escape from the surface of the U-235 mass. Naturally occurring uranium does not undergo this chain reaction, much less detonate spontaneously, because it is only 0.72 percent U-235 and any neutrons expelled by fissioning U-235 atoms are absorbed by the U-238 nuclei that predominate. To have a self-sustaining reaction, therefore, it is necessary to have a sample of U-235 that is free from impurities and is large enough that only a small proportion of the neutrons escape from the surface. The minimum quantity that meets this last requirement is called the *critical mass*.

Only a few nuclei are known to undergo fission reactions. Two others that do are U-233 and plutonium-239, both of which are produced in *breeder reactors* (see p. 825). In each case the fission process is basically the same.

We now have the means for controlling the fission process to extract the energy released when each nucleus fragments. The atomic power reactor (Figure 25.9) consists of the fissionable material or fuel, a means of controlling the chain reaction, and a coolant. The fuel can be in any physical form or chemical combination, but most commercial reactors employ solid UO_2 since it is easy to handle. Control over the chain reaction is maintained by capturing some of the ejected neutrons with neutron-absorbing *control rods*, which can be made of materials such as cadmium or boron. The position of the control rods determines the rate of the fission process, and they can be inserted far enough into the fuel elements to "shut down" the reaction effectively. The coolant removes the heat generated by the reaction and usually transfers it to a second system, which in turn is used to drive steam turbines. Coolants can be water, heavy water (deuterium oxide, D_2O), various gases, or liquid metals. Modern reactors also require

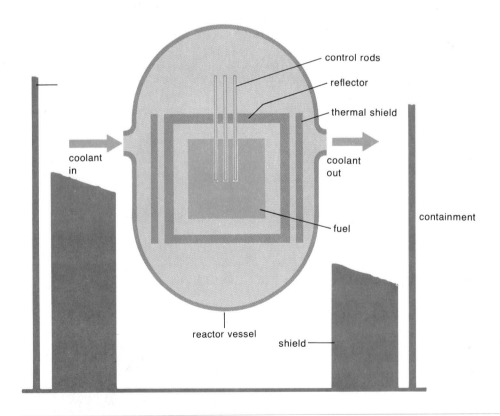

Figure 25.9 Schematic diagram of an atomic reactor.

a *moderator* (which may serve as the coolant as well), which slows down the ejected neutrons so that they can be more easily captured by fissionable nuclei. The moderator is usually water or heavy water.

There continue to be many serious questions about the practical use of nuclear energy, among them what do we do when the relatively small supply of U-235 is gone, and where do we store the dangerous and long-lived radioactive wastes of the fission process? There is also the danger that the integrity of an operating reactor can be breached, say by sabotage or earthquake. But there is no danger of an atomic explosion. The concentration of fissionable U-235 in the fuel is too low to permit an explosive reaction; should a reactor overheat, the loss of the moderator would ensure that fewer slow neutrons were available to perpetuate an accelerating reaction.

25.12 Nuclear Fusion

In nuclear fusion, light nuclei are combined to form heavier and more stable nuclei. The mass defect per nucleon increases and huge amounts of energy are released. However, to initiate nuclear fusion, the lighter nuclei must collide with tremendous force, sufficient to bring the nuclei into such close contact ($<10^{-12}$ cm) that the short-range nuclear forces come into play. (Recall the calculation of nuclear density in Example 25.1.) The activation energy barrier for fusion reactions is so great that ignition temperatures greater than 100 million kelvins are required.

One place these temperatures are available is the sun, where the following overall reaction probably takes place:

$$4\,{}^{1}_{1}H \longrightarrow {}^{4}_{2}He + 2\,{}^{0}_{+1}\beta$$

This reaction is likely the sum of three separate fusion reactions:

$${}^{1}_{1}H + {}^{1}_{1}H \longrightarrow {}^{2}_{1}H + {}^{0}_{+1}\beta$$

$${}^{2}_{1}H + {}^{1}_{1}H \longrightarrow {}^{3}_{2}He$$

$${}^{3}_{2}He + {}^{3}_{2}He \longrightarrow {}^{4}_{2}He + 2\,{}^{1}_{1}H$$

The requisite ignition temperatures are also available in atomic bombs, and this fact is the basis of the thermonuclear, or hydrogen, bomb. An atomic bomb provides the temperatures and pressures to trigger a fusion reaction.

Current research is aimed at producing these same awesome temperatures under controlled conditions so that fusion can be harnessed to do useful work. As if the problems of producing such temperatures were not great enough, a way must be devised to hold the fuel together long enough (the *containment time*) to undergo fusion. Efforts now center on two approaches: the "magnetic bottle" and laser fusion. In the first, complex magnetic fields are used to confine a hot *plasma* (matter at such high temperatures that it consists of positive ions and very mobile electrons). So far, fusionable plasmas at temperatures of 10 millions of kelvins have been contained for only fractions of a second. In laser fusion, extremely powerful and precisely timed pulses of monochromatic laser light converge simultaneously on a small solid pellet of fusionable material. A portion of the pellet implodes, forming a hot, dense mass in which fusion can take place. Containment time under these conditions of extreme density need not exceed a few nanoseconds.

Neither of these methods has yet produced energy approaching the demands of the process itself. However, research continues on a large scale, largely because the fusion reactor is such an attractive goal, free from so many of the problems that plague fission reactors. Products of the fusion reaction are generally not radioactive, and the fuel supply, deuterium and tritium from water, is practically unlimited.

25.13 Energy from the Nucleus

To a world that has finally been jolted into an awareness of the inadequacy of its conventional fuel reserves, the nucleus of the atom is a tantalizing source of energy. But its use is so fraught with problems that its promise may never be realized. When founded in 1946, the U.S. Atomic Energy Commission fully expected that within 30 years nuclear fission reactors would be providing almost as much energy as the fossil fuels. For many reasons, these predictions have not been borne out: Fission is currently supplying something over 10 percent of this nation's electrical energy.

Chief among the reasons for this slowdown has been public concern about the dangers inherent in a process that harnesses a fission reaction. Some of the fears stem from ignorance, but others reflect critical safety and environmental problems of nuclear power generation. The possibility of accidents or malfunctions at a reactor presents a terrifying picture of radioactive substances strewn around the countryside—a picture difficult for scientists and engineers to erase no matter how carefully laid their plans may be. Also, whereas it is true that atomic energy production is free from the atmospheric pollution of fossil fuel combustion, it does present other pollution problems. Waste heat, for example, must be absorbed, and rivers and lakes can tolerate only a limited temperature increase. (An innovative "floating reactor plant" is being developed that some hope will solve this problem. Situated on the ocean, 2 or 3 miles offshore, it is enclosed in a protective breakwater. The plant's thermal effects will be minimized because it will be able to discharge waste heat into the ocean.)

A much larger problem associated with the production of nuclear power is how to properly dispose of spent fuel. The original plans were that these nuclear wastes, containing many radioactive isotopes, would first be stored underwater at the power plant for several months to permit isotopes with short half-lives to decay to manageable levels. The waste would then be sent to fuel-reprocessing plants, where useful materials, including uranium and plutonium, would be removed chemically. The future of such fuel reprocessing, however, is being brought into serious question, largely because it produces fuels so rich in fissionable materials that they could be used in weapons. In any case, nobody has devised a risk-free plan for storing radioactive wastes, whether reprocessed or not. Currently, growing amounts of unreprocessed spent fuel are being stored in huge underground tanks with arrangements for constant cooling and stirring. Radioactive decay in these wastes generates enough heat to cause them to boil irregularly if they are not constantly cooled.

The severest problem of nuclear waste disposal is presented by the radioactive isotopes with half-lives greater than 20,000 years. According to one critic, safeguarding the long-lived wastes of a nuclear industry big enough to supply most of the world's energy would require a kind of "nuclear priesthood" to watch over the radioactive legacy.

A final problem of nuclear power is the supply of fuel. Only a very small amount of U-235 is consumed to produce large amounts of energy, but uranium is a scarce element and U-235 constitutes only 0.7 percent of natural uranium. As a consequence, our supplies of uranium might be exhausted even before we run out of more conventional energy sources, especially coal. To meet this problem, a new generation of nuclear reactors, called *breeder reactors*, have been designed. Several have been built, but their development in the United States has been hampered by environmental and political concerns.

Breeder reactors have the special property that they produce more fuel than they consume! This is not to say that the *law of conservation of mass* has been repealed but rather that in the fission of U-235 some of the ejected neutrons are used to transmute other elements into fissionable fuel. Thorium-232 and uranium-238 (the bulk of natural uranium) are called *fertile isotopes* because they can be converted to fissionable isotopes when bombarded with neutrons. The Th-232 is transmuted to fissionable U-233 by a series of three reactions:

$$^{232}_{90}\text{Th} + ^{1}_{0}n \longrightarrow ^{233}_{90}\text{Th}$$

$$^{233}_{90}\text{Th} \longrightarrow ^{0}_{-1}\beta + ^{233}_{91}\text{Pa}$$

$$^{233}_{91}\text{Pa} \longrightarrow ^{0}_{-1}\beta + ^{233}_{92}\text{U}$$

The U-238 isotope is transmuted to fissionable plutonium-239 by a second series of reactions:

$$^{238}_{92}\text{U} + ^{1}_{0}n \longrightarrow ^{239}_{92}\text{U}$$

$$^{239}_{92}\text{U} \longrightarrow ^{0}_{-1}\beta + ^{239}_{93}\text{Np}$$

$$^{239}_{93}\text{Np} \longrightarrow ^{0}_{-1}\beta + ^{239}_{94}\text{Pu}$$

These fissionable products can then be recycled to the reactor.

25.14 Applications of Nuclear Chemistry

Medicine The earliest application of nuclear transformations to medicine was the use of radium in the treatment of cancer. Malignant tissue was exposed to gamma rays, which retard the growth of cancerous cells. Radium has subsequently been replaced for this purpose by less expensive gamma emitters, such as cobalt-60.

A second medical application exploits the tendencies of some elements to accumulate in certain parts of the body. For example, iodine (whether stable or radioactive) will rapidly concentrate in the thyroid gland. Thus in the treatment of an overly active thyroid, sodium iodide containing radioactive iodine-131 is introduced orally. The gamma rays emitted from the radioactive isotope, which concentrate in the thyroid, destroy a portion of the gland and cure the hyperthyroidism.

Radioactive tracer studies now are often a part of the diagnostic treatment during a stay in a hospital. Because the radiation from isotopes can be monitored outside the body, physicians often can follow the movement of certain materials through the body by incorporating a radioactive *tracer* in them.

Dating An interesting application of the radioactive properties of uranium has been the determination of the earth's age. The end product of the U-238 series is Pb-206. Thus it is assumed that whenever U-238 and Pb-206 (and little or no other Pb isotopes) are found together, the Pb-206 is a product of radioactive decay. From the relative amounts of U-238 and Pb-206 in a single sample, then, the number of U-238 half-lives that have elapsed can be calculated. The half-lives of all the isotopes between U-238 and Pb-206 are very small compared to the half-life of U-238. Even the half-life of U-234, 2.7×10^5 yr, is only 1/17,000 of the half-life of U-238, 4.51×10^9 yr. Thus only the decay of the parent isotope need be considered. The result is a good estimate of the age of the sample, which can be taken as the minimum age of the earth. Calculations of this sort have established the age of the earth as between 4 and 6 billion years.

To date younger materials, especially the remains of living things, radioactive isotopes with much shorter half-lives are necessary. Furthermore, the isotopes of the natural radioactivity decay series are useless since they have been decaying since the origin of the earth. Fortunately, naturally occurring radioactive isotopes of lighter elements are available to date materials of recent periods. Among them are H-3, C-14, and K-40. The most widespread use of these lighter isotopes is in *radiocarbon dating*, a technique developed by Willard F. Libby (b. 1908) and for which he received the Nobel Prize in 1960. The concentration of C-14, among the more abundant isotopes of carbon, is exceedingly small, but it is constant due to equal rates of production and decay. It is produced in the upper atmosphere:

$$^{14}_{7}\text{N} + ^{1}_{0}n \longrightarrow ^{14}_{6}\text{C} + ^{1}_{1}p$$

And it decays with a half-life of 5568 yr:

$$^{14}_{6}C \longrightarrow ^{14}_{7}N + ^{0}_{-1}\beta$$

Carbon-14 is therefore present in the atmosphere as CO_2, which is assimilated by living plants and by the animals that consume the plants. As a consequence, all living matter contains a constant proportion of C-14. When a plant or animal dies, however, the metabolic processes that replenish the supply of C-14 cease, and the concentration begins to decline. The amount of C-14 left in the remains of a plant or animal is thus a direct measure of the time that has elapsed since its death.

Neutron Activation Analysis An analysis method that requires only small samples is especially useful when rare or precious objects must be analyzed. In neutron activation analysis, minute samples are bombarded with neutrons and then the elements present are identified by the characteristic emissions of the radioactive isotopes produced. Archeologists can thus identify trace impurities in an artifact by using only a few particles scraped from the object. The analysis can then be compared with geological samples from many areas to determine the origin of the object. Neutron activation analysis also provides a means of estimating the concentration of particular elements in the body—especially trace elements—more accurately than is possible with conventional microchemical methods.

Sterilization Radioactive emissions are used to sterilize materials that cannot be subjected to high temperatures. Many foods, pharmaceuticals, and medical supplies (such as sutures) benefit from this process, called *cold sterilization*. Its basis is the lethal effect of radiation on small organisms. For example, the radiation from Cs-137 destroys trichinae in pork. An additional advantage of cold sterilization is that materials can be sterilized after they are packaged.

Materials Testing Because of its penetrating power, radiation can be used to inspect solid materials. The radiation is sent through the material to a photographic film. By examining the shadow pattern on the film, inspectors can then locate flaws, poor welds, embedded foreign matter, and nonuniformities in density.

Sidelight
Chemistry and Forgery

One of the most successful forgeries of history was finally exposed by the work of radiochemists. Hans Van Meegeren was an artist born in the Netherlands in 1889. He studied the work of Vermeer, the great seventeenth-century Dutch master, until he could copy it with all its subtleties. He obtained old canvas and prepared paints similar to those used by Vermeer. His first successful forgery, *Christ and His Disciples at Emmaus*, was sold for $280,000! He painted and sold five more fake Vermeers, netting an estimated fortune of nearly $3,000,000.

In 1945, while being tried as a Fascist collaborator, Van Meegeren told the world that he had forged these six Vermeers. Art experts refused to believe him, and he died in 1947 with the art world still convinced of the paintings' authenticity. Not until 1968 did radiochemical evidence establish that the works were indeed modern forgeries and not seventeenth-century masterpieces.

White lead pigment contains two members of the $4n + 2$ decay series (p. 799), radium-226 and lead-210. Because of the complex relationship of half-lives in this series, the concentrations of these two isotopes remain about equal until the lead ore is refined. When the ore is converted to a pigment, most of the radium is removed. Scientists calculated, however, that within 150 years, certainly within the 300 years since Vermeer painted, radioactive decay should have restored the Ra-226 concentration to its equilibrium level. But analysis showed that a great imbalance still existed between the lead and radium isotopes in the *Emmaus*, and the art world finally had to accept that it had been deceived.

Questions and Problems

25.1 Theories of Nuclear Structure

15. How does the mass of the nucleus compare with the mass of the atom? How do the volumes compare?

16. Experiments have shown that the radius of the nucleus is directly proportional to the cube root of the number of nucleons. How would the radius of the C-12 atom compare with that of the U-238 atom?

25.2 Natural Radioactivity

17. How do nuclear reactions differ from chemical reactions?

25.4 Nuclear Equations

18. Supply the missing numbers and symbols in the following nuclear equations:

a. $^{28}_{13}\text{Al} \longrightarrow \underline{\hspace{1cm}} + ^{0}_{-1}\beta$

b. $^{239}_{93}\text{Np} \longrightarrow ^{239}_{94}\text{Pu} + \underline{\hspace{1cm}}$

c. $^{212}_{84}\text{Po} \longrightarrow ^{208}_{82}\text{Pb} + \underline{\hspace{1cm}}$

d. $^{64}_{29}\text{Cu} \longrightarrow ^{0}_{-1}\beta + \underline{\hspace{1cm}}$

25.5 Nuclear Stability

19. Calculate the binding energy of the C-13 nucleus. The atomic mass is 13.00335. Compare it with that of C-12.

20. Calculate the binding energy of the B-11 nucleus. The atomic mass is 11.00931. Compare it with that of B-12.

21. Calculate the mass equivalence of the heat of combustion of a mole of carbon. What percent of the mass has been converted to heat?

25.6 Neutron-to-Proton Ratio

22. What is the parent element of the radioactive series to which each of the following isotopes belongs?
  a. thorium-229
  b. radon-219
  c. astatine-215
  d. radium-226
  e. radon-220
  f. actinium-227
  g. thallium-206

23. The end product of the thorium-232 radioactive decay series is lead-208. How many alpha particles and how many beta particles are emitted in this series?

24. If a nucleus has too high a neutron-to-proton ratio, it may acquire greater stability by the emission of what particle?

25. A nucleus with too low a neutron-to-proton ratio might acquire greater stability in what ways?

25.7 Half-Life

26. A radioactive isotope has a half-life of 1.00 hr. How long will it take for $\frac{31}{32}$ of the original sample to decay?

27. A radioactive isotope has a half-life of 1.00 month. What portion of a sample will be left after
  a. 6 months? b. 1 yr?

25.8 Induced Nuclear Transformations

28. Supply the missing numbers and symbols in the following equations.
  a. $^{196}_{78}\text{Pt} + \underline{\qquad} \longrightarrow ^{197}_{78}\text{Pt} + ^{1}_{1}\text{H}$
  b. $^{235}_{92}\text{U} + ^{1}_{0}n \longrightarrow ^{141}_{56}\text{Ba} + 3\,^{1}_{0}n + \underline{\qquad}$
  c. $^{94}_{42}\text{Mo} + \underline{\qquad} \longrightarrow ^{95}_{43}\text{Te} + ^{1}_{0}n$

29. Write nuclear equations for the following changes:
  a. Bromine-75 emits a positron.
  b. Potassium-39 captures a proton and emits an alpha particle.
  c. Mendelevium-255 captures an alpha particle and emits two neutrons.
  d. Sodium-23 captures a deuteron and emits a proton.
  e. Curium-240 captures an alpha particle and emits two neutrons.
  f. Calcium-44 captures a proton and emits a neutron.
  g. Cobalt-59 captures a neutron and emits gamma rays.
  h. Beryllium-9 captures an alpha particle and emits a neutron.

25.10 Artificial Radioactivity

30. What is the effect upon the mass number and the atomic number when the following nuclear transformations occur?
  a. Alpha particle emitted.
  b. Beta particle emitted.
  c. Gamma ray emitted.
  d. Positron emitted.
  e. Neutrino emitted.
  f. K-electron captured.
  g. Deuteron captured, followed by proton emission.
  h. Neutron captured, followed by proton emission.
  i. Neutron captured, followed by alpha emission.
  j. Alpha particle captured, followed by proton emission.

k. Proton captured, followed by alpha emission.
l. Alpha particle captured, followed by neutron emission.

31. Give the atomic number, mass number, and symbol for the isotope produced by
   a. the loss of a beta particle by Pb-210.   b. alpha emission by Ac-225.
   c. K-electron capture by Be-7.   d. positron emission by P-30.

25.11 Nuclear Fission

32. Upon fission of a uranium-235 atom, barium-144 and two neutrons were produced. What other product was formed?

33. Why do most fission products subsequently decay by beta emission?

25.12 Nuclear Fusion

*34. How can both fission and fusion, which are essentially opposite processes, be exothermic?

35. Show that the overall solar fusion reaction (p. 823) is actually the sum of the several steps indicated.

25.13 Energy from the Nucleus

36. What is a breeder reactor?

Discussion Starters

37. Explain or account for the following:
   a. Neutrons are more effective in producing nuclear transformations if they are moving at a relatively low speed, but alpha particles are more effective if they are moving at a relatively high speed.
   b. Emission of gamma rays has no effect on either the mass number or the atomic number.
   c. Positron emission and K-electron capture have the same effect on the atomic number.
   d. A charged electroscope discharges rapidly when in the vicinity of a radioactive substance.

*38. Suggest an explanation that might be advanced for the stability of nuclei with even numbers of protons or neutrons.

Four-Place Logarithms

N	0	1	2	3	4	5	6	7	8	9
10	0000	0043	0086	0128	0170	0212	0253	0294	0334	0374
11	0414	0453	0492	0531	0569	0607	0645	0682	0719	0755
12	0792	0828	0864	0899	0934	0969	1004	0138	1072	1106
13	1139	1173	1206	1239	1271	1303	1335	1367	1399	1430
14	1461	1492	1523	1553	1584	1614	1644	1673	1703	1732
15	1761	1790	1818	1847	1875	1903	1931	1959	1987	2014
16	2041	2068	2095	2122	2148	2175	2201	2227	2253	2279
17	2304	2330	2355	2380	2405	2430	2455	2480	2504	2529
18	2553	2577	2601	2625	2648	2672	2695	2718	2742	2765
19	2788	2810	2833	2856	2878	2900	2923	2945	2967	2989
20	3010	3032	3054	3075	3096	3118	3139	3160	3181	3201
21	3222	3243	3263	3284	3304	3324	3345	3365	3385	3404
22	3424	3444	3464	3483	3502	3522	3541	3560	3579	3598
23	3617	3636	3655	3674	3692	3711	3729	3747	3766	3784
24	3802	3820	3838	3856	3874	3892	3909	3927	3945	3962
25	3979	3997	4014	4031	4048	4065	4082	4099	4116	4133
26	4150	4166	4183	4200	4216	4232	4249	4265	4281	4298
27	4314	4330	4346	4362	4378	4393	4409	4425	4440	4456
28	4472	4487	4502	4518	4533	4548	4564	4579	4594	4609
29	4624	4639	4654	4669	4683	4698	4713	4728	4742	4757
30	4771	4786	4800	4814	4829	4843	4857	4871	4886	4900
31	4914	4928	4942	4955	4969	4983	4997	5011	5024	5038
32	5051	5065	5079	5092	5105	5119	5132	5145	5159	5172
33	5185	5198	5211	5224	5237	5250	5263	5276	5289	5302
34	5315	5328	5340	5353	5366	5378	5391	5403	5416	5428
35	5441	5453	5465	5478	5490	5502	5514	5527	5539	5551
36	5563	5575	5587	5599	5611	5623	5635	5647	5658	5670
37	5682	5694	5705	5717	5729	5740	5752	5763	5775	5786
38	5798	5809	5821	5832	5843	5855	5866	5877	5888	5899
39	5911	5922	5933	5944	5955	5966	5977	5988	5999	6010
40	6021	6031	6042	6053	6064	6075	6085	6096	6107	6117
41	6128	6138	6149	6160	6170	6180	6191	6201	6212	6222
42	6232	6243	6253	6263	6274	6284	6294	6304	6314	6325
43	6335	6345	6355	6365	6375	6385	6395	6405	6415	6425
44	6435	6444	6454	6464	6474	6484	6493	6503	6513	6522
45	6532	6542	6551	6561	6571	6580	6590	6599	6609	6618
46	6628	6637	6646	6656	6665	6675	6684	6693	6702	6712
47	6721	6730	6739	6749	6758	6767	6776	6785	6794	6803
48	6812	6821	6830	6839	6848	6857	6866	6875	6884	6893
49	6902	6911	6920	6928	6937	6946	6955	6964	6972	6981
50	6990	6998	7007	7016	7024	7033	7042	7050	7059	7067
51	7076	7084	7093	7101	7110	7118	7126	7135	7143	7152
52	7160	7168	7177	7185	7193	7202	7210	7218	7226	7235
53	7243	7251	7259	7267	7275	7284	7292	7300	7308	7316
54	7324	7332	7340	7348	7356	7364	7372	7380	7388	7396
N	0	1	2	3	4	5	6	7	8	9

Four-Place Logarithms *(continued)*

N	0	1	2	3	4	5	6	7	8	9
55	7404	7412	7419	7427	7435	7443	7451	7459	7466	7474
56	7482	7490	7497	7505	7513	7520	7528	7536	7543	7551
57	7559	7566	7574	7582	7589	7597	7604	7612	7619	7627
58	7634	7642	7649	7657	7664	7672	7679	7686	7694	7701
59	7709	7716	7723	7731	7738	7745	7752	7760	7767	7774
60	7782	7789	7796	7803	7810	7818	7825	7832	7839	7846
61	7853	7860	7868	7875	7882	7889	7896	7903	7910	7917
62	7924	7931	7938	7945	7952	7959	7966	7973	7980	7987
63	7993	8000	8007	8014	8021	8028	8035	8041	8048	8055
64	8062	8069	8075	8082	8089	8096	8102	8109	8116	8122
65	8129	8136	8142	8149	8156	8162	8169	8176	8182	8189
66	8195	8202	8209	8215	8222	8228	8235	8241	8248	8254
67	8261	8267	8274	8280	8287	8293	8299	8306	8312	8319
68	8325	8331	8338	8344	8351	8357	8363	8370	8376	8382
69	8388	8395	8401	8407	8414	8420	8426	8432	8439	8445
70	8451	8457	8463	8470	8476	8482	8488	8494	8500	8506
71	8513	8519	8525	8531	8537	8543	8549	8555	8561	8567
72	8573	8579	8585	8591	8597	8603	8609	8615	8621	8627
73	8633	8639	8645	8651	8657	8663	8669	8675	8681	8686
74	8692	8698	8704	8710	8716	8722	8727	8733	8739	8745
75	8751	8756	8762	8768	8774	8779	8785	8791	8797	8802
76	8808	8814	8820	8825	8831	8837	8842	8848	8854	8859
77	8865	8871	8876	8882	8887	8893	8899	8904	8910	8915
78	8921	8927	8932	8938	8943	8949	8954	8960	8965	8971
79	8976	8982	8987	8993	8998	9004	9009	9015	9020	9025
80	9031	9036	9042	9047	9053	9058	9063	9069	9074	9079
81	9085	9090	9096	9101	9106	9112	9117	9122	9128	9133
82	9138	9143	9149	9154	9159	9165	9170	9175	9180	9186
83	9191	9196	9201	9206	9212	9217	9222	9227	9232	9238
84	9243	9248	9253	9258	9263	9269	9274	9279	9284	9289
85	9294	9299	9304	9309	9315	9320	9325	9330	9335	9340
86	9345	9350	9355	9360	9365	9370	9375	9380	9385	9390
87	9395	9400	9405	9410	9415	9420	9425	9430	9435	9440
88	9445	9450	9455	9460	9465	9469	9474	9479	9484	9489
89	9494	9499	9504	9509	9513	9518	9523	9528	9533	9538
90	9542	9547	9552	9557	9562	9566	9571	9576	9581	9586
91	9590	9595	9600	9605	9609	9614	9619	9624	9628	9633
92	9638	9643	9647	9652	9657	9661	9666	9671	9675	9680
93	9685	9689	9694	9699	9703	9708	9713	9717	9722	9727
94	9731	9736	9741	9745	9750	9754	9759	9763	9768	9773
95	9777	9782	9786	9791	9795	9800	9805	9809	9814	9818
96	9823	9827	9832	9836	9841	9845	9850	9854	9859	9863
97	9868	9872	9877	9881	9886	9890	9894	9899	9903	9908
98	9912	9917	9921	9926	9930	9934	9939	9943	9948	9952
99	9956	9961	9965	9969	9974	9978	9983	9987	9991	9996
N	0	1	2	3	4	5	6	7	8	9

Answers to Selected Questions and Problems

1 Introductory Concepts

9. a. 1×10^6 c. 3.48×10^{-5} e. 8.30×10^{-13}
     g. 1.00×10^{-1} i. 3.17×10^5

10. a. 0.00000000000000100   c. 6,110,000,000   e. 0.0000788
     g. 0.000000000000479   i. 0.000863

24. a. 6.71×10^6 c. 9.00×10^{-11} e. 9.36×10^{12} g. 9.99×10^{14}

25. a. 5.48×10^2 e. 2.92×10^{-1} g. 3.54×10^{13} i. 5.17×10^{-8}

26. a. 3.00×10^{-6} c. 1.05×10^4 e. 9.00×10^{24}
     g. 1.95×10^2 i. 3.94×10^{49}

27. a. 3.31×10^5 c. 7.00×10^{17}

28. a. 3.03×10^3

30. 5.71 km/liter

31. 4.45×10^3 m^2

33. a. 1.27×10^4 ml   b. 1.64×10^3 g or 1.64 kg

35. a. 2.00 g/liter, 10.0 g, 5.00 liters
     c. 0.420 g/ml, 5.10×10^2 g, 1212 ml
     e. 0.0721 kg/liter, 6.00×10^3 kg, 8.32×10^4 liters
     g. 1.23 lb/ft^3, 0.0387 lb, 889 cm^3

37. a. 1.00×10^3 g/liter   c. 1.00×10^{-3} kg/cm^3 e. 8.30 lb/gal

38. total volume = 1.000×10^3 cm^3; total mass = 8.19×10^3 g; average density = 8.19 g/cm^3

39. a. $-23°$C, $-10°$F, 250 K   c. $-225°$C, $-373°$F, 48 K   e. $40°$C, $104°$F, 313 K

42. $311°$F = $155°$C

44. $-40°$F

45. $160°$C, $320°$F

46. $(X°B - 0°B)\left(\dfrac{74.6°C}{100°B}\right) + 5.5°C = Y°C$

48. a. 4.9812×10^3 K

50. 1.89×10^4 cal, 1.08×10^4 cal

51. 28.1°C

2 Some Basic Concepts of Chemistry

14. a. 22

20. 52.1

21. a. 58.44   c. 100.09   e. 74.10   g. 46.08

23. a. 0.0842 gH/gC   c. 0.105 gH/gC

24. a. 2.35×10^{24} atoms F   c. 7.46×10^{21} atoms Na

25. b. 1.62×10^{-9} g at. wt. K

26. a. 1.10×10^2 g oxygen gas

27. a. 1.37×10^{-2} mol

28. a. 1.99 g at. wt. Na

29. a. 2.89×10^{22} atoms   c. 9.03×10^{22} atoms

32. 6.48×10^{18} atoms

34. 2.23×10^3 g CO_2

36. 26.8 g Al

38. a. CH_3 c. AlN_3O_9 or $Al(NO_3)_3$ e. Sb_2S_3
g. $Al_2C_3O_9$ or $Al_2(CO_3)_3$ i. Na_2HPO_4

40. $C_{17}H_{19}NO_3$

42. $(C_5H_3)_4$ or $C_{20}H_{12}$

3 Equations and Stoichiometry

7. a. 21 molecules O_2 c. 17.1 mol O_2 e. 264 g CO_2 g. 0.840 g O_2

9. 138 g As_4O_{10}

11. 4.07 mol O_2

13. C_2H_6

15. 6.38×10^{-6} g O_2

17. 4.33×10^{19} molecules N_2

19. 473 g CO_2

21. a. 275 g SO_2 b. 206 g O_2 c. 4.28 mol H_2O

23. 9.80 ml H_2O

25. 4.78×10^{-13} g O_2

27. a. 7.62×10^{24} molecules HNO_3 b. 107 g $Cu(NO_3)_2$
    c. 5.28 g NH_3 d. 4.94×10^{24} molecules NH_3

29. 93.4%

30. a. 3.33 g $CaCl_2$

31. a. 5.64×10^3 g sucrose

32. a. 29.3 ml

33. a. 12.1 M c. 18.4 M e. 17.4 M

34. 0.70 M

35. 5.6 g NaOH

37. 2.42 g H_2

39. 1.16×10^3 lb $Ca(OH)_2$; $23.20

4 Atomic Structure

11. Between 1 and 2 miles

13. 24.3

15. Rb-85, 76.5%; Rb-87, 23.5%

17. 3.95×10^{-22} g; 237 times

19. 217

5 Electron Configurations and the Periodic Table

13. 6.00×10^{14} Hz

15. 6.00×10^{-7} m per wavelength

17. a. 6.63×10^{-26} J   c. 3.6×10^{-19} J   e. 6.63×10^{-15} J

23. Element A = nitrogen; Element B = magnesium; Element C = boron

25. a. $(4, 0, 0, +\frac{1}{2}$ or $-\frac{1}{2})$ e. $(3, 2, [-2, -1, 0, +1,$ or $+2], +\frac{1}{2}$ or $-\frac{1}{2})$

37.

	i	ii	iii	iv
a.	Representative	13	3	p

46. a. :Ẍ: c. ·Ẍ:

7 Covalent Bonding—Molecular Structure

12.

Before	After
a. s-2, p-2	sp^3-4
g. s-2, p-6	sp^3d^2-8

17. a. sp^3d^2 d. sp^2

18. a. sp^3 d. sp

	Electron Formula	Hybridization	Electron Pair Geometry	Molecular Shape
20. a.	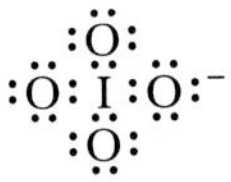 :Ö: :Ö:I:Ö:⁻ :Ö:	sp^3	tetrahedral	tetrahedral
m.	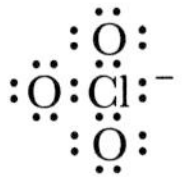 :Ö: :Ö:Cl:⁻ :Ö:	sp^3	tetrahedral	trigonal pyramidal
o.	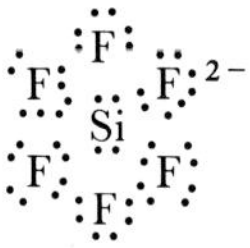 :F: :F: ²⁻ :F: Si :F: :F: :F:	sp^3d^2	octahedral	octahedral
p.	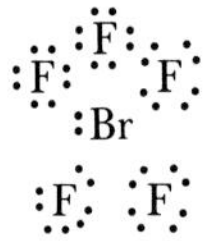 :F: :F: :F: Br :F: :F:	sp^3d^2	octahedral	square pyramidal
u.	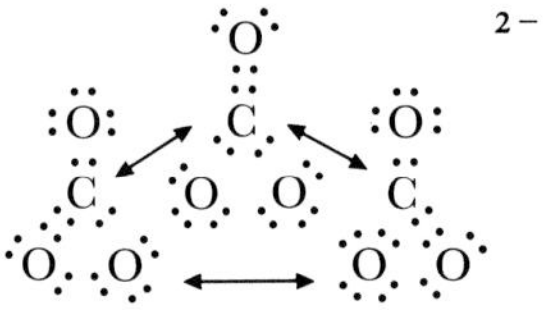 :Ö: ... :Ö: ²⁻ (resonance structures of CO_3^{2-})	sp^2	trigonal	trigonal

8 Gases

4. a. 13.7 liters   b. 7.32 liters   c. 4.29 liters

9. 9.80×10^{-1} atm

11. gas: 1.43×10^3 ml; liquid: 1.43 ml; solid: 1.43×10^{-1} ml

12. a. 1.22×10^3 cu ft   c. 5.66×10^2 cu ft

13. a. 34.0 ml

14. a. 228 K or $-45°C$

16. 1.80×10^2 ml

19. a. 1.27 atm   c. 29.0 atm

20. a. 2.00 atm   c. 0.333 atm   e. 0.250 atm

21. 4.24×10^3 liters CO_2

25. H_2: 2.50 mol; He: 1.00 mol

26. a. 1.36 mol N_2

28. a. 4.46×10^{-2} mol CO

29. 45.9 g

33. 2.70×10^{-2} mol N_2, 7.22×10^{-3} mol O_2, 3.20×10^{-4} mol Ar, 1.1×10^{-5} mol CO_2

35. 1.29 g/liter

39. $MW_A = 64.0$

41. NH_3: 5.93×10^2 cm; HCl: 4.1×10^2 cm

42. a. 1.96 g CO_2/liter

44. 37.9

46. 1.34×10^{-22} g/molecule HBr

49. a. 22.4 atm

50. a. 2.24×10^2 liters

53. diameter = 5.00 ft

55. a. 8.53×10^5 liters   **b.** 1.22×10^6 g O_2

57. C_3H_8

60. 284 K

62. 4.89 g/liter

9 Liquids and Solids—The Condensed Phases

20. 6, 12, 8

29. a. 2.5×10^4 cal

31. 90.5

33. 56.8 ml

34. a. 1.39×10^4 cal

36. 2.02×10^3 g

38. 10.9°C

43. a. contract

47. 9.15 cm^3/mol

50. 1.44×10^{-10} m

52. $E = 3.62 \times 10^{-8}$ cm; $r = 1.28 \times 10^{-10}$ m

54. a. 0.682, 6   **c.** 0.352, 3

55. corners: $8(\frac{1}{8}) = 1$ Z; faces: $6(\frac{1}{2}) = 3$ W; empirical formula: ZW_3 or W_3Z

57. $n\lambda = 2d \sin \theta$

$$d = \frac{n\lambda}{2 \sin \theta} = \frac{(1)(1.5374 \times 10^{-10} \text{ m})}{2 \sin 16°10'} = \frac{1.5374 \times 10^{-10} \text{ m}}{2(0.27843)} = 2.76 \times 10^{-10} \text{ m}$$

59. b. H_3BO_3 **d.** $Si_2O_7^{6-}$

Charge on pyrosilicate: pyro structure requires 56 electrons—2 Si's and 7 O's donate 50 electrons; thus charge = -6.

62. a. 109.5°   c. 120°

63. a. 4   b. 3

67. a. $CaCO_3$; electrostatic (ionic)   c. S_8; London dispersion forces
  f. Fe; metallic, electrostatic

10 Solutions

8. 10.0%

10. 1.77×10^{-2} mol glucose/total mol

13. $-0.93°C$

17. 9.50 mg/100 ml

19. 33.8%

21. Li_2SO_4: Solubility decreases with increasing temperature. The dissolving process is exothermic and ΔH is negative. Hydration energy is greater than lattice energy. Consider $\Delta G = \Delta H - T\Delta S$. No definite conclusion can be made about ΔS, since ΔH guarantees a negative ΔG regardless of the value of ΔS.

K_2SO_4: Solubility increases with increasing temperature. The dissolving process is endothermic and ΔH is positive. Lattice energy is greater than hydration energy. Since the process is spontaneous and thus ΔG is negative, ΔS must then be positive and sufficiently large.

24. a. 263 g NH_3/liter solution   c. 413 g NH_3/kg H_2O
  f. 0.304 mole fraction NH_3

27. 0.325 liter MeOH

29. 38.3 g NaCl

30. a. 0.885 M c. 1.45 M

32. a. 29.3%, 12.9 m; 0.189 mole fraction CH_3OH
  c. 30.0%, 4.37 m; 0.0729 mole fraction H_3PO_4

33.

	g Solute per Liter Solution	M	M	sp gr	% by Weight	Mole Fraction Solute
H_2SO_4	1.78×10^3	18.1	2.5×10^2	1.85	96.0	0.82
$C_{12}H_{22}O_{11}$	585	1.71	2.73	1.21	48.3	0.0469
C_2H_5OH	424	9.21	18.4	0.924	45.9	0.249
KI	414	2.49	2.8	1.30	31.8	0.048

35. 4.81 liters

37. 5.55×10^{-3} m

38. a. 5.06 M c. 19.4 M

39. a. 4.5×10^{-5} molecule Ar/molecule H_2O
  c. 2.2×10^{-7} ion/molecule H_2O

41. 1.8 m; MW = 3.5×10^2

43. MW = 256; S_8 is molecular formula

45. 43.4%

47. 10.4 qt

49. 26.4 mm Hg

50. b < c < a < d

51. 16.9 atm

11 Thermodynamics

3. $\Delta E = \Delta H - p\Delta V$; $p\Delta V = nRT$
  1. Assume all components are gases! 2. For exothermic reactions, $\Delta H < 0$
  a. $C_3H_8 + 5O_2 \longrightarrow 3CO_2 + 4H_2O$ $\Delta n = +1$ $|\Delta E| > |\Delta H|$

7. a. -713 kcal

8. a. 12 kcal

9. HBr: -85.0 kcal/mol; HI: -68.0 kcal/mol

16. a. $\Delta H = 15.9$ kcal; $\Delta S = -0.0290$ kcal; $\Delta G = 24.5$ kcal
  c. $\Delta H = -122.6$ kcal; $\Delta S = 0.0848$ kcal; $\Delta G = -148$ kcal
  e. $\Delta H = -1128$ kcal; $\Delta S = -0.438$ kcal; $\Delta G = -997$ kcal

19. $\Delta S_{Cl_2} = -8.90 \times 10^{-3}$ kcal/(mol)(K); $\Delta S_{H_2O} = -5.26 \times 10^{-3}$ kcal/(mol)(K)

22. System is at equilibrium, so $\Delta G = 0$.

12 Chemical Kinetics

8. a. 2nd   b. 0

18. 3.2 ml/min; 2.5 ml/min

20. V increases 5%; KE increases 10%

22. $x = 1$; $y = 2$

24. First order [A]; Second order in [B]; 0 order in [C]; Third order overall

26. $\frac{7}{8}$

28. 0.0211 g

30. $k = 3.29 \times 10^{-4}$

32. $k = 4.8 \times 10^2$

13 Molecular Equilibria

9. 1.50

11. 0.964

12. a. left   b. left   c. right   d. right   e. left   f. right

15. 0.735

17. 1.50 mol/liter

20. $K = 0.150$; $[H_2] = 2.09$; $[CO_2] = 2.09$; $[CO] = 5.91$; $[H_2O] = 4.91$

22. a. $K = 0.500$ **b.** 4.00 mol/liter $COCl_2$ removed

25. $K = 2.00$, 4.00 mol/liter SO_2 added

27. $K = 0.0643$; $[Cl_2] = 0.327$

29. $K = 0.13$

31. $[CO_2] = [H_2] = 1.33$; $[CO] = [H_2O] = 0.67$

34. $[W] = 1.23$; $[X] = 3.23$; $[Y] = 6.77$; $[Z] = 1.77$

35. a. 0.050   **b.** 0.0022

36. a. 1.9×10^{-7}

14 Acids and Bases

1. 6.0×10^{16} H^+ and OH^-/liter

8. a. ionic: $(H^+ + I^-) + (K^+ + OH^-) \longrightarrow (K^+ + I^-) + H_2O$
net ionic: $H^+ + OH^- \longrightarrow H_2O$
c. ionic: $(H^+ + NO_3^-) + NH_3 \longrightarrow (NH_4^+ + NO_3^-)$
net ionic: $H^+ + NH_3 \longrightarrow NH_4^+$
d. ionic: $HF + (Na^+ + OH^-) \longrightarrow (Na^+ + F^-) + H_2O$
net ionic: $HF + OH^- \longrightarrow F^- + H_2O$
e. ionic: $2(H^+ + ClO_4^-) + Mg \longrightarrow (Mg^{2+} + 2ClO_4^-) + H_2$
net ionic: $2H^+ + Mg \longrightarrow Mg^{2+} + H_2$

9. a. $NO_2^- + H_2O \rightleftharpoons HNO_2 + OH^-$
b. $NH_4^+ + H_2O \rightleftharpoons NH_3 + H_3O^+$
d. $Fe^{3+} + H_2O \rightleftharpoons Fe(OH)^{2+} + H^+$ or $[Fe(H_2O)_5(OH)]^{2+} + H^+$

22. 1.32 equivalents NaOH

24. 247g H_2SO_4

26. 0.018 liter

40. 79

42. 24.3 g base/eq

44. a. 14.4 eq   **b.** 2.00 eq

15 Acid-Base Equilibria

13. a. $2.0 - 4.25$ **c.** $10 - 12$

16. a. 0.010

17. a. 0.20

20. 6.5

21. a. 1.7×10^{-4}

23. 5.4×10^{-10}

25. 2.6×10^{-8}

27. $2.0 \times 10^{-1}\ M\ NH_3$

29. $86\ g\ KC_2H_3O_2$

30. a. $4.0 \times 10^{-5}\ M$ c. $8.1 \times 10^{-11}\ M$

32. a. 9.3   b. 0.0 to the nearest tenth   c. 0.3 to the nearest tenth

33. a. 2.4/1

34. a. $[S^{2-}] = 1.2 \times 10^{-22}\ M$ b. $[S^{2-}] = 2.1 \times 10^{-18}\ M$

35. a. $[CO_3^{2-}] = 5.6 \times 10^{-11}\ M; [H_2O] = 55\ M$

36. a. 5.6×10^{-10}

38. 7.3×10^{-3}

39. 8.3×10^{-11}

40. a. 8.9

41. a. $6.0 \times 10^{-4}\ M$ c. $0.10\ M$
    e. $0.015\ M$ g. $2.1 \times 10^{-2}\ M$
    i. $3.3 \times 10^{-3}\ M$ k. $1.4 \times 10^{-5}\ M$
    m. $4.5 \times 10^{-4}\ M$ o. $0.20\ M$
    q. $7.2 \times 10^{-3}\ M$ s. $5.5 \times 10^{-3}\ M$
    u. $7.3 \times 10^{-5}\ M$ w. $7.9 \times 10^{-2}\ M$
    y. $3.0 \times 10^{-5}\ M$ aa. $1.3 \times 10^{-3}\ M$
    cc. $4.8 \times 10^{-2}\ M$ ee. $0.010\ M$

42. a. $[H_3O^+] = 0.10\ M, [Cl^-] = 0.10\ M, [H_2O] = 55\ M,$
    $[OH^-] = 1.0 \times 10^{-13}\ M$
    e. $[OH^-] = [HOAc] = 9.0 \times 10^{-6}\ M, [OAc^-] = 0.15\ M, [Na^+] = 0.15\ M,$
    $[H^+] = 1.0 \times 10^{-9}\ M, [H_2O] = 55\ M$

44. blue, pH = 6.7; yellow, pH = 7.5

45. a. b

16 Precipitation and Solubility Products

1. a. $2\ NaI + Pb(NO_3)_2 \longrightarrow PbI_2 + 2\ NaNO_3$
   $2(Na^+ + I^-) + (Pb^{2+} + 2NO_3^-) \longrightarrow PbI_2 + 2(Na^+ + NO_3^-)$
   $2I^- + Pb^{2+} \longrightarrow PbI_2$
   c. $Fe_2(SO_4)_3 + 6\ NaOH \longrightarrow 2\ Fe(OH)_3 + 3\ Na_2SO_4$
   $(2Fe^{3+} + 3SO_4^{2-}) + 6(Na^+ + OH^-) \longrightarrow 2Fe(OH)_3 + 3(2Na^+ + SO_4^{2-})$
   $Fe^{3+} + 3OH^- \longrightarrow Fe(OH)_3$
   i. $3\ BaCl_2 + 2\ Na_3PO_4 \longrightarrow Ba_3(PO_4)_2 + 6\ NaCl$
   $3(Ba^{2+} + 2Cl^-) + 2(3Na^+ + PO_4^{3-}) \longrightarrow Ba_3(PO_4)_2 + 6(Na^+ + Cl^-)$
   $3Ba^{2+} + 2PO_4^{3-} \longrightarrow Ba_3(PO_4)_2$

8. 54.5 g NiS

10. 26.3 ml NaOH solution

13. $0.159\ M\ (NH_4)_2CO_3$

14. a. 7.7×10^{-13} c. 5.6×10^{-23}

15. a. 4.3×10^{-14} b. 1.6×10^{-9}

17. a. $1.0 \times 10^{-4}\ M$ c. $1.2 \times 10^{-4}\ M$

18. c. $3.3 \times 10^{-3}\ g\ Ag_2CO_3$

19. $2.0 \times 10^7\ S^{2-}$ ions/liter

23. $[Ba^{2+}] = [BaCrO_4] = 1.1 \times 10^{-5}\ M$

25. b. $[Ag^+] > 6.9 \times 10^{-5}$

27. $[Ca^{2+}] = 5.6 \times 10^{-6}\ M$

30. 3.8×10^4 kg of $Na_2CO_3 \cdot 10H_2O$

32. $[CO_3^{2-}] = 4.6 \times 10^{-4}\ M$

17 Oxidation and Reduction

6. a. -3 e. $+5$

18. a. $H = +1;\ Cl = +1;\ O = -2$
    e. $N = -3;\ H = +1$
    h. $H = +1;\ S = +7;\ O = -2$

22. a. $Cu + Cl_2 \longrightarrow (Cu^{2+} + 2\ Cl^-)$

25. a. $Na_2SO_3 + 2\ NaMnO_4 + 2\ NaOH \longrightarrow Na_2SO_4 + 2\ Na_2MnO_4 + H_2O$
    b. $KMnO_4 + 8\ HCl + 5\ FeCl_2 \longrightarrow MnCl_2 + 5\ FeCl_3 + KCl + 4\ H_2O$
    c. $24\ CuS + 64\ HNO_3 \longrightarrow 24\ Cu(NO_3)_2 + 3\ S_8 + 16\ NO + 32\ H_2O$
    d. $5\ Zn + 12\ HNO_3 \longrightarrow N_2 + 5\ Zn(NO_3)_2 + 6\ H_2O$

26. a. $3\ F_2 + 3\ H_2O \longrightarrow 6\ HF + O_3$
    b. $2\ P_4 + 5\ HClO_4 + 12\ H_2O \longrightarrow 8\ H_3PO_4 + 5\ HCl$
    c. $16\ FeCl_3 + 8\ H_2S \longrightarrow 16\ FeCl_2 + 16\ HCl + S_8$
    d. $2\ Bi(OH)_3 + 3\ Na_2SnO_2 \longrightarrow 2\ Bi + 3\ Na_2SnO_3 + 3\ H_2O$

27. (Coefficients)
    a. $1,1,4 \longrightarrow 1,2,1,2$
    b. $8,5 \longrightarrow 1,4,4,4$
    c. $6,5 \longrightarrow 3,5$
    d. $1,3 \longrightarrow 1,4$

28. a. $F_2 + Zn \longrightarrow ZnF_2$
    d. $8\ HNO_3 + 3\ Zn \longrightarrow 2\ NO + 3\ Zn(NO_3)_2 + 4\ H_2O$

29. a. $3\ OH^- + 3\ H_2O + P_4 \longrightarrow PH_3 + 3\ H_2PO_2^-$
    e. $6\ H^+ + 5\ H_2O_2 + 2\ MnO_4^- \longrightarrow 2\ Mn^{2+} + 5\ O_2 + 8\ H_2O$

30. a. 0.741 equivalent   b. 6.04 equivalents

31. a. 63.0 g/equivalent   c. 15.8 g/equivalent

33. 9.00 g/equivalent

35. 1.16 equivalents

37. 72.4 g Cl_2

38. 3.43 liters (STP)

18 Electrochemistry

9. a. 1.00 faraday   **b.** 2.00 faradays

15. a. Cell: $Mg|Mg^{2+}||Co^{2+}|Co$ $E°_{cell} = -0.28 - (-2.375) = +2.10$ V
Anode: $Mg \longrightarrow Mg^{2+} + 2e^-$
Cathode: $Co^{2+} + 2e^- \longrightarrow Co$
Overall: $Mg + Co^{2+} \longrightarrow Co + Mg^{2+}$
(+) electrode: cathode; (−) electrode: anode
Direction of electron flow: anode to cathode
... positive ion movement: toward cathode
... negative ion movement: toward anode
Change in mass of electrodes: anode decreases, cathode increases

b. Cell: $Pt|I^-|I_2||Cl^-|Cl_2|Pt$ $E°_{cell} = +1.36 - (+0.535)$
Anode: $2I^- \longrightarrow I_2 + 2e^-$ $= +0.82$ V
Cathode: $Cl_2 + 2e^- \longrightarrow 2 Cl^-$
Overall: $2I^- + Cl_2 \longrightarrow 2 Cl^- + I_2$
(+) electrode: cathode; (−) electrode: anode
Direction of electron flow: anode to cathode
... positive ion movement: toward cathode
... negative ion movement: toward anode
Change in mass of electrodes: none

21. $+0.1952$ V

23. a. $E° = 0.552; n = 2; \log K = 18.7 ; K = 10^{19}$

24. Electrolysis
a. dilute H_2SO_4 solution
Anode: $2 H_2O \longrightarrow O_2 + 4 H^+ + 4e^-$
Cathode: $2e^- + 2H^+ \longrightarrow H_2$
f. $Pb(NO_3)_2$ solution
Anode: $2 H_2O \longrightarrow O_2 + 4 H^+ + 4e^-$
Cathode: $2e^- + 2 H_2O \longrightarrow H_2 + 2 OH^-$

25. a. 2.90×10^4 coul   **e.** 1.72×10^4 coul

26. a. 86.7 g Cr; 1.67 g at. wt; 1.00×10^{24} atoms

28. 0.867 g Cr

30. 2.96 g Cu; 2.24 g O_2; 0.283 g H_2

19 The Chemistry of the Nonmetals of Groups VIA and VIIA

13. a. -2 **b.** $+4$ **c.** $+6$ **d.** $+2$

26. a. $2H^+ + CO_3^{2-} \longrightarrow H_2CO_3 [H_2CO_3 \longrightarrow H_2O + CO_2]$
c. $H^+ + OH^- \longrightarrow H_2O$
e. $2 NaCl + H_2SO_4 \longrightarrow Na_2SO_4 + 2 HCl$

g. $16 H^+ + 10 Cl^- + 2 MnO_4^- \longrightarrow 5 Cl_2 + 2 Mn^{2+} + 8 H_2O$
i. $HI + H_2O \longrightarrow H_3O^+ + I^-$
k. $Br_2 + H_2O \longrightarrow HBrO + H^+ + Br^-$
m. $PbO_2 + 4 H^+ + 2 Cl^- \longrightarrow Cl_2 + Pb^{2+} + 2 H_2O$

27. a. H^+; Cl^-; OH^-; H_2O
    b. H^+; F^-; HF; H_2O; OH^-; FHF^-
    c. Cl_2; H_2O; H^+; Cl^-; $HClO$; OH^-; ClO^-

30. c. 7.75%

32. a. H_2; H_2O b. 0.18 g H_2 c. 30.5 g H_2O

37. a. $KClO_3$ b. HgO c. KNO_3 d. MnO_2

52. 525 ml

54. 12.9 liters

57. 38.7 liters

59. 3.00×10^2 liters

20 The Chemistry of the Nonmetals of Groups IVA and VA and the Noble Gases

25. c, f, g, h, i

28. phosphide ion, P^{3-}; phosphonium ion, PH_4^+; PCl_4^+, PCl_6^- ions which compose PCl_5 in the solid state

42. b. $HCO_3^- \rightleftharpoons H^+ + CO_3^{2-}$ and $HCO_3^- + H_2O \rightleftharpoons H_2CO_3 + OH^-$ (greater)

21 The Chemistry of Metals

5. $[2 Li^+ + :\ddot{\underset{..}{O}}:^{2-}]$ $[2 Na^+ + :\ddot{\underset{..}{O}}:\ddot{\underset{..}{O}}:^{2-}]$ $[K^+ + :\overset{..}{\underset{..}{O}}:\overset{..}{\underset{..}{O}}:^-]$

23. Greater oxidation number results in greater acidity $MnO < Mn_3O_4 < MnO_2 < MnO_3 < Mn_2O_7$

29. 0.314 ton

22 Organic Chemistry

5. C—C—C—C—C—C—C
         |
         C

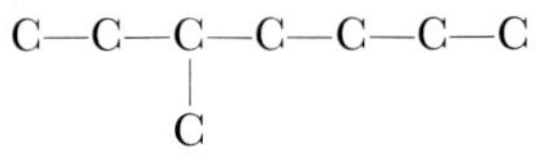

   3-methylheptane

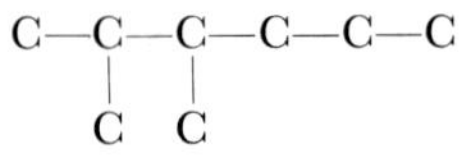

   C—C—C—C—C—C
       | |
       C   C

   2,3-dimethylhexane

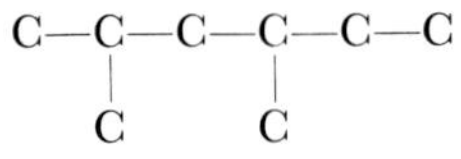

   C—C—C—C—C—C
       | |
       C   C

   2,4-dimethylhexane

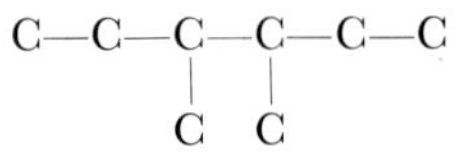

   C—C—C—C—C—C
         | |
         C   C

   3,4-dimethylhexane

                C
                |
   C—C—C—C—C
       | |
       C   C

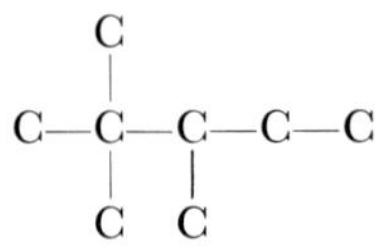

   2,2,3-trimethylpentane

7. b. 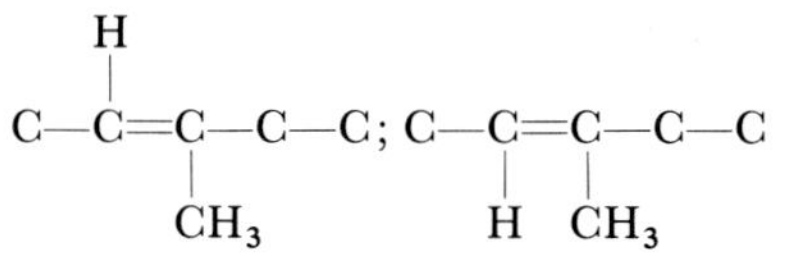
$$C-\overset{H}{\underset{CH_3}{C}}=C-C-C; \quad C-C=C-C-C$$
$$\overset{|}{H}\overset{|}{CH_3}$$

35. 0.862 g/liter

51. 64.5 liters

56. 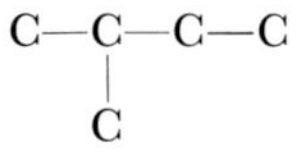 $C-C-C-C$
    $\overset{|}{C}$

59. 8.60×10^{17} molecules

78. $CH_4 \xrightarrow[light]{Cl_2} CH_3Cl \xrightarrow[aq]{NaOH} CH_3OH$

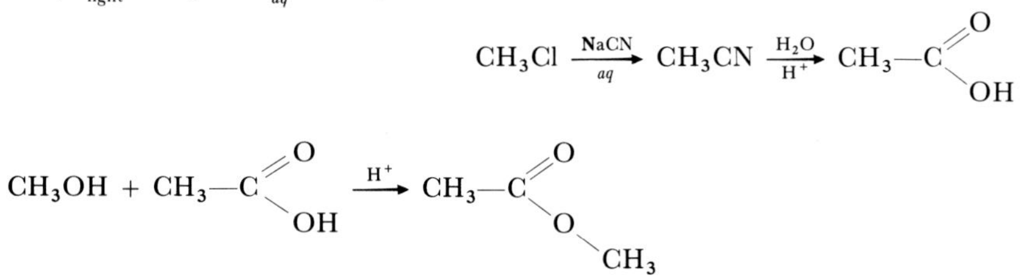

$CH_3Cl \xrightarrow[aq]{NaCN} CH_3CN \xrightarrow[H^+]{H_2O} CH_3-C\overset{O}{\underset{OH}{\diagdown}}$

$CH_3OH + CH_3-C\overset{O}{\underset{OH}{\diagdown}} \xrightarrow{H^+} CH_3-C\overset{O}{\underset{O-CH_3}{\diagdown}}$

23 The Structure and Properties of Biomolecules

19. 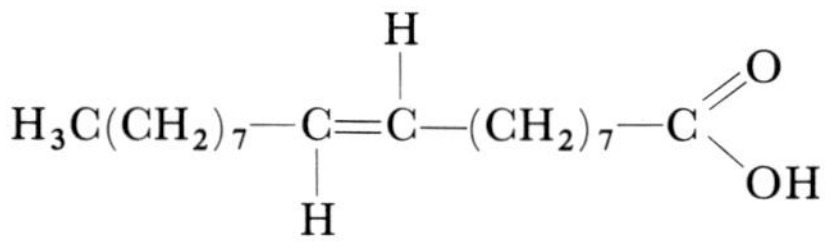 $H_3C(CH_2)_7-\overset{H}{\underset{H}{C}}=\overset{}{\underset{}{C}}-(CH_2)_7-C\overset{O}{\underset{OH}{\diagdown}}$

   elaidic acid

28. Phe Val Asp Gly His Leu Cys Gly Ser His

30. a. amino acids
    b. glucose
    c. glucose

24 Coordination Compounds

4. a. essentially ionic b. -2 c. -1 d. 4
   e. $+2$ f. Zn^{2+} g. SCN^-

5. a. tetraiodomercurate (II) ion
   b. sodium tetrahydroxoaluminate (III)
   c. pentacyanoethylcobaltate(III) ion

6. a. $[Co(NO_2)(NH_3)_5]^{2+}$
   b. $[Co(en)_3](NO_3)_3$
   c. $K_4[Fe(CN)_6]$

11. two; the H_2O ligands can occupy *cis* or *trans* positions. one; optical isomer of *cis* form. Total number is three.

14. a. $Na_3[Fe(CN)_6]$

18. a. $K[Ag(CN)_2]$
    c. $[Cr(en)_3]^{3+}$
    g. $[Fe(CO)_5]$

23. $[PtICl(NH_3)_2]$ has a coordina   number of 4.
Pt is in Group VIII B   "X" represents e^- pair from ligand

    Pt atom: [↿⇂][↿⇂][↿⇂][↿][↿] [↿⇂]   Pt(II) ion: [↿⇂][↿⇂][↿⇂][↿][↿] [ ]
                  d   s   d   s

If dsp^2-square planar: [↿⇂][↿⇂][↿⇂][↿⇂][×][×][×][×]   Diamagnetic No unpaired electrons
(must pair e^- up) dsp^2

25 Nuclear Chemistry

7. a. 0.03

16. $\dfrac{R_{U\text{-}238}}{R_{C\text{-}12}} = 2.7$

18. a. $^{28}_{14}Si$

19. Binding energies per nucleon: C-12 = 7.728 C-13 = 7.470. C-12 is more stable (has a higher binding energy/nucleon) by 1.034 times C-13.

21. $3.64 \times 10^{-8}\%$

23. 6 α particles and 4 β particles

27. a. $\frac{1}{64}$ will remain

28. a. ^2_1D

29. a. $^{75}_{35}Br \longrightarrow ^{0}_{+1}e + ^{75}_{34}Se$
    c. $^{255}_{101}Md + ^4_2He \longrightarrow 2\,^1_0n + ^{257}_{103}Lw$

31. a. $^{210}_{83}Bi$

Index

Boldface number = definition of term

A, mass number, 90
Absolute entropies, table of, 364
Absolute zero, 225, 361
Absorption, 395
Absorption spectra, 102
Accelerators, 816
Acetaldehyde, 717
Acetic acid, 459, 521, 720
  glacial, 459
Acetone, 718
Acetylene, 155, 201, 691
  production, 707
Acid, **176**, 426
  Arrhenius, 427, 430
  binary, 430
  carboxylic, 720
  concentrated, 435
  conjugate, 437
  diprotic, 431
  fatty, 735
  Lewis, 441
  monoprotic, 430
  nomenclature, 177
  nonvolatile, 434
  oxidizing, 528
  polyprotic, 431, 472
  reactions with metals, 433
  reactions with salts, 434
  strength of, 439
  strong, 430, 455
  ternary, 430
  triprotic, 431
  volatile, 434
  weak, 430, 458
Acid anhydrides, 430
Acid-base trends, 444
Acid salt, 436
Acid solution, 428
Acidic amino acids, 750
Acidic oxide, 430
Acidosis, 471
Acids, common, 430
Actinides, 128, 662
Activated complex, 389
Activation energy, 390
Active site, 759
Activity, 414, 524, 559
Activity series, 523
Actual yield, 65
Addition compounds, 764
Addition polymers, table of, 703
Addition reactions, 718
Additives, gasoline, 706
Adipic acid, 724
Age of the earth, 826
Alchemy, 34
Alcoholic beverages, 714
Alcohols, 713
  nomenclature, 713
  preparation, 712

Alcohols (continued):
  primary, 715
  properties, 715
  reactions with active metals, 713
  secondary, 715
  structure, 713
  tertiary, 715
  uses of, 715
Aldehydes, 717
  nomenclature, 717
  properties, 718
  uses, 717
Aldose, 739
Aliphatic hydrocarbons, 675
Alkali metals, 648
Alkaline cells, 549
Alkaline-earth metals, 648
Alkalosis, 472
Alkanes, 676
  nomenclature, 682
  properties, 696
  reactions of, 698
Alkenes, 675, 688
  nomenclature, 688
  properties, 696
  reactions of, 699
Alkoxide, 713
Alkyl group, 709
Alkyl halides:
  nomenclature, 709
  reactions, 712
Alkynes, 675, 690
  nomenclature, 690
Allotropes, 278
  oxygen, 592
  phosphorus, 278, 624
  sulfur, 278, 597, 600
  tin, 663
Alloys, 308, 657
Alloy system, 658
Alpha-amino acids, 747
Alpha carbon, 747
Alpha-helix, 754
Alpha particle, 86
Alpha particle scattering
  experiment, 87
Alpha rays, 86, 799
Aluminum, 648, 667
  production, 568
Aluminum chloride, 667
Aluminum fluoride, 667
Aluminum hydroxide, 667
Amalgam, dental, 308
Amino acid sequence, 752
Amino acids, 747
  acidic, 750
  basic, 749
  essential, 747
  neutral, 748
  sulfur-containing, 750

Ammines, 764, 790
Ammonia, 56, 260, 479, 613
  as a fertilizer, 479
  fountain, 616
  manufacture of, 590
  molecule, 186, 189
  properties of, 613
  synthesis, 402, 417
Ammonium chloride, 466, 765
Ammonium cyanate, 674
Ammonium dichromate, 612
Ammonium ion, 163
Ammonium nitrate, 480
Ammonium nitrate as a fertilizer, 480
Ammonium nitrite, 611
Ammonium salts, 617
Amorphous solids, 301
Amount of substance, 11
Ampere, 11, 78
Amphiboles, 295
Amphoteric hydroxide, 445, 790
Amplitude, 96
Amylopectin, 745
Analysis, 46
Anemia, sickle-cell, 752
Anesthetic, 716
Anhydride, acid, 430
Anion, **145**
Anode, 79, **542**
  electrolytic cell, 564
  galvanic cell, 541
  mud, 568
Anthracene, 695, 802
Antibodies, 733
Antibonding orbital, 209
Antiknock gasolines, 583
Antimony, 668
Antipyretic, 62
Apoenzyme, 759
Aqua regia, 656, 791
Aristotle, 33
Aromatic hydrocarbons, 675, 693
Aromatic substitution, 704
Arrhenius acids, 430
Arrhenius bases, 428
Arrhenius concept, 427
Arrhenius equation, 391
Arsenic, 668
Arsenic oxides, 668
Arsine, 668
Artificial radioactivity, 818
Aryl group, 709
Aryl halides, 712
Asbestos, 295
Associated liquids, 260
Astatine, 153, 574, 817
Aston, F. W., 85
Astringent, 62
Asymmetric carbon atom, 686
Atherosclerosis, 736

Atmosphere, 217
Atmospheric oxygen, 593
Atmospheric ozone, 593
Atmospheric pressure, 215
Atom, 33
  ground state, 105
Atomic bomb, 823
Atomic energy, 822
Atomic hydrogen torch, 591
Atomic number, **89**
Atomic properties, 133
Atomic radii, table of, 135
Atomic radius, 133
Atomic reactor, 822
Atomic structure, theory of, 87
Atomic theory, 34
Atomic volume, 117
Atomic weight, 36, 91
Atomization enthalpy, 356
Attachment isomerism, 783
Aufbau principle, **112**
Autoionization, 426, 454
Average reaction rate, 375
Avogadro, Amadeo, 42, 228
Avogadro's hypothesis, **228**
Avogadro's number, **42,** 228
Azimuthal quantum number, 107

B vitamins, 761
Bakelite, 717
Balance, 3, 34
  direct reading scale, 5
  triple-beam, 4
Balancing equations, 55
Balmer series, 104
Barium sulfate, 501
Barometer, 215
Bartlett, Neil, 640
Base, 426
  Arrhenius, 427
  conjugate, 437
  Lewis, 441
  strong, 429, 455
  weak, 429, 458
Base metals, **655**
Base units, 10
Basic amino acids, 749
Basic oxide, 429
Basic oxygen furnace, 658
Basic solution, 428
Battery, 548
  storage, 550
Bauxite, 568
Becker, H., 815
Becquerel, Antoine-Henri, 86, 797
Bent molecules, 186
Benzaldehyde, 717
Benzene, 693
  reactions, 794
  structure, 693
Benzpyrene, 695
Beryl, 293
Beryllium, 160, 667
Beryllium fluoride, 160, 191
Beryllium hydroxide, 667
Berzelius, Jöns, 39, 674
Beta-helix, 754
Beta particle, 86
Beta rays, 86, 799
Betatron, 816
Bevalac, 816
Bevatron, 816
Beverage, 310
Beverage alcohol, 714
Bicarbonates, 636
Bifluoride ions, 580
Binary acid, 430
Binary compounds:
  nomenclature, 176
Binding energy, 805
  calculation, 805
  per nucleon, 808
Biological chemistry, 733
Biomolecules, 733
Bismuth, 664

Bleaching properties of chlorine, 582
Blister copper, 567
Blood:
  buffers in, 471
  pH, 469
Body-centered cube, 267
Bohr, Niels, 102, 110
Boiling point, **253**
  hydrogen compounds, 261
  normal, 253
Boiling point elevation constant, 325
Bomb calorimeter, 347
Bond:
  covalent, 145, 149
  double, 155
  homonuclear, 598, 674
  ionic, 145
  multiple, 154, 198
  nonpolar covalent, 156
  pi, 200
  polar covalent, 151
  sigma, 200
  stability, 145
  strengths, 356
  triple, 155
Bond angle, **183**
Bond-breaking step, 357
Bond enthalpies, 356
Bonding, 185
Bonding orbital, 209
Boric acid, 47, 297
Boron, 160, 277, 295
Boron trifluoride, 160, 163, 192, 442, 766
Bothe, Walter, 815
"Bottled gas," 57
Boyle, Robert, 220
Boyle's law, **220**
Brackett series, 105
Bragg, William Henry, 287
Bragg, William Lawrence, 287
Bragg equation, 287
Branched-chain reactions, 394
Brass, 658
Breeder reactors, 822, 825
Bromate-bromide reaction, 381
Bromine, 114
  addition reaction, 700
  applications, 583
  preparation, 576
  production, 583
Bromine pentafluoride, 161
Brønsted, Johannes, 437
Brønsted-Lowry acid-base concept, 437
Bronze, 658
Brown, Robert, 265
Brownian movement, 265
Buffers, **469**
Buret, 484
Butane, 678, 681
Butenes, 689
Butter, 308
Butyne, 691

°C, 25
Calcite, 688
Calcium carbide, 707
Calcium carbonate, 634
Calcium chloride, 651, 653
Calcium hydride, 590
Calcium hydroxide, 653
Calcium sulfate, 653
Calomel, 502
"Caloric," 341
Calorie, 27
Calorie, large 27
Calorimeter, 345
Canal rays, 84
Cancer, 708, 783
Candela, 11
Carbides, 635
Carbohydrates, **733,** 739
Carbon, 633
  abundance, 633
  diamond, 292
  graphite, 298

Carbon (continued):
  in steel, 658
  properties of, 634
Carbonated drinks, 310
Carbonates, 636
  heating, 593
Carbon black, 590
Carbon dioxide, 36, 65
  automobile exhaust, 411
  crystal structure, 280
  Lewis acid, 443
  molecular structure, 200
  properties of, 636
Carbonic acid, 472, 636
Carbon monoxide, 36
  automobile exhaust, 411
  cigarette smoke, 514
  pollutant, 61, 65, 708
  properties of, 635
  reaction with hemoglobin, 635
Carbon tetrachloride, 578, 699
Carbonyl carbon:
  reactions of, 718
Carbonyl chloride, 636
Carbonyl compounds, 717
Carbonyls, 636, 764
Carboxylic acids, 720
  nomenclature, 721
  preparation, 721
  reactions, 721
Carcinogenic compounds, 695
Catalyst, **395,** 593, 594, 702
  carrier, 395
  enzymes, 758
Cathode, 79, **542**
  electrolytic cell, 564
  galvanic cell, 541
Cathode-ray tube, 79
Cathode rays, 79
Cation, **145,** 496
Cations, 496
Cavendish, Henry, 639
Cell:
  alkaline, 549
  commercial, 548
  concentration, 560
  Daniell, 540
  dry, 548
  electrolytic, 562
  fuel, 548, 552
  galvanic, 540
  lead storage, 550, 568
  Leclanché, 548
  mercury, 549
  nickel-cadmium, 551
  primary, 548
  rechargeable, 548
  secondary, 548
  silver oxide, 550
Cell reaction equation, 543
Cellulose, 184, 733, 746
Celsius, 25
Center of coordination, **767**
Centi-, 11
Centigrade, 25
Cesium, 652
Chadwick, James, 89, 815
Chain hydrocarbons, 675
Chain isomerism, 689
Chain reactions, 393, 820
Charge density, 774
Charge densities, table, 774
Charge, electrical, 77
Charge-to-mass ratio of electron, 81
Charles, Jacques, 223
Charles' law, **223**
Cheese, 308
Chelating agents, **767**
Chemical activity, **153**
Chemical analysis, 46
Chemical change, **6**
Chemical equations, **54**
Chemical kinetics, **373**
Chemical nomenclature, 172
Chemical properties, **4**

Chlorine, 114
  addition reaction, 700
  applications, 582
  bleaching properties of, 582
  as a disinfectant, 582
  and methane, 393
  occurrence, 573
  properties, 576
  substitution reaction, 699
Chlorine trifluoride, 161, 196
Chromate-dichromate
  equilibrium, 661
Chromate ion, 661
Chromic acid, 661
Chromium(II) hydroxide, 662
Chromium(III) hydroxide, 662
Chloroform, 578, 699, 711
Chloromethane, 578, 698
Chlorophyll, 787
Cis-trans isomerism, 689, 783
Classes of derivatives, 710
Closed system, 344
Closest-packed crystal lattices, 274
Coal, 591, 633, 706
Coal gas, 707
Coal tar, 707
Cockcroft, John, 816
Coefficients, 55, 495
Coenzyme, 759
Coke, 707
Colligative properties, 323
Collision:
  molecular, 377
Colloidal suspension, **307**
Colloids, types, 308
Color:
  of compounds, 652, 657
  of coordination compounds, 786
  of flames, 652
  of ions, 665
Combining capacity, 146
Combining volumes of gases, 227
Combustion, 513, 697
Commercial cells, 548
Common-ion effect, **466**, 506
Complexes:
  geometries of, 776
  inert, 774
  inner, 778
  labile, 774
  outer, 778
Complex ion, **767**
Compounds, 9
  intermetallic, 657
Compression of gases, 219
Compression ratio, 705
Concentration cell, 560
Concentrations, 67, 316, 380
  calculations, 317
  formality, 316
  molarity, 316
  mole fraction, 316
  normality, 447
Concentrations and equilibrium, 400, 404
Condensation, 244
Condensed phases, 214, 244
Condensed structural formulas, 677
Conductor, electrical, 426
Conjugate acid, 437
Conjugate base, 437
Conservation of mass, law of, 34, 54
Constant-volume method, 345, 347
Constructive interference, 285
Contact catalysis, 395
Containment time for fusion, 823
Continuous spectrum, 102
Control rods, 822
Conversion equation, 19
Conversion factor, 20
Conversions, temperature, 25
Coordinate covalent bond, 163, 765
Coordination complexes, 764, **767**
  color, 786
  geometry, 776
  hybridization, 777

Coordination complexes (continued):
  magnetic properties, 777
  stability, 773
Coordination compounds,
  nomenclature, 769
Coordination isomerism, 782
Coordination number, 269, **767**
Copper, 114
  conductivity, 568
  electrodes, 567
  reaction with nitric acid, 619
Copper(II) chloride, 659, 790
Copper(II) ion, 442
Copper sulfate pentahydrate, 766
Copper(II) sulfate solution,
  electrolysis, 567
Corrosion, 569, 669
Cortisone, 48
Cosmic rays, 99
Coulomb, 78
Covalence, 149
  in halides, 659
Covalent bonds, **145**, 149
Critical mass, 822
Critical pressure, **244**
Critical temperature, **244**
Crookes, Sir William, 402
Crookes tube, 79
Cryolite, 568, 574
Crystal lattice, 266
  closest-packed, 274
Crystal systems, 274
Cube, geometry of, 267
Curie, Marie, 797
Curie, Pierre, 797
Curium, 817
Current electricity, 77
Cyanide ion, 766
Cyclic process, 361
Cycloalkanes, 676, 691
  nomenclature, 691
Cycloalkenes, 676, 692
  nomenclature, 692
Cycloalkynes, 676
Cyclobutane, 691, 699
Cyclohexane, 691
Cyclohexene, 692
Cyclopentane, 691
Cyclopentene, 692
Cyclopropane, 691, 699
Cyclotron, 816
Cysteine, 753
Cystine, 753

d^n ions, 158
d^{10} ions, 157
$d^{10}s^2$ ions, 157, 158
d-orbital shapes, 779
d-orbital splitting, 779
Dacron, 722, 724
Dalton, John 34, 232
Daniell cell, 540
DDT, 711
Decay series, 799
Definite proportions, law of, 34
Deliquescence, **651**
Delocalized electrons, 203, 282
Democritus, 33
Denaturant, 714
Density, **22**
  of gases, 219
  of nuclear material, 796
Dental amalgam, 308
Deoxyribose, 743
Derivative classes, 710
Derivatives, hydrocarbon, 709
Desalination, 329
Destructive interference, 285
Detergent, 737, 762
Deuterium, 587, 824
Deuterium oxide, 822
Deuteron, 818
Deviations from the gas laws, 237
Diabetes, 718
Diabetes mellitus, 740

Diamagnetic, 160
Diamond, 292, 633
Diatomic molecules, 40, 149
Dichloromethane, 578, 698
Dichromate ion, 661
Diethyl ether, 716
Differentiating electron, 119
Diffraction, 284
Diffraction, x-ray, 284
Diffusion of gases, 219
Dilution problems, 320
Dimer, 619
Dimerization of NO_2, 619
Dimethyl ether, 766
Dimethyl mercury, 192
Dimethyl terephthalate, 724
Dinitrogen pentaoxide, 385
Diphosphine, 626
Dipolar ion, 747
Dipole, 151
Dipole attractions, 258
Dipole moment, 151, 184
  induced, 258
Diprotic acid, 431
Diradicals, 600
Disaccharides, 744
Discharge, 77
Disorder, 362
Disproportionation, 577
Dissociation enthalpy, 356
Distillation, 329
Distorted tetrahedral shape, 196
Donor atom, **767**
Double bond, 155
Double salts, 764
Dry cell, 548
dsp^2 hybridization, 777
d^2sp^3 hybridization, 778
Ductility, 130
Dulong and Petit, law of, 37
Duralumin, 652

$E°$, 556
EDTA, 775, 788
Efficiency, 362
Efflorescence, **651**
Einstein, Albert, 110, 805
Eka-aluminum, 115
Eka-boron, 115
Eka-silicon, 115, 668
Elaidic acid, 735
Elastic collision, 234
Elastomers, 702
Electric current, unit of, 11
Electrical charge, 77
Electrical conductor, 426
Electrical discharge, 79
Electrical energy, 7, 76
Electrical potential, 78
Electricity, 76
  current, 77
  static, 77
Electrochemical series, **523**
Electrochemistry, 540
Electrodes, 79
  inert, 545
Electrolysis, 77, 562
  of copper(II) sulfate solution, 567
  of hydrochloric acid, 565
  of molten sodium chloride, 563
  of sodium chloride solution, 564
  of sulfuric acid, 565
Electrolyte, strong, **184**
Electrolyte, weak, **184**
Electrolytic cell, 562
Electrolytic reaction, 540
Electromagnetic radiation, 96
Electromagnetic spectrum, 98, 99
Electromotive force, 556
Electron, 77, 79
  charge-to-mass ratio, 81
  differentiating, 119
  excitation, 103
  oil-drop experiment, 81

Electron (continued):
  properties of, 91
  velocity, 106
Electron cloud, 109
Electron configuration, 119
  and group number, 128
  and periodic table, 123
  and stability, 127, 145
  four quantum numbers, 120
  noble gas, 145
  one quantum number, 120
  two quantum numbers, 120
Electron configurations, table of, 124
Electron formula, **138**, 165
Electron-pair acceptor, 441
Electron-pair donor, 441, 766
Electron-pair geometry, 189
Electron promotion, 188
Electron-transfer method, 532
Electronegativity scale, 151, 152
Electrons:
  delocalized, 203, 282
  unpaired, 159
  valence, 146
Electrophile, 699
Electroplating, 568
Electropositive elements, 151
Electroscope, 800
Elemental gases, 228
Elementary particles, **796**
Elementary reactions, 387
Elements, Latin names, 39
  names of, 38
  second period, 578
e/m ratio, 81
Emission spectra, 102
Emission spectrum, hydrogen atom, 104
Empirical formula, 46, 49, 676
Enantiomers, 685
End point, 484
Endothermic reaction, **256**
Energy, 76, 340
  activation, 390
  electrical, 7, 76
  heat, 7, 25, 76
  hydration, 314
  kinetic, 7, **76**
  lattice, 313
  nuclear, 76, 822, 824
  potential, 7, **76**
  radiant, 7, 96
  solvation, 314, 493
  sound, 76
  transformations of, 7
Energy levels, 103
  diagram, 105
Engine knock, 706
Enthalpy, **348**
  absolute, 351
  of atomization, 356
  of combustion, 351
  of dissociation, 356
  of formation, 351
  relative, 351
Entropy, 359, **361**
  absolute, table of, 364
Enzyme, 395, 733, 758
Equation:
  balancing, 55, 529
  cell-reaction, 543
  half-reaction, 543
  ionic, 432
  net ionic, 432
  nuclear, 802
  overall, 432
  precipitation reaction, 493
  thermochemical, 350
Equilibrium, 400
  acid-base, 454
  calculations, 407
  chromate-dichromate, 661
  and concentration, 400, 404
  and free energy change, 414
  heterogeneous, **500**
  liquid-vapor, 247

Equilibrium (continued):
  physical, **247**
  and pressure, 405
  and reaction rates, 416
  solid-liquid, 253
  solubility, 499
  and temperature, 405
  water, 454
Equilibrium constant, 400, **403**
  and $E°$, 561
  units for, 408
Equivalence point, 484
Equivalent weight, 77
  in acid-base reactions, **447**
  in oxidation-reduction reactions, **533**
Essential amino acids, 747
Essential fatty acids, 736
Esters, 721, 722
  hydrolysis of, 723
Ethanal, 717
Ethane, 676, 679
Ethanoic acid, 720
Ethanol, 713
Ethene, 155, 199, 689
Ethers, 716
  nomenclature, 716
Ethyl alcohol, 714
Ethyl butyrate, 723
Ethyl formate, 722
Ethyl methanoate, 722
Ethylene, 155
Ethylene bromide, 711
Ethylenediamine, 768
Ethylene glycol, 714
Ethylene series, 675
Ethyne, 155, 691
  production, 707
Evaporation, 245, 324
Exact numbers, 17
Excess of a reagent, 63
Exothermic reaction, **256**
Expansion of the octet, 161
Exponential numbers, 12
  addition, 13
  division, 14
  multiplication, 14
  roots and powers, 14
  subtraction, 13

$°F$, 25
Face-centered cube, 267
Fahrenheit scale, 25
Faraday, 554
Faraday, Michael, 77, 569
Faraday's laws, 569
Fast reaction, 373
Fats, 722, 734
Fatty acids, 735
  common, 735
  essential, 736
Feldspars, 295
Ferromagnetic, 160
Ferrous metals, 658
Fertile isotopes, 825
Fertilizer, 479, 631
  ammonia, 479
  ammonium nitrate, 480
Fibers, 724
Filling order of electrons, 126
Filtrate, 493
Fireworks, 655
First-order reactions, 385
Fischer, Emil, 740
Fission, 819
Flame colors, 652
Flavoring agents, 717
Fluorescent minerals, 86
Fluoride ion, 147
Fluorine:
  applications, 583
  comparison with other halogens, 684
  and electronegativity scale, 151
  natural occurrence, 574
  and periodic classification, 114
  properties, 576

Fluorocarbons, 583, 594
Fluorspar, 574
Flux, 466
Foam, polyurethane, 308
Fog, 308
Fogtracks, 801
Folic acid, 761
Force, 215
Forgery, 827
Formaldehyde, 717
Formalin, 717
Formality, 316
Formic acid, 462, 720, 721
Formula, **40**
  closed form, 432
  condensed structural, 677
  empirical, 42, 46, 676
  molecular, 676
  open form, 432
  structural, 676
Formula unit, 42, 147
Formula weight, 41
Forward reaction, 401
Francium, 153, 648, 817
Frasch process, 598
Free energy, **365**, 413
  and $E°$, 562
  and equilibrium, 414
  of formation, 368
Freezing point, 254
  normal, 254
Freezing point depression constant, 326
Freon, 711
Frequency, 96
Fresh water, 328
Fructose, 742
Fruit odors and flavors, 723
Fuel cell, 548, 552
Fuels, 63, 362, 698
Functional group, 709
Functional-group isomerism, 716
Furanose, 743
Fusible alloys, 664
Fusion, 809, 823

G, 365
ΔG, 365
$\Delta G°$, 365
$\Delta G°_f$, 368
galactose, 742
Galilei, Galileo, 215
Gallium, 663
Galvanic cells, 540
Galvanic reaction, **540**
Galvanized iron, 670
Gamma rays, 86, 799
Gases, 214
  collecting, 251
  colors, 219
  compression, 219
  densities, 219
  diffusion, 219
  elemental, 228
  monoatomic, 40
  noble, 39
  odors, 218
  physiological effects, 218
  properties, 218
  solubility, 309
Gas laws, 220
  deviations from, 237
Gas law constant, 229, 332
Gasoline, 591, 698, 705
  additives, 706
  antiknock, 583
  combustion, 65
  octane number, 706
  straight-run, 705
  tetraethyl lead, 41, 708
  unleaded, 706
Gay-Lussac, Joseph, 223, 227
Geiger, Hans, 87
Geiger, Müller counter, 800
Gelatin, 308
Gems, 295

General gas law equation, 229
General rate-law equation, 380
Geneva system, 683
Geometric isomerism, **685**, 689, 783
Geometry:
    molecular, 183
    of the cube, 267
    of some complexes, 776
Gerlach, W., 107
Germanium, 668
Gibbs, Willard, 365
Glacial acetic acid, 459
Glucose, 739
Glycerine, 714, 715
Glycerol, 714, 715
Glycine, 747
Glycogen, 745
Goldstein, Eugen, 84
Graham's law, **235**
Grain alcohol, 714
Gram atomic weight, 42
Gram formula weight, 42
Gram molecular weight, 42
Graphite, 298, 633
Gravitational force, 1
Greek philosophers, 33
Ground state of the atom, 105
Group IA elements, 648
    chemical properties, 648
    comparison with Group IIA metals, 654
    hydrates of, 650
    physical properties, 650
    uses, 652
Group IIA elements, 648
    chemical properties, 648
    comparison with Group IA metals, 654
    hydrates of, 650
    physical properties, 650
    uses, 652
Group VA elements:
    properties of, 612
Group VII A elements:
    properties of, 575
Group number and electron configuration, 128
Gypsum, 653

H, 348
ΔH, 309, 348
$\Delta H°$, 351
$\Delta H_f°$, 351
Haber, Fritz, 402
Haber process, 402, 590
Hahn, Otto, 819
Half-cells, 542
Half-life, **385**, 812
Half-reaction equation, 543
Half-reactions, **521**
Halide ions, polarization, 659
Halides, covalence in, 659
Halides, organic, 709
Hall, Charles, 568
Halogens, 114, 149, 573
    applications, 582
    comparison with hydrogen, 591
    preparation, 576
    properties, 575, 576
    reaction with water, 577
Hard water, 497
Heat, 7, 25, 27, 76, 341
Heat of fusion, 254
Heat of solution, **309**
Heat of sublimation, 256
Heat of vaporization, **246**
Heavy metal ions, 762
Heavy water, 822
Heisenberg uncertainty principle, **109**
Heisenberg, Werner, 110
Helium, 640
Helix, 753
Hematite, 658
Hemes, 788
Hemiacetals, 719
Hemiketals, 719
Hemoglobin, 635, 733, 757, 788

Henry's law, **310**
Hertz, Heinrich, 100
Hess's law, 354
Heterocyclic compounds, 743
Heterogeneous mixture, **8**, 307
Hexagonal lattice, 274
Hg_2^{2+} ion, 157
Homogeneous materials, **8**
Homogeneous mixture, 8, 307
Homolog, 676
Homologous series, 676
Homonuclear bonds, 598, 674
Hormones, 733
Hund's rule, **112**
Hybridization, **189**
    and shape of complexes, 777
Hybrid orbitals, 188
    sp, 191
    sp^2, 192
    sp^3, 189
    sp^3d, 194
    sp^3d^2, 197, 777
    dsp^2, 777
    d^2sp^3, 778
Hydrated proton, 426
Hydrates, 650, 764
    vapor pressure, 650
Hydration, 314
Hydrazine, 617
Hydride ion, 146, 589
Hydrides, 516, 589
Hydrocarbons, **674**
    aliphatic, 675
    aromatic, 675, 693
    chain, 675
    classification, 675
    combustion, 697
    derivatives of, 675
    physical properties, 696
    reactions of aromatic, 704
    reactions of saturated, 698
    reactions of unsaturated, 699
    saturated, 675
    unsaturated, 675
Hydrocarbon derivatives, 707, 709
Hydrocarbon fuels, 698, 707
Hydrocarbon halides, table of, 711
Hydrocarbon residue, 709
Hydrochloric acid, 316
    electrolysis, 565
Hydrofluoric acid, 430, 580
Hydrogen, 586
    abundance, 586
    addition reaction, 700
    applications, 590
    comparison with halogens, 591
    industrial production, 410, 590
    isotopes, 587
    preparation, 526
    properties, 587
Hydrogen atom, emission spectrum, 104
Hydrogen bomb, 823
Hydrogen bonding, 258
Hydrogen bromide:
    addition reaction, 700
    formation, 383
Hydrogen chloride:
    addition reaction, 700
    fountain, 581
Hydrogen fluoride, 260, 580
    molecule, 186
Hydrogen fountain, 588
Hydrogen halides, 578
    preparation, 578
    properties, 580
Hydrogen iodide, formation, 389, 400
Hydrogen ion, 454
Hydrogen molecule, 186
Hydrogen peroxide, 69, 374, 392, 701
Hydrogen sulfide, 63, 601
Hydrogenation, 591, 735
Hydrolysis, **435**, 476
    of metallic ions, 438
    of phosphorus halides, 626
Hydrolysis constant, 477

Hydronium ion, 427
Hydrosulfuric acid, 472
Hydroxide ion, 454
Hydroxides, 428
    amphoteric, 445
    of IA, IIA metals, 651
Hydroxo-complexes, 790
Hygroscopy, **651**
Hypertonic solutions, **335**
Hypohalous acids, 577
Hypotonic solutions, **335**

Ice, 301
Iceberg, 329
Ideal gas, 221
Ideal gas constant, 332
Ideal gas law equation, 229
Ideal solutions, 323
Indicators, 480, **481**
Indium, 663
Induced dipole moment, 258
Inert complexes, 774
Inert-pair ions, 157, 664
Infrared radiation, 99
Initial reaction rate, 376
Inner complex, 778
Inner transition elements, 128, 662
Inorganic compounds, 634
Insecticides, 712
Instability constants, table, 773
Instantaneous rate of reaction, 376
Insulin, 751
Interference, 284
Interhalogen compounds, 580
Interionic forces, 493
Intermediate compound, 395
Intermetallic compounds, 657
Intermolecular forces:
    dipole attractions, 258
    hydrogen bonding, 258
    London dispersion forces, 258
    van der Waals forces, 258
Internal energy, 345
International Bureau of Weights and Measures, 10
International System of Units, 10
International Union of Pure and Applied Chemistry, 683
Interstitial solid, 657
Iodine, 114, 583
Iodine pentafluoride, 196
Ion, 134
Ion-electron method, 529
Ion exchangers, 497
Ionic bond, **145**
Ionic compounds, nomenclature, 175
Ionic equation, 432
Ionic radii, 148, 288, 290
Ionic solids, 274, 283
Ionization, 134, 426, 455
    degree of, 429
Ionization constant, 459
    table of, 464
Ionization energy, 134
    table of, 136
Ionization isomerism, 782
Ionizing effect of radioactivity, 801
Ion product, 507
Ion-product constant, 455
Ions, **145**
    d^n, 158
    d^{10}, 157
    $d^{10}s^2$, 157, 158
    inert-pair, 157
    migration of, 54
    monoatomic types, 156
    nomenclature, 172
    polyatomic, 163, 164
    pseudo-noble gas, 157
    spectator, 494, 543
    transition metal, 159
Ion clusters, 330
Ion separations, 495
Ionosphere, 134
Iron, 658

Iron(III) chloride, 659, 791
Iron ore, 658
Iron(II) oxide, 669
Iron(III) oxide, 593, 669
Irreversible process, 360
Isoamyl acetate, 723
Isoelectronic species, 147
Isolated system, 344
Isomerism:
  attachment, 783
  chain, 689
  cis-trans, 689, 783
  coordination, 782
  functional-group, 716
  geometric, **685,** 689, 783
  ionization, 782
  ligand, 783
  linkage, 783
  optical, **685,** 785
  position, 689
  structural, 681
Isooctane, 706
Isoprene, 702
Isothermal process, 360
Isotonic solutions, 335
Isotopes, 36, 38, **86,** 90, 91
  fertile, 825
  natural abundance, 91
IUPAC system, 683

Joliot, Frederick, 815, 818
Joliot-Curie, Irene, 815, 818
Joule, 27, 100
Joule, James, 341

K (equilibrium constant), 403
K (kelvin), 25
K_a, 459
$K_{b.p.}$, 325
$K_{f.p.}$, 326
K_h, 477
K_{ins}, 773
K_p, 413
K_{sp}, 501
K_w, 455
K-electron capture, 819
Kelvin, 11, 25
Kelvin temperature scale, 225
Ketones, 717
  nomenclature, 717
Ketose, 739
Ketosis, 718
Kilo-, 11
Kilocalorie, 27
Kilogram, 11
Kinetic energy, 7, **76,** 378
Kinetic molecular theory, 233
Kinetic theory of liquids, 264
Knock, engine, 706

Labile complexes, 774
Lactose, 744
Lanthanides, 128, 662
Laser fusion, 824
Lattice energy, 313
Lavoisier, Antoine, 34, 341
Law:
  Boyle's, **220**
  Charles', **223**
  combining volumes, **228**
  conservation of mass, **34,** 54
  definite proportions, **34**
  Dulong and Petit, **37**
  Faraday's, 77, 569
  first law of thermodynamics, **345**
  Graham's, **235**
  Henry's, **310**
  Hess's, **354**
  ideal gas, 229
  multiple proportions, **35**
  natural, 34
  Octaves, 115
  partial pressures, **232**
  periodic, 115

Law (continued):
  Raoult's, **323**
  second law of thermodynamics, **361**
  third law of thermodynamics, **363**
Lead, 664
  additives, 664
  reaction with sulfuric acid, 666
Lead battery, 550
Lead oxide, red, 36
Lead oxide, yellow, 36
Lead storage cell, 568
Leaving group, 712
Le Chatelier's principle, **249,** 309, 404
Leclanché dry cell, 548
Length, unit of, 11
Leveling effect, 439
Levulose, 742
Lewis acid, 441, 699, 767
Lewis acid-base concept, 441
Lewis base, 441, 699, 767
Lewis, Gilbert N., 441
Lewis structures, 138
Libby, Willard F., 826
Ligands, **767**
  isomerism, 783
  polydentate, 767, 775
  unidentate, 767
  volatile, 775
Light, particle nature of, 96
  plane-polarized, 686
  speed of, 97
  visible, 96
  wave nature of, 96
Light metals, 114
"Like dissolves like" rule, 312
Lime, 707
Limestone, 634
Limiting reagent, 63, 66
Line spectrum, 102
Linear accelerators, 816
Linkage isomerism, 783
Linoleic acid, 735
Linolenic acid, 735
Lipids, **733,** 734
Liquid air, distillation, 597, 611
Liquid oxygen, 597
Liquids, 264
  solubility, 312
Liquid-vapor equilibrium, 247
Lithium aluminum hydride, 721
London dispersion forces, 258
Lowry, T. M., 437
Lox, liquid oxygen, 597
Luminous intensity, unit of, 11
Luster, 130
Lyman series, 105

M, 67, 316
m, 316
Macromolecules, 733
Magnesium, 114, 652
Magnesium nitride, 515
Magnetic bottle, 823
Magnetic moment, 777
Magnetic oxide, 595
Magnetic quantum number, 107
Magnetite, 658
Main reaction, 65
Malleability, 22, 130
Maltose, 744
Manganate ion, 661
Manganese(IV) oxide, 593
Marsden, Ernest, 87
Mass, **1,** 22
  unit of, 11
Mass action, principle of, **387**
Mass of an atom, calculation, 43
Mass defect, 805
Mass-energy equivalence, 805
Mass number, **89**
Mass spectrograph, 85
Mass spectrometer, 38
Matter, 1
  classification of, 9
Maximum density temperature, 262

Measurement:
  of osmotic pressure, 333
  of vapor pressure, 249
Mechanical equivalent of heat, 342
Mechanical work, **342**
Mechanism, **387**
Medical applications, nuclear reactions, 826
Medicinal chemistry, 758
Medicines, 761
Melting, 254
Melting points, 266, 281
Melting points, table, 277
Membranes, 331, 334
Mendeleev, Dmitri, 115
Mercuric oxide, 34
Mercury, 34, 114
Mercury cell, 549
Mercury(I) chloride, 502
Mercury(I) ion, 157, 660
Mercury(II) chloride, 656
Mercury-mercury bond, 502
Meta-, 696
Metal carbonyls, 764
Metal halides, 577
Metal hydroxides, 428
Metal-ion activator, 759
Metal oxides, 429, 514
Metallic solids, 280
Metalloids, 133, 666
Metals, 130, 646
  abundance, 647
  base, **655**
  noble, **655**
  Group IA properties, 648, 650
  Group IIA properties, 648, 650
  properties, 574
  reduction potentials, 649
Metaphosphoric acids, 295, 629
Metaphosphorous acid, 629
Meter, 11
Methanal, 717
Methane, 393, 676
Methane molecule, 188, 189
Methane series, 675
Methanoic acid, 720, 721
Methanol, 713, 714
Methyl acetate, 722
Methyl alcohol, 714
Methylbenzene, 695
Methyl butyrate, 723
Methyl chloride, 578, 698
Methylene chloride, 578, 699
Methyl ethanoate, 722
Methyl group, 679
Metric system, 10
Meyer, J. Lothar, 117
Micelle, 737
Micro-, 11
Migration of ions, 541
Milk, 308
Milli-, 11
Millikan, Robert A., 81
Minerals, 514
  fluorescent, 86
  phosphorescent, 86
  silicates, 295
Mirror image, 685
Miscibility, 312
Mixture, 8
Moderator, 823
Molar heat of fusion, 254
Molar heat of vaporization, **246**
Molar volume, **228,** 332
Molarity, 67, 316
Mole, 11, 42
Molecular collision, 377
Molecular formula, 49, 676
Molecular orbitals, 206
Molecular shape, 183, 189
Molecular velocities, 235, 378
  distribution, 246
  and evaporation, 245
Molecularity, 387
Molecules, 35, 40
  diatomic, 40, 149

Molecules (continued):
   kinetic energy of, 378
   monoatomic, 39
Mole fraction, 316
Monatomic gases, 40
Monatomic ions, types, 156
Monatomic molecules, 39
Monazite sand, 662
Monomer, 619, 701
Monoprotic acids, 430
Monosaccharides, 739
Moseley, H. G. J., 89
Multiple bonds, 154, 198
Multiple proportions, law of, 35
Muscalure, 690

Names of the elements, 38
Naphthalene, 695
Natural law, 34
Natural radioactivity, 797
Negative charge, 77
Neptunium, 817
Nernst equation, **559**
Net ionic equation, 432
Network solids, 276, 292
Neutral amino acids, 748
Neutral solution, 428
Neutralization, **432**
Neutrino, 819
Neutron, **89**, 815
   properties of, 91
Neutron activation analysis, 827
Neutron-to-proton ratio, 809
Neutron stars, 797
Newlands, J. A. R., 114
Newton, 216
Newton, Isaac, 99
Nickel-cadmium cell, 551
Nitrates, 620
   heating, 593, 621
Nitric acid, 430, 527, 620
   reaction with copper, 619
Nitric oxide, 618
   molecule, 212
Nitride ion, 147
Nitrites, 620
Nitrito ligand, 783
Nitro ligand, 783
Nitrogen, 114, 611
   abundance, 611
   comparison with phosphorus, 632
   compounds with hydrogen, 613
   fixation, 402
   inactivity, 613
   isotopes, 611
   oxidation numbers, 615
   preparation, 611
   properties, 613
Nitrogen dioxide, 619
   dimerization, 619
Nitrogen molecule, 154, 614
Nitrogen oxides, 49, 618, 708
   electron formulas, 618
Nitrous acid, 430, 620
Nobel prize, 83, 110, 702, 740, 799, 815, 826
Noble gas electron configuration, 145
Noble gases, 39, **128**, 132, 151, 639
   compounds, 641
   properties, 641
Noble metals, 114, **655**
NO$^+$ ion, 212
Nomenclature, 172
   acids, 177
   alcohols, 713
   aldehydes, 717
   alkanes, 682
   alkenes, 688
   alkyl halides, 709
   alkynes, 690
   binary compounds, 176
   carboxylic acids, 721
   coordination compounds, 769
   cycloalkanes, 691
   cycloalkenes, 692
   ethers, 716

Nomenclature (continued):
   Geneva system, 683
   ionic compounds 175
   ions, 172
   IUPAC system, 683
   ketones, 717
   organic halides, 709
   ternary acids, 177
Nonelectrolyte, 184
Nonmetal halides, 577
Nonmetal oxides, 430
Nonmetals, 132
   properties, 574
Nonpolar covalent bonds, 156
Nonpolar covalent radius, 134
Nonvolatile acids, 434
Normal boiling points, 253
Normal freezing point, 254
Normality, **447**
Nuclear chemistry, 796
Nuclear densities, 796
Nuclear energy, 76
Nuclear equations, 802
Nuclear fission, 809, 819
Nuclear forces, 797
Nuclear fusion, 809, 823
Nuclear reactions, 345
   medical applications, 826
Nuclear stability, 804
Nuclear structure, 796
Nucleon, 89
Nucleophile, 699
Nucleophilic substitution, 712
Nucleus, atomic, **89**
Nylon, 724

Octahedral geometry, 197, 784
Octane number, 706
Octanes, 698
Octet rule, 145
Odd molecules, 207
Odors, 218
Ohm, 78
Oil-drop experiment, 81
Oils, 591, 734
Olefins, 675
Oleic acid, 735
Opposing reactions, 400
Optical activity, 688
Optical isomerism, **685**, 785
   biological significance, 686
Orbital shape, 109
Orbitals, **108**, 185
   antibonding, 209
   boundary surfaces, d orbitals, 112
   boundary surfaces, s and p orbitals, 110
   bonding, 209
Order, 362
Order of a reaction, 380
Ordered quadruple, 110
Ores, 514
Organic chemistry, definition of, 674
Organic compounds, 634
Organic halides, nomenclature, 709
Orlon, 701, 703
Ortho-, 696
Orthophosphoric acid, 629
Orthophosphorous acid, 629
Osmosis, 330
Osmotic pressure, 331
   measurement, 333
Ostwald process for nitric acid, 620
Outer complex, 778
Overall equation, 432
Overvoltage, 566
Oxidation, 513, **515**
Oxidation number, 170, 171, **516**
   and acidity, 664
   rules for, 516
Oxide ion, 594
Oxides, 518, 651
   acidic, 430
   basic, 429
   heating, 593
   metallic, 429

Oxides (continued):
   nonmetal, 430
   of arsenic, 668
   of nitrogen, 618
   of phosphorus, 627
Oxidizing acids, 528
Oxidizing agent, **515**
   table of, 520
   strengths, 522
Oxyacids of phosphorus, 629
Oxygen, 34, 114, 592
   abundance, 592
   comparison with sulfur, 605
   industrial aspects, 596
   in lakes and streams, 311
   preparation, 593
   properties, 593
Oxygen difluoride, 517, 576, 596
Oxygen furnace, 658
Oxygen molecule, 212
Oxyhemoglobin, 788
Oxyhydrogen torch, 591
Ozone, 592, 594, 708

Pacemaker, 549
Palmitic, 735
Para-, 696
Paradichlorobenzene, 711
Paraffins, 675
Parallel electron spin, 112
Paramagnetism, 160, 207
Partial pressures, law of, **232**
Particle nature of light, 96
Pascal, 216
Pascal, Blaise, 215
Passive metals, 670
Pauli exclusion principle, **110**
Pauling, Linus, 151, 640
Pentane, 682
Peptide linkage, 725, 751
Peptides, 752
Percent composition, 45
Percent ionization, 462
Percent yield, 65
Perchloric acid, 456
Perfumes, 717
Period number and electron
   configuration, 123
Periodic classification of elements, 114
Periodic law, **115**
Periodic table, 115
   complete, 118
   with electron configurations, 129
   groups, 119
   Mendeleev, 116
   periods, 119
Periodic trends, 154
Permanganate ion, 661
Peroxide ion, 212, 594
Peroxides, 517, 518, 651
Petroleum, 704
   composition, 705
   fractions, 705
   refining, 705
pH, 457
Phase diagram, 255
   solution, 328
   water, 257
Phenanthrene, 695
Phenol, 715
Phenolphthalein, 65
Phenyl group, 695
Pheromones, 690
Phosgene, 636
Phosphate fertilizers, 631
Phosphine, 626
Phosphonium compounds, 626
Phosphorescent minerals, 86
Phosphoric acid, 62, 630
Phosphors, 802
Phosphorus, 114, 278, 623
   acids of, 628
   allotropes, 624
   comparison with nitrogen, 632
   oxidation numbers, 625

Phosphorus (continued):
　oxyacids of, 629
　preparation, 625
　properties, 624, 625
　red, 624
　white, 624
Phosphorus halides, 626
　hydrolysis of, 579, 626
Phosphorus oxides, 627
Phosphorus pentachloride, 194, 627
Photoelectric effect, 100, 101
Photography, 660
Photon, **100**
Physical change, **6**
Physical properties, **4**
　of solutions, 323
Physical states, **214**
　in equations, 57
Pi bond, 200
Pi cloud, 202
Pig iron, 658
Planck, Max, 100, 110
Planck's constant, 100
Plane-polarized light, 686
Plasma, 824
Plaster of Paris, 653
Plastics, 702
Plato, 33
Plutonium, 817
pOH, 457
Poisons, 761
Polar covalent bonds, 151
Polarity, 183
Polarization, 258
Pollution:
　carbon monoxide, 61, 65
　fertilizers, 631
　gasoline, 41, 707, 708
　lead, 41
　mercury, 192
　nonmetallic oxides, 430
　ozone, 594
　smog, 46
　sulfur dioxide, 63, 602
　waste heat, 824
Polonium, 668, 799
Polyatomic ions, 163, 164
Polydentate ligands, 767, 775
Polyethylene, 155, 701
Polyhydric alcohols, 714
Polymers, 701
　addition, 701, 703
　condensation, 701, 724
　natural, 701
　synthetic, 701
Polyprotic acids, 431, 472
Polysaccharides, 745
Polysulfides, 601
Polyurethane foam, 308
Polyvinyl acetate, 703
Porphyrins, 787
Position isomerism, 689
Positive charge, 77
Positive rays, 84
Positron emission, 819
Potassium, 114
Potassium bromate, 381
Potassium chlorate, 593
Potassium permanganate, 701
Potassium superoxide, 595
Potential, 78
Potential difference, 556
Potential energy, 7, **76**
Precipitate, 493, 507
Precipitation reactions, 498
Predicting molecular geometries, 204
Prefixes, SI, 11
Pressure, 215
　atmospheric, 215
　critical, **244**
　and equilibrium, 405
　osmotic, 331
　standard, 217
Pressure units, 217
Pressure-volume work, **342**

Primary cells, 548
Primary valence, 765
Principal quantum number, 106
Principle of Le Chatelier, **249**, 309, 404
Principle of mass action, **387**
Probability, thermodynamic, 363
Probability density pattern, 109
Products, 34, 54
Progress of the reaction, 384
Promethium, 817
Promotion of electrons, 188
Propane, 676, 679
　combustion, 57
Propene, 689
Properties, 132
　atomic, 133
　of carbon, 634
　chemical, **4**, 132
　colligative, 323
　of gases, 218
　of halogens, 575, 576
　of hydrogen, 587
　of metals, 574
　of nitrogen, 613
　of noble gases, 641
　of nonmetals, 574
　of an object, 22
　of oxygen, 593
　of phosphorus, 624, 625
　physical, **4**, 130, 132
　of solids, 265
　of solutions, 323
　of a substance, 22
　of sulfur, 601
Propyne, 691
Protein molecules, 263
Proteins, **733**, 751
　denaturation of, 755
　fibrous, 755
　globular, 755
　pleated-sheet structure, 754
　primary structure, 753
　quaternary structure, 757
　secondary structure, 753
　tertiary structure, 753
　transport, 733
Protium, 587
Proton, 85
　hydrated, 426
　properties of, 91
Proton-acceptor, 437
Proton-donor, 437
Pseudo-noble gas ions, 157
Pseudosolids, 301
Pure substances, 8
Pyranose, 743
Pyrophosphoric acid, 629
Pyroxenes, 295

Q, 409
q, 345
q_p, 348
q_v, 347
Qualitative analysis, 46, 496
Quanta, 100
Quantitative analysis, 46
Quantum number:
　allowed values, 108
　l, azimuthal quantum number, 107
　m, magnetic quantum number, 107
　n, principal quantum number, 106
　s, spin quantum number, 107
Quantum theory, **102**
Quartz, 295

R, 229, 332
Racemic mixture, 688
Radiant energy, 7, 96
Radiation:
　detection, 799
　infrared, 99
　ultraviolet, 99
Radical, 708

Radii, ionic, 288, 290
Radioactivity, 86
　artificial, 818
　decay processes, 385
　decay series, 799
　natural, 797
　tracers, 826
　wastes, 825
Radiocarbon dating, 826
Radium, 652, 799, 815
Radius ratio, 288
Ramsey, William, 265, 640
Randomness, 362
Raoult's law, **323**
Rare earths, 128, 662
Rate constant, 380
Rate-determining step, 389
Rate-law equation, 380
Rates:
　factors affecting, 377
　of reactions, **373**
　and temperature, 377
Rayleigh, Lord, 640
Rays:
　alpha, 86, 799
　beta, 86, 799
　cathode, 79
　cosmic, 99
　gamma, 799
Reactants, 34, 54
Reaction quotient, 409
Reaction rate, 373
　average, 375
　and $E°$, 566
　and equilibrium, 416
　initial, 376
　instantaneous, 376
Reactions:
　addition, 718
　of alkyl halides, 712
　branched-chain, 394
　of carboxylic acids, 721
　chain, 393, 820
　conditions of, 377
　copper and nitric acid, 619
　electrolytic, **540**
　endothermic, 256
　exothermic, 256
　fast, 373
　galvanic, **540**
　mechanisms, 387
　metal and acid, 433
　neutralization, 432
　nuclear, 345
　opposing, 400
　order of, 380
　precipitation, 498
　progress, 384
　redox, 521, 540
　salt and acid, 434
　self-propagating, 394
　substitution, 698, 712
　of unsaturated hydrocarbons, 699
Reactor, atomic, 822
Rechargeable cells, 548
Red lead oxide, 517
Red phosphorus, 625
Redox reactions, 521, 540
Reducing agent, **515**
　table of, 520
Reducing sugars, 745
Reduction, 513, **515**
Reference standard, atomic weights, 38
Refining, 514, 567
Relative solubility, 496
Representative elements, **128**
Representative metals:
　comparison with transition metals, 665
Resistance, electrical, 78
Resonance, 162, 201
Resonance hybrid, 163
Restoration, oil paintings, 69
Reverse reaction, 401
Reversible process, 360
Ribose, 743

Rocket fuel, 49
Roentgen, Conrad, 83
Rotation about a bond, 200
Rounding off, 17
Rubber, 702
Ruby, 667
Rumford, Count, 341
Rusting, 669
Rutherford, Ernest, 87, 815

S, 361
$S°$, 363
ΔS, 361
Sacrificial metal, 670
Salt, **432**
  double, 764
Salt bridge, 543
Saltpeter, 611
Salt water, 328
Saponification, 723, 737
Sapphire, 667
Saran, 703
Saturated hydrocarbons, 675
Saturation, **322**
Scavengers, 652
Scintillation counter, 802
Seawater, 573
Second, 11
Secondary cells, 548
Secondary valence, 765
Second-order reactions, 388
Seed crystal, 323
Semipermeable membranes, 331, 334
Separation of ions, 495
Sequencers, amino acid, 753
Shielding effect, 134
Shifting an equilibrium, 405
Shortening, 591
SI, 10
SI prefixes, 11
Sickle-cell anemia, 752
Side reactions, 65
Sigma bond, 200
Significant figures, 16
Silanes, 639
Silicates, 295
Silicon, 637
Silicon dioxide, 55, 635
Silicon tetrachloride, 638
Silicon tetrafluoride, 55
Silver bromide, 659
Silver chloride, 659
Silver iodide, 659
Silver oxide cell, 550
Silver sulfide, 655
Simple cubic lattice, 267
Slope of a curve, 376
Smog, 46
Smoke, 218, 308
Soap, 497, 737
Sodium, 114, 652
Sodium acetate, 322, 477
Sodium bicarbonate, 653
Sodium carbonate, 653
Sodium chloride, 9, 582, 652
  electrolysis, 563
  electrolysis of aqueous solution, 564
Sodium hydroxide, 653
Sodium iodide 802
Sodium ion, 147
Sodium nitrate, 611
Sodium stearate, 737
Sodium thiosulfate, 603
Soft drinks, 310
Solder, 664
Solid-liquid equilibrium, 253
Solids: 265
  amorphous, 301
  characteristics, 299
  interstitial, 657
  ionic, 274, 283
  metallic, 280
  molecular, 274
  network, 276, 292
  properties, 265

Solids (continued):
  pseudo-, 30
  solubility, 313
  substitutional, 657
Solid solutions, 657
Solubility, 322, 429
  equilibria, 499
  of gases, 309
  of liquids, 312
  relative, 496
  rules, 494
  of solids, 313
Solubility product, **501**
Solubility products, table of, 504
Solutes, **67**
Solutions, 8, 67, 307
  acid, 428
  basic, 428
  charge neutrality, 542
  concentrated, 67
  dilute, 67
  heat of, **309**
  hypertonic, 335
  hypotonic, 335
  ideal, 323
  isotonic, 335
  neutral, 428
  physical properties, 323
  true solution, **307**
  types, 308
Solvated ions, 314
Solvation, 314
Solvation energy, 314, 493
Solvent, **67**
Sound energy, 76
Spectator ions, 494, 543
sp hybrid orbitals, 191
sp^2 hybridization, 199
sp^2 hybrid orbitals, 192
sp^3 hybrid orbitals, 189
sp^3d hybrid orbitals, 194
sp^3d^2 hybrid orbitals, 197
Specific gravity, **23**
Specific heat, 27, 37
  and entropy, 363
Spectra, 102
  absorption, 102
  emission, 102
Spectrometer, 102
Spectrum:
  continuous, 102
  electromagnetic, 98, 99
  line, 102
  visible, 98
  white light, 101
Speed of light, 97
Spin quantum number, 107
Spontaneity, 366
Spontaneous processes, 340, 361, 373
Spontaneous reaction, 373
Spring scale, 2
Square planar geometry, 197
Square pyramidal geometry, 197
Stability and electron configuration, 127, 145
Stability, nuclear, 804
Stable nuclei, 810
Standard conditions, 225
Standard enthalpy of combustion, 351
  table of, 353
Standard enthalpy of formation, 351
  table of, 352
Standard free energy of formation, 368
  table of, 369
Standard hydrogen electrode, **545**
Standard pressure, 217
Standard reagent, 484
Standard reduction potential, **556**
  table of, 557
Standard states, 351
Standard temperature, 225
Starch, 184, 745
State functions, **344**
State of a system, 344
State, thermodynamic, 344

Static electricity, 77
Stearic acid, 735
Steel, 657, 658
Stereoisomerism, **685**
Steric effects, 767
Sterilization, 827
Stern, Otto, 107
Stibine, 668
Stock, Alfred, 172
Stock system of nomenclature, 172
Stoichiometry, **54,** 231
  precipitation reactions, 498
Storage battery, 550
STP, 225
Strassman, Fritz, 819
Stress on an equilibrium, 247, 404
Strong acid, 430, 455
Strong base, 429, 455
Strong electrolyte, **184**
Structural formula, 676
Structural isomerism, 681
Styrofoam, 703
Sublevels, 106
Sublimation, 256
Subscripts, 40, 55
Substitution, aromatic, 704
Substitution reactions, 698, 712
Substitutional solid, 657
Substrate, 759
Sucrose, 744
Sugars, properties of, 744
Sulfa drugs, 761
Sulfanilamide, 761
Sulfide ion, 147
Sulfides, 601
Sulfur, 63, 114, 278, 597
  abundance, 598
  allotropes, 600
  comparison with oxygen, 605
  mining, 599
  -2 oxidation state, 601
  $+4$ oxidation state, 601
  $+6$ oxidation state, 602
  properties, 601
Sulfur-containing amino acids, 750
Sulfur dioxide, 63, 162, 202, 408, 602
Sulfur hexafluoride, 161, 196
Sulfur trioxide, 163, 203, 294, 444, 602
Sulfuric acid, 431, 475, 527, 550, 602
  electrolysis, 565
  reaction with lead, 666
Sulfurous acid, 601
Supercooled water, 257
Super-HILAC, 816
Superoxide, 517, 518, 651
Superoxide ion, 212, 594
Supersaturation, 322
Surface tension, 263
Surfactant, 737
Surroundings, thermodynamic, 344
Symbols, 39
Synchrotron, 816
Syndets, 737
System:
  closed, 344
  isolated, 344
  thermodynamic, 344

Tarnish, 655
Tears, pH of, 457
Technetium, 817
Teflon, 583, 703
Telluric acid, 668
Tellurium, 668
Tellurium tetrachloride, 196
Temperature, 25
  conversions, 25
  critical, **244**
  and equilibrium, 405
  and reaction rate, 377
  standard, 225
  unit of, 11
Ternary acids, 177, 430
  nomenclature, 177
Tetrachloromethane, 578, 699

Tetraethyl lead, 41, 706, 708
Tetrahedral geometry, 189
Tetraphosphorus decaoxide, 628
Tetraphosphorus hexaoxide, 627
Thallium, 663
Theoretical yield, 65
Thermal expansion, 219
Thermodynamics, 340
    equations, 350
    first law of, 345
    probability, 363
    second law of, 361
    third law of, 363
Thermometers, 25
Thermonuclear bomb, 823
Thermoplastic substances, 702
Thermosetting plastics, 702
Thiosulfates, 603
Thomson, Benjamin, Count Rumford, 341
Thomson, J. J., 81, 85, 87
Threshold frequency, 100
Time, unit of, 11
Tin, 663
"Tin disease," 663
Tin(IV) chloride, 443
Tinplate, 664, 670
Titrant, 484
Titration, **482**
Titration curve, 484
Tollen's test, 719
Toluene, 695
Torch, atomic hydrogen, 591
Torch, oxyhydrogen, 591
Torr, 217
Torricelli, Evangelista, 215
Tourmaline, 688
Tracers, radioactive, 826
Transition elements, **128**, 655
    anions, 661
    comparison with representative
        metals, 665
    halides, 659
    ions, 159
Transition state, 389
Transmutations, 815
Transport proteins, 733
Transuranic elements, 817
Trichloromethane, 518, 699
Triglycerides, 734
Trigonal bipyramidal geometry, 194
Trigonal planar geometry, 192
Trigonal pyramidal geometry, 189
Triple bond, 155
Triple point, 256
Triprotic acid, 431
Tristearin, 737
Tritium, 587, 824

True solution, **307**
T-shaped geometry, 196
Turnover number, 759

U, 345
ΔU, 345
Ultraviolet radiation, 99
Uncertainty in measurement, 16
Unidentate ligands, 767
UNILAC, 817
Unit cell, 266
Unit conversion, 19
Units, 341
    of pressure, 217
Unleaded gasoline, 706
Unpaired electrons, 159
Unsaturated hydrocarbons, 675
Unsaturated shortenings, 736
Unsaturation, test for, 701
Unsymmetrical dimethylhydrazine, 50
Uranium, 819
Uranium-238 decay series, 798
Uranium(VI) fluoride, 583
Urea, 674
Urine, pH of, 457

$\bar{V}$, 332
Vacuum, 215
Vacuum pump, 215
Valence, **146**
Valence electrons, 146
    primary, 765
    secondary, 765
Valence energy level, 146
Valence shell electron repulsion theory, 189
Van der Waals forces, 258
Vanillin, 717
Van't Hoff equation, 332
Vapor pressure, 247, 323
    of hydrates, 650
    lowering, 323
    measurement of, 249
    of water, 248
Vegetable oils, 591
Velocity, 96
Velocity of electrons, 106
Velocity, molecular, 378
Vinegar, 720
Vinyl chloride, 711
Vinylite, 703
Visible light, 96
Visible spectrum, 98
"Vital force" theory, 674
Vitamins, 759
Volatile acids, 434
Volatile ligands, 775
Volatility, 323

Volt, 78
Voltage multipliers, 816
Von Guericke, Otto, 215
VSEPR theory, 189
Vulcanization, 702

w, 345
Walton, Ernest, 816
Wastes, radioactive, 825
Water, 260, 426
    density of, 23
    equilibrium, 454
    of hydration, 459
    supercooled, 257
    vapor pressure, 248
Water gas, 410, 590
Water molecule, 186, 189
Water-softener, 497, 775
Wavelength, 96
Wave nature of light, 96
    constructive interference, 285
    destructive interference, 285
Waxes, 736
Weak acids, 430, 458
Weak bases, 429, 458
Weak electrolyte, **184**
Weight, **1**
Weightlessness, 2
Werner, Alfred, 765
Whipped cream, 308
White phosphorus, 625
Wilson cloud chamber, 801
Wöhler, Friedrich, 674
Wood alcohol, 714
Woods metal, 664
Work, 341
    mechanical, **342**
    pressure-volume, **342**
    useful, 366

Xenon difluoride, 196
Xenon tetrafluoride, 161, 197, 641
X-ray diffraction, 284, 289
X rays, 83, 284
Xylenes, 696

Yield, stoichiometric, 65

Z, atomic number, 90
Zeolites, 497
Zero, absolute, 225, 361
Zeros in a number, 16
Ziegler, Karl, 702
Ziegler process, 702
Zinc chloride preparation, 585
Zinc sulfide, 802
Zwitterion, 747

To the owner of this book:

We hope that you have enjoyed *General Chemistry* as much as we enjoyed writing it. We'd like to know as much about your experiences with the book as you care to offer. Only through your comments and the comments of others can we learn how to make *General Chemistry* a better book for future readers.

School _____ Your Instructor's Name _____

1. What did you like *most* about *General Chemistry?* _____

_____

2. What did you like *least* about the book? _____

_____

3. Were all of the chapters of the book assigned for you to read? _____

(If not, list the chapter numbers that were not.) _____

_____

4. How would you compare the graphics in *General Chemistry* to those in other college textbooks you have read? _____

_____

5. Were there any unclear chapters? If yes, please list the chapter numbers. _____

_____

_____

6. Were there any unclear illustrations? If yes, list the figure numbers. _____

_____

7. In the space below or in a separate letter, please let us know what other comments about the book you'd like to make. (For example, were any chapters *or* concepts particularly difficult?) We'd be delighted to hear from you!

Optional:

Your Name _____ Date _____

May Wadsworth quote you, either in promotion for *General Chemistry* or in future publishing ventures?

Yes _____ No _____

FOLD HERE

FIRST CLASS
PERMIT NO. 34
BELMONT, CA.

BUSINESS REPLY MAIL
No Postage Necessary if Mailed in United States

Dr. B. Richard Siebring/Dr. Mary Ellen Schaff

Wadsworth Publishing Company
10 Davis Drive
Belmont, CA 94002

CUT PAGE OUT

Chemical Elements Alphabetically by Symbol

Symbol	Element	Atomic Number	Symbol	Element	Atomic Number
Ac	Actinium	89	Mg	Magnesium	12
Ag	Silver	47	Mn	Manganese	25
Al	Aluminum	13	Mo	Molybdenum	42
Am	Americium	95	N	Nitrogen	7
Ar	Argon	18	Na	Sodium	11
As	Arsenic	33	Nb	Niobium	41
At	Astatine	85	Nd	Neodymium	60
Au	Gold	79	Ne	Neon	10
B	Boron	5	Ni	Nickel	28
Ba	Barium	56	No	Nobelium	102
Be	Beryllium	4	Np	Neptunium	93
Bi	Bismuth	83	O	Oxygen	8
Bk	Berkelium	97	Os	Osmium	76
Br	Bromine	35	P	Phosphorus	15
C	Carbon	6	Pa	Protactinium	91
Ca	Calcium	20	Pb	Lead	82
Cd	Cadmium	48	Pd	Palladium	46
Ce	Cerium	58	Pm	Promethium	61
Cf	Californium	98	Po	Polonium	84
Cl	Chlorine	17	Pr	Praseodymium	59
Cm	Curium	96	Pt	Platinum	78
Co	Cobalt	27	Pu	Plutonium	94
Cr	Chromium	24	Ra	Radium	88
Cs	Cesium	55	Rb	Rubidium	37
Cu	Copper	29	Re	Rhenium	75
Dy	Dysprosium	66	Rh	Rhodium	45
Er	Erbium	68	Rn	Radon	86
Es	Einsteinium	99	Ru	Ruthenium	44
Eu	Europium	63	S	Sulfur	16
F	Fluorine	9	Sb	Antimony	51
Fe	Iron	26	Sc	Scandium	21
Fm	Fermium	100	Se	Selenium	34
Fr	Francium	87	Si	Silicon	14
Ga	Gallium	31	Sm	Samarium	62
Gd	Gadolinium	64	Sn	Tin	50
Ge	Germanium	32	Sr	Strontium	38
H	Hydrogen	1	Ta	Tantalum	73
Ha	Hahnium*	105	Tb	Terbium	65
He	Helium	2	Tc	Technetium	43
Hf	Hafnium	72	Te	Tellurium	52
Hg	Mercury	80	Th	Thorium	90
Ho	Holmium	67	Ti	Titanium	22
I	Iodine	53	Tl	Thallium	81
In	Indium	49	Tm	Thulium	69
Ir	Iridium	77	U	Uranium	92
K	Potassium	19	V	Vanadium	23
Kr	Krypton	36	W	Tungsten	74
Ku	Kurchatovium*	104	Xe	Xenon	54
La	Lanthanum	57	Y	Yttrium	39
Li	Lithium	3	Yb	Ytterbium	70
Lr	Lawrencium	103	Zn	Zinc	30
Lu	Lutetium	71	Zr	Zirconium	40
Md	Mendelevium	101			

*Names not yet officially approved.